NINTH TEXAS SYMPOSIUM ON RELATIVISTIC ASTROPHYSICS

ALBERT EINSTEIN

Photograph Courtesy of Lotte Jacobi

ANNALS OF THE NEW YORK ACADEMY OF SCIENCES
Volume 336

NINTH TEXAS SYMPOSIUM ON RELATIVISTIC ASTROPHYSICS

Edited by Jürgen Ehlers, Judith J. Perry, and Martin Walker

The New York Academy of Sciences
New York, New York
1980

Library of Congress Cataloging in Publication Data

Texas Symposium on Relativistic Astrophysics,
9th, Munich, 1978.
Ninth Texas Symposium on Relativistic
Astrophysics.

(Annals of the New York Academy of Sciences;
v. 336)
1. Astrophysics—Congresses. 2. General
relativity (Physics)—Congresses. I. Ehlers,
Jürgen, 1929– II. Perry, Judith J.
III. Walker, Martin. IV. Series: New York
Academy of Sciences. Annals; v. 336.
Q11.N5 vol. 336 [QB460] 500s 80–11614
[523.01]

SP

Printed in the United States of America

ISBN 0-89766-045-5

Dedicated to

ALFRED SCHILD

ONE OF THE FOUNDERS OF THE TEXAS SYMPOSIA
ON RELATIVISTIC ASTROPHYSICS
PHYSICIST, GEOMETER, SCHOLAR, HUMANIST
1921–1977

ANNALS OF THE NEW YORK ACADEMY OF SCIENCES

VOLUME 336

February 15, 1980

NINTH TEXAS SYMPOSIUM ON RELATIVISTIC ASTROPHYSICS*

Editors

JÜRGEN EHLERS, JUDITH J. PERRY, AND MARTIN WALKER

International Organizing Committee

P. G. BERGMANN, A. G. W. CAMERON, J. EHLERS,
E. J. FENYVES, R. GIACCONI, S. A. KORFF,
Y. V. NE'EMAN, I. OZSVATH, I. ROBINSON,
A. SALAM, E. L. SCHÜCKING, M. M. SHAPIRO,
L. C. SHEPLEY, H. J. SMITH, J. STACHEL,
W. WALLACE, J. A. WHEELER, AND L. WOLTJER

Local Organizing Committee

J. EHLERS *(Chairman)*, RH. LÜST, J. PERRY,
M. WALKER, AND O. VON KÖLLER

CONTENTS

*This series of papers is the result of a conference entitled Ninth Texas Symposium on Relativistic Astrophysics held on December 14, 15, 16, 17, 18, and 19, 1978 in Munich, West Germany, sponsored by the Deutsche Forschungsgemeinschaft, the Max-Planck-Institut für Physik und Astrophysik, the State of Bavaria, the City of Munich, and the New York Academy of Sciences.

Part II. Cosmology and Galaxy Formation

Part III. Gamma-Ray Astronomy

Part IV. Gravitation Theory and Fundamental Interactions

Part V. Supernovae and Collapse

Part VI. Neutron Stars, Pulsars, and Supernova Remnants

Part VII. X-Ray Astronomy

PREFACE

Judith J. Perry and Martin Walker
Max-Planck-Institut für Physik und Astrophysik
8000 Munich 40
Federal Republic of Germany

For the first time in the history of this illustrious sequence of meetings, a Texas Symposium on Relativistic Astrophysics was held not just outside Texas, but also outside the continental United States. Among Germans, Bavarians are regarded in much the same way Texans are by Americans, and indeed comments on the similarity between Texas and Bavaria were rife at the meeting. Being amateurs at their trade, the local organizing committee made nearly all possible organizational blunders, but nevertheless it seems that a good time was had by most.

From the 13th to the 19th of December, 1978, astronomers, astrophysicists, and physicists having an interest in the latest developments in relativistic astrophysics gathered in Munich for their biennial confabulation. The symposium proceeded at its usual breakneck pace but with one free day thrown in for the relief of the jet-lagged North American contingent. The daytime sessions, consisting solely of invited talks, covered extragalactic sources, galaxy formation and cosmology, gamma-ray astronomy, gravitational theory, matter and the universe, supernovae and collapse, neutron stars, pulsars and supernovae remnants, and x-ray astronomy. For those still unsaturated at the end of a day, evening workshops covered galaxy formation, quasars and galactic nuclei, quantum gravity, gauge theories of gravitation, supernovae and collapse, the Hubble diagram and the expansion of the universe, neutron stars and pulsars, gravitational radiation, and accretion in astrophysics.

Due to circumstances beyond their control, three of our invited speakers were unable to attend. We are deeply grateful to R. A. Muller for springing in at almost no notice with an excellent talk on the microwave background, and to Walter Lewin for extending his talk on x-ray sources considerably beyond his prepared material.

This volume consists of papers based on the invited talks, nearly all in the order in which they were presented. As a departure from previous practice, we asked the chairmen of the workshops to present brief summaries of the discussion in the session they chaired. Most chairmen did so, and their reports have been incorporated into the proceedings in the relevant places. Joseph Taylor and Peter McCulloch, with their report of a measured decrease in the binary period of PSR 1913+16 consistent with gravitational radiation damping, provided a tribute to the prestige of these meetings as occasions for the announcement of dramatic new results.

The well-known scientific editor of the newspaper in which a number of publicity-conscious scientists have announced their latest hot news in recent years, Walter Sullivan, of *The New York Times*, enlivened the banquet with his account of historical sidelights on the evolution of our subject.

As members of the local organizing committee, we cannot claim, as have our illustrious predecessors, to have worked on the meeting for a year or more. Indeed, work was only begun in earnest in the latter part of July, five months before the symposium. We do not recommend this procedure to those of our successors who wish to prevent their hair from turning grey.

Our chairman, Jürgen Ehlers, was a pillar of strength through all tribulations, and carries the bulk of responsibility for the success of the symposium. We also owe a vote of thanks to Rhea Lüst of the Max-Planck-Institut for Astrophysics and Otto von Köller from the central administration of the Max-Planck-Gesellschaft, who completed the local committee.

Our colleague Martin Sohnius of the MPI for Physics assisted most ably throughout. We are particularly grateful to Gerhard Riedel of our institute's administration for making bureaucracy appear transparent.

Rosita Strasser handled the day-to-day secretarial side of the meeting so well that we feared the worst when she went on a month's holiday in September, but Cornelia Rickl jumped in to hold us above water. Helen Northrop assisted at various stages. The symposium could not possibly have taken place without the generous financial support of the Max-Planck-Gesellschaft through its President Reimar Lüst, and its general administration, in particular Dr. Edmund Marsch and his collaborators, of the Deutsche Forschungsgemeinschaft, and of our own institute through its director Rudolf Kippenhahn. We also gratefully acknowledge financial support from the Bavarian government and the city of Munich.

The personnel of the Munich American Express office made travel and accomodation arrangements for many conferees and negotiated special rates for us with the Munich Sheraton Hotel. Their assistance, as well as that of the Munich Sheraton staff, was a great help to us.

SCIENTIFIC OVERVIEW

Jürgen Ehlers

Max-Planck-Institut für Physik und Astrophysik
8000 Munich 40
Federal Republic of Germany

In view of the fact that specialists from many different fields are present, and since many scientists are attending a Texas symposium for the first time due to the changed location, it may be useful to preview briefly the scientific program and to relate this meeting to previous ones.

The first Texas symposium was conceived in 1963 by Alfred Schild and his friends Ivor Robinson and Engelbert Schücking, as a forum for the discussion of topics which are in the intersection of the domains of interest of astrophysicists and relativists. It was Alfred Schild who invented the name "relativistic astrophysics." Throughout the years he played a leading part in the organization of Texas Symposia which have focused the attention of many scientists on this fascinating field. The present meeting is the ninth in the series he founded, and the first to be held without him. The symposium is devoted to his memory.

The central theme of the first Texas conference was the nature of the strong radio sources, the possible connection of which with general relativity was established by the Hoyle–Fowler-hypothesis according to which the energy of these sources was due to the gravitational collapse of a supermassive star of between 10^5 and 10^8 solar masses.

The first topic to be discussed at this meeting is closely related to the principal theme of the first Texas symposium. It concerns those relatively small substructures of galaxies and other compact sources which emit large amounts of nonthermal radiation, mostly as synchrotron radiation from relativistic electrons. Typically such regions are comparable in size with the solar system and radiate as much power optically as a whole galaxy. The information obtainable about such compact, active objects from observations of radio, infrared, visible, and x-radiation is discussed in turn. This part of the meeting will in particular be concerned with the most powerful species of compact sources, the quasars.

Whereas the first part of this symposium is devoted to localized sources of radiation on the scale of galactic nuclei, the next session begins with a discussion of the highly isotropic, thermal microwave-background radiation attributed in the framework of the standard expanding universe models to the dense, hot initial stage of the universe. The discussion of this cosmological "primeval fireball," predicted theoretically by Gamov, Alpher, and Herman in 1948 and first observed by Penzias and Wilson in 1965, forms a natural prelude for the next major topic, the problem of the formation of galaxies and related material structures from a more or less structureless cosmic substratum or "ylem." During this meeting an extensive discussion of theories of galaxy formation, to be augmented by a workshop, will be the main cosmological topic, rather than the standard subject of H_0, q_0, etc. These latter subjects will, however, be discussed in a workshop.

Next on the program follows a survey of γ-ray sources and what information it provides on the overall structure and dynamics of galaxies as well as its substructures, especially the central regions of our own and that of active galaxies; here a link is established with the first session.

The topics mentioned so far are largely observationally oriented. On Saturday, attention will be directed to some theoretical themes. Detailed quantitative tests of gravitation theories so far have been based on the dynamics of the solar system, and more precise and new kinds of tests are expected from studies of the binary system containing the radio pulsar PSR 1913+16. To understand the motions of such systems one needs a theory of isolated systems of gravitationally interacting bodies. The controversial question, discussed at the previous Texas symposium, of how well developed and trustworthy the existing parts of such a theory are, will be discussed. Moreover, the part which computer-work is playing and might play in the future in gravitational theories, its theoretical foundation and prospects will be taken up, in continuation of talks some of us have heard at previous meetings. In connection with these talks on "classical GR" the present state and future prospects of comparisons between gravity theories and observations will be considered.

Ultimately one wishes to understand astrophysical processes in terms of elementary particles and their interactions. Particularly the behavior, perhaps even the creation, of matter in the very early universe and in collapsing stars requires a description of matter, fields, and geometry under extreme conditions of density, temperature, curvature, etc. Apart from astrophysical contexts, one would like to understand how Einstein's geometric theory of space-time and gravity fits together with theories of electromagnetic, strong, and weak interactions. Such fundamental theoretical issues will be discussed on Saturday as well as in evening workshops on quantum gravity and on gauge theories of gravity. The theory session on Saturday will also be concerned with some very general questions of cosmology, that is, how the matter and energy in the universe are distributed among various components, and the relation between observations and theoretical assumptions in cosmology will be discussed in a critical way.

After these general questions of principle the pendulum swings back to more specific topics, now on the scale not of galaxies, but of individual stars. Late stages of stellar evolution, explosions of massive stars, and observational and theoretical aspects of supernovae will be discussed, followed by considerations of neutron stars, pulsar radiation mechanisms, and the processes in supernova remnants. The presently available timing observations of the binary pulsar, so dear to relativists as a possibly clean, moderately relativistic 2-body system, will be reported.

The last day of the symposium will be devoted to various theoretical and observational aspects of x-ray astronomy. This topic, which was already extensively discussed at the two preceding Texas symposia, is related to late stages of stellar evolution, particularly in binary star systems.

I wish to end this Introduction with some remarks about Albert Einstein. As you know, there will be many celebrations next year of the one hundredth anniversary of the birth of the founder of the theories of relativity. Because of their subject, the Texas symposia may all be considered as a tribute to Einstein. In view of the closeness of this symposium to the hundredth birthday of Albert Einstein, we, i.e., the organizers, thought it appropriate to remind us all of the enormous extent to which Einstein's thoughts pervade the subjects of our discussions, by having his picture near to the speaker.* The photograph shows Einstein during the time (1902–1909) when he worked as a technical expert at the patent office in Bern. Let us honor Einstein by striving for a better understanding of nature during the talks and discussions to follow. I hope we shall have a stimulating conference.

*The photograph is reproduced in this volume as the frontispiece.

Part I. **Extragalactic Sources**

RADIO GALAXIES AND QUASARS

K. I. Kellermann*

Max-Planck-Institut für Radioastronomie
D-5300 Bonn 1
Federal Republic of Germany

Observations made with a variety of instruments during the past few years show that the radio emission from galaxies and quasars originates from a wide range of dimensions, extending from a small fraction of a parsec to more than 3 million parsecs. The smallest sources found in the nuclei of nearby galaxies such as M81, M82, and M104 are only about 0.01 pc in extent, and the weak radio source at the nucleus of our own Galaxy is a factor of 10^3 smaller still and comparable in size and luminosity with a variety of stellar radio sources.[1] At the other extreme, the giant radio galaxies such as 3C 236, DA 240, and NGC 315 are the largest objects known in the universe, and observations of the polarized radio emission indicate large scale ordered magnetic fields extending over millions of parsecs.[2]

The Giant Radio Galaxies

For nearly two decades of study of radio galaxies, the giant radio galaxies had gone unnoticed due to their low surface brightness and to the poor sensitivity of radio source surveys to objects of this type. The total energy contained in these extensive clouds of plasma in the form of relativistic particles and magnetic field is large, but perhaps not as spectacularly so as is generally assumed. Minimum energy (equipartition) estimates of the magnetic field strength required to interpret the observed synchrotron emission from radio galaxies lie in the range $10^{-6} \lesssim B \lesssim 10^{-4}$ G and so the magnetic (and presumably particle) energy density may not differ substantially from that found in our own quite "normal" Galaxy. On the other hand, the volume of the radiating region is very large, so the corresponding total energy appears overwhelmingly great. High resolution observations, however, show that these extensive volumes are not uniformly filled with plasma; instead there appears to be considerable fine structure in the form of knots and filaments, so that the total energy estimates are reduced by perhaps three orders of magnitude compared with that estimated earlier on the assumption that the entire volume is uniformly filled by relativistic particles and magnetic field. In addition, some of the earlier energy estimates assumed that most of the energy was contained in heavy relativistic particles which increased the required energy by yet another order of magnitude. But it is now generally accepted that the relativistic electrons are not secondary particles resulting from the interaction of the more numerous heavy particles; the acceleration of the relativistic particles appears to

*On leave from the National Radio Astronomy Observatory, Green Bank, West Virginia 24944

0077–8923/80/0336–0001 $1.75/1 © 1980, NYAS

be the result of a more efficient direct process, and there is no evidence of a hidden reservoir of heavy particles.

Nevertheless, the minimum energy content in radio galaxies and quasars is still large by any standards, perhaps 10^{58} or 10^{59} erg, and several generations of astrophysicists have been kept busy trying to understand the source of energy. Early speculations centered around the concepts of colliding galaxies, and later, exploding galaxies became fashionable. But with any of these "catastrophic" models the subsequent energy losses due to expansion are very great[3] and in order to account for 10^{58} or 10^{59} erg throughout the large observed volume of the radio galaxies, the energy of the initial explosion would have to exceed the rest energy of the galaxy by many orders of magnitude. In addition, the discovery of continuing activity, including the compact variable radio sources, in the nuclei in many radio galaxies and quasars, indicates a continuous process which generates the relativistic electrons, rather than a single cataclysmic explosion. Current discussions of the ultimate energy source are now centered around black holes, and it is speculated that the efficient transport of energy from the black hole to the extended radio lobes occurs by what is commonly referred to as a "relativistic beam or jet." Evidence for these beams of relativistic plasma streaming from the nuclei to the outer radio lobes is seen in recent high-resolution radio maps (see reference 4). Moreover the various emitting regions generally show a remarkable degree of alignment ranging from dimensions less than a parsec to the order of a Megaparsec, implying a common directive mechanism which operates over a wide range of dimensions and time (see reference 1).

The Compact Sources and Superluminal Motion

The compact radio sources themselves do not present any particular energetic problem. Assuming that they are opaque synchrotron sources, measurement of the self-absorption cutoff frequency, angular size, and peak flux density can be used to estimate the magnetic field strength and total energy. But because the compact components appear to be quite inhomogenous, these quantities are not well defined. Estimates of the magnetic field strengths lie in the range $10^{-5\pm2}$ G and the total energy is $\sim 10^{54\pm3}$ erg (see reference 1).

A serious challenge to the conventional picture of incoherent synchrotron radiation has come from the observations of very rapid variations in flux density and size, which have led to speculations that the compact radio sources are either the result of some coherent emission process or that they do not lie at the commonly assumed cosmological distances.[5-7] The problem is simply that if the source dimensions are assumed to be less than the distance traveled by light (or radio signals) during a typical radio outburst, then the corresponding brightness temperature, particularly at long wavelengths,[7] exceeds the normal synchrotron emission limit of 10^{12}K. Ironically, the first detection of variability in extragalactic radio sources by Sholomitsky in the USSR[8] was for the most part rejected by Western scientists, because this naive light-travel-time argument showed that the surprisingly rapid variations seen by Sholomitsky were "theoretically impossible" (see reference 9).

The observed variations of flux density with frequency and time are generally interpreted in terms of the expanding source model originally discussed by Shklovsky[10] and further developed by Pauliny-Toth and Kellermann[11] and van der Laan;[12] but it is now realized that the simple light-travel-time arguments are inadequate if the radiating plasma is moving with highly relativistic velocity, because signals from different parts of the source which arrive at the same time were not emitted at the same time.[13,14]

Indeed, synchrotron emission at centimeter wavelengths in typical cosmic magnetic fields of $\sim 10^{-4}$ Gauss requires microscopic relativistic motion with $\gamma \sim 10^3$. Since there appears to be little thermal plasma associated with variable radio sources[15,16] it is perhaps not unreasonable that 1% of the energy could be in the form of collective relativistic motion. The presence of absorption lines of different redshift in the spectra of some quasars also suggest that matter is ejected with a large but well-defined velocity. Other indirect evidence for relativistic motion comes from the absence of interstellar scintillations implying dimensions greater than that given by the light-travel-time limits independent of the emission mechanism.[17]

More directly the high resolution VLBI observations of galactic nuclei and quasars show rapid changes in the angular structure which, when combined with distances based on the conventional cosmological redshift interpretation, imply apparent linear motion well in excess of the velocity of light. We commonly refer to this as "superluminal" motion and a variety of clever phenomenological interpretations have been proposed based on phase velocity effect, reflections, varying opacity, gravitational lenses, and specialized geometries.[18-22] Most of these models, however, do not stand up to detailed comparison with the observations.

Of the ten or so strong extragalactic sources that have been studied in some detail, four show definite evidence of superluminal expansion.[1,23-25,27,28] Three of these, 3C 273, 3C 279, and 3C 345 are quasars, while 3C 120 is identified with a Seyfert type galaxy. Observations at short centimeter wavelengths made during the period 1970 to 1977 show that in each of these sources there are two or more components which appear to separate from each other with apparent velocities in the range $5 < v/c < 30$.

3C 120

In 3C 120 there have been multiple events, each of which has produced a pair of components which separate along a constant preferred position angle, but with different velocities.[29]

3C 345

In 1974, 3C 345 appeared to be a simple double source. Between roughly 1970 and 1976, the components apparently separated at a rate of 0.17 milliarcsec yr^{-1} ($7c$) with no evidence of any change in component size.[23,24] Since 1977 the structure has become more complex, and there has been no evidence for any further increase in the component separation. However, during this period one component has become rather diffuse and elongated so it is now difficult to unambiguously determine the separation.[25,40,44]

3C 273

In 3C 273 the situation is even more complex. An extended source (3C 273A) some 20 kpc in size lies along the well known jet about 50 kpc away from the quasar.[26] The high resolution VLBI observations of the quasar component (3C 273B) show three compact components and a somewhat larger "jet-like" extension about 50 parsecs away and curving toward the extended jet component.[25,27] When examined with a resolution just sufficient to resolve the triplet, the components appear to separate with superluminal velocity, although the data are not sufficient to determine unambiguously the individual component motions.[24,25]

3C 279

For 3C 279 the observational data are less complete, particularly during the past few years although there appears to have been at least two outbursts, each of which has resulted in a pair of rapidly separating components. The most recent of these events holds the current speed record with an apparent velocity of about $30c$.[27,28]

Stationary Sources

Unlike the rapid motions found in 3C 120, 3C 273, 3C 279, and 3C 345, some objects, such as the nucleus of NGC 1275 (3C 84) and the quasar 4C 39.25, show no evidence for any motion at all. Indeed, there is a rather curious distribution of observed velocities. About half of the well-observed sources have apparent velocities in the range $5 < v/c < 30$, while the other half have $v/c < 1$. There appear to be no sources with velocities $v/c \sim 1$. Moreover, there is no obvious way that the superluminal sources are distinguished from the stationary ones. 3C 345 and 4C 39.25 each have a similar flux density and redshift, both show the familiar double component structure, and both exhibit flux density variations on time scales of the order of a year. Yet in 3C 345 the components have expanded at a rate of 0.17 milli arc sec per year ($7c$) while in 4C 39.25 the motion is less than 0.01 milli arc sec per year ($0.4c$).[29]

NGC 1275

NGC 1275 is a well-known active galaxy and has been studied in great detail over a wide range of frequencies and angular resolutions. Figure 1 shows schematically the distribution of radio emission, on scales ranging from a few parsecs to a few hundred kiloparsecs. Intercontinental VLBI observations made at 21 cm show a strong source in the nucleus with overall dimensions of about 12 pc which is partially obscured by a cloud of HI in front of the North-East part of the source.[31] The 2.8 cm observations, illustrated in Figure 2 show that the nucleus contains three main emission regions, each of which may in turn contain multiple components.[45] The three major components show no relative motion with an upper limit of 10,000 km sec^{-1}, but the individual components appear to expand with an average velocity of about 30,000 km sec^{-1}. The rate of increase of component dimensions as well as the increase in the total

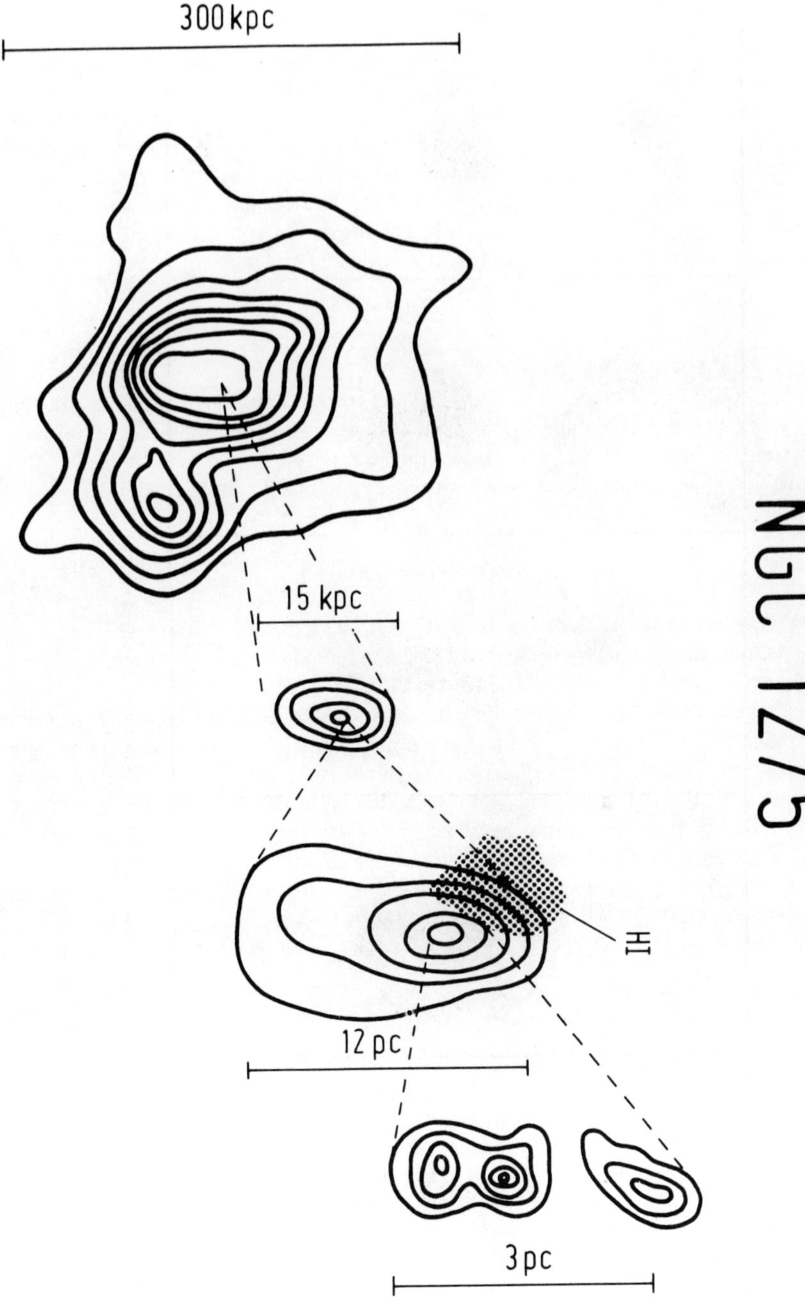

FIGURE 1. Schematic illustration showing the radio structure of NGC 1275 on a scale from 3 psc to 300 kpc.

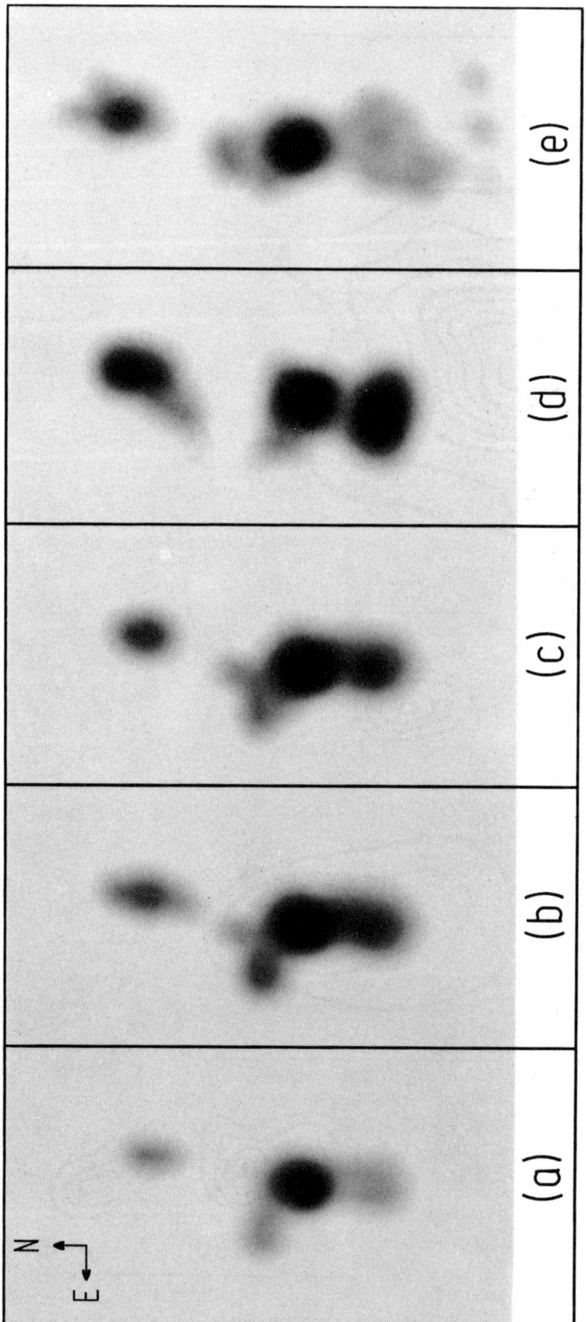

FIGURE 2. Radiophotos of the nucleus of NGC 1275 made at 2.8 cm at epochs 1972.3 (a), 1972.8 (b), 1973.2 (c), 1974.5 (d), 1976.6 (e). The direction lines are each 0.001 arc seconds (0.5 psc) in extent.

flux density (FIGURE 3) suggest a continual acceleration or injection of relativistic particles powered by an event which began about 20 years ago. The overall source extent of ~10 light years, the small upper limit to any component motion (<.03c), and the maximum age of only 20 years imply distinct stationary regions where relativistic particles are accelerated or injected into a magnetic field. On the other hand, the overall linear structure and apparently similar component ages deduced from the component expansion and flux density increase suggest that the different emission regions are not independent, and may be excited by an electromagnetic wave or relativistic shock which propagates from a common origin.[32] Alternatively, the components may have initially had a very high velocity, but then stopped when they reached their present position.

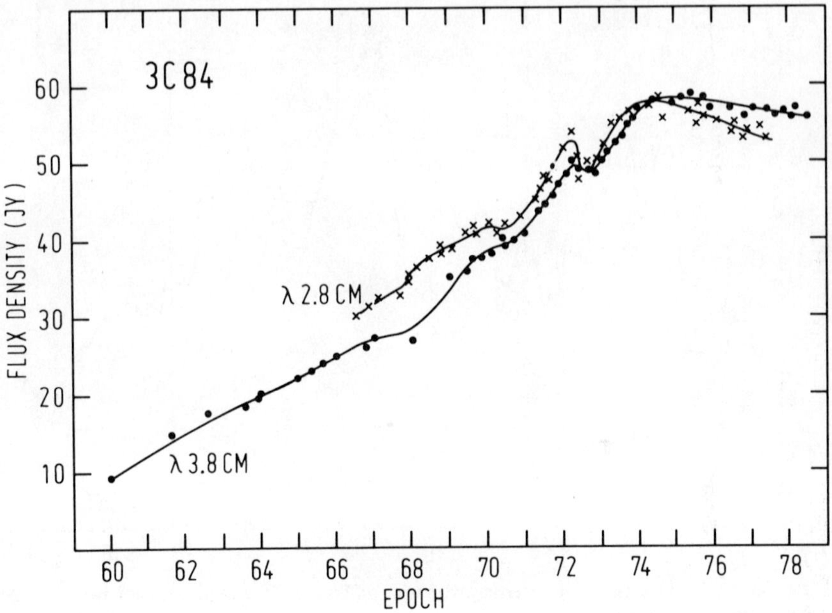

FIGURE 3. The variation of flux density with time of the 2.8 and 3.8 cm emission from NGC 1275. Taken from reference 32.

The behavior of NGC 1275, which has fixed expanding components, is in contrast to 3C 345, where the components are rapidly separating, but have not changed much in size.

REALITY OF SUPERLUMINAL MOTION

When evidence for superluminal motion was first presented in 1971, only single baseline interferometers were used, and alternate interpretations in terms of more complex fixed geometries were discussed.[33,34] Later observations, made with multielement interferometer systems and including more sophisticated analysis techniques, appeared to have confirmed the superluminal double picture.[23,24] But recently the situation has again become clouded. In at least two sources, 3C 120 (FIGURE 4) and 3C 345, there has clearly been a decline in activity and the flux density of these sources has decayed significantly in the past few years.[35,36] As a result, the component separation in these sources is less well defined[25] and it has been speculated by some that the earlier reports of superluminal motion were due to incomplete data and oversimplified analysis.[37]. This, however, is unfounded; multielement interferometers which use phase as well as amplitude information have been available for five years. It has been shown that earlier 3C 120 and 3C 345 were indeed simple equal component

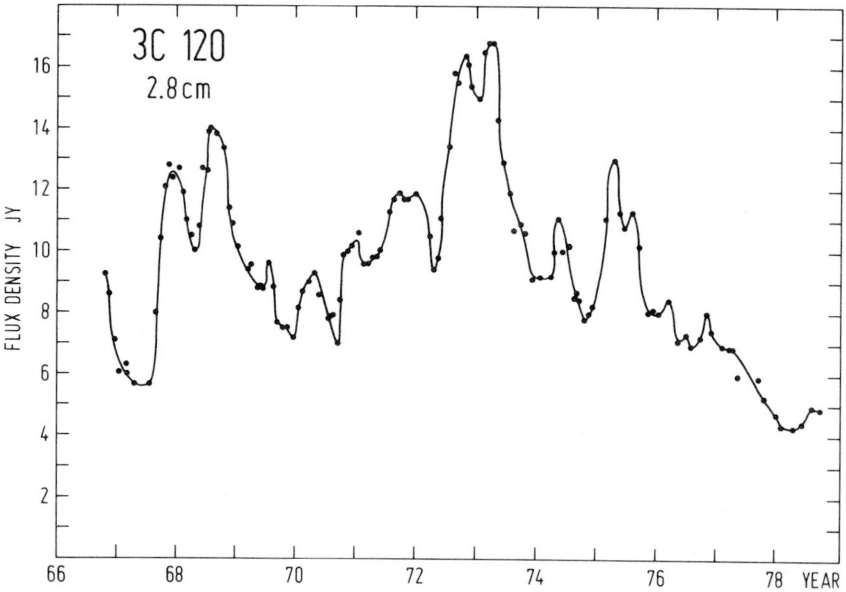

FIGURE 4. The variation of flux density with time at 2.8 cm of the radio emission from 3C 120. Taken from reference 36.

double sources,[29,38,39] although a simple inspection of the more recent data[40] shows that the structures have become asymmetric, the component intensities very different, and the dimensions less well defined. The nature of these sources *has* changed apparently due to the fading of one component, but there is no evidence that the earlier interpretation was incorrect.

INTERPRETATIONS OF SUPERLUMINAL MOTIONS

While the interpretation of superluminal motion is still open to speculation, we must emphasize that the observation of rapid angular motion is not by itself remarkable; nor, indeed, with the aid of some hindsight, even unexpected. It is a consequence of the rapid flux density variations which, at the usually assumed cosmological distances, suggests relativistic motion, if the observations are to be interpreted in terms of the usual expanding synchroton source model.

In order to explain the rapid flux density variations and the apparently excessive brightness temperatures that were implied, Rees initially postulated relativistic component expansion.[13] So far there is no observational evidence for any superluminal component expansion. Moreover it appears that these relativistic models may not be valid for simple source expansion,[41] but the observation of apparent superluminal transverse motion as well as the rapid flux density variations is still most readily interpreted as direct experimental evidence of highly relativistic bulk motions oriented nearly along the line of sight. Attempted interpretations in terms of noncosmological redshifts or nonconventional emission mechanisms are probably premature.

This is not to imply, however, that the interpretation in terms of relativistic motion is without difficulty. A commonly discussed configuration consists of two blobs of emitting plasma ejected from a common origin each with a velocity v.[47] The apparent transverse velocity V of separation of the two components is

$$V = \frac{2v \sin\theta}{1 - \beta^2 \cos^2\theta}$$

which is maximized when $\sin\theta = 1/\gamma$, $\beta = v/c$, $\gamma = (1 - \beta^2)^{-1/2}$.

In this case the apparent transverse velocity of the approaching component is $\sim \gamma c$ while the receding component "barely moves" with a velocity $\sim c/2$. In triple component sources, such as 3C 273, it is possible, therefore, in principle to distinguish between models where two components are ejected in opposite directions from a stationary center, from models where both components are ejected towards the observer at different times or with different velocities.

The required large, intrinsic, bulk velocities of $\sim 0.99c$ do, however, introduce several formidable problems:

a. Kinetic energy of the order of γ times the component mass is tied up in the motion of the components.

b. A relatively rapid deceleration is expected since all of the activity occurs well within the relatively dense galactic nuclei or quasar emission-line regions. There is no evidence that this happens.

c. The differential Doppler shift causes the intensity of the approaching component to be very much greater than that of the receding or stationary component by a factor ranging from 30 to 10.[6] But the observations typically show nearly equal component intensities.

d. The motion must be aligned within a cone $\sim$10 degrees along the line of sight, which has an *a priori* probability of less than one percent; yet, about half of all observed compact sources show apparent super-light motion. Although these statistics are based on small numbers (only about 10 sources have been adequately observed), a comparable fraction of a larger sample show the rapid flux density variations, from which superluminal motion may be indirectly inferred.[7,46]

It is possible that the necessary preferred alignment along the line of sight is the result of selecting only those components which are moving nearly towards the observer, and which thus have enhanced emission due to the Doppler beaming. The stationary or receding components would then be much weaker and probably would be undetected with the relatively poor dynamic range of current VLBI maps. It is unlikely, however, that within individual objects there could be a more or less isotropic or random ejection of radiation blobs, with only the approaching ones being strong enough to observe. This would imply that successive events within a given object should appear with random orientation in the plane of the sky. But in both 3C 120[24] and 3C 279,[27] where multiple outbursts have occurred, the position angle remains invariant, although the velocities of the various events differ. This implies a preferred orientation in the individual objects, with the Doppler enhancement selecting only those sources which are beamed towards the earth.[42,43] It has been suggested that such a model would also account for the approximately 90% of all quasars which are radio quiet, as these would then represent the large fraction of objects which are beamed away from the observer.[43] However, it is not clear how apparently stationary sources such as NGC 1275 and 4C 39.25 fit into such a picture, since the slowest motions expected are where the motion is perpendicular to the line of sight, and which should show the intrinsic relative velocity of either c or $2c$ depending on whether both components are moving or one is stationary.

Also in some superluminal sources such as 3C 273 and 3C 345, the more remote clouds of emission (and in the case of 3C 273, the optically radiating jet) appear to lie at the end of a jet-like bridge whose origin and orientation is coincident with the superluminal compact components. This could imply that these much larger and more massive clouds also move with high relativistic velocity.

It is also possible that there is no actual matter moving with very high velocity, but only an electromagnetic wave which "ignites" stationary matter. The problems associated with the absence of measurable deceleration, the large kinetic energy of bulk motion, and the relative component intensities are eliminated; but then it is not possible to interpret the apparently fortuitous orientation as a result of Doppler enhancement and selection.

Summary

There is a growing consensus that the sources of energy for radio galaxies and quasars lies in condensed massive objects, possibly black holes, and that the energy is efficiently transported hundreds of kiloparsecs or megaparsecs away by well colimated relativistic jets or beams. There is no evidence that the radio emission from either the extended or compact components is due to any mechanism other than the conventional incoherent synchrotron process. The time scales of the flux density variations and the VLBI observations of changes in the source brightness distributions suggest relativistic motion aligned nearly along the line of sight, although it is not easy to understand the observed equal component amplitudes, the consistency of the position angles, over a wide range of space and time, the absence of deceleration, and the apparently fortuitous geometry. Future refinements of the models may well lead to a better agreement between theory and observation, but this nearly satisfactory scenario may well suffer the same fate as the colliding and exploding galaxies which were equally as popular in the previous decades!

Acknowledgments

I am greatly indebted to many colleagues who have provided data in advance of publication, and for stimulating and informative discussions, and particularly to B.

Geldzahler, I. Pauliny-Toth, R. Porcas, and J. Romney for their helpful comments on the manuscript.

References

1. Kellermann, K. I. 1978. Physica Scripta **17:** 257.
2. Willis, A. G. 1978 Physica Scripta **17:** 243.
3. Scheuer, P. A. G. 1974. Mon. Not. R. Astron. Soc. **166:** 513.
4. Fomalont, E. B. *et al.* 1979. Astron. & Astrophys. In press.
5. Jones, T. W. & G. R. Burbidge. 1973. Ap. J. **186:** 791.
6. Colgate, S. A. & A. G. Petschek. 1979. BL Lac Objects. A. Wolfe, Ed. In press.
7. Fanti, R., A. Ficarra, F. Mantovani, L. Padrielli & K. Weiler. 1978. Astron. & Astrophys. Suppl. In press.
8. Sholomitsky, G. B. 1965. Soviet Astronomy AJ. **9:** 516.
9. Maltby, P. & A. T. Moffet. 1965. Astrophys. J. **142:** 409.
10. Shklovsky, J. 1966. Soviet Astron. - AJ **9:** 683.
11. Pauliny-Toth I. I. K. & K. I. Kellermann. 1966 Astrophys. J. **146:** 643.
12. van der Laan, H. 1966. Nature **211:** 1131.
13. Rees, M. J. 1967, Mon. Not. R. Astron. Soc. **135:** 345.
14. van der Laan, H. 1971. "The Nuclei of Galaxies," :245. North Holland Press.
15. Wardle, J. F. C. 1977. Nature **269:** 563.
16. Jones, T. W. & S. L. O'Dell. 1977. Astron. & Astrophys. **61:** 291.
17. Condon, J. J. & B. Dennison. 1978. Astrophys. J. **224:** 835.
18. Blandford, R. D., C. D. McKee & M. F. Rees. 1977. Nature, **267:** 211.
19. Milgrom, M. & J. N. Bahcall. 1978. Nature **274:** 349.
20. Lynden-Bell, D. 1977. Nature **270:** 396.
21. Epstein, R. I. & M. J. Geller. 1977. Nature **265:** 219.
22. Cavaliere, A. *et al.* 1971. Science **173:** 525.
23. Shaffer, D. B. & K. I. Kellermann. 1977. "L' Evolution des Galaxies et ses Implications Cosmologiques," :347. CNRS, France.
24. Cohen, M. H. *et al.* 1977. Nature **268:** 405.
25. Cohen, M. H. *et al.* 1979. Astrophys. J. In press.
26. Conway, R. G. & D. Stannard. 1975. Nature **255:** 310.
27. Pauliny-Toth, I. I. K. *et al.* 1979. Astron. & Astrophys. In press.
28. Cotton, W. *et al.* 1979. Astrophys. J. Lett. In press.
29. Shaffer, D. B. *et al.* 1977. Astrophys. J. **218:** 353.
30. Miley, G. K. & G. C. Perola. 1975. Astron. & Astrophys. **45:** 223.
31. Romney, J. 1978. Ph.D. Thesis, Calif. Inst. of Tech.
32. Preuss, E. *et al.* 1979. Astron. & Astrophys. In press.
33. Dent, W. A. 1972. Science **175:** 1105.
34. Kellermann, K. I. *et al.* 1974. Astrophys. J. Lett. **189:** L19.
35. Dent, W. A. Private communication.
36. Andrew, B. Private communication.
37. Discussion at VLBI Symposium, Heidelberg, FRG Aug 1978.
38. Seielstad, G. A., *et al.* 1979. Astrophys. J. **229:** 53.
39. Wittels, J. J. *et al.* 1976. Astrophys. J. Lett. **260:** L75.
40. Readhead, A. C. S. *et al.* 1979. Astrophys. J. In press.
41. Terrell, J. T. 1977. Astrophys. J. Lett. **213:** L91.
42. van den Bergh, S. 1978. Private communication.
43. Scheuer, P. A. T. & A. C. R. Readhead. 1979. Nature **227:** 182.
44. Unpublished Observations at 1.3, 2.8, and 6 cm.
45. Pauliny-Toth, I. I. K. *et al.* 1978. Pisma v Astronomicheski Jornal. **4:** 51.
46. Condon, J. J. *et al.* 1979. Astron. J. **81:** 1.
47. Ozernoi, L. M. & V. N. Sazinov. 1969. Astrophys. Space Sci. **3:** 395.

PROPERTIES OF BL LAC OBJECTS

Arthur M. Wolfe

Department of Physics
University of California at San Diego
La Jolla, California 92093
and
Department of Physics and Astronomy
University of Pittsburgh
Pittsburgh, Pennsylvania 15260

INTRODUCTION

BL Lac objects have only recently come into their own. Once considered to be the exclusive domain of continuum radio astronomers and optical photometrists, BL Lac objects are now studied by optical and radio spectroscopists, x-ray astronomers, etc. That is, they are currently observed with the full technology of modern observational astronomy. Interest in these enigmatic objects has increased so much that they were the sole topic of a recent conference at the University of Pittsburgh.

This dramatic change in status can be attributed to two realizations. First, the distinctly nonthermal energy source in a BL Lac is extremely compact and extremely well ordered, even for a QSO-like object: as we shall see, there is *no* direct evidence that optical continuum radiation, the dominant form of energy loss, emitted by most QSOs is in fact nonthermal. Therefore, BL Lac objects provide the most direct link we have to the primary "machine" which has eluded astrophysicists since the first of these Texas symposia. Second, the recent discovery of a galactic component in the optical radiation emitted by many BL Lacs provides another link: that is, between a QSO-like object and a galaxy of stars. Besides establishing distances, observations of the galactic component may clarify whether the BL Lac phenomenon is an evolutionary stage through which all galaxies must pass, or whether it is restricted to galaxies belonging to a special class.

In this paper, I shall (1) examine the criteria proposed to define the BL Lac class and show how methods used to find BL Lacs influence properties thought to be intrinsic; (2) describe evidence that BL Lacs belong to a wider class of objects that include the optically violent variable (OVV) QSOs; and (3) examine the connection between BL Lacs and the galaxies in which they are embedded.

DEFINING CHARACTERISTICS

The following are criteria typically used to define a BL Lac object:

(A) A radio source with a flat spectrum at cm wavelengths.
(B) A star-like object that radiates nonthermal optical continuum radiation, and is possibly associated with a nebulosity.
(C) An object with variable optical and radio flux densities.
(D) An object with strong and rapidly variable optical polarization.
(E) An object that radiates *no* emission lines.

Let us examine each property in turn.

0077-8923/80/0336-0012 $1.75/1 © 1980, NYAS

Without exception, all known BL Lac objects were discovered because of their radio emission. Not only are they radio emitters, but almost all known BL Lacs have flat or inverted radio spectra at cm wavelengths.[1] Since a large fraction of BL Lacs were discovered in high frequency surveys, it is possible that BL Lacs with steep-spectra were missed. However, the few BL Lacs found in low-frequency surveys also have flat spectra at cm wavelengths, so the latter property appears to be genuine and not an artifact of observational bias.[2]

But the reliance on radio emission still introduces important selection effects. The most obvious of these is the bias against detecting radio quiet—that is, optically selected—BL Lacs. These might dominate the BL Lac population in the same way that optically-selected QSOs dominate the QSO population. Another bias arises from inaccurate radio positions ($\Delta\theta \sim 60''$) that forced early workers in the field to seek peculiarities in color or optical morphology for optical identifications. Later on when more precise ($\Delta\theta \sim 1''$) radio positions were available, BL Lac objects with "neutral" colors and uninteresting morphology, such as PKS 0735 + 178, were discovered: objects like this would surely have been missed previously.[3] Also it is important to note that many BL Lacs were found accidentally in searches for high-redshift QSOs and in other surveys not devoted to finding BL Lacs. As a result, the *known* population of BL Lacs comprises a mixed sample drawn from heterogeneous selection criteria. It is entirely possible that these may be a subset of a much larger population.[3]

Higher up in frequency we find that BL Lacs tend to have steep power-law spectra at optical wavelengths. While the median spectral index of QSOs $<\alpha_{opt}> \approx -1$, that for BL Lacs is given by $<\alpha_{opt}> \approx -2$, where the flux density $S_\nu \propto \nu^\alpha$. When the optical emission of BL Lacs and QSOs drawn from the same cm survey is compared to the radio emission, one finds an interesting correlation: on the average, the presence of a bright flat-spectrum radio source usually implies optical emission from a compact object that radiates a power-law continuum. Condon[3] quantifies this statement by noting that the optical magnitude can be predicted from the radio flux by extending an $\alpha = -0.7$ power-law from that frequency at which the radio source is optically thin, to IR and optical wavelengths. Since BL Lacs tend to be self-absorbed down to millimeter wavelengths rather than centimeter wavelengths as for QSOs, a BL Lac with a 1 Jy flux density at millimeter wavelengths is predicted to be $1 \rightarrow 2$ mag brighter than a QSO which is 1 Jy at say 6 cm. This has been confirmed by observation. Note too that these BL Lacs should be fainter than the QSOs at centimeter wavelengths. This might explain why QSOs outnumber BL Lacs, by a factor of $10 \rightarrow 20$, in centimeter surveys limitied to fairly high flux densitites, say $S_\nu >$ 0.5 Jy. Future centimeter surveys that are more sensitive may show that the BL Lacs are relatively numerous.[3]

The variability of BL Lacs is an important property that sets them apart from most QSOs. The shorter time-scales for BL Lac variability further suggests that they are more intimately connected to the primary "machine" than is the average QSO. At optical wavelengths the distinction is especially clear cut. Whereas the average QSO varies by $\Delta B \approx 0.3 \rightarrow 0.5$ mag in a time-scale $\Delta t \approx$ few years, most BL Lacs vary by $\Delta B > 1$ mag during $\Delta t \sim 1$ day $\rightarrow$ 1 week. More extreme examples of variability can be found in the 1969 outburst of BL Lac itself when $\Delta B \approx 1$ mag during $\Delta t \sim 1$ hr,[5] and in the 1975 outburst of AO 0235 + 164 when $\Delta B \sim 5$ mag and $\Delta t \approx$ wks.[6] In this respect, BL Lacs are similar to a subclass of radio QSOs of which 3C 446,454.3, and

279 are famous examples. Not only do these optically violent variables (OVVs) undergo strong outbursts of short duration, but they also have relatively steep optical spectral indices. In comparison, violent outbursts like these do not occur in optically-selected QSOs. For some reason, the presence of radio emission, presumably in the compact core, indicates emission from a region closely associated with the primary "machine."

The distinction between QSOs and BL Lacs is less clear with respect to radio variability. A recent study[7] of a sample of flat-spectrum QSOs and BL Lacs reveals that the incidence of variability in both classes is about the same. However, the *degree* of variability is higher and the duration of the outbursts shorter in the BL Lacs. Thus σ/S, the median r.m.s. scatter in 3.7 cm flux density relative to the mean flux density, is larger than 30% for 5 out of 20 BL Lacs, whereas none of the 14 QSOs varied to this degree. Moreover, Δt for the BL Lacs was on the order of months, and on the order of years for the QSOs. Significantly, radio outbursts of the OVV class resemble the BL Lac variations in that they are relatively strong. But their time-scales tend to be longer.

These considerations support the idea that BL Lacs and OVVs are the most compact and strongly variable sources among the general class of QSOs. Within the BL Lac-OVV subclass there are further distinctions which are important. For example, a comparison between optical and radio variability shows that the fractional change of optical flux density is at least an order of magnitude larger than in the radio. BL Lac itself was the strongest known radio variable in the early 1970s,[8] and even then $(\Delta S_\nu/S_\nu)_{radio} \ll (\Delta S_\nu/S_\nu)_{optical}$. The strongest outbursts at these two wavelengths can be represented as follows:

$$\text{Radio:} \quad \Delta S_\nu \sim (2 \rightarrow 5) \cdot (S_\nu)_{initial},$$

$$\text{Optical:} \quad \Delta S_\nu \sim (30 \rightarrow 100) \cdot (S_\nu)_{initial}.$$

The second distinction, already alluded to above, is that the BL Lac-OVV outbursts typically last a week or less in the optical, and months in the radio. Extreme examples of this contrast in time-scales are BL Lac objects like 1727 + 50 (IZw 186) and 2335 + 03 (4C 03.59) which vary optically, but not at radio wavelengths.[7]

The difference between radio and optical time-scales reflects the larger size of the radio source. The optical source, which has to be no larger than the size of the solar system, is probably linked more closely to the primary "machine." But the reasons for the difference between fractional changes in flux are less clear. It is possible that the *actual* fractional change of the variable radio component is as large as that detected optically. However, the *observed* fractional change would be small if there existed a constant radio component with flux density comparable to the maximum of the variable component.[9] An example of this may be a constant component in BL Lac with $S_\nu \approx 2$ Jy at $\lambda = 3$ cm., since this radio source has never decreased below 2 Jy throughout its erratic history.[8] On the other hand, the fractional variabilities may be intrinsically different. If the energy reservoir needed to drive the radio source were larger (because of the larger volume?), then a smaller fraction of energy would be tapped during a radio outburst, resulting in a lower fractionaly variability.[10] In the future, VLB maps made during an outburst should be compared with those of the pre-outburst source. This should help to determine the fractional change of the variable component instead of the entire source.

The frequent incidence of radio and optical variability in BL Lacs raises the question of whether the two types of outburst are related? Although the obervational parameters describing radio and optical outbursts may be different for a given object, the two events could still be causally connected. Indeed, the $\nu^{-0.7}$ extrapolation discussed above implies that the same distribution of particles is responsible for both radio and optical emission, indicating a causal link between both types of source.

The 1975 outburst of the BL Lac AO 0235 + 164 is the clearest example of such a link. In this case, both radio[11] and optical[6] fluxes peaked during November of 1975, and were relatively constant before and after this time. While the duration of the optical burst was ~1 → 2 wks, as compared to ~2 months for the radio burst, a superposition of the two light curves reveals an unambiguous correlation. However, cross-correlation analyses of optical and radio light curves in most BL Lacs do not reveal any correlations. Aside from OJ 287 and NRAO 512, there is little evidence to support a causal link in a large sample of BL Lacs.[12] But this neglects the effects of synchrotron self-absorption. If an expanding shell of synchroton electrons emits radio and optical photons, the radio flux density would increse at wavelengths where the shell is optically thick, while the optical flux density always decreases. Another effect that would mask synchronous outbursts is the occurrence of a large number of overlapping events. Both effects were presumably absent during the isolated event in AO 0235 + 164, but may determine the appearance of outbursts in most BL Lacs and OVVs. A good example of this type of ambiguity can be found in the recent outburst of 3C 446. This well known OVV underwent a tremendous optical burst during the second half of 1977.[13] Since that time, repeated VLB observations reveal that the radio structure (at $\lambda = 2.85$ cm) of 3C 446 which was initially constant has now changed.[14] It is tempting to attribute the radio changes to the optical burst. On the other hand, the historical optical and radio light curves are not correlated,[12] so it is possible that the coincidence of radio and optical changes in this object is random in origin.

Of the criteria listed above, strong optical polarization is the best indication that BL Lacs and OVVs form a well-defined subclass of QSOs. In their survey of 102 optical- and radio-selected QSOs, Stockman and Angel[15] find no objects with optical linear polarization $p > 3\%$ and an average (r.m.s.) value $\langle p \rangle = 0.6\%$, when BL Lacs and OVVs are excluded. The latter, however, have polarizations in the range $p \approx$ 5–30%. Stated differently, the distribution of optical polarization is bimodal, with the BL Lacs and OVVs alone occupying the peak at strong polarization, and the remaining QSOs concentrated at essentially zero polarization. This is the strongest evidence we have that (i) the BL Lac phenomenon is intrinsically nonthermal, and (ii) the OVVs and BL Lacs are the same "beast."

Stockman and Angel further find that if BL Lacs and OVVs are excluded, neither the optically-selected nor radio-selected QSOs show significant optical polarization. Thus there is no evidence for nonthermal emission from optically-selected QSOs. Not only does this set optically-selected QSOs apart from the BL Lacs and OVVs, but it also implies that they are different from *all* radio-selected QSOs. This is because even radio-selected objects that are not in the BL Lac-OVV class contain non-thermal "machines," as evidenced by their radio synchrotron-emission.[16] Because the BL Lac phenomenon is essentially non-thermal, it follows that radio-quiet BL Lacs may not exist. If these ideas are correct, then sensitive radio surveys of all types of QSOs should reveal a bimodal distribution with two well defined peaks: one at faint or zero luminosities for optically-selected objects, and the other at strong luminosities for radio-selected objects. Preliminary studies with the VLA do not reveal such a bimodal distribution.[17] Note that another way to check whether optically- and radio-selected QSOs are different is to measure IR polarization in a sample of QSOs equally divided into both classes. Whereas the optical continuum radiated by most QSOs may be depolarized by an opaque atmosphere,[18] the IR emission may come from optically thin regions associated with high-frequency radio emission. It would be interesting if significant IR polarization (say at $\lambda = 10\mu$) were found in optically-selected objects that are truly radio-quiet; i.e., no centimeter flux at the $\approx$0.1 mJy level. This would argue *against* intrinsic differences between the two classes. Moreover, a very efficient source of opacity at $\lambda > 10\mu$ would have to be present since a low-energy cutoff to the synchroton electron distribution would yield too much radio flux as a result of the $\nu^{1/3}$ spectral behavior longward of 10 μ.

Let us now turn to criterion (E), the absence of emission lines. This widely publicized characteristic of BL Lacs has generated a lot of discussion. Considerable effort has gone into finding ways of suppressing emission lines; e.g., no gas, no ionizing photons, etc. The problem, however, has now been removed largely due to the work of J. S. Miller.[19] He finds that 6 out of the 12 BL Lacs in his sample radiate weak and narrow [OIII] (and in some cases [OII] and Hβ) emission lines characteristic of a low density gas. The lines appear weak because they are superposed on a strong continuum, and as a result the energy emitted per second, L([OIII]), is typically ten times stronger than that radiated by normal elliptical galaxies. Because of the large luminosity and presumably low density n_e, the gaseous mass, $M_{gas} \propto L([OIII])/n_e$, will probably range from $10^4 \rightarrow 10^5\ M_\odot$. This is large compared to the 10–$10^3 M_\odot$ that radiates the broad emission lines detected in QSOs. Therefore, the question concerning gas in BL Lacs no longer is "where is it?", but rather "where is the dense, turbulent component?" This question is answered by including OVVs in the BL Lac class. The extended class contains objects with similar continuum properties, but widely different emission-line properties. At one end we have AO 0235 + 164 which has very little, if any gas, as evidenced by the lack of spectral features at minimum light. Then, there is BL Lac with low-density gas, and finally 3C 446 with dense and turbulent gas. The close similarity between the nonthermal emission of 3C 446 and AO 0235 + 164 implies that emission-line properties are peripheral to the energetics underlying the primary "machine."

The Galaxy Connection

The first indication that BL Lacs are embedded in galaxies stems from their frequent association with extended images on photographic plates. The nebulosity surrounding the star-like object in PKS 0548-322 is a good example of such an association.[5] Some low-redshift QSOs are also attached to nebulosities which appear different from those around BL Lacs. While the QSO nebulosities typically have angular diameters $\theta \sim 3 \rightarrow 4''$, the BL Lac images are in the range $\theta \approx 20 \rightarrow 40''$. Furthermore, the QSO nebulosities are found in $\lesssim$1% of all QSOs, while ~30% of the known BL Lacs have nebulosities. This suggests that BL Lacs are in relatively nearby, say $z < 0.1$, galaxies, and that QSOs are in galaxies that are far away, say $z \gtrsim 1$.[20]

Spectroscopic observations of the nebulosities confirm this conjecture for the BL

Lacs, but not the QSOs. The QSO nebulosities are comprised of giant regions of ionized gas responsible for [OIII], Hβ, and other nebular emission lines.[21] In no case is there evidence for absorption lines indicating the stellar population of a galaxy. However, stellar absorption lines are found in BL Lac itself,[22] and in 7 other BL Lacs.[19] TABLE 1 lists the redshifts and absolute magnitudes of the galaxy component, as well as the absolute magnitude of the central nonthermal object. Although these magnitudes correspond to an aperture with a diameter of only ~3″, the galaxy components are still detectable. When these magnitudes are corrected to the entire galaxy, one finds a magnitude range of $-22.5 < M_v < -21$. So far, these galaxies have been identified as giant ellipticals on the basis of the scale-height and form of the surface-brightness distribution.[5] Note that the nuclear bulges of early-type spirals could reproduce the observed stellar spectrum, but not the extent of the light distribution.

It is important to emphasize that the association of BL Lacs with giant ellipticals has been demonstrated only for objects with relatively low redshifts. High-redshift BL

TABLE 1

REDSHIFTS AND ABSOLUTE MAGNITUDES OF GALAXIES SURROUNDING BL LACS*

		M_v†	
Object	Z	Galaxy	Central Object
Mk 421	0.030	−20.2	−22.1
Mk 180	0.044	−19.3	−19.7
AP Lib	0.049	−20.6	−20.0
Mk 501	0.034	−19.3	−20.1
1727 + 50	0.055	−19.0	≳ −19.9
3C 371	0.050	−20.0	−20.7
BL Lac	0.0695	−19.8	−23

*Data taken from Miller *et al.* (ref. 19).
†Absolute magnitudes in ~ 3″ diameter aperture.

Lacs and OVVs like AO 0235 + 164, PKS 0735 + 178, and 3C 446 would be orders of magnitude more luminous, if they are at cosmological distances, and thus could be intrinsically different. By extending the BL Lac-elliptical connection to large z's we would, among other things, gain insight into their evolutionary history. Moreover, we would establish unambiguous distances to objects that may be among the most luminous in the universe, and at the same time learn something about the relationship between the central object and the galaxy that surrounds it. The problems involved in finding high-redshift galaxies superposed on luminous continuum sources are illustrated in FIGURE 1. Here I have plotted redshift vs M_v for each BL Lac and, when known, for the galaxy component detected through ~3″ of aperture. A comparison between the predicted curve for first-ranked giant ellipticals and absolute magnitudes of high-z BL Lacs shows that 1400 + 162 and possibly AO 0235 + 164 (at minimum light) are promising candidates for revealing surrounding galaxies.

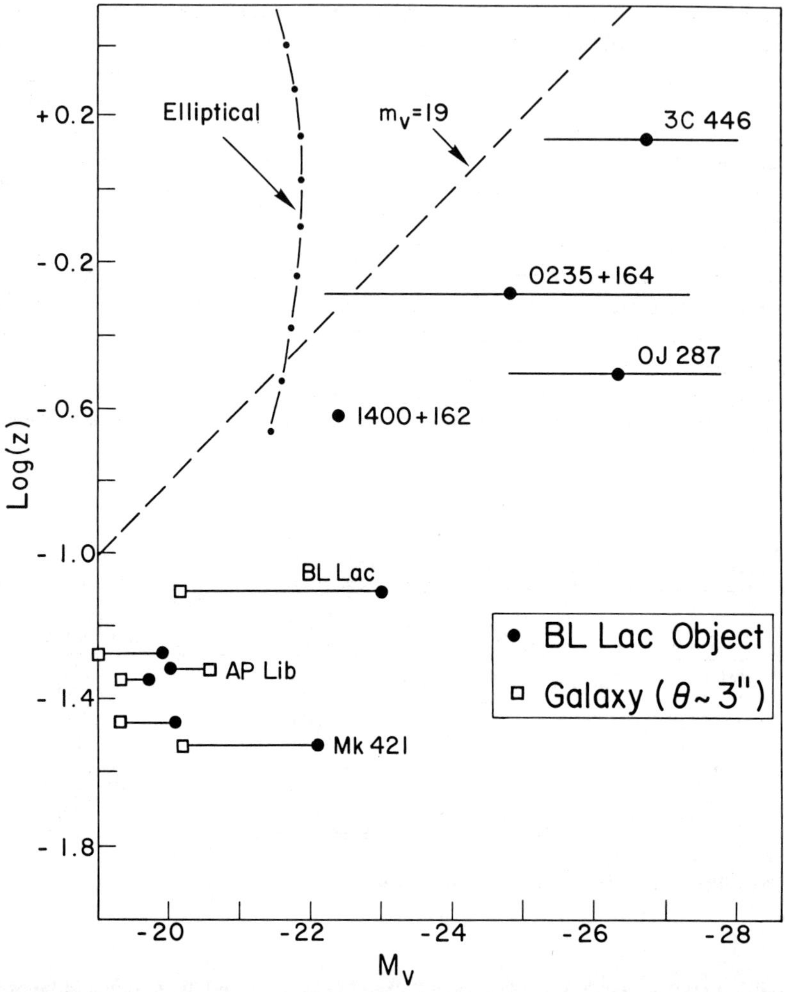

FIGURE 1. Redshift (z) vs absolute magnitude (M_v) of BL Lacs and associated galaxies. Galaxy magnitudes obtained from reference 19 where aperture sizes are given. Curve labeled "elliptical" is M_v of first-ranked giant elliptical galaxy observed through aperture with angular diameter, θ = 3″: results were obtained through use of Sandage's curve of growth.[27] Curve labeled m_v = 19 is line of that constant apparent visual magnitude. M_v for nonthermal component obtained from reference 19.

PROSPECTS

TABLE 2 compares BL Lacs, OVVs, and "average" QSOs with respect to properties usually assumed to define the BL Lac class. Even a crude division of properties such as this distinguishes BL Lacs and OVVs as a separate class of QSO. This is especially true with regard to optical polarization, the strength and duration of optical outbursts, and the mean optical spectral-index. Variability of optical polarization, which also distinguishes this subclass, is discussed by Dr. R. Angel elsewhere in this volume. The radio properties appear to be less fundamental, but could be important if systematic probes for radio-quiet BL Lacs are unsuccessful. While emission lines are no longer essential in defining BL Lac objects, they may be important in distinguishing nebulosities surrounding QSOs from those surroundong BL Lacs. Indeed, the failure to find first-ranked giant ellipticals around a sample of 15 low-luminosity QSOs[23] supports there being intrinsic differences between the two types of nebulosity.

Finally, there are a number of observations that may shed light on the problems discussed above, and some are listed as follows:

1. *Detection of nonthermal optical objects in elliptical galaxies with flat-spectrum radio sources:* Quite a few elliptical galaxies contain compact radio sources resembling those in BL Lacs.[24] If these are low-luminosity BL Lacs, then the optical or IR flux may be predicted from the radio flux by the $\nu^{-0.7}$ extrapolation discussed earlier. In the case of NGC 1052, for example, the predicted visual magnitude, $m_v \approx 16$, would be superposed on a galaxy contribution of $m_v \approx 13$–14 in a 3″ aperture. The best method to sort out the nonthermal contribution is to look for polarized optical or IR radiation. Detection of these components would aid in constructing a luminosity function for BL Lacs, and clarify the relationship between BL Lacs and galaxies.

2. *Detection of radio-quiet BL Lacs:* Discovery of radio-quiet BL Lacs would be very important as discussed above. It is not clear at present how this could best be done, but the technique of photographic polarimetry[25] appears promising. Another

TABLE 2

PROPERTIES OF QSOs, BL LACS, AND OVVs

Property	Class of Objects: "Average" QSO	BL Lac	OVV
Optical spectral index	$-1.6 \rightarrow 0.0$	$-2.0 \rightarrow -0.8$	$-2.0 \rightarrow -0.8$
Optical outburst			
Amplitude (ΔB)	$0.3 \rightarrow 0.5$ mag	$1.0 \rightarrow 5.0$ mag	$1.0 \rightarrow 5.0$ mag
Duration	few years	days to weeks	days to weeks
Optical polarization (p)	<3%	$5 \rightarrow 30\%$	$5 \rightarrow 25\%$
Radio source has flat spectrum at cm wavelengths	Significant fraction	most	most
Radio outburst			
Amplitude (ΔS/S)	≤30%	≤60%	≤60%
Duration	years	months	months to years
Radio quiet	~95%	none	none
Emission lines	Broad permitted, narrow forbidden	Narrow forbidden	Same as QSOs
Nebulosity	Ionized gas	Elliptical galaxy	Not observed

method now in use entails comparison of Schmidt plates taken at monthly intervals to search for optical outbursts.[26]

3. *Search for BL Lacs in sensitive cm surveys:* BL Lacs selected from existing high-frequency radio surveys tend to be significantly brighter in the optical than the plate limit of the Palomar Sky Survey. More sensitive radio surveys may result in the detection of optically faint BL Lacs which should help in estimating their true space density.

Summary

BL Lac objects are examined in the context of their relationship to QSOs, and to galaxies. BL Lacs and optically violent variable (OVV) QSOs are radio-selected objects that form a homogeneous subset of nonthermal sources that are exceedingly compact and well ordered. Optically-selected QSOs, however, show no evidence for non-thermal emission, indicating that they might be intrinsically different from the radio emitters. Consequently, radio-quiet BL Lacs may not exist.

Accurate spectroscopy reveals the presence of (i) narrow emission lines and (ii) stellar absorption lines in a large fraction of known BL Lacs. The emission lines indicate that BL Lacs contain large amounts ($M \sim 10^6\ M_\odot$) of low-density gas, contrary to previous assertions. The absorption lines arise in luminous elliptical galaxies in which the BL Lacs are embedded. However, there is no evidence that QSOs are embedded in galaxies.

Future observations of optical or IR polarization in elliptical galaxies with compact radio sources will help to elucidate the faint end of the BL Lac luminosity function. A search for BL Lacs in sensitive radio surveys at cm wavelengths should provide an unbiased estimate of their properties.

Acknowledgments

I wish to thank the participants of the Pittsburgh Conference on BL Lac objects for stimulating many of the ideas expressed in this paper, and F. H. Briggs for helpful comments.

References

1. Stein, W. A., S. L. O'Dell & P. A. Strittmatter. 1976. Ann. Rev. Astron. and Ap. **14:** 173.
2. Hazard, C. 1978. *In* Pittsburgh Conference on BL Lac Objects. A. M. Wolfe, Ed.: 395. University of Pittsburgh. Pittsburgh, Penn.
3. Condon, J. J. 1978. *In* Pittsburgh Conference on BL Lac Objects. A. M. Wolfe, Ed.: 21. University of Pittsburgh. Pittsburgh, Penn.
4. O'Dell, S. L. 1978. *In* Pittsburgh Conference on BL Lac Objects. A. M. Wolfe, Ed.: 312. University of Pittsburgh. Pittsburgh, Penn.
5. Kinman, T. D. 1978. *In* Pittsburgh Conference on BL Lac Objects. A. M. Wolfe, Ed.: 82. University of Pittsburgh. Pittsburgh, Penn.
6. Rieke, G. H., G. L. Grasdalen, T. D. Kinman, P. Hintzen, B. J. Wills & D. Wills. 1976. Nature. **260:** 754.

7. WARDLE, J. F. C. 1978. *In* Pittsburgh Conference on BL Lac Objects. A. M. Wolfe, Ed.: 39. University of Pittsburgh. Pittsburgh, Penn.
8. DENT, W. A. 1978. *In* Pittsburgh Conference on BL Lac Objects. A. M. Wolfe, Ed.: 66. University of Pittsburgh. Pittsburgh, Penn.
9. DENT, W. A. 1978. *In* Pittsburgh Conference on BL Lac Objects. A. M. Wolfe, Ed.: 407. University of Pittsburgh. Pittsburgh, Penn.
10. REES, M. J. 1978. *In* Pittsburgh Conference on BL Lac Objects. A. M. Wolfe, Ed.: 406. University of Pittsburgh. Pittsburgh, Penn.
11. LEDDEN, J. E., H. D. ALLER & W. A. DENT. 1976. Nature. **260:** 752.
12. POMPHREY, R. B., A. G. SMITH, R. J. LEACOCK, C. N. OLSSON, R. L. SCOTT, J. T. POLLOCK, P. EDWARDS & W. A. DENT. 1976. Astron. J. **81:** 489.
13. MILLER, J. S. & H. B. FRENCH. *In* Pittsburgh Conference on BL Lac Objects. A. M. Wolfe, Ed.: 228. University of Pittsburgh. Pittsburgh, Penn.
14. BRIGGS, F. H., R. L. BROWN, K. J. JOHNSTON & A. M. WOLFE. 1978. In preparation.
15. STOCKMAN, H. S. 1978. *In* Pittsburgh Conference on BL Lac Objects. A. M. Wolfe, Ed.: 149. University of Pittsburgh. Pittsburgh, Penn.
16. HAZARD, C. 1978. *In* Pittsburgh Conference on BL Lac Objects. A. M. Wolfe, Ed.: 393. University of Pittsburgh. Pittsburgh, Penn.
17. CONDON, J. J., J. E. LEDDEN, S. L. O'DELL & B. DENNISON. 1979. In preparation.
18. BLANDFORD, R. D. & M. J. REES 1978. *In* Pittsburgh Conference on BL Lac Objects. A. M. Wolfe, Ed.: 328. University of Pittsburgh. Pittsburgh, Penn.
19. MILLER, J. S., H. B. FRENCH & S. A. HAWLEY 1978. *In* Pittsburgh Conference on BL Lac Objects. A. M. Wolfe, Ed.: 176. University of Pittsburgh. Pittsburgh, Penn.
20. KRISTIAN, J. 1973. Astrophys. J Lett **179:** L61.
21. WAMPLER, E. J., L. B. ROBINSON, E. M. BURBIDGE & J. A. BALDWIN. 1975. Astrophys. J Lett. **198:** L49.
22. MILLER, J. S., H. B. FRENCH & S. A. HAWLEY. 1978. Astrophys. J. Lett. **219:** L85.
23. MILLER, J. S. 1978. *In* Pittsburgh Conference on BL Lac Objects. A. M. Wolfe, Ed.: 413. University of Pittsburgh. Pittsburgh, Penn.
24. EKERS, R. D. 1978. Physica. Scripta. **17:** 171.
25. CRAINE, E. R., R. DUERR & S. TAPIA. 1978. *In* Pittsburgh Conference on BL Lac Objects. A. M. Wolfe, Ed.: 99. University of Pittsburgh. Pittsburgh, Penn.
26. HAZARD, C. 1978. *In* Pittsburgh Conference on BL Lac Objects. A. M. Wolfe, Ed.: 390. University of Pittsburgh. Pittsburgh, Penn.
27. SANDAGE, A. 1972. Astrophys, J. **173:** 485.

EMISSION-LINE SPECTRA OF ACTIVE GALACTIC NUCLEI AND QUASARS*

Donald E. Osterbrock

Lick Observatory
Board of Studies in Astronomy and Astrophysics
University of California
Santa Cruz, California 95064

Active galactic nuclei and quasars, because of their large luminosities, apparently larger than can be accounted for by nuclear reactions in the stars involved, are generally considered subjects within the domain of relativistic astrophysics.[1] In the present paper I shall review the most recent observational results in the subject, particularly the work done at Lick Observatory with which I am most familiar. We have concentrated on a spectrophotometric survey of radio galaxies and Seyfert galaxies, which seem from their spectra to be closely related to quasars, but which from their forms clearly are galaxies. Hence, we are reasonably sure that the observed redshift is cosmological in origin, and gives the distance of the object and therefore (from the measured magnitude) its luminosity. I shall review this work, and also in less detail, similar measurements of the emission-line spectra of quasars, attempting to summarize the main observational results, and as much of the physical interpretation as we know. It must be realized that a full theoretical picture of these objects does not exist at the present time, although we do have some clues. My hope is that this paper will stimulate further thinking by theoretical astrophysicists and relativists, leading to further understanding of the active nuclei of galaxies and quasars.

A large fraction of radio galaxies (galaxies that are strong radio sources) have relatively strong emission lines in their optical spectra.[2] Very often the radio sources are large double lobes, symmetrically located far from the optical galaxy (to which all our spectra refer). From our survey at Lick Observatory, most emission-line radio galaxies are narrow-line radio galaxies[3] (NLRG), with lines having half-widths of order 500 km/sec, much wider than gaseous nebulae, significantly wider than most radio-quiet galaxies with emission lines, but much narrower than the other class of radio galaxies to be discussed later, broad-line radio galaxies (BLRG).

The archtypical NLRG is Cyg A, one of the first radio galaxies identified by Baade and Minkowski. A Lick Observatory scan of its spectrum[4] is shown in FIGURE 1. The strong emission lines of [O III], [N II], and H I are also strong in planetary nebulae and H II regions, and indicate the presence of ionized gas. However, Cyg A in particular and NLRG in general show a wider range of ionization in their spectra than planetary nebulae or H II regions, having strong [O I], [S II] and [Fe VII], ranging in many cases up to [Fe X] and even [Fe XI]. There is a strong featureless continuum in Cyg A that shows no absorption lines at all; it has an approximately power-law form with slope $\upsilon^{-3.3}$ (observed) or $\upsilon^{-1.6}$ (corrected for interstellar extinction of the same amount determined from the reddening of the H I lines). Most other NLRG have

*This research was supported in part by a grant from the National Science Foundation AST 76-18440.

0077–8923/80/0336–0022 $1.75/1 © 1980, NYAS

similar emission-line spectra, but usually the absorption lines of the underlying galaxy are visible.[5] An example with a quite strong stellar continuum, 3C 98, is shown in FIGURE 2. By comparing the depth of the absorption lines in the spectrum of a radio galaxy with the depth of the same lines in a "normal" galaxy without emission lines it is possible to decompose the continuous spectrum into an integrated stellar galaxy contribution and a featureless contribution.[6]

Seyfert galaxies are objects with spectra showing strong "broad" emission lines covering a wide range of ionization.[7] Thus their spectra are very similar to the optical spectra of radio galaxies. Most Seyfert galaxies are spirals, to the extent that a form classification can be given. On the basis of their spectra, they can be divided into two classes, and of these the Seyfert 2's are the objects with emission lines having widths similar to those in NLRG. The spectrum of Akn 347, a fairly typical member of this

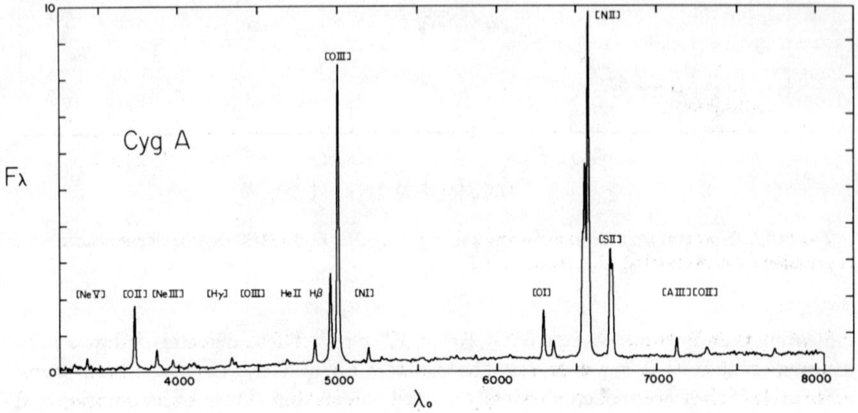

FIGURE 1. Spectral scan of narrow-line radio galaxy Cyg A, λλ3300–8100. This and all other scans in this paper show relative flux per unit wavelength interval versus wavelength in the rest system of the emitting object.

class, is shown in FIGURE 3. A detailed spectrophotometric study by Koski led to the conclusion that the optical spectra of Seyfert 2 galaxies and NLRG are indistinguishable.[6] The relative intensity ratios of the H I lines in these galaxies indicate extinction by dust, presumably within or near the objects themselves, since the amount of extinction is often considerably larger than can be accounted for by dust within our galaxy. The relative intensities of the [N II] and [O III] lines from different excited levels indicate electron temperatures $T \approx 1$ to 2 eV in many of these objects. This probably indicates that the energy input to the ionized gas occurs by photoionization by ultraviolet and soft x-ray radiation. The ionizing radiation cannot come from normal stars, because the range of ionization is so great, but qualitatively could come from the featureless continuum with an approximate power-law υ^{-n} with n approximately equal to 1.5. The electron density N_e in the ionized gas in many NLRG and Seyfert 2's is $\lesssim 10^4$ cm^{-3}. Others have smaller [O III] intensity ratios, probably

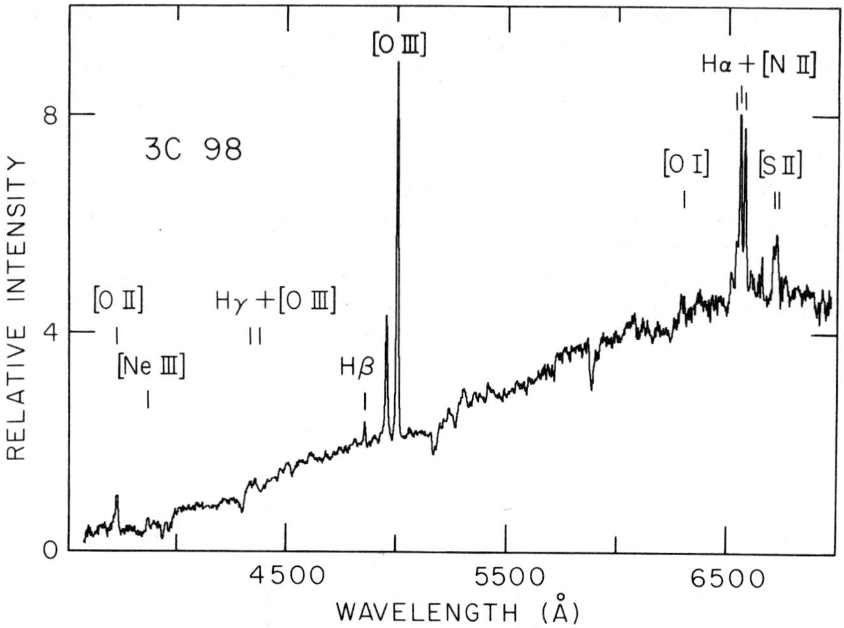

FIGURE 2. Spectral scan of narrow-line radio galaxy 3C 98, λλ3600–6900. (Reproduced with permission from Astrophys. J., reference 5.)

indicating that in these objects $N_e \approx 10^5$ or 10^6 cm^{-3}. These densities follow on the assumption that T is more or less the same in all of these objects, which seems reasonable if they are indeed physically related objects, but this is an assumption and not a rigorously proved conclusion. The abundances of the elements H, He, O, Ne, N, S, etc., are approximately normal. Photoionization models calculated on this basis fit qualitatively the mean relative intensities of Seyfert 2's and NLRG. The observed spectrum of the Crab Nebula, known to be photoionized by a power-law spectrum, fits them better, and a combination of the Crab Nebula plus a high-ionization planetary nebula fits best of all.[6]

A minority of the radio galaxies with emission lines in their spectra, about one-third of the total, have very broad H I emission lines. The H I profiles are irregular, are different in different galaxies, and in some cases are different from one H I line to another in the same galaxy.[8] These are the objects we call BLRG, spectra of four of which are shown in FIGURE 4. Note that the full widths at zero-intensity of the broad H I lines range up to nearly 0.1 c. Only the H I, He I, and He II lines have broad profiles; all the forbidden lines have narrow widths and are much the same as the forbidden lines in NLRG. In form most of the BLRG are N galaxies, objects with a bright nearly stellar nucleus containing most of the light of the system, thus nearly quasars, but with a small amount of faint low-surface-brightness nebulosity around them.[9] The BLRG often have extremely large H I decrements. In several, including two of the galaxies in FIGURE 4, absorption lines of the integrated stellar galaxy continuum can be seen, but on the average the featureless continuum is considerably stronger in BLRG than in NLRG.

The Seyfert galaxies with spectra most nearly similar to BLRG are Seyfert 1 galaxies, defined as having broad H I, He I, and He II lines and much narrower forbidden lines in their spectra. An example, Akn 120 is shown in FIGURE 5. These Seyfert 1 galaxies often have strong permitted Fe II emission, which is not observed in ordinary planetary nebulae and H II regions. There is a wide range of intensity of Fe II emission in otherwise similar Seyfert 1 galaxies. The Fe II emission features are blends of many lines, so their individual line profiles cannot be determined accurately, but they are clearly broad and have approximately the same profiles as the H I lines, in contrast to the narrow forbidden lines in these galaxies.

The absence of any forbidden lines in the broad-line region shows that the electron

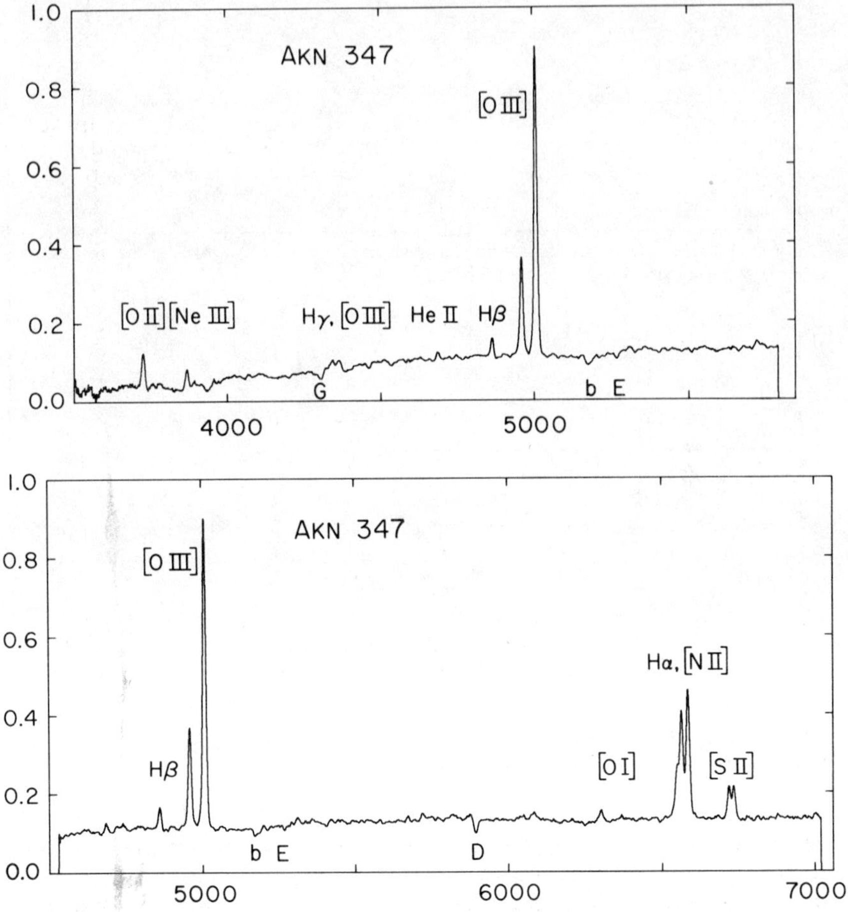

FIGURE 3. Spectral scans of Seyfert 2 galaxy Akn 347, λλ3500–5800 and λλ4600–7000.

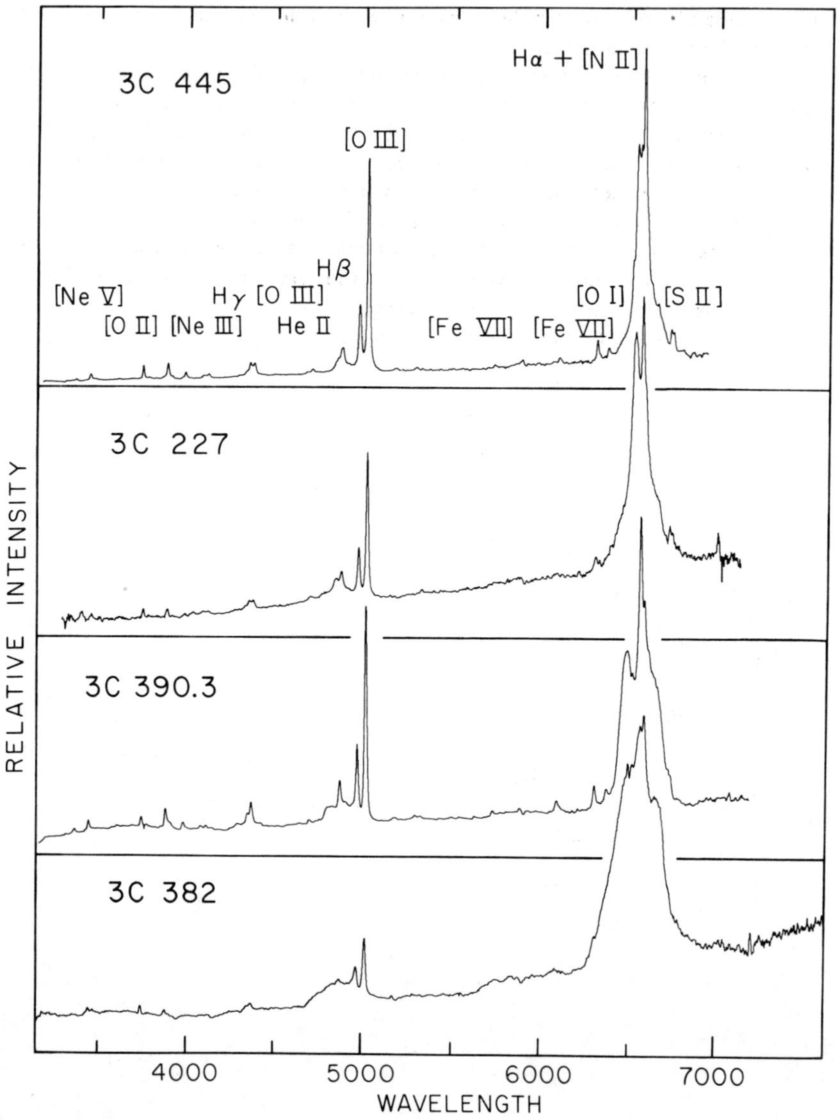

FIGURE 4. Spectral scans of four broad-line radio galaxies as indicated, λλ3500–7000. (Reproduced with permission from Astrophys. J., reference 8).

density must be sufficiently high in this region so that all metastable levels are chiefly collisionally deexcited. This sets a lower limit to the electron density at 10^8 or 10^9 cm^{-3}. The temperature is not known, because only the H and He lines are observed, but is probably of order 1 eV, perhaps slightly lower. From the fact that H and He lines are observed it also appears that the abundances, at least of the most abundant elements, are not strikingly different from the abundances in normal stars and nebulae. In some cases the ionization is quite high in the narrow-line regions of Seyfert 1 galaxies. An outstanding example is III Zw 77, shown in FIGURE 6, which has quite strong [Fe X] and [Fe XI] in its spectrum.

The H I intensity ratios in Seyfert 1 galaxies and BLRG do not agree with those expected from recombination plus reddening by interstellar extinction alone. This indicates that other physical processes must also be effective in the line-emission

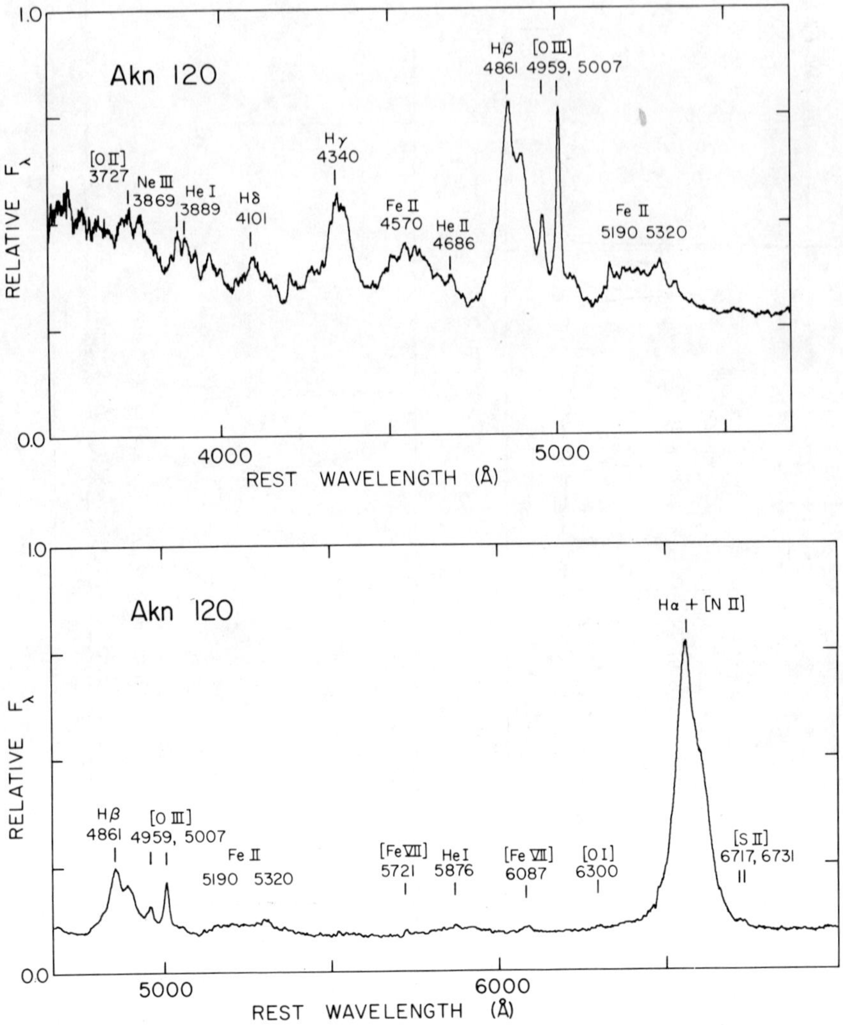

FIGURE 5. Spectral scans of Seyfert 1 galaxy Akn 120, λλ3500–5700 and λλ4700–7000. (Reproduced with permission from D. E. Osterbrock & M. M. Phillips, Publ. Astron. Soc. Pac. **89:** 251, 1977.)

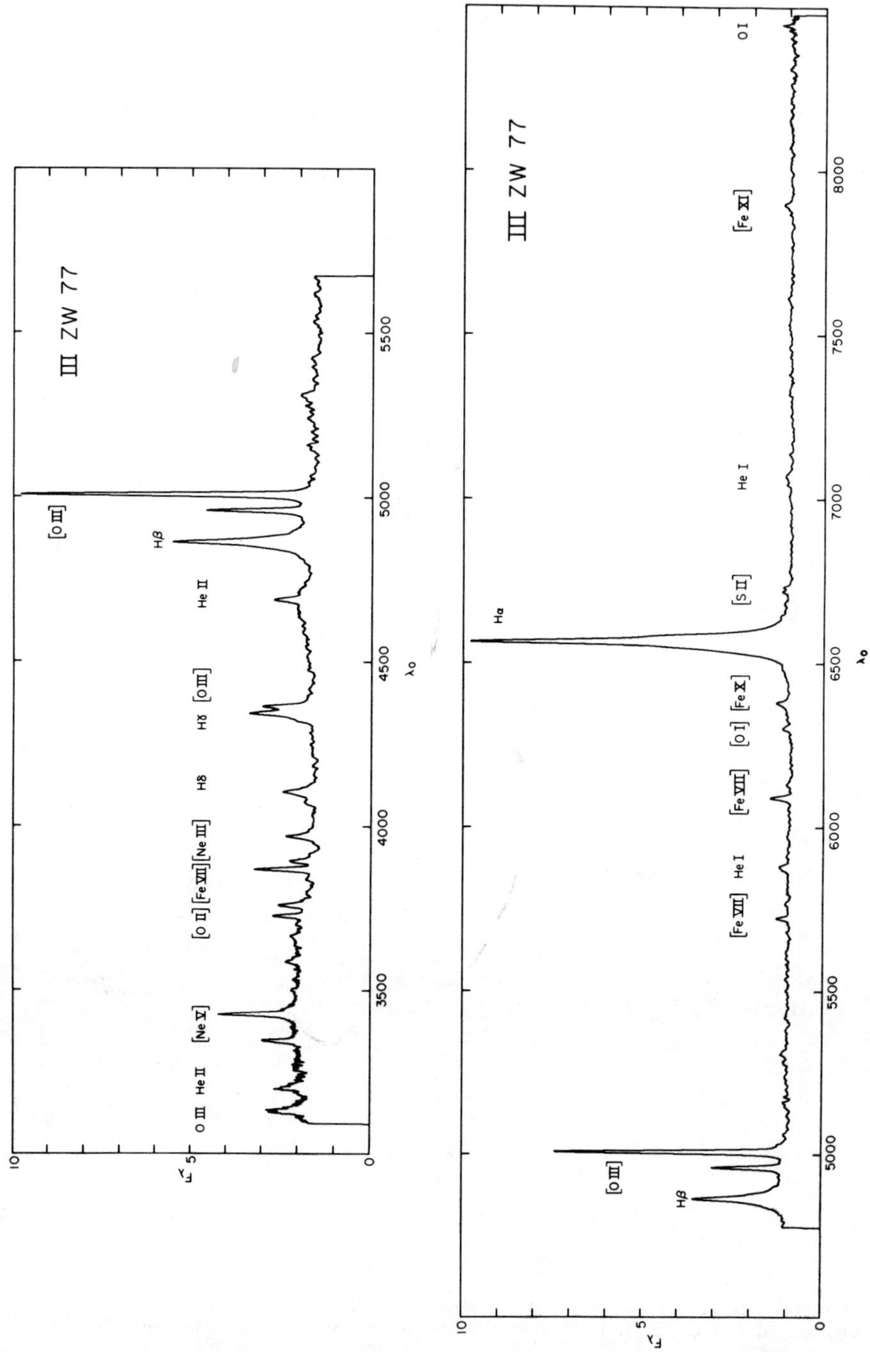

FIGURE 6. Spectral scans of Seyfert 1 galaxy III Zw 77, λλ3150–5700 and λλ4800–8500.

process.[8,11] Because of the high density, these processes are probably self-absorption and collisional excitation from the 2 ^{2}S and 2 ^{2}P levels of H I.

TABLE 1 summarizes the similarities and differences among Seyfert galaxies and radio galaxies. The number of Seyfert galaxies is small (apparently in the volume of space around our galaxy about 1 percent of all spirals are Seyfert galaxies) but nevertheless large in comparison with the number of radio galaxies, which are rare objects indeed in space.

Furthermore, Seyfert 1 galaxies and BLRG are not identical in spectra, although they are quite similar. This is shown in FIGURE 7, where it can be seen that on the average BLRG have large $H\alpha/H\beta$ ratios while Seyfert 1 galaxies have considerably stronger Fe II emission. These optical characteristics must be connected with the physical structure of the objects. They thus present a clue to the conditions under which radio emission does or does not occur in connection with active galactic nuclei, although this clue is not completely understood.[9]

There are Seyfert galaxies intermediate between types 1 and 2, objects whose H I profiles combine a broad-line profile with a narrow-line profile. An example is NGC

TABLE 1

RADIO- AND SEYFERT-GALAXY COMPARISON

Optical Emission Lines	Seyfert Galaxies (radio quiet)	Radio Galaxies
Broad H I, He I, He II	Seyfert 1 ~ 2/3	BLRG ~ 1/3
	Fe II strong	Fe II weak
Narrow forbidden lines	S	N
Narrow H I, He I, He II	Seyfert 1 ~ 1/3	BLRG ~ 2/3
	Spectra indistinguishable	
Narrow forbidden lines	S	cD, D, E

985, shown in FIGURE 8, which might be called a Seyfert 1.5 galaxy. In fact there is a continuous sequence of objects with all possible relative amounts of broad- and narrow-line profiles, indicating that in these galaxies all possible relative amounts of dense gas with a large velocity distribution, and lower-density gas with a considerably lower velocity distribution, can exist.[12] The BLRG are more like these intermediate Seyfert 1.5 galaxies, having strong narrow components of their H I lines and strong forbidden lines, than they are like typical Seyfert 1 galaxies. Thus, more radio galaxies are NLRG, and those that are not NLRG still have strong narrow-line components.

One further observational result, due to Sramek and Tovmassian,[13] is that a certain number of Seyfert galaxies are weak radio emitters, much fainter in the radio region than typical radio galaxies. The Seyfert galaxies that are detectable radio sources nearly all are Seyfert 2 galaxies; the ones that are not Seyfert 2's are mostly Seyfert 1.5 objects, and almost none of them are typical Seyfert 1 galaxies.

Low-redshift quasars have in their spectra the same emission lines that are observed in Seyfert and radio galaxies.[14] All the well-observed low-redshift quasars have broad H I emission lines in their spectra; no quasars are known with spectra like

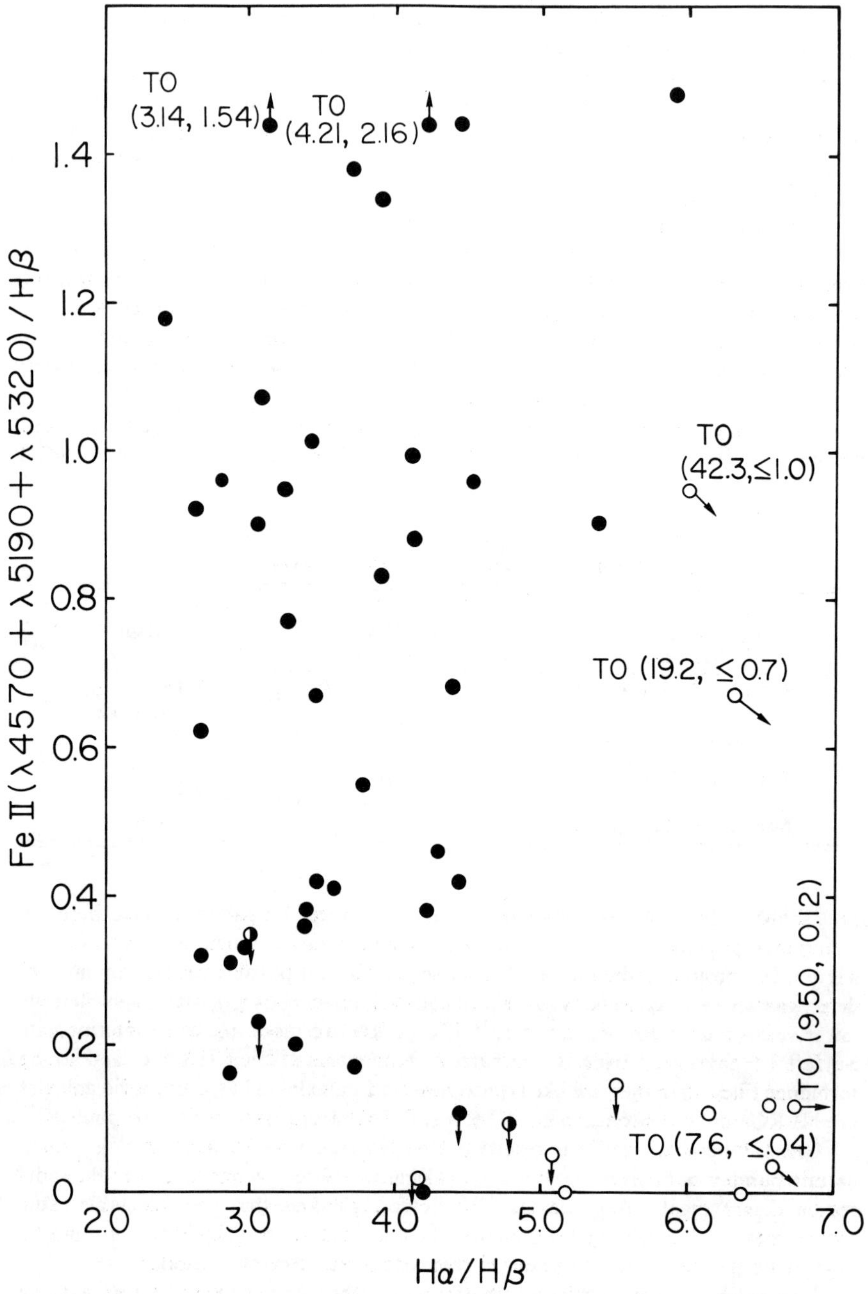

FIGURE 7. Measured Fe II/Hβ vs. Hα/Hβ intensity ratios for Seyfert 1 galaxies (solid circles), broad-line radio galaxies (open circles), and two radio-quiet N galaxies (half-filled circles). (Reproduced with permission from Astrophys. J., reference 9.)

NLRG or Seyfert 2 galaxies. Most quasars do not have Fe II emission in their spectra and thus are more similar to BLRG than to Seyfert 1 galaxies, although a few quasars do have Fe II emission. Spectra of several quasars, taken by Baldwin at Lick Observatory with the same scanner as the Seyfert- and radio-galaxy spectra of this paper, are shown in FIGURE 9.

Quasars with larger redshifts provide the opportunity for us to observe their ultraviolet spectra from earth.[15] They generally have broad emission lines of C IV λ1549, C III] λ1909, O III], N III], N IV], N V, O VI, and so on, in other words all the lines expected from the highly ionized nebulae with $T \approx 1$ to 2 eV as deduced from the optical spectra of low-redshift BLRG, Seyfert 1 galaxies, and quasars. The presence of C III] λ1909 in nearly all quasars in which this spectral region can be observed, and the absence of C III λ977, show that the C^{++} 2 ^3P level is not greatly collisionally deexcited, and thus that the electron density is $\lesssim 10^{10}$ cm^{-3}. To the extent that the

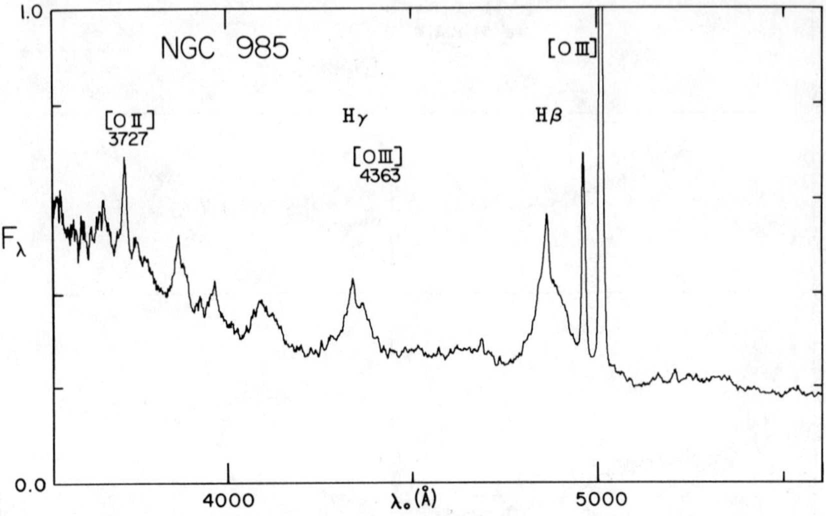

FIGURE 8. Spectral scan of intermediate Seyfert 1.5-type galaxy NGC 985.

broad-line regions of BLRG and Seyfert 1 galaxies are the same as quasars this electron density limit presumably applies to them. Very recently C III] λ1909 has been observed in the Seyfert 1 galaxy NCC 4151 from the IUE satellite above the atmosphere, confirming that at least this one Seyfert 1 galaxy is like the quasars in this respect.[16] This observation very clearly emphasizes the importance of ultraviolet observations of active nuclei of galaxies from space.

In quasars with sufficiently large redshift, H I Lα is observed in emission in every case. Photometry over a long baseline linking the optical and ultraviolet spectral regions is difficult, but the results to date by Baldwin,[17] Wu,[18] and Davidsen, Hartig and Fastie[19] give line ratios of order $L\alpha/H\beta \approx 3$–6. This is much smaller than the value of order 25–40 predicted by pure recombination spectra. Undoubtedly this low ratio results from the effects of dust and collisional excitation from the 2 ^2S and 2 ^2P

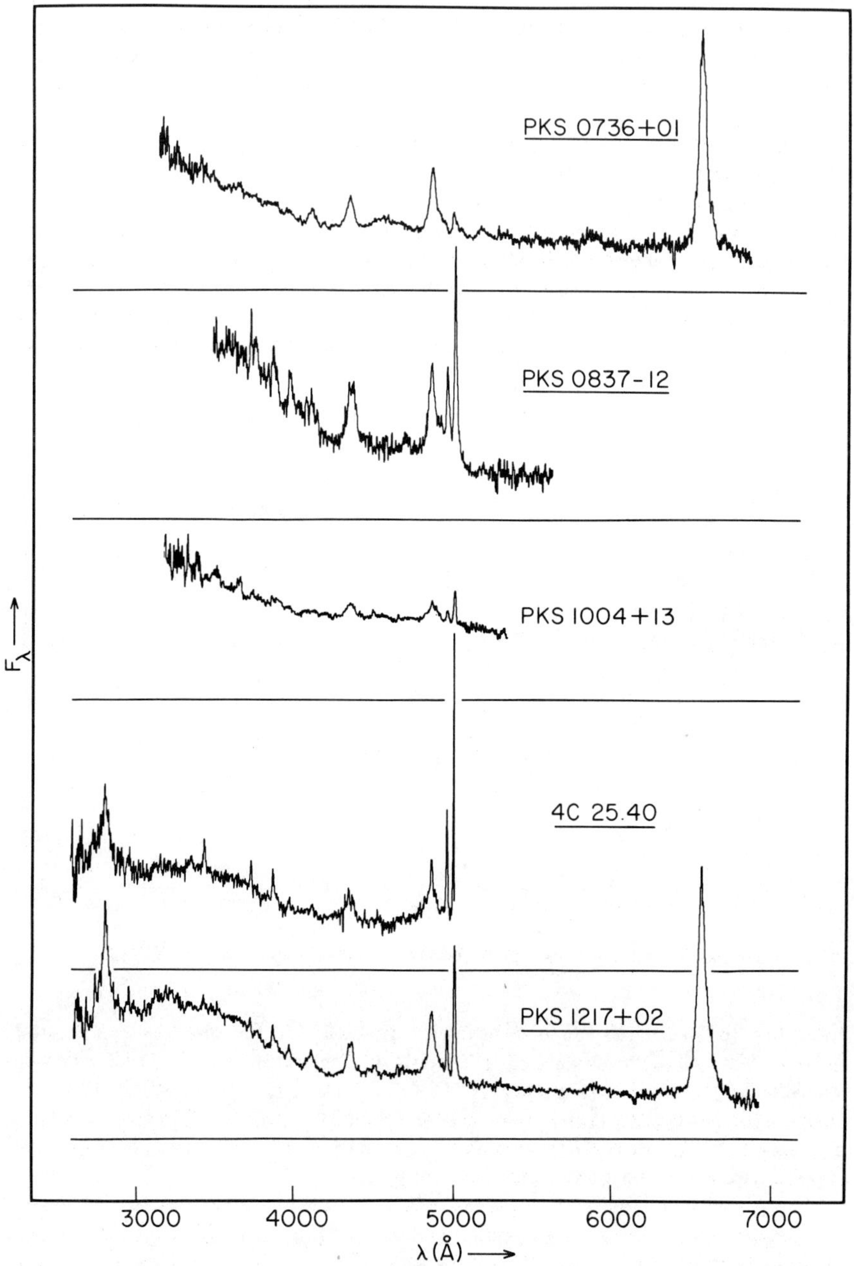

FIGURE 9. Spectral scans of five low-redshift quasars, as indicated, λλ2700–7000. (Reproduced with permission from Astrophys. J., reference 14.)

levels of H^0, both of which have been deduced to be important in Seyfert galaxies and BLRG from analyses of optical spectra, as described above. Further observational data of this kind, not only for H I but also for He II λλ1640,3203, which should not be so greatly affected by collisional excitation and self-absorption, plus further theoretical analysis will undoubtedly give more information on the density and structure of the ionized gas clouds in quasars. There is some suggestion that the ratio $L\alpha/H\beta$ is more or less the same in all quasars.[20] The observations are still at an early stage and it would be wise to confirm this result observationally before erecting too great a theoretical structure on it. If it is observationally confirmed it may indicate a general property of the ionization structure of clouds in motion in quasars.

Finally I should like to discuss an overall physical model, based on the spectra of Seyfert and radio galaxies, but presumably applying also to quasars, to the extent that they are similar to the other objects. Photoionization seems clearly to fit all the observational data on the narrow lines, and it seems most likely that the input of energy to the narrow-line gas is from a source emitting a featureless continuum, approximately power-law in form, extending to high energies, at least 200–300 eV.[21] Qualitatively this source can be understood as resulting from a rotating accretion disk about a black hole,[22] and the luminosity required fits approximately with a central mass $\approx 10^7$ or 10^8 $M_\odot$. The dimensions of the narrow-line region, judged from the density and from marginal space resolution in a few of the very nearby objects, is of order 10^3 pc.

The next question is what the energy input mechanism is to the broad-line dense gas. From its density and in some cases light variation, its typical dimension is of order 0.1 pc and velocities range up to 0.05 c, although 5×10^3 km/sec half-width would be more typical. Mechanical input of energy[23] (dissipation of shock waves, collisions between clouds, etc.), slowing down of keV ions, chiefly protons,[24] and photoionization by the same featureless continuum that photoionizes the narrow-line gas have all been suggested as possibilities. Some recent observational evidence seems to favor the photoionization picture.[25]

If photoionization is the energy input mechanism and if the photoionized region absorbs all the available photons, then it is easy to see that there is a simple relationship between the strength of an H I recombination line, say Hβ, and the total number of photoionizing photons. Put another way, this means that the equivalent width of the Hβ emission line depends upon the form of the ionizing continuum and the fraction of available photons that are absorbed, as pointed out by Searle and Sargent.[26] To test this idea observationally, the distribution functions of the equivalent widths of the narrow-line components of the emission lines in Seyfert 1, Seyfert 1.5 (objects with a composite profile containing both broad and narrow parts), and Seyfert 2 galaxies have been assembled in FIGURE 10. Note that the statistics actually pertain to [O III], but as the ratio of narrow-line Hβ/[O III] appears to be nearly the same in all these objects, ≈0.1, this graph is, except for a scale factor, very similar to the distribution function of narrow-line components of Hβ. These distribution functions show that the equivalent widths of the narrow-line components of Hβ are on the average considerably weaker in the Seyfert 1 and the Seyfert 1.5 galaxies, the objects with dense gas in their nuclei, than in the Seyfert 2 galaxies, which do not have dense gas in their nuclei. Thus the presence of the dense broad-line gas apparently tends to inhibit or prevent the photoionization of the narrow-line gas by radiation in the

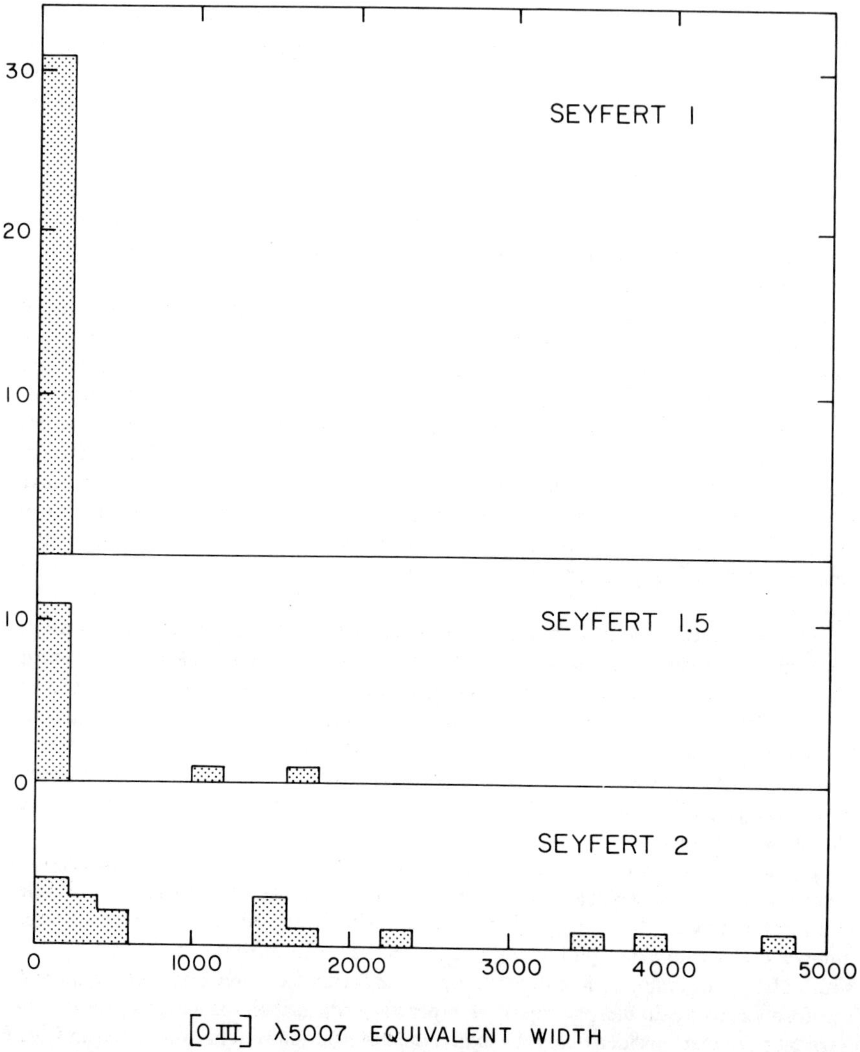

FIGURE 10. Distribution functions of [O III] λ5007 emission-line equivalent widths expressed in terms of featureless continuum for Seyfert 1, 1.5, and 2 galaxies separately.

featureless continuum. It seems reasonable to imagine that this inhibition actually occurs through absorption of the ionizing photons in the dense gas itself. To test this hypothesis the statistics of the total equivalent widths of Hβ, broad and narrow lines together, have also been assembled and shown in FIGURE 11. Here it can be seen that the total Hβ equivalent width has much more nearly similar distribution functions in the three types of galaxies, suggesting that photoionization is indeed the energy input mechanism to the broad-line gas as well as to the narrow-line gas.[25]

Such photoionization models have been calculated for quasars by MacAlpine,[27] Shields,[28] and Netzer,[29] and seem to fit their broad-line spectra, including the ultraviolet lines of C III], C IV, etc. A problem, emphasized by Baldwin and Netzer,[30] is that the line profiles of different stages of ionization appear to be more nearly similar than would be expected, for ions such as C^{++} with lower ionization potentials would be expected on the average to be further from the source of ionizing photons than C^{+++}, a higher-ionization-potential ion, and thus to have different velocity profiles. Further work on this problem would help in understanding the structure of quasars.

A geometrical model that reproduces the main observed features of active nuclei of galaxies is shown in FIGURE 12. The most natural assumption to explain the large internal velocities shown by the line widths is that they are due to a combination of turbulence and rotation. This is an assumption, not a unique deduction from the data, but it seems difficult to understand how the material making up the dense gas could have exactly zero angular momentum, and at the small dimensions of the dense gas any angular momentum implies a large rotational velocity. Furthermore, the statistics of line widths can be fitted by an assumed combination of turbulence and rotation.[11] Whatever photons are not absorbed in the dense gas penetrate through to the outer lower-density narrow-line gas and photoionize it. On this model a galaxy with a large amount of dense broad-line gas would be an almost pure Seyfert 2. A galaxy with an intermediate amount of dense gas is shown in the figure; in this situation ionizing photons escape from the broad-line disk near the poles but not at the equator. This object would then be observed as a Seyfert 1.5 galaxy with a combined broad- and narrow-line profile. Within each density regime, the highest ionization occurs closest

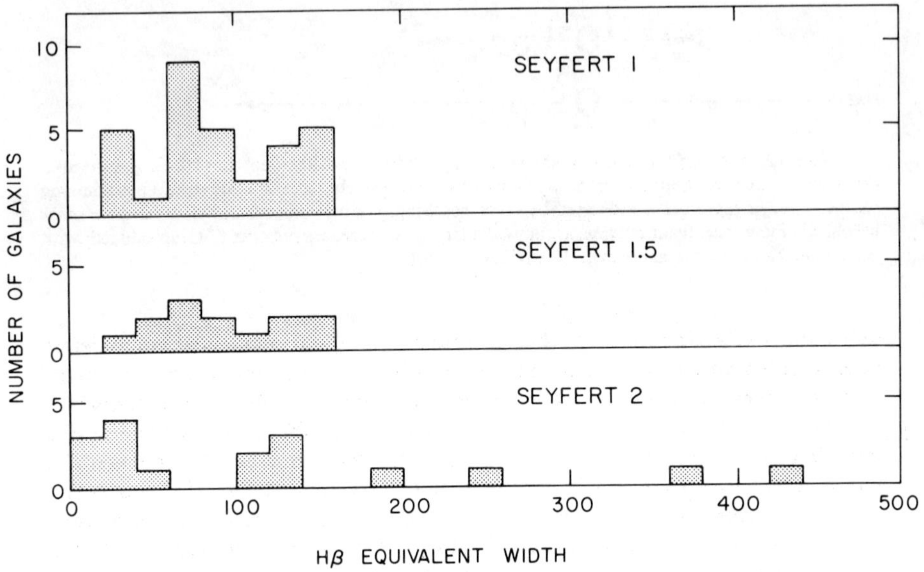

FIGURE 11. Distribution functions of Hβ total (broad + narrow components) emission-line equivalent widths expressed in terms of featureless continuum for Seyfert 1, 1.5, and 2 galaxies separately.

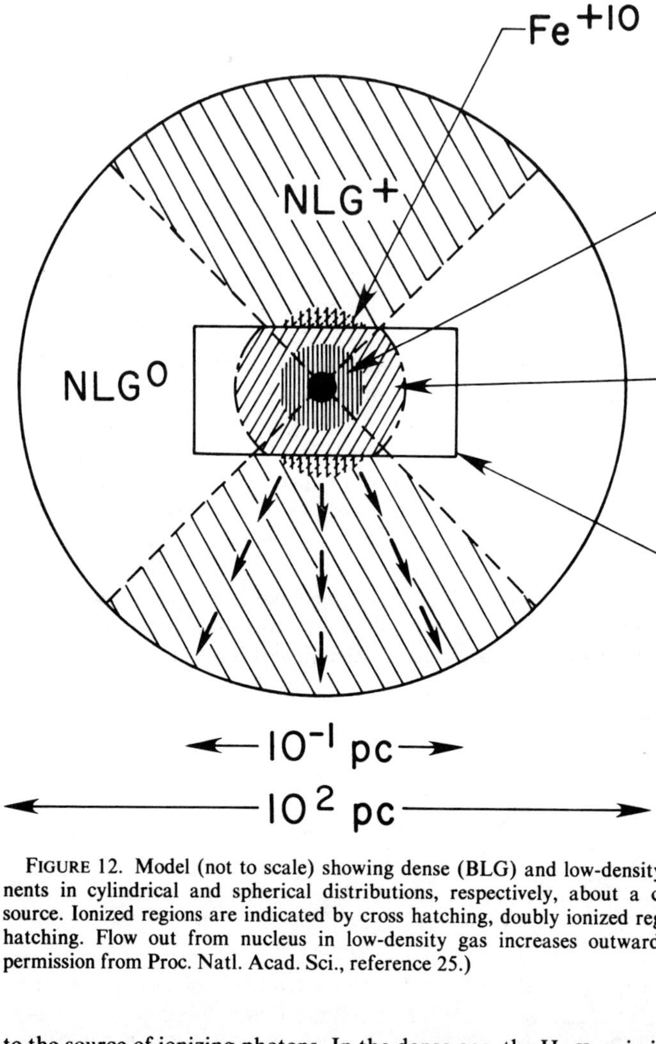

FIGURE 12. Model (not to scale) showing dense (BLG) and low-density (NLG) gas components in cylindrical and spherical distributions, respectively, about a central photoionizing source. Ionized regions are indicated by cross hatching, doubly ionized regions by double cross hatching. Flow out from nucleus in low-density gas increases outward. (Reproduced with permission from Proc. Natl. Acad. Sci., reference 25.)

to the source of ionizing photons. In the dense gas, the He II emission would therefore be expected to arise closer to the center than He I or H I, and therefore to have on the average higher rotational velocity and hence a broader profile, a deduction which seems to be confirmed by the scanty observational data. Also in the narrow-line gas, the [Fe X] and [Fe XI] lines would be expected to occur closest to the photoionizing source. These lines are observed to have small but detectable blue shifts, presumably indicating currents or flow velocities in the inner part of the narrow-line region away from the source of photoionization, which, together with the presence of dust absorbing radiation from the further side of the narrow-line region, could explain this observation.

Furthermore, the correlation between radio emission and the strength of the narrow optical components of the emission lines can perhaps be understood on this model. The narrow lines are strong if ionizing photons can penetrate through the dense gas near the nucleus, particularly at the poles. Radio emission occurs when radio plasma, generated in the central source, can escape, presumably along the axis of rotation, and get far enough away from the galaxy to form radio emission lobes. Thus, according to the model, escape of ionizing photons and escape of relativistic plasma both occur in the direction of the poles in those galaxies in which the amount of dense gas in these directions is small. This model is by no means proved, but it does agree qualitatively with a large amount of observational evidence and should be tested against further observational evidence.

Within the past few years x-ray observations have shown that nearly every bright Seyfert 1 galaxy is an x-ray source.[31,32] The x-rays seem to be an extension of the featureless continuum into the keV region. They presumably are emitted by the central source, and there must be channels or tunnels through the dense gas, which must be clumped into condensations, since the x-rays could not escape through a homogeneous medium of the size and density required by the optical observations. The observed strengths of the highest-ionization optical lines observed, [Fe X] and [Fe XI], are not particularly well correlated with the observed strength of the x-rays, but the data are still scanty.

No Seyfert 2 galaxies have been detected as x-ray sources, but if the x-ray luminosity is proportional to the featureless continuum luminosity, none of them would be expected to be detected (except possibly NGC 1068) because of the relative weakness of the featureless continuum in Seyfert 2's. NGC 1068, the only bright classical Seyfert 2 galaxy, is a somewhat peculiar object with a large amount of observed dust[33] and is not representative of typical objects of this class. If the x-ray measurements can be made just a little more sensitive, several more Seyfert 2 galaxies should become detectable. These measurements would thus provide a good test of the idea that the central source in Seyfert 2 galaxies is basically similar to the central source in Seyfert 1 galaxies.

Dust is probably present in the dense regions of Seyfert 1 galaxies and BLRG.[21] Its presence cannot be established with certainty from the $H\alpha/H\beta$ ratios alone because of the competing effects of collisional excitation and self-absorption. Nearly all Seyfert galaxies have infrared excesses, and the infrared spectra of Seyfert 2 galaxies show that their excesses are clearly due to dust.[34] In Seyfert 1's the featureless continuum itself contributes at least part of the excess, and it is not clear whether or not dust is also present.[35] The fragmentary data I have been able to collect on the correlation between $H\alpha/H\beta$ ratios and infrared excesses is not conclusive, but at this conference very recent results by Rieke show there is a correlation of the type expected from absorption of photons by dust.

Acknowledgments

I am particularly grateful to my colleagues G. R. Blumenthal, W. G. Mathews, and J. S. Miller, and also to all the participants in the UCSC Summer Workshop in 1978 on Quasars and Active Galactic Nuclei, for many interesting and productive discussions of the subjects reviewed in this paper.

REFERENCES

1. Quasi-stellar sources and gravitational collapse. Proc. 1st Texas Conference. I. ROBINSON, A. SCHILD & E. L. SCHUCKING, Eds. University of Chicago Press. Chicago, Ill.
2. SMITH H. E., H. SPINRAD & E. O. SMITH. 1976. Publ. Astron. Soc. Pac. **88:** 621.
3. OSTERBROCK, D. E. 1977. *In* Radio Astronomy and Cosmology. D. L. Jauncey, Ed. D. Reidel Pub. Co. Dordrecht, Holland.
4. OSTERBROCK, D. E. & J. S. MILLER. 1975. Astrophys. J. **197:** 535.
5. COSTERO, R. & D. E. OSTERBROCK. 1977. Astrophys. J. **211:** 675.
6. KOSKI, A. T. 1978. Astrophys. J **223:** 56.
7. WEEDMAN, D. W. 1977. Annu. Rev. Astron. Astrophys. **15:** 69.
8. OSTERBROCK, D. E., A. T. KOSKI & M. M. PHILLIPS. 1976. Astrophys. J. **206:** 898.
9. GRANDI, S. A. & D. E. OSTERBROCK. 1978. Astrophys. J. **220:** 783.
10. PHILLIPS, M. M. 1978. Astrophys. J. Supp. **38:** 187.
11. OSTERBROCK, D. E. 1977. Astrophys. J. **215:** 773.
12. OSTERBROCK, D. E. & A. T. KOSKI. 1976. Mon. Not. R. Astron. Soc. **176:** 61p.
13. SRAMEK, R. A. & H. M. TOVMASSIAN. 1975. Astrophys. J. **196:** 339.
14. BALDWIN, J. A. 1975. Astrophys. J. **201:** 26.
15. STRITTMATTER, P. A. & R. E. WILLIAMS. 1976. Annu. Rev. Astron. Astrophys. **14:** 307.
16. BOKSENBERG, A. *et. al.* 1978. Nature **275:** 34.
17. BALDWIN, J. A. 1977. Mon. Not. R. Astron. Soc. **178:** 67p.
18. WU C.-C. 1977. Astrophys. J. Lett. **217:** L117.
19. DAVIDSEN, A. F., G. HARTIG & W. FASTIE. 1977. Nature **275:** 404.
20. ZIRIN, H. 1978. Astrophys. J. **222:** L105.
21. OSTERBROCK, D. E. 1978. Physica Scripta **17:** 285.
22. REES, M. J. 1977. Ann. N. Y. Acad. Sci. **302:** 613.
23. DALTABUIT, E. & D. COX. 1972. Astrophys. J. Lett. **173:** L13.
24. PTAK, R. & R. E. STONER. 1973. Astrophys. J. Lett. **179:** L89.
25. OSTERBROCK, D. E. 1978. Proc. Natl. Acad. Sci. USA. **75:** 540.
26. SEARLE, L. & W. L. W. SARGENT. 1968. Astrophys. J. **153:** 1003.
27. MAC ALPINE, G. M. 1971. Astrophys. J. **175:** 11.
28. SHIELDS, G. A. 1976. Astrophys. J. **204:** 330.
29. NETZER, H. 1976. Mon. Not. R. Astron. Soc. **117:** 473.
30. BALDWIN, J. A. & H. NETZER. 1978. Astrophys. J. **213:** 1.
31. ELVIS, M., T. MACCACARO, A. S. WILSON, M. J. WARD, M. V. PENSTON, R. A. E. FOSBURY & G. C. PEROLA. 1978. Mon. Not. R. Astron. Soc. **183:** 129.
32. TANNANBAUM, H., G. PETERS, W. FORMAN, R. GIACCONI, C. JONES & Y. AVNI. 1978. Astrophys. J. **223:** 74.
33. HILDEBRAND, R., S. E. WHITCOMB, R. WINSTON, R. F. STIENING, D. A. HARPER & S. H. MOSELEY. 1977. Astrophys. J. **216:** 698.
34. ALLEN, D. A. 1976. Astrophys. J. **207:** 367.
35. STEIN, W. A. & D. W. WEEDMAN. 1976. Astrophys. J. **205:** 44.

ABSORPTION-LINE SPECTRA OF QUASI-STELLAR OBJECTS: PHYSICAL PROPERTIES OF ABSORBING GAS*

E. Margaret Burbidge

Department of Physics
University of California, San Diego
La Jolla, California 92093

Historical Background

Absorption lines were discovered in the spectrum of the strong radio-emitting QSO 3C 191 in 1966 (Burbidge, Lynds, and Burbidge, 1966; Stockton and Lynds, 1966). Prior to this, emission lines in QSOs had been studied in some detail and redshifts up to slightly larger than $z = 2$ had been found. The absorption lines in 3C 191 were easily identified on low-resolution spectra with resonance transitions of H, C, N, O, Mg, Si at z_{abs} slightly less than z_{em}; the lines are much narrower than the emission lines and, despite the presence of C IV and N V, are obviously produced in less hot gas in front of the energy source and emission-line region.

The second QSO (also a radio source) in which absorption lines were found is PKS 0237 − 23, and in this object the identification of the very large number of absorption lines proved difficult. The answer (Burbidge, Lynds, and Stockton, 1968; Bahcall, Greenstein, and Sargent, 1968) was that the lines are due to familiar transitions in the common elements at many different redshifts, some very much less than z_{em}.

In the 10 years since the deciphering of the spectra of PKS 0237 − 23, many more QSOs with absorption lines have been discovered and there is a large body of literature on the subject. A review with extensive bibliography has been published by Perry, Burbidge, and Burbidge (1978). As of December 1978, absorption lines have been definitely seen and confirmed in 60 QSOs out of some 800 that have been studied, and are tentatively noted in a further 60 objects. Hydrogen 21-cm absorption has been detected in a few QSOs and BL Lac objects, but I shall concern myself with the optical absorption lines.

TABLE 1 lists the principal absorption features that have been identified. It should be remembered that the atmospheric cutoff from $\lambda < 3000$ Å means that, depending on the redshifts, a selection of these lines are available from ground for $z > 0.2$ up to the (so far) largest known z_{em}, 3.53, while the International Ultraviolet Explorer and the forthcoming Space Telescope extend the range for low redshifts. For $z = 2$, Lyα is at 3648 Å, and for $z = 3.5$, at 5472 Å.

The absorption lines, which in most (but not all) cases have $z_{abs} < z_{em}$, may arise in gas clouds or filaments ejected from the QSO, or in intergalactic clouds, galactic halos, or the disks of spiral galaxies lying between us and the QSO. Much has been written on the controversial question of which of these origins predominates; if the QSOs are at the cosmological distances given by their redshifts one would expect some

*Extragalactic research at UCSD is supported by the NSF under grant AST 77-22560 and by the NASA under grant NSG 7377.

0077-8923/80/0336-0039 $1.75/1 © 1980, NYAS

intervening galaxies (Wagoner, 1967), but not the very large number of different z_{abs} found in PKS 0237 − 23, for example, nor the patchy distribution of absorption redshifts in which some QSOs have few or even no well defined sets of absorption lines, while others have more than a dozen in addition to hundreds of lines shortward of Lyα emission. It is now generally accepted that the latter appear to be Lyα absorption in cloudlets with insufficient optical depth to show any of the heavier elements. Burbidge *et al.* (1977) showed that in order to account for the absorptions in PKS 0237 − 23 by intervening galaxies, it would be required that such galaxies are surrounded by gaseous halos up to 0.5 Mpc in diameter, to provide the necessary cross sections. On the other hand, for the large values found for $\Delta z = z_{em} - z_{abs}$, giving velocity differences

$$\frac{v}{c} = \beta = \frac{(1 + z_{em})^2 - (1 + z_{abs})^2}{(1 + z_{em})^2 + (1 + z_{abs})^2},$$

we find that values of β up to 0.5 or 0.6 are sometimes found; in a large number of cases $\beta \approx 0.1$. Ejection at speeds of 1/10 to 1/2 the velocity of light creates stringent demands on theories of possible ejection processes (explosion, radiation pressure, particle pressure).

TABLE 1

PRINCIPAL LINES SEEN IN ABSORPTION

Line	Rest λ (Å)	Line	Rest λ (Å)
Mg I	2852	Si II	1260, 1193
Mg II	2803, 2796	C II	1335
Mn II	2606, 2594, 2576	Lyα	1216
Fe II	2600, 2586, 2383	Si III	1206
C IV	1551, 1548	Lyβ	1025
Si IV	1403, 1396	O VI	1038, 1032

The existence of cases where $z_{abs} > z_{em}$ creates difficulties for all present hypotheses, because the velocity differences indicated can be as large as 3500 km/sec or more and it is difficult to model the requisite infall of gas. On the other hand, positive $z_{abs} - z_{em}$ might be explained as being due to intervening galaxies in unusual clusters of galaxies around the QSOs, but these random motions would then exceed those found in well-studied typical galaxy clusters out to (say) $z = 0.04$ or so.

TYPES OF ABSORPTION OBSERVED

Whatever the origin of the absorption lines, they are extremely interesting because those that are intrinsic to the QSOs should throw light on the physical processes going on in these enigmatic objects, while for those that arise in intervening material, if the QSOs are at their cosmological distances, the QSOs can serve as proof of unseen matter in the universe at look-back times which are a large fraction of the age of the universe. Thus it is important to assemble the facts about absorption lines and separate the types of lines seen into physically meaningful classes.

The following facts are now apparent:

1. Most of the absorption lines are very narrow; 10–20 km s^{-1} wide in many cases that have been studied in detail for optical absorption lines and narrower, say 4 km s^{-1} or less, for a few cases where 21-cm absorptions have been studied.
2. What appears to be a single, narrow, incompletely saturated absorption line on moderate dispersion spectrograms may break up into a number of individual lines, very narrow, some heavily saturated and others less so. This complicates any analysis of the line strengths by conventional curve-of-growth techniques.
3. Some lines may be completely "black," i.e., very highly saturated, while others are not.
4. Abundances of the elements, in the few cases where they have been studied, and subject to the inaccuracies due to (2), appear to be either near "normal," i.e., solar, or somewhat underabundant in elements heavier than hydrogen.
5. Some QSOs have very wide trough-like absorptions, resembling exotic supernovae.
6. Both high- and low-ionization lines are found.
7. The excitation is commonly very low, and cases of absorption out of excited levels are rare.
8. All QSOs with $z_{em} \gtrsim 2.6$ or so, for which adequate studies have been made, exhibit a large discontinuity in the density of lines (number per wavelength interval) at the position of Lyα emission. Shortward of Lyα the line density is very large, and it is now commonly accepted that these lines are all Lyman absorption lines, mostly Lyα but sometimes higher Lyman series transitions.

Let us consider two main categories: the broad lines and the narrow lines.

Broad Lines

While some QSOs have a very broad Lyα absorption and relatively narrow lines of other elements (e.g., PHL 957, 1331 + 170), the prototype of the kind of absorption I want to discuss here is PHL 5200. Data obtained at Lick Observatory with the Wampler-Robinson IDS spectrophotometer are summed and shown in Figure 1. In this object $z_{em} = 1.98$; the absorptions are adjacent on the short wavelength side to the emissions, identifications are obvious, the state of ionization is high, and the absorptions are so strong that much of the C IV emission is cut away and most of the Lyα emission is absorbed by N V absorption. The widths of the lines give Δv up to about 2×10^4 km s^{-1}. Sharp absorption lines are weak or absent; however, there is well-defined structure in the bottom of the absorption troughs (particularly well seen in C IV).

C III] λ 1909 is not a resonance transition, and does not have an absorption, but there is a short-wavelength shoulder extending into the velocity domain where the resonance absorptions appear. The obvious interpretation is that an optically thick gas outflow is occurring; clearly the problem of detailed interpretation of the absorption profiles must involve solutions of the radiative transfer equation in outward-moving

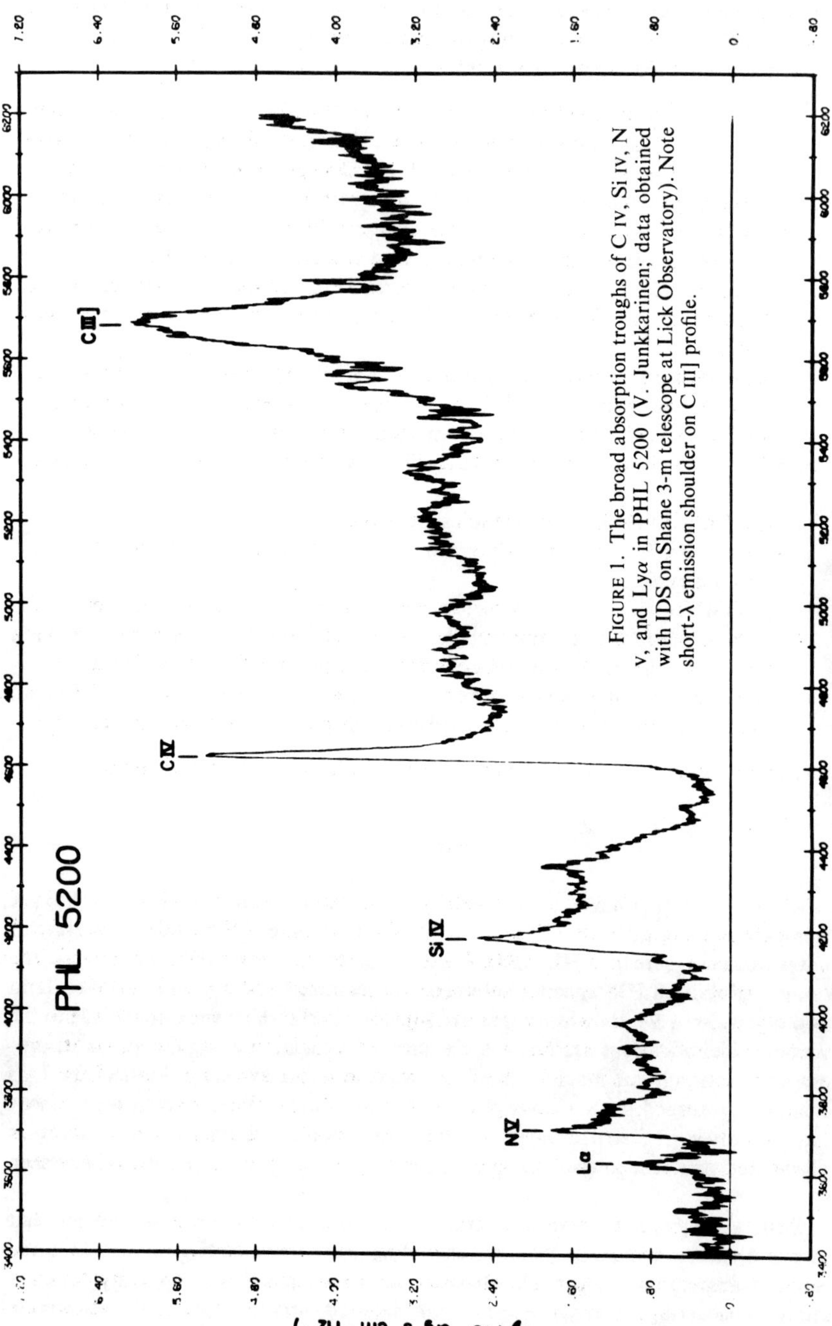

FIGURE 1. The broad absorption troughs of C IV, Si IV, N V, and Lyα in PHL 5200 (V. Junkkarinen; data obtained with IDS on Shane 3-m telescope at Lick Observatory). Note short-λ emission shoulder on C III] profile.

gas. This problem is under detailed study by a graduate student at the University of California, San Diego, V. Junkkarinen.

Another very interesting object is Q1246 − 057, with $z_{em} = 2.21$, which is under study by several people. It has broad absorptions *displaced* from the emission, with Δv up to 35,000 km s^{-1} or 0.1 c, and has narrow absorptions as well.

At present, some dozen or so broad-lined objects of the type of PHL 5200 and Q1246 − 057 are known. New samples of radio-quiet QSOs, particularly the University of Michigan survey by MacAlpine and colleagues, have yielded several cases. A number of these are being studied in detail by Junkkarinen. It appears, as far as our present data go, that all the "supernova-like" objects are radio-quiet, and this has important implications for their interpretation. One explanation of the radio-quiet QSOs is that they are highly compact objects, so that radio self-absorption processes inhibit the appearance of any low-frequency radiation. It is not known whether any of these objects have strong infrared radiation, nor whether any are variable. It does appear that PHL 5200 has not been subject to violent luminosity fluctuations during the period covered by the Harvard plate stacks (W. Liller, private communication).

An interesting feature of PHL 5200 is the long persistence of its broad absorptions—as the first such object to be discovered, it has now been observed (admittedly with ever-increasing signal/noise) over some 11 years. The expanding shells of supernovae, of course, break up; in galactic novae it is usual for the originally broad-line profiles produced by optically thick expanding gas shells to break up into sets of narrow absorption lines.

Narrow Lines

The majority of the absorption lines in QSO spectra are narrow; how narrow is usually not known because in the majority of cases the observed line profiles are determined by the instrumental resolution. A typical case, which was discussed by Burbidge, Smith, Weymann, and Williams (1977) is the radio QSO PKS 0454 + 039; one of their spectrograms is reproduced in Figure 2. There are two important questions that can be addressed in discussing the narrow lines: the physical conditions in the absorbing gas, and the distribution both in $\Delta z = z_{em} - z_{abs}$ and with respect to z_{em}. A full discussion of these issues is addressed in the review by Perry *et al.* (1978); we give a brief summary here.

Physical Conditions. There are no well-established differences between systems with $z_{abs} \approx z_{em}$ and those with z_{abs} very different from z_{em}. There exist systems that can be categorized as pure low-ionization systems, in which the predominant features are transitions of Lyα, C II, Si II, Fe II, Mg II, and Mg I. There are also pure high-ionization systems, showing Lyα, O VI, N V, Si IV, and C IV. In addition, there are mixed-ionization systems, with various combinations of these transitions. It is interesting that these three categories, together with one of the broad supernova-like systems, are exemplified by four QSOs which are adjacent to each other in a list ordered by redshift of the QSOs known to have absorptions. These objects are PHL 938 (z_{em} = 1.955, low ionization), PKS 0119 − 04 (z_{em} = 1.955, high ionization), 3C 191

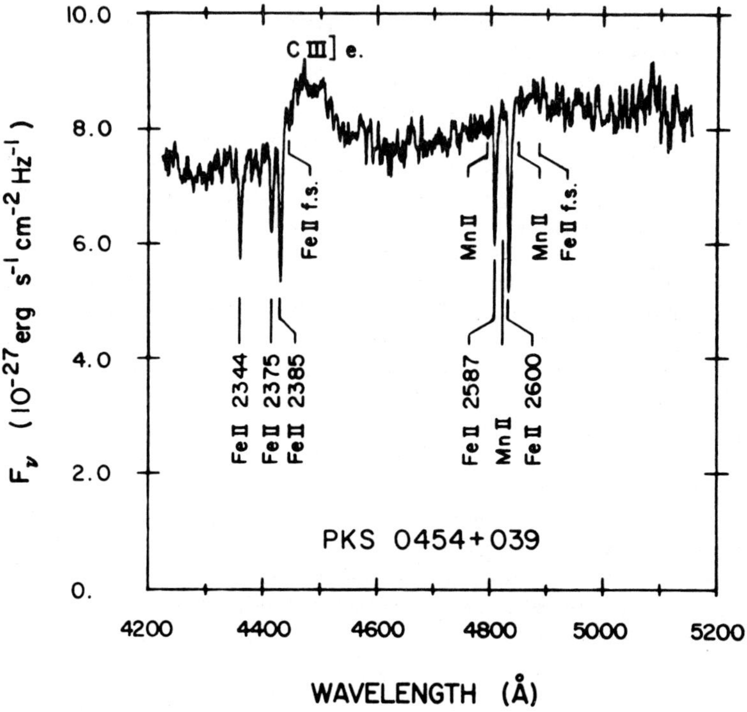

FIGURE 2. Strong narrow absorption lines of Mg II and Fe II in PKS 0454 + 039 at $z_{abs} << z_{em}$ (Burbidge, Smith, Weymann, and Williams, 1977).

(z_{em} = 1.956, mixed ionization), and PHL 5200 (z_{em} = 1.98, broad absorptions, as discussed in the section entitled "Broad Lines").

The absence of lines arising from excited fine-structure states has been used, since the early work by Bahcall and Wolf (1968), to set minimum, and quite large, distances of the absorbing material from the source of radiant energy, and maximum (low) limits on the electron density in the absorbing gas. Recent work in progress by C. L. Sarazin, B. P. Flannery, and by A. M. Wolfe, however, has shown that these distances of the absorbing gas from the energy source may actually be much less than earlier calculations suggested. Nevertheless, the large values of Δv, and the large numbers of clouds or filaments, imply that a large energy supply is needed for conversion to the requisite amount of kinetic energy, if the clouds are ejected from the QSO.

Distribution. The distribution of absorption-line redshifts with respect to both z_{em} and $\Delta z = z_{em} - z_{abs}$ is interesting. Here one must consider the rest wavelengths of the absorption lines in TABLE 1. At low z_{abs} only lines of the less abundant species, Mg, Fe, Al, can be seen. However, surveys extending over the region 3100–8000 Å are becoming available and these give good coverage of the stronger lines; C IV to Mg II, for example, have a good range of redshifts. A problem in the past has been to obtain good resolution over a wide wavelength region; the highest resolution (that obtained by A. Boksenberg and colleagues) has been restricted to limited spectral regions, while the wider coverage (that obtained at Steward Observatory, Lick Observatory, and Kitt Peak National Observatory) does not achieve resolutions greater than about 1–2 Å.

Despite these limitations, some tentative conclusions can be drawn:

1. Absorption-line systems are not very frequent for $z_{em} \leq 1.5$.
2. The number of absorption-line systems varies markedly for $1.5 \leq z_{em} \leq 2.2$.
3. Absorption becomes frequent for $z_{em} > 2.3$.
4. There is a very rich absorption spectrum for $2.8 < z_{em} < 3.5$.

Surveys underway at present by M. Burbidge, R. Weymann, and A. Boksenberg and their colleagues will, it is hoped, provide firmer data and determine the extent, if

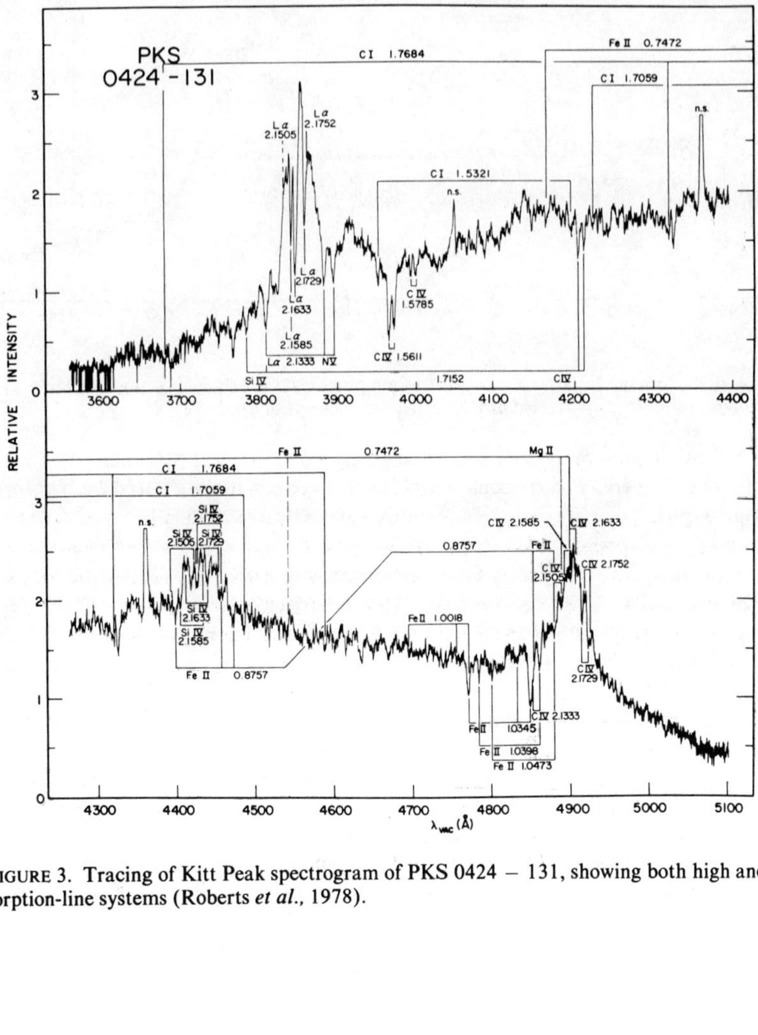

FIGURE 3. Tracing of Kitt Peak spectrogram of PKS 0424 – 131, showing both high and low absorption-line systems (Roberts *et al.*, 1978).

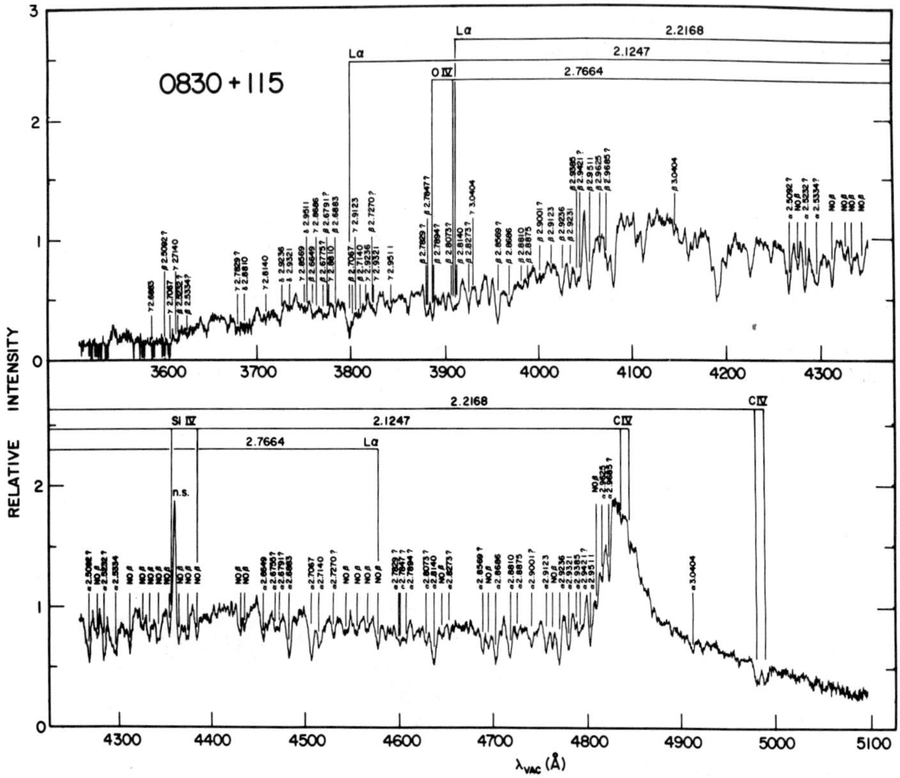

FIGURE 4. Tracing the Kitt Peak spectrogram of PKS 0830+115, showing very large number of Lyman absorptions shortward of Lyα emission (Roberts *et al.*, 1978).

any, to which present results are affected by observational selection (Perry *et al.*, 1978). The University of Arizona search by B. Peterson for Mg II and Fe II absorption in objects with $0.4 < z_{em} < 1.5$, for example, has already yielded the interesting result that absorption appears to be rare in the range $0.4 < z_{em} < 1.0$, although systems with z_{abs} in this range are frequently found in objects with higher z_{em} (Roberts *et al.*, 1978). The objects 0424 − 131 and 0830 + 115 are particularly interesting in this respect; Kitt Peak spectrograms of these used by Roberts *et al.* are reproduced in FIGURES 3 and 4, and it can be seen that interesting candidate low-z_{abs} systems may be present, especially in 0424 − 131.

Column Densities and Abundances. Generally the data have been insufficient for good studies, particularly in view of recent 21-cm work by A. M. Wolfe, which shows that apparently single optical absorption lines may actually be a composite of a few separate lines with varying degree of saturation. Objects for which studies have been made include 0735 + 178 (Carswell *et al.*, 1974), 0235 + 164 (Wolfe and Wills, 1977), 0528 − 250 (Smith, Jura, and Margon, 1979), and PHL 938 and Ton 1530 (Chan and Burbidge, 1971). The spectrogram of Smith *et al.* is reproduced in FIGURE 5.

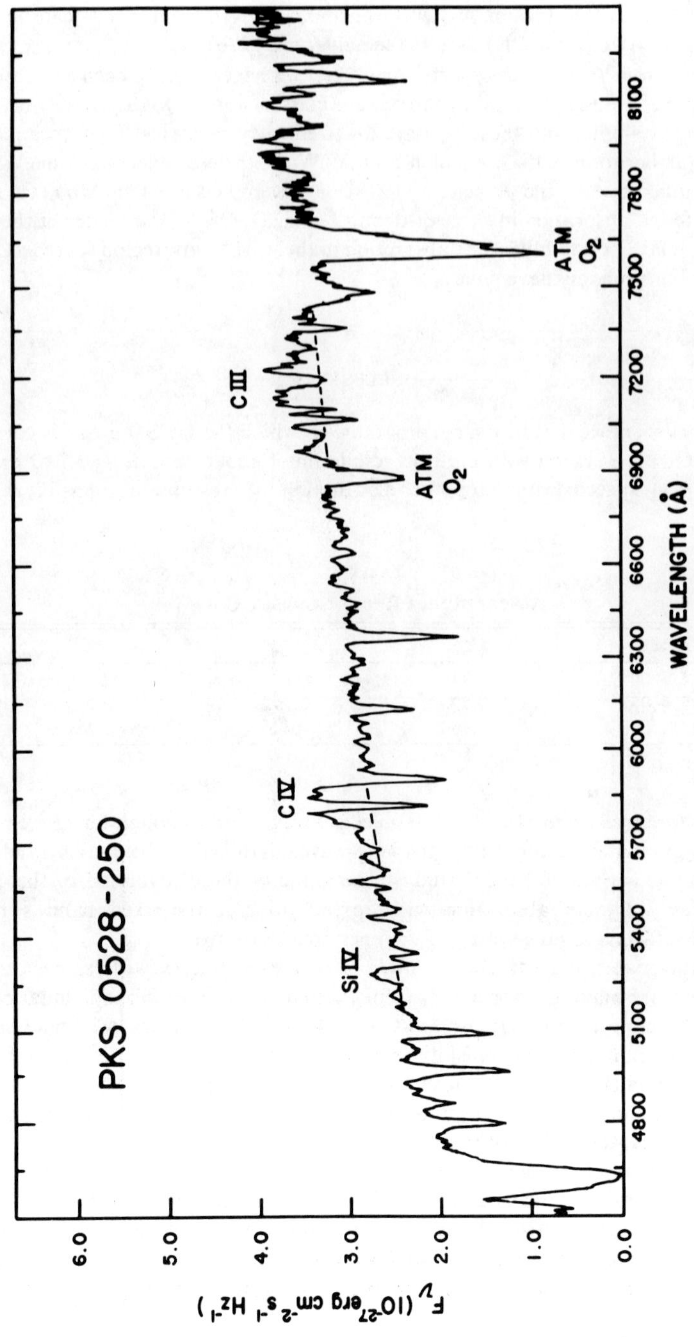

FIGURE 5. Strong absorption lines in PKS 0528 − 250 at $z_{abs} > z_{em}$; emission lines are very weak (Smith, Jura, and Margon, 1979).

In general, for lines of any species (e.g., MgII) to be seen, the column density will be of order 10^{14} cm^{-2}. Values of N_{HI} will in general be ~10^{18} cm^{-2}, but can be much higher, e.g. ~10^{21} cm^{-2} in PHL 957 (Coleman *et al.*, 1976).

Absorption in Particular Objects. At UCSD we have recently been searching for Mg II and Fe II absorption in an interesting class of radio QSOs: those with large VLBI fringe visibility and those that are low-frequency variables. This work is being done in collaboration with J. Condon and A. Wolfe. These objects are likely to be compact. Junkkarinen and Burbidge (1978) have searched eight such objects so far and have found absorption in three of them (TABLE 2). Of the remainder of the eight objects, one fairly certainly has no absorption in the wavelength region searched, while the other four probably have none.

CONCLUSIONS

In conclusion, we describe some programs on which the UCSD group is currently working. One is a search with uniform resolution of about 2 Å, using Kitt Peak and Cerro Tololo spectrograms, for the distribution of absorption lines in *a fixed*

TABLE 2
ABSORPTION IN RADIO COMPACT QSOs

Object	z_{em}	z_{abs}	Ions for z_{abs}
2134 + 004	1.936	0.629	Mg II, Fe II
2145 + 067	0.990	0.792	Mg II
2223 + 210	1.960	0.605	Mg II

wavelength range (about 300 Å) in all objects with z_{em} large enough for Lyα emission to be at a greater wavelength than the long-wavelength end of the chosen band. This should enable a good statistical study to be made of the distribution of the clouds showing only Lyman absorption with respect to z_{em}; the fixed band searched corresponds to a fixed range of z_{abs} if the lines are mostly Lyα.

Secondly, we are continuing a study of low-z_{abs} systems with respect to z_{em}. Thirdly, we are making a special study of particularly interesting individual objects such as the close pair of QSOs, 1548 + 114a and b, in which we found several absorption lines in the larger-redshift object. Another particularly interesting object is Hoag No. 48 in the Hercules Cluster region (Hoag, Burbidge, and Smith, 1977). A fairly low-resolution spectrogram of this, showing its very strong and complex absorption lines, is reproduced in FIGURE 6.

Of the many theoretical programs underway, I will mention only a few: the population of excited fine-structure levels and its theoretical implications is under study at the Center for Astrophysics, Harvard; hydrodynamic flow in ambient gas set up by "winds" from a QSO is being studied here in Munich at the Max-Planck-Institut and also at the Universities of Leeds and Manchester; and a new attack on the

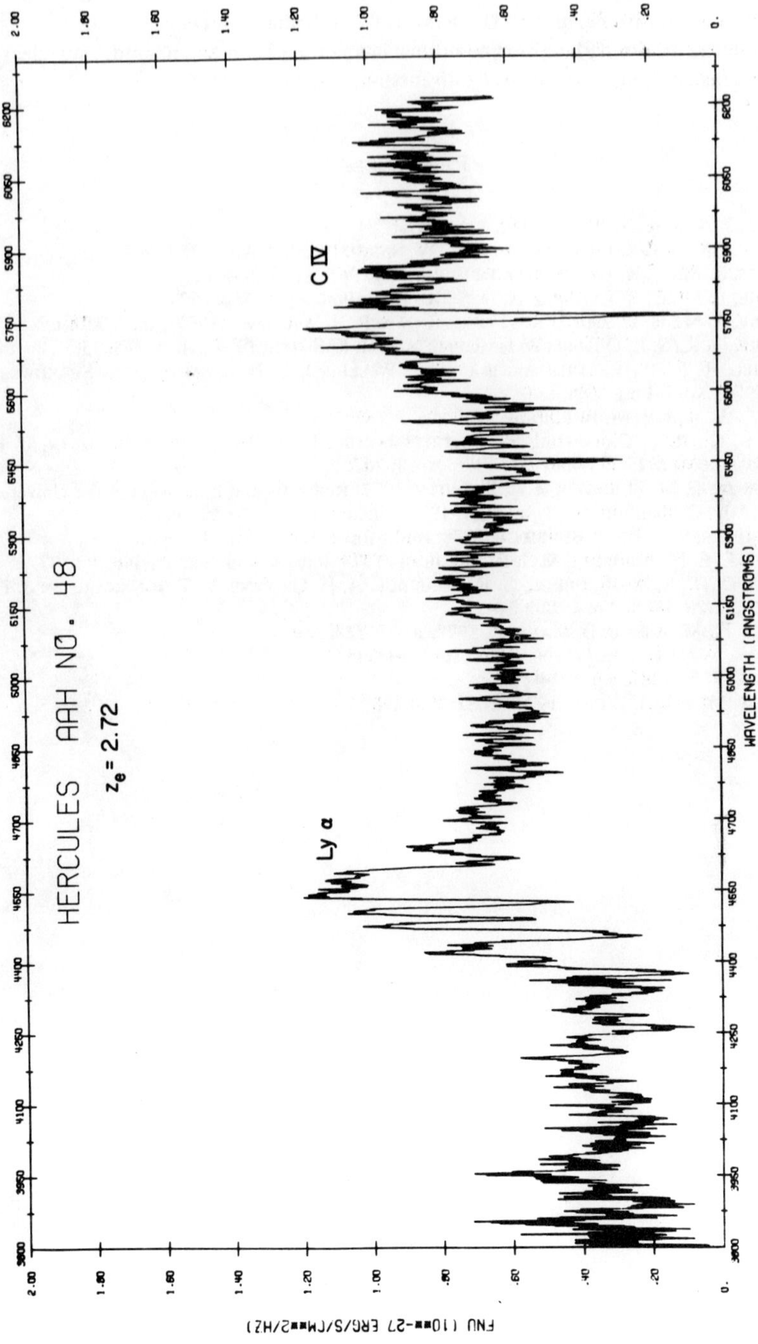

FIGURE 6. One of the radio-quiet QSOs in the Hercules cluster region found by Hoag (Hoag *et al.*, 1977). Note the great strength of the absorptions even at low resolution.

problem of curves of growth, by a synthesis of optical and 21-cm data, is underway at the University of Pittsburgh and University of California, San Diego.

While the origin of the absorption lines has not yet been deciphered, they clearly provide a fascinating topic for study, discussion, and controversy!

References

BAHCALL, J. N. & R. A. WOLF. 1968. Ap. J. **152:** 701.

BAHCALL, J. N., J. L. GREENSTEIN & W. L. W. SARGENT. 1968. Ap. J. **153:** 689.

BURBIDGE, E. M., C. R. LYNDS & G. R. BURBIDGE. 1966. Ap. J. **144:** 447.

BURBIDGE, E. M., C. R. LYNDS & A. N. STOCKTON. 1968. Ap. J. **152:** 1077.

BURBIDGE, E. M., H. E. SMITH, R. J. WEYMANN & R. E. WILLIAMS. 1977. Ap. J. **218:** 1.

BURBIDGE, G. R., S. L. O'DELL, D. H. ROBERTS & H. E. SMITH. 1977. Ap. J. **218:** 33.

CARSWELL, R. F., P. A. STRITTMATTER, R. E. WILLIAMS, T. D. KINMAN & K. SERKOWSKI. 1974. Ap. J. Lett. **190:** L101.

CHAN, Y.-W. T. & E. M. BURBIDGE. 1971. Ap. J. **167:** 213.

COLEMAN, G., R. F. CARSWELL, P. A. STRITTMATTER, R. E. WILLIAMS, J. BALDWIN, L. B. ROBINSON & E. J. WAMPLER. 1976. Ap. J. **207:** 1.

HOAG, A. A., E. M. BURBIDGE & H. E. SMITH. 1977. Redshifts and Expansion of the Universe. I.A.U. Colloquium No. 37 and C.N.R.S. Colloquium No. 263:521 Paris.

JUNKKARINEN, V. & E. M. BURBIDGE. 1978. Bull. Amer. Astron. Soc. **10:** 688.

PERRY, J. J., E. M. BURBIDGE & G. R. BURBIDGE. 1978. Pub. Astron. Soc. Pacific. **90:** 337.

ROBERTS, D. H., E. M. BURBIDGE, G. R. BURBIDGE, A. H. CROWNE, V. T. JUNKKARINEN & H. E. SMITH. 1978. Ap. J. **224:** 344.

SMITH, H. E., M. JURA & B. MARGON. 1979. Ap. J. **228:** 369.

STOCKTON, A. N. & C. R. LYNDS. 1966. Ap. J. **144:** 451.

WAGONER, R. V. 1967. Ap. J. **149:** 465.

WOLFE, A. M. & B. J. WILLS. 1977. Ap. J. **218:** 39.

INFRARED OBSERVATIONS OF EXTRAGALACTIC SOURCES*

G. H. Rieke

Steward Observatory and Lunar and Planetary Laboratory
University of Arizona
Tucson, Arizona 85721

INTRODUCTION

Infrared work has generally been considered somewhat alien and perhaps even irrelevant to students of extragalactic sources. It has been pursued primarily by a small group of astronomers who did relatively little work in other spectral regions and were motivated in part by different problems and questions from those occupying the attention of other astronomers. Now the infrared research has produced important conclusions and hypotheses that need to be brought forcefully to the attention of other astronomers, both to test these conclusions against other pieces of knowledge and to improve our understanding of the sources. As preparations were being made for this paper, an extensive review of the infrared observations was being finished[1]; some of the highlights of this review will be discussed below.

"NORMAL" GALACTIC NUCLEI

Over the years, a number of galactic nuclei that were not thought to be particularly extraordinary have been found to produce enormous infrared fluxes that place their luminosities in the range of the nuclei of Seyfert galaxies and require their discussion as "active" galaxies. Examples include NGC 253, NGC 2903, M81, M51, and NGC 6946.

Recently, a small but statistically homogeneous sample of bright spiral galaxies was studied at $10\mu m$.[2] Combined with spectral information from far-infrared measurements of an even more limited sample,[3] these observations indicate an average $M/L \lesssim 0.2$ (in solar units) for the nuclei of these galaxies. Thus in these regions matter is being converted to energy an order of magnitude more rapidly than was thought previously and at a rate that is difficult to sustain over a cosmic time scale with stellar processes as we understand them. A number of possibilities need to be explored:

1. The observational foundation of this work needs to be expanded, to refine estimates of the average M/L of spiral galaxy nuclei and thus to define better the implications described above.
2. The original source of the energy that is degraded into the detected infrared fluxes needs to be determined.

*This work was supported in part by a grant from the National Science Foundation, No. AST-7822482.

0077–8923/80/0336–0051 $01.75/1 © 1980, NYAS

3. Theories for the evolution of spiral galaxy nuclei and ideas regarding their relation to other types of active extragalactic sources need to be reevaluated.

The dominant luminosity sources in many spiral galaxy nuclei have escaped detection until now because they are embedded in dense clouds of dust that absorb their visual and ultraviolet emission and reradiate it in the infrared. Direct evidence for very strong extinction has been found in a number of the infrared-emitting nuclei by observing the absorption feature at 10μm produced by silicates in interstellar dust.[4] A particularly dramatic example is NGC 6946, a nearby, nearly face-on Sc galaxy listed by Keel and Weedman[5] among the 448 galaxies with the brightest nuclei in the optical. However, the strength of the 10-μm feature indicates that the infrared source is behind an obscuration of $A_v \gtrsim 30$ magnitudes, even though it lies very near the nucleus. In addition, the true nucleus is sufficiently obscured that the optical bright spot identified as the nucleus is displaced by $2'' \approx 100$ pc from the true nucleus of the galaxy.[4]

The original source of the energy seen in the infrared could be of some exotic type, or it might be that stars are forming at an exceptionally high rate. The second of these hypotheses has the advantage that its consequences can be explored in considerable detail, given the wealth of information available about the evolution of massive stars. Therefore, M. J. Lebofsky and I have been considering a model of the infrared source in M82 in which the luminosity is a consequence of a powerful burst of star formation. M82 is usually considered to be a very exceptional galaxy, but its infrared source is very similar in all measured parameters to those in the more powerful of the "normal" spirals, such as NGC 253 and NGC 6946. We have begun with M82 because somewhat more information is available about it; however, the conclusions drawn will almost certainly be applicable to the other galaxies with similar infrared properties.

The model is constrained by:

1. The total luminosity of 3×10^{10} $L_\odot$;[3]
2. The total mass in the source region of $\sim 6 \times 10^8$ $M_\odot$;
3. The ionizing flux, estimated from the Bα and Bγ strengths;
4. The flux generated by red giants in the nucleus, estimated from near-infrared photometry corrected for reddening (it is clear from the strength of the CO absorption bands at 2.3μm that red giants dominate the near-infrared emission), and;
5. The supernova rate, estimated from the nonthermal radio flux.

It appears to be just possible to meet all of these constraints under three fundamental conditions:

1. A very large percentage ($\geq$50%) of the mass in the nuclear region participates in the star burst;
2. The formation of stars less massive than 4 $M_\odot$ is strongly inhibited, and;
3. The luminosity outburst has a lifetime $<10^8$ years.

The first two of these conditions indicate that the process of star formation is significantly modified by conditions in the nucleus of M82 compared with those in the solar neighborhood. The first and third conditions, combined with the low average value of $M/L \lesssim 0.2$ found in the survey of spiral galaxy nuclei, indicates that the activity in these sources is sporadic and must be refueled by replenishment of the interstellar medium, probably by material from outside the nuclear region. These conclusions seem plausible enough that the stellar burst model can be accepted tentatively as the explanation for the observed behavior of spiral galaxy nuclei, although it is by no means established as the only possibility.

Active Galaxies and QSOs

Seyfert galaxies are characterized by strong infrared fluxes and have therefore been studied extensively in that spectral region. For those of type 2 (permitted line-widths of the same order as forbidden ones), the infrared emission is generally agreed to arise from thermal reradiation by dust around the nucleus. In the case of NGC 1068, the infrared measurements show the presence of enough dust to produce a very large optical depth in the optical—$A_v \sim 4$ to 10.[6,7] This result needs to be reconciled with what is known about the optical-to-ultraviolet continuum and spectrum. For the type-1 Seyfert galaxies (permitted lines much broader than forbidden ones), there has been substantial debate concerning the origin of the infrared fluxes. Recent evidence indicating that thermal radiation plays an important role includes:

1. The presence of 10-μm silicate absorption in the spectra of some of these galaxies;[4]
2. An anticorrelation between the strengths of the ultraviolet and infrared excesses,[8] and;
3. A correlation between the reddening deduced from Balmer line ratios and the strength of the infrared excess.[8]

The nonthermal nuclear source probably also plays a significant role, at least in the near infrared. Higher-quality observations of infrared variability could clarify this role substantially. It is now established that some of these galaxies vary in the near infrared,[1] but so far the observations cannot exclude that the varying source may be dust heated by the nonthermal nuclear source.

Since dust probably plays an important role in many type-1 Seyfert galaxy nuclei, its effects must be allowed for carefully in theoretical studies of these sources. For example, at least part of the observed departures from line ratios predicted by recombination theory undoubtedly arises from extinction. In addition, the ultraviolet continua appear to be significantly reddened, so that extrapolation of the observed slope is a hazardous procedure for estimating the ionizing flux.

To the extent that QSOs are related to type-1 Seyfert galaxies, the cautions in the preceding paragraph must be applied to them also. The analogy between these two types of objects is strengthened by the observed infrared properties of the non-optically-violently-variable QSOs.

Since most Seyfert galaxies have been identified by their ultraviolet excesses, primarily from the work of Markarian, the anticorrelation between the strengths of the infrared and ultraviolet excesses will result in a bias against infrared-bright examples of this class of galaxy. We have recently found that a similar bias operates for QSOs, even for samples that are not selected on the basis of color.[10] We observed the positions of flat spectrum radio sources that were empty fields to the limit of the Palomar survey. In six cases we detected a source at 2.2μm. These six sources represent 40% of the entire sample that had radio positions sufficiently accurate to guarantee that the source lay within the beam used in the infrared. Spectral information was obtained for five of the sources; all five can be characterized by power laws of index −2.5 to −3.3 between 0.6 and 2.2μm and therefore *all five* are redder in this spectral range than the reddest QSOs known previously (e.g., AO 0235 + 164, 3C 68.1). At least two of the sources have varied on a time scale of a month. These objects therefore appear to be related to optically violently variable QSOs, but this tentative conclusion requires further observational and theoretical investigation.

REFERENCES

1. RIEKE G. H. & M. J. LEBOFSKY. 1979. Annu. Rev. Astron. Astrophys. In press.
2. RIEKE G. H. & M. J. LEBOFSKY. 1978. Ap.J. Lett. **220:** L37.
3. TELESCO, C. M. 1977. Galaxies as Far-Infrared Sources. Ph.D. Thesis. Univ. of Chicago.
4. LEBOFSKY M. J. & G. H. RIEKE. 1979. Ap.J. **229:** 111.
5. KEEL, W. C. & D. W. WEEDMAN. 1978. Ap.J. **83:** 1.
6. JONES, T. W., C. M. LEUNG, R. J. GOULD & W. A. STEIN. 1977. Ap.J. **212:** 52.
7. LEBOFSKY, M. J., G. H. RIEKE & J. C. KEMP. 1978. Ap.J. **222:** 95.
8. RIEKE, G. H. 1978. Ap.J. **226:** 550.
9. NEUGEBAUER, G., J.B. OKE, E. E. BECKLIN & K. MATTHEWS. 1979. Ap.J. **230:** 79.
10. RIEKE, G. H., M. J. LEBOFSKY & T. D. KINMAN. 1979. Submitted to Ap.J. Lett.

IMPLICATIONS OF THE VARIABILITY AND POLARIZATION OF BL LAC OBJECTS*

J. R. P. Angel and R. L. Moore

Steward Observatory
University of Arizona
Tuscon, Arizona 85721

We discuss here the results and interpretation of recent optical polarimetric observations obtained at Steward Observatory of BL Lac objects and violent quasar outbursts. The rapid variability and strong polarization of these objects gives information about the emission processes at their very centers, where it seems likely that there are massive black holes. Our program is to try to characterize these properties and their differences to aid in understanding these remarkable objects. We note at the outset that, in contrast to the objects considered here, the large majority of QSOs (virtually all radio quiet and most radio loud) show only weak polarization and little if any short-term variability.[1]

A program of polarimetric monitoring of 12 BL Lac objects with $V \lesssim 16$ was carried out for several months at the 36″ telescope of Steward Observatory.[2] Half of this sample was found to vary with a characteristic timescale of one day, that is to say the variations from hour to hour in strength and angle of polarization were monotonic and not large, but from night to night the changes were erratic and quite substantial (up to 20° angle, 5% in strength of polarization). The remainder of the sample did not show detectable nightly variations, and in general had a preferred orientation of the position angle, preserved over years of observation. The objects of known redshift showing the characteristic one-day variability (BL Lac and NGC 1275) have luminosities (0.3–10μ) of $\sim 10^{46}$ erg/sec.

The outburst of B2 1308 + 326 in the spring of 1978 gave us a chance to compare the polarimetric properties of a much more luminous source. The redshift determined by Miller (1978) is 0.996, yielding from the observed outburst flux in the 0.3–10μ band (0.15μ–5μ in the rest frame) a luminosity of 10^{48} erg/sec, again assuming isotropic emission. This is about the same as that of AO 0235 + 164.[3] Repeated measurements of the flux and polarization of B2 1308 + 326, in both the optical and infrared, are reported by Moore *et al.*[4] These data are remarkably similar in character to those obtained for the objects 100 times fainter. Again there is little power in the variations from hour to hour, but the polarization varies erratically by up to $\sim$40° and 7% with a timescale of two days (one day in the rest frame). In this object we can show that the variability is associated with a substantial fraction of the energy output, since the infrared polarimetric variations at 2.2μ track those in the visible, and the optical photometry showed changes of up to 50% in one day (rest frame).

We have recently made extended observations of the short-term variability of BL Lac with continuous polarization measurements over periods of up to eight hours.

*This work was supported in part by the National Science Foundation under Grant AST 7822714.

0077–8923/80/0336–0055 $1.75/1 © 1980, NYAS

These measurements, and those made during the 36″ survey, clearly show variability on a timescale of ~1 day. In addition, the measurements of position angle through any given night show a fairly smooth rotation in angle of 1–2°/hr. The direction of rotation may change from night to night, and longer term trends are superimposed. However, linear fits to the position angle variations on nine nights yield rotation rates of 0.8–2.5°/hr; there is only one other night where four or more observations are available and the rate was 7°/hr. These rates are derived from data taken in different months and different years. The remarkable consistency with which we observe this rotation suggests some underlying driving mechanism.

We now consider how these results can be interpreted, making the assumption that the emission is indeed isotropic, and thus the total luminosity very large. Because of the strong similarity in the variability and polarization of BL Lac objects and the very luminous outbursts of AO 0235 + 164 and B2 1308 + 326, it seems likely that the same engine is at work, and that it can operate up to a maximum luminosity of $\sim 10^{48}$ erg/sec. The corresponding mass for emission at the Eddington limit is 10^{10} $M_\odot$, so one can argue that the engine mass will not be very much less than this. The maximum mass that can be expected to vary strongly in one day is found by increasing the mass within a one light-day region until the event horizon is out to one light day. More realistically, we set one day equal to $5R_s/c$, since strong emission is not expected within 5 Schwarzschild radii R_s. The limiting mass is then 2×10^9 $M_\odot$. It follows that in order to emit such high luminosity without greatly exceeding the Eddington limit, we are driven to consider radiation emitted from close to a black hole of this mass. The fact that the timescale of one day is found over a wide range of luminosity points to the same mass of about 2×10^9 $M_\odot$ being present in all these objects. A much smaller mass of 3×10^7 $M_\odot$ was suggested for BL Lac by Elliot and Shapiro,[5] based on the same type of arguments from the Eddington limit and timescale of variability. The discrepancy arises because of their use of shorter timescales and lower luminosity, and because of different assumption about the closeness of the emission to the black hole. We believe the higher mass to be more consistent with the observational data now available.

The argument that the emission comes from close to a black hole is based only on containing a huge radiation density in a very small volume. The close proximity to the black hole is, of course, consistent with the source of energy being gravitational and the concept that infalling material radiates some of its energy before disappearing into the hole. We suggest that the continuous rotation of position angle as observed in BL Lac is a consequence of the emission regions being in rotation about the black hole.

The possibility that synchrotron radiation could, under such extreme conditions, be the optical emission mechanism has recently been considered in detail by Blandford and Rees.[6] It could account for the very high polarization, up to 30% seen at times in some sources. The main constraint for synchrotron radiation is that the magnetic field strength must be very high (~kG) to avoid severe inverse Compton losses; and that there also must be very few nonrelativistic electrons in the source to avoid depolarization by Faraday rotation and electron scattering. This in turn requires continuous acceleration of the electrons, or very many repeated accelerations within one day. For the extreme conditions of B2 1308 + 326 Moore *et al.* (reference 4) find that at any given time, 3/4 of all the electrons must be relativistic.

We conclude from the above considerations that a black hole of mass a few times 10^9 $M_\odot$, as indicated by the timescales of variability, may be present in BL Lac objects and violently variable quasars. The wide range of luminosities can be accounted for by different accretion rates onto the black holes. We cannot exclude larger masses in those objects where rapid variability is not detected, but the apparent limit on the maximum luminosity of 10^{48} erg/sec for any QSO suggests that these may not occur. An upper limit to the mass is quite interesting since it implies that either a black hole stops accreting material at some point, or that a more massive black hole is unable to produce observable emission. One cutoff mechanism was proposed by Hills[7] who pointed out that when a black hole reaches $M \sim 3 \times 10^8$ $M_\odot$, solar-type stars would reach the event horizon before being tidally disrupted, thus cutting off tidal breakup as the source of material for an accretion disc. The fact that we derive a larger mass for BL Lac objects (in particular B2 1308 + 326, with very high luminosity) indicates that there is some other source for the infalling material. Growth of very massive black holes could be limited simply by the absence of material which can be readily accreted, i.e., the inner nucleus of the host galaxy has been swept clean. The nucleus of M87, which apparently has a central mass $\sim 5 \times 10^9$ $M_\odot$,[8] yet shows no BL Lac characteristics, is perhaps an example of a black hole with no significant supply of material to be accreted.

The similarity of the mass of bright QSOs derived from the Eddington limit, and of BL Lacs and violent QSOs from their variability, suggest that BL Lac objects may be burned-out bright quasars. This identification is supported by the similarity in space density of quasars with luminosity 10^{47}–10^{48} erg/sec. at $z \sim 2$ and the observed local space density of BL Lac objects.[9] It is tempting to imagine an evolution in which early massive galaxies process some hydrogen by thermonuclear burning in massive stars, then accrete this material preferentially (perhaps as gas lost from the stars) to build the central black hole in the period of high QSO luminosity. The violent variable QSOs could then be a short-lived phase when the main accretion supply was running out and the objects would then die down to become BL Lac objects or optically inactive objects like the M87 nucleus.

In conclusion we recognize that the arguments made above are based on the assumption that emission is isotropic and the rest frame timescales are as observed. Blandford and Rees[6] suggest that the radiation we see from BL Lac objects may be that from jets with relativistic speeds directed towards the observer, greatly enhancing the apparent luminosity and reducing the observed timescales. We lean towards the power heavy scenario, because it has fewer degrees of freedom and seems to make a coherent picture of QSO evolution.

References

1. Stockman, H. S. 1978. *In* Pittsburgh Conference on BL Lac Objects. A. M. Wolfe, Ed. University of Pittsburgh. Pittsburgh, Penn.
2. Angel, J. R. P., T. A. Boroson, M. T. Adams, R. E. Duerr, M. S. Giampara, M. S. Gresham, P. S. Gural, E. N. Hubbard, D. A. Kopriva, R. L. Moore, B. M. Peterson, G. D. Schmidt, D. A. Turnshek, M. S. Wilkerson, N. V. Zotov, J. Maza

& T. D. KINMAN. 1978. *In* Pittsburgh Conference on BL Lac Objects. A. M. Wolfe, Ed. University of Pittsburgh. Pittsburgh, Penn.

3. RIEKE, G. H., G. L. GRASDALEN, T. D. KINMAN, P. HINTZEN, B. J. WILLS & D. WILLS. 1976. Nature **260:** 754.
4. MOORE, R. L., J. R. P. ANGEL, M. J. LEBOFSKY, G. H. RIEKE, W. WIESNIEWSKI, S. L. MUFSON, H. R. MILLER, B. Q. MCGIMSEY, R. M. WILLIAMON & F. J. VRBA. 1979. Ap.J. In press.
5. ELLIOT, J. L. & S. L. SHAPIRO. 1974. Ap.J. Lett. **192:** L3.
6. BLANDFORD, R. D. & M. J. REES. 1978. Preprint.
7. HILLS, J. G. 1975. Nature **254:** 295.
8. SARGENT, W. L. W., P. J. YOUNG, A. BOKSENBERG, K. SHORTRIDGE, C. R. LYNDS & F. D. A. HARTWICK. 1978. Ap.J. **221:** 731.
9. SETTI, G. & L. WOLTJER. 1977. Ap.J. Lett. **218:** L33.

X-RAY QUASARS*

Hale V. Bradt

Department of Physics and
Center for Space Research
Massachusetts Institute of Technology
Cambridge, Massachusetts 02139

SUMMARY

Seven x-ray sources have been identified with quasars or Seyfert-quasar transition objects as classified by optical observers: IIIZw2, Fairall-9, 0241+622, 3C273, IC4329A, Mk509, and MR2251−178. In all but one case (Fairall-9) the identification is based upon an x-ray position precise to ~30″. The x-ray luminosities range from $10^{44.0}$ to $10^{46.1}$ ergs s^{-1} (H_0 = 50 km s^{-1} Mpc^{-1}), and in two cases (3C273 and MR2251−178) x-ray variability by factors of 2–5 in months has been detected.

Each of these optical objects is quite luminous ($-27.3 < M_v < -23.8$), is stellar in appearance, and exhibits an associated nebulosity. The latter range from the edge-on spiral galaxy IC4329A to the nonthermal jet of 3C273. Five of the objects are known to be radio emitters, three with 5 GHz flux densities in excess of 200 mJy, and two (IIIZw2 and 3C273) demonstrably variable. In this paper we review the observational results for these seven objects at radio, infrared, optical, ultraviolet, x-ray and gamma-ray frequencies. The x-ray results all precede those forthcoming from the recently launched Einstein Observatory.

These objects are examples of the most extreme Seyferts or of close quasars (D = 95–950 Mpc). Their x-ray properties (i.e. variability and spectra) do not differentiate them from Seyfert nuclei. 3C273 is more luminous in the optical and x-ray than the most luminous of the other six objects but only by a factor ~6. Its lack of an observable low-frequency cutoff ($N_H < 4.5 \times 10^{21}$) in x-rays and its variability are also similar to other Seyferts and BL Lac objects. Thus these objects help establish the similarity between the several types of active extragalactic nuclei.

The space density of the six objects (excluding 3C273) within 600 Mpc (~2×10^{-8} Mpc^{-3} mag^{-1} at $M_v \sim -24.3$) is comparable to that inferred from other studies of Seyferts and quasars. This suggests that these seven x-ray emitting objects constitute a substantial fraction of the luminous transition objects in the neighborhood of the Galaxy. The proximity of these objects makes possible VLBI studies with good spatial resolution, to ~1 pc, as well as optical and x-ray studies of the nebulosities to ~1 kpc.

The x-ray variability of objects such as these (e.g., NGC 4151) indicates that the x-rays derive from compact regions of size <0.01 pc, comparable to that of the optical continuum and permitted line regions. Thus, correlated optical and x-ray observations (spectrum and intensity) show particular promise for the further understanding of the emission processes.

*This work was supported in part by NASA contract numbers NAS5-24441 and NAS8-27972.

0077–8923/80/0336–0059 $1.75/1 © 1980, NYAS

INTRODUCTION

X-rays are known to emanate from the active nuclei of Seyfert type 1 and Seyfert type 2 galaxies, narrow-emission-line galaxies, *N* galaxies, and BL Lac objects. (See reference 44† for a review.) Three objects with quasar characteristics have also been found to emit x-rays: 3C273,[18,80,97,113,128,127,22,114,156] MR2251 – 178[117,26] and 0241 + 622.[6,96] Two of these (MR2251 – 178, 0241 + 662) were located through their x-ray emission, much as 3C273 was located through its radio emission.[63,131] In addition there are several x-ray sources which have been identified with very luminous Seyfert nuclei ($M_v < 23.8$) that have been described as quasars or quasar-like: Fairall–9 (F–9),[46,115,149,145,100,60,32] Mk509,[43,41,84] IC4329A,[43,36,39,99,153] and IIIZw2.[132,9,59,55] The x-ray positions for six of these seven objects have been measured with modulation collimators to precisions of 20″–40″ (FIGURE 1). The exception is Fairall – 9 with an x-ray precision of ~15′,[31] but as discussed below its identification as an x-ray source is quite secure.

The 2–10 keV (0.5–2.4 EHz)‡ x-ray luminosities of the seven objects range from $10^{44.0}$ to $10^{46.1}$ erg s^{-1} which are comparable to the optical and infrared luminosities and to the radio luminosities of IIIZw2 and 3C273. The x-ray fluxes from a number of active galactic nuclei are variable on time scales reaching down to at least ~1 d, e.g., NGC4151,[105] Cen A,[154,106] and Mk421,[118] indicating the compact nature of the x-ray sources. Thus the character of the x-ray emission provides direct information about the inner portion (<0.01 pc) of the emission regions and clues to the nature of the power source. Among the seven objects discussed here, 3C273[113,22,156] and MR2251 – 178[31,116] have already been shown to exhibit x-ray variability. One of the quasars (3C273) is probably a gamma-ray (~100 MeV) emitter[141] and another (0241 + 622) may be.[66] The x-ray emission process may be closely related to the gamma-ray emission process.[74,140]

The three objects previously listed[6,150] as quasars (3C273, 0241 + 622, MR2251 – 178) are the optically brightest[6] (apparent magnitude after correction for extinction in the Galaxy, FIGURE 2a) of all the 633 quasars[23] in a recent catalog. This

†References are in alphabetical order in reference list.
‡In this paper, we make use of the following:

$H_0 = 50$ km s^{-1} Mpc^{-1}
1.0THz ≡ 10^{12} Hz
1.0PHz ≡ 10^{15} Hz → 0.30μm
1.0EHz ≡ 10^{18} Hz → 4.14 keV
1.0μJy ≡ 10^{-29} W Hz^{-1} m^{-2}
≡ 0.242 × 10^{-11} erg cm^{-2} s^{-1} keV^{-1}

Average flux densities for broad-band measurements: $S(\nu)_{av} = \int S(\nu)d\nu/\Delta\nu$.
1.0 Uhuru count s^{-1} ≃ 1.2 μJy (0.5–2.4 EHz, 2–10 keV).
1.0 Ariel-5 SSI count s^{-1} ≃ 2.6 μJy (0.5–2.4 EHz, 2–10 keV).

FIGURE 1. Precise (<1′) x-ray error regions obtained with the SAS-3 (circles) and HEAO-1 (parallelograms) modulation collimators for 6 of the 7 objects: IIIZw2 (SCHNOPPER *et al.*[132]), 0241 + 622 (APPARAO *et al.*[6]), 3C273 (BRADT *et al.*[22]), IC4329A (DELVAILLE *et al.*[36]), Markarian 509 (DOWER *et al.*[41]), MR2251 – 178 (RICKER *et al.*[116]). See these references for photo credits. The x-ray region for the seventh object, Fairall – 9, has an area of 0.21 deg² (COOKE *et al.*[31]).

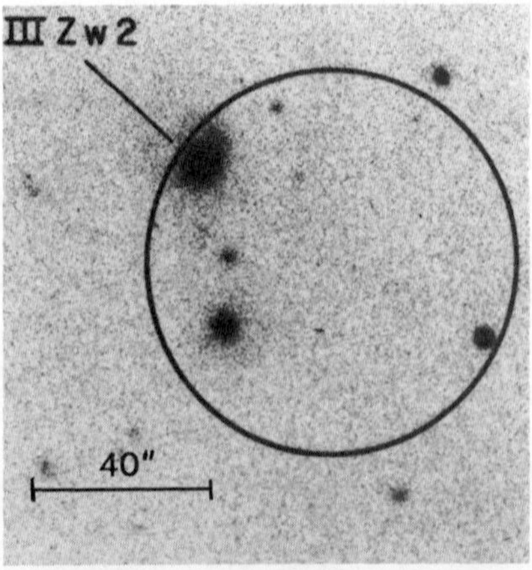
III Zw 2
40″

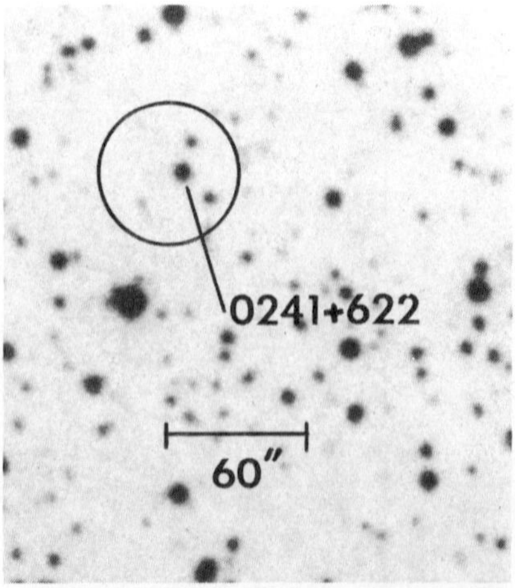
0241+622
60″

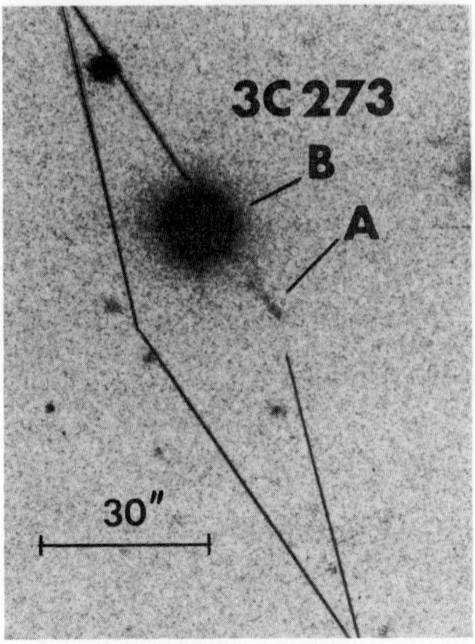
3C 273
B
A
30″

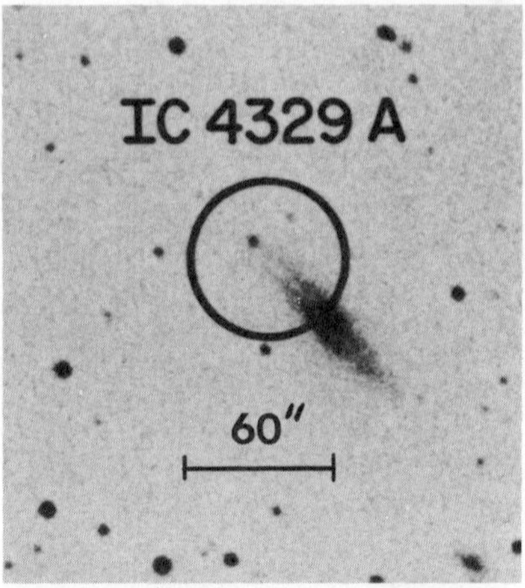
IC 4329 A
60″

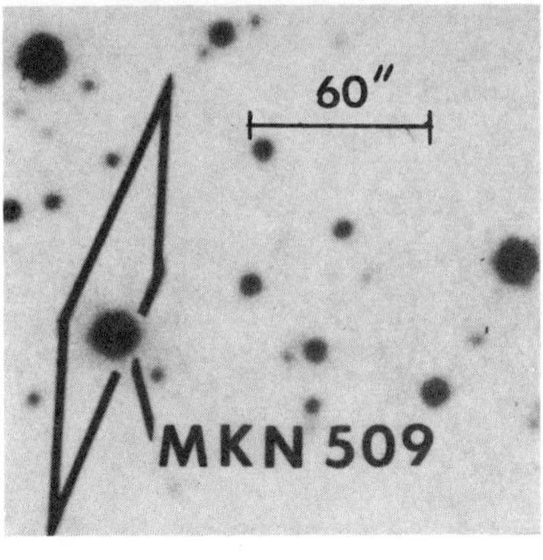
60″
MKN 509

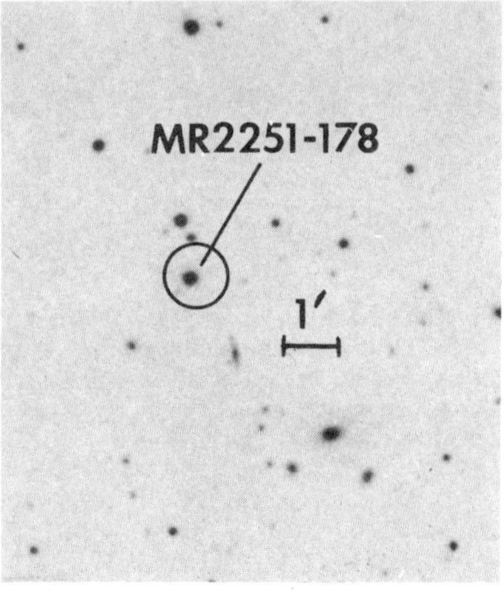
MR2251-178
1′

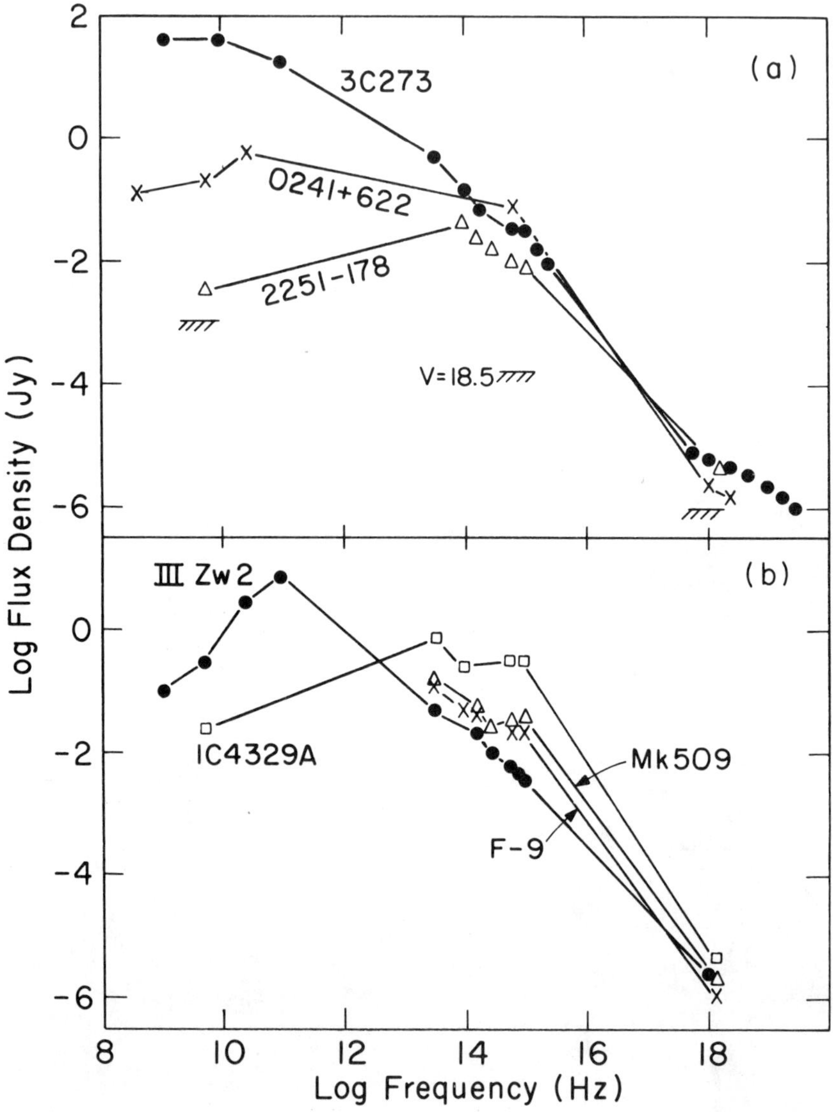

FIGURE 2. Spectra of the seven objects. See TABLE 1 for fluxes and references. For variable sources, the brightest of the flux densities $S(\nu)$ given in TABLE 1 are shown. Note that these are not necessarily the historical maxima (cf. IIIZw2, FIGURE 4 and 3C273, FIGURE 8). The approximate radio, optical, and x-ray detection thresholds (pre HEAO-2) are indicated in the upper figure. The COS-B γ-ray fluxes are shown in FIGURE 5 and FIGURE 11.

is because they are among the closest to the sun, i.e., at distances comparable to the more distant Seyfert type 1 (Sy 1) galaxies. The ratios of x-ray to optical luminosities for these three objects are comparable to one another. This indicates that increased x-ray sensitivity will lead to the detection of the (apparently) fainter and presently cataloged optical quasars. These three objects as a group are essentially indistinguishable in their x-ray and optical characteristics from the extreme Sy 1 nuclei also discussed here. They thus help to establish the common nature of Sy 1 nuclei and quasars.

The purpose of this paper is to review the observational data for these seven exceptionally luminous, compact, extragalactic objects. Their characteristics at radio, infrared, optical, ultraviolet, x-ray and γ-ray frequencies are presented in TABLES 1a–g. We will discuss each of the objects in the following sections (in order of increasing right ascension) and will conclude with a brief discussion of the objects as a class.

IIIZw2 (0007+106)

The x-ray emission from IIIZw2 was discovered in modulation collimator data from SAS-3 that was accumulated during 17–22 August 1977.[132] The data from the two channels (2–6, 6–11 keV; 0.5–1.5, 1.5–2.7 EHz) are consistent with an energy spectrum of the form $S(\nu) \propto \nu^{-0.3}$, with a flux density of 2.4 μJy at 10^{18} Hz (1.0 EHz). The source does not appear in either the 2A or 4U catalogs.[31,48] Further analyses of the Ariel-5 and Uhuru data yield 3σ upper limits[43,132] which do not convincingly indicate variability.[132] A positive detection with the HEAO-1 GSFC experiment yields an intensity[98] which is consistent with the SAS-3 intensity.

The optical object was first reported as a compact galaxy[159,160] because of its slightly nonstellar appearance; a faint nebulosity extends 10″ to the north-northwest (FIGURE 1). Spectra taken by a number of observers[9,129,81,109,55] all show the classic Sy 1 or quasar emission features (FIGURE 3a). The apparent visual magnitude $V = 14.8$[53] and redshift ($z = 0.09$) lead to $M_v = -24.1$. The object is considered to be intermediate between the Seyfert and quasar classes[9,59] or to be a quasar.[55] The latter group[55] obtained the spectrum of the nebulosity and found emission lines with the same redshift as the stellar object. In 1976, the object was reported as possibly variable in the optical.[53] During the recent radio outburst described below, the optical continuum increased substantially and became bluer,[102] but the IR flux[120] remained constant[121] as expected for thermal radiation from a dust cloud.[121]

Radio emission was first reported in 1968.[17] Variability (13% in one day, and also on time scales of months) at 2.7 GHz, was discovered in 1973.[72] A dramatic radio event occurred in 1977 wherein the 14.5 GHz flux rose from 100 mJy in January 1975, to 350 mJy in June 1977, and to 1050 mJy in November 1977.[157] Observations with the VLA and the NRAO 36-ft Kitt Peak antenna in December 1977 and January 1978[132] showed a spectrum rising from 0.09 to 6.9 Jy over the frequency range 1.4 to 90 Ghz (FIGURE 4), indicative of an extremely compact and rapidly evolving source.[132] A VLBI study[155] at 10.6 GHZ in March 1978, shows 85% of the flux to come from an unresolved region of diameter <1.0 pc.

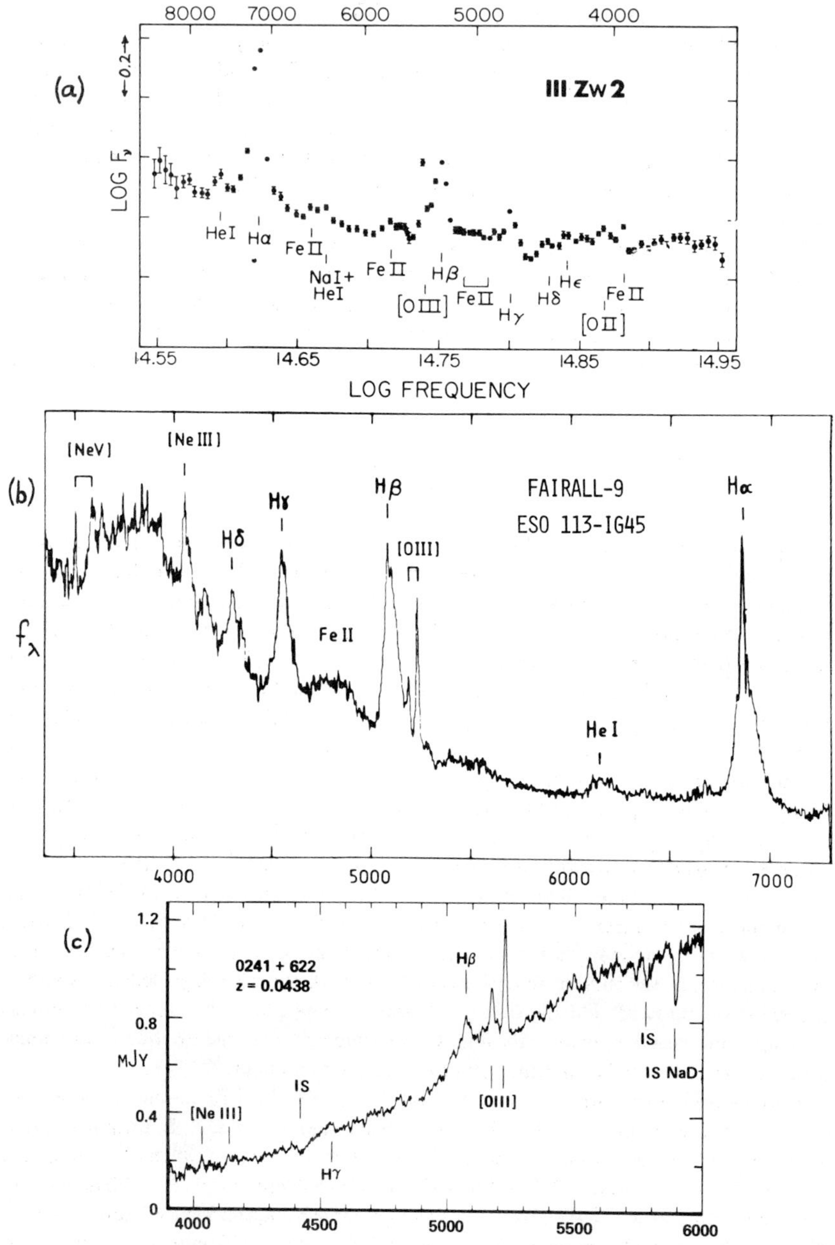

FIGURE 3. Optical spectra of the seven nuclear objects with apertures ranging from 3″ to 7″. (a) IIIZw2 (GREEN *et al.*[55]), (b) Fairall–9 (WARD *et al.*[145]), (c) 0241 + 622 (MARGON and KWITTER[96]), (d) 3C273 (BOKSENBERG[15]), (e) IC4329A (WILSON and PENSTON[153]), (f) Mrk 509 (OSTERBROCK and KOSKI[110]), (g) MR2251 – 178 (CANIZARES *et al.*[26]).

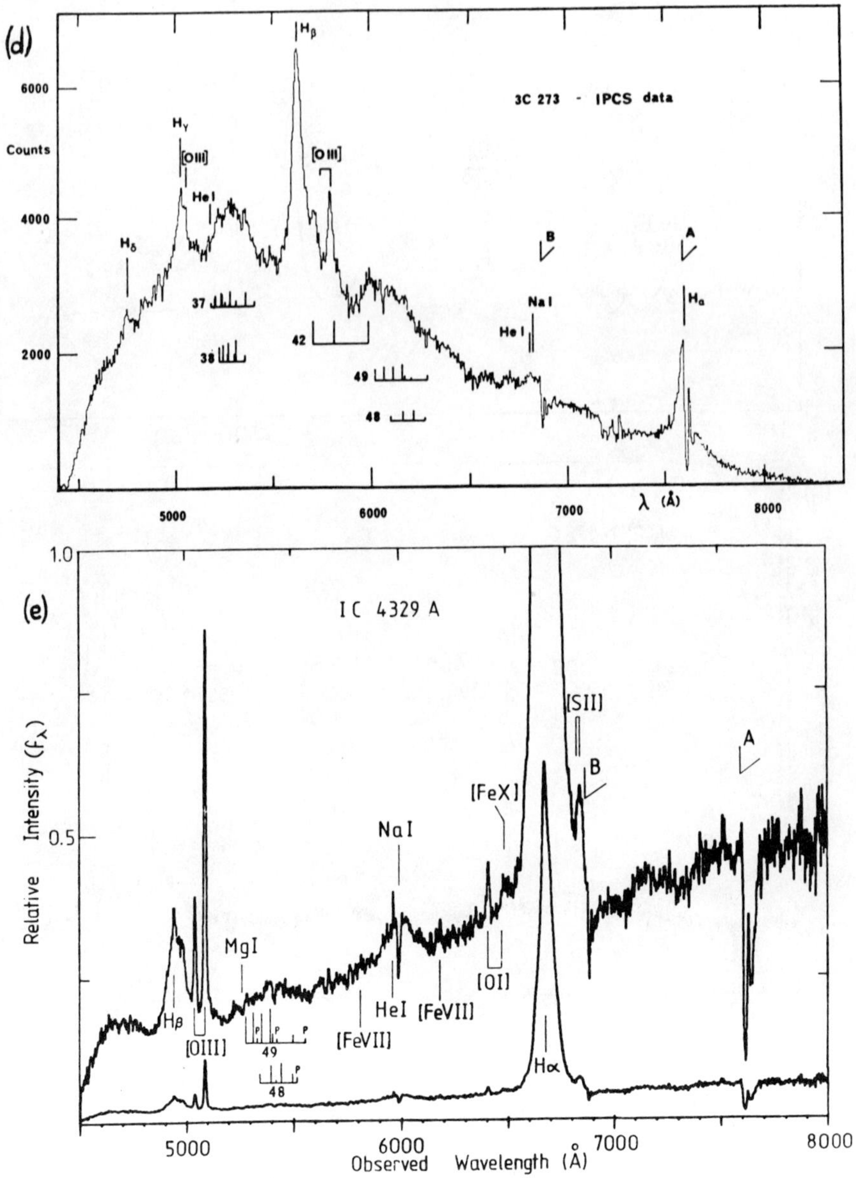

FIGURE 3. (*continued*)

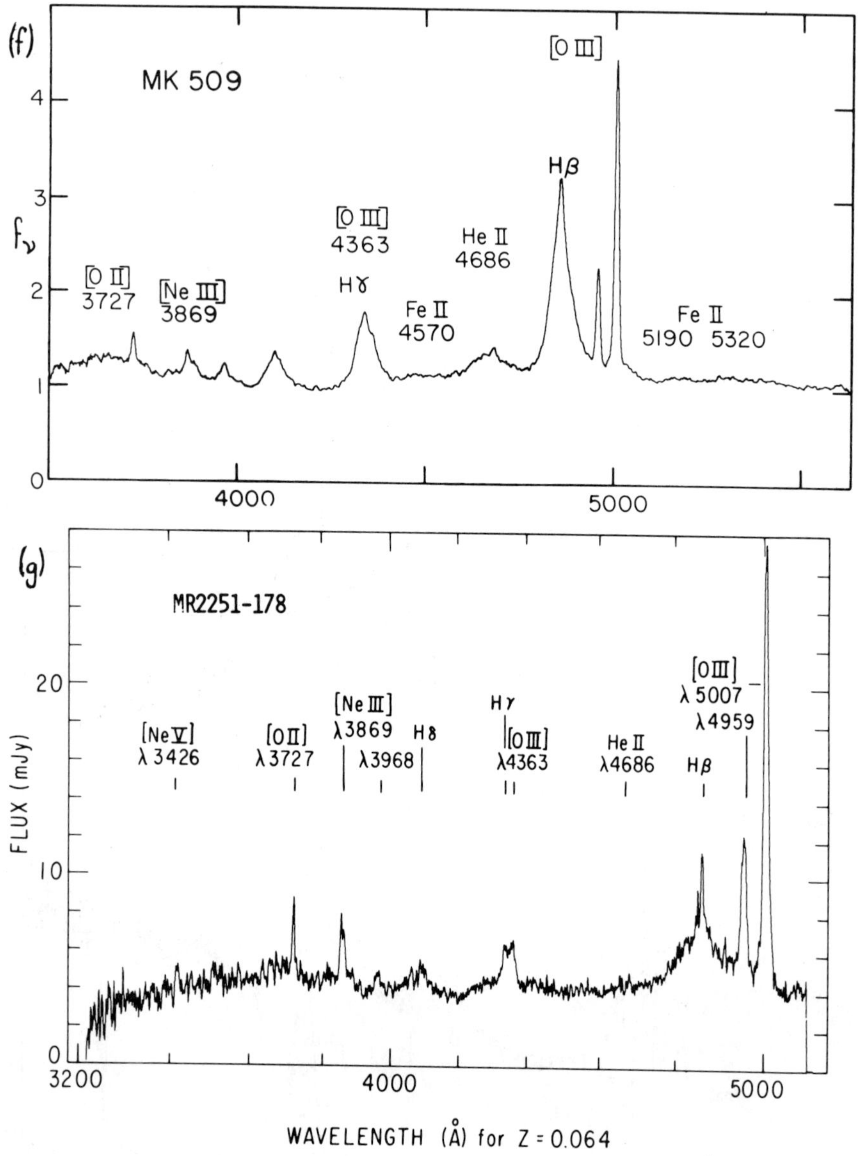

FIGURE 3. (*continued*)

Fairall–9 (0121–590)

This galaxy was located through the ESO quick blue survey,[46] the ESO/Uppsala survey,[70,149] and the Ariel-5 x-ray survey.[31,145] It was identified as an extreme Seyfert type 1[46,149,145] and associated with the x-ray source,[115,13,145] in each case independently by several groups. The association with the x-ray source 2A0120−591 is based upon an x-ray error region[31] of area 0.2 deg^2 since there have been no modulation collimator detections of this weak (1.3 ± 0.3 μJy; 0.5–2.4 EHz) source. However, precise measurements with modulation collimators have confirmed that essentially all of the Seyferts proposed to be x-ray sources on the basis of comparable positions, e.g., reference 43, are correct.[57,41] Nevertheless, a confirmation of the identification would be comforting.

The x-ray data are very sparse. The Ariel-5 detection was at a significance of only 5σ. An Uhuru source (4U0106 − 59) with a 3 deg^2 error region and flux density of 3.3 ± 0.5 μJy may also be this source.[48] Neither the Ariel-5 nor the Uhuru data provide any information on variability.

Optical observations[46,149,145,32,60] all show this to be an extreme Seyfert 1 galaxy

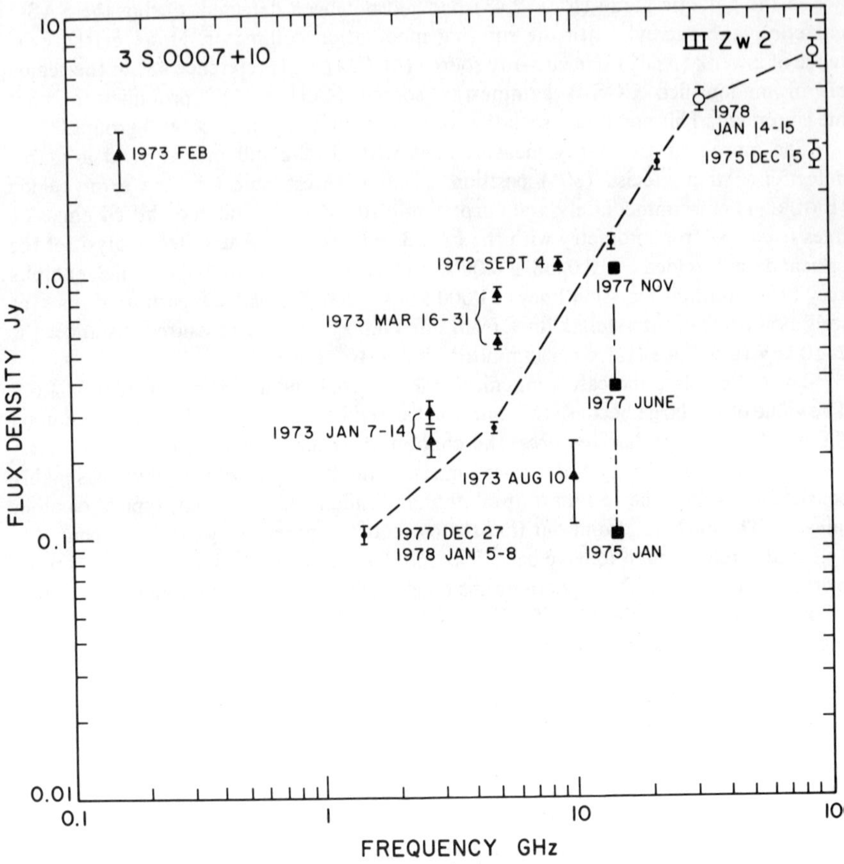

FIGURE 4. Radio spectrum of IIIZw2 in 1978 January (dashed line, SCHNOPPER *et al.*[132]) and previous results: open (1975) circle (B. ULICH in reference 132); squares (WRIGHT *et al.*[157]); and triangles (HUCHTMEIER and WRIGHT[72]). Taken from SCHNOPPER *et al.*[132].

which at $V = 13.2$ is one of the brightest Seyferts in the south.[145] With $z = 0.045$, its distance is almost exactly that of 0241+622. Its absolute magnitude is $M_v - 24.2$[149,145] (for $H_0 = 50$) and the dominance of the stellar nucleus ($L_{nucl}/L_{gal} \sim 3 - 10$; reference 149) place this in the Seyfert-quasar transition region as pointed out by all the authors referenced above. Its spectrum (FIGURE 3b) shows high excitation features with unusually weak [OIII] compared to Hβ. In this respect it is similar[145] to 3C273.[15] The envelope shows well-defined "lanes" or "spiral arms."[149]

In the infrared,[100,51,32] it is very luminous, 2×10^{45} erg s^{-1}, and variable ($\Delta K = 0.7$ mag., $\Delta L = 1.2$ mag. in 2 months, not typical of Seyferts[32]). A small and slow visual variability ($\Delta V = 0.2$ mag.) did not correlate with the IR variability.[32]

0241+622

The x-ray quasar, 0241+622, was located through its x-ray emission.[6] The object lies in the galactic plane ($b^{II} = 2.4°$) but had not been detected[7] during the SAS-3 galactic-plane survey[19] with the rotating modulation collimator. However the existence of a weak (3 μJy) Uhuru x-ray source (4U0241+61; reference 48) in the region of an unidentified COS-B gamma-ray source (CG135+1)[66] prompted another measurement. This possible association was commented upon by several groups.[66,92,75]

The more sensitive x-ray measurement with SAS-3 did indeed yield a source detection and a precise (30″) position.[6] The brightest object in the error region (FIGURE 1) was immediately and surprisingly found to exhibit red-shifted emission lines from spectrophotometry with the Lick 3-m telescope.[6] A detailed analysis of the optical data[96] yielded $z = 0.0438 \pm 0.0001$ and a spectrum (FIGURE 3c) which exhibits very broad Balmer emission lines (24,000 km s^{-1} FWZI), narrow permitted lines, as well as unshifted interstellar lines from our Galaxy. At the measured distance, the 2–10 keV (0.5–2.4 EHz) x-ray luminosity is 3×10^{44} erg s^{-1}.

The optical data indicate a magnitude of $V = 16.4$ and an extinction of $A_v \sim 4.6$.[96] The value of A_v, large because the source is located in the galactic plane, is based upon the width of interstellar features, the continuum spectral index, the Balmer decrement, the Hβ luminosity, and 21-cm measurements, all of which yield reasonably consistent results. The resultant absolute magnitude is $M_v \sim -25.3$, typical of other quasars. The authors[96] point out that the corrected apparent magnitude is about $V =$ 11.8 and bright enough to have been discovered years earlier had it not been obscured by the galactic plane. Deep photographs taken with the ISIT video camera on the Kitt Peak 4-m telescope indicate a faint ($V \sim 17.5$) nebulosity around 0241+622[47], and subsequent spectroscopic observations confirm its existence.[95]

A radio search was carried out at the VLA, and a relatively intense radio source within 2″ of the stellar-appearing object was discovered.[6] The radio spectrum is typical of Seyfert nuclei of quasars, decreasing from 0.29 to 0.20 Jy at low frequencies (1.5 to 4.9 GHz) and thereafter increasing up to 0.53 Jy at 22 GHz.[6] Interferometry[143] with baselines of 5 and 127 km has resolved the source into an extended component (~3 kpc; $S(\nu) \propto \nu^{-0.7}$) and a point-like component (<0.4 kpc) with a rising spectrum (0.4–2.7 GHz). This source is bright in the radio and is closer[6] to the sun than all the quasars cataloged in reference 23, excepting one very low luminosity object ($M_v = -19$). It is an excellent candidate for VLBI measurements. The red-shift distance (270 Mpc for $H_0 = 50$ km s^{-1} Mpc^{-1}) yields high spatial resolution (1 milliarcsec = 1.3 pc).

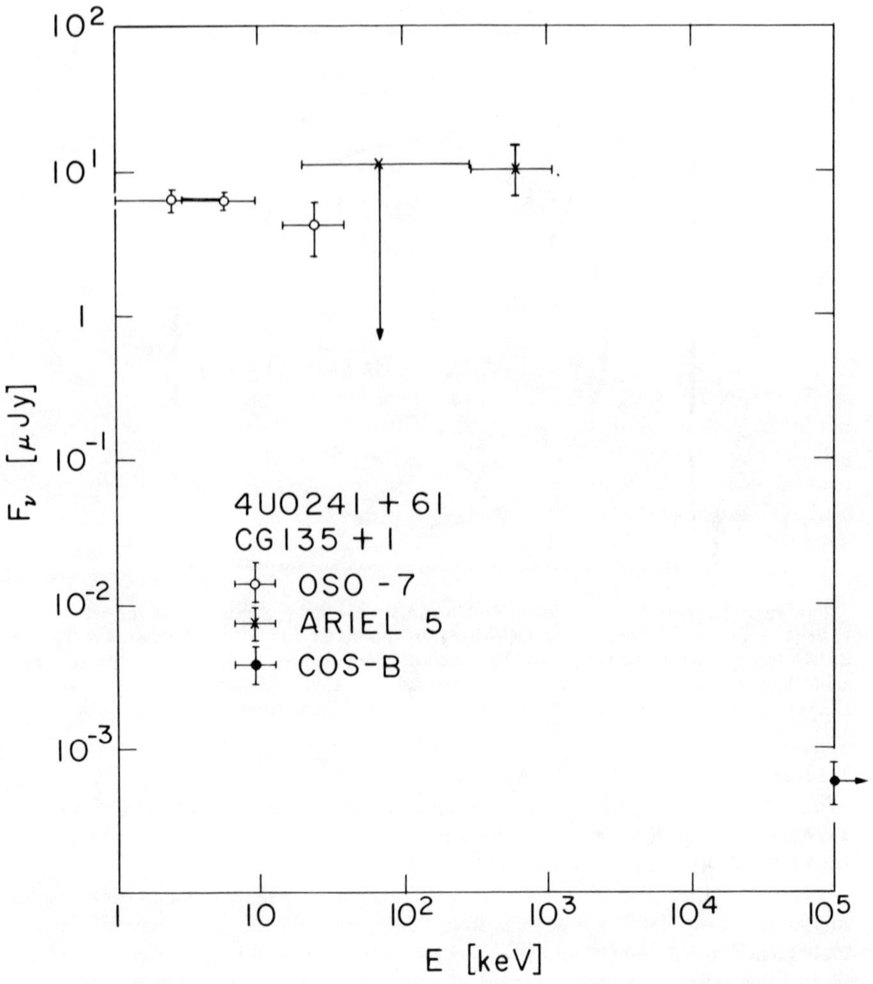

FIGURE 5. High-energy spectrum from region of 0241+622. OSO-7 (MARASCHI *et al.*[92]), Ariel 5 (COE *et al.*[28]), and COS-B (HERMSEN *et al.*[66]). These data may represent fluxes from one or more sources other than quasar 0241+622 due to the relatively large angular resolutions of the instruments (3°, 8°, and 2.5° FWHM for OSO-7, Ariel-5, and COS-B, respectively) and the source densities in the galactic plane ($b^{II} = 2°$; see text and FIGURE 6). Adapted from MARASCHI *et al.*[92]

A limit on the short term x-ray variability for 0241+622 was obtained from the NRL HEAO-1 data.[135] A six-day observation yielded a limit of <10%. The flux densities measured with the Uhuru,[48] SAS-3,[6] and HEAO-1[135] satellites are all consistent with a nonvarying flux of ~2 μJy (2–11 keV; 0.5–2.7 EHz). In contrast, the 3–10 keV (0.7–2.4 EHz) OSO-7 flux density (FIGURE 5) was 6 μJy,[92] indicating variability if source confusion is not present in the 3° (FWHM) field of view. The OSO-7 data also indicate a relatively hard spectrum with a substantial 15–40 keV (3.6–9.7 EHz) flux density of 4.2 μJy (FIGURE 5). A 0.3–1.2 MeV (70–290 EHz)

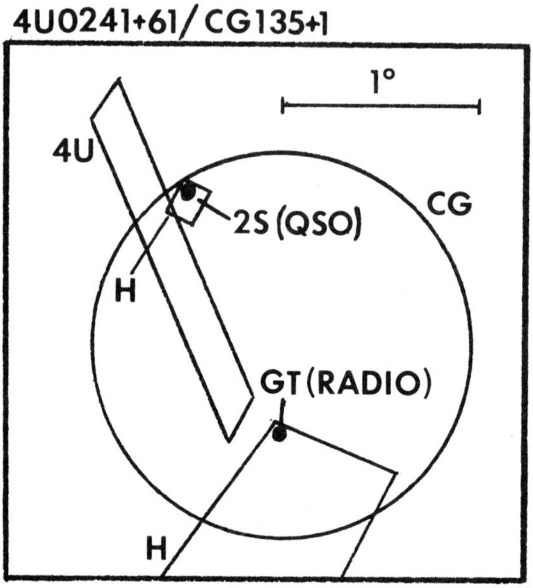

FIGURE 6. The position of the quasar and the SAS-3 position (2S, Apparao *et al.*,[6] see FIGURE 1) relative to the locations of the COS-B γ-ray source CG135+1 (HERMSEN *et al.*[66]), the variable-radio and OB star GT0236+610 (GREGORY and TAYLOR,[56] SANDULEAK,[126] HJELLMING *et al.*[68]), and the x-ray source error regions obtained from Uhuru (4U; FORMAN *et al.*[48]) and the HEAO-1 A-1 experiment (H; SHARE *et al.*[135]). Adapted from APPARAO *et al.*[6] and SHARE *et al.*[135]

source has also been detected in this region with the 8° FWHM scintillation telescope on Ariel-5, at a flux density of 10 μJy.[28] Finally, a search for 100–1000 GeV [(0.24–2.4) × 10^{26} Hz] radiation yields only upper limits.[65]

The possibility of source confusion in this region is real. A highly variable radio source, GT0236+610[56], 1.3 degrees from 0241+622 (FIGURE 6), has been convincingly identified with an OB star.[126,68] Since variable radio-emitting stellar systems are frequently associated with x-ray sources,[67,21] it is not unlikely that this particular radio source is also a (variable) x-ray source. The Uhuru source could be the GT source (FIGURE 6) as could the OSO-7 source. The NRL group[135] report a weak source in their HEAO-1 data of strength about 0.7 μJy with a position consistent with the GT source. On the other hand an upper limit of < 0.4 μJy (95% confidence) was obtained at another time with SAS-3.[6]

The association of the quasar and CG135+1, the COS-B gamma-ray source, remains in doubt. The extragalactic nature of the identification together with the low-latitude galactic distribution of the gamma-ray sources[66] argues against the association. However, the subsequent COS-B discovery[141] of a 100 MeV source in the vicinity of 3C273, another of the close and bright quasars, keeps the issue alive. The 1–40 keV, ~1 MeV, and ~100 MeV flux densities[92,28,66] from this region are summarized in FIGURE 5 where we stress that these fluxes could well arise from different objects. The GT stellar radio source has also been proposed as the counterpart of CG135 + 1,[56] where the authors point out the similarity of the radio source to the radio, x-ray, and gamma-ray emitter, Cyg X-3 (see references 56, 21).

3C273 (1226+023)

The first report of x-ray emission from 3C273 was based upon data obtained during a sounding rocket flight.[49] The statistical significance of this result was questioned[8] and responded to by the original authors who presented the raw counting rate data for the 160 s celestial scan.[50] We find that these data yield an intensity of 43 ± 42 counts for the 17 s when 3C273 was within the central 2/3 of the collimator field-of-view, i.e., the data which, a priori, should yield the greatest signal-to-noise ratio. A subsequent rocket flight by the Berkeley group[18] yielded a 3 standard deviation signal for a source in this region, and early Uhuru data[80] confirmed this result at a level of 6 standard deviations. The 2A and 4U catalogs[31,48] list detections at

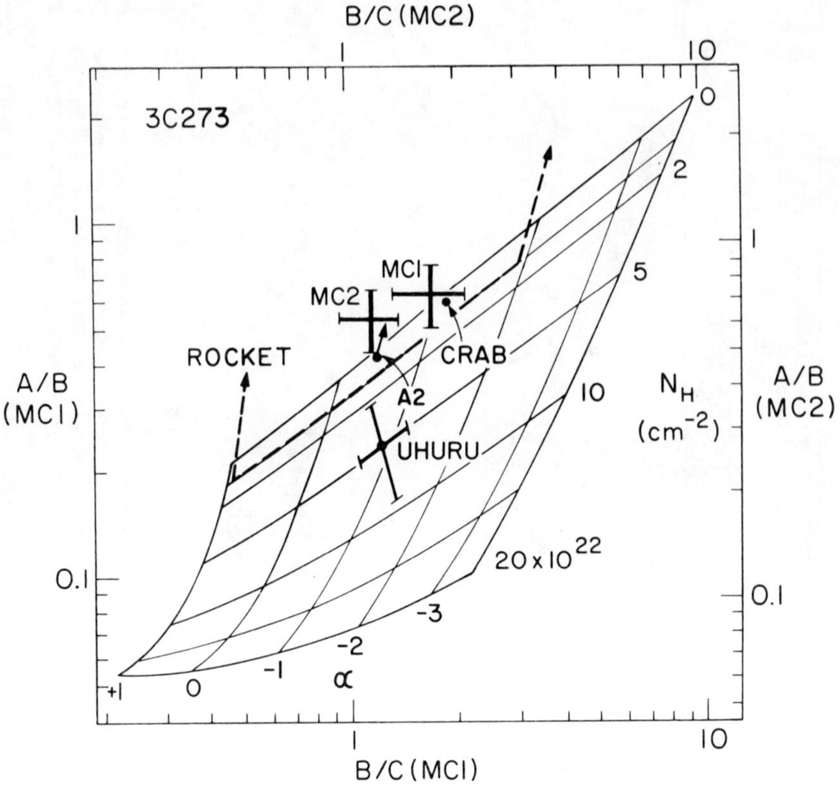

FIGURE 7. Plot of 3C273 for a power-law x-ray energy spectrum: 1.5–13.5 keV, $S(\nu) \propto \nu^{\alpha} \exp[-\sigma(\nu)\, N_H]$. UHURU (KELLOGG[79]), ROCKET (MARGON *et al.*[97]), A2 (HEAO-1 A2 experiment, WORRALL *et al.*[156]), MC1 and MC2 (HEAO-1 modulation collimators, BRADT *et al.*[22]). Adapted from reference 22.

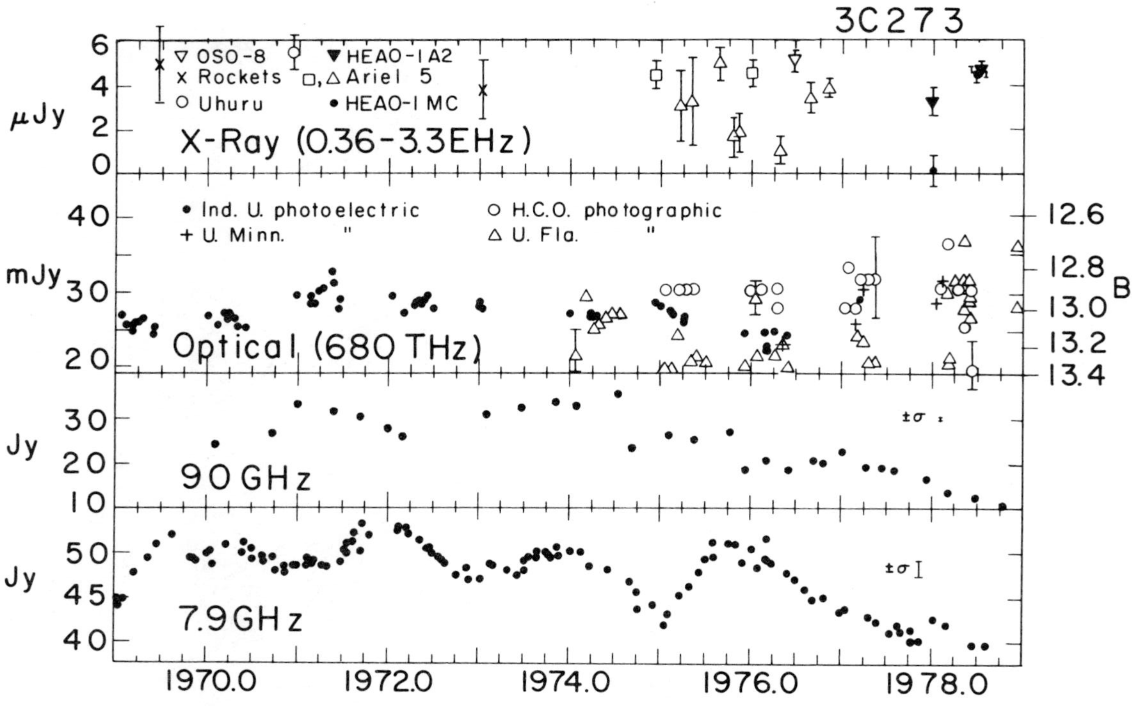

FIGURE 8. X-ray, optical, and radio flux densities from 3C273 versus time. The x-ray error bars do not include possible systematic offsets between the several experiments. The optical error bars do attempt to take systematic errors into account. The 1.0 GHz flux for 3C273 contains a substantial contribution from the jet 3C273A.[61,62] *X-RAY (1.5–13.5 keV):* (x) BOWYER *et al.*,[18] MARGON *et al.*,[94] MARGON *et al.*[97]; (○) KELLOGG *et al.*[80]; (□) SANFORD[127]; (△) POUNDS[113]; (●) BRADT *et al.*[22]; (▿, ▾) WORRALL *et al.*[156] *OPTICAL (B band):* (○) BURKHEAD and HILL,[24] previous work referenced therein, and reference 22 after 1975.5; (+) O'DELL *et al.*[107]; (○) reference 22 from the Harvard patrol plates; (△) reference 22. Typical 1 σ errors are ~0.04 mag for the photoelectric measurements. *RADIO: 90 GHz:* HOBBS and DENT,[69] reference 22 after 1977.0, typical error ±0.4 Jy. *7.9GHz:* DENT and KAPITZKY,[38] reference 22 after 1974.5, typical error $\pm 0.\bar{8}$ Jy. Adapted from BRADT *et al.*[22].

~10 standard deviations of 3 to 5 μJy, (2–10 keV) and positional error radii of ~8′ (FIGURE 9).

Spectral information from the Berkeley rocket flights[94,97] indicated the existence of a significant 1–2 keV flux whereas the Uhuru data[79] indicated a substantial low energy cutoff (E_a = 2.8 keV) corresponding to 5 × 10^{22} H atoms cm^{-2}. This would

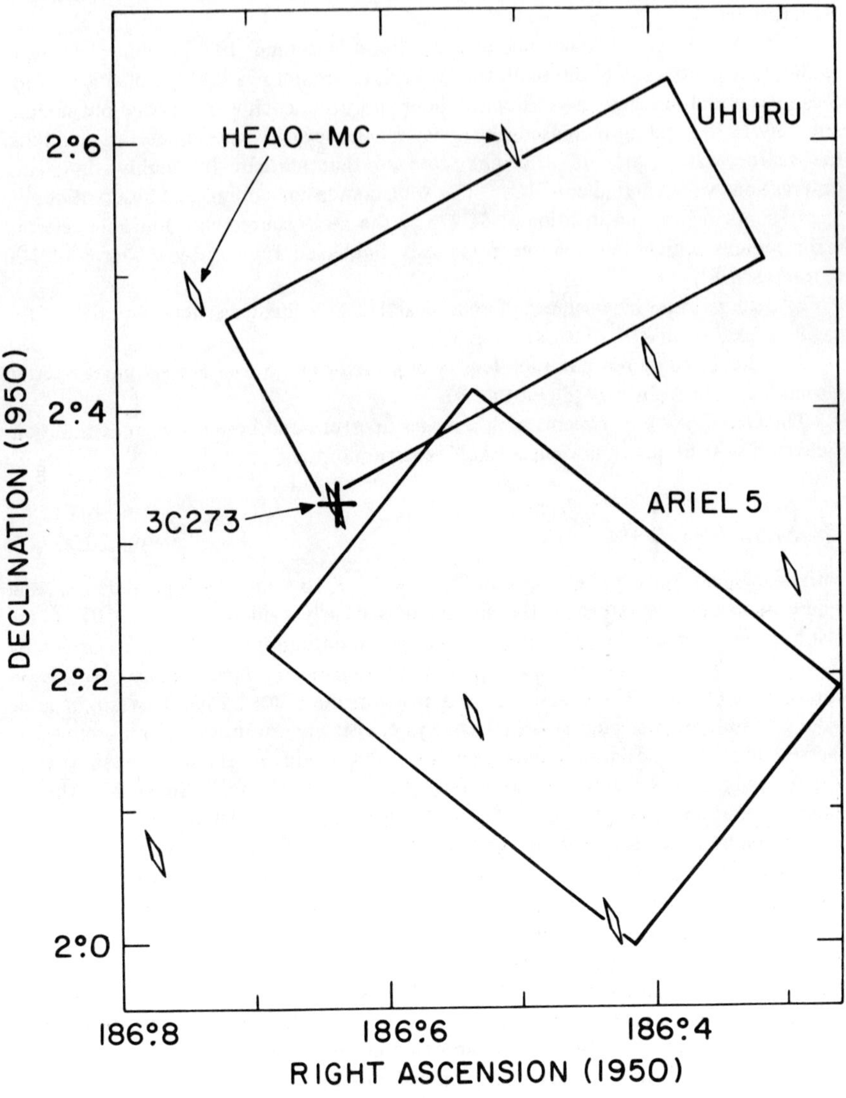

FIGURE 9. Multiple x-ray error regions obtained for 3C273 from the HEAO-1 modulation collimators (BRADT *et al.*[22]). Also shown are the Uhuru (FORMAN *et al.*[48]) and Ariel-5 (COOKE *et al.*[31]) error rectangles. From reference 22.

seem to indicate a variable cutoff. However, the results could well be self-consistent because the quoted errors (shown in FIGURE 7) for the Uhuru spectral parameters were underestimated.[97] The x-ray source intensity has been reported to be variable in the Ariel-5 SSI data.[113,31] It has also been detected with the MSSL Ariel-5 experiment[127,128] and the GSFC OSO-8 experiment.[156] The flux density of the source from the several detections ranged from 1 to 5 μJy (FIGURE 8). All in all, the x-ray data on this faint source have been very sparse, at least until the recent HEAO-1 observations.

The HEAO-1 satellite scanned over 3C273 in December 1977, in June 1978, and again in December 1978, although the latter data are not available as of this writing. Several pointed observations of about 6 hours duration each were carried out in June and July 1978 yielding increased sensitivity for the several experiments. At this time, the source was clearly in a bright state so that statistically highly significant observations were carried out.[114,22,156] The A3 (modulation collimator) observations[22]:

1. confirm the identification of 3C273 as the x-ray source with multiple celestial error regions which reduced the previously published regions by a factor of 500 (FIGURES 1, 9);

2. demonstrate the existence of an intense 1–2 keV flux consistent with that of the earlier rocket results[94,97] (FIGURE 7), and;

3. indicate an increase in flux density of a factor of $\gtrsim 3$ over the six month period December 1977 to July 1978 (FIGURE 8).

The GSFC A2 experiment, with a larger aperture and better spectral resolution, yields[156] a best-fit power-law energy keV spectrum of the form:

$$S(E) = 0.016\ \mathrm{E}^{-0.41}\ \mathrm{keV\ cm^{-2}\ s^{-1}\ keV^{-1}} \qquad (2\text{–}60\ \mathrm{keV})$$
$$S(\nu) = 5.9\ (\nu/10^{18})^{-0.41}\ \mu\mathrm{Jy} \qquad (0.5\text{–}14.5\ \mathrm{EHz})$$

with no evidence for a low-energy cutoff: $N_H < 4.5 \times 10^{21}$ cm^{-2} (90% confidence), and where additional structure in the spectrum is clearly evident (FIGURE 10). These results also indicate an increase in flux over the 6-month period, but by a factor of only ~1.4. The difference arises from an apparent discrepancy in the December 1977 fluxes obtained from the A2 and A3 HEAO-1 instruments (FIGURE 8). This might arise from a combination of poor statistics and systematic uncertainties.[22] Both groups find that the high statistics data of June and July 1978 provide no evidence for short term variability on time scales which range from 1.5 hr to 30 d. The 2σ limits from the A3 experiment are 30% and 15% for 1.5 and 17 d variability, respectively.[22]

The high-energy spectrum of 3C273 was measured from 13 to 120 keV (3.1–29 EHz) with the HEAO-1 A4 experiment[114] (FIGURES 10 and 11). A power-law fit to the data yields a logarithmic slope of -0.67 ± 0.14, again for the energy spectrum. Gamma-ray emission (~100 MeV) has been reported for a 7 deg^2 region which includes 3C273 by COS-B observers.[141] This source was the only significant peak in the 20° radius (1250 deg^2) field of view. The assurance that this source is indeed 3C273 rests upon the unusual nature of 3C273 and the relatively low probability (~2%) of finding a cataloged x-ray source in a given gamma-ray error region at high galactic latitude.[141] In the absence of a refined gamma-ray position, spectral points at energies intermediate between a few KeV and 100 MeV can help solidify the identification. The spectral data of the A4 experiment begin to fill this need. The A4

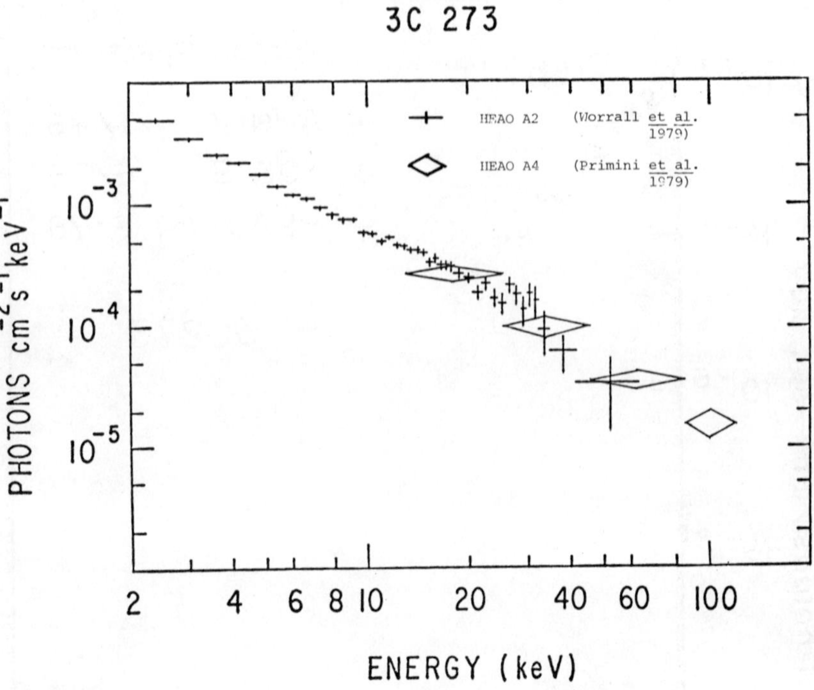

FIGURE 10. X-ray spectrum of 3C273 from the HEAO-1 A2 experiment (WORRALL *et al.*[156]) and HEAO-1 A4 experiment (PRIMINI *et al.*[44]). Adapted from reference 156 by PRIMINI and LEVINE.

energy spectrum can be extrapolated to the gamma-ray spectral points with a logarithmic slope[114] of -1.05 ± 0.03 (FIGURE 11).

The radio emission from 3C273 is intense and variable.[37,4,77,38,69,112,5] In 1964–1966, the 15.5 GHz flux density rose[4] from ~23 Jy to ~50 Jy, at which level it has remained since, albeit with ~ ±20% variations on time scales of months (see FIGURE 8 herein and also Figure 1(o) of reference 112). A lunar occultation of the radio flux[63] led to the optical identification.[131]

VLBI measurements show several components with sizes and spacings of order 1 milliarcsec (5 pc) and with apparent velocities greater than the speed of light (reference 78 and references therein; also see reference 86). These are located in the stellar optical component 3C273B. The source 3C273A at the end of the jet (FIGURE 1) is much larger. It is not detected in VLBI observations[27,73] (size $\gtrsim$200 pc at 600 MHz; i.e., unresolved at $4.6 \times 10^6\lambda$)[73] and is resolved at 1.5″ with lunar occultations[61,62] and at ~5″ (with a possible point component) in 1.5″ interferometry.[30] Thus, even though the HEAO-1 position measurement could not distinguish the two radio components[22] (FIGURE 1), it is likely that the variable x-ray source is associated with the radio and optical activity of 3C273B.

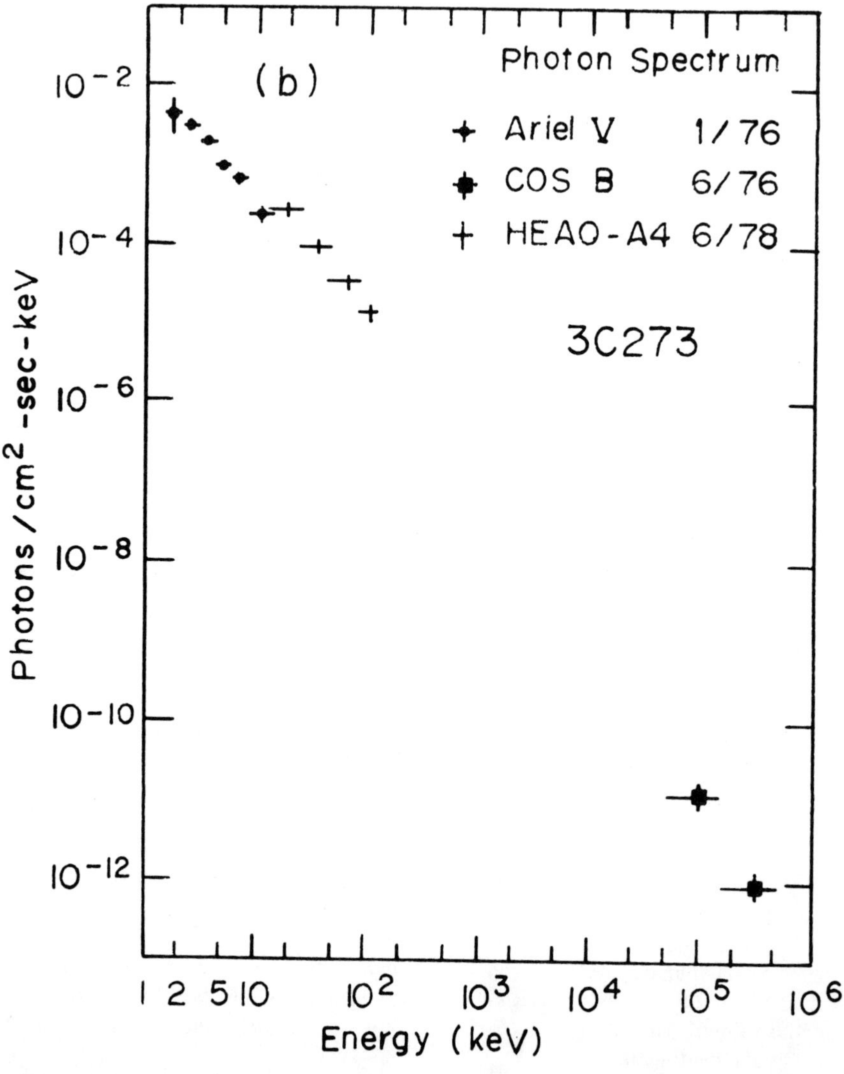

FIGURE 11. Number spectrum of 3C273: Ariel-5 (SANFORD[127]), HEAO-1 A4 (PRIMINI *et al.*[114]), COS-B (SWANENBURG *et al.*[141]). It is possible, but unlikely, that the COS-B source is not 3C273 (see text). Taken from PRIMINI *et al.*[114]

A compilation of radio, optical, and x-ray fluxes for 3C273 is presented in reference 22 (FIGURE 8). The 1978 x-ray increase was accompanied by a decrease in flux density at high radio frequencies (90 GHz) and probably by increased optical activity (see below). The x-ray data show no systematic trends due to the infrequent samplings. Several possible correlations between the radio and x-ray data are clearly not reproducible (e.g., the high x-ray/radio flux ratio in mid-1978 and the converse in mid-1976; and the approximately coincident downward 90 GHz and x-ray transitions at ~1975.8). The data demonstrate that the x-ray flux is not easily correlated with the radio, especially if possible delays are taken into account. A systematic search for optical–radio correlations has been carried out with negative results.[112]

The optical flux was relatively quiet from 1973 through 1977. Due to the absence of good photoelectric coverage in 1978, the photographic data for the later years are also presented in FIGURE 8. The work of each of the several groups exhibits good internal consistency but also systematic offsets relative to the other sets of data. A well documented (2 plates) flare at 1976.0 and the extreme variations reported by the University of Florida and Harvard College Observatory (H.C.O.) groups in 1978[22] appear to indicate real variability. The short term variability of the optical [e.g., ΔB = 0.09 mag. in 40 min[25] and $\Delta B \simeq 0.3$ mag. in 1d (1978 Dec 9–10; reference 137 and FIGURE 8)] may lend itself more easily to correlated studies with the x-ray intensity, although greatly increased frequency of observations in both wavelength domains would be required.

3C273 has been studied extensively through optical spectroscopy (e.g., references 131, 108, 11, 15) as well as at infrared[122,3] and ultraviolet[158,33,16] wavelengths. Its optical spectrum is notable for its low [OIII]/Hβ ratio and strong FeII emission. Studies of the jet (color and structure) indicate nonthermal emission and show no evidence of bright spots.[85]

IC4329A (1346–300)

This galaxy, an edge-on spiral, is an emission-line object in a richness class 0 cluster, 3′ from the giant elliptical IC4329.[124,36] Optical studies[39,99,153] (FIGURE 3e) show the characteristic broad Balmer lines (14,600 km s^{-1} FWZI) and narrow forbidden lines of a Sy 1. Photometry yielded V = 14.4 mag (10″ diaphragm), and the Balmer decrement indicated extinction of $A_v \simeq 4.2$.[39] A later study[153] yielded values of A_v ranging from 2.5–4.8, leading to absolute visual magnitudes of −22.9 to −25.2 for the red-shift distance of 94 Mpc. It is the nearest object in the present sample and has been called "the nearest quasar?"[153] It exhibits [FeX][153] and peculiar double-peaked Hβ emission.[39,99,153] (The latter feature appears as an asymmetric line shape in the red spectrum of FIGURE 3e.)

The 10 μm flux density (0.8 ± 0.07 Jy) leads to a 7.9 – 13 μm (23–38 PHz) luminosity of 8×10^{43} erg s^{-1}.[82,119] A radio flux of 24 ± 1 mJy at 5 GHz has been detected from IC4329A with a 4′ beam.[40]

The galaxy was tentatively identified with the 4.2 ± 0.5 μJy (0.5–2.4 EHz) x-ray source 2A1347–300 by the Ariel-5 group[31,43] on the basis of the 0.1 deg^2 x-ray error box that also includes the giant elliptical IC4329. The Uhuru data did not yield a significant detection.[142] Subsequent detections at 10–20 standard deviations with the SAS-3 modulation collimators yielded a position precise to 30″ which established IC4329A as the x-ray source[36] (FIGURE 1). The latter measurement yielded a flux density comparable to that from Ariel-5, and the two energy channels (2–6, 6–11 keV) indicated a flux spectrum somewhat harder than the Crab Nebula [$S(\nu) \propto \nu^{-0.3}$].

Mk 509 (2041 – 109)

Mk 509 was classified[84] as a Seyfert type 1 similar to a quasar because of its broad Hα-emission line, compact appearance, and high optical luminosity (M_v = –23.9; references 139, 43). Its spectrum[110,3,109] shows a FWZI for Hβ of 15,100 km/s.[109] The surrounding nebulosity is faint and apparently structureless.[1] It is luminous in the infrared[139,3,120] but does not show the strong broad permitted FeII lines which are typical of several other infrared objects, e.g., 3C273.[3]

The object lies in the 0.40 deg^2 celestial error region of the Ariel-5 x-ray source 2A 2040 – 115,[31,43] and was identified as the counterpart by that group. A recent precise position (FIGURE 1) from the HEAO-1 modulation collimator confirms the identification.[41] The Ariel-5 flux density[31] of 2.4 ± 0.3 μJy (2–10 keV; 0.5–2.4 EHz) corresponds to a luminosity of 2.5 × 10^{44} erg s^{-1} at the red-shift distance[84,3] of 210 Mpc. A search for x-ray emission with the Uhuru data base did not yield a statistically significant detection.[142] Comparison of the results yields no evidence for x-ray variability.

We are unaware of a radio detection or limit.

MR2251 – 178

This was the first quasar to be located through its x-ray emission.[117] The SAS-3 rotating modulation collimator was used to reduce the celestial positional error of the variable Ariel-5 x-ray source 2A2251 – 179[31] from 0.2 deg^2 to 1.4 arcmin^2. An examination of the objects within the x-ray error circle on the Tololo-Michigan Objective Prism Survey immediately showed that the indicated object (FIGURE 1) exhibited bright red-shifted emission lines. A search for radio emission with the VLA at 5 GHz showed emission at a flux density of 3.2 ± 0.3 mJy, from within ~1″ of the optical object. Another bright object, to the north in the original x-ray error circle, lies outside a refined error circle of 30″ radius[116] (FIGURE 1). It appears to be a compact elliptical galaxy without emission features.[117] For the distance quoted below, the Ariel-5 x-ray flux gives an x-ray luminosity which varies over the range <2 – 16 × 10^{44} erg s^{-1}. At the time of the SAS-3 observation (August 1977), the luminosity was about 5 × 10^{44} erg s^{-1}.[117]

A subsequent optical study[26] yielded V = 14.1, B = 14.6, Z = 0.064, and a Balmer decrement consistent with the assumed galactic extinction of A_v = 0.31. These values yield M_v = –24.1 and an optical UBV (λ = 3000–7000Å) luminosity of 1.1 × 10^{45} erg s^{-1}. Spectrophotometry of the object taken at the McGraw Hill Observatory with the Mark 1 and Mark 2 scanners (FIGURE 3g) show broad (18,000 km s^{-1}) Balmer emission, and lines due to He I, He II, [Ne II], [OI], [OII], [OIII], [Ne III], and [Ne V]. The authors conclude that the spectrum is typical of other quasars, and hence either there is no optical signature of the prodigious x-ray emission, or *all* quasars are x-ray emitters.

A recent deep plate obtained with the CTIO 4-m telescope by J. Graham[116] appears to show a faint nebulosity associated with the object. Also, a survey of the Harvard plate stacks[116] shows optical variability of ~2 magnitudes, B ~ 14–16.

DISCUSSION

The radio to x-ray spectra of the seven objects discussed herein are presented in FIGURE 2. The plotted values are for the selected frequencies, extinction factors, and times presented in TABLE 1. The seven are grouped according to their earlier classification by some authors as quasars (upper figure) and extreme Seyferts (lower figure). The x-ray fluxes are all very close to the detection limits of ~1 μJy (prior to the launch of the Einstein Observatory) whereas the optical fluxes are well above the detection threshold of $V \sim 18.5$ (FIGURE 2). It has been noted previously[43,6] that the *apparent* optical fluxes of the x-ray emitting Seyferts and quasars are preferentially among the most intense of the cataloged objects[148,23] in the two classes. Thus, more sensitive x-ray studies should lead to the detection of many presently cataloged quasars and Seyferts which optically are fainter.

The fact that the present x-ray studies lead to the detection of emission from the apparently brightest of the optical objects indicates at least a rough proportionality between the x-ray and optical fluxes for the objects presented here. [The more general question of whether the x-ray flux is always proportional to the optical continuum is very dependent upon selection factors and is still uncertain.[42] See discussion and FIGURE 4 of reference 43 (note the three open circles to the left in this figure with $V = 12.9$–13.6).] For the three objects in FIGURE 2a, the x-ray to optical ratios are the same within a factor of about 15. In contrast, the ratios of radio to x-ray fluxes differ by a factor of 10^4. This suggests that the optical and x-ray generation processes are more closely related to one another than are the x-ray and radio processes. The fluxes from the four objects in FIGURE 2b are also consistent with this pattern, where the extreme radio variability of IIIZw2 (FIGURE 4) and the large optical extinction correction applied to IC4329A must be kept in mind.

The x-ray flux from Seyfert galaxies has been reported to be correlated[43] with the parameters, e.g., Balmer line widths, IR flux, and possibly the optical continuum (but see comment above), that are often associated with a small (<0.01 pc) emission region (reference 150 and references therein) in the nucleus. The rapid variability of these components (e.g., one hour for the optical continuum of 3C273[25] and one day for the x-ray flux from NGC4151[105] and two months for the IR flux from F-9[32]) establishes their small extent. (Note, however, the lack of IR variability during the radio-optical outburst in IIIZw2, which suggests thermal radiation from a dust cloud of size >1 pc.[121]). The association of the x-ray flux with these small regions of high plasma densities ($\gtrsim 10^8$ cm^{-3}) and broad Balmer lines, due presumably to fast-moving particle streams,[90,111] is strengthened by a recent compilation[42] of x-ray luminosities and Balmer linewidths (FIGURE 12), which shows a rather convincing correlation. This set of parameters is relatively free of selection effects because the measured linewidths are independent of the distance to the source.[43]

In contrast, the narrow permitted lines and radio fluxes are believed to derive from larger regions. For example, VLBI measurements of 3C273B[78] resolve components of size down to ~3 pc, but note the ~1 month variability at 90 GHz (FIGURE 8). Sizes of ~1 kpc for both the permitted optical lines and the radio flux from Sy 2 nuclei have been directly resolved (e.g., references 34, 35, 150).

Optical nebulosities have been detected at each of the seven objects. The photographic appearance of IC4329A[39] and F-9[149] and the detection of interstellar absorption in IC4329A[153] indicate that the nebulosities of these two objects are stellar. The nonthermal jet[85] of 3C273 is the only reported nonstellar nebulosity in this group. The nature of the nebulosities of IIIZw2, 0241 + 622, Mk 509, and MR2251 − 178 are as yet undetermined, although spectroscopic studies of the envelope of IIIZw2 are consistent with a stellar component.[55] Despite these extended stellar components, each of the nuclear objects has been considered, in the literature, to fall within the

TABLE 1a*

NAMES, POSITIONS, RADIO

Names[a,b]	Position[a,c]		Radio[a]		
	α(1950) δ(1950) $\pm 2\sigma$	l^{II} b^{II}	ν GHz	S_ν^d mJy	Log $L^{e,f}$ 1–90 GHz erg s^{-1}
0007 + 106 IIIZw2 S (132) H (98)	$00^h\ 07^m\ 56\overset{s}{.}77$ 10° 41′ 48″ ± 1″ radio(132). cf. opt.(152)	107 − 51	1.5 4.9 22.5 90.0 Jan 78, during outburst; cf. (17,72,157);VLBI(155).	100 ± 10 270 ± 10 2600 ± 200 6900 ± 200(132)	 44.14 1–90GHz
0121 − 590 Fairall − 9 ESO113-IG45 2A,4U	01 21 51.22 − 59 03 58.9 ±0.4 opt.(149)	295 − 58			
0241 + 622 4U,S(6) H(135)	02 41 00.68 62 15 27.5 ± 1″ radio(6)	136 2	1.5 4.9 22.5 Dec. 1977.cf.(143).	290 ± 20 200 ± 10 530 ± 50(6)	42.44 for $\alpha = -0.5$ from 22–90 GHz
1226 + 023 3C273 4U,2A, Rockets (18,97), OSO(156), H(22,114,156).	12 26 33.29 2° 19′ 42.0″ ± 1″ radio, 3C273B(63)	290 64	1.0† 7.9‡ 90‡ variable; see FIGURE 8. †1970; includes 3C273A. ‡Jan. 1978. cf. (4,5,37,38,69). Lunar Occult. (61–63). VLBI (78,86). 3C273A(30,62).	40,000(77) 42,500 ± 900(22) 16,100 ± 200(22)	45.36
1346 − 300 IC4329A 2A,S(36)	13 46 27.87 − 30 03 40.6 ± 1″ opt. (152)	318 31	5	24 ± 1(40)	39.98 for $\alpha = -0.5$ 1–90 GHz
2041 − 109 Mk509 2A,H(41)	20 41 26.25 − 10 54 17.4 ± 1″ opt. (152)	36 − 30			
MR2251 − 178 2A,S(117)	22 51 25.89 − 17 50 54.2 ± 1″ opt. (117)	46 − 61	4.9	3.2 ± 0.3 (117)	40.33 for $\alpha = -0.5$ 1–90 GHz

*All lettered footnotes follow TABLE 1g.

TABLE 1b

INFRARED

Source	ν THz[g]	λ μm	$S_\nu^{a,d}$ mJy	α^h	Log $L^e_{1\text{-}30\mu m}$ 10–300 THz erg s^{-1}
IIIZw2	28	10.6	44 ± 9	−0.4	45.23(120)
	136	2.2	21 ± 2	(120)	
	240	1.25	9.3 ± 1(120)		
			not variable (121)		
Fairall−9	28	10.6	130 ± 40(51)	−1.0	45.3(32)
	88	3.4	51 ± 10*(32)	(32)	
	136	2.2	46 ± 5*(32)		
			*Jan. 1978; variable (32); cf. (100).		
0241+622					
3C273	28	10.6	240 – 590†(122)		46.6(122)
	88	3.4	143 ± 6(3)		
	136	2.2	83 ± 3(3)		
			†variable		
IC4329A	28	10.6	770 ± 70		44.81(120)
	83	3.6	250 ± 40 (120)		
			cf. (82, 119)		
Mk509	28	10.6	140 ± 20	−0.6	44.93(120)
	136	2.2	56 ± 5	(120)	
	240	1.25	28 ± 3(120)		
			cf. (3, 139)		
MR2251−178	88	3.4	44‡	∼−1.2	45.4
	136	2.2	22‡	(116)	
	240	1.25	18‡(83)		
			‡from FIGURE 4 of (116)		

TABLE 1c

OPTICAL

Source	$z_0^{a,i}$	D^j Mpc	V B–V U–B	A_v^k	ν THz[g]	S_ν^m mJy	M_v^n	Log $L^e_{0.3\text{-}0.7\,\mu m}$ 430–1000THz erg s^{-1}
IIIZw2	0.090	535	14.80	0.23	545	5.7	−24.1	45.00
	(9,55,129)		+0.44		682	4.5		
			−0.73(53).		822	4.1		
			variable		cf. (55)			
			(53,102)					
Fairall−9	0.0455	275	13.23	0.21	545	24	−24.2	45.14
	(149)		0.20		682	23		
	cf.(60,145)							
		−0.9						
		5(149)		822	26			
			cf.(46,100)		cf. (32)			

TABLE 1c (*continued*)

Source	$z_0^{a,i}$	D^j Mpc	V B–V U–B	A_v^k	ν THz[g]	S_ν^m mJy	M_v^n	Log $L^e_{0.3\text{–}0.7\,\mu m}$ 430–1000THz erg s^{-1}
0241+622	0.0445 (96)	265	16.4 1.4 — (96)	~4.6 (96) cf.(47)	545 682 822	73 80 —	−25.3	~45.66 for U−B = 0
3C273	0.158 (15,131)	945	12.80 0.21 −0.82(24) variable m_{pg}:13.0–11.5 (138) m_v:12.93–12.66 (24) cf. (22,112)	0.20	545 682 822	35 34 34	−27.3	46.36
IC4329A	0.0157 (153) cf. (124)	95	14.44 0.96 0.51(39)	~4.2 (39) 2.5–4.8 (153)	545 682 822	310 450 315	−24.6 for A_v = 4.2	45.39
Mk509	0.0351 (3) cf. (84)	210	13.12 0.23 −0.93(139)	0.36	545 682 822	30 30 34	−23.9	45.02
MR2251−178	0.064 (26)	385	14.11 0.52 −0.86 var. (116)	0.31 (26)	545 682 822	11.6 8.7 9.1	−24.1	45.03

TABLE 1d

OPTICAL (*continued*)

Source	Log L (Hβ)[p] erg s^{-1}	Hβ FWZI km s^{-1}	[OIII]/Hβ (4959 + 5070)/Hβ	FeII	Comments
IIIZw2	43.44(109) cf.(147)	15,800 (109) cf. (9)	0.48 (55) cf.(109)	Present (55,109)	Nearby galaxies have same z (55,129). Nebulosity emission at same z(55). Broad doubled Balmer lines (9).
Fairall−9	43.36 (60,145)	10650 (145)	0.16(60) cf. (145)	Strong (60,145)	Hβ asymmetric (100). L_{nucl}/L_{gal} ~3–10 (149). Nearby galaxy has same red shift (60).

TABLE 1d (*continued*)

Source	Log L (Hβ)[p] erg s^{-1}	Hβ FWZI km s^{-1}	[OIII]/Hβ (4959 + 5070)/Hβ	FeII	Comments
0241+622	43.38(96)	24,000 Hα(96)	Moderate(96)	Possibly present (96)	Nebulosity ~1 mag fainter than nucleus (47).
3C273	44.39(15) cf.(147)	20,000 Hα(15) cf. (11,108)	0.10(15)	Strong (15)	Structure and colors of jet (85,131)
IC4329A	43.45(39)	14,600 (153) cf. (39,99)	0.86(39) cf.(99,153)	Present (153)	Double peaked Hβ profile (39,99,153). Continuum prob. variable (153). [Fex] emission (153).
Mk509	43.44(109) cf.(84,110,147)	15,100 (109)	0.32(109)	Weak (3,43) cf. (109)	
MR2251−178	43.15(26)	18,000 (26)	0.88(26)	Not reported (26)	Fuzzy patch SW of object (116). Continum slope $\alpha = -1.4$ (26)

TABLE 1e
X-RAY

Source	ν EHz[g]	E keV	$S_\nu^{a,d,q}$ μJy	α^h 2–10 keV 0.5–2.4 EHz	Log L^r 2–10 keV 0.5–2.4 EHz erg s^{-1}	Positional Uncertainty[s] (″,90%)
IIIZw2	1.0	4.14	2.4 ± 0.3 (132) cf. (98)	−0.3 ± 0.3(132)	45.17	40″(132)
Fairall−9	1.3	5.3	1.3 ± 0.3 (31)		44.36	15′(31)
0241+622	0.91 2.0	3.8 8.3	2.2 ± 0.2 1.6 ± 0.2(6) cf.(48,92)	−0.4 ± 0.3(6)	44.54	30″(6)
3C273	0.5 1.0 2.0 4.0	2.1 4.1 8.3 16.6	7.9 ± 0.5 5.9 + 0.4 4.5 ± 0.3 3.4 ± 0.2 (156)	$\alpha = -0.4 \pm 0.2$, 2–60 keV, spect. shows structure(156).	46.06	22″(22)
	8.3 15.3 24.6	34 63 102	2.3 ± 0.5 1.5 ± 0.3 1.0 ± 0.3 (114)	$N_H < 4.5 \times 10^{21}$ cm^{-2} (156). $\alpha = -0.67 \pm 0.14$ 13–120 keV (114).	46.76 (2–100 keV, bright state)	
	June–July 1978; variable; cf. (18,22,80,94,97,113,127)					

TABLE 1e (*continued*)

Source	ν EHz[g]	E keV	$S_\nu^{a,d,q}$ μJy	α^h 2–10 keV 0.5–2.4 EHz	Log L^r 2–10 keV 0.5–2.4 EHz erg s^{-1}	Positional Uncertainty[s] (",90%)
IC4329A	1.3	5.3	4.5 ± 0.5 (31)	−0.3(36)	43.95	30"(36)
Mk509	1.3	5.3	2.4 ± 0.3 (31)		44.39 cf. (43)	25"(41)
MR2251−178	1.3*	5.3	4.5 ± 1.0 (31)	−0.5 ± 0.5(117)	44.76–45.20	30"(116)
	0.9†	3.8	2.0 ± 0.3 (117)			
	2.0†	8.3	1.4 ± 0.3 (117)			
	*1975; †1977; variable					

TABLE 1f

ULTRAVIOLET

Source	ν PHz[g]	$S_\nu^{a,d}$ mJy	α^h	Log L^e 1.0–2.5 PHz erg s^{-1}
3C273	0.97	25.1	$\alpha \simeq -1.0$(16)	46.35
	1.4	15.8		
	2.2	9.1(16) continuum. Line spectrum(16) cf.(33,158)		

TABLE 1g

GAMMA-RAY

Source	ν 10^{21} Hz	E MeV	$S_\nu^{a,d}$ nJy	α^h	Log $L^e_{50\text{–}500MeV}$ 12–120 × 10^{21} Hz erg s^{-1}
3C273	24	100	0.8 ± 0.3	−1.3	46.51
	79	325	0.18 ± 0.06(141)		
0241+622	See text and FIGURE 5 for discussion of gamma-ray sources in region of 0241+622: 0.300–1.2 MeV (28); >100 MeV (66); 100–1000 GeV upper limit (65)				

NOTES TO TABLES 1a–1g

[a]Reference numbers are given in parentheses. Numbering is the same as in reference list for this paper. References after *cf.* contain similar data or relevant comments. The references in the table (and text) for the seven sources are listed here.

IIIZw2
9 Arp 1968
17 Bolton 1968
43 Elvis *et al.* 1978
53 Green 1976
55 Green *et al.* 1978
59 Gunn 1971
72 Huchtmeier & Wright 1973
81 Khachikian & Weedman 1974
98 Marshall *et al.* 1979
102 Miller 1979
109 Osterbrock 1977
120 Rieke 1978
121 Rieke & Lebofsky 1979
129 Sargent 1970
132 Schnopper *et al.* 1978
147 Weedman 1976
148 Weedman 1977
152 Wilson & Meurs 1978
155 Wittels *et al.* 1978
157 Wright *et al.* 1977
159 Zwicky 1967
160 Zwicky 1971

IC4329A
31 Cooke *et al.* 1978
36 Delvaille *et al.* 1978
39 Disney 1973
40 Disney 1976
43 Elvis *et al.* 1978
82 Kleinmann & Wright 1974
99 Martin 1974
119 Rieke 1976
120 Rieke 1978
124 Sandage 1975
142 Tananbaum *et al.* 1978
148 Weedman 1977
152 Wilson & Meurs 1978
153 Wilson & Penston 1979

MR2251-178
26 Canizares *et al.* 1978
31 Cooke *et al.* 1978
83 Kleinmann *et al.* 1979
116 Ricker *et al.* 1979
117 Ricker *et al.* 1978

Fairall-9
13 Bergeron 1978
31 Cooke *et al.* 1978
32 Danks *et al.* 1979
46 Fairall 1977
48 Forman *et al.* 1978
51 Frogel 1978
60 Hawley & Phillips 1978
70 Holmberg *et al.* 1978
100 Martin *et al.* 1978
115 Ricker 1978
145 Ward *et al.* 1978
149 West *et al.* 1978

0241 + 6226
6 Apparao *et al.* 1978
7 Apparao *et al.* 1978
28 Coe *et al.* 1978
47 Ford 1978
48 Forman *et al.* 1978
56 Gregory & Taylor 1978
65 Helmken & Weekes 1979
66 Hermsen *et al.* 1977
68 Hjellming *et al.* 1978
75 Julien & Helmken 1978
92 Maraschi *et al.* 1978
95 Margon & Ford 1978
96 Margon & Kwitter 1978
126 Sanduleak 1978
135 Share *et al.* 1979
143 Tzanetakis *et al.* 1978

Mk 509
1 Adams 1977
3 Allen 1976
31 Cooke *et al.* 1978
41 Dower *et al.* 1980
43 Elvis *et al.* 1978
84 Kopylov *et al.* 1974
109 Osterbrock 1977
110 Osterbrock & Koski 1976
120 Rieke 1978
139 Stein & Weedman 1976
142 Tananbaum *et al.* 1978
147 Weedman 1976
148 Weedman 1977
152 Wilson & Meurs 1978

3C273
3 Allen 1976
4 Allen *et al.* 1968
5 Andrew *et al.* 1978
8 Argyle 1968
11 Baldwin 1975
15 Boksenberg *et al.* 1975
16 Boksenberg *et al.* 1978
18 Bowyer *et al.* 1970
22 Bradt *et al.* 1979
24 Burkhead & Hill 1975
25 Burkhead & Stein 1971
27 Clarke *et al.* 1969
30 Conway & Stannard 1975
31 Cooke *et al.* 1978
33 Davidsen *et al.* 1977
37 Dent 1965
38 Dent & Kapitzky 1976
48 Forman *et al.* 1978
49 Friedman & Byram 1967
50 Friedman *et al.* 1968
61 Hazard 1965
62 Hazard *et al.* 1966
63 Hazard *et al.* 1963
69 Hobbs & Dent 1977
73 Jauncey *et al.* 1971
74 Jones 1979
77 Kellerman & Pauliny-Toth 1971
78 Kellermann *et al.* 1977
79 Kellogg 1973
80 Kellogg *et al.* 1971
85 Kron *et al.* 1972
86 Legg *et al.* 1977
93 Margon 1978
94 Margon *et al.* 1972
97 Margon *et al.* 1975
101 McBreen 1979
107 O'Dell *et al.* 1978
108 Oke 1965
112 Pomphrey *et al.* 1976
113 Pounds 1977
114 Primini *et al.* 1979
122 Rieke & Low 1972
127 Sanford 1977
128 Sanford & Ives 1976
131 Schmidt 1963
137 Smith 1979
138 Smith & Hoffleit 1963
141 Swanenburg *et al.* 1978
147 Weedman 1976
156 Worrall *et al.* 1979
158 Wu 1977

[b]Source names. X-ray detections are also indicated: Rockets, Uhuru (4U: refs. 48,80), Ariel-5 (2A: refs. 31, 127), OSO-8, SAS-3(S), and HEAO-1 (H).

[c]Best available position from radio or optical work, as indicated. Errors are ~90% confidence.

[d]Sample flux densities. See quoted references for complete data. 1.0 Jy = 10^{-26} W m^{-2} Hz^{-1}; 1.0 nJy = 10^{-9} Jy.

[e]Luminosities are from the integration of fluxes S(ν) and slopes (α) in the previous columns, unless referenced. The luminosity distance, $D = cz_0/H_0$, $H_0 = 50$ km s^{-1} Mpc^{-1} is used. See note i below.

[f]The spectral shape $S(\nu) \propto \nu^{\alpha}$, $\alpha = -0.5$, is adopted where measurements are not available within the 1–90 GHz range.

[g]1.0 THz ≡ 10^{12} Hz; 1.0 PHz ≡ 10^{15} Hz ⟶ 0.30 μm; 1.0 EHz = 10^{18} Hz ⟶ 4.14 keV.

[h]α is the logarithmic slope: $S_{\nu} \propto \nu^{\alpha}$.

[i]z_0 is the galactocentric red shift, corrected in quoted work or here according to $\Delta V = 300 \sin l^{II} \cos b^{II}$; errors where quoted are about 1 place in the last digit.

[j]Luminosity distance, $D = cz_0/H_0$; $H_0 = 50$ km s^{-1} Mpc^{-1}.

[k]A_v is the interstellar absorption in the Galaxy as reported by the observers or according to $A_v = 0.18 \csc|b^{II}|$mag. For IC4329A, A_v includes absorption in the edge-on spiral.

[m]The three optical flux densities, $S(\nu)$, are derived respectively from V, B, and U. These magnitudes are corrected for extinction and then converted directly to flux densities with the typical conversions given in ALLEN.[2] Thus they include contributions from spectral line features.

[n]M_v, the absolute visual magnitude, is derived from values of V, A_v, and D in previous columns.

[p]The values of Hβ luminosity given for IIIZw2 and Mk509 are estimated from the equivalent widths given in ref. 109. $H_0 = 50$ km s^{-1} Mpc^{-1}.

[q]The flux density $S(\nu)$ for the Crab Nebula is 1000 μJy at 1.33 EHz (5.5 keV).

[r]X-ray luminosity for the indicated (previous column) power-law spectrum, or if not given, from the rate to flux conversion for the 2A catalog (ref. 31).

[s]Radius r_e of circle with area A of quoted 90% confidence error region: $A = \pi r_e^2$.

definitions of a quasar according to one or more of the criteria given in references 10, 125, and 130.

The seven "quasars" discussed in this paper thus appear to be both extreme examples of the Seyfert phenomenon and close examples of the quasar phenomenon. The distribution in FIGURE 12, which includes the other detected x-ray Seyferts, shows a continuum of points as does the distribution of x-ray and optical luminosities (FIGURE 13). In these distributions, only 3C273 lies slightly apart from the other objects. However the factor of 3 to 10 which separates it from the other objects does not seem sufficient to distinguish it as a different type of object. Similar arguments, based solely on optical properties, have been made by others (e.g., reference 146).

The similarity of the x-ray characteristics of 3C273 to those of other x-ray emitting galactic nuclei further strengthens the quasar-Seyfert-BL Lac link. The demonstrated x-ray variability of 3C273 (FIGURE 8) and MR2251 – 178[31,117] is typical of other known Seyferts[142,43] and the BL Lac object Mk421.[118] The 1–2 keV x-ray flux from 3C273[97,22,156] is also found to emanate from other Seyferts[103] and from BL Lac objects.[104,64,134] This absence of a low frequency cutoff indicates an absence of intervening matter, $<4.5 \times 10^{21}$ H atoms cm^{-2} for 3C273.[156] Thus the x-ray spectra do not lend independent support to the view that the presence of optical emission lines in quasars, compared to their absence in BL Lac objects, is due simply to an excess of nuclear gas in the former.[93]

The low-energy fluxes will provide significant constraints on the conditions, e.g., the filling factor, within the small broad-Balmer-line region from which the x-rays are presumed to emanate.[151,153] The typical conditions for this region ($D \sim 10^{-2}$pc; $n \sim 10^8$

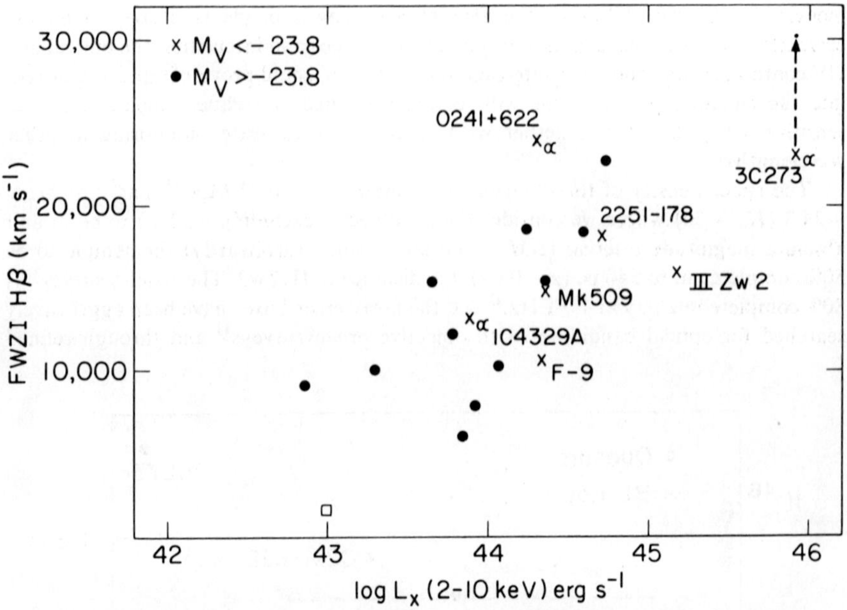

FIGURE 12. The Balmer line width (Hβ FWZI; or Hα FWZI = α) versus the x-ray luminosity. Crosses: the seven objects discussed here. Circles: other x-ray emitting Sy 1 nuclei. Square: a Sy 2 nucleus (NGC 5506 = 2A1410−029; RUBIN,[123] GRIFFITHS *et al.*,[58] and references therein). Provided by M. Elvis, private communication.[42]

cm^{-3}) clearly could give rise to substantial x-ray absorption. Recent calculations[136] address this issue and yield self-consistent results for 3C273.[156]

Models for the x-ray generation have been discussed and summarized by a number of authors (e.g., references, 14, 29, 76, 136, 45, 87, 88, 91, 150, 43, 105, 156). These include the synchrotron self-Compton process, thermal/Compton models wherein the low-frequency photons are Compton scattered to high energies by the thermal electrons, and thermal emission possibly arising from shocks caused by the high-velocity streams known to exist in these objects. The gamma-ray flux from 3C273 has been discussed in terms of second-order inverse self-Compton scattering,[74] synchrotron radiation derived from a pair-producing $>10^{14}$ eV γ-beam,[140] and its interaction with the x-ray photons.[101] Scenarios and detailed models to explain the x-ray variability are now forthcoming (references 52, 87, 105, 89, and references therein).

It has been stressed by several of these authors that studies of correlated variability (e.g., x-ray/optical or x-ray/radio) can provide critical information about the nature of the energy generation processes. Scenarios have been invoked to explain noncoincident (but possibly related and delayed) x-ray/radio events in 3C120[52] and Cen A.[12] However, the limited coverage and the limited number of detected events make it difficult to associate the events with confidence.

The data for 3C273 (FIGURE 8), with a high frequency of radio events and limited x-ray coverage, illustrate the difficulties inherent in such studies. However, the behavior of 3C273 at the several wavelengths already places constraints upon models.[156] The variability in the several wavelength bands is clearly rich, and correlated observations, e.g., of the Balmer line shapes and intensities, the optical and UV continuum, and the x-ray intensity and spectrum, could provide important insight into the emission process. This will require sustained or frequent monitoring with sensitive x-ray detectors together with routine and accurate monitoring at other wavelengths.

The space density of these objects is of order $\sim 2 \times 10^{-8}$ Mpc^{-3} mag^{-1} at $M_p \simeq -24.3$ ($H_0 = 50$) where we consider the six objects (excluding 3C273) to be in unit absolute magnitude interval ($\Delta M_\nu = 1$), and assume (arbitrarily) the sample to be 50% complete out to 540 pc ($z = 0.09$), the distance to IIIZw2. The Ariel-5 survey[31] is 80% complete to 2 μJy at 1.3 EHz,[43] and the x-ray error boxes have been aggressively searched for optical candidates with objective prism surveys[43] and through refined

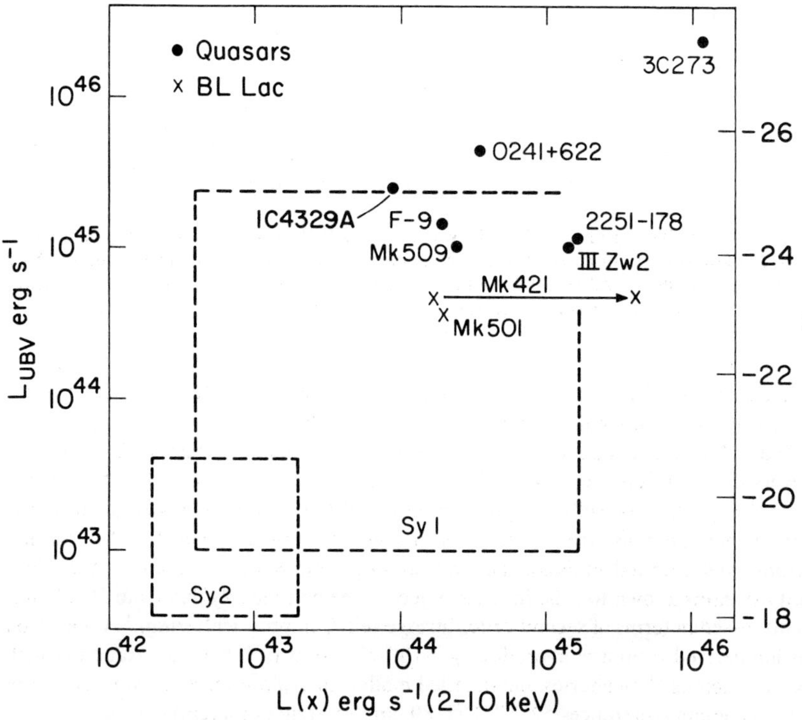

FIGURE 13. Optical (3000–7000 A) versus x-ray (2–10 keV) luminosity for the seven quasars and for two x-ray emitting BL Lac Objects: Mk421 (ULRICH *et al.*,[144] COOKE *et al.*,[31] RICKETTS *et al.*,[118] HEARN *et al.*[64]), and Mk501 (ULRICH *et al.*,[144] FORMAN *et al.*,[48] SCHWARTZ *et al.*[133]). Mk421 flared by a factor of $\sim$25 in several days.[118] The known x-ray variables 3C273 and MR2251$-$178 are shown at their maximum x-ray luminosities. The optical extinctions for 0241+622 and IC4329A are rather large ($A_\nu \sim 4$) and uncertain by $\sim \pm 1$.[96,153] The scale for the absolute magnitude, M_ν, is set for the colors of IIIZw2. The approximate extremes of L_x and L_{UBV} for x-ray emitting Seyfert nuclei are also shown: Sy 1 (ELVIS *et al.*,[43] TANANBAUM *et al.*,[142] MARSHALL *et al.*,[98]); Sy 2 (WARD *et al.*,[145] BRADT *et al.*[20]). The apparent proportionality between L_{UBV} and L_x is in part a selection effect.[43]

SAS-3 and HEAO-1 x-ray positions followed by spectroscopic observations of individual objects. On the other hand, the flux densities of these objects are all very close to the x-ray threshold and, for instance, IC4329A at 94 Mpc would not have been detected in x-rays at 550 Mpc. Also there are at least two other objects, with $D < 600$ Mpc ($z < 0.1$) and $M_v < -23.8$, viz., IZw1 ($D = 365$ Mpc, $M_v = -24.1$)[129,148] and PG1351+64 ($D = 530$ Mpc, $M_v = -23.8$),[53] which have not yet been detected at x-ray wavelengths ($L_x < 5 \times 10^{44}$ erg s^{-1} for IZw1[43]). [Other somewhat more distant transition objects ($z < 0.17$, $M_v < -23.7$) which have not yet been reported as x-ray sources are IIZw136;[129,148] PG0026+12, PG1001+05, PG2349−01;[53,54] 1612+262 (TON 256) and 1803+676.[23]]

The above space density is comparable to that obtained from a modest extrapolation of the luminosity function for Markarian Seyferts,[71] to $M_p = -24.3$ ($\sim 1.5 \times 10^{-8}$ Mpc^{-3} mag^{-1} corrected to $H_0 = 50$ from $H_0 = 75$) and that inferred by the present author from early results[54] of the Palomar Bright Quasar Survey, i.e., two objects at $M_B = 24.5 \pm 0.5$ within 1000 Mpc in 1424 deg^2 ($\sim 1 \times 10^{-8}$ Mpc^{-3} mag^{-1}). Thus, the seven such objects discussed herein may constitute a substantial portion of the Seyfert–quasar transition objects in our immediate neighborhood.

These x-ray transition objects may be studied with particularly high sensitivity compared to other quasars because they are close (95–950 Mpc) and hence quite bright at x-ray and optical wavelengths. They also can be studied with relatively high spatial resolution, e.g., with VLBI techniques to ~1 pc or with optical techniques to ~1 kpc for the study of the associated nebulosities. However, it is clear that variability studies will continue to provide the information about the smallest spatial scales in these compact objects (<0.01 pc).

ACKNOWLEDGMENTS

The author is grateful to a number of colleagues for helpful comments and suggestions, including J. Delvaille, M. Elvis, P. Joss, B. Margon, R. Mushotzky, F. Primini, G. Ricker, D. Schwartz, M. Ward, A. S. Wilson, D. Worrall, and his colleagues on the SAS-3 and HEAO-A3 programs which led to some of the results herein. He thanks T. Armstrong and T. Welch for assistance in the preparation of this paper.

REFERENCES

1. ADAMS, T. F. 1977. Astrophys. J. Supp. **33:** 19.
2. ALLEN, C. W. 1976. Astrophysical Quantities, 3rd Edition. The Athlone Press. London, England.
3. ALLEN, D. A. 1976. Astrophys. J. **207:** 367.
4. ALLEN, R. J., A. H. BARRETT & P. P. CROWTHER. 1968. Astrophys. J. **151:** 43.
5. ANDREW, B. H., J. M. MACLEOD, G. A. HARVEY & W. J. MEDD. 1978. Astron. J. **83:** 863.
6. APPARAO, K. M. V., G. F. BIGNAMI, L. MARASCHI, H. HELMKEN, B. MARGON, R. HJELLMING, H. V. BRADT & R. G. DOWER. 1978. Nature **273:** 450.
7. APPARAO, K. M. V., H. V. BRADT, R. G. DOWER, R. E. DOXSEY, J. G. JERNIGAN & F. LI. 1978. Nature **271:** 225.
8. ARGYLE, E. 1968. Science **159:** 747.
9. ARP, H. C. 1968. Astrophys. J. **152:** 1101.

10. BAHCALL, J. N. 1971. Astron. J. **76:** 283.
11. BALDWIN, J. A. 1975. Astrophys. J. Lett. **196:** L91.
12. BEALL, J. H., W. K. ROSE, W. GRAF, K. M. PRICE, W. A. DENT, R. W. HOBBS, E. K. CONKLIN, B. K. ULICH, B. R. DENNIS, C. J. CRANNELL, J. F. DOLAN, K. J. FROST & L. E. ORWIG. 1978. Astrophys. J. **219:** 836.
13. BERGERON, J. 1978. Private communication in reference 149.
14. BERGERON, J. & E. E. SALPETER. 1973. Astron. Astrophys. **22:** 385.
15. BOKSENBERG, A., K. SHORTRIDGE, R. A. E. FOSBURY, M. V. PENSTON & A. SAVAGE. 1975. Mon. Not. R. Astron. Soc. **172:** 289.
16. BOKSENBERG, A. *et al.* 1978. Nature **275:** 404.
17. BOLTON, J. G. 1968. Private communication in reference 9.
18. BOWYER, C. S., M. LAMPTON, J. MACK & F. DE MENDONCA. 1970. Astrophys. J. Lett. **161:** L1.
19. BRADT, H. V., K. M. V. APPARAO, G. W. CLARK, R. DOWER, R. E. DOXSEY, D. R. HEARN, J. G. JERNIGAN, P. C. JOSS, W. MAYER, J. MCCLINTOCK & F. WALTER. 1977. Nature **269:** 21.
20. BRADT, H. V., B. F. BURKE, C. R. CANIZARES, P. E. GREENFIELD, R. L. KELLEY, J. E. MCCLINTOCK & J. VAN PARADJIS. 1978. Astrophys. J. Lett. **226:** L111.
21. BRADT, H. V., R. E. DOXSEY & J. G. JERNIGAN. 1979. *In* X-ray Astronomy. Adv. in Space Exploration. Vol. 3. W. Baity and L. Peterson, Eds. :3 Pergamon Press. Oxford, England.
22. BRADT, H. V., R. E. DOXSEY, M. D. JOHNSTON, D. A. SCHWARTZ, M. S. BURKHEAD, W. A. DENT, W. LILLER & A. G. SMITH. 1979. Astrphys. J. Lett. **230:** L5.
23. BURBIDGE, G. R., A. H. CROWNE & H. E. SMITH. 1977. Astrophys. J. Supp. **33:** 113.
24. BURKHEAD, M. S. & R. K. HILL. 1975. Pub. Astron. Soc. Pacific. **87:** 821, and previous work referenced therein.
25. BURKHEAD, M. S. & W. L. STEIN. 1971. Pub. Astron. Soc. Pacific. 1971. **83:** 830.
26. CANIZARES, C. R., J. E. MCCLINTOCK & G. R. RICKER. 1978. Astrophys. J. Lett. **226:** L1.
27. CLARKE, R. W., N. W. BROTEN, T. H. LEGG, J. L. LOCKE & J. L. YEN. 1969. Mon. Not. R. Astron. Soc. **146:** 381.
28. COE, M. J., J. J. QUENBY & A. R. ENGEL. 1978. Nature **274:** 343; see also I.A.U. Circular 3003.
29. COLGATE, S. A., J. D. COLVIN & A. G. PETSCHEK. 1975. Astrophys. J. Lett. **197:** L105.
30. CONWAY, R. G. & D. STANNARD. 1975. Nature **255:** 310.
31. COOKE, B. A., M. J. RICKETTS, T. MACCACARO, J. P. PYE, M. ELVIS, M. G. WATSON, R. E. GRIFFITHS, K. A. POUNDS, I. MCHARDY, D. MACCAGNI, F. D. SEWARD, C. G. PAGE & J. J. L. TURNER. 1978. Mon. Not. R. Astron. Soc. **182:** 489.
32. DANKS, A. C., W. WAMSTEKER, N. VOGT, P. SALINARI, M. TARENGHI & H. W. DUERBECK. 1979. Astrophys. J. Lett. **227:** L59.
33. DAVIDSEN, A. F., G. F. HARTIG & W. G. FASTIE. 1977. Nature. **269:** 203.
34. DE BRUYN, A. G. & A. S. WILSON. 1976. Astron. Astrophys. **53:** 93.
35. DE BRUYN, A. G. & A. S. WILSON. 1978. Astron. Astrophys. **64:** 433.
36. DELVAILLE, J. P., M. J. GELLER & H. W. SCHNOPPER. 1978. Astrophys. J. Lett. **226:** L69.
37. DENT, W. A. 1965. Science **148:** 1458.
38. DENT, W. A. & J. E. KAPITZKY. 1976. Astron. J. **81:** 1053.
39. DISNEY, M. J. 1973. Astrophys. J. Lett. **181:** L55.
40. DISNEY, M. J. 1976. Quoted in reference 34.
41. DOWER, R., R. E. GRIFFITHS, H. V. BRADT, R. E. DOXSEY & M. D. JOHNSTON. 1980. Astrophys. J. **235.** In press.
42. ELVIS, M. 1979. Private communication.
43. ELVIS, M., T. MACCACARO, A. S. WILSON, M. J. WARD, M. V. PENSTON, R. A. E. FOSBURY & G. C. PEROLA. 1978. Mon. Not. R. Astron. Soc. **183:** 129.
44. FABIAN, A. C. 1979. This volume.
45. FABIAN, A. C. & M. J. REES. 1979. *In* X-ray Astronomy. Adv. in Space Exploration. Vol. 3. W. Baity & L. Peterson, Eds. :381. Pergamon Press. Oxford, England.

46. FAIRALL, A. P. 1977. Mon. Not. R. Astron. Soc. **180:** 391.
47. FORD, H. C. 1978. Bull. Am. Astr. Soc. **10:** 450.
48. FORMAN, W., C. JONES, L. COMINSKY, P. JULIEN, S. MURRAY, G. PETERS, H. TANANBAUM & R. GIACCONI. 1978. Astrophys. J. Supp. **38:** 357.
49. FRIEDMAN, H. & E. T. BYRAM. 1967. Science **158:** 257.
50. FRIEDMAN, H., E. T. BYRAM & T. A. CHUBB. 1968. Science **159:** 747.
51. FROGEL, J. 1978. Private communication in reference 60.
52. GELLER, M. J., J. P. DELVAILLE, F. J. PINEDA, F. D. ROSENBERG & H. W. SCHNOPPER. 1979. In preparation.
53. GREEN, R. F. 1976. Pub. Astron. Soc. Pacific. **88:** 665.
54. GREEN, R. F. & M. SCHMIDT. 1978. Astrophys. J. Lett. **220:** L1.
55. GREEN, R. F., T. B. WILLIAMS & D. C. MORTON. 1978. Astrophys. J. **226:** 729.
56. GREGORY, P. C. & A. R. TAYLOR. 1978. Nature **272:** 704.
57. GRIFFITHS, R. E., U. BRIEL, D. A. SCHWARTZ, J. SCHWARZ, R. E. DOXSEY & M. D. JOHNSTON. 1979. Mon. Not. R. Astron. Soc. **188.** In press.
58. GRIFFITHS, R. E., R. E. DOXSEY, M. D. JOHNSTON, D. A. SCHWARTZ, J. SCHWARZ & J. C. BLADES. 1979. Astrophys. J. Lett. **230:** L21.
59. GUNN, J. E. 1971. Astrophys. J. Lett. **164:** L113.
60. HAWLEY, S. A. & M. M. PHILLIPS. 1978. Astrophys. J. **225:** 780.
61. HAZARD, C. 1965. *In* Quasi Stellar Sources and Gravitational Collapse. I. Robinson, Ed.: 135. University of Chicago Press. Chicago, Ill.
62. HAZARD, C., S. GULKIS & A. D. BRAY. 1966. Nature **212:** 461.
63. HAZARD, C., M. B. MACKEY & A. J. SHIMMINS. 1963. Nature **197:** 1037.
64. HEARN, D. R., F. J. MARSHALL & J. G. JERNIGAN. 1979. Astrophys. J. Lett. **227:** L63.
65. HELMKEN, H. F. & T. C. WEEKES. 1979. Astrophys. J. **228:** 531.
66. HERMSEN, W., B. N. SWANENBURG, G. F. BIGNAMI, G. BOELLA, R. BUCCHERI, L. SCARSI, G. KANBACH, H. A. MAYER-HASSELWANDER, J. L. MASNOU, J. A. PAUL, K. BENNETT, J. C. HIGDON, G. G. LICHTI, B. G. TAYLOR & R. D. WILLS. 1977. Nature **269:** 494.
67. HJELLMING, R. M. 1976. *In* The Physics of Nonthermal Radio Sources. G. Setti, Ed.: 203. Reidel. Dordrecht, Holland.
68. HJELLMING, R., D. HOGG, H. HVATUM, P. GREGORY & R. TAYLOR. 1978. IAU Circular 3180.
69. HOBBS, R. W. & W. A. DENT. 1977. Astron. J. **82:** 257.
70. HOLMBERG, E. B., A. LAUBERTS, H. E. SCHUSTER, & R. M. WEST. 1978. Astron. Astrophys. Supp. **31:** 15.
71. HUCHRA, J. & W. L. W. SARGENT. 1973. Astrophys. J. **186:** 433.
72. HUCHTMEIER, W. & A. E. WRIGHT. 1973. Astrophys. Lett. **15:** 209.
73. JAUNCEY, D. L. *et al.* 1971; quoted in Kellermann *et al.* 1971. Astrophys. J. **169:** 1.
74. JONES, T. W. 1979. Astrophys. J. In press.
75. JULIEN, P. F. & H. F. HELMKEN. 1978. Nature **272:** 699.
76. KATZ, J. I. 1976. Astrophys. J. **206:** 910.
77. KELLERMANN, K. I. & I. I. K. PAULINY-TOTH. 1971. Astrophys. Lett. **8:** 153.
78. KELLERMANN, K., D. SCHAFFER, G. PURCELL, I. PAULINY-TOTH, E. PREUSS, A. WITZEL, D. GRAHAM, R. SCHILIZZI, M. COHEN, A. MOFFET, J. ROMNEY & A. NIELL. 1977. Astrophys. J. **211:** 658.
79. KELLOGG, E. M. 1973. *In* X- and Gamma-Ray Astronomy. I.A.U. Symposium 55, H. Bradt & R. Giacconi, Eds.: 171. Reidel. Dordrecht, Holland.
80. KELLOGG, E., H. GURSKY, C. LEONG, E. SCHREIER, H. TANANBAUM & R. GIACCONI. 1971. Astrophys. J. Lett. **165:** L49.
81. KHACHIKIAN, E. YE. & D. W. WEEDMAN. 1974. Astrophys. J. **192:** 581.
82. KLEINMANN, D. W. & E. L. WRIGHT. 1974. Astrophys. J. Lett. **191:** L19.
83. KLEINMANN, S., S. SARGENT & R. JOYCE. 1979. Private communication in reference 116.
84. KOPYLOV, I. M., V. A. LIPOVETSKII, V. I. PRONIK & K. K. CHUVAEV. 1974. Astrofizika. **10:** 483.
85. KRON, G. E., H. D. ABLES & A. V. HEWITT. 1972. Pub. Astron. Soc. Pacific **84:** 303.

86. LEGG, T. H., N. W. BROTEN, D. N. FORT, M. J. S. QUIGLEY, F. V. BALE, P. C. BARBER & J. L. JEN. 1977. Astrophys. J. **211:** 21.
87. LIGHTMAN, A. P., R. GIACCONI & H. TANANBAUM. 1978. Astrophys. J. **224:** 375.
88. LIGHTMAN, A. P. & G. B. RYBICKI. 1979. Astrophys. J. Lett. **229:** 15.
89. LIGHTMAN, A. P. & G. B. RYBICKI. 1979. Astrophys. J. **232:** 882.
90. MACALPINE, G. M. 1974. Astrophys. J. **193:** 37.
91. MANCHANDA, R. K., E. COSTA, B. M. BELLI & P. UBERTINI. 1978. Astrophys. Space Sci. **53:** 231.
92. MARASCHI, L., T. MARKERT, K. M. V. APPARAO, H. BRADT, H. HELMKEN, W. WHEATON, W. A. BAITY & L. E. PETERSON. 1978. Nature **272:** 679.
93. MARGON, B. 1978. Private communication.
94. MARGON, B., S. BOWYER & M. LAMPTON. 1972. Astrophys. J. **174:** 471.
95. MARGON, B. & H. FORD. 1978. Private communication.
96. MARGON, B. & K. B. KWITTER. 1978. Astrophys. J. Lett. **224:** L43.
97. MARGON, B., M. LAMPTON, S. BOWYER & R. CRUDDACE. 1975. Astrophys. J. **197:** 25.
98. MARSHALL, F. E., E. A. BOLDT, S. S. HOLT, R. F. MUSHOTZKY, S. H. PRAVDO, R. E. ROTHSCHILD & P. J. SERLEMITSOS. 1979. Astrophys. J. Supp. **40:** 657.
99. MARTIN, W. L. 1974. Mon. Not. R. Astron. Soc. **168:** 109.
100. MARTIN, W. L., J. E. PENFOLD & I. S. GLASS. 1978. Mon. Not. R. Astron. Soc. **184:** 15p.
101. MCBREEN, B. 1979. Astron. Astrophys. **71:** L19.
102. MILLER, J. 1979. Private communication.
103. MUSHOTZKY, R. F., F. E. MARSHALL, E. A. BOLDT, S. S. HOLT & P. J. SERLEMITSOS. 1979. Astrophys. J. In press.
104. MUSHOTZKY, R. F., E. A. BOLDT, S. S. HOLT, S. H. PRAVDO, P. J. SERLEMITSOS, J. H. SWANK & R. H. ROTHSCHILD. 1978. Astrophys. J. Lett. **226:** L65.
105. MUSHOTZKY, R. F., S. S. HOLT & P. J. SERLEMITSOS. 1978. Astrophys. J. Lett. **225:** L115.
106. MUSHOTZKY, R. F., P. J. SERLEMITSOS, R. H. BECKER, E. A. BOLDT & S. S. HOLT. 1978. Astrophys. J. **220:** 790.
107. O'DELL, S. L., J. J. PUSCHELL, W. A. STEIN & J. W. WARNER. 1978. Astrophys. J. Supp. **38:** 267.
108. OKE, J. B. 1965. Astrophys. J. **141:** 6.
109. OSTERBROCK, D. E. 1977. Astrophys. J. **215:** 733.
110. OSTERBROCK, D. E. & A. T. KOSKI. 1976. Mon. Not. R. Astron. Soc. **176:** 61p.
111. OSTERBROCK, D. E., A. T. KOSKI & M. M. PHILLIPS. 1976. Astrophys. J. **206:** 898.
112. POMPHREY, R. B., A. G. SMITH, R. J. LEACOCK, C. N. OLSSON, R. L. SCOTT, J. T. POLLOCK, P. EDWARDS & W. A. DENT. 1976. Astron. J. **81:** 489.
113. POUNDS, K. 1977. Ann. N.Y. Acad. Sci. **302:** 361.
114. PRIMINI, F. A., B. A. COOKE, C. A. DOBSON, S. K. HOWE, A. SCHEEPMAKER, W. A. WHEATON, W. H. G. LEWIN, W. A. BAITY, D. E. GRUBER, J. L. MATTESON & L. E. PETERSON. 1979. Nature. **278:** 234.
115. RICKER, G. R. 1978. Nature **271:** 334.
116. RICKER, G. R., G. W. CLARK, R. E. DOXSEY, R. G. DOWER, J. G. JERNIGAN, C. R. CANIZARES, J. P. DELVAILLE, G. M. MACALPINE & R. M. HJELLMING. 1979. *In* X-ray Astronomy. Adv. in Space Exploration. Vol. 3. W. Baity & L. Peterson, Eds. :281. Pergamon Press. Oxford, England.
117. RICKER, G. R., G. W. CLARK, R. E. DOXSEY, R. G. DOWER, J. G. JERNIGAN, J. P. DELVAILLE, G. M. MACALPINE & R. M. HJELLMING. 1978. Nature **271:** 35.
118. RICKETTS, M. J., B. A. COOKE & K. A. POUNDS. 1976. Nature **259:** 546.
119. RIEKE, G. H. 1976. Astrophys. J. Lett. **206:** L15.
120. RIEKE, G. H. 1978. Astrophys. J. **226:** 550.
121. RIEKE, G. H. & M. J. LEBOFSKY. 1979. Astrophys. J. **227:** 710.
122. RIEKE, G. H. & F. J. LOW. 1972. Astrophys. J. Lett. **176:** L95.
123. RUBIN, V. C. 1978. Astrophys. J. Lett. **224:** L55.
124. SANDAGE, A. 1975. Astrophys. J. **202:** 563.
125. SANDAGE, A. 1972. Astrophys. J. **178:** 25.

126. Sanduleak, N. 1978. IAU Circular 3170.
127. Sanford, P. W. 1977. Ann. N. Y. Acad. Sci. **302:** 386.
128. Sanford, P. W. & J. C. Ives. 1976. Proc. R. Soc. Lond. A. **350:** 491.
129. Sargent, W. L. W. 1970. Astrophys. J. **160:** 405.
130. Schmidt, M. 1970. Astrophys. J. **162:** 371.
131. Schmidt, M. 1963. Nature **197:** 1040.
132. Schnopper, H. W., J. P. Delvaille, A. Epstein, W. Cash, P. Charles, S. Bowyer, R. M. Hjellming, F. N. Owen & W. D. Cotton. 1978. Astrophys. J. Lett. **222:** L91.
133. Schwartz, D. A., H. V. Bradt, R. E. Doxsey, R. E. Griffiths, H. Gursky, M.D. Johnston & J. Schwarz. 1978. Astrophys. J. Lett. **224:** L103.
134. Schwartz, D. A., R. E. Doxsey, R. E. Griffiths, M.D. Johnston & J. Schwarz. 1979. Astrophys. J. Lett. **229:** L53.
135. Share, G. H., K. Wood, R. L. Kinzer, S. Shulman, J. Samimi, W. N. Johnson, D. Yentis, J. Meekins, W. Evans, D. McNutt, E. Byram, T. Chubb & H. Friedman. 1979. *In* X-ray Astronomy. Advances in Space Exploration. Vol. 3. W. Baity & L. Peterson, Eds. :535. Pergamon Press. Oxford, England.
136. Shields, G. A. & R. F. Mushotzky. 1979. Astron. Astrophys. In press.
137. Smith, A. G. 1979. Private communication.
138. Smith, H. J. & D. Hoffleit. 1963. Nature **198:** 650.
139. Stein, W. A. & D. W. Weedman. 1976. Astrophys. J. **205:** 44.
140. Swanenburg, B. N. 1978. Astron. Astrophys. **70:** L71.
141. Swanenburg, B. N., K. Bennett, G. F. Bignami, P. Caraveo, W. Hermsen, G. Kanbach, J. L. Masnou, H. A. Mayer-Hasselwander, J. A. Paul, B. Sacco, L. Scarsi & R. D. Wills. 1978. Nature **275:** 298.
142. Tananbaum, H., G. Peters, W. Forman, R. Giacconi, C. Jones & Y. Avni. 1978. Astrophys. J. **223:** 74.
143. Tzanetakis, A., R. E. Spencer, C. R. Masson & J. E. Baldwin. 1978. Mon. Not. R. Astron. Soc. **185:** 63p.
144. Ulrich, M-H., T. D. Kinman, C. R. Lynds, G. H. Rieke & R. D. Ekers. 1975. Astrophys. J. **198:** 261.
145. Ward, M. J., A. S. Wilson, M. V. Penston, M. Elvis, T. Maccacaro & K. P. Tritton. 1978. Astrophys. J. **223:** 788.
146. Weedman, D. W., 1976. Q. J. R. Astr. Soc. **17:** 227.
147. Weedman, D. W. 1976. Astrophys. J. **208:** 30.
148. Weedman, D. W. 1977. Ann. Rev. Astron. Astrophys. **15:** 69.
149. West, R. M., A. C. Danks & G. Alcaino. 1978. Astron. Astrophys. Lett. **62:** L13.
150. Wilson, A. S. 1979. Proc. R. Soc. London Ser. A. **366:** 461.
151. Wilson, A. S. 1979. Private communication.
152. Wilson, A. S. & E. J. A. Meurs. 1978. Astron. Astrophys. Supp. **33:** 407.
153. Wilson, A. S. & M. V. Penston, 1979. Astrophys. J. In Press.
154. Winkler, P. F. & A. E. White. 1975. Astrophys. J. Lett. **199:** L139.
155. Wittels, J. J., W. D. Cotton & I. I. Shapiro. 1978. Astrophys. J. Lett. **225:** L47.
156. Worrall, D., R. F. Mushotzky, E. A. Boldt, S. S. Holt & P. J. Serlemitsos. 1979. Astrophys. J. In press.
157. Wright, A. E., D. A. Allen, P. A. Krug, D. C. Morton & M. G. Smith. 1977. IAU Circular 3145.
158. Wu, C-C. 1977. Astrophys. J. Lett. **217:** L117.
159. Zwicky, F. 1967. Adv. Astron. Astrophys. **5:** 267.
160. Zwicky, F. 1971. Catalog of Selected Compact Galaxies and of Post-Eruptive Galaxies (Guemligen:Zwicky).

QUASARS NEAR COMPANION GALAXIES AND THE PROBLEM OF THE ABSOLUTE MAGNITUDES OF QUASARS

Halton Arp

*Hale Observatories**
Carnegie Institution of Washington
Pasadena, California 91101

INTRODUCTION

The first quasars that were associated with low redshift galaxies were associated with eruptive and disturbed galaxies [1,2] (Arp 1966a, b, 1973, 1976). It was concluded that these quasars had been ejected in some kind of explosive process. Later it was concluded that smaller companion galaxies in the vicinity of large galaxies were more often in early stages of evolution characterized by secondary ejection, fission, and disruption (Arp 1972). Quasars were discovered so close to these companions that their chance of being accidently projected background quasars seemed negligible (Arp, Baldwin, and Wampler, 1975). The numbers of such quasars associated with low redshift galaxies, and particularly with companions to larger galaxies, has continued to grow over the years. Discoveries up to the year 1976 were summarized in Colloques Internationaux du C.N.R.S. No. 263 (Arp 1976). In the two years since then to the present 9th Texas Symposium on Relativistic Astrophysics, many more high probability associations of quasars have been discovered. The number has roughly doubled in these last two years and now enables us to study in more detail the properties of these quasars which are apparently associated with low redshift galaxies.

In particular, it is interesting, and crucial, to compare the probability that these quasars are in fact much closer than their redshift distance to the probability that some other quasars are at their redshift distances. As we leave the early, simple concepts about quasars and gain more information, we must address the question whether there is only one kind, two or more distinct kinds of quasars, or a wide range of characteristics. From the increased number of associations reported here, it is possible to make a comparison of the absolute magnitudes of the associated quasars to absolute magnitudes of other quasars derived in the cosmological assumption. Some progress on answering this question about possible different kinds of quasars can then be made.

Finally, we utilize the existing data on all sides of this question to propose several possible hypotheses about quasars which fit all the major, current observational requirements.

*Operated jointly by the Carnegie Institution of Washington and the California Institute of Technology.

0077–8923/80/0336–0094 $1.75/1 © 1980, NYAS

Some New Associations

We discuss below some of the more interesting associations found since 1976. The complete, up to date listing of all the highest probability of associated quasars with the presently available parameters of the association are listed in Table 1.

a) NGC 5055

This spiral galaxy of bright apparent magnitude and large apparent diameter has two apparent companions aligned roughly 30 arc min on either side. The NW companion has a very bright emission region in it. The SE companion has a quasar very close to it. Figure 1 shows the configuration of NGC 5055 and its companions. Only the area around the companions has been inspected for quasar candidates.

b) NGC 5669

The asymmetrical, disturbed appearance of the spiral NGC 5669 is apparent from the photograph in Figure 2. Close by to the NW is an apparent companion. This companion has no defined form but, it is blue and has emission lines at about the same redshift as NGC 5669. Close to this companion is a bright (V_{5000} = 17.7 mag) quasar. The chance of such a bright quasar being an accidental projection of unrelated field quasar is only 0.003.

c) NGC 5982

Figure 3 shows three galaxies close together and aligned. If the central one, NGC 5982 is an E galaxy, then the ones on either side, being spirals, should have lesser masses even though one is brighter in apparent magnitude. In that case we would interpret them as having been ejected from the E galaxy. But the brightest galaxy, NGC 5985 is also the one with the lowest redshift and therefore may be the origin of ejection.

In any case, NGC 5981 is a considerably fainter galaxy and is obviously a companion in that sense. It is also the one to which a quasar, UB1, falls very close. Another quasar, UB2, has very much the same redshift but falls farther away from NGC 5981. A third quasar falls still further away and has considerably different redshift (probably $\sim z = 1.03$).

These quasars were discovered by J. W. Sulentic and H. Arp (unpublished). On the same 10 × 10 inch dual exposure UB plate taken with the 48-inch schmidt, they searched two circular regions displaced well away from NGC 5982 of the same size as searched around NGC 5982, namely 1.5 deg^2. In this total of 3.0 deg^2 of control field, Sulentic and Arp found two quasars close together of magnitude V_{5000} = 18.0 mag and 18.3 mag and one of V_{5000}=20.0 mag. Compared to the average density of quasars (Sandage & Luyten 1969, Setti & Woltjer 1973, and Arp, Sulentic and di Tullio 1978), this is more bright quasars than would be expected (two instead of one) and far fewer faint quasars than would be expected (one instead of 30). This perhaps emphasizes two points: one is that the quasar background is variable from region to

TABLE 1

QUASARS NEAR COMPANION GALAXIES

Main Galaxy			Companion Galaxy				Quasar				
	m_c	z_0 (km/sec)		dist	mag	z (km/sec)		dist	V_{5000}	z (km/sec)	probability
				A Probability ≤ 0.01							
NGC 157	11.1 mag	1749	pec sp	30′			BSO 1	119″	19.0 mag	.756	.006
NGC 1068	9.7	1134	NGC 1073	90′	11.3	1,245	BSO 1	104″	19.6–20.1	1.94	.01
							BSO 2	117″	18.9	.598	.006
							RSO	84″	20.1	1.404	.02
			NGC 1087	120′	11.3	1,844	UB 1	170″	19.1	2.147	.01
NGC 1595		4360	NGC 1598	2.′8		4,790	BSO 1	119″	(20)		.03
							BSO 2	222″	(19.5)		.03
			Carafe gal	6.′8		4,605	BSO 1	124″	(19)		.008
NGC 3726	11.1	818	pec gal	15′	15.1		BSO 1	100″	18.4	1.13	.002
NGC 4395	10.3	307	pec ring	70′		6,600*	UB 1	145″	18.7	1.265	.01
							B 6	329″	18.4	1.038	.03
							3C 270.1	303″	18.7	1.521	[.04]†
							U 3	946″	18.7	(1.677)	.3
NGC 5055	9.6	587	pec sp	32′	15.5		UB 1	315″	18.3	.910	.01
NGC 5297	12.4	2500	NGC 5296	1.′5	15.0		BSO 1	55″	19.3	.963	.003
NGC 5669	12.0	1372	blue em	6′	15.3	1,392	UB 1	70″	17.7	.766	.0003
NGC 5689	12.4	2352	NGC 5682	8.′8	14.5	2,413	BSO 1	95″	19.2	1.94	.005
NGC 5982	12.1	3083	NGC 5981	6.′1	14.2	2,836	UB 1	107″	19.0	2.132	.006
			NGC 5985	7.′3	11.6	2,671	UB 2	740″	19.0	1.968	.3
NGC 7171	12.6	2770	IC 1417	12.′2			BSO 1	76″	17.8	.73	.0003
NGC 7448	12.1	2431	NGC 7463	26′	13.5	2,676					
			NGC 64	26′	14.5	2,108					
			65	26′	13.3	2,190	UB 1	128″	19.2	1.66	.01

TABLE 1 (*continued*)

QUASARS NEAR COMPANION GALAXIES

	Main Galaxy		Companion Galaxy				Quasar				
	m_c	z_0 (km/sec)		dist	mag	z (km/sec)		dist	V_{5000}	z (km/sec)	probability
B Probability ≥ 0.01 and Special Cases											
NGC 2681	11.2	760	I/A dbl	35′	2 × 16	21,300	0846 + 51W1	12″	19–20	1.860	[.001]‡
M81	7.9	95	M82	38′	9.1	388	Hoag #1	382″	20	1.92	
							2	512″	21	1.92	
							3	585″	21	1.92	
NGC 3338	11.5	1191	dstb sp	20′SE	13.8		UB 2	251″	19.7	2.14	.15
			pec sp	16′NW	15.5		UB 1	218″	20.4	(2.04)	.27
NGC 3379	10.5	756	NGC 3384	7′.2	10.8	642	UB 13	149″	20.6	.50	.13
			NGC 3389	6′.3	12.6	1,138	UB 1	265″	19.4	1.11	.07
							UB 2	679″	19.8	1.28	1.1
							UB 4	646″	19.9	1.11	1.1
							UB 8	795″	18.7	1.13	.2
							UB 5	1036″	19.2	1.19	.8
							UB 15	1574″	19.7	1.13	3.6
NGC 3435	14.0		NGC 3407	18′	15.0		4C 61.2	173″	16.3	.43	[.0002]§
NGC 3992	10.6	1149	dstb sp	14′	14.1		W1	280″	20.3	1.75	[.3]§
NGC 7585	12.6	3504	NGC 7592	14′.8		7,446	UB 1	415″	18.7	1.410	.04
			NGC 7576	10′.7		3,729					
NGC 520	12.0	2272							(18)	.77	
									(17.5)	.67	
									18.4	2.11	
									17.3	.72	
NGC 622	14.0	5483					UB 1	71″	18.5	.91	.001
							BSO 1	73″	20.2	1.46	.02
NGC 4395	10.3	307					UB 1	370″	19.2	.77	.07

*Em line z = 6600 km/sec, abs line z = 5400 km/sec.

†Presence of quasars known first from radio detection.

‡Probability to centroid of interacting double.

§Presence of quasar known from radio detection.

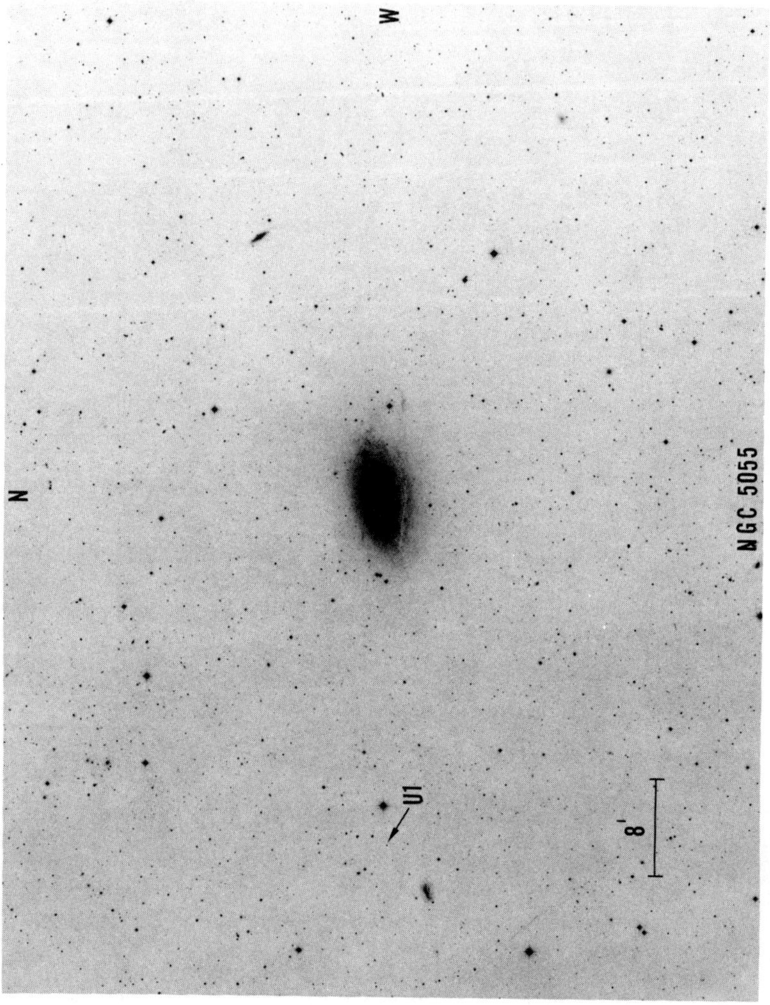

FIGURE 1. NGC 5055 and companion galaxies. Quasar near SE companion is marked UB1.

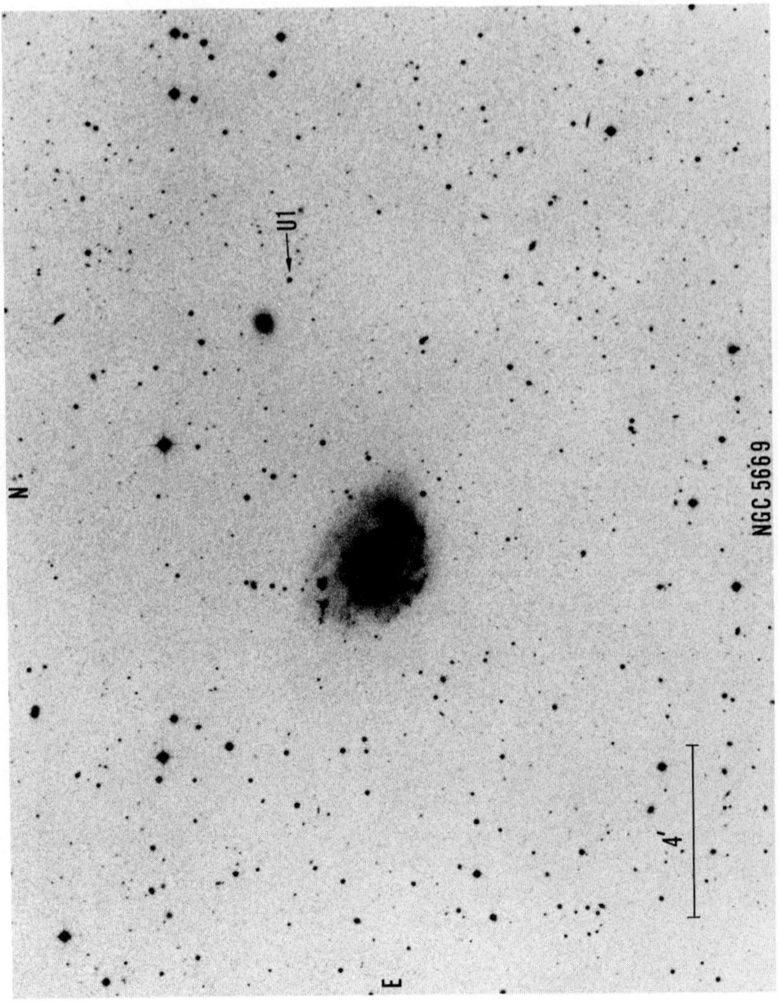

FIGURE 2. NGC 5669 and its blue, emission line, companion galaxy. Quasar marked near companion has $V_{5000} = 17.7$ mag.

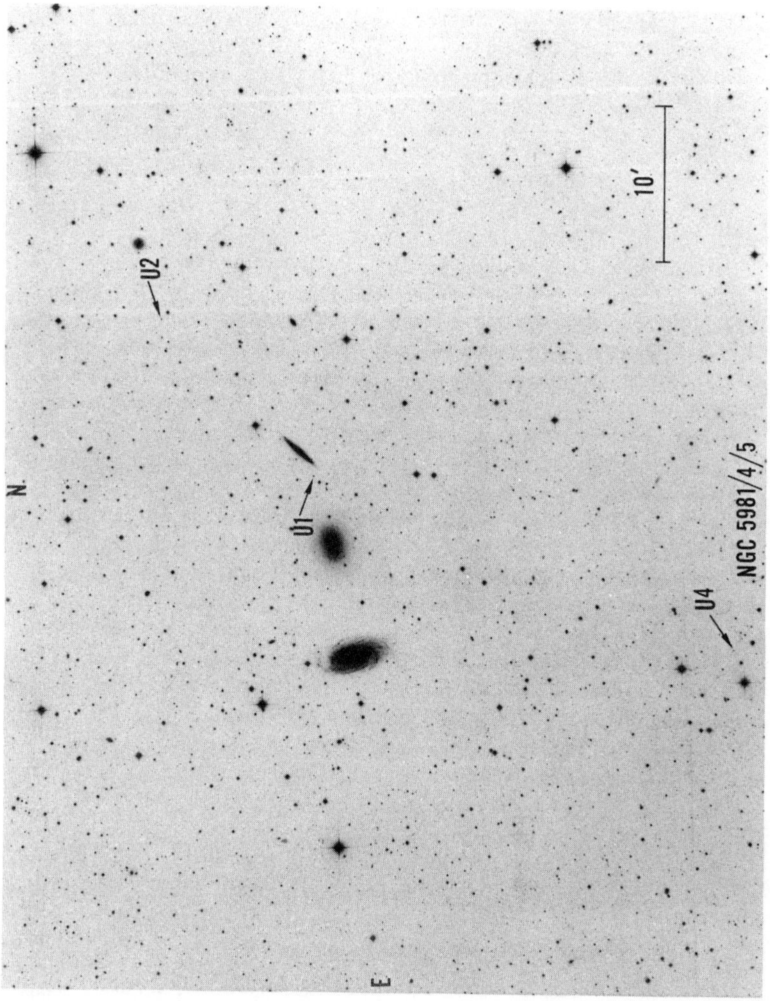

FIGURE 3. NGC 5981, NGC 5982 and NGC 5985. Quasars marked are from a complete search of 1.5 deg^2 around triplet of galaxies. Search of two additional control fields of same size 10 × 10 inch P48 plate yields two quasars of V_{5000} = 18.0 and 18.3 mag and one of V_{5000} = 20.0 mag. Data from Arp and Sulentic unpublished.

region in the sky, and two is that the number of fainter quasars is perhaps not as great as claimed in Sandage and Luyten (1969).

As for the three quasars in the NGC 5981 field all around $V = 19.0$ mag, the statistics are small, but reasonably we could expect UB1 and possibly UB2 might belong to the triplet of galaxies and UB4 to be an unrelated field quasar. The above discussion is meant to be illustrative of the kind of numbers we typically expect to deal with in galaxy and control fields and to show why it is principally the quasars close to the associated galaxies which are capable of demonstrating an association if found close enough and in sufficient numbers. It also demonstrates why average densities of field quasars must be drawn from careful studies of large areas of the sky.

d) NGC 7448

East of the bright spiral NGC 7448 is an interacting triplet of galaxies NGC 7463/4/5. About in the middle of this group is a quasar as shown in FIGURE 4. Deep superposition photographs and spectroscopic study of the members of this group might give additional evidence for the association of this quasar beyond the statistical probability of chance of about 0.01 quoted here.

e) NGC 1068

The Seyfert galaxy NGC 1068 has a number of apparent companion galaxies including NGC 1073 and NGC 1087. Arp and Sulentic (1979) showed three quasars fell in the outskirts of the later type barred spiral, NGC 1073. FIGURE 5 shows these quasars identified. Subsequently a quasar was identified close to NGC 1087. Data on these quasars are summarized in TABLE 1. The total probability of finding four quasars brighter than 20.0 mag and closer than 170 arc sec to two companion of NGC 1068 is $\sim 10^{-5}$.

f) NGC 622

This galaxy is not at present known to be a companion galaxy to any larger galaxy. But recently Arp has discovered two quasars which are in close proximity to it and resemble very much the situation in NGC 1073. The brighter quasar shown in FIGURE 6 has a $z = 0.91$ and the fainter $z = 1.46$. These redshifts are similar to the redshifts of two of the NGC 1073 quasars.

The brighter, lower redshift quasar is extremely red. The continuum to the short wavelength side of Mg II 2798 is so steep in this quasar that it could not be a normally flat, lower z quasar seen at a higher cosmological redshift. The argument that material from the galaxy is reddening UB1 could not apply to UB2 which is very blue.

STATISTICS OF ASSOCIATIONS

In all the cases listed in TABLE 1 the probability of chance association has been computed in the following way: at the observed distance of the quasar from the associated galaxy, we compute the amount of area encompassed within this radius

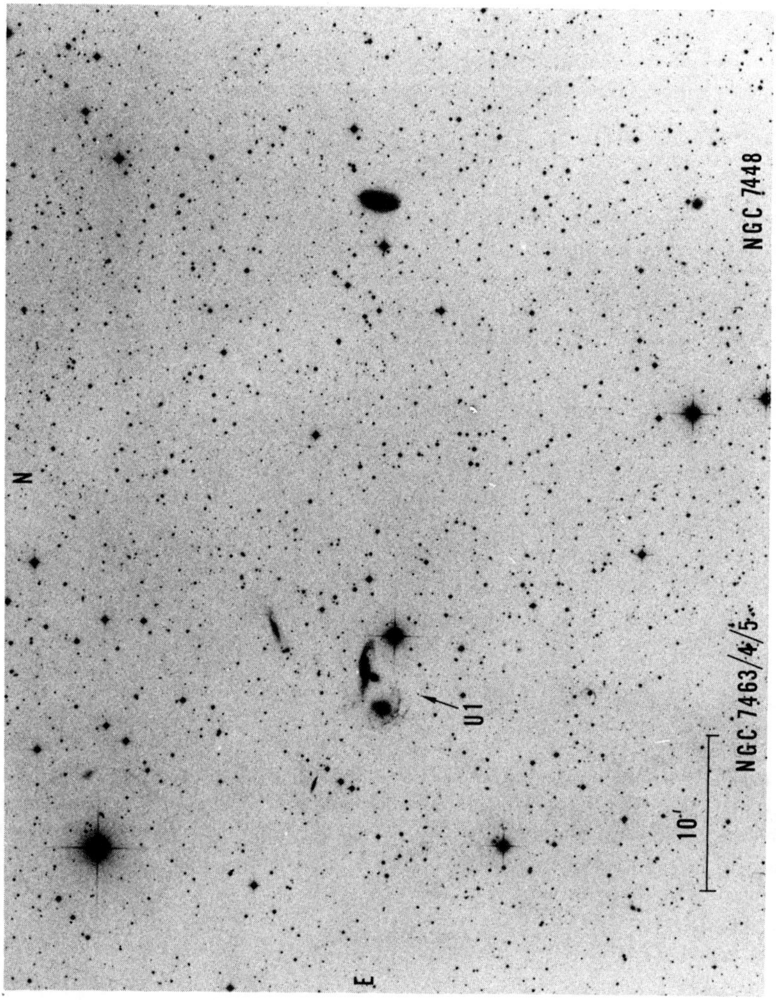

FIGURE 4. East of bright spiral NGC 7448 is an interacting triplet NGC 7463/4/5. Newly discovered quasar in tripled marked U1.

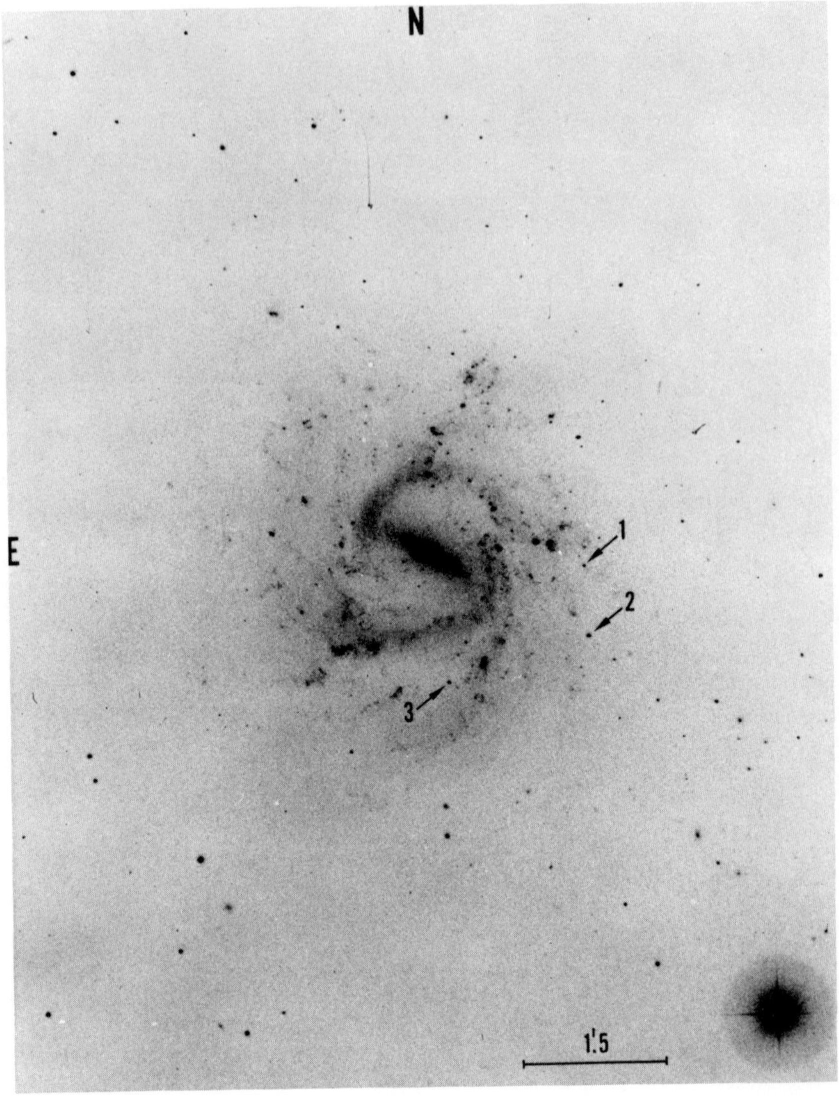

FIGURE 5. Three quasars near edge of NGC 1073. From Arp and Sulentic (1979). Quasars confirmed by E. M. Burbidge, Junkkarinen and Kosky (unpublished).

from the galaxy. From the average areal density of quasars of that brightness we compute the chance that we include one quasar within our area around the galaxy.

The average density of quasars over the sky is taken from Sandage and Luyten (1969). This density has been checked by Arp, Sulentic, and di Tullio (1979) and other workers in the field. It is clear that the average density of quasars agrees within a factor of two as determined by different investigators in different areas of the sky. Therefore the probabilities of association computed in TABLE 1 must be reliable.

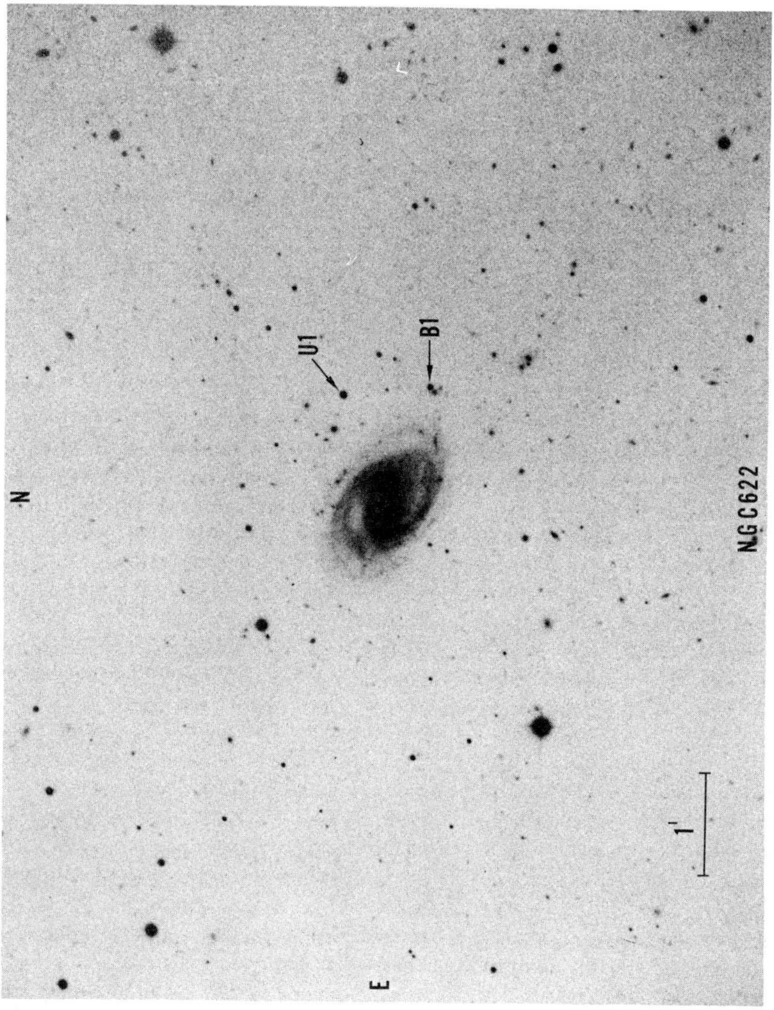

FIGURE 6. Newly discovered quasars near NGC 622. The quasar marked in U1 is $z = 0.91$ and very red; the quasar market B1 is $z = 1.46$ and very blue.

Finding the companion galaxies in TABLE 1 related with background quasars by chance is therefore in each case very small, between 1 in 100 to 1 in 1000.

The only question that remains is how many galaxies have been searched in total for such quasar associations. In fact, only one or two areas around companions of larger galaxies have been searched with UB two color plates which *did not* reveal a quasar. Moreover, more intensive search may reveal more quasars around these objects. No more than the order of the 14 successful galaxy-companion associations in TABLE 1 have been cursorily searched for blue stellar objects.

But the main galaxies in TABLE 1 are all brighter than $m_c = 12.6$ mag (essentially Shapley-Ames galaxies). They excluded the crowded region of the Virgo Cluster; they include only one large southern declination object. Therefore there are only a restricted number of galaxies which qualify as main galaxies, and only a fraction of such galaxies have proper companions to qualify in TABLE 1. Therefore even if *all* bright galaxies with companions in the regions covered by TABLE 1 had been inspected, there would not have been nearly enough to produce the cases in TABLE 1 by accident with known background densities of quasars.

To illustrate, if we take all northern Shapley-Ames galaxies between the right ascension of NGC 5055 and NGC 7448, we have a total of 151 galaxies north of Dec $= -14$ degrees. Some of these are *E* galaxies or spirals without obvious companions. Only about ⅓ are candidate spirals with companions of 15.5 mag or brighter. As TABLE 1 shows, in this interval of Shapley-Ames numbers I have found seven quasars near companions each with a probability of $p \leq 0.01$ of being accidental. If no more of the remaining 50 candidates give successful identifications, we still have seven successes out of 50 trials, and the combined probability of the seven being an accident is $P = 10^{-6}$.

Since I have not yet investigated the other 50 candidates, however, I have really had 7 successes out of 7 trials at $p = \leq .01$ in this interval of Shapley-Ames galaxies. The probability of that being chance is $P = 10^{-14}$. It appears, therefore, that the association of some high redshift quasars with companions of low redshift galaxies has a probability of being chance of only between 10^{-6} and 10^{-14} at this stage.

Now that a sufficient number of associations with companions have been found, we can use the examples in TABLE 1 to define a set of criteria for further candidate systems. By inspection of TABLE 1 we can see that the bright galaxy should be a Shapley-Ames galaxy with corrected apparent magnitude brighter than about 12.6 mag. The companion galaxy should be brighter than 15.5 apparent magnitude, and its distance from the bright galaxy should be less than about 30 arc min. It is seen that the brightest, nearest systems like NGC 5055 have companions which are the most distantly spread. Most of the companions in TABLE 1 are within 15 arc min, but for the sake of a starting definition, we adopt 30 arc min. (NGC 1068 and its companions are a special case where an investigation of all Shapley-Ames galaxies in a fairly sparse area of the sky shows a strong, isolated grouping within about 120 arc min of the central, very bright, Seyfert NGC 1068.)

EXPLOSIONS IN COMPANION GALAXIES —M82

The majority of the companion galaxies listed in TABLE 1 have disturbances and perturbations which could only reasonably be attributed to violent internal activity.

For example, the ring galaxy 70 arc min NW of NGC 4395 is isolated but shows morphological evidence of activity and has an emission line redshift of $z = 6600$ km/sec as opposed to an absorption line redshift of $z = 5400$ km/sec. The extension of the galaxy through the ring prints roughly in the direction of the four quasars which are, uncannily, all between $V_{5000} = 18.3$ and 18.7 mag. It is concluded that these quasars have been ejected more or less in a line by an explosion in the associated galaxy. In this respect they resemble the chain of quasars from NGC 520 (Arp 1970) and those across NGC 3384 (Arp, Sulentic, and di Tullio 1979).

It would come as striking confirmation of these previous observations then, if one of the most active companion galaxies to one of the nearest spiral galaxies turned out to have quasars associated with it. But this is exactly what has happened in the case of M82. A. Hoag (in preparation) has discovered three quasars SE of M82. He reports they represent a concentration in order of magnitude closer than found in other regions of the sky. Moreover, all three have redshifts of about $z = 1.92$. Obviously these quasars are associated. FIGURE 7 shows how these three quasars can be traced back in a cone of ejection to M82. This cone originates in M82 just along the most conspicuous, linear absorption feature that traverses M82. Interference filter photographs of M82 (Lynds and Sandage 1963) show gas emerging from the same side of M82 as the quasars. Recently spectroscopic observations by Axon and Taylor (1978) revive the explosion interpretation of M82, and HI mapping (Crutcher, Rogstad, and Chu 1978) show a pair of concentrations across the center along the region of the quasar cone (FIGURE 8). The coincidence of these alignments and features argues strongly, in my opinion, that the three Hoag quasars emerged recently from the exploding M82.

STOCKTON'S OBSERVATIONS OF OBJECTS NEAR LOW REDSHIFT QUASARS

Recently spectra were obtained for nonstellar objects in the vicinity of 27 quasars between redshifts of $z = 0.142$ and $z = 0.450$ (Stockton 1978). In 8 cases some of these presumed galaxies turned out to have within 1000 km/sec the same redshift as the quasar.

There are several questions about Stockton's work that present themselves at this point. They are: 1. Why do these quasars appear in small groups of presumed galaxies and not in rich clusters? 2. Why do not galaxies appear around the other 19 quasars in Stockton's investigation? 3. Why is the percentage of his successful companion galaxies independent of distance from the quasar? This last aspect is the most serious since it implies that if the radius of search were enlarged, he would continue to go on getting associated galaxies, and there would be no "grouping" around the quasar.

The most serious interpretative criticism which can be levelled at this work, however, is Stockton's rejection of the possibility that only the steep (radio) spectrum quasars might be associated with galaxies at the same redshift. From 10 steep and 7 flat spectrum quasars he only associates one flat spectrum with a galaxy of the same redshift. But even that one is not a quasar—as can be seen by consulting Stockton's own photograph in the paper which clearly shows a nebulous, compact galaxy image. So he is left with no flat spectrum quasars associated with galaxies of the same redshift. This should be contrasted to the original quasars which Arp (1966) associated with low redshift disturbed galaxies—all of which turned out to be compact

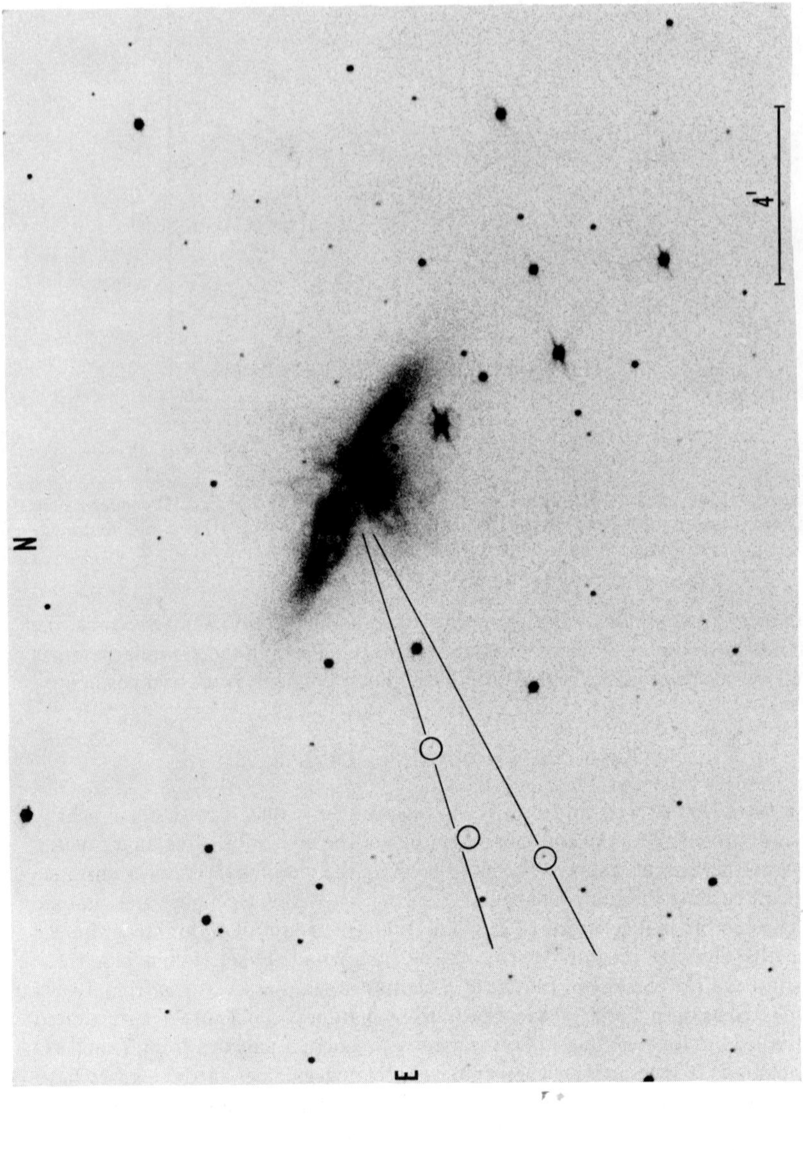

FIGURE 7. Photograph of M82 taken in line of [O II] 3727 A gaseous emission line with Palomar 48-inch schmidt. Lines connecting the three Hoag quasars back to center of M82 are drawn in. Linear absorption feature traversing M82 and aligned with quasar line is seen well in original Halpha photographs of Lynds and Sandage (1963).

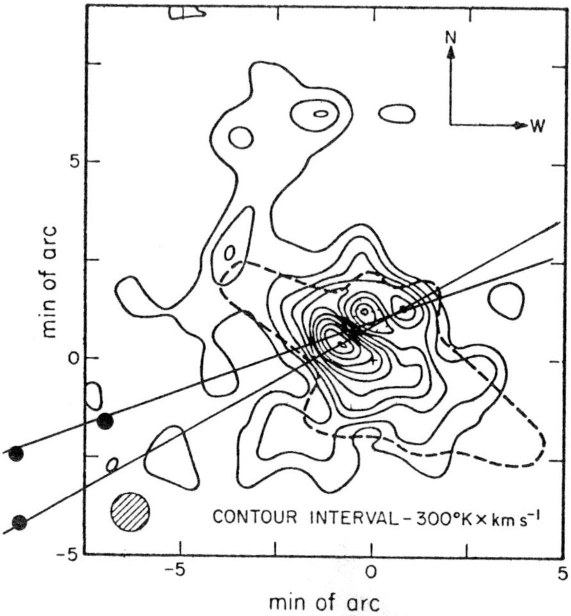

FIGURE 8. HI map of M82 from Crutcher, Rogstad and Chu (1978). Two HI condensations are seen across center of M82 on line with proposed quasar ejection cone HI is cleared out along line of cone to the southeast. Position of quasars indicated by filled circles.

radio sources, hence probably flat spectrum (Rowan-Robinson 1972). Of course, the difference of behavior of the flat and steep spectrum quasars in the z, m diagram has been well established by Setti and Woltjer (1973) at a previous Texas Symposium.

ABSOLUTE MAGNITUDES OF QUASARS

By associating at high probabililty the quasars in TABLE 1 with their nearby galaxies, we can compute the absolute magnitude of the quasar by using an estimated modulus to the parent galaxy. We always take the main galaxy as a distance determinant because the companion galaxies have usually slightly higher and, in some cases, considerably higher redshifts. Using a Hubble constant of 50 km/sec/Mpc we compute the absolute magnitudes for all the quasars in TABLE 1 and plot these luminosities against redshift in FIGURE 9. The most certain associations from TABLE 1A are plotted as open circles; the less certain associations from TABLE 1B are plotted as small filled circles. The quasars associated with isolated galaxies from TABLE 1C are not plotted. It is less clear whether the associated galaxies in these cases have redshifts indicative of distances. (The NGC 520 quasars fit well in the general sum of FIGURE 9, but the NGC 622 quasars fall high, and the NGC 4395 quasar falls low.)

It is apparent that the quasars near companion galaxies have a peculiar redshift distribution. Most of them are between $z = 0.6$ and 1.4 and between $z = 1.8$ and 2.1.

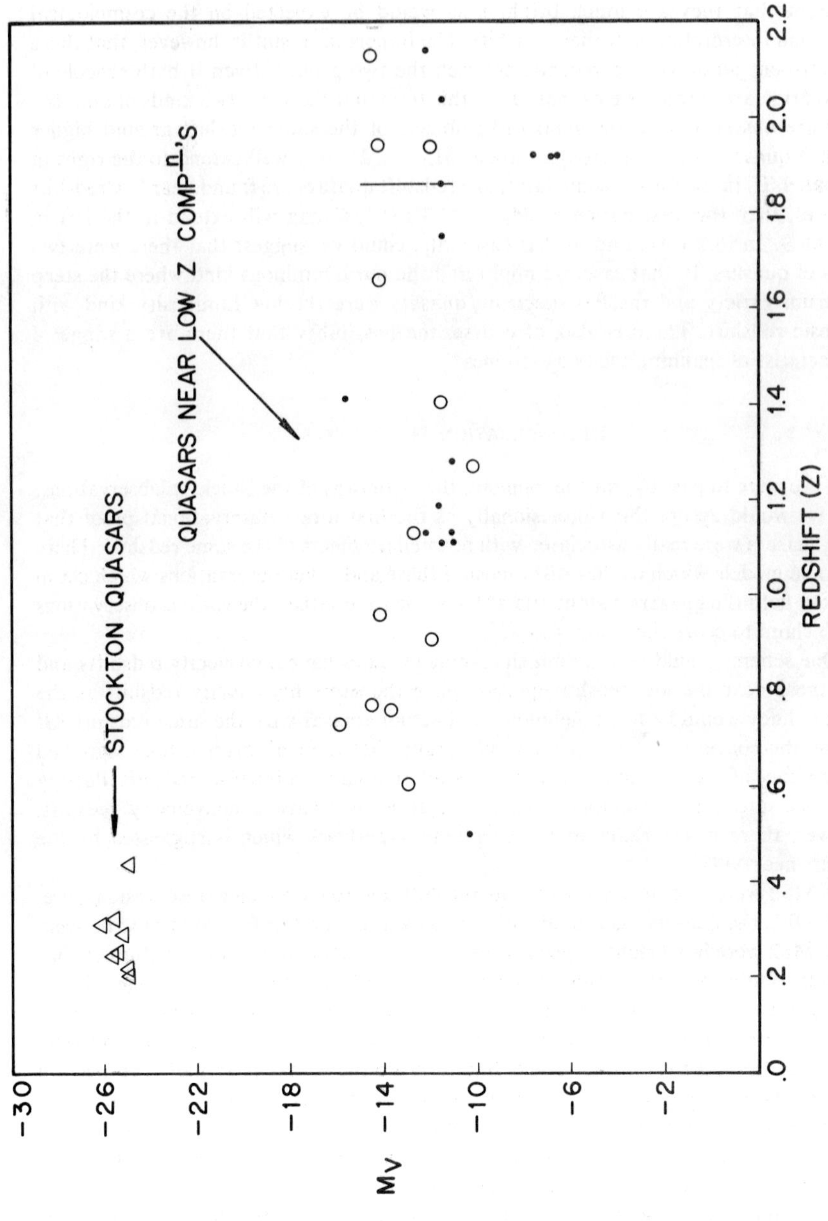

FIGURE 9. Absolute magnitudes of quasars using redshift of associated, main galaxy to determine distance modulus. Open circles are associations with chance probability ≤0.01, TABLE 1A; small filled circles are remainder of quasars in TABLE 1B. Open triangles are quasars at their redshift distances from Stockton (1978).

This redshift distribution is unlike that of the 3CR quasars or that of other groups of quasars (see Arp, Wills and de Ruiter 1979).

In FIGURE 9 the eight Stockton quasars are plotted with open triangles. It is apparent that they are much brighter as would be expected on the cosmological expansion interpretation of their redshift. The important result is, however, that there is at present no overlap in redshift between the two groups. Even if both schools of observation are correct, we cannot say at this time that there are two kinds of quasars. If future observations show galaxy-like objects of the same redshift around higher redshift quasars, then the group at about $M_V = -25$ may well extend to the right in FIGURE 9. If, in the future, some faint, low redshift quasars are found near low redshift galaxies, then the distribution at $M_V = -13$ to -14 mag will extend to the left in FIGURE 9. In that case, and in that case only, could we suggest that there were two kinds of quasars. In that case, we might find the more luminous kind where the steep spectrum variety and the flat spectrum quasars were the low luminosity kind with intrinsic redshift. There is also, of course, the possibility that there are a range of characteristics spanning the two extremes.

RECONCILIATION HYPOTHESES

If we were to pass by, for the moment, the criticisms of the Stockton observations, then we would accept them provisionally as the first direct observational proof that some quasars were really associated with nonstellar objects of the same redshift. There are some models which could satisfy most of these and other observations which claim to prove both the quasars distant and at the same time satisfy the various observations which claim to prove the quasars local.

One scheme would be to admit that some galaxies have nonvelocity redshifts and that those near the low redshift quasars have the same nonvelocity redshift as the quasar. They would be like a nebulous, companion quasar with the same redshift. Of course, the conventional viewpoint would argue that their absorption lines signalled the presence of a stellar component which could not have an intrinsic redshift. Putting aside the question of whether a galaxy of stars can have a nonvelocity redshift, however, there is a perhaps more acceptable hypothesis which is suggested by the quasars near M82.

If M82 were at a distance where its redshift due to cosmological expansion were, say $z = 0.2$, the quasars associated with it in TABLE 1 would be far too faint to be seen. But if M82 were in a bright stage of its explosion, at that distance we hypothesize that it, itself, would look like a quasar. Then M81 would be detected and measured as a nearby galaxy of the same redshift as the quasar. The conventional explanation of quasars is thay they are events that go off in the nuclei of distant galaxies. We only add to that hypothesis that the event that lights up the galaxy is actually the ejection of some faint, high intrinsic redshift, compact objects. FIGURE 10 illustrates a schematic of this proposal. Then there would be a class of objects that behaved like conventional quasars and another class of objects that were much less luminous but became visible when ejected from nearby galaxies.

This hypothesis also has features which would be helpful in explaining other currently puzzling observational restrictions. For example, the flaring nuclei might be more associated with steep radio spectrum quasars. Steep radio spectra are usually double radio sources such as associated with classical giant *E* galaxy ejection events. Such kinds of galaxies are most likely to have approximately conventional redshift distances. The high *z*, low luminosity objects which are ejected might then be associated with flat spectra, since it is well known that the flat spectra quasars show much more scatter in the *z*, *m* diagram than do the steep spectrum quasars. As mentioned earlier, there is a strong tendency for the quasars of anomalous redshifts in association with low *z* galaxies to be flat spectrum. Secondly, if it is principally the companion galaxies which are lighting up to form the conventional kinds of quasars, this would explain why some of the low redshift quasars are found in small groups but never in rich clusters. Occurrence in a fairly small companion galaxy might also explain the puzzling observation that many low redshifts quasars, if events in normal

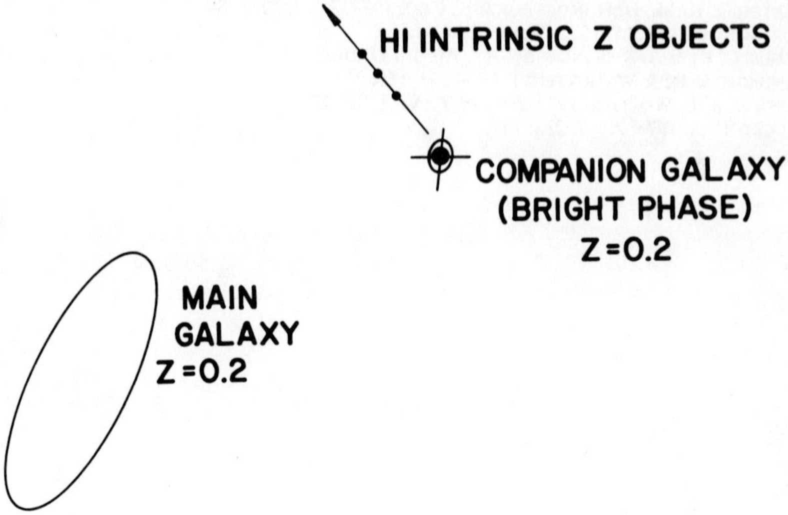

FIGURE 10. Schematic of hypothesis that companion galaxies undergo bright phase which appears stellar or nearly stellar when seen at a distance corresponding to a cosmological redshift of $z = 0.2$ or greater. Hypothesized ejection of high intrinsic redshift objects from this companion would appear at apparent magnitude $V = 26$ mag or fainter.

galaxies, should show disks of these galaxies extending out beyond their stellar images. There are many that show no indication at all of underlying galaxy, although, as has been claimed now, there are sometimes larger galaxies of the same redshift lying in the vicinity.

It is perhaps characteristic of a subject that when we first learn of it, it seems very simple. As we continue to gather data and learn to distinguish classes and differences, the phenomena can become eventually very complex. The many differences now appearing in radio quasars of different radio characteristics, radio quiet quasars, BL Lac objects and their different colors and eruptive behavior attests to this growing complexity of information. Perhaps we may look back some day to our early picture of "a quasar" as very primitive and crude.

References

ARP, H. 1966a. Science **151:** 1214.
ARP, H. 1966b. Ap. J. **148:** 321.
ARP, H. 1970. Ap. J. **75:** 1.
ARP, H. 1972. Bull. Amer. Astr. Soc. **4:** 397.
ARP, H. 1973. The Redshift Controversy. G. G. Field, Ed. W. A. Benjamin, Inc. Reading, Mass.
ARP, H. 1975. L' Evolution des Galaxies et ses Implications Cosmologique. Colloque I.A.U. No. 37. Colloque C.N.R.S. 263. Paris, September 1976.
ARP, H., J. A. BALDWIN & E. J. WAMPLER. 1975. Ap. J. **198:** L3.
ARP, H., SULENTIC, J. W. & di TULLIO, G. 1979. Ap. J. In press.
ARP, H. & J. W. SULENTIC. 1979. Ap. J. In press.
ARP, H., A. G. WILLIS, & H. DE RUITER. 1979. Astron. & Astrophys. In press.
AXON, D. J. & K. TAYLOR. 1978. Nature **274:** 37.
CRUTCHER, R. M., D. H. ROGSTAD & K. CHU. 1978. Ap. J. **225:** 784.
ROWAN-ROBINSON, M. 1972. Nature **236:** 112.
LYNDS, C. R. & A. R. SANDAGE. 1963. Ap. J. **137:** 1005.
SANDAGE, A. R. & W. LUYTEN. 1969. Ap. J. **155:** 913.
SETTI, G. & L. WOLTJER. 1973. Ann. N.Y. Acad. Sci. **224:** 8.
STOCKTON, A. 1978. Ap. J. **223:** 747.

LIST OF PAPERS PRESENTED AT THE WORKSHOP ON QUASARS AND GALACTIC NUCLEI*

F. Pacini

Laboratorio di Astrofiscia Spaziale
Frascati, Italy

MITCHELL C. BEGELMAN, MARTIN J. REES (Institute of Astronomy, Cambridge), AND ROGER BLANDFORD (California Institute of Technology, Pasadena): A twin-jet model for radio trails.

MONICA BELTRAMETTI (Inst. für Theoretische Astrophysik, Universität Heidelberg): Model-dependent expanding quasar envelopes.

P. BENVENUTI (ESA-IUE Observatory, Madrid): Recent IUE data.

G. F. BIGNAMI, C. E. FICHTEL, R. C. HARTMAN, AND D. J. THOMPSON (NASA/Goodard Space Flight Center, Greenbelt): Galaxies and gamma ray astronomy.

A. BRACCESI, V. ZITELLI, F. BONOLI (Istituto di Astronomia dell'Università, Bologna), AND L. FORMIGGINI (Laboratorio di Radioastronomia del C.N.R., Bologna): A sample of very faint ultraviolet excess objects in the 13^h+36 field-II: A discussion of the number–magnitude relation of optically selected quasars and a new determination of the luminosity function and of the characteristics of its cosmological evolution.

A. CAVALIERE (University of Rome) AND P. MORRISON (M.I.T., Cambridge, Mass): Extreme non-thermal emission from active galactic nuclei.

W. KUNDT (Inst. für Astrophysik, Universität Bonn): A model for all the violent phenomena occurring in nuclei of galaxies.

H. VAN DER LAAN (Leiden): Recent results on radiogalaxies.

DANIEL A. SCHWARZ (Center for Astrophysics, Cambridge, Mass.): X-ray emitting BL Lac objects located by the scanning modulation collimator on HEAO-1.

B. N. SWANENBURG (Huygens Laboratory, Leiden): Gamma rays, the missing link.

M. TARENGHI (ESO/Cern): IUE results on the jet of M87.

P. A. WEHINGER (MPI für Astronomie, Heidelberg), S. WYCKOFF (Ohio State University), AND A. BOKSENBERG (University College London): Spectra of fuzz around quasars.

K. W. WEILER (MPI für Radioastronomie, Bonn), R. FANTI, A. FICARRA, F. MANTOVANI, AND L. PADRIELLI (Laboratorio di Radioastronomia del C.N.R., Istituto di Fisica "A. Righi," Bologna): Superluminal intensity variations in extragalactic radio sources.

J. C. WHEELER, G. A. SHIELDS, AND M. J. DUNCAN (Center for Theor. Physics, Univ. of Texas at Austin, Austin): The sustenance of black holes in galactic nuclei.

D. M. WORRALL (Laboratory for High Energy Astrophysics, NASA/Greenbelt and Dept. of Physics and Astrophysics, Univ. of Maryland), R. F. MUSHOTZKY, E. A. BOLDT, S. S. HOLT, AND P. J. SERLEMITSOS (Laboratory for High Energy Astrophysics, NASA/Greenbelt): The X-ray Spectrum of 3C273.

*This workshop covered so many different and partly controversial subjects that it seemed impossible to the chairman to prepare a short and adequate summary. Only those contributions for which abstracts were submitted to the chairman are listed here. Interested readers can approach the authors directly.

THE SONG OF CREATION

This is the signal primeval. The murmuring
    microwave background,
Hidden in noise and galactic emission, indistinct
    in the starlight,
Cast by ylem of eld, with message old and prophetic,
Cast at 3 degrees Kel, and spectral shape blackbody.
High in his cabled laboratory, the perplexed
    astrophysicist
Speaks, and in accents disconsolate answers the
    wail of the background.

This is the signal primeval; but where are the shapes
    that within it
Huddled to galaxies as they felt the pull of their masses?
Where are the superclusters, and horizons of causality—
Smoothed by neutrino viscosity, but reflecting an image
    of creation?

Ye who believe in experiment that hopes, and endures,
    and is patient,
Ye who believe in the beauty of strength of scientists'
    striving,
Fund the experiment to study the cosmic background!
Fund the difficult project to hear the song of creation.

(to be submitted as
a proposal to NASA)

(with apologies to
H. W. Longfellow)

Richard A. Muller

COSMIC MICROWAVE BACKGROUND—PRESENT STATUS AND FUTURE PROSPECTS

Richard A. Muller

Department of Physics and
Lawrence Berkeley Laboratory
University of California
Berkeley, California 94720

This is the signal primeval.
The murmuring microwave background,
Hidden in noise and galactic emission,
*Indistinct in the starlight.**

Bigbang! Bigbang! burning bright
In the darkness of the night!
What immortal hand or eye,
*Could frame thy fearful symmetry?**

It has been 13 years since the discovery of the cosmic microwave radiation by Penzias and Wilson of Bell Laboratories,[1] but it is only recently that the explanation put forth by Dicke, Peebles, Roll, and Wilkinson[2] has been nearly universally accepted by the astophysics community. Only with the now clear observation of the peak of the Planckian spectrum and its Wien tail have the doubts of many skeptics been answered. The correct prediction of the entire spectrum after the observation of the intensity at only one frequency must certainly be considered one of the great triumphs of cosmological theory. One of the indications of the impact of their discovery has been the award of the Nobel Prize in physics to Penzias and Wilson this year (1978). The cosmic microwave radiation is not only the strongest evidence we have for the hot, compressed early universe, but it provides one of the most direct means for studying that universe. After a brief review of the origin of the radiation according to the "standard model," I shall discuss the status of present measurements of the spectrum and the large-angular-scale isotropy, and finally indicate what the prospects are for improved measurements in the next decade.

ORIGIN OF THE RADIATION

The very existence now of a "standard model" for the origin of the universe is a reflection of the present acceptance of the theory.[3] In this model, half a million years after the big bang the universe was uniformly filled with a hot plasma of photons, electrons, protons, and heavier atoms. The expansion of the universe cooled the plasma

*Apologies to H.W. Longfellow and W. Blake.

0077–8923/80/0336–0116 $1.75/1 © 1980, NYAS

to the point that the electrons and protons combined to form neutral hydrogen. The previously opaque universe suddenly became transparent, since photons are much less strongly coupled to neutral hydrogen than to the plasma. From that period until now the cosmic blackbody photons have been traveling virtually unscattered,[4] carrying information about the nature of the universe at the time of the decoupling (FIGURE 1). In the Lorentz frame moving with the plasma, the photons have a blackbody spectrum

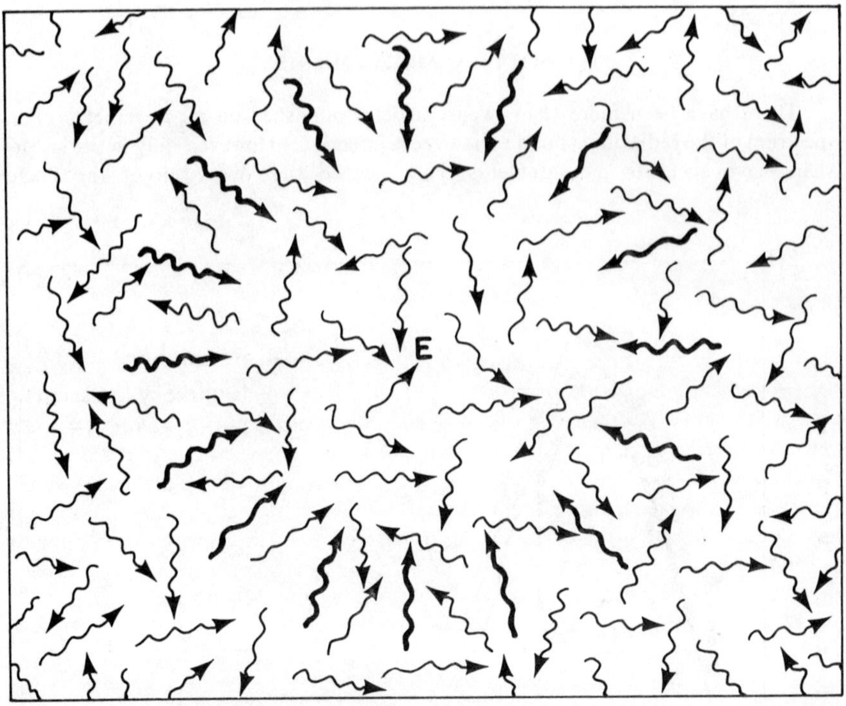

FIGURE 1. Origin of the cosmic background radiation. About 0.5 million years after the big bang, the free electrons and protons in the plasma combine to form neutral hydrogen, and the mean-free-path of the blackbody photons becomes comparable to the size of the universe. Those photons that will reach the earth in A.D. 1978 are drawn with the heavier lines; the future position of the earth is indicated with the letter E. In a comoving frame the photons have a temperature of about 4500 K; in the earth frame their temperature is Doppler shifted to about 3 K.

with a characteristic temperature of about 4500 K; however, the Hubble expansion gives this frame a velocity, with respect to ours, large enough to redshift the radiation by a factor of 1500 to the observed temperature of 3 K. (The fact that a Doppler shift preserves the Planckian form of the spectrum was first derived by Peebles and Wilkinson.[5]) Study of the radiation is therefore a study of the shell of matter that last scattered the radiation.

The cosmic blackbody photons are coming from the most distant region of space ever observed, and were emitted earlier in time than any other cosmological signal. Although the "background" radiation was originally given that name because of its potential interference with satellite communications, the word has taken on a new and vivid meaning: the radiating shell of matter forms the spatial background in front of which all other astrophysical objects, such as the quasars, lie. Until methods are devised to detect the neutrinos or gravitons that decoupled earlier, we will have no direct means of viewing beyond this background.

Spectrum Measurements

There have been more than twenty articles published on measurements of the spectrum of the radiation. (For a review see reference 6.) However, only in the last few years have accurate measurements truly verified the Wien tail of the Planck

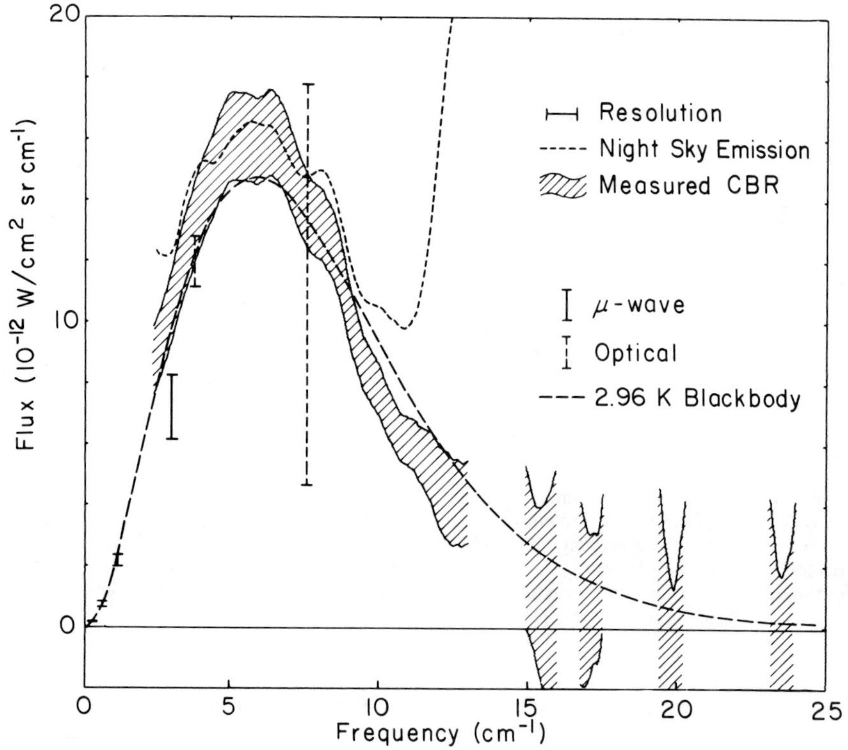

FIGURE 2. Measured spectrum of the cosmic background radiation, from the experiment of Woody and Richards (reference 7). The shaded region corresponds to the $\pm$ 1σ limits. The gaps in the data are at frequencies at which strong atmospheric emission lines cause the errors to become very large.

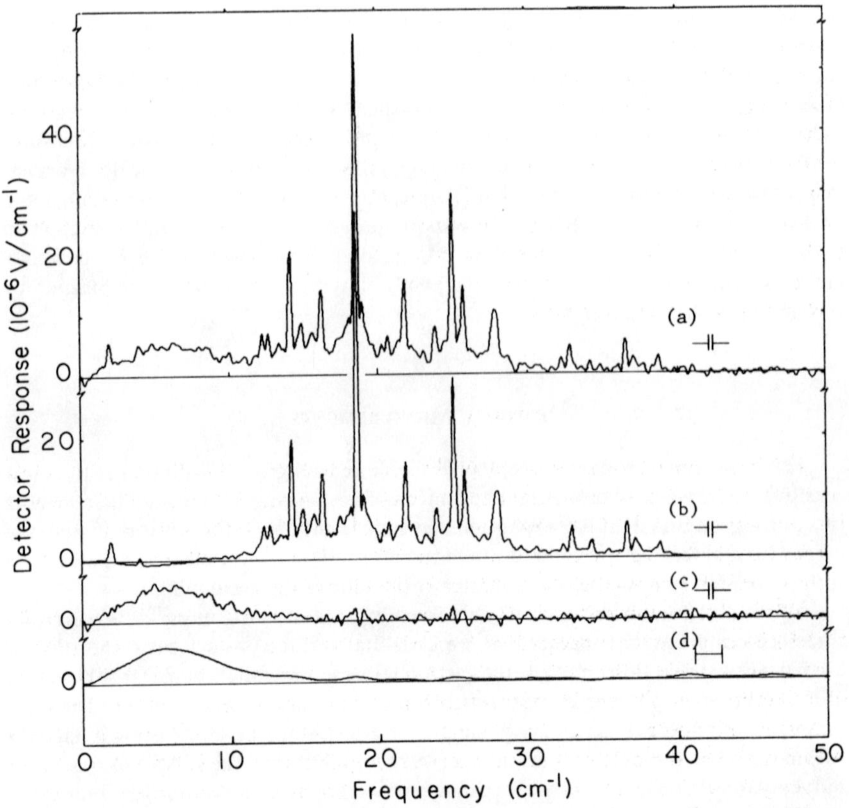

FIGURE 3. The "raw data" from the Woody–Richards experiment is shown in (a), adapted from reference 7. Plotted is the Fourier transform of the instrument output, corrected for instrument response. Note that the Wien tail of the spectrum can be seen without correction for atmospheric emission. The calculated atmospheric emission is shown in (b), and then the net cosmic background is shown in (c) and (d) at two different spectral resolutions.

distribution. Particularly impressive are the recent results of D. Woody and P.L. Richards,[7] shown in FIGURE 2. These results are plotted on linear–linear axes rather than the traditional log–log in order to emphasize the fact that most of the energy of the background radiation is in the range of frequencies measured in this experiment, a region that had few previous measurements because of intense atmospheric emission. Woody and Richards used a balloon-borne Fourier spectrometer, which is sensitive to the entire range of frequencies simultaneously. The spectrum is obtained by Fourier transform of the output signal, with correction for atmospheric emission and detector response. Their "raw" spectrum with instrumental correction but no atmospheric subtraction is shown in FIGURE 3. Note that even without atmospheric correction the Wien tail of the radiation is clearly seen in the raw data.

The striking confirmation of the Planckian shape is the most important result of the Woody–Richards experiment. However, they point out that they do see a statistically significant deviation from the Planckian shape: their spectrum is high in the region of 3 to 7 cm^{-1} and low from 9 to 12 cm^{-1}. The probability of a fluctuation accounting for this effect is about 10^{-6}, corresponding to about 5 standard deviations. (The spectrometer resolution is shown in FIGURE 2 and is about 1 cm^{-1}. Measurements of the spectrum separated by this interval are effectively statistically independent.) Great care was taken by the experimenters to determine if the deviation could be instrumental; this is particularly important to do since an overall reduction in normalization by 22% brings the data into good agreement with a 2.8 K blackbody curve. If the deviation from the blackbody curve is cosmological in origin, its interpretation is of great interest.

ISOTROPY MEASUREMENTS

The large-angular-scale isotropy of the cosmic background radiation is the most sensitive probe we have of several phenomena of cosmological interest. These include the isotropy of the Hubble expansion, possible rotation of the universe,[8] and the existence of very long wavelength gravitational radiation.[9] Anisotropy can also arise from a nonuniform distribution of matter at the time of the decoupling.

Anistrophy has now been clearly observed by at least two groups;[10,11] however, its smooth cos θ behavior suggests that we should attribute it to a local rather than a cosmological effect: the motion of the earth relative to the radiation. After subtraction of the cosine term, there is no statistically significant anisotropy remaining. The most sensitive experiment so far is that done by our group at Berkeley[11,12] using a 33-GHz twin-horn Dicke radiometer flown in a series of flights aboard the NASA-Ames Earth Survey Aircraft (U-2). In this experiment we were able to put an upper limit (90% confidence level) on the existence of a second-order (quadrupole) component to the background radiation of 1.1×10^{-3} K or about 4 parts in 10,000 of a 2.8 K signal.[12] This limit refers to four of the five possible quadrupole components; the fifth component is indistinguishable in the data from the dipole (cosine) component due to our lack of data in the southern hemisphere. The ambiguity will be resolved with a few flights in the southern hemisphere, scheduled to take place early in 1979. In addition to the quadrupole limit, we place an upper limit on the sky "roughness" (i.e., the statistical significance of the scatter of the data points around the cosine fit) of 0.60×10^{-3} K at the 95% confidence level.[12]

Although my original interest in the experiment was due to the cosmological phenomena, perhaps the most interesting result turned out to be the local velocity of the earth. Like Michelson and Morley, we were surprised by the result, but for the opposite reason: we obtained a velocity for the Milky Way (and the local group of galaxies) much higher than we had expected. The data for flights through early 1978 are shown in FIGURE 4, along with the cosine fit. The amplitude of the cosine is $3.61 \pm 0.54 \times 10^{-3}$ K, in a direction of $\alpha = 11.23 \pm 0.46$ hr, $\delta = 19 \pm 7.5$ deg (reference 12). For a cosmic background temperature of T, this implies a velocity for the earth of $361 \times (3\text{ K}/T)$ km/sec. When this is folded with the known velocity of the sun relative to the Milky Way galaxy (from galactic rotation) one finds a net velocity for the local

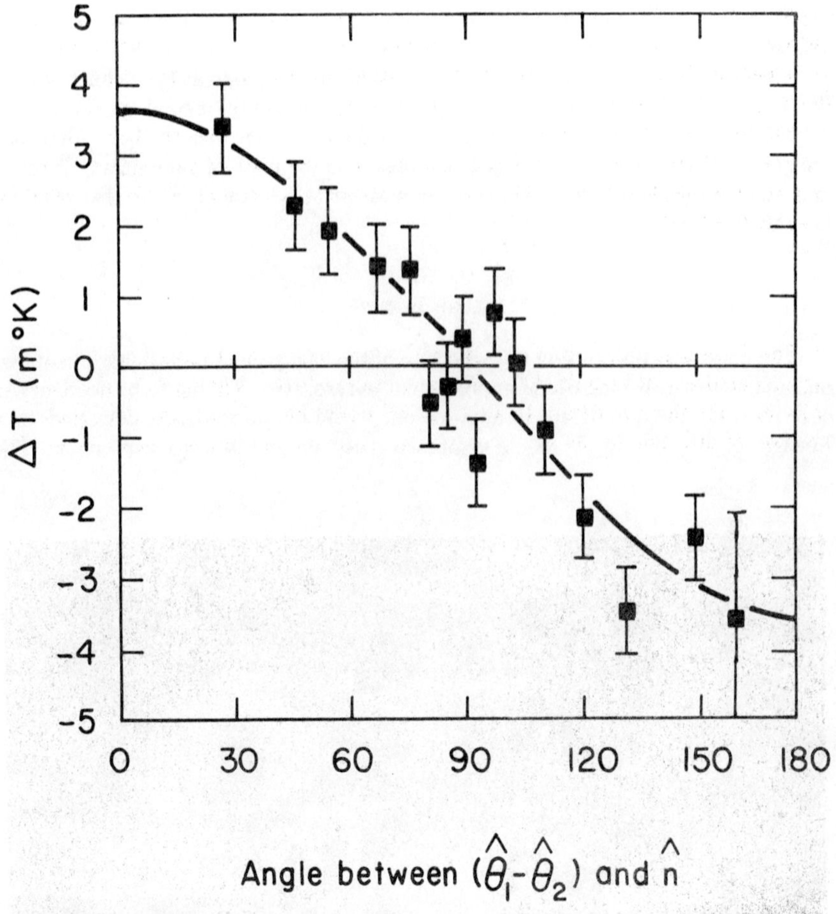

FIGURE 4. Anisotropy data from the 33-GHz U-2 experiment, adapted from reference 12. The measured anisotropy ΔT in millidegrees Kelvin is plotted versus the angle between the instrument axis and the observed direction of maximum temperature $\hat{n}$ = (11.23 hours R.A., 19° Dec.). The data shown are for the seven flights between April 1977 and May 1978. Each point is an average of approximately five 20-minute "legs" of straight and level flight.

group of galaxies (assuming $T = 2.8$ K) of about 550 ± 75 km/sec in the direction α = 10.5 hr, $\delta = -11$ deg.

It is the large magnitude of this velocity that is particularly disturbing, especially since peculiar velocities for nearby galaxies are small. The relative velocity of the Milky Way and the Andromeda galaxy is only 80 km/sec, and Peebles has concluded[13] that the peculiar (non-Hubble) velocity of the Virgo cluster is similarly small. But large peculiar velocities seem to be cropping up recently in diverse areas of cosmology. V. C. Rubin reported at the last Texas symposium on a high velocity

(3500 < ca < 6500 km/sec) for the Milky Way measured with respect to nearby galaxies[14] (but in a different direction than ours; see references 11 and 12). And in this conference J. Peebles is reporting the recent results of his group at Princeton that sees similarly high velocity differences in pairs of (presumably) bound galaxies. It is difficult to reconcile the high kinetic energy of these galaxies with the belief that they are gravitationally bound unless one supposes that the mass of each galaxy is much greater than the visible mass. The missing mass must be comparable to that required to close the universe.

The Future

The deviations observed in the spectrum of the background radiation are tantalizing, and an order-of-magnitude improvement in sensitivity is going to be necessary in order to study them in detail. In addition one would like to study the deviations as a function of direction in the sky, a combined spectrum and isotropy experiment. The

FIGURE 5. The planned COBE satellite; an artist's conception. The apertures at the center are for the 3.3 to 0.33 mm spectrum radiometer and for the diffuse infrared photometer; the four pairs of microwave horns are for the isotropy radiometers. The radiometers are shielded from the sun and earth by the large conical shade.

isotropy measurements have already yielded an unexpected discovery in the high peculiar velocity of the local group of galaxies, but as of yet no intrinsic anisotropy has been observed. Again one would like to see an order-of-magnitude improvement. Fortunately just such an improvement in both the spectrum and isotropy is on the near horizon: COBE.

COBE stands for COsmic Background Explorer, a satellite which may be flown by NASA as early as 1983. The COBE scientific team includes many of the scientists in the United States who have made the best state-of-the-art measurements from rooftops, balloons, and airplanes. The planned configuration is shown in FIGURE 5, and includes the following experiments:

1. Spectrum measurement from 3.3 to 0.33 mm, based on differential comparison with an adjustable temperature blackbody, with a sensitivity of 2×10^{-14} w/cm^2sr cm^{-1} for each 7 degree diameter field of view. This instrument should be capable of measuring the spectrum to 0.1% of the peak flux for each field of view on the sky.

2. Anisotropy measurements at four frequencies: 23.5, 31.4, 53, and 90 GHz, again with a 7 degree field of view. The differential microwave radiometers should be able to detect deviations from isotropy to 10^{-4} of the background intensity, and should be limited only by the galactic synchrotron emission. Inclusion of four frequencies should allow subtraction of galactic emission.

3. Diffuse infrared background photometer to measure the infrared region from 8 to 300 microns in six octave bands, with a sensitivity of 10^{-13} w/cm^2sr for each 1 degree field of view.

As a physicist who is not actively working on COBE, I feel free to express my delight at the beautiful design of the experiments and to express unbiased optimism for the expected data. Without a satellite an order-of-magnitude improvement over the existing data would be nearly impossible; with COBE we can expect to have it in the next few years. The advantages of COBE come not only from the sensitivity, but also from the complete sky coverage afforded by its polar orbit. The COBE sensitivity is such that the uncertainty in the galactic synchrotron emission may set the limit to the accuracy of its measurements. If this is the case, then COBE may also provide the ultimate limit to the knowledge that we can derive from measurements of the cosmic background.

REFERENCES AND NOTES

1. PENZIAS, A. & R. WILSON. 1965. Astrophys. J. **142:** 419.
2. DICKE, R. H., P. J. E. PEEBLES, P. G. ROLL & D. T. WILKINSON. 1965. Astrophys. J. **142:** 141.
3. HARRISON, E. R. 1973. Annu. Rev. Astron. Astrophys. **11:** 155. Also P. J. E. PEBBLES. 1971. Physical Cosmology. Princeton University Press. Princeton, N.J.
4. It is possible that the radiation was last scattered at a z of about 6 when the formation of stars could have reionized intergalactic matter. Such a rescattering affects our interpretation of the background measurements but does not alter their importance; $z = 6$ is still quite remote.
5. PEEBLES, P. J. E. & D. T. WILKINSON. 1968. Phys. Rev. **174:** 2168.
6. DANESE, L. & G. DEZOTTI. 1977. Rivista del Nuovo Cimento **7:** 277.
7. WOODY, D. & P. L. RICHARDS. 1979. Phys. Rev. Lett. **42:** 925.

8. HAWKING, S. W. 1967. Mon. Not. R. Astron. Soc. **135:** 413. Also C. B. COLLINS & S. W. HAWKING. 1973. Mon. Not. R. Astron. Soc. **162:** 207–320.
9. BURKE, W. L. 1975. Astrophys. J. **196:** 329–334.
10. COREY, B. E. & D. T. WILKINSON. 1976. Bull. Astron. Astrophys. Soc. **8:** 351.
11. SMOOT, G., F. GORENSTEIN & R. A. MULLER. 1977. Phys. Rev. Lett. **39:** 898. M. V. GORENSTEIN, R. A. MULLER, G. F. SMOOT, J. A. TYSON. 1978. Rev. Sci. Inst. **49:** 440. R. A. MULLER. 1978. Sci. Am. **238:** 64–74.
12. GORENSTEIN, M. V. Ph.D. Thesis. University of California, Berkeley. November 1978. Unpublished.
13. PEEBLES, P. J. E. 1978. Comments Astrophys. Space Phys. **7:** 197–205.
14. RUBIN V. C., N. THONNARD, W. K. FORD, Jr. & M. S. ROBERTS. Astron. J. **81:** 687.

THE ENERGY SPECTRUM OF THE UNIVERSE

Geoffrey Burbidge

*Kitt Peak National Observatory**
P.O. Box 26732
Tucson, Arizona 85726
and
Department of Physics
University of California
San Diego, California 92037

Introduction

The universe is made up of matter and energy. Some components have been detected, but there are others which are speculated on for which we have no observational evidence. The discussion here will be confined mostly to what we have observational evidence for.

A review of this kind is not new. What is summarized here was described extensively by me in two reviews prepared for IAU Symposia in 1970 and 1972, respectively (Burbidge, 1972; Burbidge, 1974) and since then little in our knowledge based on observation has changed.

The major constituents of the universe which are easily detected are the discrete objects which emit large fluxes of electromagnetic radiation—galaxies, quasi-stellar objects, and objects related to them. Less easily detected but now known to be present are a number of smooth fields of electromagnetic radiation.

From an energetic standpoint, the two major components are the discrete objects and the microwave background radiation: all other detected components probably arise from the discrete objects, or are a result of the interaction of the radiation emitted by the discrete objects with the microwave background.

The constituents can be summarized as follows:

a. Mass condensations which are observed;
b. Diffuse gas which is observed in limited volumes;
c. Fields of electromagnetic radiations which are observed;
d. Relativistic particles which are observed in limited volumes;
e. Gravitational radiation which has not been observed;
f. Neutrinos which have not been observed;

We will discuss each of them in turn.

Galaxies—Mass Condensations

We can estimate the mass density in galaxies (ρ_g) if we can obtain the number density and the average mass per galaxy. Mass measurements for galaxies can be obtained only for those comparatively nearby. Thus the average masses of galaxies are

*Operated by the Association of Universities for Research in Astronomy, Inc., under contract with the National Science Foundation.

0077–8923/80/0336–0125 $1.75/1 © 1980, NYAS

obtained either from individual mass densities or by using the luminosities and mass-to-light ratios. The mass density obtained is proportional to the square of the Hubble constant H_0. Throughout the discussion I have assumed that $H_0 = 50$ km sec^{-1} Mpc^{-1}.

The mass density has been estimated many times going back to Hubble in 1936 and the changes in the value obtained have been due to revisions in the distance scale and also changes in the estimates of galaxy masses. The most conservative estimate based on visible mass in galaxies is

$$\rho_g \simeq 10^{-31} \text{ gm cm}^{-3}.$$

If one takes seriously the idea that all galaxies have heavy halos made up of unseen matter, a theoretical argument that has been pushed too far by Ostriker and his collaborators, we can only guess at the increase in the value of ρ_g. However, Rubin and her collaborators have begun to find that many spiral galaxies have flat rotation curves extending very far from the nucleus and this certainly suggests that the masses of the spirals have been underestimated. Thus, optimistically we may conclude that the average masses have been underestimated by a factor ~ 3, i.e.,

$$\rho_g \simeq 3 \times 10^{-31} \text{ gm cm}^{-3}.$$

If further we assert without proof, as is often done, that the virial holds in general for groups and clusters of galaxies, we may be able to increase ρ_g further so that perhaps

$$\rho_g \simeq 3\text{–}10 \times 10^{-31} \text{ gm/cm}^{-3}$$

The critical density needed to close the universe is

$$\rho_c = 5 \times 10^{-30} \text{ gm/cm}^3.$$

Thus the observed matter fails by a factor of 50 to close the universe and even if the actual mass density is greater than this it appears unlikely that it is large enough to reach the critical value.

Diffuse Gas

Many attempts have been made to detect diffuse intergalactic gas. Attempts to detect cold gas through its absorbing properties have failed unless one chooses to interpret the observed absorption lines in the spectra of QSOs as due to intergalactic clouds at large redshifts. I consider that this is highly unlikely, but if it is assumed it is possible to estimate within wide limits the mean density, and it turns out to be very low:

$$10^{-39} < \rho_{ig} < 10^{-34} \text{ gm/cm}^3.$$

The other possibility which has been widely canvassed for many years is that large amounts of hot gas are present and it is detectable from its x-ray emission.

X-rays almost certainly due to hot gas with temperatures in the range 10^7–10^8 degrees Kelvin have now been detected from a number of rich clusters of galaxies. However this gas does not make up a large part of the mass of the clusters, and in fact it is not adequate to provide the unseen mass which must be present if the clusters are bound stable systems. Thus:

$$\rho \text{ (gas in clusters)} < \rho_g.$$

Apart from the discrete extragalactic x-ray sources there is a diffuse background x-ray flux with a spectrum ranging from a fraction of an electron volt to ~ 100 MeV. This background flux may be due to discrete sources, to diffuse hot gas spread throughout the universe, to interactions between relativistic electrons and microwave photons in large volumes, or to a combination of these. Some attempts are still being made by Perrenod and Field and others to argue that hot gas is the dominant source and may be present at closure density ρ_c. However, in view of the fact that many discrete sources are radiating x-rays (Seyfert galaxies, QSOs, and clusters have all been identified) such sources may make a large contribution to the background, and there is no certainty that any significant part of the diffuse x-ray flux comes from hot gas.

Fields of Electromagnetic Radiation

Energy is carried by diffuse radiation fields covering a wide range of frequencies from radio frequencies (10^7–10^{10} Hz) to hard γ rays (~ 10^{20}–10^{21} Hz). The radiation

Table 1
Energy Density

Type	(erg cm^{-3})	Origin	Comments
Radio	~10^{-18}	Discrete sources	Detected
Microwave	~10^{-12}	Big bang	Detected
Infrared	~10^{-14}–10^{-15}	Discrete sources	Not yet detected
Optical	~10^{-14}	Discrete sources	Detected
UV	~10^{-14}	Discrete sources	Not yet detected
X-ray	~10^{-15}	Not known	Detected
γ-ray	~10^{-17}	Not known	Detected

fields which can be separately identified are described in Table 1. Among these, the microwave background radiation of cosmological origin contributes the greatest energy mass density but is still very small (~ 10^{-33} gm/cm^3) compared with the mass in visible galaxies. Of course in the conventional expanding universe this radiation field was dominant at early epochs. All of the other radiation fields almost certainly derive their energy ultimately from nuclear or gravitational energy in discrete objects, in some cases in interaction with the microwave background radiation.

Relativistic Particles

Relativistic protons and electrons with a small admixture of heavier ions make up the cosmic ray flux in our own galaxy with an energy density ~10^{-12} erg/cm^3. Radio observations show that relativistic electrons exist in similar quantities in other spiral galaxies and it is reasonable to assume that relativistic nuclei are present in them also.

The evidence from the powerful but rare nonthermal radio sources indicates that immense fluxes of relativistic electrons exist in them in volumes which can range from cubic parsecs up to cubic megaparsecs in a few very large sources. The sources of these particles are thought to be supernovae and explosive processes in galactic nuclei. Relativistic electrons dissipate their energy through synchrotron radiation and Compton scattering and by escape, while relativistic nuclei dissipate energy by electromagnetic processes as they slow down in matter and then by nuclear collisions or escape.

The overall energy density of relativistic particles in the universe is not known. If they are largely confined to galaxies and radio sources and do not gradually fill up larger volumes, the mean energy density averaged over space is probably no greater than 10^{-16} erg/cm^3. If they are able to fill up large volumes like clusters of galaxies to the energy density found in our own galaxy, the mean energy density may be as great as 10^{-14} erg/cm^3.

Gravitational Radiation

Despite many strenuous efforts, no gravitational radiation has yet been detected. The results on the binary pulsar, though, reported at this meeting by Taylor and his associates, suggest that gravitational radiation is radiated as was predicted by the Einstein theory.

Theoretical speculations have been made that suggest that a large amount of energy may be carried by gravitational waves. However, without direct detection we cannot put meaningful limits on the energy that is carried in this mode.

Neutrinos

While there may be a large flux of cosmic neutrinos and antineutrinos, none have been detected as yet.

Those neutrinos which arise in thermonuclear processes in stars only contribute a few percent of the flux in optical radiations, i.e., this component of neutrino energy density is of order 10^{-16} erg/cm^3. However, if much of the energy released in the gravitational collapse of stars comes out in the form of neutrinos, the energy density could be as high as 10^{-12} erg/cm^3.

High energy neutrinos are presumably released in the violent events in the nuclei of galaxies which give rise to high-energy charged particles, but the contribution of these neutrinos to the total flux is not known.

The experimental limits to the cosmic flux of neutrinos are set by the cutoff in the β spectrum and are still considerably above ρ_c. Thus most of the mass-energy in the universe could be in the form of neutrinos, though from a theoretical standpoint this appears to be very unlikely.

Conclusion

The two major components of mass and energy that are known to exist in the universe are the discrete radiating objects—galaxies—whose origin is really not known, and the microwave background radiation whose origin is thought to be cosmological. Probably all of the other components arise from these two. They are generated directly by the conversion of thermonuclear or gravitational energy, or through interaction between the microwave photons and the components generated by the discrete sources.

These observational results have been described within the framework of a conventional cosmological model. However, the only quantity that really enters in when energy densities are derived is the distance parameter, and this only in cases where the components are not directly measured.

REFERENCES

BURBIDGE, G. R. 1972. IAU Symposium No. 44: External Galaxies in Quasi-Stellar Objects. D. S. Evans, Ed.: 493–517. D. Reidel Publishing Company. Dordrecht, Holland.

BURBIDGE, G. R., 1974. IAU Symposium No. 58: The Formation and Dynamics of Galaxies. J. R. Shakeshaft, Ed.: 93–108. D. Reidel Publishing Company. Dordrecht, Holland.

LIMITS TO VERIFICATION IN COSMOLOGY

G. F. R. Ellis

Department of Physics
University of Alberta
*Edmonton, Alberta, Canada T6G 2J1**

Introduction

This paper does not aim to present a general theory of limits to verification in cosmology, for the very good reason that such a general theory does not yet exist. Rather it considers some specific cases and raises some particular questions relating to the theme.

To give the paper a specific focus, I will consider the following question: How can we *observationally* prove the universe is well-described by a Friedmann-Robertson-Walker (FRW) universe model? There are two aspects to the question: First, how can we prove it is described by *some* FRW model? Second, how can we find the best *particular* FRW description? The importance of the first of these questions is undoubted, as the most fundamental features of our present understanding of the universe are its uniformity (spatial homogeneity and isotropy) and its expansion from a hot big bang. Accordingly, these features need to be subjected to searching observational tests to make sure that *all* of the rather meager data available *consistently* lends support to this interpretation. The significance of the conclusion means that it should be checked in all the ways available to us. The second question is then essentially that of the value of the Hubble constant H_0 and deceleration parameter q_0, determining the size and nature of the best-fit FRW universe model. Both of these aspects will be considered in the body of the paper.

Assumptions

The basic assumptions we make are, firstly, that *ordinary local physics holds everywhere* in the observed region of the universe, i.e., we try to interpret the observations without introduction of "exotic" physical processes. Some support for this supposition is provided, for example, by observations of the spectra of distant galaxies. Cosmography—that is, the question of what can be determined without the introduction of a specific dynamical theory for space-time curvature—will be briefly considered; but in the main we shall deal with cosmology, where we assume such a theory and, for want of solid proof that any other theory is needed, take that theory to be general relativity. The second assumption is that the *"local" astronomy of the universe is reasonably well understood* (that is, the organization of matter into stars, star clusters, galaxies, clusters of galaxies, and so on) and that *essentially the same structures occur at large distances.* This again is supported by observations of distant galaxies.

*Permanent address: Department of Applied Mathematics, University of Cape Town, Cape Town, South Africa.

0077-8923/80/0336-0130 $1.75/1 © 1980, NYAS

The Basic Problem

It follows from these assumptions that *the universe is large*—we cannot travel to any appreciable portion of it in the time available to us. (This is a quite nontrivial feature—we can imagine universes small enough that one could traverse the whole universe in the course of a lifetime.) In addition, no signal can bear information faster than light. Accordingly, on the scale involved in cosmological models, *we can only view the universe from one space-time point,* and the only information we receive about the universe is that arriving at this space-time point ("here and now") on timelike or null curves. This is the fundamental feature underlying the problem of verification in cosmology.*

The Evidence

Four kinds of evidence are available that bear to some degree on the cosmological question. These are (i) local physical experiments; (ii) local geophysical and astronomical observations; (iii) extragalactic high-energy massive particle ("cosmic-ray") observations; and (iv) astronomical observations of distant objects and background radiation. The latter might in the future include observations made through gravitational waves and neutrinos, but currently and in the foreseeable future, the only high resolution observations of distant objects will be astronomical observations made by electromagnetic radiation. Accordingly, it is appropriate to consider in detail what these observations tell us.

IDEAL GALACTIC OBSERVATIONS

We accept the usual assumption that the geometrical optics approximation gives a good description of the propagation of electromagnetic radiation to us from distant galaxies. Then in a particular cosmology, the *space-time geometry* determines the null geodesics which are the light rays along which we receive information, and the specific *matter distribution* determines the potential observations (e.g., redshifts, observed magnitudes, number counts) that can be made. The way in which these observations could in principle lead to a determination of the local space-time geometry has been explained in a beautiful paper by Kristian and Sachs;[2] we here outline further development of the theory by Ellis, Maartens, and Nel.[3]

The clearest way to consider the situation is to set up *observational coordinates* $x^a = \{w, y, \theta, \phi\}$ in the following way (FIGURE 1): let w be the time of observation of a galaxy as measured by an observer O; y be a measure of distance to the galaxy (e.g., an affine parameter distance down the null geodesic from the observer to the galaxy); and θ, ϕ the direction of observation in ordinary spherical coordinates. Then the surface $\{w = \text{constant}\}$ is the past light cone of the observer O at time w; the null geodesics generating this light cone have the tangent vector field $k_a = w_{,a}$ and are given by $\{w, \theta, \phi \text{ constant}\}$, and y measures distance down the null cone (y represents both spatial distance from O and the time lag from the time of light emission to the time of

*See reference 1. The only cosmological quantity whose rate of change seems remotely measurable is source redshifts (see reference 66).

light reception). Choosing the world line of the observer O to correspond to $y = 0$, certain central conditions have to be satisfied† by these coordinates at $y = 0$. It now follows† that the space-time metric takes the form

$$g_{ab} = \begin{pmatrix} a & b & v_2 & v_3 \\ b & 0 & 0 & 0 \\ v_2 & 0 & h_{22} & h_{23} \\ v_3 & 0 & h_{23} & h_{33} \end{pmatrix}, \quad (1)$$

where one component may be normalized by choice of y (e.g., $b = 1$ if y is chosen as an affine parameter for the null geodesic vector field **k**). Observations from our position

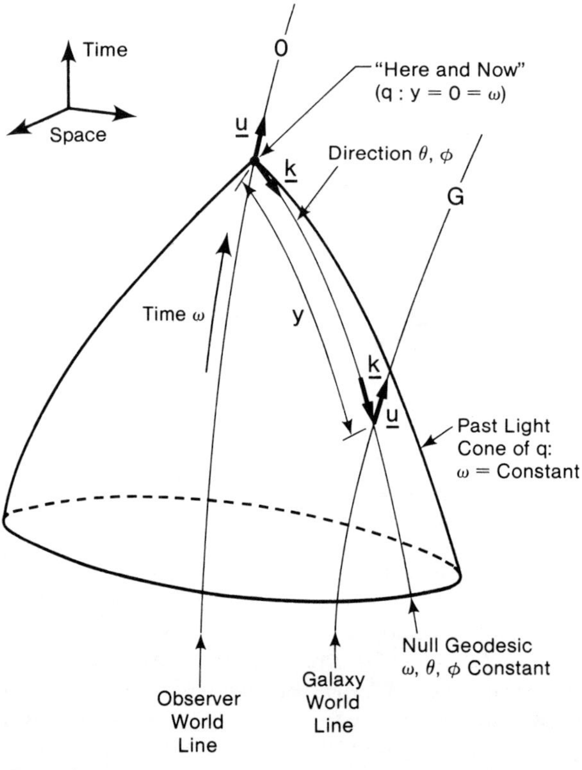

FIGURE 1. Observational coordinates based on the world line of an observer O characterize the time of observation w; the position in the sky θ, ϕ; and the distance y to the galaxy observed.

in space and time ("here and now") may be taken to correspond to observations made from the point $\{w = r = 0\}$; they enable various features to be determined on our past light cone (the surface $w = 0$).

In principle, one can carry out the following program to determine the space-time geometry from observations made on this light cone.

†Derivations and further details of the following results will be given in reference 3.

1. The metric components h_{AB} $(A,B = 2,3)$ are directly observable, if the intrinsic size and shape of the observed object is known. One can measure the size and shape of the image and compare this with the known size and shape of the object to obtain the metric components h_{AB} directly; for these just relate the actual size of the object to the corresponding observed angles through the relation $dl^2 = g_{ab}\, dx^a dx^b$, $dy = dw = 0$. In this way, one can determine the *distortion effect* caused by anisotropic curvature (the image at the observer is distorted relative to the apparent shape near the object, e.g., a spherical galaxy might look like an elliptical galaxy) and the *focusing effect* caused by isotropic curvature. The most convenient way to express the latter is through the "observer area distance" r_0, defined‡ by the relation $dS_0 = r_0^{\,2}\, d\Omega_0$, where $d\Omega_0$ is the solid angle subtended by the object and dS_0 its cross-sectional area perpendicular to the line of sight. In terms of the coordinates introduced above, $r_0^{\,4} \sin^2\theta = \det(h_{AB})$; the curvature of space-time results in r_0 being related in a non-Minkowskian way to other measures of distance to the galaxy. Knowing h_{AB}, one knows the intrinsic geometry of the null cones $\{w = \text{constant}\}$.

2. *The matter 4-velocity components $u^a (a = 0, 2, 3)$ are observable.* Let τ be proper time measured along the world line of the galaxy; then its 4-velocity is $u^a = dx^a/d\tau$. The components u^θ, u^ϕ can be determined directly from the observed proper motions of the galaxy, while the component u^w may be obtained directly from the observed galaxy redshift[2,4]: $1 + z = (u^a k_a)_{gal}/(u^b k_b)_{obs} = u^w$ (we use the normalization $\mathbf{u} \cdot \mathbf{k} = 1$ at $y = 0$ for the null-geodesic tangent vector $\mathbf{k}$). The final component u^y is not directly observable; however, it can be found in terms of the metric tensor components (**1**) from the normalization condition $u^a u_a = -1$ (which follows from the definition of u^a).

3. *The matter density is deducible from observations if the metric component b is known.* The number count dN of galaxies for some increment $y \rightarrow y + dy$ in distance down the null geodesics determines the product of the metric component b with the number density n of sources at y, and is observable if y is an observable quantity (e.g., the observer area distance r_0). Hence if b is known, n can be found, and if the mass present per source is known, then the mass density ρ can be found. Basically ρ is proportional to dn, with the ratio being determined by the mass-to-light ratio and the observational selection function f (f is the fraction of sources in $y \rightarrow y + dy$ which are detected by the measurements made).

The *only* detailed features of the cosmological model that are accessible to direct observation are the quantities just listed (namely, $\{h_{AB};\, u^a(a = 0, 2, 3);\, [bn]\}$), on the past null cone $C^-(q)$ of our space-time position q.

4. *The metric components a, b, v_A are not measurable directly* through any null-geodesic observations made by an observer at q (these quantities, which determine the imbedding of the light cone, are not reflected directly in galaxy shape or size measurements, redshifts, or number counts). Thus in general, one

‡See reference 4. This is the "corrected luminosity distance" of reference 2. It can also be determined, in principle, from measurements of the apparent magnitude m of a source of known intrinsic luminosity.[2,4]

will only know their limiting values at q (where they are determined by the central conditions at $y = 0$).

Consequently, *the cosmographic idea does not work:* one cannot determine the space-time metric or the matter distribution on the past null cone $C^{-}(q)$ by cosmological observations alone. As a particular example, when no dynamical equations are imposed, the fact that one found all cosmological observations to be exactly spherically symmetric would *not* imply that the space-time was spherically symmetric about 0.

5. Given the general relativity field equations and a suitable equation of state for the matter, one can integrate down the null cone $w = 0$ from the central world line to determine $(a, b, v_A; n; u^y)$ on this null cone from the known quantities $\{h_{AB}(0, y, \theta, \phi);\ u^a(0, y, \theta, \phi),\ (a = 0, 2, 3);\ [\text{nb}](0, y, \theta, \phi)\}$ and the central conditions on the metric components. Thus *one can in principle determine the metric tensor, and the matter distribution and velocity, on this null cone from cosmological observations, if the field equations and a suitable equation of state are assumed.* As a particular example, given the field equations and assuming that the equation of state is that of a pressure-free perfect fluid ("dust"; this is likely to be a good description at recent times), spherically symmetric observations *do* imply that the space-time is spherically symmetric on this null cone.

6. Because of the hyperbolic nature of the field equations, unless the universe has finite space sections and we have already seen around it, [1] *one cannot predict off the null cone to the future without assuming the values of unobservable initial data.* While one has sufficient data to completely determine the solution in the interior of the null cone (i.e., to the past), except in the finite universe case mentioned, new information can come from outside to upset our expectations (based on analytic continuation of present conditions) of the future; in order to obtain unique predictions from the field equations, one must assume that a "no-interference" data condition holds on a suitable surface in the past (FIGURE 2). However, given such a condition, one can in principle carry out a complete observational cosmology program: *One can in principle determine the metric and matter distribution on the past null cone $w = 0$, and then extend the solution to a space-time region around this null cone, if no incoming radiation interferes with the predicted solution.* It is clear on physical grounds that one must assume such a no-interference condition if we cannot already see all round the universe, for we cannot be sure that the sun will rise tomorrow unless we can be sure, for example, that a strong incoming gravitational field will not interfere with its stability and motion. On the other hand it is also clear that this assumption is a reasonable one to make, at the present time; for we can regard it as a continuity assumption, extrapolating from present circumstances where this condition does indeed appear to be satisfied.

FRW Observational Conditions

Within the framework of this "observational cosmology" program, the space-time is locally determined uniquely by cosmological observations, so the question posed in

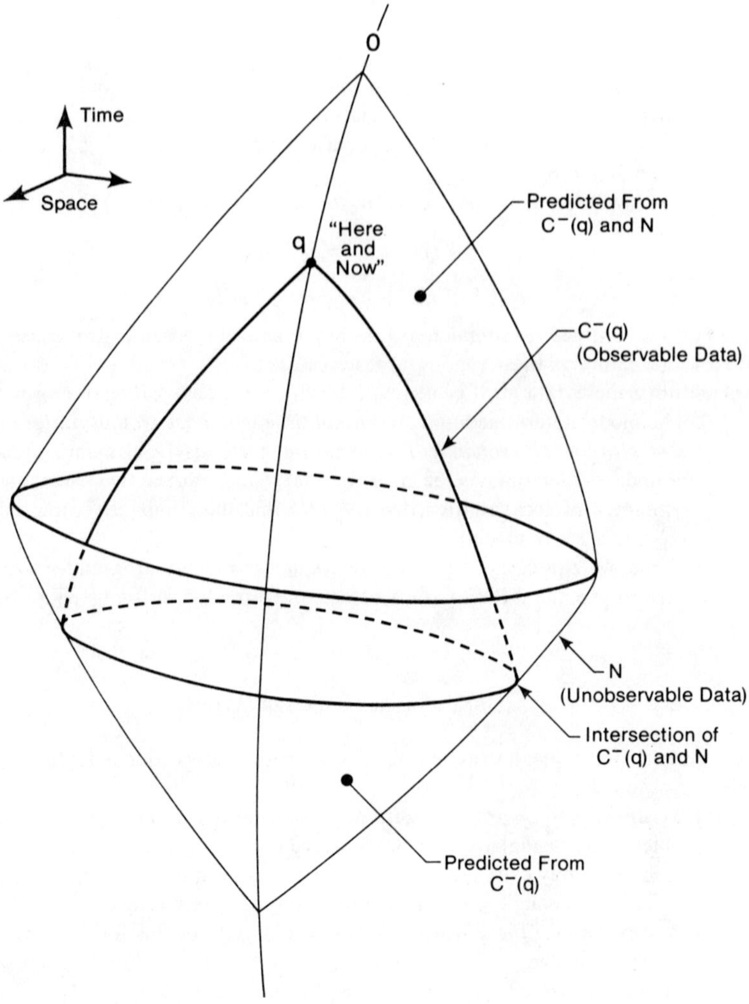

FIGURE 2. Observational data on the past null cone of the point q has to be supplemented by data on a surface such as the surface N shown, in order to make predictions into the future (unless the past null cone already circles the whole universe, as it does in the Einstein static universe).

this paper now becomes: *What observational data will imply that the universe is a FRW universe model?* The situation is as follows[3]: the space-time determined by astronomical observations is FRW (spatially homogeneous and isotropic) if and only if:

(i) $u^\theta = u^\phi = 0$, i.e., there are *no proper motions;*

(ii) there is *no distortion effect;* then h_{AB} takes the form

$$h_{AB} = \begin{pmatrix} r_0^2 & 0 \\ 0 & r_0^2 \sin^2\theta \end{pmatrix}; \qquad (2)$$

(iii) *all relations between observable quantities* [e.g., [m, z) and dN, z) relations] *are isotropic,* so that the relations between the fundamental quantities r_0, z, and ρ are isotropic [e.g., if we use z for the coordinate y, $r_0\,(z, \theta, \phi) = r_0(z)$ and $\rho(z, \theta, \phi) = \rho(z)$];

(iv) the derived relations *take the FRW form* (see references 4,7)

$$r_0(z) = H_0^{-1} q_0^{-2} (1 + z)^{-2} \{q_0 z + (q_0 - 1)(\sqrt{1 + 2q_0 z} - 1)\}, \qquad (3a)$$

$$\rho(z) = (1 + z)^{-3} \rho_0, \quad \rho_0 = 6 q_0 H_0^2. \qquad (3b)$$

(Here we have assumed for simplicity that the pressure and cosmological constant are both zero; one can relax these conditions if necessary).

In addition, these data must be compatible with other observational information:

(v) The model determined must be *compatible with background radiation data and with local astronomical information,* particularly, element and particle abundances, estimates of ages of stars and radioactive elements, and estimates of local matter density. *All* this data must be compatible to confirm a FRW model!

In principle, we can carry out an observational cosmology program in which we systematically prove the FRW nature of the universe by verifying points (i)–(v) above.

Realistic Galactic Observations

In contrast to the ideal situation sketched above, observational reality is quite different. Specifically,

(a) *We cannot detect proper motions in the cosmological context.* The real problem here is that we can only observe on one light cone; such transverse velocities as may exist are too small to be measured directly, but nevertheless may be large enough to be important dynamically (see reference 2).

(b) *We cannot detect the distortion effect.* The reason for this is that there do not exist in the universe objects that have clearly defined "standard shapes" so that observation of their apparent shapes could be unambiguously interpreted as due to the distortion effect. For example, an apparently elliptical galaxy cannot be proven (e.g., by a distinguishing spectral feature) to be intrinsically spherical; an apparent alignment of galactic axes could as well be explained by local astrophysical effects related to galaxy formation, as by the distortion effect; and so on. Even if one could find one or two objects, or even directions in the sky, where the effect could be clearly established, this would be very far from what is actually needed: to carry out an observational cosmology program in full, one needs (both for proper motions and the distortion effect) the effect to be determined for *all distances* (e.g., for all redshifts) *and for every direction in the sky.* This is clearly quite unattainable in practice.

(c) While *we can in principle check observational isotropy,* both for objects at specific distances (e.g., for particular ranges of z) and overall, difficulties arise because of problems of statistics and definition. Relatively few objects may have been observed, or may even be available for observation (at small z); at what level of fluctuation would we accept that there was clearly a departure from isotropy? No clear criteria for isotropy or lack of isotropy have been laid down, so far.

(d) In determining $r_0(z)$ and $\rho(z)$ from observations, among the problems we run into are *observational limits, astrophysical uncertainty, statistical problems,* and *redshift and screening dilution* of information. We shall discuss these problems in the succeeding sections.

The reality of the situation, therefore, is that although in principle the cosmology is completely determinable from galactic observations, in practice one cannot obtain significant parts of the necessary data directly. We have to reply on indirect evidence to set good limits on the proper motions and distortion; in particular, to deduce a FRW description of space-time, we have to either *assume* that these quantities are zero *a priori,* or deduce this indirectly from other measurements. We lack the ability to directly check these features of the standard uniform cosmological models.

Ideal Observations in a Spherically Symmetric Cosmology

Let us now assume that we somehow have decided that an accurate cosmological model *is isotropic:* there are no proper motions, no distortion effect, and all the observations are isotropic (that is, (i)–(iii) are satisfied). The space-time metric is then given by (**1**) with $v_A = 0$ and h_{AB} as in (**2**); for a suitable choice of the coordinate y, the remaining nontrivial quantities a, b, r_0, z, and ρ will depend only on w and y. In principle, $r_0(z)$ is directly observationally measurable (by observing apparent magnitudes or angles as a function of redshift);$[nb](z)$ can be found directly from number counts, and knowledge of b will determine $\rho(z)$ if the mass-to-light ratio is known. Assuming this ratio and a suitable equation of state, one can integrate up the field equations to determine the metric components a, b, and the matter stress tensor $T_{\alpha\beta}$ on the null cone $w = 0$ and in a region around this null cone (the "no-interference" condition is automatically satisfied in a space-time that is isotropic *at all times* about the world line of the observer O). To show that space-time is a specific FRW space-time now requires *verifying* the FRW forms (**3**) for $r_0(z)$ and $\rho(z)$ for particular values of H_0 and q_0, and *checking* their compatibility with other observational data [that is, (iv) and (v) above remain to be checked]. In principle one can carry this out, but in practice a series of problems arise which we now discuss.

Observational Methods

In order to make clear statements relating to these problems, one needs to consider the methods by which the observations are made. It is *not* sufficient merely to state that one can measure galactic magnitudes; indeed, the bolometric magnitude m in terms of which the theory is conventionally stated is not really an observationally determinable quantity at all.[5] The situation is reflected in Baum's observation[6] that between the years 1920 and 1960 the observed diameter of the galaxy M32 increased exponentially with time, from 3 to 30 seconds of arc; Baum comments "the data tell us more about the threshold of detection than the object involved." This implies similar problems may arise in measuring galactic magnitudes, for these depend in an essential way on an angular diameter estimation.

To look at this question, we use here a simple model[5] of the way observations of extended objects are made in practice. Let the galaxy be at a distance characterized by an observer area distance r_0; then a distance r at the object from its center, perpendicular to the line of sight, corresponds to an observed angle α from the center of the image, where $r = \alpha r_0(z)$. Suppose the brightness distribution across the surface of the galaxy perpendicular to the line of sight is given by

$$B = B_0 \mathcal{J}(\nu) f(r/a),$$

where B_0 *is the central surface brightness,* $\mathcal{J}$ *is the spectrum, and* f is the radial brightness function, chosen so that $f(0) = 1$ and scaled by the length a [for example, f might be the Hubble profile $f(r/a) = (1 + r/a)^{-2}$ or the de Vaucouleurs profile $f(r/a) = \exp\{-(r/a)^{1/x}\}$, with $x = 1$ for spirals and $x = 4$ for ellipticals]. Then conservation of photons implies that the observed specific intensity of radiation at an angle α from the center of the image is given by [5,4,7]

$$I_\nu(\alpha) = \{[B_0/(1+z)^3] \cdot \mathcal{J}(\nu(1+z)) \cdot f(\alpha r_0/a)\}. \tag{4}$$

Accordingly (if we for the moment ignore galactic and intergalactic absorption, and "seeing" effects) one can determine the following measurable quantities[5]:

1. The *galactic magnitude* $m_\nu(A)$ *measured through an aperture of semi-angle* A centered on the galactic image obtained by integrating (**4**) up to the angle A and changing to a "magnitude" scale; one finds

$$m_\nu(A) = -2.5 \log_{10}\{[B_0\mathcal{J}(\nu(1+z))/(1+z)^3] \cdot (a/r_0)^2 \cdot g(Ar_0/a)\} + \text{const}, \tag{5}$$

where $g(\beta) \equiv \int_0^\beta 2\pi\alpha f(\alpha)\, d\alpha$.

2. In any particular observational situation, there will be a detection threshold determining the limiting specific intensity $S_L(\nu)$ to which the galactic image can be detected (see references 8,14). The angle from the center at which the observed brightness profile drops to this limit is then the *apparent angle* A_{ap} measured for the galaxy; it is determined, on using equation (**4**), by the relation

$$I_\nu(A_{ap}) = S_L(\nu). \tag{6}$$

The corresponding magnitude measured by using an aperture of this value is given [from (**5**)] by

$$m_{ap} \equiv m_\nu(A_{ap}); \tag{7}$$

this is the *apparent magnitude* of the galaxy. If the observed specific intensity never exceeds the detection limit $S_L(\nu)$, then $A_{ap} = m_{ap} = 0$ and the galaxy cannot be detected by the apparatus in question at the set threshold (i.e., assuming the exposure time and other limiting factors remain constant).

The quantities A_{ap} and m_{ap} are the 'natural' measures of an image that arise in observations of distant objects, e.g., in deep sky plates; however, when detectors are used in detailed observations to determine the magnitude of a galaxy, the quantity

$m_\nu(A)$ may be a better description of the measurements made, for suitable choice of aperture value A.

Observational Limits

It is commonly assumed that galactic observations are magnitude limited, i.e., that simply m_{ap} is the limiting factor in detection. However, this is not so; in fact it is clear that both average image brightness and apparent angle play an important role in detection limits and in selection effects.[9–15] In essence, if we consider[9–11] a set of sources with the same intrinsic magnitude but various diameters, then for very large diameters the objects will not be detected because their average surface brightness will drop below the detection limit, while if the diameter is very small the apparent angle will be very small, and the object will not be easily distinguished from a star (it will appear "quasi-stellar").

However, it is clear that we cannot discuss these effects in terms of the *object* size and luminosity alone, for the actual selection effects are determined by the *image* size and intensity.[9,12–14] Accordingly, one should clearly separate out those aspects which are intrinsic to the objects observed, and those which are intrinsic to the measuring situation (in particular, the distance from the object to the observer). This can be done in the following way: define, for a given class of objects, an *object plane* whose axes are[10] the object's intrinsic magnitude and its diameter, and an *image plane* whose axes are the apparent magnitude and apparent angle of the given class of objects. Then, for objects of the kind considered which are situated at a specific distance from the observer in a given space-time, Equations 4–7 determine the *observational map,* which maps each point in the object plane into a corresponding observed point in the image plane.§ The determinants of the map are (i) the intrinsic properties of the class of objects considered, namely, their spectrum $\mathcal{J}$ and radial brightness distribution function f (assumed to be the same for all objects in the class); (ii) the detection limit $S_L(\nu)$ for the detection apparatus in question; (iii) the space-time properties as specified by the observer area distance $r_0(z)$ in terms of the redshift z; and (iv) the distance between object and observer as measured by the redshift z. FIGURE 3 shows this map for a FRW universe with $q_0 = +1$, for galaxies with an $x = 1$ de Vaucouleurs profile and with spectra described by a spectral index of 0.7. One can see how the images of particular objects (marked as points #1–4 in the object plane) move in the image plane, if these objects are considered at different values of redshift z. (Remember that "seeing" and absorption effects have not been included in this simple model; they would further displace the points in the image plane.)

Now the crucial feature is that *observational limits are limits placed in the image plane.* Typical kinds of limits that might occur are shown in FIGURE 3: images to the right of the brightness limit would be difficult to detect because of their faintness, optical images to the left of the angle limit would easily be confused with stars, and in addition the magnitude limit might play a role (particularly in radio or x-ray observations, where the minimum angle limit would not be directly significant). One

§Note that the source luminosity L is given by $L = D^2 B_0 G$, where D is its diameter and G is a constant depending on the radial function f. The magnitude is given by $M = -2.5\log_{10}L +$ const.

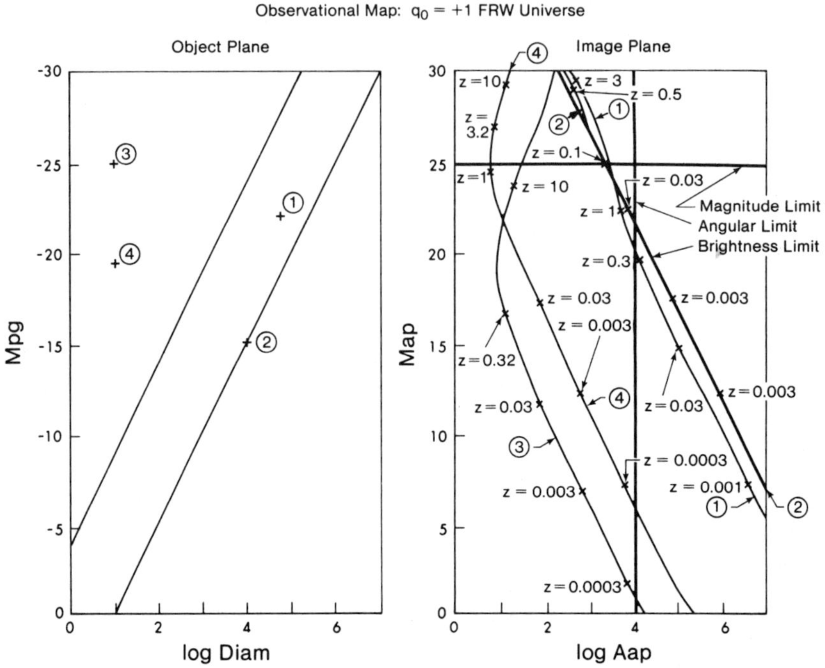

FIGURE 3. The observational map for galaxies with a de Vaucouleurs profile ($x = 1$) and spectral index of 0.7, in a FRW universe with $q_0 = +1$. The image points of galaxies #1–4 are shown for various values of the redshift z of the galaxy.

can see that as the redshift increases the image of each object moves into an image region where it is unlikely to be observed, or may not even be detected.

Various interesting features can be deduced. Firstly, one can obtain the *inverse map* from the image plane to the object plane, deducing the diameter and magnitude of the object from its image size and apparent magnitude (assuming the other features determining the map are known). In particular, one can map back any particular set of observational limits from the image plane into the object plane, showing (see FIGURE 4) which objects are likely—at a given redshift—to be easily detectable, and which are not; and showing how as the redshift increases, the "observable" region in the object plane steadily decreases (FIGURE 5). This makes clear the nature of possible populations of objects which might be present in a particular space-time without having been detected[15] by particular observational techniques, because they lie outside the observational limits for particular redshift values.

Secondly, one can project the map to obtain, for particular objects in the class, (apparent angle, redshift) and (apparent magnitude, redshift) relations. An important feature of these relations¶ is the existence of a *limiting redshift* for observation of any

¶In contrast to the (m, z) relation, which is usually plotted.

particular object; by (**4**), for high enough redshift, the specific intensity received will drop below the detection limit. Accordingly, in contrast to the theoretical situation where this detection limit is not taken into account, the (A_{ap}, z) and (m_{ap}, z) curves asymptotically approach this limiting redshift value for small angles and large magnitudes (FIGURES 6,7).

Astrophysical Uncertainty

The curves in FIGURE 3 are plotted for objects that maintain the same magnitude and diameter, i.e., which remain fixed in the object plane as the redshift increases. However, one certainly has to take evolution (that is, change in intrinsic source properties with z) into account, for studies of galactic evolution (see references 16,17) show that both luminosity and spectral evolution may be expected in a FRW universe at high z, and studies of "galactic cannibalism"[18] show that radial evolution may also be appreciable. Accordingly[18,67] one needs to consider the evolution of individual objects of a given kind, represented by a movement of the corresponding representative point in the object plane as z increases (i.e., as we look back to greater distances

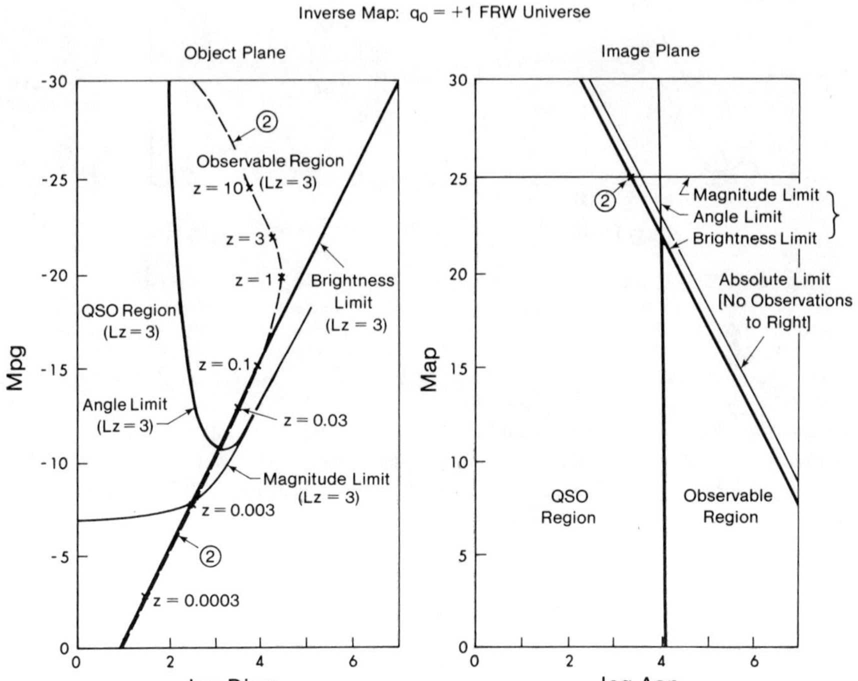

FIGURE 4. The inverse map, for the same conditions as in FIGURE 3. The image #2 corresponds to the various positions shown in the object plane, for different values of z. Also shown are the maps of the various observational limit lines into the image plane, for the case when $\log_{10}cz = 3$.

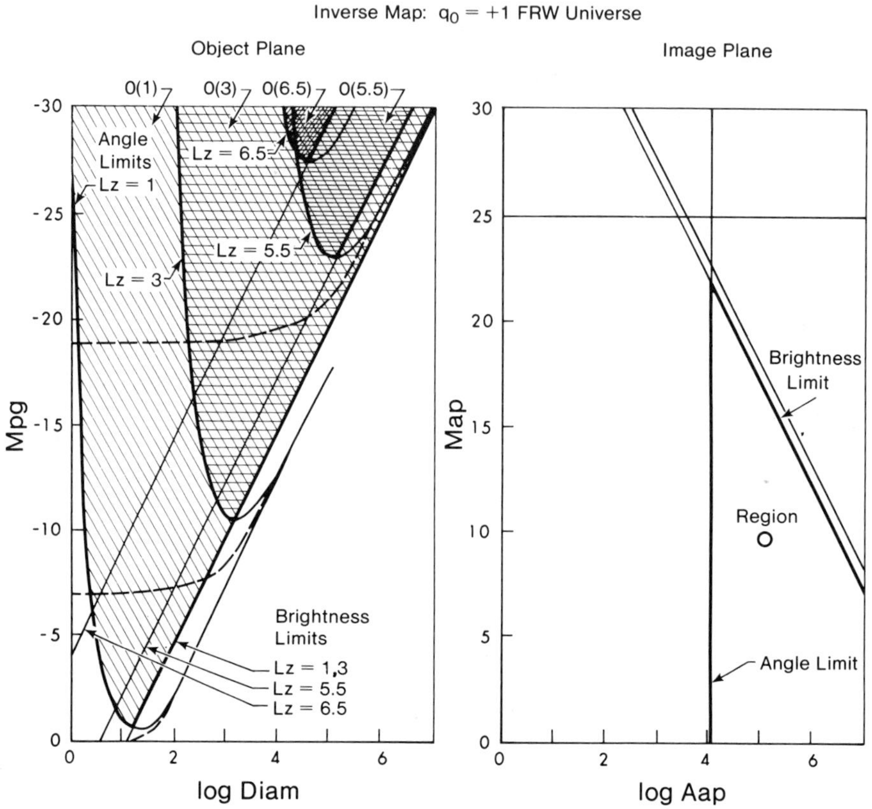

FIGURE 5. Taking the observable region in the image plane as O, the corresponding set of points in the object plane, representing galaxies which can be detected because their images lie in O, move up to include only larger and more luminous galaxies as z increases (shown for log cz = 1, 3, 5.5, and 6.5).

and earlier times). FIGURES 6 and 7 show the effect of radial evolution $a(z)$ and luminosity evolution $L(z)$ on the observed apparent magnitudes and apparent angles, determined by Equations 4–7. The most important feature is that such evolution changes the limiting redshift value, thereby giving the curves considerable spread (the most extreme cases occurring when the evolution causes the maximum difference in intrinsic source brightness, e.g., when the luminosity L increases with z, but the radius a decreases with z). In contrast to this, if no evolution occurs, but the value of q_0 in a FRW universe is varied, there is very little spread in the curves so obtained (FIGURE 8)—for they all have to agree at small z values, and they also have to end at the same limiting value of z because changing q_0 does not change this limit. The (A_{ap}, z) and (m_{ap}, z) curves can therefore only differ for a small range of values of z, as q_0 varies.

One might imagine that other measurement techniques [using[19] specified aperture angles $A(z)$ depending on z, and using Equation 5 to obtain $m(z) \equiv m_v(A(z))$, for example] would get around the problem of needing to know radial evolution, but this is

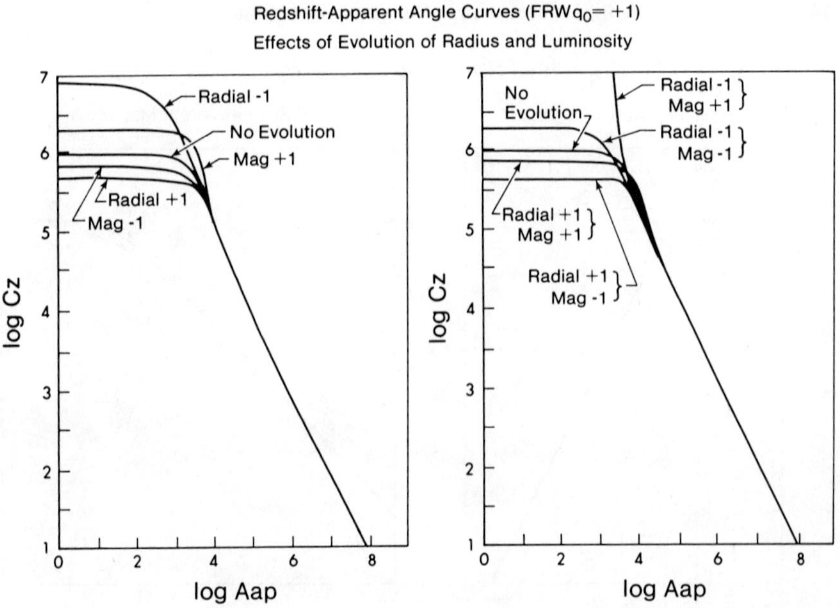

FIGURE 6. The (redshift, apparent angle) graphs for galaxies as in FIGURE 3 and various evolutionary conditions. The evolution is taken as proportional to $(1 + z)$ or $(1 + z)^{-1}$, and is applied to the galaxy intrinsic luminosity and/or radius. [Thus, "radial −1, mag +1" means a radial evolution in which a length scale a is proportional to $(1 + z)^{-1}$ as one goes back in the past, and an evolution of galaxy intrinsic magnitude in which the luminosity L is proportional to $(1 + z)$ as one goes back into the past.]

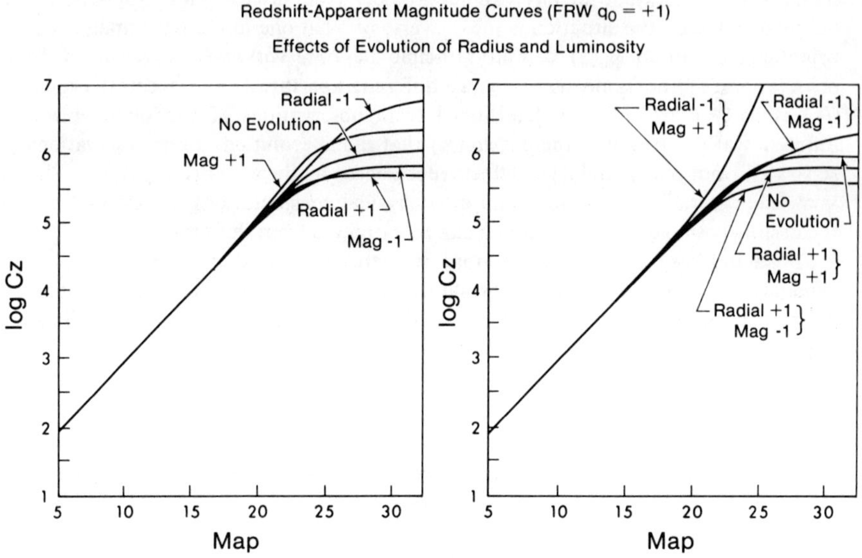

FIGURE 7. The (redshift, apparent magnitude) graphs for galaxies as in FIGURE 6.

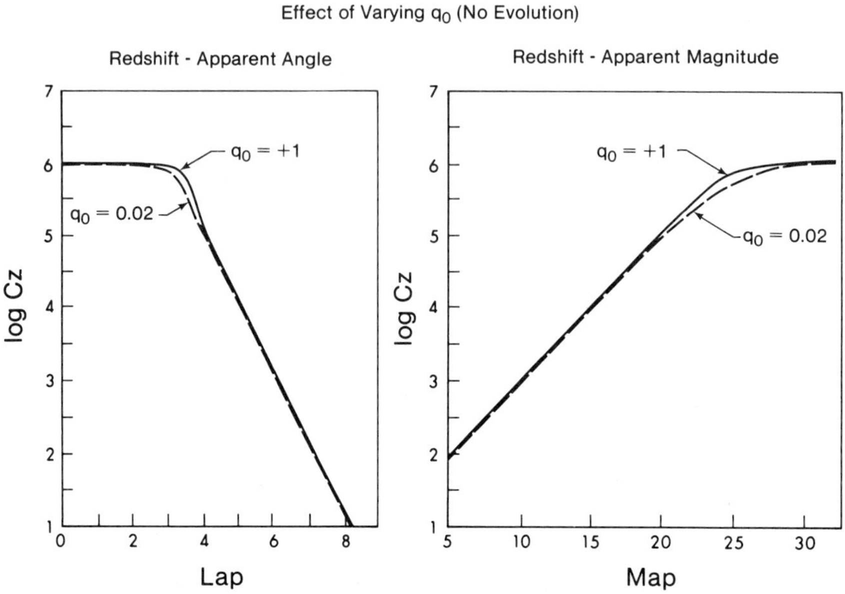

FIGURE 8. The effect of different FRW models on the apparent magnitude and apparent angle curves (no intrinsic evolution of the galaxies has been assumed).

not so—any method of estimating magnitudes that relies on the use of apertures must depend on this knowledge, and apertures *have* to be used in an essential way to measure the magnitude of any extended object (the situation for QSOs is rather different). In fact, the situation is the converse of what one might have imagined: the brightness evolution $B_0(z)$ can in principle be *observationally determined* from observations of the same object using different apertures‖; the spectral evolution $\mathcal{J}(\nu, z)$ can in principle be determined from observations of the same object at different wavelengths; but[5] there is *no way* that radial evolution can be observationally separated from the cosmological effects represented by the area distance $r_0(z)$. Thus a knowledge of radial evolution, from astrophysical considerations, is essential before the desired cosmological information can be extracted from the data.

It should also be emphasized that interpretation of the *number counts* also demands knowledge of the radial, brightness, and spectral evolution; for all enter the determination of the galaxy selection function in an essential way (as they all affect the limiting redshift to which a particular class of objects is observable[12]).

Selection Effects

In considering either number counts with distance or the luminosity function of a cluster of galaxies, it is clear that the observed curve does not necessarily represent all

‖See reference 5. In principle, one can deal with extra complications arising from such features as a color gradient caused by two different stellar populations in the source.

the objects "out there"; rather it represents the separation between those objects which are seen and counted, and those which are either not seen or not counted. Thus the observed curve is determined by the objects that are both detected and selected for inclusion in a catalogue.

As an example, consider[11] a line of constant magnitude in the object plane. This is mapped into a line in the image plane that represents approximately constant observed magnitude for small angles, but turns over sharply at large angles where the surface brightness drops towards the limiting value (FIGURE 9). This suggests[11] that of all the galaxies with a given constant magnitude at a given redshift, those which are most likely to be observed are those presenting the largest apparent angle, i.e., which lie farthest to the right in the image plane (corresponding to the turnaround of the curve); in the object plane, these are situated in the region where the line of constant magnitude approaches the right-hand brightness limit. Disney suggests[11] that selection based on this effect may explain the observed distribution of central brightnesses in many clusters of galaxies. While this interpretation is controversial,[14] it is clear that many galaxies of the same magnitude could be present and still go undetected.

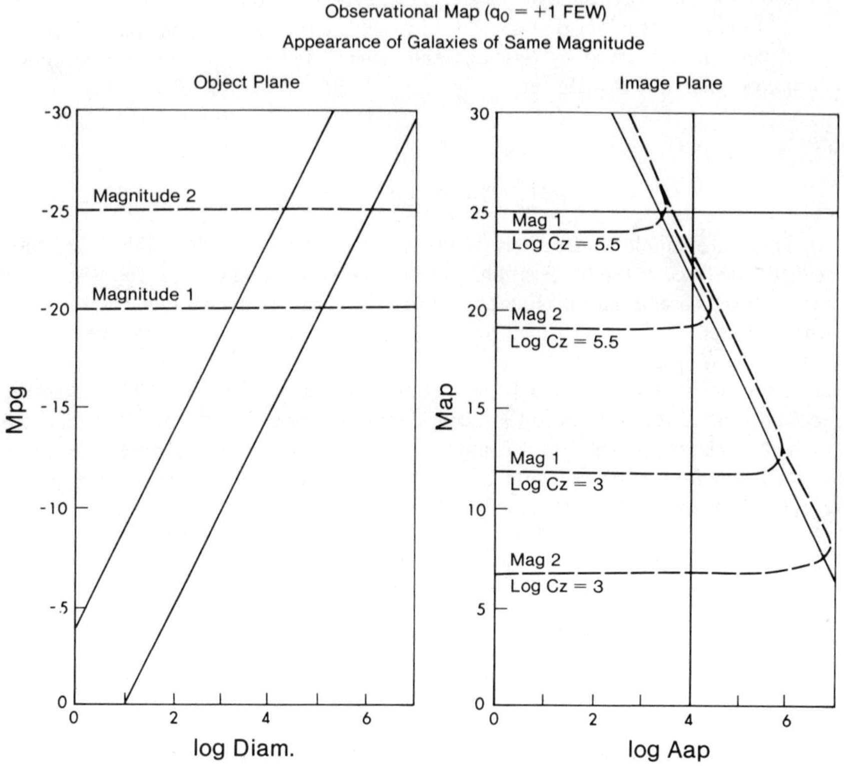

FIGURE 9. The image of families of galaxies which are all intrinsically of the same magnitude (but of varying diameters), for log $cz = 3$ and 5.5.

As a further example, consider a cluster of galaxies in which the galaxies obey a linear (magnitude, log diameter) relation, i.e., which lie on a line in the object plane (see references 20–22). The crucial question (FIGURE 10a) is whether this line has a slope greater or less than that of a line of constant brightness. If the slope is equal to or steeper than lines of constant brightness, then (FIGURE 10b) the observed points lie on a line that is close to a line of constant average observed brightness; at any particular redshift, one will fail to observe some of the larger galaxies with larger radii, and as the redshift increases the proportion of galaxies lying beyond the observational limit increases. However, if the slope of the line in the object plane is less than that of constant brightness, the galaxies at both the bottom and the top end of the line are undetected, the first because they exceed both angular and apparent magnitude limits, and the second because they exceed brightness limits. The observed curve has a sharp turnover at a well-defined apparent magnitude and angle, and this turnover "rolls up" as z increases (FIGURE 10c). If one were observing a cluster of galaxies with such a (magnitude, diameter) relation, an effect similar to that proposed by Disney could occur: of all the galaxies observed in the cluster, it is those near the turnover in the observed curve (corresponding to where the line on the object plane crosses the brightness limit) that would be regarded as the most luminous cluster galaxies. Accordingly the existence of such a sharp turnover of the observed curve in the image plane is the kind of effect that could lead to selection effects responsible for the very small deviation observed by Sandage and others[23] for the apparent magnitudes of "brightest galaxies" in a cluster.

The Observational Situation

These are simple illustrations of the kinds of effects which could occur. More realistic analyses of the observations can be carried out, and in fact[5] the best way to carry out the observations is probably to use a two-dimensional photodetector together with a computer program which produces analyses similar to those above but with more realistic spectral and surface brightness assumptions (a program of this kind has been written by Butchins and Ripp[24]). Interesting new effects could be taken into account if more realistic spectra are considered, for example, "banding" could occur: for a given class of objects, there could be a visibility limit z_1 so that these objects were undetectable for greater redshifts, because of low apparent brightness, until a redshift z_2 is reached where a steep step in the spectrum could bring the observed brightness up above the detection threshold again. Thus the class of objects considered could be visible in several redshift bands, but invisible for values of z between these bands (for a given set of detectors). Again, in analyzing selection effects, one would want to consider the brightness line determining saturation, as well as the limiting brightness.[13]

We do not aim to pursue these features further here; rather, for present purposes,

---

FIGURE 10. The images of families of galaxies which are nearly of the same intrinsic brightness (but varying magnitude and diameter). Case (b): the image of families whose slope is greater than a family of constant brightness. Case (c): the image of families whose slope is less than that of a family of constant brightness.

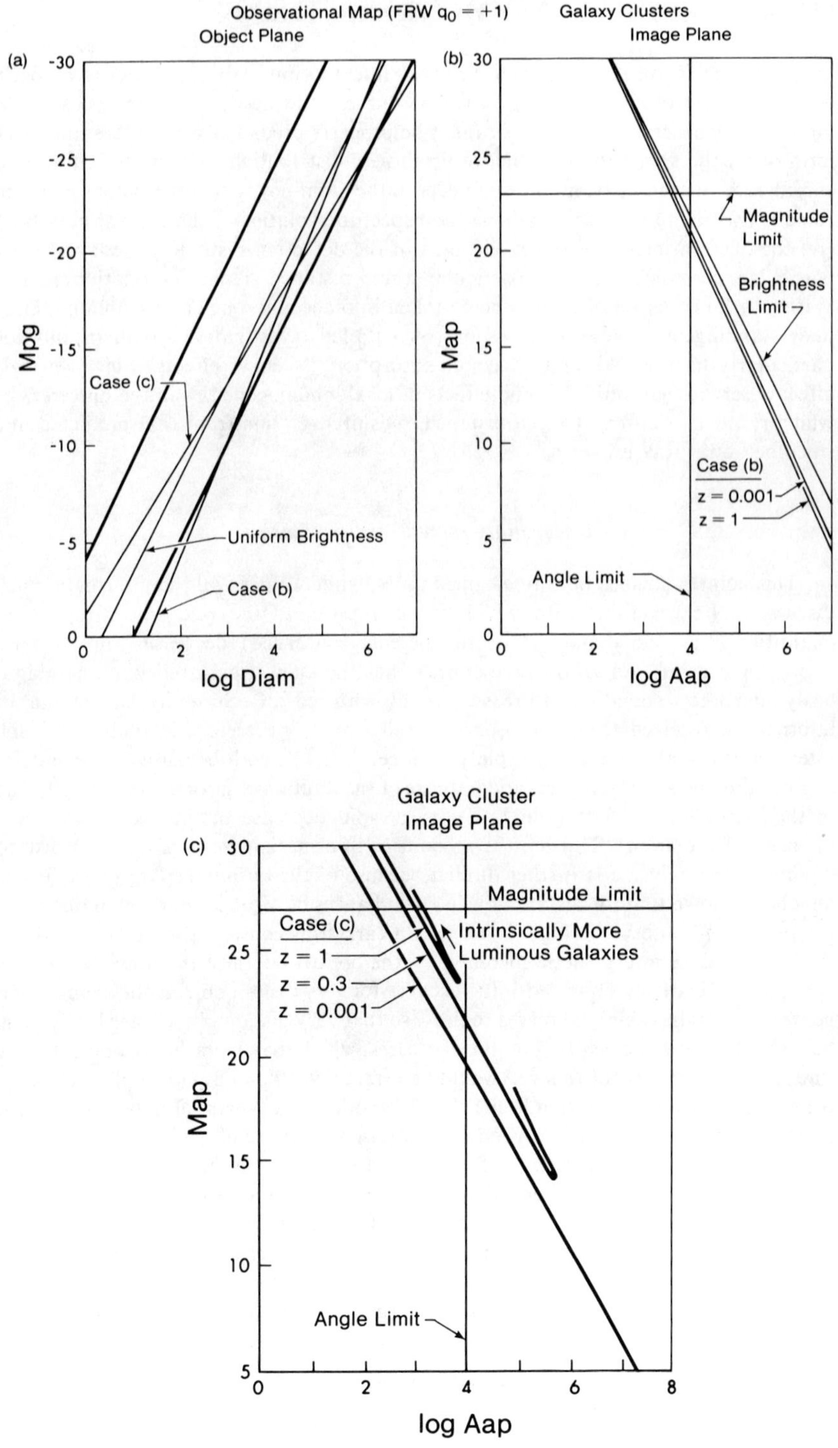
Observational Map (FRW q0 = +1)
Galaxy Clusters
Object Plane
Image Plane
(a)
(b)
Mpg
Case (c)
Uniform Brightness
Case (b)
log Diam
Map
Magnitude Limit
Brightness Limit
Case (b)
z = 0.001
z = 1
Angle Limit
log Aap
Galaxy Cluster
Image Plane
(c)
Magnitude Limit
Case (c)
z = 1
z = 0.3
z = 0.001
Intrinsically More Luminous Galaxies
Angle Limit
Map
log Aap

the important feature that needs to be taken into account is that, in order to properly understand the observed (m, z) or (N, z) curves, one needs *an entire package of information:* observational curves cannot be properly predicted without assuming the form of (i) the spectrum and surface brightness distribution for objects in the class considered; (ii) the distribution of objects in the (luminosity, log diameter) plane, at some z; (iii) the luminosity, diameter, and spectral evolution of the class of objects††; (iv) the observational and selection limits of the detection system(s) used; and (v) a particular cosmology, i.e., a particular (area-distance, redshift) relation r_0 (z). Without *all* of this set of factors being taken into account, one cannot obtain a clear understanding of galactic observations. A detailed investigation would in addition particularly have to take into account absorption; "seeing" effects, which strongly affect observational limits[12,25]; and effects of local inhomogeneities in the universe,[69,70] which result in a different (area-distance, redshift) relation from that predicted in a smoothed-out FRW universe.

Uncertainty Increases with Redshift

This point is obvious but important. Firstly, while we need all the information just discussed, the certainty with which we can estimate the relevant astrophysical quantities (e.g., the diameters of the objects concerned) decreases rapidly with increasing redshift. As *all* of these entities must be known, our ability to unambiguously interpret observations decreases rapidly with redshift. Secondly, the amount of information received from each source—and which is therefore available for such interpretation—also decreases rapidly with redshift,[26–28] both because of the smaller angles subtended by the objects and because of the dilution of information directly due to the increasing redshift (which causes a rapid decrease in the received specific intensity of radiation). Thirdly, the amount of information we receive about distant regions in the universe is further diluted because of the *cosmic veiling effect.* It has long been known that for distant objects, the chances of a galaxy intervening between the source and observer is high.[29–31] One can carry this consideration further: even if there is no dense intergalactic medium,[32,33] the optical depth of the universe may be very high at relatively small redshifts, because for very distant objects the chances of a *galactic core* intervening increases rapidly, so that very distant objects are likely to lie behind a cosmic veil consisting of galactic cores, which may be rather opaque at early times; but rather little of this veil would be directly visible to us (most of it would lie below the brightness detection limit). In effect, one has a "forest of galaxies" at least partially obscuring the view beyond (cf. Harrison's picture of a "forest of stars"[34]). For example, if we consider cores of diameter 4 Kpc, at a redshift of 10 they obscure over 40% of the sky in a high-density universe (60% in a low-density universe), and at a redshift of 20, 70% and 90% respectively. Radiation from higher redshifts must come through such galactic cores, posing problems of interpretation if they are very opaque (e.g., if there are large 'active cores') in many galaxies at early times.

The screening due to galaxies and intergalactic matter is essentially complete

††Luminosity and spectral evolution can in principle be *determined* from the observations, providing the morphological evolution is known.

when the optical depth reaches unity; this will happen at different distances for different wavelengths, but for each wavelength there will be some maximal distance y^* such that *no* detailed information reaches us from $y > y^*$ (we simply receive integrated background radiation from beyond). The very existence of background radiation with a nearly blackbody spectrum is evidence that the optical depth *does* become unity at a finite distance[35]; in a FRW universe the maximal distance at which this would happen is about $z^* \simeq 1000$, but it might well happen at much smaller redshift values in high-density universes.[36,37]

In summary, at larger redshifts we rapidly receive less data and have greater problems in interpreting uniquely the data we do receive. Therefore the hope of obtaining direct observational evidence of the space-time curvature fades out with increasing distance and growing observational and interpretational uncertainties. At some finite redshift value z^* the universe becomes opaque to electromagnetic radiation, and we cannot directly observe its nature at greater distances.

Observationally Verifying the FRW Universes

Returning now to the question posed at the beginning of this article, the situation may be summarized as in Table 1. To observationally verify a particular FRW model, we need to confirm the isotropy of the universe about us; the FRW nature of the space-time (within the family of isotropic space-times); and the values of q_0 and H_0 (within the family of FRW space-times). We consider these in turn.

Isotropy

(i) *Direct tests.* To show isotropy directly, we need to show that proper motions and distortions are zero, and that the observable relationships [the (area distance, redshift) relation—which is essentially the (m, z) relation—and the (number count, redshift) or (number count, magnitude) relation] are isotropic. In principle one can carry out these checks, but in practice the proper motions and distortions can only be checked at the point q ("here and now"), and even that estimation is difficult.[2]

The observable relationships must be isotropic *independent* of all problems of interpretation (e.g., the nature of source evolution or the nature of selection effects) as long as observational techniques and equipment are constant in the sample used to check isotropy. However, statistical problems arise in these checks. As an example, there is some observational evidence for anisotropy in the (m, z) relation,[38,39] and this can be interpreted as due to motion of our galaxy relative to the mean motion in our neighborhood, or due to clumpiness in the distribution of galaxies together with selection effects, or to an anisotropic velocity field, or in other ways. As far as testing the FRW models is concerned, which interpretation is chosen is a secondary consideration: the FRW prediction is uniquely that the observations should be isotropic. The problem is rather a statistical one: is the observed deviation from isotropy (assuming it is confirmed by further observations) *large enough* to disprove the presumed isotropy of the universe? The answer to this may depend on the interpretation put on the cause of anisotropy, but currently we are unable to answer the question, for we do not have any statistically usable test which will either confirm or deny isotropy: we do not have a valid operational criterion. Until there is some agreed criterion for isotropy ("we will reject the FRW hypothesis if the anistropy in test X at redshift y is above $z\%$ in a sample of size S"), the question is unanswerable.

In consequence, direct tests of isotropy pose problems both in that observational and interpretation problems prevent the motion and distortion investigations being carried out to the accuracy needed; and statistical decision problems arise in verifying the isotropy of the observational relationships. The latter problem is exacerbated by the problems of galactic (and possible intergalactic) extinction, and by the small body of data available for examination (basically, because of the time and effort involved in making the necessary measurements). Even given that this situation will improve with time, the question of a decision criterion may well continue to pose significant problems.

TABLE 1

THE WAY IN WHICH VARIOUS OBSERVATIONAL DATA ENABLE ONE TO VERIFY THE FRIEDMAN-ROBERTSON-WALKER (FRW) NATURE OF THE UNIVERSE

	Check space-time isotropy? (within general space-times)		Check FRW form? (within isotropic space-times)		Check H_0q_0? (within FRW space-times)	
Observational Data	at $z = 0$?	away from $z = 0$?	at $z = 0$?	away from $z = 0$?	H_0	q_0
Proper motions	?	x	x	x	x	x
Distortions	?	x	x	x	x	x
(Number count, redshift) relation	✓	✓	✓	?	x	?
(Area distance, redshift) relation	✓	✓	✓	?	✓	✓
Background radiation intensity	x	✓	x	✓	x	x
Consistency with local data:						
local matter density	x	x	x	x	x	?
element abundances	?	x	?	x	✓	✓
star and element ages	x	x	x	x	✓	✓
galaxy existence and properties	x	x	x	x	?	?

(ii) *Indirect tests.* Further evidence on isotropy comes from the near-isotropy of the microwave background radiation. However, one must approach this with some caution, particularly because the anisotropy in that radiation that has been measured implies large-scale motions of matter in our vicinity[40]; this again begins to raise the problem of definition of isotropy (how large a motion of our galaxy, or of local groups of galaxies, relative to the microwave background radiation will we accept as consistent with the isotropy demanded by the FRW universe models?). In fact, one has no direct evidence on the redshift of emission of the radiation, and there is a wide range of possibilities for this emission redshift.[36,37] Without additional evidence, the observations do not preclude that this emission redshift could vary with the direction of observation. By itself, the isotropy of the radiation does not provide evidence directly for isotropy either of the matter emitting the radiation or of the space-time, for one can arrange anisotropic universes and matter distributions to produce isotropic radiation observations on a particular world line. Even if the redshift of emission of the radiation is the same in all directions, its isotropy does not directly prove that the space-time is isotropic about us.[41] While use of additional assumptions, such as the Copernican principle, may let us give strong arguments for the FRW models,[35] we are here engaged in examining what may be inferred directly from the observations, without the addition of such extra assumptions that are not based on observational data; in this context, the radiation isotropy—a single integrated measurement in each direction—has limited inferential power.

Suppose one puts aside these cautions and accepts the isotropy of the radiation as relating directly to isotropy of the space-time: even then one must not overemphasize the conclusion, for the measurement is, of course, an *upper limit* on the anisotropy observed from a suitable 4-velocity, rather than a statement of *zero* anisotropy; and this limit only applies since the radiation was emitted, i.e., since the value y^* of y corresponding to the distance where the universe became opaque. A very small anisotropy at that time could be below our signal detection limit but could imply very large departures from isotropy at earlier times. For example, arbitrarily small departures from isotropy at the time of decoupling could occur in tilted Bianchi V universes, in which the early behavior is totally different from that in a FRW universe; there could exist *two* separate singularities in such a universe but the density could *remain finite* everywhere (see references 42,43).

Limits on the anisotropy of the early universe have also been deduced from element abundance measurements[44–46]; this should also be treated with caution. We shall shortly discuss the element formation question further, but for the present the point to be noticed is that the evidence can directly relate only to the *near vicinity* of our world line in the past; where this evidence has been taken to imply major statements on the isotropy of the universe over large distance scales, this has been done on the basis of *a priori* assumptions about the homogeneity of the universe, which we are not making in the present analysis of the observations.

(iii) *Considered overall,* the evidence for isotropy will necessarily remain basically incomplete because of the problems in determining the proper motions and distortions. The evidence available may turn out to be compatible with an isotropic model at late times (i.e., since the decoupling of the background radiation), but because of the missing evidence it will *not* be possible to *uniquely* establish isotropy directly from the available data. On the other hand, if *all* the items of evidence that are available consistently point to isotropy, then (despite the problems that may arise in interpreting each item separately) the combined weight of evidence can make it reasonable to infer isotropy at late times.‡‡ It will still be unreasonable to extend this assumption back to very early times because the observational limits on isotropy can

‡‡There is a possibility that a detailed examination of the integrability conditions for the field equations would show that isotropic (m, z) and (dN, z) curves for all z would *imply* vanishing proper motions and distortions. Such a result would considerably strengthen the possibility of directly verifying isotropy.

only be upper limits back to some value y^*, and the very early universe could be very different indeed from a FRW universe and still be compatible with any given set of limits on this anisotropy.

FRW Form (Assuming Isotropy)

(i) *Direct tests.* In order to characterize the universe as a FRW universe, within the class of spherically symmetric universes, we have to show it is spatially homogeneous. One can do this by showing the (r_0, z) and (ρ, z) relations have the FRW form (3), and in principle this can be verified from the (r_0, z) and (dN, z) or (dN, r_0) relations.§§ However the problems discussed above, relating to observational limits, selection effects, and the need to know the source "evolution" (i.e., the variation of source characteristics with z) have made it difficult for us so far to obtain the quadratic term in the power series expansion of $r_0(z)$, let alone any more detailed information about this function, and calibration problems make it difficult even to determine H_0 with accuracy. Further while we have quite detailed information on the number counts of galaxies and radio sources, the conversion of this into information about $\rho(z)$ is proving very difficult: basically, we have been unable to directly determine the mass-to-light ratio at our own position q, let alone at any other position (where evolutionary effects play a significant role). Thus while one can hope to determine the FRW form at q (where the usual ambiguity as to whether an observed variation is a time variation, or a spatial variation, can be uniquely resolved), even verification of the relation $\rho_0 = 6\ q_0 H_0^2$ at this point is very difficult because of the calibration problems, and it does not appear possible in practice to directly verify the FRW form at any substantial value of z.

In fact, the current situation is quite the converse[1]: instead of using the number count observations to verify the FRW form of the universe, the FRW assumption is used[52] to reduce the large uncertainty concerning physical processes in evolving radio sources. The hope of resolving this astrophysical uncertainty and so using the number counts to prove the FRW form of $\rho(z)$ at large z is remote. In principle the (r_0, z) relation is more hopeful, but a full check of the FRW geometry by this means depends on detailed knowledge of source brightness, radial and spectral evolution at large z; it is therefore unlikely that direct observations can ever produce a definitive answer.¶¶ In fact at large distances, we have (essentially because of this uncertainty) no way of directly determining whether an *apparent* variation from homogeneity (see reference 47) is due to a variation of source properties with time in a FRW universe, or due to a variation of source properties with spatial distance from us (contrary to the FRW assumption). As particular examples, we may find it difficult to disprove hierarchical models of the universe,[48,49] and it seems that we cannot observationally distinguish between a FRW universe and a spatially inhomogeneous universe that is like a FRW universe near us, but very inhomogeneous spatially at large distances from us, with our

§§As has been remarked, r_0 is determined if m can be inferred for galaxies of known intrinsic luminosity.

¶¶Particularly in view of recent observations that suggest that source *morphology* may change rather rapidly with redshift (S. van den Bergh, private communication).

local group of galaxies surrounded at all times by a singularity rather than having the singularity occur in the past in a spatially homogeneous way (FIGURE 11, and see Appendix B of reference 41). Thus direct observations can verify the FRW form at q, but not at large distances.‖‖

(ii) *Indirect tests.* Indirect information comes from the background radiation spectrum, which is certainly consistent with the FRW form, but not uniquely so[41]; all that is required to explain the major component of the radiation is a source of blackbody radiation. Even in the FRW scheme, there are several alternative possible sources of this radiation[37]; it can be plausibly explained also in other models (such as the inhomogeneous universe of FIGURE 11). Other indirect information comes from the element abundances, but (as mentioned above) while this may provide evidence on the homogeneity of our universe in our own history near to our galaxy world line, it does not directly do so for distant regions (this situation would change if we could directly provide convincing evidence on element abundances in distant galaxies; however, it seems unlikely that we can do so). Thus indirect observational information on the homogeneity of the universe does not seem able to point uniquely to a FRW form for the universe, at large distances from us.

(iii) *Considered overall,* the observational verification of spatial homogeneity at large distances does not seem to be possible either by direct or indirect observational data.

Determining q_0 and H_0 (within the FRW family)

(i) *Direct tests.* In principle we can determine q_0 from the relation

$$q_0 = \rho_0/6H_0^2, \tag{8}$$

by measuring the local number density of galaxies, and determining the local rest-mass density ρ_0 from this. However, because of the uncertainty in the mass-to-light ratio (see references 8,50,51) referred to before, we cannot determine ρ_0 from number counts at present. In fact, relation (**8**) is currently *run the other way* to place limits on the value of ρ_0 from q_0, and hence observed values of q_0 are used to put limits on the mass-to-light ratio.

The major problem here relates to the unobservability of many kinds of matter; while dynamical methods can give some indication of galactic masses, they do not[50,52] determine the density of a uniform intergalactic distribution of mass (due to gas, black holes, massive neutrinos, rocks, etc.). Consequently, determination of q_0 via the local mass density will always be open to some questioning.†††

The (r_0, z) relations, on the other hand, can in principle[2] give us H_0 and q_0 directly.

‖‖Unless we can prove that the universe is a spatially bounded (finite) FRW universe (note that this can happen if $k = -1$ or 0, see reference 68), and *we have already seen round it.* (This could be shown if we could provide convincing evidence that any galaxy cluster we see in some particular direction has also been seen by us in another direction). In that case, by observing nearby galaxies we effectively see to large distances because of the spatial topology.

†††Additionally, it is here that the exact validity of Einstein's equations becomes important; a cosmological constant, for example, would invalidate this method of estimating q_0. One could also consider the consequences of some of the other available gravitational theories, e.g., those of Dicke, Rosen, Canuto, Moffatt, and Rastall.

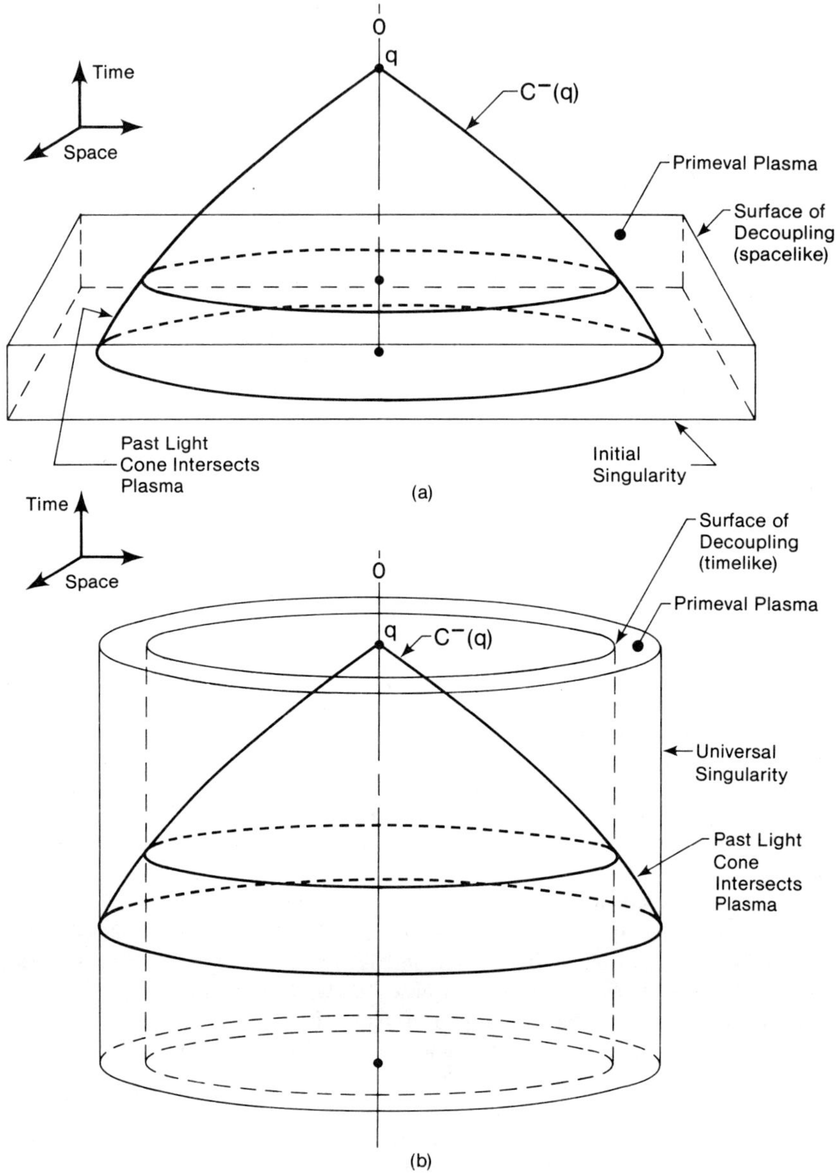

FIGURE 11. Two observationally similar situations: (a) a FRW universe with a spatially homogeneous primeval fireball, and (b) a spherically symmetric inhomogeneous universe, nearly spatially homogeneous near O's worldline, but surrounded by a timelike fireball and singularity.

Calibration problems play a considerable role in causing uncertainty in determining H_0, these calibration problems being related to and worsened by statistical problems[52,53] (particularly the paucity of calibration objects at convenient distances to complete the distance chain). The evolution problem is already serious at distances corresponding to the quadratic term, making the direct determination of q_0 problematic. However, there does not appear to be any reason why this question will not be resolved when sufficiently detailed studies of galaxy diameter and luminosity distributions are available, together with detailed estimations of the evolution of these quantities with z. The main theoretical problem here is the need to estimate the morphological and radial evolution of the sources. While the brightness evolution can in principle be observationally determined,[5] there is no way of separating radial evolution effects from the curvature effects which are represented by the function $r_0(z)$. The morphological evolution must be known because the map from the object plane to the image plane depends on the form of the function f expressing the radial brightness variation of the source; given this knowledge, the problem of determining q_0 depends on resolution of the problem of radial evolution. This will apply to any new method of determining q_0 on the basis of direct astronomical observation (see reference 52 for some of the ingenious methods currently proposed), because each such method can only determine r_0 relative to some length scale a; and the astrophysical problem then is to determine the change of that length scale a with z.

Any remaining uncertainty concerning the evolution of the effective size a of the object looked at must be reflected in a corresponding uncertainty in q_0. In addition, full analysis would require consideration of absorption, "seeing", and "lumpiness" effects in the real universe, as each of these can significantly affect the observations.

(ii) *Indirect tests.* Consistency of the FRW models with local data, particularly the element abundances and the star and element ages, can be used to infer quite strong limits on acceptable values of q_0 and H_0 (see references 52, 54, 55). However, the certainty which can be attributed to these estimates should not be overestimated. Consider the helium and deuterium abundances, for example. What one may note here is that the theory of element formation in FRW universes together with the present observations of light element abundances, have been used, in interesting recent papers, to give limits on the variation of the gravitational constant G, the lifetimes of neutrinos associated with heavy leptons, the time variation of Planck's constant, and the number of types of heavy leptons and their masses (see references 56–59), and these element abundances would also be sensitive to variations in the weak-coupling constant, the neutron–proton mass difference, and the strong-coupling constant.[57] The fact that it has been found worthwhile to make estimates of these limits from the observations of light element abundances (plus the FRW assumption) shows that there is sufficient uncertainty in each of these quantities and their variation over very long timescales that these estimates significantly improve our knowledge of the relevant physical behavior. But of course this implies that the converse is also true: precisely because of the uncertainty concerning local physical laws reflected in this series of papers, we cannot with any great certainty use the element abundance observations to place limits on the FRW universe parameters. The point is that we cannot have things both ways: if we know the universe is a FRW universe, we can use these observations to improve our knowledge of local physical laws, and if we know the local physical laws, we can use the observations to check on the FRW universe, but we cannot use the *same* limited set of numbers (the He and D abundances) to simultaneously settle *both* sets of questions. One data point is one data point and, try as we may, we cannot squeeze more than one piece of information from it. In fact, these papers show that the uncertainty in the local physics involved is such that we *cannot unambiguously* use the element abundance data to restrict the FRW models, for in so doing we are extrapolating local physical laws way beyond the bounds of their experimental verification.

Similar comments may apply even to the determination of the local mass density ρ_0; for estimates of H_0 and q_0 have been used[60] to put limits on heavy neutrino masses and lifetimes. Thus part of the uncertainty in ρ_0 is also a reflection of uncertainty in local physical laws. This is related to, and reflected in, uncertainty about the equation of state of the dominant matter component in the universe.

Because a significant part of the age of the FRW universe models has elapsed since a "low density and temperature" state has been attained (i.e., since $z = 1000$), the age estimates are less prone to this kind of criticism, for the physics involved at these times is better verified than that involved at the early times of element formation. However, even the limits derived from age estimates should be treated with a bit of caution, because of the possibility that some of the "constants" involved in determining stellar and element ages might vary significantly in the time periods involved.

Finally, it is certainly also possible that significant limits on H_0 and q_0 may be derived from studies of galaxy formation and clustering (see reference 61). It is clear that a thoroughly convincing analysis of the astrophysical and dynamical processes involved will be necessary to make a good case here; but it is certainly possible that this may succeed.

(iii) *Considered overall,* it is likely (as in the case of the isotropy measurements) that the separate arguments for particular values of H_0 and q_0 may be open to some scepticism, but that eventually we shall be able to build up an overall picture that makes a good case for particular values because of the combined weight of evidence.

Summary

There are three steps involved in directly observationally verifying a particular FRW model universe. The first step (checking isotropy) cannot be carried out directly, but a sufficient part of the requirements for isotropy can be statistically tested that a good—but not watertight—argument from the observations may be possible. The second step (verifying spatial homogeneity) can be carried out "here and now," but evolutionary and information problems prevent a unique answer from being arrived at, at large distances from us.‡‡‡ The third step (isolating a particular FRW model) is also made difficult by these problems; the indirect evidence—while it may help verify a particular set of values of q_0 and H_0—should be treated with caution because an extrapolation of physical theory beyond the bounds where it has been verified is involved. Nevertheless the weight of direct and indirect evidence, interpreted in the light of a plausible theory of galaxy morphological, brightness, spectral, and radial

‡‡‡Unless the universe has finite spatial sections (if this is true, it would probably be rather difficult *proving* this was so, if $k = 0$ or -1).

evolution, may enable us to isolate particular values of H_0 and q_0 with reasonable certainty. However, this step would *not* prove the universe is a FRW universe; it would rather determine H_0 and q_0 *if* the universe has a FRW form, which might not be true.

Because of the problems in proving homogeneity, therefore, important features of the universe may be undecidable on the basis of available direct and indirect evidence. Use of extra assumptions, such as a homogeneity or uniformity principle, may enable us to make a good case for a particular FRW universe. But in the end the FRW nature of the resulting form (specifically, the assumption of homogeneity at large distances from us) will almost certainly be a result of this (unverifiable) homogeneity assumption, rather than a consequence of observational evidence.

Uncertainty in Cosmology

A full cosmological theory should relate the following elements:

Philosophical Principles		Cosmological Models		Observations
principle 1	→	~~model 1~~	← Limits to Verification	observation 1
principle 2		model 2		observation 2
....		~~model 3~~		....

One may have various models, some of which are eliminated on the basis of observational evidence (e.g., they might predict very anisotropic microwave background radiation), and some on the basis of philosophical principles, such as the cosmological principle, Mach's principle, an anthropic principle, and so on[62]; in general these principles are unverified and may indeed be unverifiable. In the usual approach, a heavy emphasis is laid on philosophical principles rather than observational tests, in the sense that the homogeneity assumption (the cosmological or Copernican principle) is laid down first, and only then are theoretical observational relations derived for comparison with the data; thus any model which does not satisfy the cosmological principle is eliminated *a priori*. The homogeneity principle itself is not subjected to searching observational tests. However, the intention underlying the usual approach to cosmology is to view it as an *experimental* subject, open to contradiction by observations; and so a good case can be made that there should be a better balance between observations and unverified *a priori* principles, than at present. Verification of models should, where possible, take precedence over the imposition of unverified principles, which should not be introduced into the theory until this becomes unavoidable (at present, the cosmological principle is built into the structure of the theory and analysis of the observations so thoroughly that it is very difficult interpreting the data presented in a way that is unbiased by this assumption).

In placing stronger emphasis on verification rather than on *a priori* principles, it is clear that *limits* to verification should then be taken seriously. Our analysis of the difficulties of verification lend support to McCrea's suggestion[63] that a mature cosmological theory should include as a basic element, an *uncertainty principle* explicitly acknowledging our limited ability to observationally verify our models, and accord this uncertainty a fundamental status in our understanding of observational cosmology. Such an understanding would, for example, recognize the feature that while we can in principle completely verify the FRW nature of the universe by galactic observations, in practice observational limits and astrophysical uncertainty strongly restrict the extent to which such a program can actually be carried out. These uncertainties in cosmology are partially related to uncertainties in microphysics (for any uncertainties regarding fundamental physics restricts the degree to which the predictions of physical theory can be used to check cosmological models, and there is always uncertainty in microphysics when the limits of laboratory verification are reached), and to uncertainties in the application of microphysics to macrophysics.§§§

The contention is that *practical limits to verification should be accorded a fundamental status in our view of the nature of cosmology and in our assessment of the picture of the universe we derive from our observations,* for in the end they limit the extent to which cosmology can be considered to be an observationally based subject.

In addition, even within the framework of a cosmological theory that is content to rely from the outset on unverifiable principles to delimit the universe to be a FRW universe, a greater emphasis on the nature of the measurements made and their limitations, may lead to greater clarity of analysis of the evidence used to choose particular values of q_0 and H_0, and so ultimately to a better understanding of the observational basis of the models put forth.

Acknowledgments

I wish to thank the Max Planck Institute for Physics and Astrophysics, Munich, for hospitality and support, and the President of the University of Alberta for support through a NRC grant. Discussions with many colleagues, particularly R. Maartens, S. Nel, J. Perry, K. Lake, and W. H. McCrea have been valuable in developing these ideas.

References

1. Ellis, G. F. R. 1975. Cosmology and verifiability. Q. J. R. Astr. Soc. **16:** 245–264.
2. Kristian, J. & R. K. Sachs. 1966. Observations in Cosmology. Astrophys. J. **143:** 379–399.
3. Ellis, G. F. R., R. Maartens & S. D. Nel. 1979. Observational Cosmology. In preparation.
4. Ellis, G. F. R. 1971. Relativistic cosmology. *In* General Relativity and Cosmology. R. K. Sachs, Ed. Rendiconti Scuolo Enrico Fermi, XLVIII Corso. Academic Press. New York.
5. Ellis, G. F. R. & J. Perry. 1979. Towards a 'correctionless' observational cosmology. Mon. Not. R. Ast. Soc. **187:** 357–370.
6. Baum, W. A. 1962. The detection and measurement of faint astronomical sources. *In* Astronomical Techniques. W. Hiltner, Ed. University of Chicago Press. Chicago.
7. Ehlers, J. & R. K. Sachs. 1971. Kinetic Theory and Cosmology. *In* Astrophysics and General Relativity. Vol. 2. Chretien *et al,* Eds. Gordon and Breach. New York.
8. Spinrad, H., J. P. Ostriker, R. P. S. Stone, L-T. G. Chiu & G. Bruzual A. 1978. Halos of spiral galaxies: Photometry and mass-to-light ratios. Astrophys. J. **225:** 56–66.

§§§For example, it is conceivable that gravity is *not* the dominant force in the early universe,[64] or that some astrophysical bodies are almost-transitive systems.[65]

9. Zwicky, F. 1957. Morphological Astronomy. 112–114. Springer, Berlin.
10. Arp, H. 1965. A very small, condensed galaxy. Astrophys. J. **142:** 402–406.
11. Disney, M. 1976. Visibility of galaxies. Nature. **263:** 573–575.
12. Ellis, R. S., R. Fong & S. Phillipps. 1977. Angular diameter counts of galaxies: a method for determining the selection criteria for deep Schmidt plates. Mon. Not. R. Astron. Soc. **181:** 163–182.
13. Fairall, A. P. 1978. What are Zwicky's compact galaxies? Observatory. **98:** 1–7.
14. Allen, R. J. & F. H. Shu. 1979. The extrapolated central surface brightness of galaxies. Astrophys. J. **227:** 67–72.
15. Harwit, M. 1975. The number of class A phenomena characterizing the universe. Q. J. R Astr. Soc. **16:** 378–409.
16. Tinsley, B. M. 1973. Analytical approximations to the evolution of galaxies. Astrophys. J. **186:** 35–49.
17. Tinsley, B. M. 1977. Galaxy counts, color–redshift relations, and related quantities as probes of cosmology and galactic evolution. Astrophys. J. **211:** 621–637; **216:** 349–350.
18. Hausman, M. A. & J. P. Ostriker. 1978. Galactic cannibalism. III. The morphological evolution of galaxies and clusters. Astrophys. J. **224:** 320–336.
19. Gunn, J. E. & J. B. Oke. 1975. Spectrophotometry of faint cluster galaxies and the Hubble diagram: an approach to cosmology. Astrophys. J. **195:** 255–268.
20. Freeman, K. C. 1970. On the disks of spirals and SO galaxies. Astrophys. J. **160:** 811–830.
21. Gudehus, D. H. 1973. Radius-parameter and surface brightness as a function of galaxy total magnitude for clusters of galaxies. Astronom. J. **78:** 583–593.
22. Strom, S. E. & K. M. Strom. 1978. Gravitational encounters in rich clusters of galaxies. Astrophys. J. **225:** L93–L97.
23. Sandage, A., J. Kristian & J. A. Westphal. 1976. The extension of the Hubble diagram. Astrophys. J. **205:** 688–695.
24. Butchins, A. & M. Ripp. 1978. Astronomy Department, University of Cape Town. Unpublished manuscript.
25. Perry, J. & G. F. R. Ellis: Cosmological observations. In preparation.
26. Metzner, A. W. K. & P. Morrison. 1959. The flow of information in cosmological models. Mon. Not. R. Astron. Soc. **119:** 657–664.
27. Whitrow, G. J. & B. D. Yallup. 1964. Background radiation in isotropic world models. I. Mon. Not. R. Astron. Soc. **127:** 301–318.
28. Gudehus, D. H. 1972. Cosmology and information. Publ. Astron. Soc. Pac. **84:** 818–822.
29. Wagoner, R. V. 1967. Some effects of an intervening galaxy on the radiation from very distant objects. Astrophys. J. **149:** 465–479.
30. Bahcall, J. N. & E. E. Salpeter. 1966. Absorbtion lines in the spectra of distant sources. Astrophys. J. **144:** 847–852.
31. Burbidge, G., S. L. O'Dell, D. H. Roberts & H. E. Smith. 1977. On the origin of the absorption spectra of quasi-stellar and BL Lacertae objects. Astrophys. J. **218:** 33–38.
32. Gunn, J. E. & B. A. Peterson. 1965. On the density of neutral hydrogen in intergalactic space. Astrophys. J. **142:** 1633–1636.
33. Bahcall, J. N. & R. M. May. 1968. Electron scattering and tests of cosmological models. Astrophys. J. **152:** 37–42.
34. Harrison, E. R. 1977. The dark night sky paradox. Am. J. Phys. **45:** 119–124.
35. Hawking, S. W. & G. F. R. Ellis. 1973. The Large Scale Structure of Space Time. Cambridge University Press. Cambridge. Section 10.1.
36. Rees, M. J. 1971. Effects of very long wavelength primordial gravitational radiation. Mon. Not. R. Astron. Soc. **154:** 187–195.
37. Rees, M. J. 1978. Origin of pregalactic microwave background radiation. Nature **275:** 35–37.
38. Rubin, V. C., N. Thonnard, W. K. Ford & M. S. Roberts. 1976. Motion of the galaxy and the local supercluster determined from the velocity anisotropy of distant Sc I galaxies. II. Astron. J. **81:** 719–737.
39. Jaakhola, T., H. Karoji, G. Le Denmart, M. Moles, L. Nottale & J.-P. Vigier. 1976. Additional evidence and possible interpretation of angular redshift anisotropy. Mon. Not. R. Astron. Soc. **177:** 191–213.

40. ROWAN-ROBINSON, M. 1977. Aether drift detected at last. Nature **270:** 9–10.
41. ELLIS, G. F. R., R. MAARTENS & S. NEL. 1978. The expansion of the universe. Mon. Not. R. Astron. Soc. **184:** 439–465.
42. COLLINS, C. B. & G. F. R. ELLIS. 1979. Singularities in Bianchi cosmologies. To appear in Physics Reports.
43. ELLIS, G. F. R. 1975. Singularities and cosmology. Ann. N. Y. Acad. Sci. **262:** 231.
44. BARROW, J. 1976. Light elements and the isotropy of the universe. Mon. Not. R. Astron. Soc. **175:** 359–370.
45. BARROW, J. D. & R. MATZNER. 1977. The homogeneity and isotropy of the universe. Mon. Not. R. Astron. Soc. **181:** 719–727.
46. OLSON, D. W. 1978. Helium production and limits on the anisotropy of the universe. Astrophys. J. **219:** 777–780.
47. GREEN, R. F. & M. SCHMIDT. 1978. Evidence for non-uniform radial distribution of quasars regardless of the nature of their redshifts. Astrophys J. **220:** L1–L4.
48. DE VAUCOULEURS, G. 1970. The case for a hierarchical cosmology. Science **167:** 1203–1213.
49. HAGGERTY, M. J. & J. R. WERTZ. 1972. On the redshift–magnitude relation in hierarchical cosmologies. Astrophys. J. **155:** 495–503.
50. OSTRIKER, J. P., P. J. E. PEEBLES & A. YAHIL. 1974. The size and mass of galaxies and the mass of the universe. Astrophys. J. **193:** L1–L4.
51. WHITE, S. D. M. & N. A. SHARP. 1977. Tidal interactions and the massive halo hypothesis. Nature **269:** 395–396.
52. GUNN, J. E., M. S. LONGAIR & M. J. REES. 1978. Observational Cosmology. 8th Advanced Course of the Swiss Society of Astronomy and Astrophysics. Geneva.
53. PEEBLES, P. J. E. 1978. The distance and redshift of the Virgo cluster and Hubble's constant. Comments Astrophys. Space Phys. **7:** 197–205.
54. TINSLEY, B. M. 1978. Accelerating universe revisited. Nature **273:** 208–211.
55. KAZANAS, D., D. N. SCHRAMM & K. HEINEBACH. 1978. A consistent age for the universe. Nature **274:** 672–673.
56. ZELDOVICH, Y. B. 1970. The universe as a hot laboratory for the nuclear and particle physicist. Comments Astrophys. Space Phys. **2:** 12–17.
57. BARROW, J. D. 1978. A cosmological limit on the possible variation of *G*. Mon. Not. R. Astron. Soc. **184:** 677–682.
58. TAYLER, R. J. 1978. Neutrino stability and cosmological helium production. Nature **274:** 232–233.
59. WASSERMAN, I. & K. BRECHER. 1978. Time invariance of Planck's constant. Phys. Rev. Lett. **41:** 920–923.
60. LEE, B. W. & S. WEINBERG. 1977. Cosmological bound on heavy neutrino masses. Phys. Rev. Lett. **39:** 165–168.
61. SASLAW, W. C. 1979. Galaxy clustering and anti-clustering. Nature **277:** 266–267. See also PEEBLES, P. J. E. This volume.
62. ELLIS, G. F. R. & E. R. HARRISON. 1974. Cosmological principles I. Symmetry principles. Comments Astrophys. Space Phys. **6:** 23–27. See also HARRISON, E. R. 1974. Cosmological principles II. Physical principles. Comments Astrophys. Space Phys. **6:** 29–35.
63. MCCREA, W. H. 1960. The interpretation of cosmology. Nature **186:** 1035.
64. HARTLE, J. B. 1970. Long range weak forces and cosmology. Phys. Rev. D. **1:** 394–399.
65. TAKAVOL, R. K. 1978. Is the sun almost transitive? Nature **276:** 802–803.
66. DAVIS, M. M. & L. M. MAY. 1978. New observation of radio absorbtion line in 3C286, with potential application to the direct measurement of cosmic deceleration. Astrophys. J. **219:** 1–4.
67. GUNN, J. E. & B. M. TINSLEY. 1976. Dynamical friction: the Hubble diagram as a cosmological test. Astrophys. J. **210:** 1–6.
68. ELLIS, G. F. R. 1971. Topology and cosmology. J. Gen. Rel. Grav. **2:** 7–21.
69. DYER, C. C. & R. C. ROEDER. 1974. Observations in locally inhomogeneous cosmological models. Astrophys. J. **189:** 167–175.
70. ROEDER, R. C. 1975. Apparent magnitudes, redshifts, and inhomogeneities in the universe. Astrophys. J. **196:** 671–673.
71. ROEDER, R. C. 1975. Significance of the angular diameter–redshift relation. Nature **255:** 124–125.

STATISTICS OF THE DISTRIBUTION OF GALAXIES*

P. J. E. Peebles

Joseph Henry Laboratories
Princeton University
Princeton, New Jersey 08544

INTRODUCTION

Hubble[1,2] introduced the variation of galaxy counts with apparent magnitude as a measure of the large-scale matter distribution and a test of cosmological models. Shapley[3] pioneered the study of the angular distribution of galaxy counts at fixed apparent magnitude as a measure of galaxy clustering. More recently people have turned to the joint distribution in angular position and redshift: this gives a much more detailed picture of the galaxy distribution but it demands considerably more observational work. All three traditions have been active and profitable. I summarize here a few of the current developments and discussions as they apply to statistical measures of the galaxy distribution.

THE LARGE-SCALE DISTRIBUTION OF MATTER

FIGURE 1 shows the variation of differential galaxy counts with apparent magnitude. The filled triangles are based on the Zwicky catalog of galaxies,[4] at galactic latitude $b > +40°$. The circles at the bright end show the recent results of Kirshner, Oemler, and Schechter,[5] who surveyed galaxy magnitudes and redshifts in eight small fields, each about 4° by 4° square. The open circles are based on the fields in the North Galactic hemisphere, the half-filled circles, the South hemisphere. I have used their transformations of the Zwicky magnitudes to the J system.

There has been considerable recent work on deep galaxy counts. The data shown here are taken from Kron's[6] discussion and use his transformations to J magnitudes and differential counts. The filled circles show Kron's counts, the open triangles, Rainey's counts,[7] the open squares, Brown's counts,[8] and the filled squares, the count in the Lick catalog of galaxies.[9] Not shown are the deep counts by the Durham group (reported at this conference by R. Ellis) and by Tyson and Jarvis.[10] For a graph similar to FIGURE 1 and showing more data see reference 10: the overall shape of their graph is quite close to FIGURE 1 except that Tyson and Jarvis find a somewhat lower count at $J \sim 23$.

If the space distribution of galaxies is homogeneous, and if redshift and space curvature can be neglected, the counts should vary as[1,2]

$$\frac{d\mathcal{N}}{dm} \propto 10^{0.6m}. \tag{1}$$

The solid line in FIGURE 1 has this slope: one sees that the data do not follow it very well. At $m \lesssim 13$ the counts are increased by the local concentration in and around the

*This research was supported in part by the National Science Foundation.

0077–8923/80/0336–0161 $1.75/1 © 1980, NYAS

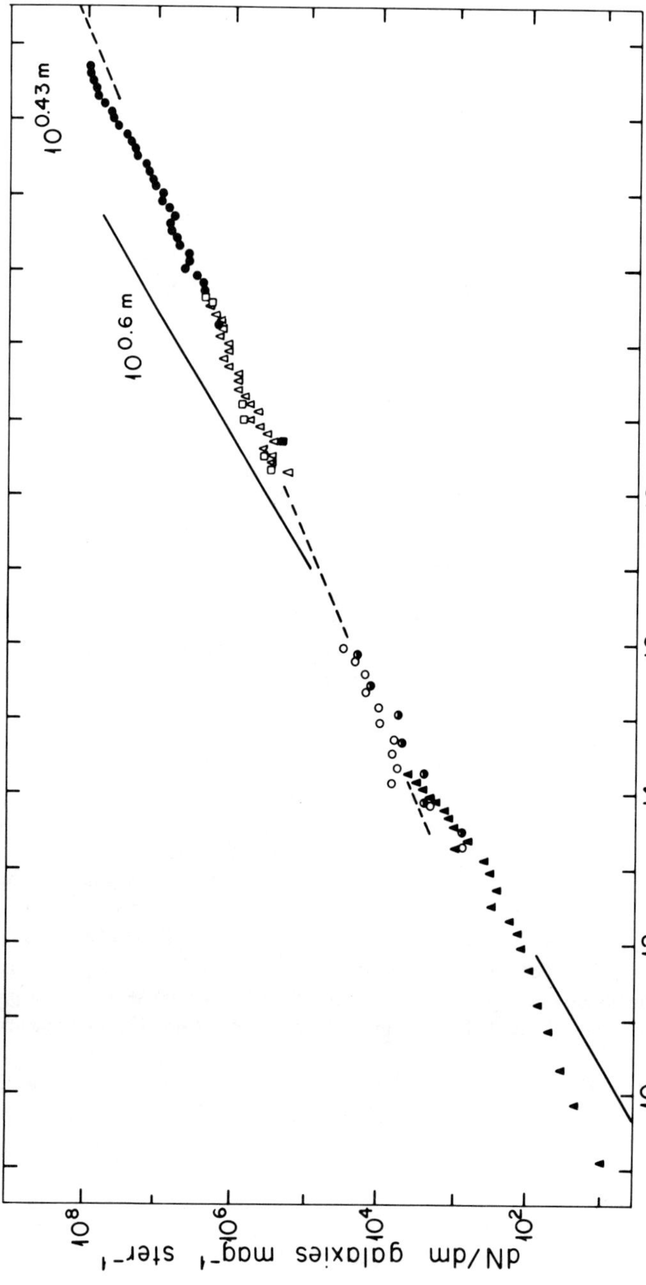

FIGURE 1. Galaxy counts (after Kron[6] and Kirshner, Oemler, and Schechter[5]). The symbols are explained in the text.

Local Supercluster. Kirshner, Oemler, and Schechter[5] have argued that there are excess counts at $m \sim 15$ in the North hemisphere, reflecting a galaxy cloud at distance $\sim 100\ h^{-1}$ Mpc.† This effect aside, one might say that the counts at $m \sim 13$ to 16 follow Equation 1 reasonably well. The counts of these bright galaxies together with redshift samples can be used to estimate the mean luminosity per unit volume due to galaxies. Kirshner, Oemler, and Schechter[5] find

$$j \simeq 2.0 \times 10^8\ h\ \mathcal{L}_\odot/\mathrm{Mpc}^3, \qquad \textbf{(2)}$$

at the J band.

The deep counts reach redshifts Z on the order of unity at magnitude $J \sim 23$. This means the magnitude measurements in the visible are detecting the blue end of the galaxy spectrum, where galaxies are not so bright. The K-corrections to the magnitudes to take account of this spectrum shift are substantial and not very easy to compute: they depend on the shapes of galaxy uv spectra and on the mix of galaxy redshifts and spectrum shapes at a given apparent magnitude, all of which are uncertain. There also is an important correction to the $10^{0.6m}$ relation for the effect of the general expansion of the universe. (For example, in an Einstein–de Sitter model the number of galaxies at $Z \leq 1$ is one-fifth of the number expected from an extrapolation of $\mathcal{N} \propto Z^3$ from the counts at low redshifts.) There may also be a significant correction for galaxy evolution, depending on when vigorous star formation terminated. It is generally thought that these three corrections can account for the discrepancy between the solid line and the counts at the faint end, but the calculations of the corrections still are tentative and actively debated.

The degree of isotropy of the deep counts has not yet been studied in detail, but we have some useful limits. For example, Kron[6] studied two fields, one in the North galactic hemisphere, the other in the South. The differential counts at $m \sim 21$ to 24 differ by no more than $\sim 30\%$.

I would list as the first major result from the deep counts the observation that galaxies roughly like the local ones exist at $Z \sim 1$ at about the local abundance—if there is an edge to our universe of galaxies it is quite close to or beyond the horizon. If we are near the center of a spherically symmetric universe the present data with reasonable K-correction would allow a global density gradient perhaps amounting to a factor ~ 3 variation in mean density from distance $\sim 100\ h^{-1}$ Mpc to $cH^{-1} \sim 3000\ h^{-1}$ Mpc. If our position is "random," the isotropy of the galaxy counts implies a much stronger limit on density variations on the scale of cH^{-1}, perhaps $\lesssim 50\%$. I have the impression that the K-correction and evolution corrections can be estimated well enough to establish whether, under the homogeneity assumption, the relativistic Friedmann–Lemaître models are needed to account for the galaxy counts. If that could be done, it would be a second considerable triumph. Of course, distinguishing among Friedmann–Lemaître models is a still more distant goal.

Shape of the Galaxy Two-Point Correlation Function

The galaxy n-point correlation functions, which are statistics taken from the theory of nonideal gases, have proved to be useful in the description of galaxy

†I write Hubble's constant as $H = 100\ h\ \mathrm{km\ s^{-1}\ Mpc^{-1}}$.

clustering. Perhaps the main reason is that there is a linear integral relation between each angular n-point function and the corresponding spatial function, so it is easy to derive the latter from the former. Use of the two-point function was pioneered by Neyman,[11] Scott,[12] Limber,[13] Rubin,[14] and Totsuji and Kihara.[15] The available data have been studied in some detail, and the statistics extended to the three- and four-point correlation functions, by the Princeton group.[16]

Another currently popular statistic is the multiplicity function introduced by Gott and Turner.[17] Designed to be analogous to the galaxy luminosity function, it is the frequency distribution of the number of galaxies found in localized concentrations whose mean densities exceed the background by some chosen factor. The problem with the application of this as a descriptive statistic is that there is no simple relation between the wanted spatial function and the observable angular function. Fall[18] describes at this conference some interesting theoretical properties of the multiplicity function.

I shall discuss only one topic, the shape of the two-point galaxy correlation function and how it might have gotten that way. The spatial two-point correlation function $\xi(r)$ is defined by the joint probability of finding galaxies centered in the volume elements dV_1 and dV_2 at separation r,

$$dP = n^2[1 + \xi(r)]\,dV_1 dV_2, \tag{3}$$

where n is the mean-space number density. The correlation function $w(\theta)$ for the angular distribution at fixed apparent magnitude m is defined by a similar equation. If expansion and redshift corrections may be ignored, and if the space distribution of galaxies is a homogeneous random process, then w scales with the depth of the sample as

$$w(\theta) = D^{-1}f(\theta D), \qquad D \propto 10^{0.2m}. \tag{4}$$

FIGURE 2 shows scaled estimates of $w(\theta)$ for three samples, with triangles for the Zwicky catalog,[4] circles for the Lick catalog,[9] and squares for the catalog of the Jagellonian field.[19] The correlation functions have been scaled to the depth of the Zwicky sample by using Equation 4 supplemented by a K-correction.[20] This correction is quite uncertain but not very large at the depths of these samples.

The fact that the scaled w estimates come out fairly close to each other is positive evidence that they are reliable measures of the space clustering and that they are not seriously affected by systematic errors in the catalogs or large-scale fluctuations in the space distribution. Thus we have a new addition to the rather short list of fundamental data for cosmology.

The points in FIGURE 2 describe a fairly good approximation to a straight line, or power law,[20]

$$w(\theta) \propto \theta^{1-\gamma}, \qquad \gamma = 1.77 \pm 0.04. \tag{5}$$

This form is produced by a power law model for the spatial function,

$$\xi = (r_0/r)^{\gamma}. \tag{6}$$

The amplitude r_0^{γ} has been quite uncertain because it depends on the galaxy

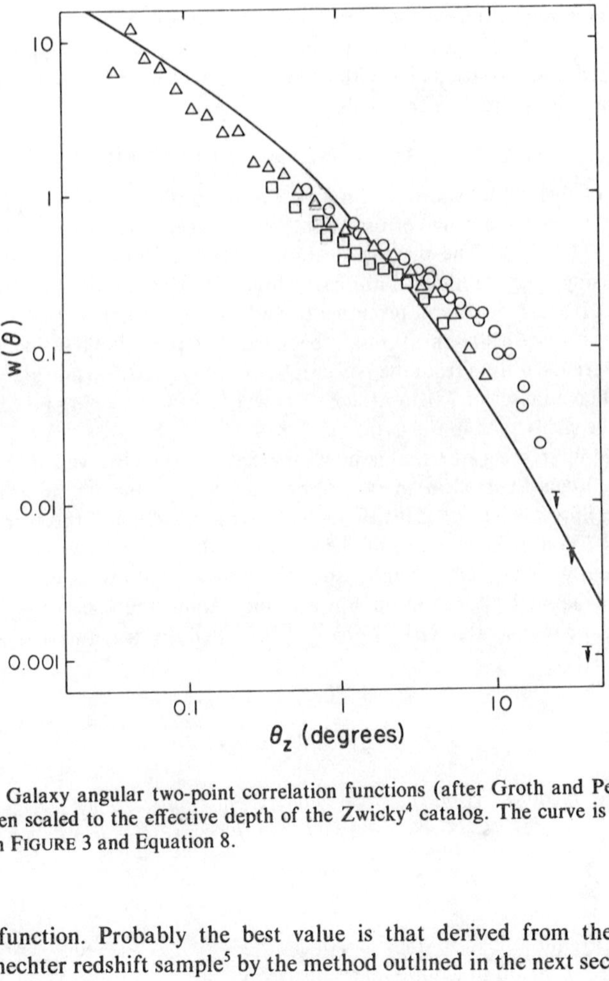

FIGURE 2. Galaxy angular two-point correlation functions (after Groth and Peebles[20]). The data have been scaled to the effective depth of the Zwicky[4] catalog. The curve is based on the model for ξ in FIGURE 3 and Equation 8.

luminosity function. Probably the best value is that derived from the Kirshner–Oemler–Schechter redshift sample[5] by the method outlined in the next section[21];

$$r_0 = 4.2 \pm 0.3 \text{ h}^{-1} \text{ Mpc}. \quad (7)$$

In the Lick sample the correlation function seems to curve downward from the power law at $r \sim r_b \sim 2r_0$.[20] This is close to the noise for the sample, however, so the effect is tentative.

Equations 6 and 7 say that if the galaxy number density is averaged through a spherical window of radius ~ 4 h^{-1} Mpc, the rms fluctuation in density after averaging is about equal to the mean density. Of course there are spots where this averaged density is a good deal larger than twice the mean, and there are spots where the clustering can be traced to scales much larger than r_b, but it appears that such fluctuations are rare enough that they do not appreciably affect $\xi(r)$.

Returning to the shape of this function, let us consider a test of possible systematic deviation from the power law. The angular function $w(\theta)$ measures the clustering projected onto the sky, and the projection smooths any rapid variation of ξ with r. But the situation is very different for a large-scale change of shape because a logarithmic derivative $-\gamma$ for ξ becomes a logarithmic derivative $1-\gamma$ for w (Equations 5 and 6). Since $\gamma \sim 2$, a slow variation of γ with r in fact appears more pronounced in $w(\theta)$ than in $\xi(r)$. This is illustrated by the model

$$\xi(r) = [(r/13.6\ \mathrm{h}^{-1}\ \mathrm{Mpc})^{1.5} + (r/4.15\ \mathrm{h}^{-1}\ \mathrm{Mpc})^{2.75}]^{-1}. \qquad (8)$$

The graph of this model shown in FIGURE 3 appears quite close to a power law with slope $\gamma \sim 1.77$ over the range of useful measures of the correlation function, $r \sim 100$ h^{-1} kpc to 10 h^{-1} Mpc. The curve in FIGURE 2 is the result of projecting this model to get $w(\theta)$, using a conventional luminosity function (the one used in Reference 20): here the curvature is more prominent, and the systematic deviation from the observations is quite apparent. It might be noted that at $r \lesssim 0.3\ \mathrm{h}^{-1}$ kpc, the model for $\xi(r)$ in Equation 8 differs from the power law $\propto r^{-1.5}$ by $\lesssim 20\%$, while at $r = 4\ \mathrm{h}^{-1}$ Mpc, $\xi(r)$ has fallen a factor of 7 below the $r^{-1.5}$ power law. A "drooping power law" of this sort would be easily seen in the data.

Now the interesting question is how (or whether) the observed simple power-law shape of $\xi(r)$ could persist in an expanding universe. I suspect[22] the problem is most severe if the universe has low density, say $\langle\rho\rangle \sim 0.1\ \rho_c$,‡ which is the currently popular estimate. If ξ is a good measure of the mass distribution then we conclude that the mass distribution averaged through a smoothing window of radius $r \sim r_0 = 4\ \mathrm{h}^{-1}$ Mpc fluctuates by about 100%, so in the upward fluctuations the mass density typically is about twice the mean value $\langle\rho\rangle$. If $\langle\rho\rangle \sim 0.1\ \rho_c$, then the gravitational energy within

FIGURE 3. Bent power-law model for $\xi(r)$ (Equation 8). The logarithmic derivative of $\xi(r)$ is marked at spots along the curve. This model yields the angular correlation function shown as the curve in FIGURE 2.

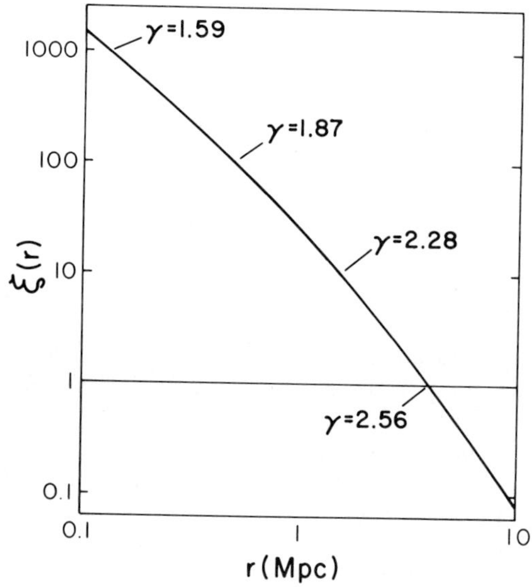

‡ρ_c is the density in an Einstein–de Sitter universe with the observed Hubble constant.

the dense patches still is well below the kinetic energy associated with the general expansion, so it is hard to see why the clustering pattern should not be expanding with the general expansion. On the other hand, if we reduce the smoothing radius to ~ 400 h^{-1} kpc, we find the dense patches have density ~ 100 $\langle \rho \rangle$ ~ 10 ρ_c if $\langle \rho \rangle$ ~ 0.1 ρ_c. Here gravity is important and we must expect that the dense patches are stable or perhaps collapsing. The problem with this is that with the clustering pattern stable on small scales and freely expanding on large scales a power law form for $\xi(r)$ could not persist very long. For example, consider what happens when the universe expands by a factor of two. If the small-scale clustering is stable the typical density within clusters does not change, while the background density drops by a factor of 8. Since ξ is fixed by the ratio of these densities we conclude $\xi(r)$ has increased by a factor of 8 at small r. If no new generations of clustering develop to fill in the power law at $r > r_0$, then $\xi(r)$ already droops at the large r end by an amount perhaps comparable to the model in Equation 8 and FIGURE 2.

What are we to make of this? It is fair to ask, with Gott and Rees,[23] whether a modest droop from the power-law model for $\xi(r)$ might not be allowed by the observations? However, as I have described, it seems to me the constraints are quite strong. It may well be that the galaxy distribution has nothing to do with the mass distribution, for it seems clear that there is a considerable amount of dark mass around the visible parts of galaxies. Of course that does not remove the puzzle of why $\xi(r)$ is such a good approximation to a power law but it does obscure the physical nature and significance of the puzzle. On the theoretical side, it has proved difficult to make the heuristic argument given above for a low-density universe more precise and quantitative, or to demonstrate that in a cosmologically flat universe $\xi(r)$ would remain an adequate approximation to a power law. Some recent developments are summarized in this volume by Fall.[18]

Very likely an important empirical diagnostic of the puzzle will be the statistical pattern of the galaxy peculiar velocities associated with the spatial clustering. A few preliminary results are discussed next.

JOINT DISTRIBUTION IN REDSHIFT AND ANGULAR POSITION

Three-dimensional maps based on galaxy redshifts and angular positions provide a much clearer and more sensitive picture of the galaxy clustering because redshift is much better than apparent magnitude as a distance measure. A dramatic example was Chincarini and Rood's map of the outskirts of the Coma cluster.[24] More recent results are described by Tifft and Gregory[25] and Einasto.[26]

The distribution of galaxies in the three-dimensional "space" with polar coordinates θ, ϕ, and cZ/H is distorted from the true configuration by peculiar redshifts, but it does represent the significant data in familiar terms. And it is interesting to think of possible statistical measures of the distribution in this redshift space as n-point correlation functions. The two-point function $\xi_v(\sigma, \pi)$ would be defined as in Equation 3 but, since redshift is not exactly proportional to distance, ξ_v would have to be written as a function of two variables, the components π and σ of the separations parallel and perpendicular to the line of sight. If peculiar redshifts were negligible, ξ_v would be isotropic and would agree with the usual spatial function $\xi((\pi^2 + \sigma^2)^{1/2})$. If each galaxy had a randomly assigned peculiar velocity, ξ_v would be elongated along the line of sight. If the mean peculiar velocity difference of pairs of galaxies were antiparallel to the separation of the pair, as would be the case if new generations of clusters were forming now, it would tend to make ξ_v flattened along the line of sight.

Statistics of redshifts and clustering have been discussed by several authors.[21,27-32] The scheme outlined above is from Reference 21.

The best available approximation to a fair sample of the distribution in redshift space comes from the survey by Kirshner, Oemler, and Schechter.[5] They measured redshifts of almost all galaxies brighter than $J = 15$ in eight fields, each $\sim 4°$ by $4°$, totaling 166 galaxies in all. The sample goes deep enough that it is not likely to be contaminated by the Local Supercluster, and because the fields are well separated it reduces the chance that a sample this small may be unduly affected by local peculiarities, like a galaxy cloud in the field. Nevertheless, the sample is small, so statistical uncertainties in the ξ_v are large and the results are therefore only a preliminary indication.

FIGURE 4 shows estimates of ξ_v: the solid histograms show ξ_v as a function of σ at fixed π, the dashed histograms show ξ_v as a function of π at fixed σ. In graph (a) the fixed variable is in the range 0 to 0.5 h^{-1} Mpc, in (b) it is in the range 0.5 to 1.0 h^{-1} Mpc, and in (c) it is in the range 1 to 2 h^{-1} Mpc. If ξ_v were isotropic, the solid and dashed histograms would agree within the statistical fluctuations. The elongation of ξ_v along the line of sight is quite pronounced.

The curves in FIGURE 4 come from the following model. Suppose the redshift difference of each galaxy pair is the sum of the cosmological part and a peculiar part w drawn at random from the distribution $f(w)$. Then

$$\xi_v(\sigma, \pi) = \int dw\, f(w)\, \xi((\sigma^2 + (\pi - w/H)^2)^{1/2}). \qquad \textbf{(9)}$$

A model for f that seems to work fairly well is[30]

$$f \propto \exp(-2^{1/2}|w|/\langle w^2\rangle^{1/2}). \qquad \textbf{(10)}$$

The curves in FIGURE 4 are based on these two equations with ξ given by Equations 6 and 7 and with the peculiar velocity dispersion

$$\langle w^2\rangle^{1/2} = 500 \text{ km s}^{-1}. \qquad \textbf{(11)}$$

The fluctuations in the ξ_v estimates are large, but it does seem that the model gives a good description of the behavior of the data. Further details on the estimation of ξ_v and the numerical fit to $\langle w^2\rangle$ are given in Reference 21.

If galaxy-peculiar velocities are isotropic and uncorrelated, Equation 11 corresponds to rms peculiar spatial velocity

$$\langle v^2\rangle^{1/2} = (3/2)^{1/2} \langle w^2\rangle^{1/2} \simeq 600 \text{ km s}^{-1} \qquad \textbf{(12)}$$

This might be considered surprisingly high for the "field" galaxies outside the great clusters. Similar large velocities are found in the redshift sample at $m < 13$,[29,31] but here it can be argued that the sample has been unduly affected by motions around the outskirts of the Virgo cluster. In the "close-isolated" pairs of galaxies observed by Turner[33] and Peterson[34] the relative line-of-sight velocities are ~ 200 km s^{-1} for spiral + SO galaxies, but perhaps more nearly comparable to Equation 11 for E pairs. The bound on peculiar motions of nearby galaxies found by Sandage and Tammann[35] is a factor of ~ 10 lower than in Equation 11. However, the results cannot be directly compared because Sandage and Tammann considered the redshifts of fairly well-isolated galaxies or else the means within groups, while Equation 11 refers to peculiar motions of galaxies and is weighted by the abundance of neighbors around a galaxy. It is not clear whether these points can account for the difference between Equation 11 and the Sandage–Tammann limit.

It is of course possible that Equation 11 just results from an accidental fluctuation in the small sample. That should be tested in the next few years as larger redshift samples become available. If $\langle w^2 \rangle$ is about right it means that a large amount of mass

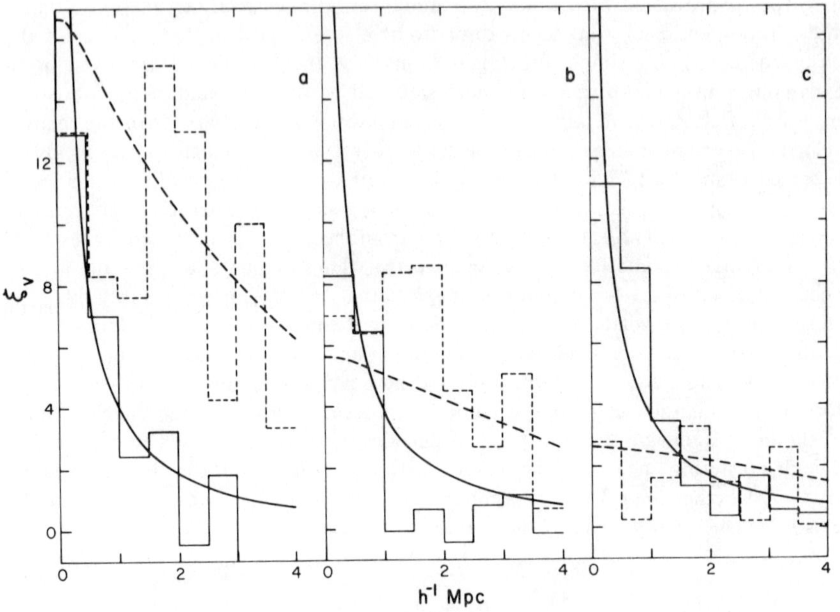

FIGURE 4. Estimates of the two-point correlation function in redshift space.[21] The curves are given by Equations 9–11.

in and around galaxies would be needed to make the general clustering pattern on scales $\lesssim r_0$ statistically stable. One estimate[21] of the wanted mean density is $\langle \rho \rangle \sim 0.4\ \rho_c$ (independent of H). With Equation 2 this would say that the mass-to-light ratio in the clustered matter is

$$M/L \sim 500 \text{ h } M_\odot/L_\odot \tag{13}$$

This is quite similar to the results from the central parts of the great clusters, but about two orders of magnitude larger than the mass-to-light ratios seen in the central parts of galaxies.

Summary

There has been considerable recent progress in measuring the variation of the deep galaxy counts with apparent magnitude, and, to a lesser extent, the possible large-scale angular variations of the counts at fixed apparent magnitude. A key problem in the interpretation of the counts is the correction for redshifts of the spectra (effective *K*-correction at given apparent magnitude). It is my impression that the counts and *K*-corrections approach the reliability needed to provide strong limits on relativistic cosmologies with appreciable density fluctuations on the scale of cH^{-1}, and to test homogeneous tired light cosmologies (with space curvature constrained by other measures, such as angular sizes of galaxies as a function of redshift).

I think the most reliable statistical measures of galaxy clustering on scales $\lesssim 10$ h^{-1} Mpc are provided by the galaxy two- and three-point correlation functions ξ and ζ, and I have mentioned what seems to me to be a serious puzzle, that is, whether the observed remarkably simple functional forms of ξ and ζ would be preserved in an expanding universe. An important next step will be the extension of the statistical measures of clustering to joint distributions in position and velocity. Does the relative velocity dispersion of galaxy pairs scale with separation as $r^{2-\gamma}$ at small r, as would be expected from ξ and ζ? Is there evidence of anticorrelation of peculiar velocities of galaxy pairs at $r \lesssim r_0$, as would be expected if new generations of clusters were forming to match the growth of ξ at small r resulting from the expansion of the universe?

Preliminary results for the two-point correlation function ξ_v suggest the galaxy peculiar velocities are large, and this suggests that the mass in groups of galaxies is considerably above what one would reckon from the mass-to-light ratios found in the central parts of galaxies. That may be helpful in understanding how $\xi(r)$ preserves its power-law shape, for in a dense universe new generations of clustering could be forming to fill in the clustering spectrum. Of course it leads us to a new puzzle: if most of the mass is outside the seen parts of galaxies, is there any reason to believe that galaxy clustering measures the mass clustering? The point can be checked by studying whether the pattern of peculiar velocities is consistent with dynamics in the clustering pattern described by the spatial correlation functions.

References

1. Hubble, E. 1926. Astrophys. J. **64:** 321.
2. Hubble, E. 1936. Astrophys. J. **84:** 517.
3. Shapley, H. 1935. Proc. Natl. Acad. Sci. **21:** 587.
4. Zwicky, F., E. Herzog, P. Wild, M. Karpowicz & C. T. Kowal. 1961–1968. Catalogue of Galaxies and Clusters of Galaxies. California Institute of Technology. Pasadena, California.
5. Kirshner, R. P., A. Oemler & P. L. Schechter. 1969. Astron. J. **83:** 1549.
6. Kron, R. G. 1978. Doctoral dissertation. University of California, Berkeley.
7. Rainey, G. W. 1977. Doctoral dissertation. University of California, Los Angeles.
8. Brown, G. S. 1974. Doctoral dissertation. University of Texas, Austin.
9. Shane, C. D. & C. A. Wirtanen. 1967. Publ. Lick Obs. **22,** part 1.
10. Tyson, J. A. & J. F. Jarvis. Astrophys. J. Lett. **230:** L153.
11. Neyman, J. 1962. *In* Problems of Extragalactic Research. G. C. McVittie, Ed.: 294. Macmillan. New York, New York.

12. Scott, E. L. 1962. *In* Problems of Extragalactic Research. G. C. McVittie, Ed.: 269. Macmillan. New York, New York.
13. Limber, D. N. 1954. Astrophys. J. **119:** 655.
14. Rubin, V. C. 1954. Proc. Natl. Acad. Sci. **40:** 541.
15. Totsuji, H. & T. Kihara. 1969. Publ. Astron. Soc. Jpn **21:** 221.
16. Peebles, P. J. E. *In* The Large-Scale Structure of the Universe. M. S. Longair and J. Einasto, Eds.: 217. Reidel. Dordrecht, Holland.
17. Gott, J. R. & E. L. Turner. 1977. Astrophys. J. **216:** 357.
18. Fall, S. M. 1979. This volume.
19. Rudnicki, K., T. Z. Dworak, P. Flin, B. Baranowski & A. Sendrakowski. 1973. Acta. Cosmologica **1:** 7.
20. Groth, E. J. & P. J. E. Peebles. 1977. Astrophys. J. **217:** 385.
21. Peebles, P. J. E. 1979. Astron. J. **84:** 730.
22. Peebles, P. J. E. 1974. Astrophys. J. Lett. **189:** L51.
23. Gott, J. R. & M. J. Rees. 1975. Astron. Astrophys. **45:** 365.
24. Chincarini, G. & H. J. Rood. 1975. Nature **257:** 294.
25. Tifft, W. & S. A. Gregory. 1979. *In* The Large-Scale Structure of the Universe. M. S. Longair and J. Einasto, Eds.: 51. Reidel. Dordrecht, Holland.
26. Einasto, J. 1979. *In* The Large-Scale Structure of the Universe. M. S. Longair and J. Einasto, Eds. Reidel. Dordrecht, Holland.
27. Jackson, J. C. 1972. Mon. Not. R. Astron. Soc. **156:** 1p.
28. Sargent, W. L. W. & E. L. Turner. 1977. Astrophys. J. Lett. **212:** L3.
29. Geller, M. J. & P. J. E. Peebles. 1973. Astrophys. J. **184:** 329.
30. Peebles, P. J. E. 1976. Astrophys. Space Sci. **45:** 3.
31. Davis, M., M. J. Geller & J. Huchra. 1978. Astrophys. J. **221:** 1.
32. Geller, M. J. & M. Davis. 1978. Astrophys. J. **225:** 1.
33. Turner, E. L. 1976. Astrophys. J. **208:** 304.
34. Peterson, S. 1978. Doctoral Dissertation. Cornell University. Ithaca, New York.
35. Sandage, A. & G. A. Tammann. 1975. Astrophys. J. **196:** 313.

GRAVITATIONAL INSTABILITY THEORY OF GALAXY FORMATION AND CLUSTERING: SOME RECENT DEVELOPMENTS

S. Michael Fall

Institute of Astronomy
Cambridge, England
and
Harvard-Smithsonian Center for Astrophysics
Cambridge, Massachusetts 02138

Introduction

A nearly uniform gravitational system is unstable to the growth of inhomogeneities because overdense regions exert a stronger gravitational attraction than underdense regions. In some form or other, this simple mechanism has played an important role in attempts to understand the cosmological development of structure on scales ranging from individual galaxies up to the largest observable aggregates of galaxy clusters. The idea is an appealing one because the physics is relatively simple and because it can explain in a natural way what might otherwise seem a paradoxical situation: the universe appears to be more homogeneous on large scales than it does on small scales. Variations on the basic gravitational instability theme are mainly the inclusion of gas-dynamical effects such as those that might set the characteristic scales of individual galaxies and distinguish them in some way from the cluster environments in which they are often found.

The purpose of this review is to discuss some recent developments in the gravitational instability theory of galaxy formation and clustering, including a confrontation with observational data. On the theoretical side of the subject, *N*-body computer simulations of several varieties have helped to sharpen the predictions of the theory and several new ideas have emerged on the roles of dissipation in protogalactic fragmentation and in galaxy collisions. On the observational side, the clustering properties of galaxies have been analyzed in some new and fruitful ways that demand a detailed comparison with theory. In addition, more and better measurements of the sizes, masses, and rotations of galaxies continue to accumulate. What follows is a collection of topics that have been chosen somewhat arbitrarily to illustrate these recent developments and their interconnections. More thorough introductions to the subject can be found in references 1–5.

Some Preliminaries

For the sake of definiteness, the "standard" big bang cosmological model will be adopted throughout this discussion. That is, it will be assumed that the universe is homogeneous in the large, that (apart from clustering effects) the redshifts of galaxies are the result of a general expansion, and that the microwave background radiation is extragalactic and originated in a dense fireball phase of the universe. Furthermore, it

0077–8923/80/0336–0172 $1.75/1 © 1980, NYAS

will be assumed that the expansion is governed by Friedmann's equations with the following present values of the defining parameters:

$$0.5 \lesssim h \lesssim 1.0, \qquad T = 2.7 \text{ K},$$
$$0.02 \lesssim \Omega \lesssim 1.0, \qquad \Lambda = 0, \tag{1}$$

all in standard notation (with $h \equiv H/100 \text{ km s}^{-1} \text{ Mpc}^{-1}$). The range of values above is intended to reflect empirical uncertainties, except for the "cosmological constant" Λ, which is almost completely unknown.

In order to avoid discussion of the highly uncertain primordial origins of structure and its evolution in the fireball, the (re)combination epoch ($z_r \simeq 1500$) will be taken as the starting point. Moreover, it will usually be assumed that on mass-scales larger than individual galaxies ($\sim 10^{12}$ $M_\odot$), the fluctuations present at that time were isothermal and had a power spectrum of the form

$$\langle |\Delta_k|^2 \rangle_r \propto k^n \qquad (-3 < n \leqslant 4), \tag{2}$$

where Δ_k is the Fourier transform of $\Delta(\mathbf{x})$, the local density contrast at the point $\mathbf{x}$, and n is an exponent to be determined empirically. For some purposes, an alternative expression in terms of mass scales is useful,

$$\Delta_r(M) \propto M^{-1/2-n/6}, \tag{3}$$

where the density contrast must now be interpreted in the root-mean-square sense when averages are taken over length scales proportional to $M^{1/3}$. Of course, the real fluctuation spectrum at recombination may not have been a pure power law on all scales but (2) and (3) are at least convenient one parameter forms to work with and seem to give reasonably good agreement with the data on galaxy clustering for n in the approximate range -2 to 0 (depending on the value of Ω; see below).

As is well known, the growth rate of the gravitational instability in an expanding medium is slow. In the absence of pressure forces, the growing mode of linear theory is approximately

$$\Delta^{(+)} \propto \begin{cases} t^{2/3} & (z \gtrsim z_f) \\ \text{const} & (z \lesssim z_f), \end{cases} \tag{4}$$

for $\Omega \lesssim 1$, where t denotes proper time from the big bang and $z_f \simeq \Omega^{-1} - 1$ is the redshift when the universe begins undecelerated expansion. Thus, the mass scale which is just beginning nonlinear condensation varies with t as

$$M_c(t) \propto \begin{cases} t^{4/(3+n)} & (z \gtrsim z_f) \\ \text{const} & (z \lesssim z_f). \end{cases} \tag{5}$$

A nearly spherical inhomogeneity which grows to $\Delta \simeq 1$ before z_f continues to expand until $\Delta \simeq 5.5$, after which time it collapses and then oscillates in a manner something like that depicted in FIGURE 1 until the equilibrium configuration is reached. An inhomogeneity which does not reach unit amplitude before z_f grows very little thereafter.

Ignoring dissipation, the above formulae and simple energy considerations lead to some useful relationships for the equilibrium configurations of objects formed by the

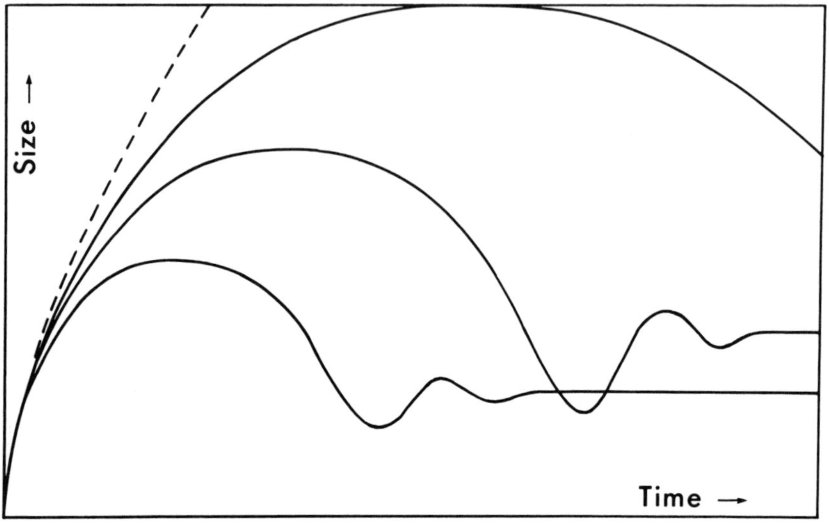

FIGURE 1. Evolution of nearly spherical inhomogeneities of different densities.

gravitational instability mechanism. They depend on the "initial" spectrum of fluctuations through the parameter n and, in terms of the "final" characteristic densities, masses, and sizes, take the form

$$\rho \propto M^{-3/2-n/2} \propto R^{-(9+3n)/(5+n)} \qquad \textbf{(6)}$$

Another prediction of the theory is that the objects ought to be spinning. The idea here is simply that many inhomogeneities will have, or will develop, some asphericity and hence a quadrupole moment as they grow. The tidal torques exerted by neighboring inhomogeneities will then result in a gain of angular momentum by each object, with most of the gain occurring near the time of maximum expansion.[6,7] The result is most conveniently expressed in terms of the dimensionless parameter

$$\lambda \equiv J|E|^{1/2}G^{-1}M^{-5/2}, \qquad (7)$$

where J is the angular momentum of the object and E is its total energy. In terms of this parameter, and ignoring dissipation once again, the theoretical prediction of Peebles[6] is

$$\lambda \simeq 0.08 \; (n \simeq -0.5), \qquad \textbf{(8)}$$

a value independent of M and well-confirmed by numerical experiment.[8,9]

Correlation Functions

An important method for describing the large-scale matter distribution, and one that has been especially stressed by Peebles and his associates, is in terms of correlations between the positions of galaxies.[4,5,10] In particular, the pair-correlation function ξ has received the most attention because of the relative ease with which it can be estimated empirically. This function is defined such that $\bar{n}^2[1 + \xi(r)]\,\delta v_1 \delta v_2$ is the joint probability of finding galaxies in the elemental volumes δv_1 and δv_2 separated by the distance r where $\bar{n}$ is the mean space density of galaxies. It is a measure of the relative strength of pairwise clustering on various scales and is known to have the approximate power-law form

$$\xi(r) \simeq (r_0/r)^\gamma \qquad (0.1\, h^{-1}\,\text{Mpc} \lesssim r \lesssim 10\, h^{-1}\,\text{Mpc}),$$
$$\gamma \simeq 1.8, \qquad r_0 \simeq (4 \pm 1) h^{-1}\,\text{Mpc}, \tag{9}$$

except that it may also have a downward bend at $r \simeq 2r_0$. A few of the higher-order correlation functions have also been estimated and they appear to have forms that are consistent with the notion of a hierarchical distribution of matter (FIGURE 2). In this case, it can be shown that the characteristic density of aggregates of galaxies varies with the clustering scale as

$$\rho(R) \propto R^{-\gamma}, \tag{10}$$

up to about 10 h^{-1} Mpc and perhaps more.[11]

The goal now is to relate these functions to the "initial" spectrum of fluctuations. With $\Omega \simeq 1$, this is relatively straightforward because most of the aggregates to which they refer can be presumed to have already, or to have nearly, reached the "final"

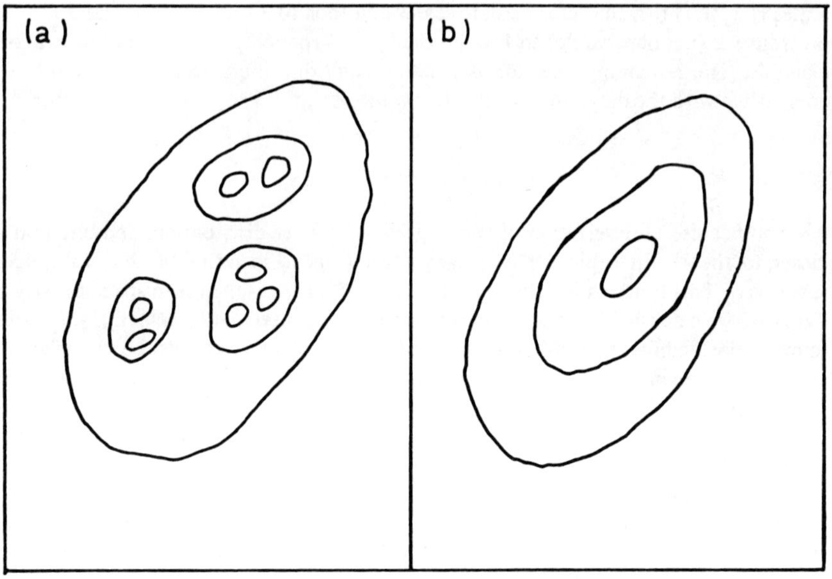

FIGURE 2. Group catalogues at different density contrasts: (a) hierarchical distribution; (b) nonhierarchical distribution. If clustering is stable, the distribution will maintain a form like (a), but if merging and disruption are important, the distribution may evolve to a form like (b).

equilibrium configurations. In this case, the comparison of (**6**) with (**10**) gives

$$\gamma \simeq (9 + 3n)/(5 + n) \qquad (\Omega \simeq 1.0), \tag{11}$$

and $\gamma \simeq 1.8$ therefore implies $n \simeq 0$, a "white noise" spectrum at recombination.[12] With $\Omega \ll 1$ the problem is not nearly so simple because aggregates on scales over most of the range that ξ has been reliably estimated must be presumed to be in the expansion or collapse phases of their evolution. Indeed, fluctuations on scales larger than about r_0 should be expanding with very little deceleration, their growth having stopped at a redshift of about z_f.

In general, the effects of low Ω should manifest themselves in ξ as a progressive steepening of this function, but the exact form this should take has been a matter of some debate.[12–14] Clearly, the smaller Ω is, the more large-scale structure is required in the initial matter distribution to compensate for its slower growth rate. A simple dynamical argument that accounts for the steepening in terms of an Ω-dependent "effective" power-law index gives[4]

$$\gamma_e \simeq \frac{-3 \log \Omega + \log \xi_v}{-\log \Omega + \gamma_f^{-1} \log \xi_v} \qquad (1 \lesssim \xi \lesssim \Omega^{-3}), \tag{12}$$

where γ_f is given approximately by (**11**) and $\xi_v \simeq 300$ is roughly the density contrast bound aggregates have when they first reach their equilibrium ("virialized") state. Although it has been tested in only a few cases, this expression is in reasonable agreement with expanding N-body experiments (see below) and, as an example, predicts $n \simeq -1.6$ for $\gamma_e \simeq 1.8$ and $\Omega \simeq 0.1$. It is not expected, however, that ξ should have perfect power-law form for $\Omega \ll 1$ because its index must vary from γ_f on small scales ($\xi \gtrsim \Omega^{-3}$) to γ_e on intermediate scales and then to $3 + n$ on large scales ($\xi \lesssim 1$) or, from 1.2 (not observable) to 1.8 and then to 1.4 (possibly observable) in the above example. The remaining question is whether such deviations from a power law are compatible with the data, and hence a low value of Ω; this is where opinions differ.[10]

MULTIPLICITY FUNCTIONS

Another useful description of the large-scale matter distribution, and one that is closer to the intuitive picture of galaxy clustering, is in terms of the multiplicity function η. This function is defined such that $\eta(M, \delta)dM$ is the mean space density of "groups" with density contrast δ and total mass in the interval $(M, M + dM)$. In order to make this definition precise, the notion of a group of galaxies must be specified and this is best done in terms of a "group catalogue" $\mathcal{C}(\delta)$, defined here to be all regions of space within which the mean density contrast has the fixed value δ. Individual regions are then the groups of the catalogue $\mathcal{C}(\delta)$. The boundaries of a family of catalogues at different density contrasts make a contour map of the matter distribution and the topology indicates how hierarchical the distribution is (FIGURE 2).

In the gravitational instability theory, the dependence of $\eta(M, \delta)$ on M at fixed δ should directly reflect the initial mass spectrum of fluctuations, independent of Ω, because Ω merely controls the relative growth rates of inhomogeneities at different densities. The dependence of $\eta(M, \delta)$ on δ at fixed M, however, should reflect both the initial spectrum and Ω in essentially the same way that they are reflected in the index of ξ. A simple set of formulae that generalizes the original theory of Press and Schechter[15] and incorporate these effects into η with the same assumptions used in the derivation of (**12**) is the following:[16]

$$\eta(M,\delta) = M^{-2}G_n\,[(M/M_1)\,\delta^{-1/\epsilon_e}],$$
$$G_n(x) \propto x^{1/2+n/6}\exp\,(-x^{1+n/3}),$$
$$\epsilon_e = -\,\gamma_e/(3-\gamma_e). \qquad \textbf{(13)}$$

Here M_1 is the current, marginally nonlinear mass scale given by Equation 23 below and, with $\gamma_e \simeq 1.8$, the required exponent is $\epsilon_e \simeq -1.5$.

Estimating the multiplicity function empirically from a large sample of galaxies is not nearly as straightforward as estimating correlation functions unless complete redshift information is available. This is because the relation between η and the projected distribution of galaxies on the sky involves several selection effects such as the apparent richness–distance correlation of groups.[16] The estimates of Gott and Turner are based on the identification of groups as surface density enhancements greater than 8 in a 14th magnitude sample of galaxies with some supplementary redshift information for distance estimates.[17] After various corrections, the result is[18]

$$\eta(M,\delta_e) \underset{\sim}{\propto} \begin{cases} M^{-1.0} & (M \lesssim 8\times 10^9 h^{-2}\mu_c L_\odot) \\ M^{-2.3} & (M \gtrsim 8\times 10^9 h^{-2}\mu_c L_\odot), \end{cases} \qquad \textbf{(14)}$$

where $\delta_e \simeq 500$ is the effective luminosity contrast of groups in the sample and $\mu_c \simeq 1600\Omega hM_\odot/L_\odot$ is the cosmic mass-to-light ratio. The estimates (**14**) do not have the functional form predicted by (**13**) but they do have a feature at the predicted mass scale $\delta^{1/\epsilon_e}M_1 \sim 8\times 10^9 h^{-2}\mu_c L_\odot$. When complete redshift samples become available, a more detailed comparison between theory and observation should be possible.

Clustering Experiments

In principle, the ideas of the previous sections can be tested against the clustering that develops in expanding *N*-body computer experiments. This is especially important because one might expect that various disruptive processes within condensed aggregates of galaxies would erase some of the hierarchical structure formed by the simple gravitational clustering mechanism discussed above (Figure 2). For example, the encounters of small substructures within large aggregates might cause them to merge during direct encounters or to suffer some rearrangement of material during near encounters and thus to alter the relations between the present clustering pattern and the initial fluctuation spectrum of the simple theory.[19] It is almost certain that some of these effects have been important in the development of the densest clusters of galaxies,[20] such as Coma, but they are rare and the question of interest here is whether disruptive effects are important in the evolution of statistical measures of clustering like ξ and η.

In their simplest form, the *N*-body experiments involve 10^3 or so particles, interacting by Newtonian gravity, with each particle interpreted as representing a single bright galaxy. The entire system is usually enclosed within a spherical reflecting

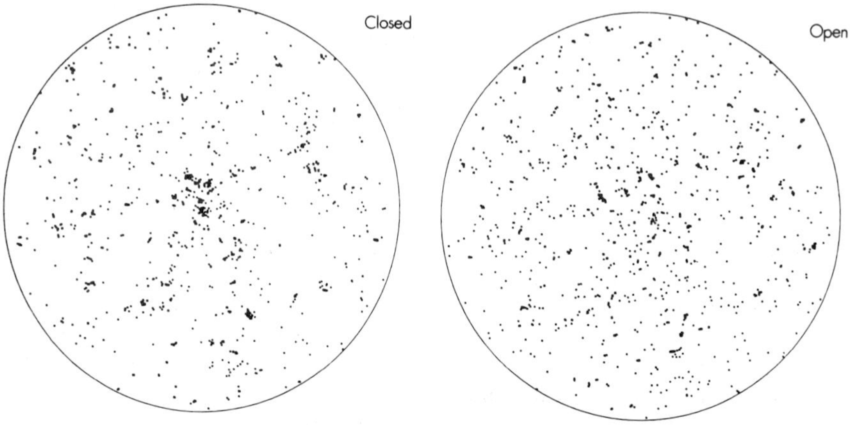

FIGURE 3. Results of two 1000-body experiments with different density parameters[4]: left, $\Omega = 1.0$; right, $\Omega = 0.26$ (final value). The initial distributions were identical (Poisson, $n = 0$) and crosses mark the positions of particles, projected onto the plane of the page, after each system had expanded by a factor of 9.0. Notice that large-scale clustering is stronger in the high-density system.

boundary, the radius of which moves according to the Friedmann equations, and initially, all of the particles are given Hubble-like expansion velocities. The initial conditions can be adjusted to simulate different values of Ω and different initial fluctuation spectra, "white noise" being the easiest and most common. Many of the experiments done to date have made use of some variant of Aarseth's integration code,[16,21–24] which is known to have high numerical accuracy. When interpreted liberally, the results of the experiments generally show agreement with both observation and theory and this, of course, is support for the gravitational instability picture (FIGURE 3).

There are, however, some technical difficulties in the interpretation of the N-body experiments,[22] most of which are connected with the relatively small number of particles involved ($N \simeq 10^3$). This results in a limited range of scales over which to measure clustering and noisy clustering estimates that show considerable experiment-by-experiment variation with statistically similar initial conditions. Perhaps even more important is the fact that fluctuations on mass scales of about 1–4 (particles), corresponding to a nonnegligible range of the scales of interest in the final distributions, are nonlinear from the start of the calculations. It is probably for these reasons that the first N-body experiments gave rise to some controversy.[21] In order to overcome some of the difficulties, several new kinds of clustering experiments are also in progress.[25,26] My own impression of the results of many experiments, taken as a whole, is that they are in qualitative agreement and, in some cases, quantitative

FIGURE 4. Averaged pair-correlation estimates from six 1000-body experiments[16] with $n = 0$ and $\Omega = 1.0$. The expansion parameter at which the estimates were made is denoted by a and the error bars on the last set of points are one standard deviation from the mean. The dashed line is a power-law with index 1.8.

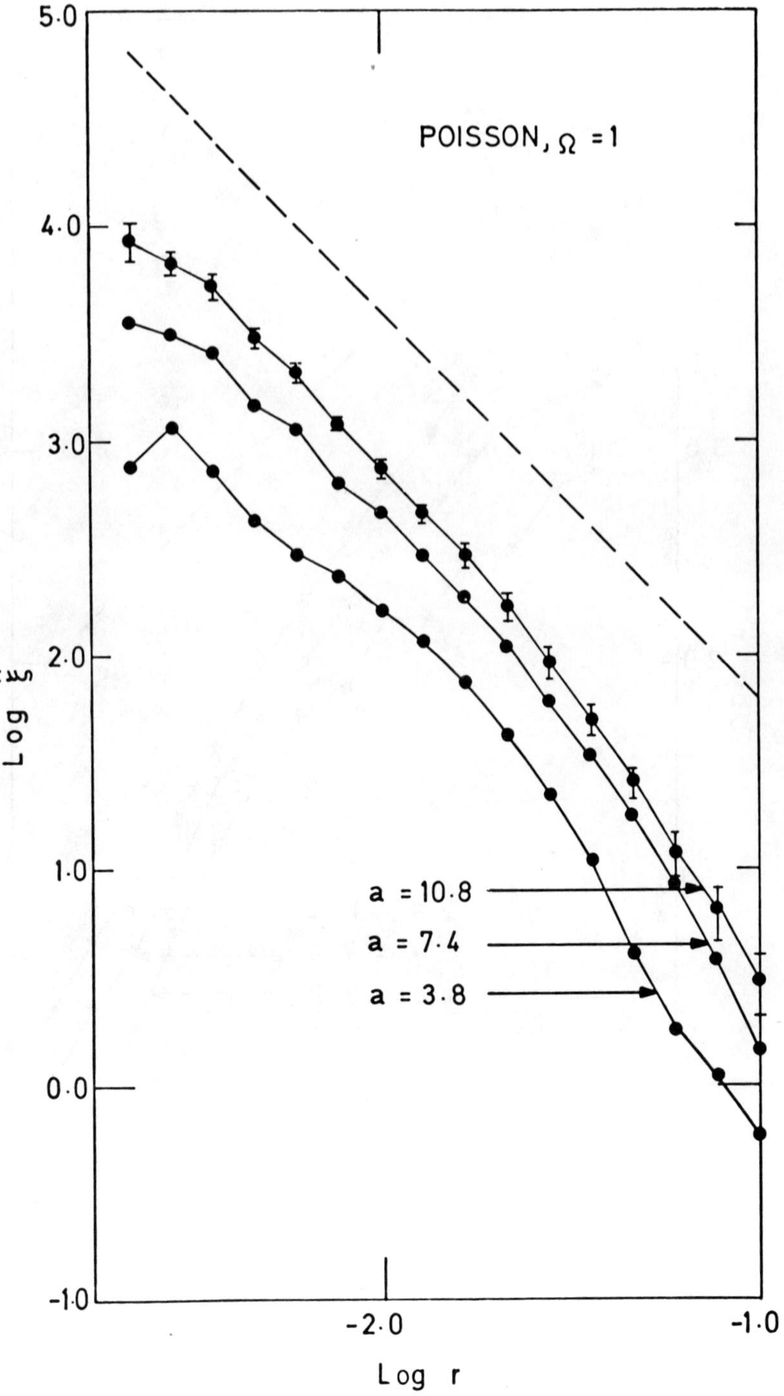

POISSON, Ω = 1
a = 10·8
a = 7·4
a = 3·8
5·0
4·0
3·0
2·0
1·0
0·0
-1·0
-2·0
-1·0
Log ξ̂
Log r

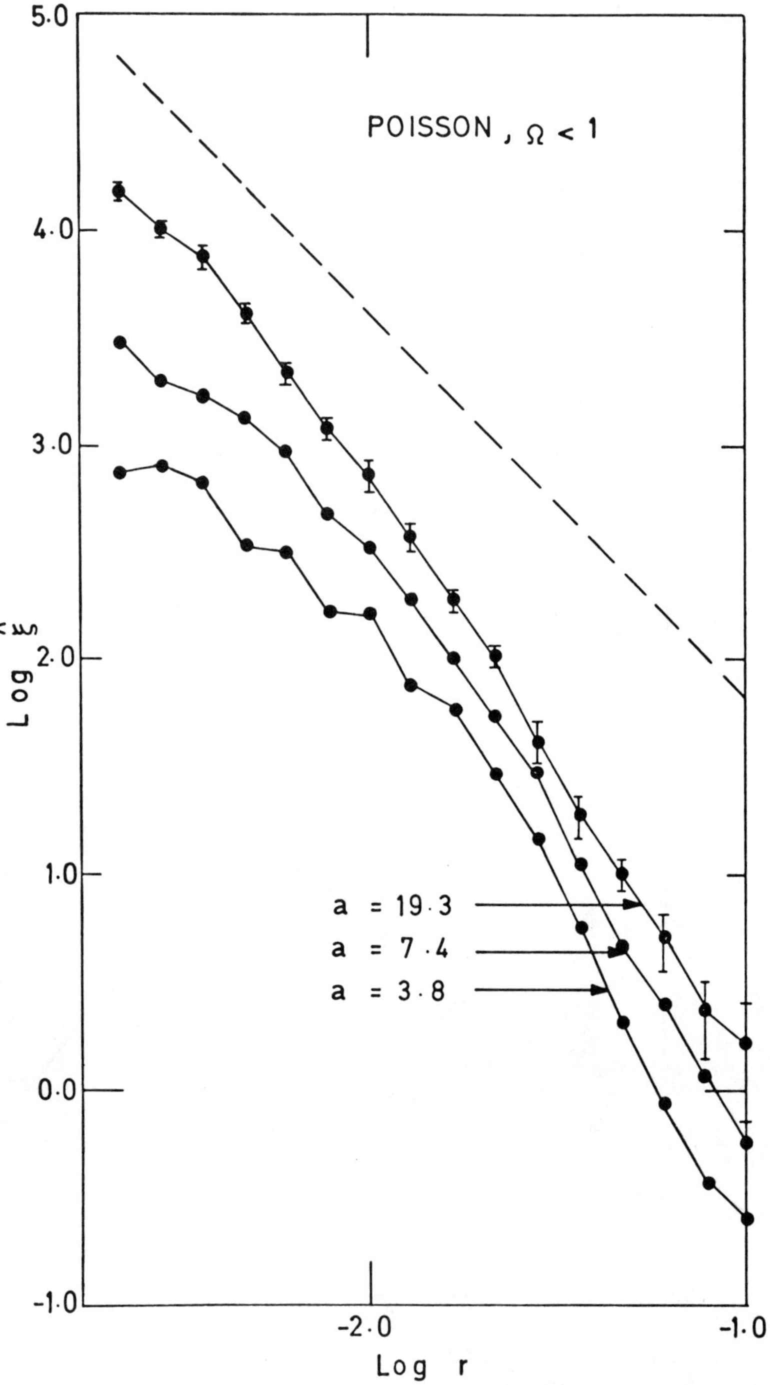
5.0
4.0
3.0
2.0
1.0
0.0
-1.0
POISSON , Ω < 1
Log ξ̂
a = 19.3
a = 7.4
a = 3.8
-2.0
-1.0
Log r

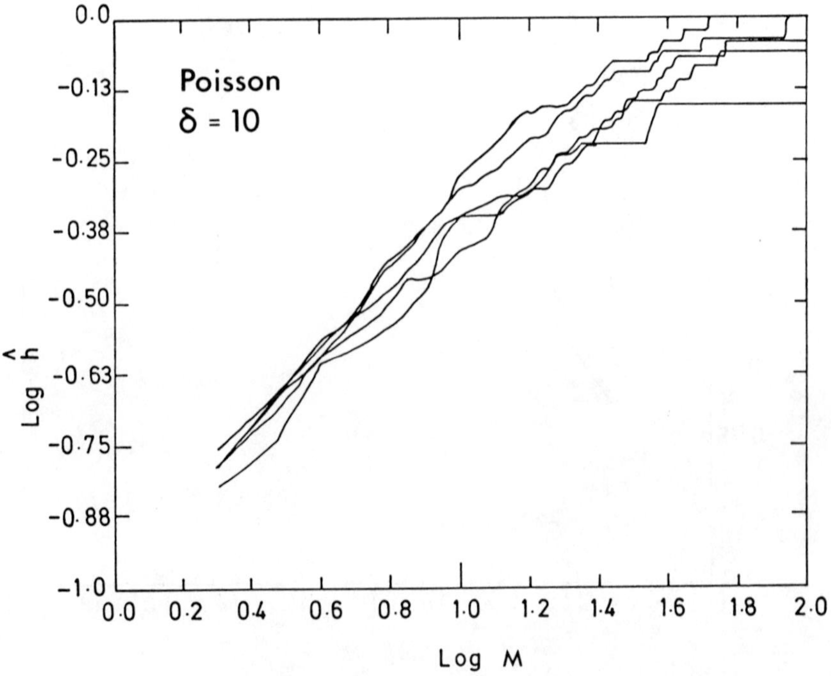

FIGURE 6. Cumulative multiplicity functions at $\delta = 10$ and $a = 10.8$ for the six 1000-body experiments[16] of FIGURE 4 with $n = 0$ and $\Omega = 1.0$. The results vary considerably between experiments but the average function is consistent with the theory of Equation 13.

agreement with the simple theory (FIGURES 4–7). That is, ξ and η show some dependence on Ω and n in the expected sense, with little or no evidence for disruptive and "bootstrap" effects, except perhaps at the highest densities ($\delta \gtrsim 500$) where the experimental statistics are poor.

DISSIPATIONLESS GALAXY FORMATION

The gravitational instability mechanism without dissipation seems to offer a satisfactory explanation of galaxy clustering and it is therefore tempting to apply its predictions to the structure of galaxies themselves. The presence of a gaseous disk in spiral galaxies is usually taken as evidence that they have suffered at least some dissipation so elliptical galaxies and perhaps also the spheroidal components of spiral galaxies are often taken as the prototypes for dissipationless galaxy formation

---

FIGURE 5. Averaged pair-correlation estimates from three 1000-body experiments[16] with $n = 0$ and $\Omega = 0.15$ (final value). The symbols have the same meaning as in FIGURE 4. Notice the progressive steepening of the correlation function as Ω decreases and the lack of a strong feature near $\xi \simeq \Omega^{-3}$.

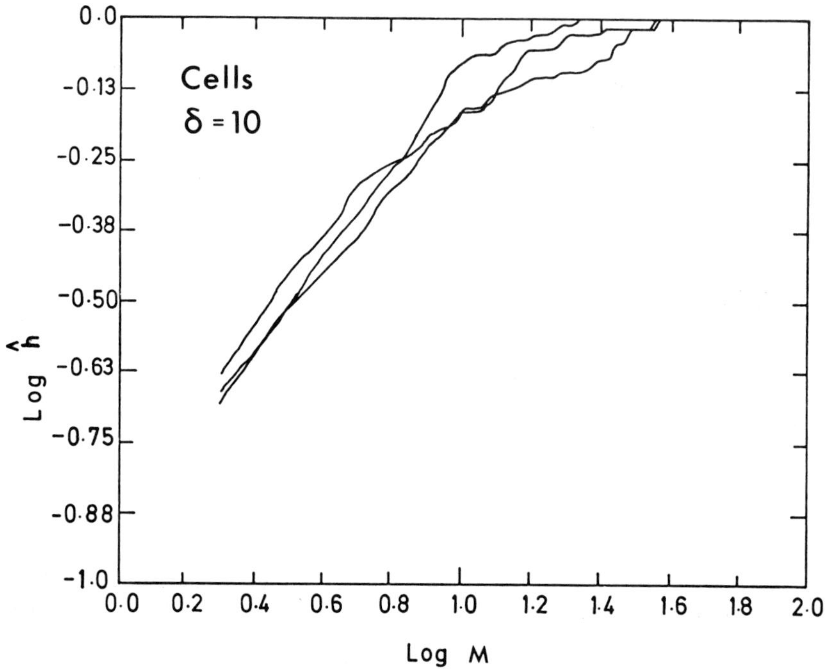

FIGURE 7. Cumulative multiplicity functions at $\delta = 10$ and $a = 15.4$ for three 1000-body experiments[16] with $n \simeq +1$ and $\Omega = 1.0$. The difference between these results and those of FIGURE 6 is significant at about the 2σ level.

models.[27] In view of the mounting evidence for heavy halos and galaxy mergers and the current ideas on the role of gas-dynamical effects in setting the characteristic masses and sizes of galaxies, it is becoming more popular to accept the notion that dissipation of some sort was involved in the formation of all galaxies. Nevertheless, it is instructive to confront the data on elliptical galaxies with the simple dissipationless theory, if only for the purposes of comparison with more realistic theories.

In terms of average quantities, the relevant observational material can be summarized as follows:

$$R_e \propto L^p \qquad p = 0.90 \pm 0.35 \qquad \text{(references 28, 29)}, \tag{15}$$

$$\sigma_0 \propto L^q \qquad q = 0.25 \pm 0.05 \qquad \text{(references 30, 31)}, \tag{16}$$

$$\langle v_c/\sigma_0 \rangle = 0.20 \pm 0.03 \qquad \text{(references 31–34)}. \tag{17}$$

Here L is the total luminosity, R_e is the half-light radius, σ_0 is the central (three-dimensional) velocity dispersion and v_c is the peak rotation velocity along the major axis of the galaxy (corrected upwards from the observed value by the mean oblate projection factor $4/\pi$). The uncertainties indicated above are entirely subjective except for those of (**17**). From the dynamical relation $\sigma_0{}^2 \propto M/R_e$, one now has

$$M/L \propto L^{p+2q-1} \propto L^{0.4\pm0.4}, \tag{18}$$

$$M \propto R_e^{1+2q/p} \propto R_e^{1.6\pm0.3}, \tag{19}$$

where, in view of the probable variation of mass-to-light ratio with galactocentric distance, M must be interpreted as the characteristic mass associated with the luminous parts of a galaxy. The first of the expressions above indicates that the corresponding mean mass-to-light ratios increase with galactic luminosity and the comparison of the second expression with (**6**) suggests $n = (p - 10q)/(p + 2q)$, or $n = -1.1 \pm 0.5$, on the dissipationless model.[31]

The recently measured internal velocities of elliptical galaxies[31-34] indicate that their flattening cannot be the result of rotation alone as, for example, in the Maclaurin spheroid models. As Binney has emphasized, this means that the galaxies are either prolate or that their velocity distributions are anisotropic,[35] although the observations now seem to point away from the prolate model in most cases. He has further pointed out that the necessary anisotropies would follow simply from highly flattened initial configurations because violent relaxation is not completely efficient at rearranging motions in the outer parts of a collapsing stellar system.[36] In the dissipationless model, the required flattening might result from a combination of the tidal stretching by neighboring protogalaxies and the propensity for even an isolated inhomogeneity to increase its asphericity.[37] In any case, the low rotation velocities are probably good news for the tidal torques theory. For example, assuming a flat rotation curve and constant velocity dispersion, one has [38]

$$v_c/\sigma_0 = 3.1\lambda \quad \text{(Plummer model)}, \tag{20}$$

with the exact value of the coefficient on the right-hand side depending on the galactic density profile. Comparing (**20**) with the empirical estimates (**17**) gives $\lambda = 0.07 \pm 0.02$, a value that is in remarkably good agreement with the theoretical prediction, although somewhat uncertain observationally. Against the tidal torques hypothesis, however, is the lack of any correlation between the orientations of close binary galaxies, so the empirical basis for the theory is ambiguous.[39]

A final point about the dissipationless model of galaxy formation concerns the degree to which individual galaxies join onto the large-scale clustering hierarchy. Very roughly, the mass-density contrast associated with the inner 10 h^{-1} kpc of a bright galaxy is $\delta_g \simeq 10^5\Omega^{-1}$, whereas an extrapolation of the correlation function (**9**) to this scale gives $(5 \pm 2) \times 10^4\, h^{-1.8}$. Thus, the clustering hierarchy may almost match for $\Omega \simeq 1$, (reference 12), but $\Omega \ll 1$ implies a poor match and hence some dissipation in the development of small-scale structure. Another way to see this is in terms of the galaxy luminosity function ϕ, which is defined such that $\phi(L)\mathrm{d}L$ is the mean space density of galaxies with luminosities in the interval $(L, L + \mathrm{d}L)$. A group catalogue $\mathcal{C}(\delta)$ at a sufficiently high density contrast ($\delta \gtrsim \delta_g$) will include only individual galaxies and the galaxy luminosity function and the group multiplicity function should be equal.[16] Thus, in any theory of galaxy formation and clustering, one has $\phi(L) = \eta(\mu_g L, \delta_g)$ where μ_g is the galactic value of M/L (roughly equal to 20 h $M_\odot/L_\odot$ for the inner 10 h^{-1} kpc of bright galaxies). In the dissipationless model, ϕ should have the same shape as η for all $\delta \lesssim \delta_g$ and a feature at the luminosity $(\delta_g\mu_c/\mu_g)^{1/\epsilon_e}\mu_c^{-1}M_1 \simeq 1 \times 10^7\, h^{-2}L_\odot$. Now Schechter[40] has shown that the luminosities of galaxies in a 13th magnitude sample can be fitted by a function of the form

$$\phi(L) \propto L^{-\alpha}\exp(-L/L^*), \tag{21}$$

with $L^* \simeq 8 \times 10^9 h^{-2} L_\odot$ and $\alpha = 1.25$ or $\alpha = 1.0$ (reference 41). The discrepancy between this result and the prediction of (**13**) and (**23**) suggests that if galaxies originally formed from a self-similar hierarchy, their luminous parts must have contracted by a factor of order 10.

Gas-Dynamical Effects

The major alternative to the simple gravitational instability theory with isothermal fluctuations at recombination is the "pancake" theory of Zeldovich and his colleagues.[42] In this theory, it is supposed that the fluctuations emerging from recombination were adiabatic and had no appreciable structure on scales smaller than the Silk mass

$$M_D \simeq 3 \times 10^{12}\,\Omega^{-5/4} h^{-5/2} M_\odot, \tag{22}$$

as the result of photon viscosity in the fireball plasma.[43] Structures on larger scales then grew by the ordinary gravitational instability mechanism while in the gaseous state into elongated figures which subsequently collapsed (at redshifts of about 3–5) to form large caustic surfaces (the "pancakes"). Once this happened, individual galaxies and other small-scale structures are presumed to have condensed out of the gas as the result of cooling behind the shocks. With the inclusion of gas-dynamical effects such as these, the theory and its confrontation with observation become difficult. The power-law form of the correlation function may, however, provide a simple constraint because the present, marginally linear mass scale is about[4]

$$M_1 \simeq 8 \times 10^{14}\,\Omega\, h^{-1}\,(r_0/4\,h^{-1}\,\mathrm{Mpc})^3\,M_\odot. \tag{23}$$

If M_D is in the range from about 10^{12} $M_\odot$ to M_1, one might reasonably expect to see some feature in ξ at the corresponding spatial scale whereas, if M_D is much larger than M_1, one would not expect to find any small-scale structure, including ourselves (FIGURE 8 and reference 44). The comparison of (**22**) with (**23**) then limits the acceptable values of the density parameter in the pancake theory to small ranges around $\Omega \simeq 0.08\ h^{-3/2}$ and $\Omega \simeq h^{-5/4}$ unless it can be shown that the nonlinear gas dynamics would have completely smeared out a feature in the spectrum at M_D.

In any theory of galaxy formation and clustering, a fundamental problem has been to explain the existence of galaxies as well-defined entities that are reasonably distinct from their surroundings. Recently, several authors[45–47] have suggested that the condition that protogalactic gas clouds be able to cool and collapse on their free-fall time scales leads naturally to the characteristic masses and radii of large galaxies. The relevant time scales are

$$t_c \simeq 3kT_v m/\rho\Lambda(T_v), \qquad t_d \simeq (24\pi G\rho)^{-1/2}, \tag{24}$$

where k is Boltzmann's constant, m is the mass of the proton, T_v is the virial temperature of the cloud, ρ is the mean density of the cloud, and Λ is the cooling function for a primordial mixture of hydrogen and helium. A cloud with $t_c \lesssim t_d$ will collapse at its free-fall velocity, cool to a temperature near 10^4K and then promptly fragment into small substructures and stars. This condition leads to the following characteristic scales:

$$M \lesssim 10^{12}\ \mathrm{M_\odot} \quad \text{for} \quad \rho \gtrsim 10^{-29}\ \mathrm{g/cm^{-3}},$$
$$R \lesssim 75\ \mathrm{kpc} \quad \text{for} \quad M \gtrsim 10^{12}\ \mathrm{M_\odot}, \tag{25}$$

depending on the geometry of the collapse (flatter in the adiabatic theory, rounder in the isothermal theory). A cloud with $M \gtrsim 10^{12}$ $\mathrm{M_\odot}$ and $R \gtrsim 75$ kpc would have contracted quasistatistically until it had cooled enough to begin free-fall and fragmentation; but the condition that it do so in the Hubble time restricts the maximum allowed masses and radii to values near 10^{12} $\mathrm{M_\odot}$ and 75 kpc and hence to values near

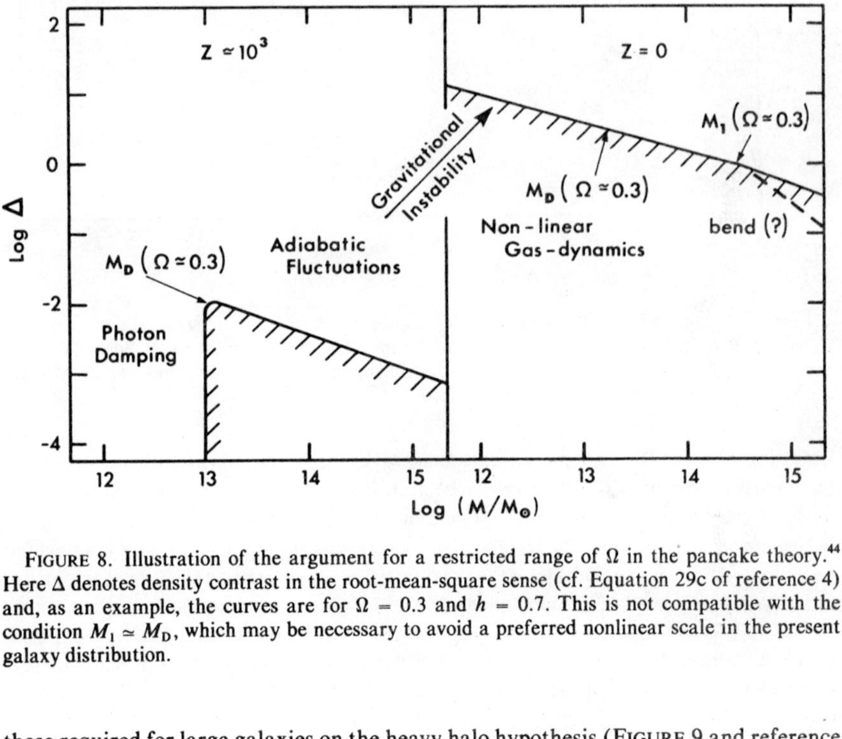

FIGURE 8. Illustration of the argument for a restricted range of Ω in the pancake theory.[44] Here Δ denotes density contrast in the root-mean-square sense (cf. Equation 29c of reference 4) and, as an example, the curves are for $\Omega = 0.3$ and $h = 0.7$. This is not compatible with the condition $M_1 \simeq M_D$, which may be necessary to avoid a preferred nonlinear scale in the present galaxy distribution.

those required for large galaxies on the heavy halo hypothesis (FIGURE 9 and reference 45).

The existence of relatively young stars in galaxies would seem to be evidence for at least some dissipation. Although probably important for our understanding of galactic evolution, the detailed connection between star formation and galaxy formation is difficult to make because of the lack of relevant observational material and the various theoretical uncertainties involved. Nevertheless, the galaxy formation models that do include star formation, such as those of Larson[48] and Tinsley,[49] can, with reasonable input physics, claim a successful comparison with the observed colors and shapes of galaxies. In general, dissipation would have permitted the formation of galaxies at

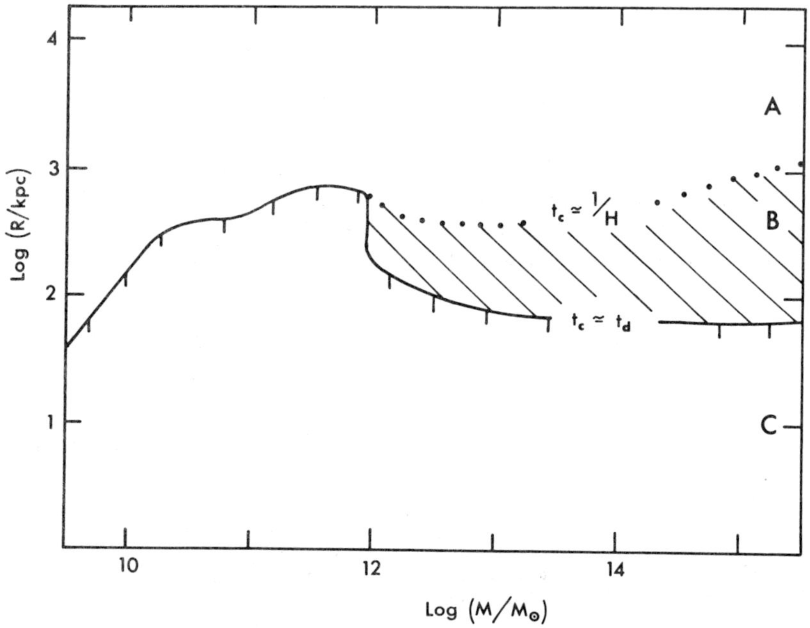

FIGURE 9. Illustration of the arguments leading to the characteristic sizes and masses of large galaxies.[45] Protogalactic gas clouds in region A cannot cool in the Hubble time; those in region B can be pressure-supported for some time but will, in some cases, evolve into region C where isothermal collapse and fragmentation can occur.

times later ($z \lesssim 10$) than the times predicted by the simple gravitational instability mechanism without dissipation ($z \gtrsim 10$) and, if the amount of dissipation was large, most of the analysis of the previous section would be invalid. Thus, the agreement between the tidal torques theory and the rotation measures of elliptical galaxies may be fortuitous unless their halos are rotating more slowly than the luminous parts or unless mergers have been important (see below). Other modifications emphasized by White and Rees[50] might be the merging and disruption of galactic halos in groups and clusters. In their variant of the hierarchical clustering model, the luminous parts of galaxies are composed of relatively young stars (populations I and II) formed from cool gas that fell to the bottoms of the potential wells of extended halos, which are presumed to be composed of pregalactic material (population III stars?).

GALAXY MERGERS

For some time now, it has been thought likely that the *cD* galaxies of rich clusters attained their large sizes by "cannabalism" and the workings of dynamical friction within the clusters.[20,51,52] Impressed by the large number of interacting galaxies in small groups and their short orbital time scales, the Toomres have gone one step further and have suggested that perhaps most large elliptical galaxies are the result of mergers, with the interacting galaxies themselves providing the necessary dynamical friction.[53,54] The idea is simply that the collision of, for example, two disk galaxies on a nearly parabolic orbit would be very sticky and that in the aftermath of the collision the structure of the protagonists would be obliterated, leaving behind a fairly amorphous system like an elliptical galaxy. Any gas in the system would have to be converted efficiently into stars in the shocked regions during the collision[55,56] or perhaps be blown out by a supernova-driven wind. The main dynamical requirement of the mergers scheme is an environment where the relative velocities of galaxies are comparable with their internal velocity dispersions ($\sigma \simeq 250$ km s^{-1}) and this has often been held against the scheme because most elliptical galaxies are found in clusters where the velocities are much higher. The expanding *N*-body experiments designed to simulate mergers in a cosmological setting show, however, that many merger remnants will reside in clusters, as a consequence of dynamical friction and merging in low-velocity subgroups that only later join to form larger, high-velocity clusters.

With isothermal fluctuations at recombination, the merger picture may also be thought of as a variation on the usual hierarchical clustering theme with the infall of material occurring in discrete lumps. Indeed, mergers may have played a role in the evolution of all galaxies and spirals may only be those galaxies that managed to retain some gas. The mergers that took place before star formation must have been highly dissipative and even those that are now occurring are inelastic in the sense that the kinetic energy in the center-of-mass frame before merging is subsequently taken up by the internal stellar motions of the product. In order to determine the range of orbital parameters over which merging occurs, a number of "cross section" experiments have been done[57–59] and the results are shown in FIGURE 10. In these experiments, each

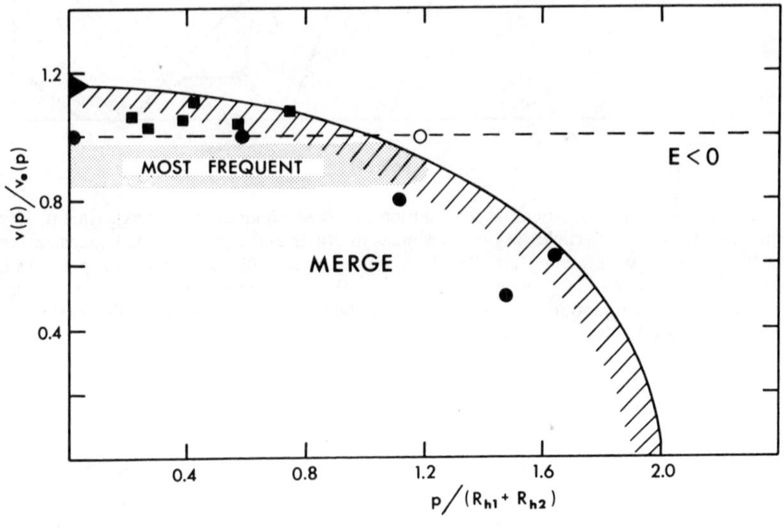

FIGURE 10. The cross section used in the cosmological merger experiments.[38] Here $v(p)$ and $v_e(p)$ denote, respectively, the encounter velocity and the escape velocity at pericenter p and R_{h1} and R_{h2} denote the half-mass radii of the protagonists. The triangle at $v(0) = 1.16\ v_e(0)$ is from the experiments of reference 57, the squares are from the experiments of reference 59, and the filled circles are from experiments E through I of reference 58, all of which merged. The open circle is from experiment D of reference 58, which did not merge after about nine crossing times. The stippled region indicates where mergers in the cosmological experiments are most frequent.

galaxy is represented by an N-body system ($N \gtrsim 30$) and pairs of "galaxies" are placed in various two-body orbits, with the result that merging takes place within one orbital time scale if the relative velocity is less than about 1.16 times the escape velocity at pericenter in the head-on collisions[57] and somewhat less in the off-axis collisions. White's experiments in particular showed that for several different initial configurations, the merger remnants would have a Hubble density profile and would be well-mixed by the time they settled into equilibrium.[58]

In order to determine which merging orbits are actually set up by the cosmological development of structure, the expanding N-body systms described in the section on clustering experiments above can be modified to include the merging of individual

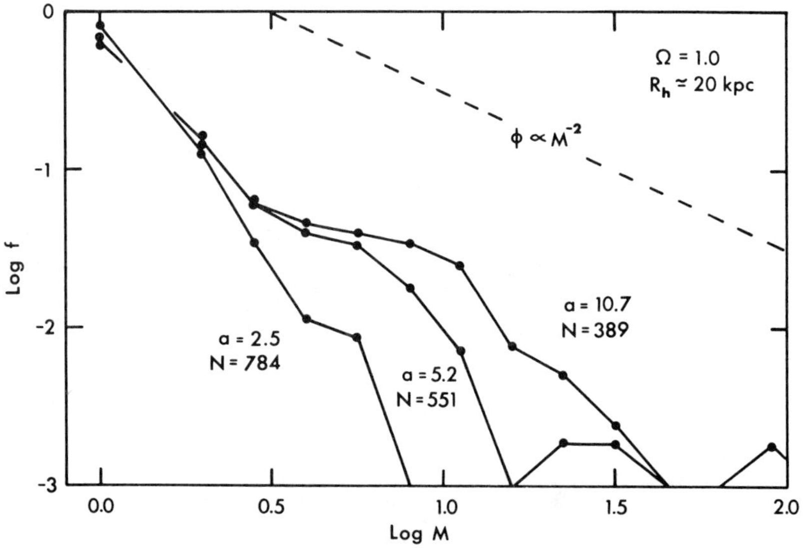

FIGURE 11. Evolution of the mass distribution in a cosmological merger experiment. Here f denotes the fraction of particles in different mass intervals and a denotes the expansion factor when the estimates were made. Initially, the system consisted of 1000 unit-mass particles in a Poisson ($n = 0$) distribution. Notice the runaway of the most massive particles; this function is not compatible with Equation 21 at the bright end but experiments with smaller particle sizes have resonably good mass functions in this region.[38]

particles. Such experiments have been done by Jones and Efstathiou[60] and by Aarseth and myself.[38] Here emphasis will be given to the latter as they are based on the cross section of FIGURE 10, use realistic particle sizes, and employ a computational scheme that conserves mass, momentum, angular momentum, and energy (in the sense that the sizes, spins, and interaction radii of merger remnants are adjusted to account for the energy and angular momentum lost from the center-of-mass frame during merging). The preliminary results of these experiments can be summarized as follows: Merging occurs hierarchically and mainly from eccentric two-body orbits of nearly zero energy. The particle mass function continually flattens as a result of a runaway effect, especially when particle sizes are large (FIGURE 11). The characteristic angular momentum acquired by remnants corresponds to $\lambda \simeq 0.07$ and the remnant mass–

radius relation is $M \propto R_h^{1.2}$. With realistic "galaxy" sizes ($R_h \simeq 2\,h^{-1}$ kpc–10 h^{-1} kpc), the fraction of merger remnants at the end of the calculations is about 0.25 for $\Omega = 0.1$ and somewhat larger for $\Omega = 1$. The large-scale clustering properties of the system (ξ and η) are always similar to those mentioned in the section on clustering experiments, but the small-scale clustering properties are sensitive to the merging process. In particular, clustering is too weak and the mass function is too flat if the sizes of particles are much larger than the luminous parts of real galaxies. Most of the results are consistent with the observations summarized in the section on dissipationless galaxy formation and lend support to the merger scheme if heavy halos are not attached to individual galaxies in groups and clusters.

Conclusions

It should be clear from what has gone before that the gravitational instability theory in one of its various forms is capable of explaining a wealth of observational data on the structure of galaxies and clusters of galaxies. Indeed, it is possible that the simple gravitational instability mechanism without dissipation can account for the large-scale matter distribution even if some dissipation—gas dynamics or mergers—is necessary to account for some aspects of small-scale structure. Nevertheless, essentially all of the observations with which the theory can now be confronted refer to the present epoch and one may well wonder if things really happened this way.

In view of the many theoretical uncertainties, it would be foolish to claim success until we have some fairly direct evidence for the cosmological development of structure. Observations from the Space Telescope, the Einstein X-Ray Satellite and the Cosmic Background Explorer, as well as from earth-based optical and radio telescopes, may soon provide us with some important clues in this regard.

Acknowledgments

I thank my collaborators Sverre Aarseth, George Efstathiou, Craig Hogan, and Silvio Bonometto for allowing me to mention our unpublished work and Jim Peebles, Richard Gott, and Paul Schechter for allowing me to mention their unpublished work.

References

1. Jones, B. J. T. 1976. Rev. Mod. Phys. **48:** 107.
2. Gott, J. R. 1977. Ann. Rev. Astron. Astrophys. **15:** 235.
3. Rees, M. J. 1978. *In* Observational Cosmology. (Saas-Fee, VIIIth Course) p. 259. Geneva Observatory. Sauverny, Switzerland.
4. Fall, S. M. 1979. Rev. Mod. Phys. **51:** 21.
5. Peebles, P. J. E. 1979. The Large-Scale Structure of the Universe. In press.
6. Peebles, P. J. E. 1969. Astrophys. J. **155:** 393.
7. Thuan, T. X. & J. R. Gott. 1977. Astrophys. J. **216:** 194.
8. Peebles, P. J. E. 1971. Astron. Astrophys. **11:** 377.
9. Efstathiou, G. & B. J. T. Jones. 1979. Mon. Not. R. Astron. Soc. **186:** 133.
10. Peebles, P. J. E. This volume.
11. Peebles, P. J. E. & E. J. Groth. 1975. Astrophys. J. **196:** 1.
12. Peebles, P. J. E. 1974. Astrophys. J. Lett. **189:** L51.
13. Gott, J. R. & M. J. Rees. 1975. Astron. Astrophys. **45:** 365.

14. Davis, M., E. J. Groth & P. J. E. Peebles. 1977. Astrophys. J. Lett. **212:** L107.
15. Press, W. H. & P. Schechter. 1974. Astrophys. J. **187:** 425.
16. Efstathiou, G., S. M. Fall & C. Hogan. 1979. Mon. Not. R. Astron. Soc. In press.
17. Turner, E. L. & J. R. Gott. 1976. Astrophys. J. Suppl. Ser. **32:** 409.
18. Gott, J. R. & E. L. Turner. 1977. Astrophys. J. **216:** 357.
19. Silk, J. & S. D. M. White. 1978. Astrophys. J. Lett. **223:** L59.
20. White, S. D. M. 1976. Mon. Not. R. Astron. Soc. **177:** 717.
21. Aarseth, S. J., J. R. Gott & E. L. Turner. 1978. *In* The Large-Scale Structure of the Universe (IAU Symposium No. 79). M. Longair and J. Einasto, Eds: 21, 63, 189. Reidel. Dordrecht, Holland.
22. Fall, S. M. 1978. Mon. Not. R. Astron. Soc. **185:** 165.
23. Efstathiou, G. 1979. Mon. Not. R. Astron. Soc. **187:** 117.
24. Gott, J. R., E. L. Turner & S. J. Aarseth. 1979. Astrophys. J. In press.
25. Efstathiou, G. & J. W. Eastwood. 1979. In preparation.
26. Fry, J. N. & P. J. E. Peebles. 1979. In preparation.
27. Gott, J. R. & T. X. Thuan. 1976. Astrophys. J. **204:** 649.
28. Kormendy, J. 1977. Astrophys. J. **218:** 333.
29. Strom, S. E. & K. M. Strom. 1978. Astron. J. **83:** 73, 732, 1293.
30. Faber, S. M. & R. E. Jackson. 1976. Astrophys. J. **204:** 668.
31. Schechter, P. L. & J. E. Gunn. 1979. Astrophys. J. **229:** 472 and private communications.
32. Bertola, F. & M. Capaccioli. 1975. Astrophys. J. **200:** 439.
33. Illingworth, G. 1977. Astrophys. J. Lett. **218:** L43.
34. Davies, R. 1978. Ph.D. Thesis. Cambridge University.
35. Binney, J. 1978. Mon. Not. R. Astron. Soc. **183:** 501.
36. Binney, J. 1976. Mon. Not. R. Astron. Soc. **177:** 19.
37. Icke, V. 1973. Astron. Astrophys. **27:** 1.
38. Aarseth, S. J. & S. M. Fall. 1979. Astrophys. J. In press.
39. Sharp, N. A., D. N. C. Lin & S. D. M. White. 1979. Mon. Not. R. Astron. Soc. **187:** 287.
40. Schechter, P. 1976. Astrophys. J. **203:** 297.
41. Turner, E. L. & J. R. Gott. 1976. Astrophys. J. **209:** 6.
42. Doroshkevich, A. G., R. A. Sunyaev & Ya. B. Zeldovich. 1974. *In* Confrontation of Cosmological Theories with Observational Data (IAU Symposium No. 63). M. Longair, Ed. p. 213. Reidel. Dordrecht, Holland.
43. Silk, J. 1968. Astrophys. J. **151:** 459.
44. Bonometto, S. & S. M. Fall. 1978. Unpublished work.
45. Rees, M. J. & J. P. Ostriker. 1977. Mon. Not. R. Astron. Soc. **179:** 541.
46. Binney, J. 1977. Astrophys. J. **215:** 483.
47. Silk, J. 1977. Astrophys. J. **211:** 638.
48. Larson, R. B. 1976. *In* Galaxies (Saas-Fee, VIth Course) p. 67. Geneva Observatory. Sauverny, Switzerland.
49. Tinsley, B. M. 1976. *In* Galaxies (Saas-Fee, VIth Course) p. 155. Geneva Observatory. Sauverny, Switzerland.
50. White, S. D. M. & M. J. Rees. 1978. Mon. Not. R. Astron. Soc. **183:** 341.
51. Ostriker, J. & S. Tremaine. 1975. Astrophys. J. Lett. **202:** L113.
52. Ostriker, J. P. & M. Hausman. 1977. Astrophys. J. Lett. **217:** L125.
53. Toomre, A. & J. Toomre. 1972. Astrophys. J. **178:** 623.
54. Toomre, A. 1977. *In* Evolution of Galaxies and Stellar Populations. B. M. Tinsley and R. B. Larson, Eds.:**401.** Yale University Observatory. New Haven, Conn.
55. Silk, J. 1978. Astrophys. J. **220:** 390.
56. Tinsley, B. M. & R. B. Larson. 1979. Mon. Not. R. Astron. Soc. **186:** 503.
57. van Albada, T. S. & J. H. van Gorkom. 1977. Astron. Astrophys. **54:** 121 and private communications.
58. White, S. D. M. 1978. Mon. Not. R. Astron. Soc. **184:** 185.
59. Roos, N. & C. A. Norman. 1979. Astron. Astrophys. **76:** 75.
60. Jones, B. J. T. & G. Efstathiou. 1979. Mon. Not. R. Astron. In press.

THEORIES OF GALAXY FORMATION

Bernard J. T. Jones

Institute of Astronomy
Cambridge, CB3 OHA England

GRAVITATIONAL INSTABILITY THEORIES: A PREVIEW

The two previous speakers in this session on galaxy formation concentrated on the results of the "gravitational instability" type of theory for the origin of galaxies. It is my aim here to explain why this theory has gained such a prominent position in current thinking, and to indicate the directions in which this theory may progress and what difficulties may have to be faced. The gravitational instability theory argues that the force of gravitation has been primarily responsible for the nature of the large-scale organization of matter in the universe. Other theories, which will be discussed below, invoke turbulence, magnetic fields, or even matter–antimatter annihilation in the early universe. In comparison with these other theories, the gravitational instability picture is easy to describe in simple terms, and it has recently been further advanced by using gravitational N-body simulations of the universe.

The gravitational instability picture comes in two basic and different forms. First there is the version that argues that primeval adiabatic fluctuations in an otherwise homogeneous and isotropic Friedmann universe are the progenitors of galaxies. Prior to the epoch of recombination, these fluctuations were rather like large-scale cosmic sound waves and were subject to the dissipative processes acting in the primeval fireball. These dissipative processes serve to define a minimum surviving mass scale, the "Silk mass":[80]

$$M_S \simeq 10^{12} (\Omega h^2)^{-5/4}\, \mathrm{M}_\odot \qquad \mathbf{(1)^*}$$

It is supposed that these perturbations are the progenitors of galaxy clusters rather than galaxies. Most of the recent work on this theory has been done by the Moscow group (Doroshkevich, Novikov, Shandarin, and Zel'dovich), who refer to this as the "pancake theory."[1]

The other version of the gravitational instability theory supposes that primeval isothermal perturbations are the progenitors of galaxies. Such perturbations are not affected by the dissipative processes of the prerecombination fireball and the spectrum of perturbations that emerges after recombination directly reflects the initial state of the universe. You get out what you put it. The important characteristic mass scale in this theory is the Jeans mass just after recombination:

$$M_J = 10^6 (\Omega h^2)^{-1/2} \delta^{-1/2}\, \mathrm{M}_\odot \qquad \mathbf{(2)}$$

where

$$\delta = \rho/\langle\rho\rangle - 1$$

*Here h denotes the present Hubble constant in units of 100 km s^{-1} Mpc^{-1}. Ω is the ratio of the present density of the universe to the density that would be required to just close the universe. Currently popular values are $h = 0.5$, $\Omega = 0.1$.

0077–8923/80/0336–0191 $1.75/1 © 1980, NYAS

is the density contrast between the fluctuation and the background. The δ factor is necessary since according to current ideas the fluctuations are probably nonlinear ($\delta \gg 1$) on the scale M_J. If the spectrum of fluctuations extends to scales as small as M_J, then it is clear that in this version of the theory, galaxies form through the gravitational agglomeration of small-scale lumps. The recent work of White and Rees[2] gives a description of this process in some detail and I shall return to that later.

Other Theories of Galaxy Formation

The ability of the cosmologist to devise new schemes for galaxy formation is quite remarkable. I shall review the current status of some of the theories that are consistent with the hot big bang origin of the universe.[3–5] Of course, if we abandon the idea that the cosmic background radiation field is born with the universe, and replace it with a scheme such as that proposed recently by Rees[6] wherein the radiation field is created relatively recently, we could consider a whole new set of theories, or revive some of the ideas of Layzer[7] on galaxy formation in cold universes. The recently reported observations of the spectrum of the cosmic background radiation by Woody and Richards[8] (see the review by Muller in these proceedings) may provide some reason to take this seriously, though it should be remembered that in doing so one has to solve the problem of making the helium in the universe.

The Cosmic Turbulence Theory

For a long time this theory was considered to be the major rival to the gravitational instability picture. The idea that galaxies are the fossilized remnants of some primordial turbulence was ingeniously set in the framework of the hot big bang theory by Ozernoi and Chernin,[9,10] and developed in considerable detail by Ozernoi and Chibisov[11] and by Kurskov and Ozernoi.[12,13] The theory sought, in terms of just one free parameter, to explain not only the characteristic masses and angular momenta of galaxies, but to describe in detail the spectrum of galaxy clustering. Superficially at least, the theory succeeded in predicting exactly the slope of the two-point correlation function describing galaxy clustering (see the reviews by Fall and Peebles in these proceedings).

A number of arguments have been put forward against the theory. Fortunately, despite the inherent complexity of the phenomenon of "turbulence," the problems can be formulated in a simple manner. A fairly detailed review of the theory is presented by Jones.[14]

The "initial condition" for the turbulence theory is that the early universe deviates only slightly from homogeneity, and is endowed with a statistically isotropic random velocity field. If the maximum velocity V_{max} is subsonic, the existence of the velocity field is in principle consistent with the assumption of homogeneity. V_{max} is the one free parameter of the theory, and it was argued by Ozernoi and Chernin[9] that a good theory of galaxy formation could be constructed with an initial V_{max} (say at $t \sim 10^{-5}$ s) that is subsonic. It was Novikov and Zel'dovich[15] who first remarked that such "initial conditions" were not as simple as they appeared. If these conditions are specified at some particular time (like $t \sim 10^{-5}$ s), then the distortions of the space-time geometry diverge as $t \rightarrow 0$. Thus:

Problem 1: The initial conditions for the cosmic turbulence theory must have been very special. Indeed, the initial conditions seem to be no less *ad hoc* in this theory than in the gravitational instability theory.

The Friedmann-like appearance of the early phase of a turbulent universe is also deceptive when one considers the cosmic synthesis of light elements. Barrow[16] has discussed this in some detail and has shown that:

Problem 2: It is difficult to reconcile the cosmic turbulence theory with the observed abundances of the light elements. The important issue here is whether or not the universe is locally isotropic at the times when the neutron–proton ratio is frozen out and when protons and neutrons combine to form deuterium. If the turbulent velocity V_{max} exceeds about one percent of the speed of light, then the cosmology is non-Friedmannian throughout the whole nucleosynthesis period, and so in those situations where the turbulence is strong enough to form galaxies, it is also strong enough to upset the standard nucleosynthesis picture.

It was argued by Ozernoi and Chernin[9] that the turbulence dynamics would establish a universal ("Kolmogorov") spectrum of turbulence during the fireball phase of the universe, and that this spectrum would be independent of the initial conditions. This inherently attractive feature of the theory has been discussed in some detail by Jones,[17] Olson and Sachs,[18] and Drury and Stewart.[19] The establishment of the universal spectrum is intimately related to the dissipation of the turbulence and it seems unlikely that the universal spectrum could in fact have been created from initially subsonic cosmic turbulence. So we have:

Problem 3: The establishment of a Kolmogorov spectrum of turbulence over a significant range of scales is unlikely unless the initial turbulence was supersonic.

It is the appearance of a universal spectrum, independent of initial conditions, that allows one to *predict* the mass–radius relations for galaxies and the two-point correlation function for the clustering of galaxies. In the gravitational instability theory the spectrum of irregularities has to be specified in an *ad hoc* manner.

Kurskov and Ozernoi[12] have attempted to model the development of the Kolmogorov spectrum during the era prior to recombination by means of an "inertial range energy transfer" model of homogeneous and isotropic turbulence, such as was popular during the 1950s. They chose the "Heisenberg" model for turbulence, modified in a somewhat arbitrary manner to describe turbulence in an expanding universe. They did indeed observe the establishment of a universal spectrum (that is hardly surprising) but it is difficult to assess the significance of this result, which is based on a very simple-minded "equilibrium" theory for the turbulence spectrum.

As the universe recombines, the photon mean free path increases rapidly and the universe becomes highly viscous, albeit for a relatively short period of time. (The recombination takes place over a period on the order of 10% of the cosmic expansion time scale.) Had the recombination been instantaneous, the turbulent motions would have been thrown into a state of supersonic chaos. Such a regime is difficult to describe in anything but qualitative terms.[3,10,20,21] The fact that the recombination is not instantaneous leads to a significant amount of damping of turbulent motions, and this in fact simplifies the description of what emerges from the recombination period. It seems that all turbulent (i.e., vortical) motions on scales less than

$$M_D \simeq 10^{12}\,(\Omega h^{2})^{-7/2}\,M_\odot$$

are damped, while the surviving eddy motions (scales $M > M_D$) generate dense lumps in strong compressive shocks. These dense lumps are far too dense to be immediately associated with any known objects (galaxies or galaxy clusters, depending on scale). Thus we have:

Problem 4: What emerges from the fireball bears no obvious relationship to any known cosmic structures.

Ironically, if we estimate the two-point correlation function $\xi(r)$ describing the resulting inhomogeneity of the universe, we find[22]

$$\xi(r) \propto r^{12/7}, \tag{3}$$

where the exponent $^{12}\!/_7$ is remarkably close to that of the observed clustering function (see the reviews by Peebles and Fall in this volume). The difficulty arises with the predicted constant of proportionality, which is vastly in excess of what is observed.[23]

It has been suggested that the cosmic turbulence theory may be constrained by observations of the spectrum and small-scale anisotropy of the cosmic microwave background radiation. Anile *et al.*[24] have made a thorough study of the question of small-scale angular variations of the microwave background radiation spectrum. An upper limit on the temperature fluctuations of 10^{-4} on angular scales of $\sim 1°$ would be evidence against there being turbulent motions at the recombination epoch. The question of spectral distortions has been investigated by Sunyaev and Zel'dovich[25] and with close attention to the mechanism of turbulent heating by Chan and Jones.[26] Unfortunately, the curious nature of the distortion reported by Woody and Richards[8] makes it difficult to interpret these distortion calculations.

Baryon-Symmetric Cosmologies

With the stimulus of the work of Omnes[27] there was a considerable amount of effort devoted a few years ago to developing a baryon symmetric cosmology. The most detailed galaxy formation model within such a cosmology was produced by Stecker and Puget.[28] These authors sought to combine various attractive features of the cosmic turbulence theory with the baryon-symmetric concept, and produced a model of annihilation-driven turbulence. At that time the arguments against the whole baryon-symmetric cosmology program were not coherently organized and it was felt that the Stecker–Puget theory should be judged on its own merits. My personal feeling was that there was a lot wrong with the paper, and it would have been a better idea to have developed a galaxy formation theory without introducing the complications of turbulence. As it happened, the underlying concept of baryon symmetry was slowly being eroded by questions of nucleo-synthesis in such a universe, the gamma-ray background, and other issues that are now succinctly put in Steigman's[29] critique of the baryon-symmetric cosmologies. Although it may be argued that no one of Steigman's points in itself constitutes a convincing argument against such cosmologies, taken as a whole they present a formidable challenge to anyone who wishes to take up the subject seriously.

Primeval Magnetic Fields

From time to time it has been suggested that a primordial chaotic magnetic field might push the matter around into lumps that then collapse to form galaxies and galaxy clusters. The general picture is reviewed by Rees and Reinhardt[30] and worked out in more detail by Wasserman.[31] Whereas there seems little doubt that a seed field can be amplified to the present value of the observed galactic magnetic field, the origin of the seed field is a controversial issue. Michel and Yahil[32] suggested that the seed field originated in stars, being blown out into the interstellar medium by mass loss processes. This hypothesis might be criticized on the grounds that, since the fields are fed into the galaxy at random, it is difficult to understand the general orientation of the galactic field as deduced from observations.

Perhaps the simpler hypothesis is to suppose that the seed field is of promordial origin, though this does not necessarily mean that the field was strong enough during pregalactic times to have had a significant dynamical effect.

Prior to recombination the magnetic pressure cannot compete with the radiation pressure for a present intergalactic magnetic field $B \sim 3 \times 10^{-6}$ G; this limit is almost certainly implied by observational evidence.[30] However, after recombination, it is possible that the magnetic pressure dominates over the gas pressure. (The small residual ionization of the post-recombination cosmic plasma is sufficient to ensure a conductivity that is sufficiently high for flux freezing.) That the magnetic pressure exceeds the gas pressure can be expressed as

$$V_A \gg c_s \tag{4}$$

where $V_A = \sqrt{B^2/8\pi\rho}$ is the Alfven speed, and $c_s = \sqrt{\gamma kT/M_p}$ is the sound speed. In this situation it is the Alfven speed that determines the "Jeans length"

$$L_B \sim V_A/\sqrt{G\rho}. \tag{5}$$

Imagine the prerecombination universe to be pervaded by an irregular magnetic field, with characteristic length scale $L_{\rm rec}$ at recombination, but with the matter distributed homogeneously. When the universe recombines, the matter is pushed around by the magnetic stresses, and it moves with speed V_A. (We shall use V_A to denote the Alfven speed at $t_{\rm rec}$.) If we take the standard gravitational instability equations[3] and solve them subject to the initial conditions that $\delta\rho/\rho = 0$ at $t_{\rm rec}$, but with the matter moving at initial velocity V_A, we find that

$$\frac{\delta\rho}{\rho} \sim \left(\frac{V_A^2 t_{\rm rec}^2}{L_{\rm rec}^2}\right)\left(\frac{t}{t_{\rm rec}}\right)^{2/3} \sim \left(\frac{L_B}{L_{\rm rec}}\right)^2\left(\frac{t}{t_{\rm rec}}\right)^{2/3}, \qquad t \gg t_{\rm rec}, L \gg L_B. \tag{6}$$

(Another way of looking at this is to note that a force V_2^A/L per unit mass can generate an "initial" density perturbation $\delta_{\rm rec} \sim V_A^2 t_{\rm rec}^2$ on the gravitational time scale, i.e., $t_{\rm rec}$, from an initial condition $\delta = 0$.) Similarly, the velocity perturbation induced by the magnetic stresses is

$$u \sim \left(\frac{V_A^2 t_{\rm rec}}{L_{\rm rec}}\right)\left(\frac{t_{\rm rec}}{t}\right)^{1/3}, \qquad t \gg t_{\rm rec}, L \gg L_B. \tag{7}$$

[These last two equations correspond to Wasserman's[31] equations (16) and (17).] Hence a prescribed $L_{\rm rec}$ defines the mass scale, and the choice of the Alfven speed on that scale V_A fixes the perturbation's history, and in particular the time of recollapse to form a bound system. In the regime $L \gg L_B$, the picture differs from the gravitational instability theory only in that the magnetic stresses have imposed a vorticity field on the cosmic flow, the velocity associated with this being given by (7). It is therefore interesting to evaluate the importance of this contribution to galactic angular momentum.

When the perturbation collapses and relaxes to an equilibrium configuration, the "thermal" velocity of the stars is on the order of

$$u_{\text{grav}} \sim \alpha \sqrt{G\rho_{\text{rec}}}\, L_{\text{rec}} \left(\frac{t_{\text{rec}}}{t_{\text{coll}}}\right)^{1/3}. \tag{8}$$

Here t_{coll} is the time at which the perturbation collapses (that is fixed by δ_{rec} and the cosmological parameter Ω,[33] and α is a numerical factor whose magnitude depends on the amount of dissipation that occurs during the collapse. In dissipationless collapse, $\alpha \sim 1$. The ratio of the rotational velocity to the "thermal" velocity is thus on the order of

$$\lambda \sim \left(\frac{L_B}{L_{\text{rec}}}\right)^2 \sim \left(\frac{M_B}{M}\right)^{2/3}, \qquad \alpha \sim 1. \tag{9}$$

Since we are specifically interested in comparing the magnetically induced spin with the tidally induced spin[34,35] (see the section entitled "The Pancake Theory" below), we can only make the comparison in the vicinity of maximum expansion and set $\alpha = 1$. The enhancement of λ by subsequent collapse and dissipation depends on the nature of the collapse, but is irrelevant as regards the comparison between the magnetic and gravitational torques.

The following remarks seem pertinent:

1. The ratio L/L_B remains constant, and hence M_B is a minimum mass for bound cosmic systems.
2. M_B is an arbitrary parameter of the theory, fixed by the assumed magnetic field strength at recombination.
3. M is fixed by the characteristic length scale of the field, and this too is an arbitrary parameter. If the field has structure on all scales, it is not clear how to define a suitable M_B.
4. $\delta \sim 1$ is achieved at a redshift (cf. Equation 6)

$$z_c \sim \left(\frac{M_B}{M}\right)^{2/3} z_{\text{rec}}. \tag{10}$$

   Hence if M is a galactic mass ($\sim 10^{11}$), M_B must be $\sim 10^3$ times smaller.
5. The λ parameter is on the order

$$\lambda \sim \left(\frac{z_c}{z_{\text{rec}}}\right) \tag{11}$$

   for dissipationless collapse at redshifts $\sim z_c$. The tidally induced λ is on the order of .05–0.1, and so for z_c sufficiently large ($\sim$ 50–100), the magnetic torque can compete with the tidal torque.
6. Whereas tidal torques are independent of mass scale, the magnetic torques depend sensitively on mass. Systems with $\lambda \sim 1$ are necessarily dense ($\sim 10^{-20}$ g cm^{-3}) and among the least massive systems ($M \sim M_B$).

"Exotic" Theories

In this category come theories invoking "lagging cores" or primordial black holes[37,38] as seeds for galaxy formation. The underlying physical mechanism is gravitational instability, only here a random distribution of black holes or lagging cores is postulated *ab initio* rather than a spectrum of density perturbations as in the conventional gravitational instability picture. There is little difference between this kind of idea and standard gravitational instability; a point that needs checking is whether the black holes that are postulated could perturb the cosmic microwave background radiation spectrum.

The Dilemma of the Initial Conditions

Putting forward a particular theory of galaxy formation is equivalent to making a statement about the nature of the initial conditions for the universe. The requirements for the initial conditions can seem quite specific; consider for example the gravitational instability theory. There, the universe is supposed to have been born in a state deviating slightly from perfect homogeneity and isotropy. The level of inhomogeneity must have been significant enough to have led to the formation of galaxies by the present epoch, but not so great that most bound systems formed as long ago as a redshift of 1000. The "pancake" theories suppose further that the perturbations were vorticity free and of constant entropy per baryon. One could of course blandly assert that galaxies are what they appear to be simply because the initial conditions were that way, and had the initial conditions been something else, we would have been asking the same question in a universe that looked somewhat different from the one we have now. The opposite point of view might be that the universe started off in a completely chaotic state, and variety of physical processes acting at different times served to single out a rather special kind of perturbation. The "chaotic cosmology" program, while more satisfying that "the universe is what it is because it was what it was" philosophy, proves quite difficult to maintain without making a strong appeal to the physics at the "Planck time", $t \sim 10^{-43}$ s after the singularity.

The regime in which it is possible to say that we really understand the physics is the history of the universe at times $t > 10^{-4}$ s. At earlier times than this, the equation of state of the cosmic plasma is a matter of speculation. We can make the following points about the era $t > 10^{-4}$ s:

1. If at the time of nucleosynthesis ($t \sim 1$ min) the universe were highly chaotic, the cosmic nucleosynthesis of light elements would lead to abundances different from those deduced on the basis of present observation.[39]

2. Neutrino viscosity can certainly damp some primordial anisotropies in the cosmic expansion, as first suggested by Misner,[40] but too much anisotropy at the time of neutrino damping might lead to too high an entropy production.[41]

3. Large amplitude adiabatic fluctuations on mass scales $\ll 10^{13}\ M_\odot$ could form shock waves and isothermal perturbations.

4. Recently Hogan[42] has displayed the enormous diversity of primordial density perturbations that is consistent with observations of the clumping of matter in the universe. Most of the observational constraints on the amplitudes apply to mass scales in excess of $10^6\ M_\odot$: these must be in the linear regime ($\delta < 1$) at the time of

recombination. Almost anything goes on mass scales smaller than 10^6 $M_\odot$. The major constraint is that a lump may have a sufficiently large value of δ to upset the standard nucleosynthesis, and yet have δ too small to prevent the lump from inflating (under the influence of its own internal pressure) and then mixing its nucleosynthesis products with visible matter.

Points 1 and 2 certainly suggest that we should appeal to physical processes in the regime $t < 10^{-4}$ s to remove any significant anisotropy. Point 3 indicates that, unless there is a process limiting the amplitudes of adiabatic perturbations in the regime $t < 10^{-4}$ s isothermal perturbations will be formed on mass scales $< 10^{13}$ $M_\odot$ by the shocking of large amplitude sound waves. The limit set on the amplitude of adiabatic perturbations by the shocking process is[20,43]

$$\frac{\delta\rho}{\rho} < \left(\frac{M}{M_J}\right)^{1/3}, \qquad M \ll M_J. \tag{12}$$

This does not help to understand the origin of a spectrum of density fluctuations that decrease with mass

$$\frac{\delta\rho}{\rho} \propto M^{-\alpha}, \qquad \alpha \sim \frac{1}{2} \text{ or } \frac{1}{3}, \tag{13}$$

such as is inferred from observation[33,44]

Turning our attention to the regime $t < 10^{-4}$ s we enter a realm of still greater uncertainty. Equations of state for this early phase of the universe have been proposed by Hagedorn,[45] Carlitz, Frautschi, and Nahm,[46] and Zel'dovich.[47] It should be remarked that in spite of the importance of these models, they have not yet been fully exploited as a basis for models of the early universe. This seems particularly true of the Hagedorn model. Thus during the brief period of state, and there may well be physical processes occurring then that could for example damp anisotropies of one kind or another and thereby create entropy. It is perhaps surprising then to find that the era around the Planck time $t_P \sim (Gh/c^5)^{1/2} \sim 10^{-43}$ s is a domain of intensive investigation. Yet a potentially important result has emerged concerning the rate of production of gravitons in anisotropic universes. Zel'dovich and Starobinskii[48] have shown that in universes with small anisotropy, the rate of graviton production is proportional to the square of the Weyl tensor C_{ijkl}:

$$\frac{dn_{\text{graviton}}}{dt} \propto C_{ijkl}C^{ijkl}. \tag{14}$$

This suggests that graviton production and particle production might occur mostly in regions of space-time where the conformal curvature is largest, and hence argues in favour of a universe that had constant curvature shortly after the Planck time. We are still some way from establishing such a result rigorously; we are too ignorant of the precise nature of the gravitational particle production process. It is investigations such as these that make one optimistic that the inhomogeneity of the universe can be fully understood in terms of physical processes.

The "Pancake" Theory

If the early universe has only small amplitude primeval adiabatic perturbations, then it is well known[49,50] that the smallest surviving scale of inhomogeneity after recombination is

$$M_S \simeq 5 \times 10^{13} (\Omega h^2)^{-1} \mathrm{M}_\odot. \quad \textbf{(15)}$$

If perturbations on this scale have amplitude $\delta \sim 0.01$, they would collapse to form bound systems at redshifts $z_c \sim 10$. It is extremely unlikely that a general lump undergoing collapse would do so spherically symetrically, and in fact the collapse of an irregularly shaped cloud will be initially towards forming a plane dense structure, a "pancake." The collapse of spheroidal clouds has been discussed in a variety of ways by Lin, Mestel, and Shu,[51] Fujimoto,[52] Icke,[53] Zel'dovich,[54] and White and Silk.[55] If the material in the universe is still gaseous at the time of collapse, the pancake will consist of a hot, shocked layer of gas, in which is embedded a dense layer of shocked gas that has had time to cool. A specific one-dimensional model was presented by Sunyaev and Zel'dovich.[56] Material continues to fall onto the pancake, squeezing what was already there, and the cool dense layer grows as more time passes for more hot shocked gas to cool.

The dense cool layer can fragment and collapse gravitationally on large enough scales. The standard Jeans criterion must be modified to take account of the fact that the layer is being squeezed by a pressure $P = \rho_0 U^2$ on either side. (ρ_0 is the density in the collapsing material, and U is the collapse velocity.) If the thickness of the layer is d and its surface mass density is σ, then the critical length scale for gravitational instability is[56]

$$l_{\text{crit}} \sim d\left(\frac{P}{G\sigma^2}\right). \quad \textbf{(16)}$$

This is a scale length measured in the plane of the pancake, and note that $l_{\text{crit}} > d$: the fragments are disks. Of course, as the layer grows, σ increases and the critical mass for instability decreases. The important mass scale is the one whose collapse time scale is shorter than the time scale for the layer to grow,[57]

$$M_t \sim 2 \times 10^8\, n_0^{-1/2}\, \mathrm{M}_\odot, \quad \textbf{(17)}$$

where n_0 is the particle number density of the collapsing system before it is shocked and heated. Typically, one is talking in terms of fragment in the mass range $2 \times 10^8 - 2 \times 10^{10}$ $\mathrm{M}_\odot$, depending on the mass and density of the collapsing system. It is clear then that we must *build* galaxies by coalescing these fragments; such a model has recently been discussed by Doroshkevich *et al.*[1]

One of the features of the pancake theory is that the planar structures that arise out of this theory provide a natural explanation for the filamentary structures that are clearly seen on the deep surveys of the galaxy distribution,[58,59] and for the "holes" in the distribution of matter that is becoming apparent in redshift surveys of particular areas of the sky (e.g., Gregory and Thompson[60]). It is of course possible that the N-body simulations of the other version of the gravitational instability picture will also show filamentary structures: At present the number of particles in these simulations are too small for this purpose ($N \sim 1000$), but even so there is a suggestion of filamentary structure in the published models of Efstathiou and Jones.[34,35] The Moscow group (Doroshkevich *et al.*[61]) have recently done some two-dimensional N-body simulations for $N \sim 4000$, and this has revealed a strong filaments–holes structure. The structure is transient, so that fixes the time scale in the simulations. Of course, in a full three-dimensional simulation we would expect the structure to be diluted by projection effects.

The following comments may suggest some future investigations on this promising theory.

1. Does the existence of a preferred mass scale M_s in the universe necessarily imply a feature in the observed galaxy clustering correlation function at some special length scale associated with M_s? Since (a) the run of density in clusters of galaxies appears to fall off roughly as r^{-2} and (b) we see the sky in projection, I suspect that there will not be a strong feature, but at worst a curving of ξ that will in any case be difficult to rule out observationally.

2. If the shocked material shrinks in its own plane with dissipation, then the λ parameter measuring the ratio of rotational velocity to velocity dispersion could become large enough to be measurable. The pancakes inevitably acquire angular momentum through tidal torques and the present observations of galaxy clusters show no detectable rotation.

3. The theory has not answered the question of how galaxies themselves form. The recent work of Doroshkevich *et al.*[61] shows how complex the galaxy building process could be in such a theory.

The Origin of Galactic Spin

If we are to argue that galaxies form from primordial density perturbations in a universe that had no local vorticity, then we face the task of explaining where galactic angular momentum comes from. In the adiabatic density perturbation theory where galaxy clusters form first there is the possibility of forming spinning fragments in the shock front. Since the fragment mass is smaller than the mass typically assigned to a large galaxy, galaxy formation must proceed via a merging of fragments, and so estimation of the angular momentum could be rather difficult. The following estimate seems to be the simplest, though not without problems! If we consider a perturbation of mass M and scale length r, which collapses to form a pancake, the typical velocity gradient along the pancake will be on the order of $\lambda_0\sqrt{G\rho_0}$, ρ_0 being the density of the perturbation just at the time of collapse, and λ_0 being a numerical factor depending on shape and other things. This of course assumes that the protocluster is somewhat irregular in shape. The relative velocity, normal to the pancake, of two points of the pancake separated by a distance l is thus $\sim \lambda_0 l\sqrt{G\rho_0}$. If a region of scale l in the pancake ends up in one galaxy, its angular momentum is on the order of

$$\mathcal{H} \sim \lambda_0 l^2 d\rho_{sh}\, l^2\sqrt{G\rho_0} \qquad \textbf{(18)}$$

where d is the thickness of the cool fragmenting layer and ρ_{sh} its density. It is customary to define a dimensionless measure of the rotation velocity by

$$\lambda = \mathcal{H}E^{1/2}G^{-1}M^{-5/2} \qquad \textbf{(19)}$$

for a mass M of total energy E. λ can be thought of as the ratio of the rotation velocity to the velocity dispersion. In the absence of dissipation during the collapse and relaxation of the pancake, we can take $E \sim GM^2/l$, whence

$$\lambda \sim \left(\frac{l}{d}\right)^{1/2}\left(\frac{\rho_0}{\rho_{sh}}\right)^{1/2}\lambda_0, \qquad \textbf{(20)}$$

d being the thickness of the layer. The quantity ρ_{sh}/ρ_0 is just the square of the shock Mach number, since the shock is strongly cooling, and must be roughly the ratio of the density in a galaxy to the typical density of a galaxy cluster (whatever that is!). At the time of fragmentation, l/d for a fragment is also approximately equal to the Mach number,[62] and so at the time the pancake fragments, $\lambda \sim \lambda_0$. In fact λ_0 measures the velocity gradient along the pancake, and since this velocity gradient is a consequence of shape irregularity and the tides from neighbouring protoclusters, this can be calculated from the tidal torques theory.[35,63] The median value of λ_0 is ~ 0.06. So we have the not too surprising result that at the time the dense layer fragments, the smallest fragments have

$$\lambda \sim 0.06. \tag{21}$$

If galaxies formed before clusters, then we can only appeal to the tidal interactions between neighboring protogalaxies as the source of galactic angular momentum. The magnitude of this nonlinear effect was estimated in the context of the gravitational instability theory by Peebles.[63] Since then it has been possible to examine the tidal transfer of angular momentum in more detail by running gravitational N-body simulations of the universe[34,35] using Aarseth's[64] Ahmad–Cohen integrational program. In terms of the λ parameter (Equation 19), the mean rotation generated by tidal torques has

$$\lambda \sim 0.06, \tag{22}$$

independent of mass. This value agrees remarkably well with Peebles' prediction.[63] Since the N-body simulations are dissipationless, this λ value is in fact a lower limit. An initial of λ of 0.06 would be consistent with the present rotation of our galaxy if the Galaxy collapsed from a radius 500 kpc. What is remarkable about this "low" value of λ is that it is consistent with observations of the rotations of elliptical galaxies.[78] It therefore implies that elliptical galaxies formed by dissipationless collapse, and that a considerable amount of dissipation has taken place during the formation of disk systems. For a full discussion of the implications of the tidal theory of angular momentum transfer see the survey by Efstathious and Jones.[34,35]

Theories Starting from Isothermal Perturbations

For small amplitude density perturbations, the Jeans mass just after recombination is

$$M_J \simeq 10^6 \, (\Omega h^2)^{-1/2} \, M_\odot. \tag{23}$$

Peebles and Dicke[65] noted the similarity of this mass to the masses typically assigned to globular clusters, and proposed a galaxy formation theory exploiting this. Fall and Rees[66] took this a stage further and discussed processes that would cull the protoglobular cluster population to that observed. Thus systems that were much more massive than M_J would be absorbed in the central regions of the galaxy by dynamical friction. Clusters that were too loosely bound would have been disrupted by tidal fields or galactic disk shocks. They were certainly successful in showing what physical processes have acted to define the parameters of the present population of galactic globular clusters, and at the same time provided a possible explanation as to why the stellar densities in the cores of elliptical galaxies and globular clusters should be so similar. The question of whether the globular cluster population is the remnant of the primeval building blocks of the galaxy is nevertheless an open one. Some tentative arguments against the Fall–Rees idea have been put forward by Gunn[67]: it is difficult to account for the observed metallicity gradients in the globular cluster population, it is difficult to understand why the density of the globular cluster component falls off as r^{-3} or $r^{-3.5}$, and the mass M_J is in fact somewhat on the large side for the globular cluster population as a whole.

White and Rees[2] have presented a theory of galaxy formation in which galaxies are gradually assembled from some basic building blocks. In this theory it is assumed that a large fraction of the mass of the universe forms into "dark material" at some early stage. The precise nature of the dark material is irrelevant; it could be black holes, "Population III" stars, or anything else that does not contribute to the present light in the universe. It is only assumed that this material is inhomogeneously distributed throughout the universe, and thus provides a set of potential wells into which the remnant gas can fall. Two physical processes operate simultaneously. The dark matter clusters under the influence of the forces of gravity: thus, neighboring lumps of dark material continually merge to form larger lumps of dark material. At the same time the remnant gas cascades into these potential wells, forming stars when its density becomes great enough for it to become self-gravitating. It is these stars that comprise the luminous component of the observed universe. An important aspect of this theory is that it provides an explanation for the apparent upper limit on galaxy masses. This upper limit arises because there is a maximum mass above which a gas cloud that is in thermal equilibrium takes longer to cool than its gravitational contraction time scale. Such a cloud can only deflate in the cooling time for the gas. This maximum mass is on the order of 10^{12} $M_\odot$, and smaller masses of gas can never be quasi-statically pressure supported unless they are cooler than 10^4 K. This important fact is due to Silk[68] and to Rees and Ostriker.[69]

There are a number of complications that must be taken into account. The most important is that in order for the gas to become self-gravitating it must dissipate energy and as it collapses it inevitably spins up. Thus the ratio of the rotation velocity to the random velocity increases above the value $\lambda \sim 0.06$, which is tidally induced (see the preceding section). This is no problem for disk galaxies where $\lambda \sim 1$, but it may be a serious difficulty if the bright parts of elliptical galaxies have formed by dissipative collapse since they are observed to have $\lambda \sim 0.06$.[34,35,70,71] Of course it might be that a substantial amount of star formation takes place before the gas has shrunk appreciably in radius: stars could form during the merging of two systems in strong shocks involving the remnant gas, or stars could form if the gas collapses to a disk first rather than undergoing a quasi-spherical collapse. However, it is difficult to argue that in some galaxies the gas undergoes a substantial reduction in radius accompanied by spin-up, whereas in others it does not do that. Although such a statistical approach to what makes the two great families of galaxies is inherently quite acceptable, it leaves unanswered the question of why elliptical galaxies are found preferentially in the cluster environment.

Recently Aarseth and Fall[72] and Jones and Efstathiou[76] have used N-body programs to simulate the hierarchical agglomeration of matter. It would be premature to discuss these results at this stage, but one feature of the simulations that is very striking is that the rate at which galaxies are built is much higher in the cluster environment than outside. Consequently, the most massive galaxies are to be found in clusters. This suggests the vague notion that "ellipticals are merged spirals," an idea that was first put forward by Toomre.[73] It is certainly true that a merger between spirals embedded in dark halos could wreck the gaseous components of the disk structures. What happens to the underlying stellar components of the disks must depend on a number of factors such as the relative directions of their rotation axes, but if the dynamics of the stellar disk is dominated by the halo rather than its own self-gravity, it seems reasonable to expect a thorough mixing of the luminous disk material as a consequence of the collision. The new galaxy will thus possess a luminous stellar system whose angular momentum derives from the relative orbital angular momentum of the merging systems and the spins of the two galaxies, together with dust and gas that will form the next generation of stars. Two points now have to be dealt with:

1. The luminous material came from two rapidly rotating disks crashing together at a random angle. Is the resulting system rapidly rotating or not?

2. If the gas that was initially in the disks is not removed it will collapse, dissipate, and form a new disk system in the center of the new galaxy.

As regards point 2, one could argue that since this is happening in the cluster environment, there may be enough cluster gas to strip the galaxy of its gas before the new disk forms. Gisler[74] has discussed this possibility in some detail. Point 1 remains to be answered, but it seems crucial if we are to maintain the idea that the luminous material of galaxies has been formed by the cascading of gas into ever-growing potential wells made up of dark material. In a different context, Brosche[75] suggested that galaxies might *acquire* angular momentum via a few mergers. His argument might be used the other way round: If N galaxies endowed with angular momenta of magnitude $\mathcal{H}$ are merged, then, for randomly oriented angular momentum vectors, the angular momentum of the new system is typically $N^{1/2}\mathcal{H}$. Hence the merging of many spiral systems could produce a low angular momentum luminous system from the stars that made up their stellar disks.

Acknowledgments

I am grateful to the Scientific Organizing Committee of the 9th Texas Symposium for the opportunity of presenting this review. I have benefitted from numerous discussions with colleagues at the Institute of Astronomy, and would particularly like to thank the research students who are working with me on this subject: George Efstathiou, Phil Palmer, and Rosie Wyse.

References

1. Doroshkevich, A. G., S. F. Shandarin & E. Saar. 1978. Mon. Not. R. Astron. Soc. **184:** 643.
2. White, S. D. M. & M. I. Rees. 1978. Mon. Not. R. Astron. Soc. **183:** 341.
3. Peebles, P. J. E. 1971. Physical Cosmology. Princeton University Press. Princeton, N.J.
4. Peebles, P. J. E. 1971. Astrophys. Space Sci. **11:**443.
5. Weinberg, S. 1972. Gravitation and Cosmology. Wiley. New York, N.Y.
6. Rees, M. J. 1978. Nature (London) **275:** 35.
7. Layzer, D. 1964. Annu. Rev. Astron. Astrophys. **2:** 341.

8. Woody, D. P. & P. L. Richards. 1979. To appear.
9. Ozernoi, L. M. & A. D. Chernin. 1968. Sov. Astron. AJ **11:** 907.
10. Ozernoi, L. M. & A. D. Chernin. 1969. Sov. Astron. AJ **12:** 901.
11. Ozernoi, L. M. & G. V. Chibisov. 1971. Sov. Astron. AJ **14:** 615.
12. Kurskov, A. A. & L. M. Ozernoi. 1974. Sov. Astron. AJ **18:** 157.
13. Kurskov, A. A. & L. M. Ozernoi. 1974. Sov. Astron. AJ **18:** 300.
14. Jones, B. J. T. 1976. Rev. Mod. Phys. **48:** 107.
15. Novikov, I. D. & Ya. B. Zel'dovich. 1970. Astrofizika. **6:** 379.
16. Barrow, J. D. 1977. Mon. Not. R. Astron. Soc. **178:** 625.
17. Jones, B. J. T. 1973. Astrophys. J. **181:** 269.
18. Olsen, D. W. & R. K. Sachs. 1973. Astrophys. J. **185:** 91.
19. Drury, L. O. C. & J. M. Stewart. 1976. Mon. Not. R. Astron. Soc. **177:** 377.
20. Jones, B. J. T. 1971. Ph.D. Thesis. Cambridge University.
21. Jones, B. J. T. & P. J. E. Peebles. 1972. Comments Astrophys. Space Phys. **4:** 121.
22. Jones, B. J. T. 1977. Mon. Not. R. Astron. Soc. **180:** 151.
23. Fall, S. M. 1979. Rev. Mod. Phys. To appear.
24. Anile, A. M., L. Danese, G. de Zotti & S. Motta. 1976. Astrophys. J. **205:** L59.
25. Sunyaev, R. A. & Ya. B. Zel'dovich. 1970. Astrophys. Space Sci. **7:**20.
26. Chan, K. L. & B. J. T. Jones. 1975. Astrophys. J. **200:** 461.
27. Omnes, R. 1971. Astron. Astrophys. **10:** 228.
28. Stecker, F. W. & J. L. Peget. 1972. Astrophys. J. **178:** 57.
29. Steigman, G. 1977. Annu. Rev. Astron. Astrophys. **14:** 339.
30. Rees, M. J. & M. Reinhardt. 1972. Astron. Astrophys. **19:** 189.
31. Wasserman, I. 1978. Astrophys. J. **224:** 337.
32. Michel, F. C. & A. Yahil. 1973. Astrophys. J. **179:** 771.
33. Gott, J. R. & M. J. Rees. 1976. Astron. Astrophys. **45:** 365.
34. Efstathiou, G. & B. J. T. Jones. 1979. Mon. Not. R. Astron. Soc. **186:** 133.
35. Efstathiou, G. & B. J. T. Jones. 1979. Comments Astron. Space Phys. In press.
36. Novikov, I. D. 1965. Sov. Astron. AJ **8:** 857.
37. Carr, B. 1975. Astrophys. J. **201:** 1.
38. Meszaros, P. 1974. Astron. Astrophys. **37:** 225.
39. Barrow, J. D. 1977. Mon. Not. R. Astron. Soc. **179:** 47P.
40. Misner, C. W. 1968. Astrophys. J. **151:** 431.
41. Caderni, N. 1979. *In* Springer Lecture Notes in Physics. M. Demianski, Ed. Springer-Verlag. Berlin.
42. Hogan, C. 1978. Mon. Not. R. Astron. Soc. **185:** 889.
43. Liang, E. P. 1977. Astrophys. J. **211:** 361.
44. Peebles, P. J. E. 1974. Astrophys. J. Lett. **189:** 51.
45. Hagedorn, R. 1970. Astron. Astrophys. **5:** 184.
46. Carlitz, R., S. Frautschi & W. Nahm. 1973. Astron. Astrophys. **26:** 171.
47. Zel'dovich, Ya. B. 1972. Mon. Not. R. Astron. Soc. **160:** 1P.
48. Zel'dovich, Ya. B. & A. A. Starobinskii. 1978. JETP Lett. **26:** 273.
49. Peebles, P. J. E. & J. T. Yu. 1970. Astrophys. J. **162:** 815.
50. Bonometto, S. A., G. F. de Zotti & F. Lucchin. 1977. Astron. Astrophys. **54:** 405.
51. Lin, C. C., I. Mestel & F. H. Shu. 1965. Astrophys. J. **142:** 1431.
52. Fujimoto, M. 1968. Astrophys. J. **152:** 523.
53. Icke, V. 1973. Astron. Astrophys. **27:** 1.
54. Zel'dovich, Ya. B. 1970. Astron. Astrophys. **5:** 84.
55. White, S. D. M. & J. I. Silk. 1979. Astrophys. J. **231:** 000.
56. Sunyaev, R. A. & Ya. B. Zel'dovich. 1972. Astron. Astrophys. **20:** 189.
57. Jones, B. J. T. & R. Wyse. 1979. In preparation.
58. Shane, C. D. & C. A. Wirtanen. 1967. Publ. Lick Obs. **22** (Part 1): 49.
59. Soneira, R. M. & P. J. E. Peebles. 1978. Astron. J. **83:** 845.
60. Gregory, S. A. & L. A. Thompson. 1978. Astrophys. J. **222:** 784.
61. Doroshkevich, A. G., E. V. Kotok, I. D. Novikov, A. N. Polyudov, S. F. Shandarin & Yu. S. Sigov. 1979. Preprint.
62. Jones, B. J. T., P. I. Palmer & R. Wyse. 1979. In preparation.
63. Peebles, P. J. E. 1969. Astrophys. J. **155:** 393.

64. AARSETH, S. J. 1979. In preparation.
65. PEEBLES, P. J. E. & R. H. DICKE. 1968. Astrophys. J. **154:** 891.
66. FALL, S. M. & M. J. REES. 1976. Mon. Not. R. Astron. Soc. **181:** 37.
67. GUNN, J. E. 1978. *In* Cambridge NATO Summer Conference on Globular Clusters. D. Hanes and B. Madore, Eds.
68. SILK, J. I. 1977. Astrophys. J. **211:** 638.
69. REES, M. J. & J. P. OSTRIKER. 1977. Mon. Not. R. Astron. Soc. **179:** 451.
70. ILLINGWORTH, G. 1977. Astrophys. J. Lett. **218:** 43.
71. GUNN, J. & SCHECHTER. 1979. Astron. J. To appear.
72. AARSETH, S. J. & S. M. FALL. 1979. In preparation.
73. TOOMRE, A. 1977. *In* Evolution of Galaxies and Stellar Populations. R. B. Larson & B. M. Tinsley, Eds. Yale University Observatory.
74. GISLER, G. 1979. Astrophys J. **228:** 385.
75. BROSCHE, P. 1970. Astron. Astrophys. **6:** 240.
76. JONES, B. J. T. & G. EFSTATHIOU. 1979. Mon. Not. R. Astron. Soc. To appear.
77. DOROSHKEVICH, A. G., E. M. SAAR & S. F. SHANDARIN. 1977. *In* The Large Scale Structure of the Universe. (IAW Symp. No. 79) M. S. Longair & J. Einasta, Eds.: 423.
78. SCHECHTER, P. & J. E. GUNN. 1979. Astron. J. In press.
79. ZEL'DOVICH, YA. B. 1977. *In* The Large Scale Structure of the Universe. (IAU Symp. No. 79) M. S. Longair & J. Einasta, Eds.: 409.
80. SILK, J. 1968. Astrophys. J. **151:** 459.

SUMMARY OF WORKSHOP ON GALAXY FORMATION

Joseph Silk

Department of Astronomy
University of California, Berkeley
Berkeley, California 94720

The theory of the formation of the galaxies is confronted by at least three outstanding problems of cosmogony. These may be summarized as (a) the origin of the fluctuations in the early universe from which galaxies arose; (b) the nature and origin of the dark matter that apparently comprises the dominant mass constituent of many galaxies, galaxy clusters, and the universe itself; and (c) the origin of the morphology and general characteristics of the galaxies. All of these problems are currently unsolved and are fundamental to any approach to cosmology, both theoretical and observational. For when we survey the distant reaches of the universe, we are looking back in time to epochs when galaxies were young, and the problems of cosmogony become inextricably coupled with the broader issues of cosmology.

A number of interesting ideas were presented at the Galaxy Formation Workshop that attempt to grapple with these vexing questions. No attempt will be made in this brief summary to present a literal transcription of the opinions presented at the Workshop, nor will all contributors and all viewpoints be covered. Rather, a subjective assessment will be given of work that the author found most relevant to the issues of galaxy formation.

ORIGIN OF THE FLUCTUATIONS

According to G.J. Lasher (I.B.M.), a quark–nucleon phase transition can occur in certain models of the very early universe. It is necessary that the initial state be that of a cold uniform quark fluid expanding according to the standard Friedmann model. Provided that the quark–nucleon phase transition is of first order, the quark phase will nucleate small bubbles of nucleon gas by quantum fluctuations. This is depicted as a tunnelling process, with the bubbles having a Poisson distribution in space and time. The bubbles grow as spherical detonation waves until the entire space is filled, there being sufficient power at long wavelengths to yield bubble mass scales of astrophysical interest. The average density of nucleation centers fixes the mass of each bubble, and this is determined according to a free parameter of the model, namely the probability per unit volume for the nucleation process. The quark fluid irreversibly converts to nucleons yielding the cosmic background radiation, and adjustment of the nucleation parameter determines the observed value for the entropy of the cosmic background radiation. This argument fixes the characteristic baryon mass per bubble at $\sim 10^6\ M_\odot$, baryon density fluctuations being generated with this characteristic mass. One interesting aspect of Lasher's result is that his characteristic mass can be expressed as $\sim (\alpha m_p/\alpha_g m_e)^{3/2}\ (\ln t_1/t_n)^3$, where α is the fine-structure constant, $\alpha_g\ (\equiv Gm_p^2/\hbar\ c)$ is the gravitational fine-structure constant, t_n is the epoch at which nuclear density is attained ($\sim 10^{-4}$ s), and t_1 is a model-dependent parameter that denotes the time when

0077-8923/80/0336-0206 $1.75/1 © 1980, NYAS

the nucleon bubbles have all coalesced ($t_1 \sim 2000$ s yields the correct entropy per baryon ratio according to Lasher's calculations). The dependence of a seed mass for galaxies on a large parameter that enters logarithmically is a definite deterrent to cosmologists who pursue the anthropic principle.

A possible difficulty with this theory for the fluctuation origin is that it relies on a specific, and still speculative and unestablished, model of strong interactions. However, it is undoubtedly premature to reject this source of density fluctuations on such grounds.

Galaxy Clustering: A Probe of the Spectrum of Primeval Density Fluctuations?

While the theory of the origin of the seed fluctuations from which galaxies sprang remains in a very unsatisfactory state, considerable progress has recently been made on defining the problem that has to be solved. Studies of galaxy clustering combined with observations of the galaxy correlation functions have been performed which provide information on the spectrum of primordial density fluctuations immediately after decoupling. Results of N-body simulations of galaxy clustering were reported by R. Gott (Princeton) and G. Efstathiou (Durham). It seems clear that the slope of the two-point correlation function is dependent on at least two parameters : the cosmological density parameter Ω and the initial conditions assumed for the density fluctuations. If Ω were known, one could therefore infer the initial fluctuation spectrum with some confidence. If $\Omega \sim 1$, the fluctuation spectrum at recombination must be of Poisson form ($\delta\rho/\rho \sim M^{-1/2}$). Gott found that $\Omega = 0.1$ is also compatible with observations provided that the initial fluctuation spectrum has more power at larger scales ($\delta\rho/\rho \sim M^{-1/3}$).

Other workers in the field have taken a somewhat different approach. S. Bonometto and F. Lucchin (Padua) have developed a hierarchical model of galaxy clustering, and they show that this model can account in a simple manner for the higher-order correlation functions. On the other hand, T. Shanks (Durham) argued that a power-law clustering model, in which clusters exist on only one large scale and possess a universal profile that declines as a power-law of the radius, can equally well simulate the observations.

Observational studies of correlation functions have now been extended to magnitudes as deep as $J = 25$. R. Ellis (Durham) has found that the galaxy angular correlation function satisfies the virial scaling relations to $J \approx 22$ to within about 30%. For a fainter sample, based on a study of a single 4-meter plate of a blank field, discrepancies are found, although the power-law form of the correlation functions holds to very small separations ($r \gtrsim 0.04\ h^{-1}$ Mpc); sampling problems cannot however yet be eliminated for the small fields studied until more 4-meter plates have been analyzed.

Galaxy Evolution: Observational Discriminants

According to P. Katgert and H. van der Laan (Leiden), faint radio galaxies show strong color evolution, being considerably bluer than nearby samples. This result has to be reconciled with the results of galaxy counts to an isophotal magnitude $J = 25$, hitherto performed on a single deep 4-meter plate and reported by R. Ellis, that reveal little or no evidence for evolution. The slope of the counts at 23.0 magnitude was found to be 0.455 ± 0.005 as compared with the predicted value of 0.40 for models with no evolution; the faint counts are not believed to be seriously affected by star/galaxy separation problems. The effect of uncertainties in K-corrections, galaxy mix, and luminosity functions has been studied from detailed calculations that allow for seeing and surface brightness effects. Since radio galaxies form a small ($\lesssim 10\%$) fraction of the total sample, one could presumably allow stronger evolution for this subset of galaxies; however, because of the strong redshift dependence of the evolution reported by the Leiden workers, one might expect the effect to show up in the faint galaxy counts, particularly when plates in two or more colors are analyzed.

One might also expect that galaxy evolution would result in a substantial energy input to intergalactic gas in the vicinity of the galaxies. C. Hogan and M. Rees (Cambridge) described a new technique for investigating nonuniform gas emission at a large redshift. Their idea is that such gas would have a unique signature because of the correlation between its spectral structure (due to line emission at 21 cm or in the Ly-α line) and angular structure (due to the clumpiness of the gas and galaxy distribution). This might enable one to distinguish it from the other sources of background radiation.

Galaxy Collisions

One of the most exciting developments in galaxy formation research has been the work on colliding galaxies. The original suggestion that such a system as the Antennae (NGC 4038/9) could be explained this way can be traced back to early papers by Zwicky (1956),[1] but the first detailed computations are due to A. and J. Toomre (1972).[2] This work emphasized parabolic encounters. Results reported at the workshop by C. Norman (Leiden) demonstrate that many ellipticals in clusters can be formed from mergers. Substantial mass ejection by halo material occurs, and the cores of the two colliding galaxies (with parabolic trajectories) merge into slowly rotating elliptical galaxies.

Another important implication of the collision hypothesis is that various other observed characteristics of galaxies can be explained. By means of successive mergers, the luminosity function of galaxies can be accounted for in a natural manner. H. Zinnecker, R. Cowsik, and S. Drapatz (Munich) reported on a Fokker-Planck approach to this problem, which results in a Gaussian function in logarithmic mass space (i.e., $\exp -[\log(M/M_0)]^2$) as a consequence of initial injection with a characteristic mass M_0 at the onset of galaxy clustering. Galaxy mergers can also account for the color distribution of cluster galaxies if one makes the simple hypothesis that each collision results in a burst of star formation and associated enrichment.[3] Galaxy collisions also provide a release mechanism for the enriched intracluster gas, which presumably is of galactic origin: supernova-driven winds evidently fail to accomplish this necessary task.[4]

A. di Fazio, A. Tornambe, and F. Vagnetti (Frascati) described a hydrodynamical treatment of the gravitational instability of a self-gravitating collapsing fluid model of the galaxy. Realistic heating and cooling functions were included. The calculations are still in progress, but one of the aims is to demonstrate that fragments can form of mass comparable to those of star clusters.

It seems clear that the halo stars must have formed in star clusters that have long since dispersed. The dark matter that apparently dominates the dynamics of galaxy halos has to be incorporated into such models. Attempts along this line have been made by J. Shaham (Jerusalem), who reported on models of extended galactic halos in which the stellar population consists of old stars, and the dark material is largely in objects that are massive planets or very low luminosity stars. According to one model, these objects formed in star clusters that subsequently merged together to form galaxies.

Several other significant problems were discussed at the workshop. The titles of these contributions are listed at the end of this report. I apologize to the authors for not covering them in these remarks, but as stated at the outset, I have only attempted to highlight what I subjectively consider to be a few of the most crucial problems in galaxy formation theory. Needless to say, the workshop resulted in no firm conclusions. Perhaps one can best summarize it by echoing Thomas Carlyle: "I don't pretend to understand the Universe—it's a great deal bigger than I am."

Papers Presented at the Workshop on Galaxy Formation

Bonometto, S. A. & F. Lucchin (Padua): A model for galaxy clustering.
Falle, S. (Leeds) & P. Mészáros (Munich): Infall of gas into galaxy clusters.
di Fazio, A., A. Tornambe & F. Vagnetti (Frascati): Hydrodynamical treatment for gravitational instabilities in a self-gravitating, collapsing fluid.
di Fazio, A., F. Vagnetti & J. R. Wilson (Frascati): Multifluid hydrodynamics for chemical evolution of EO galaxy.
Efstathiou, G. (Durham): Galaxy clustering Ω-dependence.
Ellis, R. S. (Durham): Correlation studies of deep galaxy samples.
Gerbal, D. & E. S. Solé (Meudon): Secondary maximum in density curves of clusters of galaxies.
Gott, R. (Princeton): Spectrum of primeval density perturbations.
Hogan, C. & M. J. Rees (Cambridge): New technique for detecting intergalactic gas at high redshift.
Hurwitz, H. & R. Baierlein (Middletown): Dissipation of cosmic magnetic fields.
Jehle, H. (Munich): Statistical dynamics of gravitational systems.
Katgert, P. & H. van der Laan (Leiden): Color evolution of faint radio sources.
Lasher, G. (Yorktown): New source for density perturbations: the quark–nucleon transition.
Norman, C. (Leiden): Galaxy collisions.
Nowotny, E. (Konstanz): Phase transition of second order and the beginning of the matter era.
Palmer, P. (Cambridge): Fourier filtering in numerical hydrodynamics.
Shaham, J. (Jerusalem): Galaxy halos and their formation.
Shanks, T. (Durham): Statistical measures as discriminants between theories of galaxy clustering.
Zinnecker, H., R. Cowsik & S. Drapatz (Munich): A Fokker-Planck approach to the I.M.F. of stars and galaxies.

References

1. Zwicky, F. 1956. Ergeb. Exacten Naturwiss. **19:** 344.
2. Toomre, A. & J. Toomre. 1972. Astrophys. J. **178:** 623.
3. Silk, J. 1978. Astrophys. J. **220:** 390.
4. Bregman, J. N. 1978. Astrophys. J. **224:** 768.

COS-B OBSERVATION OF THE MILKY WAY IN HIGH-ENERGY GAMMA RAYS

H. A. Mayer-Hasselwander,* K. Bennett,† G. F. Bignami,‡
R. Buccheri,* N. D'Amico,§ W. Hermsen,¶ G. Kanbach,*
F. Lebrun,‖ G. G. Lichti,† J. L. Masnou,‖ J. A. Paul,‖ K. Pinkau,*
L. Scarsi,§ B. N. Swanenburg,¶ and R. D. Wills†

**Max-Planck-Institut für Extraterrestrische Physik*
Garching bei München, Federal Republic of Germany
†Space Science Department of the European Space Agency, ESTEC
Noordwijk, the Netherlands
‡Laboratorio di Fisica Cosmica e Tecnologie Relative del C.N.R.
Università di Milano
Milan, Italy
§Laboratorio di Fisica Cosmica e Tecnologie Relative del C.N.R.
Università di Palermo
Palermo, Italy
¶Cosmic Ray Working Group, Huygens Laboratory
Leiden, the Netherlands
‖Service d'Electronique Physique
Centre d'Etudes Nucléaires de Saclay
Saclay, France

INTRODUCTION

The Caravane Collaboration's gamma-ray astronomy experiment aboard ESA's satellite COS-B has been recording celestial gamma rays in the energy range from about 50 MeV to several GeV since August 1975. The experiment features a spark chamber for gamma-ray event identification and determination of incidence direction, a CsI energy spectrometer for the measurement of gamma-ray energy and an assembly of particle counters for event selection and triggering. A description of the experiment is given by Bignami *et al.*[1] and its characteristic parameters are given by Scarsi *et al.*[2]

Most of the 37 observations, of typically one-month duration, which were performed up to the end of 1978, were devoted to the study of the galactic disc. The 20 observations listed in TABLE 1 have been used for the present analysis. These observations cover the whole range of galactic longitude, thus making it possible to present here the first complete detailed gamma-ray survey of the Milky Way with greatly improved statistical accuracy and significantly better energy measurement than in the previous survey by the SAS-2 mission.[3]

The present work concentrates on the spatial aspects of the gamma radiation, including localized sources, while the detailed analysis of spectral properties[18] and possible time variations are not sufficiently advanced to be reported here.

0077–8923/80/0336–0211 $1.75/1 © 1980, NYAS

TABLE 1

OBSERVATIONS USED IN THE ANALYSIS

Observation	Pointing l (deg)	Direction b (deg)	Epoch From	Epoch To
1	184.0	−6.1	1975—08–17	1975—09–17
2	5.2	0.0	09–17	10–20
3A	263.5	−3.3	10–20	11–08
3B	262.7	+3.5	11–08	11–28
4	73.9	+0.3	11–28	12–24
5	291.6	+0.2	12–24	1976—01–23
7	321.5	+0.3	1976—02–23	03–24
8	25.7	+1.5	03–24	04–24
9	44.9	−2.1	04–24	05–24
11	125.8	+1.1	06–24	07–24
13	352.1	+8.5	08–21	09–30
14	195.3	+4.3	09–30	11–02
16	101.0	−0.9	12–10	1977—01–17
17	155.4	0.0	1977—01–17	02–24
18	5.2	−1.0	02–24	03–07
19	219.5	0.0	03–07	04–14
21	243.3	−2.4	05–02	06–08
22	84.1	+0.5	06–08	07–15
24	345.3	−4.5	08–18	09–23
25	35.3	+0.3	09–23	11–01
26	60.0	+0.1	11–01	12–07
28	140.9	+0.1	1978—01–13	1978—02–22

METHOD OF ANALYSIS

All the gamma-ray events used have been selected according to criteria which were defined to give a good compromise between sensitive area, field of view, angular resolution, energy measurement and background rejection. The spark-chamber data were analyzed by an automatic pattern-recognition program which identified likely gamma-ray events. These were subsequently visually scanned in order to remove background and improve the accuracy of the direction determination. FIGURE 1 shows the basic experimental characteristics for paraxial incidence and the particular selection criteria adopted in the present analysis. The energy bands used in the presentation are also indicated. The band boundaries were chosen to give sufficient statistical accuracy (similar numbers of gamma rays were recorded in each range) and to allow the assumption of a constant angular resolution (point-spread function) in each of the three intervals. The field of view was limited to an 18° half-cone in order to minimize possible systematic errors in measuring gamma rays at large angles of incidence with respect to the experiment axis without significant loss of overall sensitivity.

Special investigations were conducted on individual observation-period data sets in order to assess and correct for variations of experiment sensitivity caused by aging of the spark-chamber gas and by occasional experiment malfunctions. Such corrections, of a magnitude between 5 and 20%, were applied to 10 of the observations used.

In total, about 64,000 gamma-ray events were finally accepted and processed in order to derive contour maps and profiles of the gamma-ray intensity. The data were grouped in 0.5° × 0.5° bins so as not to lose fine structure in regions of high intensity. The resulting small average number of events per bin necessitated a careful smoothing to make a clear overall presentation possible.

The smoothing procedure developed and used was required to represent as perfectly as possible a data surface in the sky map whose structure is limited by the extension of the experiment point-spread function which, in this case, is strongly energy dependent, i.e., it should not suppress the sharpest physically possible peaks. On the other hand, in order to have a smoothing effect, the procedure should not reproduce significantly narrower peaks, which evidently can only be due to statistical fluctuations.

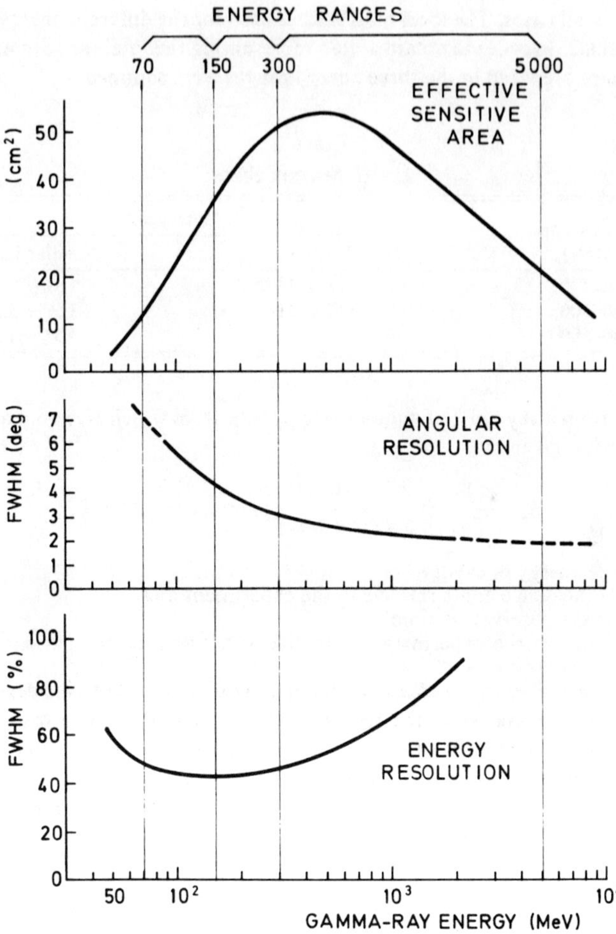

FIGURE 1. Characteristic parameters of the experiment for the data selection requirements applied in this analysis and for paraxial incidence of gamma rays. The angular resolution is indicated by the FWHM of the two-dimensional point-spread function.

It has been found that the use of two-dimensional polynomials across such a large matrix of data points leads either to flattening or to oscillation. If small regions are treated individually, the use of spline functions also results in undesired flattening of real structures. The following satisfactory method has been developed. A data matrix A, describing the sky map within each energy band, is produced. The data show statistical fluctuations. Each element of A is smoothed by the following process: a small (e.g., 9 × 9 element) matrix B, the local matrix, centered on the element of A, is extracted. This local matrix B is then approximated by a two-dimensional Legendre polynomial. The smoothed value in the central element of B is then transferred into the new matrix C, which finally replaces A. The most appropriate degree of the polynomials and the size of the local matrix have been determined by testing within each energy range. It was found that best results were achieved with third-degree polynomials in all cases. The local matrix sizes used for the different energy ranges are given in TABLE 2. In order to obtain a map representing the total range of energies, the smoothed maps prepared in the three energy ranges were summed.

TABLE 2
LOCAL MATRIX SIZES

Energy Range (MeV)	Matrix Size	
	Elements	Angular Extent
70–150	15 × 15	7.5° × 7.5°
150–300	11 × 11	5.5° × 5.5°
300–5000	9 × 9	4.5° × 4.5°

The flux from a sky bin k of angular size ω_k (sr), from which N_k gamma-ray events were recorded, is given by

$$F_k = \frac{1}{T\omega_k} \sum_{i=1,N_k} \frac{1}{\epsilon(E_i)\eta(\vartheta_i,E_i)} \text{ photon cm}^{-2}\text{s}^{-1}\text{sr}^{-1}$$

where E_i is the energy of event i
ϑ_i is its incidence angle relative to the experiment axis
T (s) is the observation time
$\epsilon(E)$ (cm^2) is the experiment's effective sensitive area for paraxial incidence (see FIGURE 1)
$\eta(\vartheta,E)$ describes the decrease of effective sensitive area with incidence angle.

Since, as a consequence of the spectrum of the galactic gamma rays, the spatial characteristics of a photon-flux map over the total energy range from 70 to 5000 MeV are dominated by the low-energy quanta which are measured with least accuracy in angular resolution, a sky map has been generated showing the parameter "on-axis counts" defined by the equation

$$G_k = \frac{1}{T\omega_k} \sum_{i=1,N_k} \frac{1}{\eta(\vartheta_i,E_i)} \text{ count s}^{-1}\text{sr}^{-1}$$

This gives directly the number of gamma-ray events which would have been recorded in the bin k if the experiment had been pointed there, observing it "on-axis."

In the presented figures a background has been subtracted or is indicated. This background has been determined from two observations at galactic latitudes around

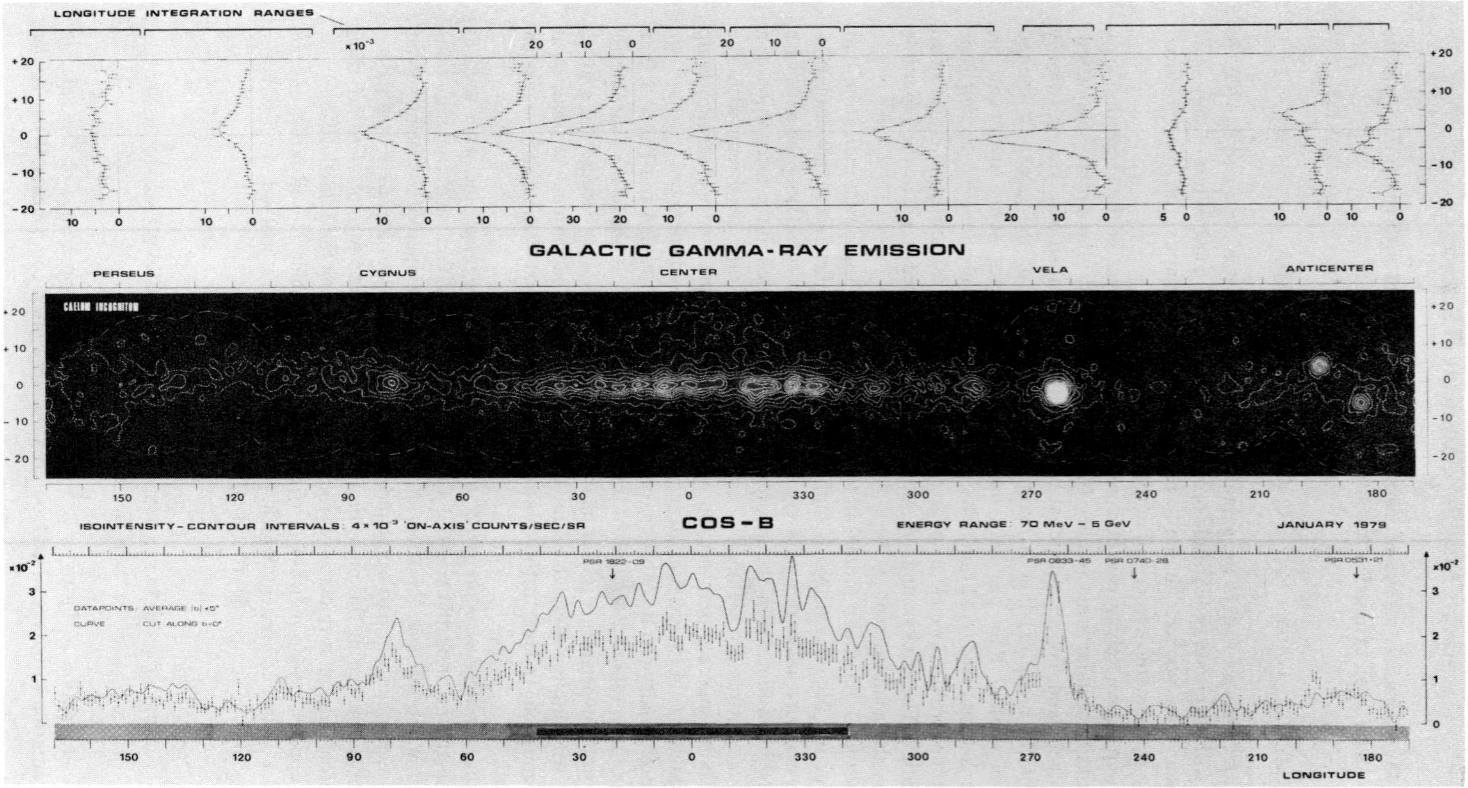

FIGURE 2. Presentation in galactic coordinates of the structure of the galactic gamma-ray emission as measured by COS-B. In the map the surface fitted to the data matrix is indicated by contour lines and a grey scale. The accepted field of view is indicated by dashed circles. The latitude profiles show the data points with errors and the fitted surface (full line). The longitude ranges over which the data are averaged are indicated by brackets. In these profiles and in the map a background is subtracted. In the longitude profile the data points with errors averaged over $\pm 5°$ are shown. The full line indicates a cross section through the fitted surface along the galactic equator. The background is indicated by the shaded area. The map and the profiles show the parameter "on-axis" counts $s^{-1}sr^{-1}$.

70° and 45°. It consists mainly of instrumental background with remaining galactic radiation and a small extragalactic component.

Galactic Gamma-Ray Emission Between 70 MeV and 5 GeV

Figure 2 shows a map of the galactic gamma-ray emission measured by COS-B, together with longitude and latitude profiles. For the map, a combination of grey scale and contour lines has been chosen to represent the smoothed data matrix, produced according to the method described in the preceding section.

The applied smoothing can be considered to be minimal in the sense that only fluctuations of angular scale smaller than the point-spread function are suppressed. On the other hand, not all individual features in the smoothed map are statistically significant. In order to allow a one-page representation of the galaxy, a limited number of contour lines has been used. As a consequence, some significant fine structure is suppressed. Additional insight into the data is provided by the longitude and latitude profiles, which also give quantitative information on the statistical uncertainties. The background, as indicated by the shaded band in the longitude profile, has been subtracted.

In Figure 3, longitude profiles in three energy ranges are shown, indicating that more structure becomes visible at higher energies, due to better angular resolution. This fine structure has been separately investigated by cross-correlating the data with the point-spread function of the instrument. This analysis led to the recognition of a number of statistically significant enhancements which are consistent with the expected point-source appearance, but which may also be due to extended features up to an angular size of two degrees, depending on intensities and energies.[4]

The preliminary result of this analysis is presented in Figure 4 and Table 3. The small flux range (about a decade) of the resolved sources[4,21] is one of the characteristics of present experiments in gamma-ray astronomy. This is so because the sources are seen with limited angular resolution against a high background of unresolved galactic emission. In fact, further granularity in the emission of the disc is suggested by the data.

Latitude profiles were prepared in the highest-energy range and were used to determine the intrinsic thickness of emissivity by folding Gaussian trial functions representing the intrinsic thickness of emissivity with the appropriate point-spread function. Only the inner part of the measured and simulated profiles were used for the fit. Examples illustrating this procedure are shown in Figure 5 together with variation of the unfolded angular thickness of the emitting region along the galactic disc.

The upper example indicates a narrow component, which is well fitted, and an asymmetric wide component which is disregarded by the method. In the lower example, only one component is present and is fitted.

Correlation between the Observed Gamma-Ray Emission and Galactic Features

The combination of Figures 2–5 gives the overall appearance of the galactic gamma radiation with its resolved (sources) and unresolved components. When trying

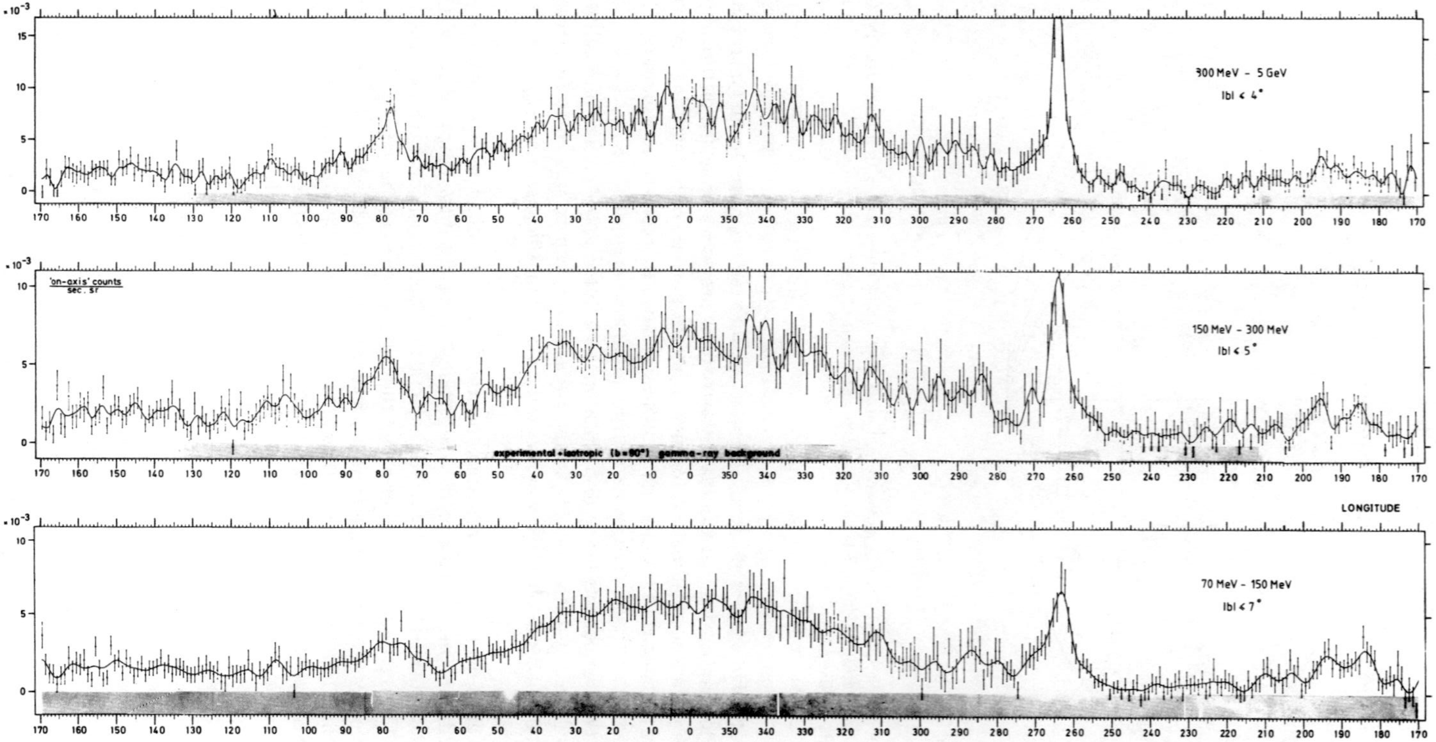

FIGURE 3. Longitude profiles of "on-axis" counts in three energy ranges. The full line indicates the fitted surface. The background is indicated.

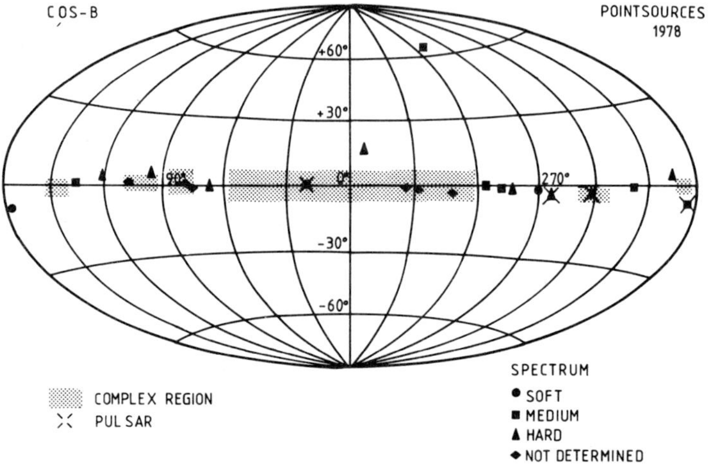

FIGURE 4. The point sources as listed in TABLE 3 are shown together with a tentative spectral hardness indication. Complex regions, where resolution of sources is difficult, and pulsars are specifically indicated.

to understand it in terms of possible correlations with the structures in our Galaxy, one has first to bear in mind that gamma-ray astronomy provides a projected picture of the sky, with no independent method for measuring the distance of the emitting features. The only exceptions to this are the identified galactic sources, the four pulsars,[17] for which distance estimates are available from radio measurements.

The two components in the disc emission, seen in the upper profiles of FIGURE 5, and noted first by Fichtel *et al.*[3] in the SAS-2 data, can be related to features at different distances in the Galaxy by assuming a constant thickness for the gamma-ray disc.

This gives the first clear indication that the data refer both to parts of the Galaxy 4 to 6 kpc distant and to much more local regions at < 1 kpc if this disc is about 200 pc thick. As for the distant features, their galactic longitude dependence is important and is related to the large scale structures of the Galaxy, whereas the local component is found to be less dependent on longitude and can be well correlated to the local matter distribution. This correlation stems from the asymmetry of the latitude dependence of the wide components, with a broad excess at positive latitudes in the central region and the corresponding (less clearly defined) excess at negative latitudes in the anticenter.

Together with possible evidence for a localized source in the Ophiucus dark cloud complex (see TABLE 3), this can be considered a good indication of gamma-ray emission from such local structures as Gould's belt, at distances varying up to 450 pc from the Sun. If such an identification, as already suggested by Fichtel *et al.*,[3] is accepted, then a production mechanism based on the interactions of cosmic rays, both protons and electrons, with nearby interstellar clouds, is in this case clearly at work (references 5–7). Without performing a correlation with the individual clouds (apart from the Ophiucus case), it is important to note that, assuming a production rate of gamma rays of $\sim 2\text{–}3 \times 10^{-25}$ photon (> 100 MeV) sec^{-1} per equivalent H atom (see references 8–19), the approximately correct flux of gamma rays can be reproduced with a mass distribution for Gould's belt as given by Lindblad,[10] Rossano,[11] and references therein. Whether this is true for the particular case of the Ophiucus complex is still open to debate.

Considering now the gamma-ray emission from more distant features, one should try to suppress from the map of FIGURE 2 the effects of the local structures described above, plus the obvious identified sources including the brightest gamma-ray emitter in the sky, the local PSR 0833 −45, which coincides by chance with the galactic disc. What remains is a mixture of resolved and unresolved emission especially bright for $55° \gtrsim l \gtrsim 285°$. This general enhancement is well established in gamma-ray astronomy[3,12] and indicates that most of the emission of gamma radiation takes place in the inner Galaxy.

The correlation of the gamma-ray emission with the large-scale features of the Galaxy can be based on the following:

(1) The small, but significant, latitude displacement of the peak emission visible in the profiles of FIGURE 2, as well as in the source distribution given in FIGURE 4,

TABLE 3

GAMMA-RAY SOURCES DETECTED BY COS-B*

Galactic l^{II}	Coordinates b^{II}	Number of Observations	Spectral Class	Remarks
21.0	+1.0	1	Soft	PSR 1822-09
65.5	0.0	3	Hard	CG 64 + 0
75.0	0.0	1	×	CG 75 + 0
78.5	+1.5	1	×	CG 78 + 1
95.5	+4.5	2	Hard	—
106.0	+1.5	1	×	—
121.0	+3.5	1	Hard	CG 121 + 3
135.5	+1.5	2	Medium	CG 135 + 1
176.0	−7.0	1	Soft	CG 176 − 7
184.5	−5.5	2	Medium	CG 185 − 5, PSR 0531 + 21
189.0	+1.0	1	×	CG 189 + 1
195.5	+4.5	2	Hard	CG 195 + 4
219.0	−0.5	1	Medium	—
243.0	−2.5	1	Hard	PSR 0740 + 28
263.5	−2.5	5	Hard	CG 263 − 2, PSR 0833 − 45
270.0	−1.0	2	Soft	—
284.0	−1.0	1	Hard	—
288.5	−0.5	1	Medium	—
291.0	+65.0	1	Medium	CG 291 + 65, 3C273
295.5	+0.5	1	Medium	—
312.0	−1.5	1	×	CG 312 − 1
327.5	−0.5	1	×	CG 327 − 0
333.5	0.0	1	×	CG 333 + 0
353.0	+16.0	1	Hard	In Ophiucus Dark Cloud Complex

*The information is preliminary; analysis continues. Error circles for positions have radii between 0.5° and 2.0°. With the exception of CG 263 − 2, source intensities (>100 MeV) are within the range 0.2 to 1.0 times that of CG 185 − 5. Spectral classification is based on the relative numbers of photons detected in the ranges 50 to 150 MeV and >150 MeV. The pulsar identifications are confirmed by timing phase analysis. Sources marked × are located in complex regions where more detailed analysis is needed to provide information additional to that given in the first COS-B catalogue.

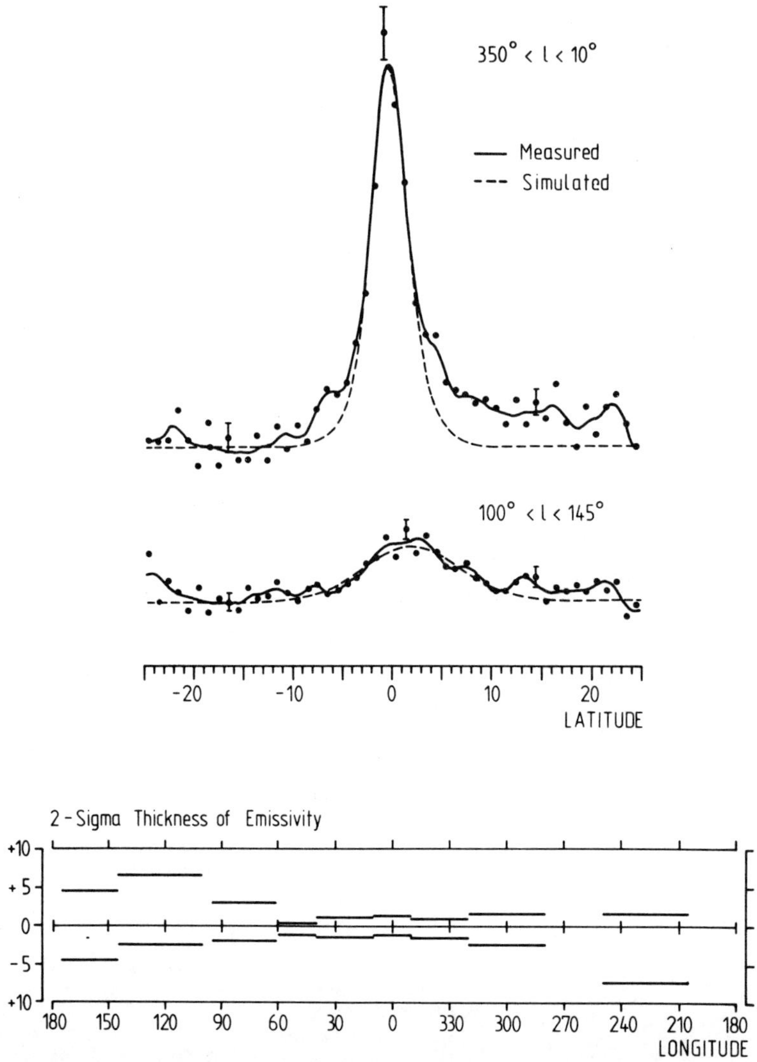

FIGURE 5. Two examples for latitude distributions ("on-axis" counts s^{-1} sr^{-1}) are given. Data points, fitted surface, and simulated distributions are shown. At the bottom the thickness of the Gaussian distributions (2σ), describing the unfolded intrinsic emissivity at varying galactic longitude, and their position in latitude are given.

corresponds well to the "hat brim" effect seen in HI (21-cm line) and considered to be a tracer of the large-scale warping of the plane.

(2) As far as the longitude variation of the intensity is concerned, the "galactic hole" at $l \sim 60°$, an empty interarm region, is present in the gamma-ray profile. Also apparent is the difference in the shape of the longitude profile in the central region, smoother from 0° to 50° and rougher from 300° to 360°. Finally, the emission from the

Cassiopeia–Perseus–Taurus ($l \sim 110°$–$170°$) region is enhanced with respect to that in the Monoceros–Puppis–Vela region ($l \sim 210°$–$250°$) which is known to be another particularly empty region of the disc.

No detailed correlation will be attempted here with particular features, nor with any model of the galactic spiral structure, as has been done by a number of authors on the basis of the SAS-2 data, which also contained similar indications (reference 13 and references therein, references 14 and 20). This is not done because such modeling implies assumptions on the production scenarios which are not unique. Moreover, assumptions are needed on whether the large scale emission is related to diffuse "elementary" processes (gas and cosmic rays) or is more granular and clumpy in nature, i.e., possibly due to the sum of unresolved sources. The results presented here leave open the problem of the relative contribution of "source" and "diffuse" processes. Certainly, sources like those already recognized in the COS-B data do contribute to the general galactic emission. The possible amount of this contribution has been discussed by Bignami *et al.*[15] and Protheroe *et al.*[16]

In conclusion, it is important to note that the data presented here refer to galactic structures on the scale of both several kpc and several hundred pc, and that they *are* related to the structure of the Galaxy as it is known from radio data. Moreover, we have evidence of at least two types of gamma-ray emitter in the Galaxy, the pulsars[17] and the local clouds in Gould's belt. No final conclusion can yet be reached on the nature of the large-scale emission because of the present experimental impossibility of unravelling its small-scale intrinsic geometry.

Note added (September 1979): Recently, radio measurements of pulsars PSR 1822 and PSR 0740 have become available which are contemporary to the COS-B gamma ray observations. The radio periods thus derived differ significantly by several hundred nanoseconds from the gamma-ray periods. While this new situation is being investigated, these identifications as gamma-ray pulsars are open to question. Please ignore these objects in TABLE 3 and FIGURE 4.

ACKNOWLEDGMENTS

We thank P. Caraveo and A. Pollock for valuable discussions. G.G.L. acknowledges receipt of an ESA fellowship.

REFERENCES

1. BIGNAMI, G. F., G. BOELLA, J. J. BURGER, P. KEIRLE, H. A. MAYER-HASSELWANDER, J. A. PAUL, E. PFEFFERMANN, L. SCARSI, B. N. SWANENBURG, B. G. TAYLOR, W. VOGES & R. D. WILLS. 1975. Space Sci. Instrum. **1:** 245.
2. SCARSI, L., K. BENNETT, G. F. BIGNAMI, G. BOELLA, R. BUCCHERI, W. HERMSEN, L. KOCH, H. A. MAYER-HASSELWANDER, J. A. PAUL, E. PFEFFERMANN, R. STIGLITZ, B. N. SWANENBURG, B. G. TAYLOR & R. D. WILLS. 1977. Proc. 12th ESLAB Symposium, Frascati, ESA SP **124:** 3.
3. FICHTEL, C. E., R. C. HARTMAN, D. A. KNIFFEN, D. J. THOMPSON, G. F. BIGNAMI, H. ÖGELMAN, M. E. ÖZEL & T. TÜMER. 1975. Astrophys. J. **198:** 163.
4. HERMSEN, W., K. BENNETT, G. F. BIGNAMI, G. BOELLA, R. BUCCHERI, J. C. HIGDON, G. KANBACH, G. G. LICHTI, J. L. MASNOU, H. A. MAYER-HASSELWANDER, J. A. PAUL, L. SCARSI, B. N. SWANENBURG, B. G. TAYLOR & R. D. WILLS. 1977. Nature **269:** 494.
5. BLACK, J. H. & G. G. FAZIO. 1973. Astrophys. J. Lett. **185:** L7.

6. PUGET, J. L., C. RYTER, G. SERRA & G. F. BIGNAMI. 1976. Astron. Astrophys. **50:** 247.
7. LEBRUN, F. & J. A. PAUL. 1978. Astron. Astrophys. **65:** 187.
8. CESARSKY, C. J., J. A. PAUL & P. G. SHUKLA. 1978. Astrophys. and Space Sci. **59:** 73.
9. FICHTEL, C. E., G. A. SIMPSON & D. J. THOMPSON. 1978. Astrophys. J. **222:** 833.
10. LINDBLAD, P. O., K. GROPE, A. SANDQVIST & J. SCHOBER. 1973. Astron. Astrophys. **24:** 309.
11. ROSSANO, G. S. 1978. Astron. J. **83:** 241.
12. BENNETT, K., G. F. BIGNAMI, R. BUCCHERI, W. HERMSEN, G. KANBACH, F. LEBRUN, H. A. MAYER-HASSELWANDER, J. A. PAUL, G. PICCINOTTI, L. SCARSI, F. SOROKA, B. N. SWANENBURG & R. D. WILLS. 1977. Proc. 12th ESLAB Symposium, Frascati. ESA SP **124:** 83.
13. BIGNAMI, G. F., C. E. FICHTEL, D. A. KNIFFEN & P. J. THOMPSON. 1975. Astrophys. J. **199:** 54.
14. PAUL, J. A., M. CASSÉ & C. J. CESARSKY. 1976. Astrophys. J. **207:** 62.
15. BIGNAMI, G. F., P. CARAVEO & L. MARASCHI. 1978. Astron. Astrophys. **67:** 149.
16. PROTHEROE, R. J., A. W. STRONG, A. W. WOLFENDALE & P. KIRALY. 1979. Nature. In press.
17. BUCCHERI, R., K. BENNETT, G. F. BIGNAMI, N. D'AMICO, W. HERMSEN, D. J. H. HUIZING, G. KANBACH, G. G. LICHTI, J. L. MASNOU, H. A. MAYER-HASSELWANDER, J. A. PAUL, B. SACCO, B. N. SWANENBURG & R. D. WILLS. 1979. Astron. Astrophys. In press.
18. PAUL, J. A., K. BENNETT, G. F. BIGNAMI, R. BUCCHERI, P. CARAVEO, W. HERMSEN, G. KANBACH, H. A. MAYER-HASSELWANDER, L. SCARSI, B. N. SWANENBURG & R. D. WILLIS. 1978. Astron. Astrophys. Lett. **63:** L31.
19. STRONG, A. W., A. W. WOLFENDALE, K. BENNETT & R. D. WILLS. 1978. Mon. Not. R. Ast. Soc. **182:** 751.
20. STECKER, F. W. 1977. Astrophys. J. **212:** 60.
21. SWANENBURG, B. N. 1979. Gamma-ray sources: A puzzle or a piece of a puzzle? *In* Stars and Stellar Systems. Astrophysics and Space Science Library. Reidel. Dordrecht. In press.

GAMMA RAYS, COSMIC RAYS, AND THE DYNAMICS OF THE GALAXY

Catherine J. Cesarsky

Section d'Astrophysique
Centre d'Etudes Nucléaires de Saclay
91 Gif sur Yvette, France

GALACTIC COSMIC RAYS

Observations from the earth, from balloons, and from satellites provide us with information on the cosmic-ray flux in the solar vicinity. The main facts are well known: the nucleon component, dominated by protons, is, at an energy of 1 GeV, one hundred times more intense than the electron component. Above a few GeV, the proton spectrum is a power law which extends over several decades; the electron spectrum is much more difficult to measure, and can only be considered well-known in the energy range from one to a few GeV.[1] At energies $\leq$ 800 MeV, the fluxes of both protons and electrons from interstellar space are very much affected by the interactions of the particles with the solar wind, and are consequently difficult to assess. In the solar vicinity, the energy density of cosmic rays, $\approx$ 1 eV/cm^3, is comparable to those of cloud motions, magnetic fields and star light.

The electron component of the cosmic-ray flux reveals its presence in other regions of the galaxy through its emission of synchrotron radiation. Studies of the radio-continuum background towards the anticenter[2] permit the establishment of the electron spectrum in the energy range 200 MeV – 2 GeV; however, these determinations are imprecise by factors of ~5, owing to a number of intervening factors that are difficult to estimate : the effects of radio-continuum emission at high frequencies and of radio-continuum absorption at low frequencies, the strength of the interstellar magnetic field, the relative distributions of magnetic fields and cosmic-ray electrons, etc.

Radio-continuum observations also tell us about the distribution of synchrotron radiation in our galaxy and in some of its neighbors. In our galaxy, this emission has been known for a long time to be partially correlated with the spiral arms[3]; this is also true of external galaxies with similar characteristics.[4] In spiral galaxies, the synchrotron radiation from the disk peaks either at the center or at a ring surrounding the center, then falls off exponentially like the blue light of the galaxy. Generally the synchrotron disk has a smaller radius than the disk of HI gas emitting 21-cm line radiation.[5] Some external galaxies, seen edge on, have more or less flat radio haloes[6]; possibly our galaxy has one too[7] — or at least a "thick disk" of synchrotron radiation.[8] These observations give us some indications on the galactic distribution of cosmic-ray electrons, but their interpretation, similar to the determination of the local energy spectrum, is heavily hindered by our lack of knowledge of the magnetic field distribution (and sometimes by the difficulty in separating the thermal and nonthermal components of the continuum radiation).

It seems frustrating to be always concentrating on a component of the cosmic radiation which, at the sun, is so insignificant in the over-all flux; means had to be

0077–8923/80/0336–0223 $1.75/1 © 1980, NYAS

found to observe the nuclear cosmic rays elsewhere. This task, it was hoped, would be fulfilled by observations of the galactic gamma radiation.

Galactic Gamma-Ray Emission

Gamma rays in the 100 MeV range are emitted through the decay of neutral pions, generated by collisions of nuclear cosmic rays of energy $\geq$ 70 MeV with interstellar-medium particles. The spectrum of the radiation thus emitted has been known for some time[9]; it exhibits a characteristic hump at 70 MeV (see Figure 1). The galactic diffuse gamma ray radiation at energy $>$ 100 MeV was expected to be dominated by this process (e.g., reference 10). Two other processes contribute to the diffuse galactic gamma-ray emission in this energy range, and they both involve cosmic-ray electrons: bremsstrahlung and inverse Compton. The bremsstrahlung emission is due to electrons of energy comparable to that of the gamma rays, and thus is subjected to the uncertainties, mentioned earlier, which hamper the determination of the low-energy electron spectrum; the gamma ray emissivity, for different spectra, is shown in Figure 1.[11] The Compton scattering of starlight photons by electrons in the GeV range and of blackbody photons by electrons in the 100 GeV range also generates high-energy gamma rays. In the galactic plane, the inverse Compton component is always expected to be the smallest of the three,[11,12] but it may play an interesting role if there is a "fat disk," a "flat halo," or a spherical halo of cosmic rays.[13]

Detailed observations of the galactic gamma radiation have now been obtained by instruments on board of two satellites : SAS-2, which operated during the period from November 1972 to June 1973, and COS-B, which has been in orbit since August 1975. (The present paper does not take into account the new COS-B data presented in the preceding paper). The SAS-2 observations have revealed the main characteristics of the gamma radiation from the galaxy.[14] The radiation is confined to a thin disk and the longitudinal distribution has a broad maximum centered on $l = 0$. Features of the interstellar gas distribution, such as the HI spiral arms, the Gould belt, and the warping of the plane, have gamma-ray counterparts. These observations of the gamma-ray distribution were interpreted by many authors in terms of diffuse processes, largely dominated at $E > 100$ MeV by the Π° decay process.[15] Essentially all the authors found that cosmic-ray gradients were necessary to explain the data, but a consensus on the scale and the magnitude of these gradients was not reached[16]; the differences are often due to the uncertainties still existing in the galactic distribution of the interstellar matter : while the HI component is relatively well determined, it is now known that the molecular hydrogen accounts for a comparable amount of interstellar matter, and that it has a different distribution.[18, 19] Unfortunately, the molecular hydrogen can only be observed indirectly through the carbon monoxide emission; the surveys are far from complete, and the interpretation of the data in terms of column and volume densities of H_2 is not straightforward.

A class of models[15,17] built to explain the observed gamma-ray distribution was based on two types of assumptions: (a) the gas in our Galaxy follows the spiral structure, as derived from HI observations[15] or HII regions data [17] and (b) the galactic magnetic field is held to the galactic disk by the interstellar matter, which is subjected to the gravitational field of the stars; in turn, it can contain the cosmic rays, allowing them to build up a pressure gradient comparable to its own.[20] These considerations prompted the assumption that the similarity between the pressures of the gas, the magnetic field, and the cosmic rays, observed in the solar vicinity, holds also elsewhere on the galactic disk, on a scale of ~1 kpc.

With these assumptions, it was possible to build detailed models that could give

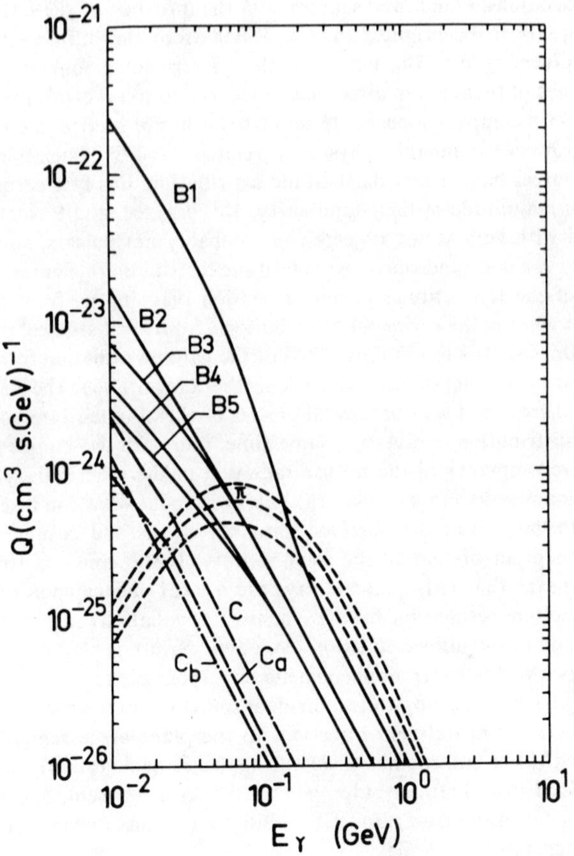

FIGURE 1. Calculated differential production rates of gamma rays, per unit volume, for a density of 1 H-atom cm^{-3}. B1 thru B5 represent the production rates of bremsstrahlung gamma rays, obtained using different estimates of the electron spectrum. π refers to gamma rays produced through the decay of neutral pions; the upper and lower limits are also represented. C_a and C_b are the inverse Compton production rates for starlight and blackbody photons; C is the total inverse Compton production rate. (Figure taken from reference 11.)

excellent fits to the gamma-ray data. As a bonus, it was possible to use the gamma-ray data to derive the distribution of interstellar gas in the galactic plane, independently of the data on CO molecular emission; the distribution of H_2 thus found[17] was flatter than that derived by the CO observers.[18, 19] Now, some or all of the conclusions gathered are put into question because of three new observational facts: (a) the extragalactic gamma-ray background is less intense than it was previously thought so that the ratio of the anticenter to the center emission was underestimated[21]; (b) the spectrum of the gamma-ray emission has been measured[22,23]; (c) the COS-B satellite has discovered that "localized" (within 2°) gamma-ray sources are common in the galactic disk. Up to now, 23 galactic sources are known[24]; four of them are pulsars : the Crab and Vela pulsars and two others which have been discovered owing to their known time variations. One of the sources is in the direction of the ρ Ophiuchis dark cloud, but appears to be brighter than expected from the diffuse emission due to cosmic-rays traversing it.[25] The nature of the 18 remaining sources is a subject of much debate; five of them are in directions tangential to spiral arms, three to six could be associated with compact supernovae remnants,[26] five or six are in directions where one finds supernovae remnants physically related to OB associations,[27] etc. The unidentified sources have a very flat latitude distribution; this fact permits a guess as to the order of magnitude of their luminosity, 10^{35} erg/sec, and to venture that they are associated with very young objects (but probably not pulsars, since these have large peculiar velocities, and thus a wide latitude distribution). Sources are responsible for most of the low latitude gamma radiation reaching us from the anticenter direction—but what is their over-all contribution? Statistical studies[28,35] suggest that they can account for 30% to virtually 100% of the gamma radiation from the galactic plane; to reach firm conclusions, more should be known about the nature of these sources—and there could well be several classes of them—their luminosity function, their galactic distribution, etc. At the same time, there are two simple arguments in favor of the predominance of the diffuse radiation in the inner galaxy. First, as we shall see, the gamma-ray luminosity of the galaxy is about what can be expected from diffuse mechanisms, given the observed amounts of gas and cosmic rays with an intensity on the order of that in the solar vicinity. If the emission from sources is important, it means that either cosmic rays are a local phenomenon (except for the high-energy electrons responsible for the synchrotron radiation) or that the sources are somewhat related to the diffuse emission too, unless nature is playing a dirty trick on us by making two widely different phenomena be responsible for comparable shares of the observed radiation. Second, while individual sources tend to have different spectra, the overall spectra from different directions in the plane are essentially identical[22]; they also match (after subtraction of an extragalactic background) the spectrum of radiation received from high latitudes, where sources are absent, and whose diffuse nature is taken for granted—again, if the radiation is not predominantly diffuse, a number of coincidences are at play.

In the next section, we briefly discuss some conclusions that can be reached from the gamma-ray data, without having to ascertain the importance of the source contribution.

Information on Cosmic Rays and the Interstellar Medium Derived from Gamma-Ray Observations

Diffuse Gamma-Ray Emissivity and the Electron Spectrum in the Local Interstellar Medium

We have seen (Figure 1) that the uncertainties in the low-energy electron spectrum made it difficult to predict the gamma-ray spectrum; Figure 2 shows three

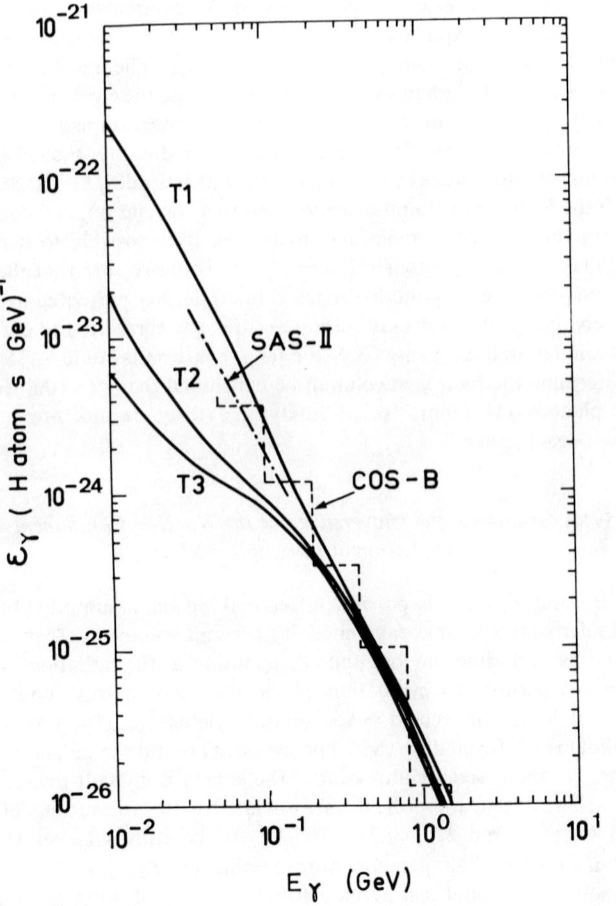

FIGURE 2. Total differential gamma-ray emissivity per H-atom for three of the electron spectra considered in Figure 1, and observed spectra. The spectra of the local gamma radiation obtained by SAS-2[23] and COS-B[22] are superimposed on the calculations, in such a way that the COS-B spectrum agrees with the calculated ones at energies > 120 MeV, and that the two observed spectra coincide in the energy range covered by both. (Figure taken from reference 11.)

possible such spectra, for three different choices of electron spectrum. In this figure, the galactic gamma-ray spectrum observed by COS-B and the spectrum of the high-latitude emission seen by SAS-2 have been superimposed and normalized in such a way as to agree with each other in the region where they overlap and with the calculated spectrum at energies > 120 MeV.

While the SAS-2 spectrum can be fitted with a power law, the COS-B spectrum is somewhat curved. Still, it is obvious that the observed spectra do not exhibit the characteristic Π° decay hump: thus, gamma-ray astronomy has failed in proving beyond doubt the presence of relativistic nuclei throughout the galactic disk. For the solar vicinity, however, where we know the flux of nuclei, we can use the gamma-ray spectrum to determine the electron spectrum and the gamma-ray emissivity $\epsilon_\odot$ of the interstellar medium, by requiring that the Π° emission be drowned by the electron emission—most likely due mainly to bremsstrahlung. The results obtained[11] are $\epsilon_\odot(>100\text{MeV}) \simeq 2.4\times10^{-25}$ photons (H atom)$^{-1}$s^{-1}, while the electron spectrum below 1 GeV is close to its upper limit as derived from radio observations.

The local emissivity $\epsilon_\odot$ can also be determined more directly. Fichtel *et al.*[22] found that the gamma-ray intensities observed by SAS-2 at latitudes $/b/>12.8°$ are linearly correlated to the hydrogen column density; from this correlation, and considering the uncertainty in the density of molecular hydrogen, it is possible to derive $\epsilon_\odot(>100$ MeV$)=(2.35\pm0.1)\times10^{-25}$ photons (H atom)$^{-1}$s^{-1}. To make sure that the correlation between gamma-ray intensity and hydrogen column density presented in reference 22 was not merely due to the fact that both quantities are correlated to csc b, Lebrun[29] performed a similar analysis, using COS-B data at constant latitude $b=15°$ and galaxy counts to determine the hydrogen column; he concluded that $\epsilon_\odot(>100$ MeV$)=(2.9\pm0.6)\times10^{-25}$ photons (H atom)$^{-1}$s^{-1}. Clearly, the three results are in very good agreement with each other.

Cosmic-Ray Gradients, the Universality of the Nucleon Component, and the Distribution of Interstellar Matter

Once $\epsilon_\odot$ is determined, it is possible, after making the assumption of cylindrical symmetry, to derive the gamma-ray emissivity per unit volume in other regions of the galactic plane by unfolding the longitude distribution of the radiation[30]; in addition, such an analysis permits an evaluation of the total gamma-ray luminosity of the galaxy. An unfolding[31] of recent SAS-2 data[21] yields $L_{galaxy}(>100$ MeV$)=(1.2–1.5)\times10^{42}$ photons s^{-1} (excluding the 2 kpc cylinder around the galactic center). The total mass M_{gas} of the gaseous component of the galaxy is difficult to ascertain, as we mentioned earlier, due to the difficulties associated with observations of H_2 clouds. Gordon and Burton[19] find $M_{gas}=4.3\times10^9\ M_\odot$. If the emissivity per H atom were constant, the corresponding galactic diffuse emission $L_{diff.\ gal.}(>100$ MeV$)\simeq(1.2–1.5)\times10^{42}$ photons s^{-1} would just account for the totality of the galactic gamma-ray flux.

A closer look at the observations shows that the hypothesis of a uniform flux of cosmic rays is not tenable, at least in the anticenter region; this can be made evident in many ways. For instance, Strong *et al.*[3] have considered the COS-B results for $160°<l<210°–20°<b<30°$. The intensities received from different latitudes, after subtracting the contributions from localized sources, are as shown in FIGURE 3. On the same figure, we have represented the emission expected from interactions of the *nuclear component* of cosmic rays with the HI and H_2 gas present in these directions. Clearly, the narrow component predicted, which would be due to cosmic rays pervading the Perseus spiral arm, is not present in the data: cosmic rays appear to die out in the first kiloparsec beyond the sun. An interesting consequence of this result is that it can close a debate which has been going on for several years[33]: we have now a proof that nuclear cosmic rays of energy in the GeV range are of galactic origin, rather than metagalactic or universal.[34] The unfolding of the SAS-2 data[31] corroborates this conclusion; even though the contribution of the unidentified sources (several of which are in the anticenter region, where, given the low background, they are relatively easy to separate out) has not been subtracted out, the authors found it necessary to let the galactic emission go to zero for galactic radii $R > 11.5$ to 12.5 kpc, while the gas is present at least up to $R = 16$ kpc.

The interpretation of the gamma-ray emission (>100 MeV) from the galactic center direction is much less straightforward. The emissivity of the galactic disk peaks at $R = 5$ to 6 kpc. As we have seen, the accepted values for the extragalactic background and for the local emissivity $\epsilon_{\odot}$ have changed since the early work; both results concur in diminishing the ratio r_{γ} = (emissivity/pc^2 at R=5 to 6 kpc)

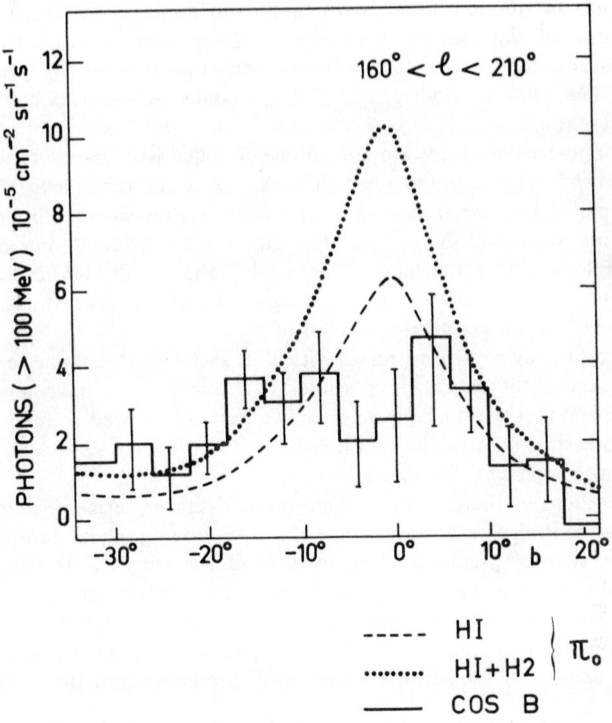

FIGURE 3. Latitude distribution of gamma rays of energy 100 MeV observed by COS-B and expected distributions for Π° decay emission for a spatially uniform cosmic-ray density. The HI estimate is based on 21-cm data; the H_2 estimate on CO data. (Figure adapted from reference 32.)

/(emissivity/pc^2 at $R = 10$ kpc) from 8 or so to only 3 to 4.[31] The galactocentric distribution of the gamma-ray emitters is very similar to that of tracers of young objects, such as supernovae remnants, pulsars, giant HII regions, or Lyman continuum photons (as derived from radio-continuum or radio-recombination line observations); it seems plausible that the gamma-ray sources have also a distribution of this type. In that case, defining $r_{\gamma \text{diff.}}$ as the equivalent of r_{γ} for the diffuse radiation, we have $r_{\gamma \text{diff.}} \simeq 3$ to 4.[35] The CO radiation also peaks at $R = 5$ to 6 kpc, but the value of the ratio r_{gas}=(gas surface density at $R = 5$ to 6 kpc)/(gas surface density at 10 kpc) is a matter of controversy. According to Gordon and Burton,[19] $r_{gas} = 3$. However, the interpretation of the CO data in reference 19 relies on the assumption that the ratio of the densities $n(CO)/n(H_2) = 6\times10^{-5}$ throughout the galaxy. In fact, in the solar neighborhood, more recent determinations yield $n(CO)/n(H_2) = 12\times10^{-5}$, leading to $r_{gas}=2$ [43]; and the possible presence of a gradient of the abundance of carbon with R may lower r_{gas} down to a value of order 1.[17] On the other hand, Solomon,[18] using his own CO data and a different procedure to derive H_2 densities, finds twice as much H_2 at $R = 5$ to 6 kp than Gordon and Burton.[19] These results have a bearing on our knowledge of the galactic distribution of cosmic radiation: $r_{\gamma diff.} = r_{gas}$ is compatible with a uniform distribution of cosmic rays in the inner galaxy, while $r_{\gamma diff.} > r_{gas}$ argues for an increase of the cosmic-ray density as R decreases from 10 to 5 kpc. The observations of synchrotron radiation from external galaxies would favor this latter possibility. The third possibility, $r_{gas} > r_{\gamma\ diff.}$, could be understood if one of the following were true:

a) Dense molecular clouds are not completely filled with cosmic rays. This could happen if cosmic rays were retained in such clouds for times long enough to be partially depleted, because of their energy losses through inelastic collisions, Coulomb collisions, and bremsstrahlung. The trapping could be due to interactions with hydromagnetic waves at the border[36] or inside[37] the clouds. However, theoretical considerations[37], as well as the detailed analysis of COS-B observations of a nearby cloud complex,[38] argue against this possibility.

b) the cosmic ray density decreases with R. A result that would be very surprising indeed! Of course, it is always possible that cosmic rays have a very patchy distribution, and that the sun happens to be in a particularly well-populated spot. This type of hypothesis is unattractive, but it is difficult to rule out—especially with regard to the nuclear component of cosmic rays.

Interpretations of the galactic gamma-ray emission in terms of purely diffusive processes had to incorporate the presence of a cosmic-ray gradient across spiral arms to reproduce some particularly sharp features of the longitude distribution of the radiation. At present, it seems very plausible that the localized sources, as most young objects, are concentrated in the spiral arms; even if their overall contribution is not overwhelming, their spatial distribution may account for the features mentioned above, thus questioning the need for a cosmic-ray gradient across the arms.

Gamma-Ray and Cosmic-Ray Halo

The existence or nonexistence of a halo, through which the galactic cosmic rays must diffuse before escaping into the intergalactic medium, has always been a subject of much debate, especially since radio-haloes were discovered on some external galaxies.[6] Since haloes are not believed to contain appreciable quantities of diffuse matter or of optical light, the only way they could be seen in gamma rays is through the inverse-Compton interaction of relativistic electrons with blackbody photons.

Spherical Gamma-Ray Haloes

Before most of the high-latitude gamma-ray observations were available, Worral and Strong[39] conjectured that a large part or all of the high-latitude radiation observed was not extragalactic, but in fact originated in a spherical halo. More recently, the SAS-2 workers[23] have divided the radiation received from latitudes $|b| > 12.8°$ into two components. The galactic component, which we mentioned earlier, is correlated with the column of matter. The residual component is found to be isotropic: in particular, the ratio of the 35–100 MeV γ-ray intensity in the directions $(300° < l < 60°)$, $(20° < |b| < 40°)$ to that from $(100° < l < 250°)$, $(20° < |b| < 40°)$ is found to be 1.10 ± 0.19. Obviously, then, there cannot be a spherical gamma-ray halo with a 15 kpc radius.

Flat Gamma-Ray Halo

Some of the radio haloes observed in external galaxies have the shape of a thin oblate spheroid; if the same phenomenon occurs in our galaxy, it could give rise to a substantial, disk-like, inverse-Compton emission.[13] This emission would also be correlated with csc b, and thus could easily be included in the "galactic component" identified by the SAS-2 workers[40]; so, the isotropy of the residual radiation does not rule out the presence of radiation from a flat halo. There are, however, two arguments against this emission being a substantial part of Fichtel *et al.*'s galactic high-latitude emission. First, we note that the gamma-ray emissivities $\epsilon_{\odot}$ (>100 MeV) determined directly from correlations with gas columns by the SAS-2 workers, as well as that derived from the spectrum,[11] both would contain a spurious component from a flat halo. The fact that they are both in excellent agreement with the determination made from COS-B data by Lebrun,[29] which is certain to consider only processes related to interstellar matter, argues against the existence of a significant flat gamma-ray halo at energies >100 MeV. Second, again, the similarity between the galactic gamma-ray spectra in the plane and at high latitudes is easier to understand if the inverse-Compton components are small in both cases. Otherwise, one would have to expect the relative contribution of inverse-Compton radiation to be the same in the galactic plane and at high latitudes. In the plane, the radiation is mostly due to the interaction of electrons with visible or infrared light, in a region coextensive with the gas. At high latitudes, the interactions arise in a volume not filled by gas and in addition a different portion of the electron spectrum is involved.

Cosmic-Ray Haloes

The "haloes" usually mentioned in the cosmic-ray literature are in fact generally nuclear cosmic-ray haloes, for it is well known that, because of the losses suffered through the synchrotron and inverse-Compton processes, the electron haloes are likely to be smaller. This is confirmed by observations of external galaxies, where the spectrum is seen to steepen as the height above the plane increases.[6] The electrons of interest here have energies in excess of 100 GeV; at these energies, the electron spectrum may have steepened considerably even at the galactic plane.[41] If the halo is of the dynamic type, and especially if it is dominated by convection, these electrons would be depleted at a small height above the plane.[42] Thus, the absence of a gamma-ray halo does not preclude the existence of a nuclear cosmic-ray halo.

Some constraints on the size of the nuclear halo can be obtained from the longitude distribution of the galactic gamma radiation, assuming that all the radiation is of diffuse origin, that the distribution of the cosmic-ray sources is known, and that the cosmic-ray propagation in the halo is purely diffusive.[44] If cosmic rays are also convected by a galactic wind (dynamic halo[42]), this method can only yield information on the diffusion coefficient of cosmic rays in the galactic disk.[45]

Summary

The galactic gamma-ray emission has recently been observed by the SAS-2 and COS-B satellites; we have attempted to summarize the impact of these results on our knowledge of galactic cosmic rays and the interstellar medium. Despite the fact that an unknown proportion of the observed gamma rays is emitted by localized sources, the present gamma-ray observations can bring information on the gamma-ray emissivity of the interstellar medium and the low-energy interstellar electron spectrum, the universality of the nuclear component of cosmic rays, and the large-scale distribution of interstellar matter. There is no firm evidence for the existence of a *gamma-ray* halo, spherical or oblate, but the existence of a *cosmic-ray* halo is not ruled out.

Acknowledgment

I thank J.A. Paul for stimulating discussions.

References

1. Buffington, A. 1978. A.P.S. Meeting at Durham, New Hampshire, September 1978.
2. Goldstein, M. L., R. Ramaty & L. A. Fisk. 1970. Phys. Rev. Lett. **24**: 1193(c).
   Cummings, A. C., E. C. Stone & R. E. Vogt. 1973. Proc. 13th Int. Cosmic Ray Conf. (Denver) **1:** 335.
   Rockstroh, J. M. & W. R. Webber. 1978. Ap.J. **224:** 677.
3. Mills, B. V. 1959. Pub. A.S.P. **71:** 627.
4. van der Kruit, P. C. & R. J. Allen. 1976. Ann. Rev. Astron. Astrophys. **14:** 417.
5. van der Kruit, P. C. 1977. I.A.U. Symposium 77.
6. Allen, R. J., J. E. Baldwin & R. Sancisi. 1978. Astron. Astrophys. **62:** 397.
7. Webster, A. 1975. Mon. Not. R. Astr. Soc. **171:** 243.
8. Baldwin, J. E. 1976. Proc. 2nd Int. Symposium Galactic Gamma Rays, NASA/GSFC X-662-76-**154:** 206.
9. Stecker, F. W. 1970. Astrophys. Space Sci. **6:** 377.
10. Fazio, G. G. 1967. Ann. Rev. Astron. Astrophys. **5:** 481.
11. Cesarsky, C. J., J. A. Paul & P. G. Shukla. 1978. Astrophys. Space Sci. **59:** 73.
12. Shukla, P. G. & J. A. Paul. 1976. Astrophys. J. **208:** 893.
    Piccinotti, G. & G. F. Bignami. 1976. Astron. Astrophys. **52:** 69.
13. Schlickeiser, R. & K. O. Thielheim. 1977. Astrophys. Space Sci. **47:** 415.
14. Fichtel, C. E., R. C. Hartman, D. A. Kniffen, D. J. Thompson, G. F. Bignami, H. B. Ogelman, H. E. Ozel & T. Tumer. 1975. Ap. J. **198:** 163.
15. Bignami, G. F. & C. E. Fichtel. 1974. Ap. J. Lett. **189:** L 65.
    Paul, J. A., M. Cassé & C. J. Cesarsky. 1976. Ap. J. **207:** 62.
    Bignami, G. F., C. E. Fichtel, D. A. Kniffen & D. J. Thompson. 1975. Ap. J. **199:** 54.
16. Stecker, F. W., P. M. Solomon, N. Z. Scoville & C. E. Ryter. 1975. Ap. J. **201:** 90.
    Puget, J. L., C. Ryter, B. Serra & G. F. Bignami. 1976. Astron. Astrophys. **50:** 247.
17. Cesarsky, C. J., M. Cassé & J. A. Paul. 1977. Astron. Astrophys. **60:** 139.

18. Scoville, N. Z. & P. M. Solomon. 1975. Ap. J. Lett. **199:** L105.
Solomon, P. H. 1978. I.A.U. Symposium No. 84.
19. Gordon, H. A. & W. B. Burton. 1976. Astrophys. J. **208:** 346.
20. Parker, E. N. 1969. Space Sci. Rev. **9:** 651.
21. Fichtel, C. E., D. A. Kniffen & D. J. Thompson. 1977, Proc. 12th ESLAB Symposium, Frascati. ESA **SP-124:** 95.
22. Fichtel, C. E., R. C. Hartman, D. A. Kniffen, D. J. Thompson, H. B. Ogelman, M. E. Ozel & T. Tumer. 1977. Ap. J. Lett. **217:** L 9.
Paul, J. A. *et al.* 1978. Astron. Astrophys. **63:** 131.
Kniffen, D. A., D. L. Bertsch, D. J. Morris, R. A. R. Palmeira & K. R. Rao. 1978. Ap.J. **225:** 591.
23. Fichtel, C. E., G. A. Simpson & D. J. Thompson. 1978. Ap. J. **222:** 833.
24. Hermsen, W. *et al.* 1977. Nature **269:** 494.
Hermsen, W. 1978, Paper presented at the 5th European Cosmic Ray Conference, Kiel.
25. Black, J. M. & G. G. Fazio. 1973. Astrophys. J. Lett. **185:** L 7.
26. Lamb, R. C. 1978. Nature **272:** 429.
Lamb, R. C. 1978. Nature **274:** 720.
27. Montmerle, T. 1979. Ap.J. **231:** 95.
28. Bignami, G. F., P. Caraveo & L. Maraschi. 1978. Astron. Astrophys. **67:** 149.
29. Lebrun, F. 1978. Thèse de 3ème cycle, Université de Paris VII.
30. Puget, J. L. & F. W. Stecker. 1974. Ap. J. **191:** 323.
Strong, A. W. & D. M. Worral. 1976. J. Phys. A: Math. Nucl. Gen. **9:** 283.
31. Caraveo, P. A. & J. A. Paul. 1979. Astron. Astrophys. **75:** 340.
32. Strong, A. W., K. Bennett, R. D. Wills & A. W. Wolfendale. 1977. Proc. 12th ESLAB Symposium. ESA SP **124:** 167.
33. Burbidge, G. & G. Brecher. 1971. Astrophys. Space Phys. **5:** 140.
34. Dodds, D., A. W. Strong & A. W. Wolfendale. 1975. M.N.R.A.S. **176:** 345.
35. Rothenflug, R. & P. Caraveo. 1979. Astron. Astrophys. In Press.
36. Skilling, J. & A. W. Strong. 1976. Astron. Astrophys. **53:** 253.
37. Cesarsky, C. J. & H. Volk. 1978. Astron. Astrophys. **70:** 367.
38. Lebrun, F. & J. A. Paul. 1978. Astron. Astrophys. **65:** 187.
39. Worral, D. M. & A. W. Strong. 1977. Astron. Astrophys. **57:** 229.
40. Stecker. F. W. 1978. I.A.U. Symposium No. 84.
41. Prince, T. A. 1979. Ap. J. **227:** 676.
42. Owens, A. J. & J. R. Jokipii. 1977. Ap. J. **215:** 685.
43. Guibert, J., J. Lequeux & F. Viallefond. 1978. Astron. Astrophys. **68:** 1.
44. Stecker, F. W. & F. C. Jones. 1977. Ap. J. **217:** 843.
45. Jones, F. C. 1978. Ap. J. **222:** 1097.

LOCALIZED SOURCES OF HIGH-ENERGY GAMMA RAYS

K. Pinkau

Max-Planck-Institut für Physik und Astrophysik
Institut für Extraterrestrische Physik
8000 Munich 40
Federal Republic of Germany

INTRODUCTION

Regions of enhanced emission are observed in the gamma-ray sky by modern telescopes. These telescopes have a limited degree of angular resolution which may eventually become 5° for a Compton telescope[1] (1–10 MeV) and is typically 5° full width half maximum (FWHM) at ~ 70 MeV for a spark chamber system like that employed by SAS-2 or COS-B,[2] decreasing to about 2° FWHM at very high energies ($\gtrsim$ 500 MeV).

Due to this limitation, it is preferable to speak of localized sources and to make no distinction *a priori* between an extended region several degrees in diameter and a true star. Identifications are possible only if spectral or temporal features suggest a correlation with observations at other wavelengths.

To date, it appears that extragalactic localized sources have been discovered in the gamma-ray domain, and they will briefly be covered first. Then, the question of gamma-ray sources in the galaxy will be dealt with in more detail.

EXTRAGALACTIC SOURCES

Observational evidence presented during the past year makes it plausible that the Seyfert galaxy NGC 4151[3] and the Quasar 3C 273[4] have been seen in gamma rays.

NGC 4151

During a balloon flight of a Compton telescope on October 1, 1977,[3] a source was observed which could not be identified by its position due to the large angular uncertainty of the Compton telescope in the coarse mode (~ 10°). The association with NGC 4151 is suggested only on account of the fact that the energy spectrum observed connects rather well with that known from x-ray observations (FIGURE 1).[5] If this identification is accepted, NGC 4151 would emit 10^{45} erg/s in the gamma-ray domain below 3 MeV; beyond 3 MeV, the spectrum sharply steepens.

NGC 4151 has been observed to be highly variable.[6] Time scales of several days for variability in the x-ray energy range have been confirmed.[6–8]

The break in the energy spectrum at 3 MeV is suggestive of Compton models[9,10] for the source. Various models may, however, fit the observed spectrum. For example, Pozdnyakov *et al.*[10] have computed Compton spectra produced by electrons with a relativistic Maxwellian distribution, and they could produce a power-law spectrum of

0077–8923/80/0336–0234 $1.75/1 © 1980, NYAS

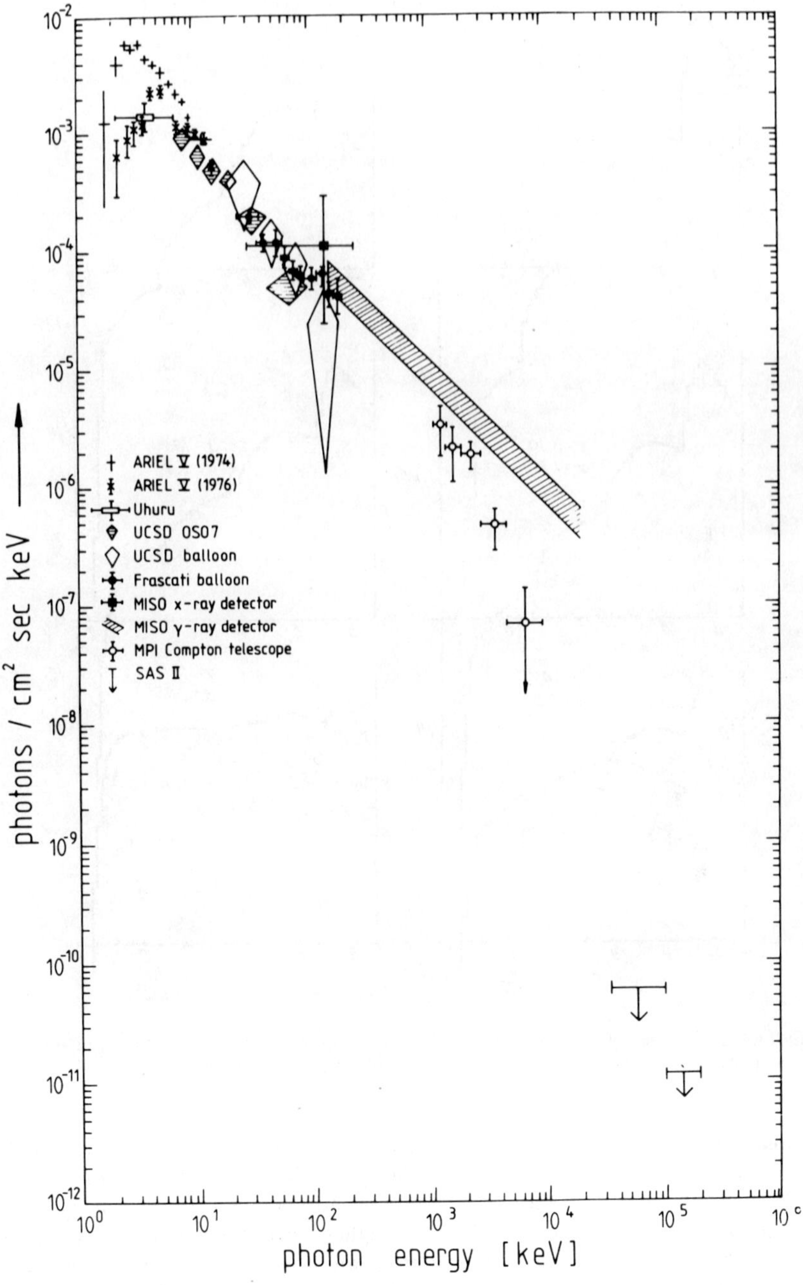

FIGURE 1. X-ray and gamma-ray intensities observed from NGC 4151 (adapted from Schönfelder).[5]

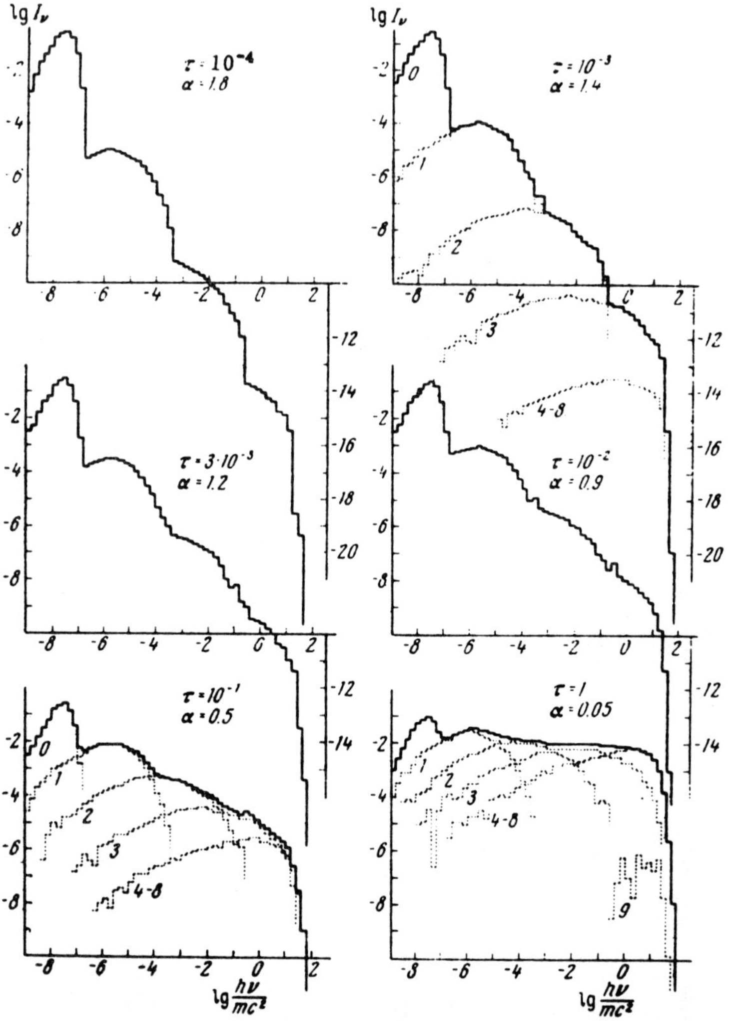

FIGURE 2. A series of emission spectra due to Compton scattering in sources of different optical depth τ. The value of α is the effective slope of the photon spectrum.[10]

photons (with exponent -1) and a cutoff depending on the thermal energy of the relativistic electrons by suitably adjusting the optical depth of the source (FIGURE 2).

In such a flat photon spectrum, however, the entire luminosity is dependent on the highest-energy photons and therefore the highest-energy electron will suffer the largest cooling rate. Also, Compton scattering will occur in the Klein–Nishina limit and this will have an impact on the lifetime of the electrons.[11,12] It is thus apparent that the real situation in NGC 4151 is likely to be more complicated than the simple mechanism used to compute the results of FIGURE 2.

3C 273

This source has been observed by COS-B.[4] Using a cross-correlation method, the authors found an excess centered on $\alpha(1950) = 12^h29 \pm 6^m$, $\delta(1950) = (3 \pm 1.5)°$ with a distribution expected for a point source. They tentatively identified this source with 3C 273, yielding then a luminosity of 2×10^{46} erg/s in the 50–500 MeV energy interval. The luminosity per decade of 3C 273 is shown in FIGURE 3.

COS-B will be devoting much of its observation time planned for next year to a study of extragalactic regions.

GALACTIC SOURCES

The total of observed localized sources in the galaxy has now been increased by COS-B to 22.[13] FIGURE 4 shows their distribution in the sky in galactic coordinates. It also indicates the positions of four pulsars found so far and of complex regions where it is difficult to isolate sources unless one is helped by the temporal structure of the sources. Away from the region of the galactic center, the gamma-ray galaxy is sufficiently simple to allow the detection of localized regions even with the limited angular resolution of $\sim 2°$.

It is significant that practically all the galactic sources are very close to the galactic plane; only 3 sources have $b^{II} > 4.5°$. The source with the largest latitude at galactic coordinates $l^{II} = 353°$ and $b^{II} = 16°$ has been associated with the Ophiucus dark cloud complex and would thus emit gamma rays because it is penetrated by cosmic rays. Since the general galactic emission is extremely narrow[13] ($|b^{II}| \lesssim 1.5°$), such a statement can be extended to include also those localized sources that have not yet been discovered.

Positive identifications by COS-B have been possible only in the case of four pulsars,[14] PSR 1822−09, PSR 0531+21, PSR 0740+28, and PSR 0833−45, and, if one accepts the two "very likely" discoveries by SAS-2,[15] of two additional pulsars,

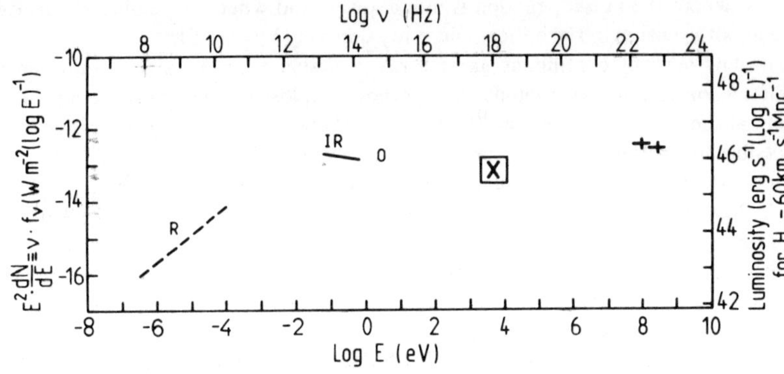

FIGURE 3. Spectrum of 3C 273 at radio (R), infrared (IR), optical (O), x-ray (X), and gamma-ray energies (the two crosses).[4]

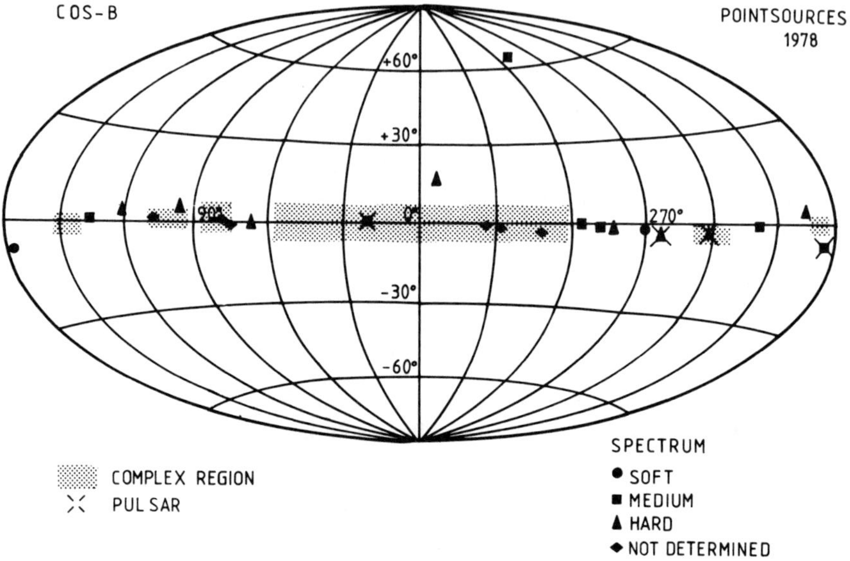

FIGURE 4. Sky map of gamma-ray sources after COS-B.[13]

PSR 1818−04 and PSR 1747−46. It is therefore suggestive to divide the discussion into two parts, pulsars and the rest.

Gamma-Ray Pulsars

The two most conspicuous pulsars are the Crab and Vela. Indeed, Vela is the strongest gamma-ray source in the sky. The radiation from Crab and Vela is entirely pulsed, the 2σ lower limit of the pulsed fraction > 50 MeV for the Crab is 78%, and for Vela 91% (FIGURE 5). One may see this in the context of the flux values at lower energies, where the pulsed fraction is much lower, and where large differences occur between both sources in the optical and x-ray domain (FIGURE 6).

The low level of continuous gamma-ray emission from Vela sets limits on the production of cosmic ray protons in supernova explosions and their storage in the vicinity of the supernova remnant.[16,17] Although these limits are not yet decisive it may be possible to see the cosmic ray protons via the π^0-decay gamma rays if gamma-ray instruments finally detect continuous emission from Vela or the Crab.

The light curves of both pulsars show remarkable differences. While the Crab is in phase throughout the entire electromagnetic spectrum (FIGURE 6), Vela is not, and it also displays quite different light curves in the radio, optical, and gamma-ray domain. However, in gamma rays, Crab and Vela have a very similar appearance showing two maxima with 0.4 phase separation.

Vela is so strong that it has been possible to derive light curves as a function of energy[18] from the COS-B results (FIGURE 7). This is quite remarkable: While the first pulse appears to be predominant until 300 MeV, it weakens above this energy and even seems to have disappeared > 1000 MeV while the second pulse is still visible.

This disappearance of pulsed gamma-ray emission with increasing energy is suggestive of theories in which gamma rays are produced by curvature radiation of electrons near the poles of the neutron star. These gamma rays may then cascade by pair production on the magnetic fields, if they have sufficiently high energies.[19–23]

COS-B has recently added two more pulsars[14] to its list of discoveries, PSR 1822−09 and PSR 0740+28. These results will be presented in detail at the pulsar workshop by the COS-B collaboration. The very interesting feature of these new discoveries is that they also show a double maximum with 0.4 phase separation, like Crab and Vela (FIGURE 8). If one goes by the COS-B results alone, one would thus conclude that one always sees the two pulses if they are visible at all.

It should be pointed out that SAS-2 has claimed[15] the discovery of two pulsars, 1747−46 at $(2.4\pm0.7)\times10^{-6}$ photons > 35 MeV/cm²s and 1818−04 at $(2.0\pm0.5)\times10^{-6}$ photons > 35 MeV/cm²s. COS-B had one look so far at 1818−04 and did not see it, while SAS-2 saw neither 1822−09 (2σ upper limit=2.7×10^{-6}) nor 0740+28 (2σ upper limit=2.1×10^{-6}). The flux values reported by COS-B for PSR 1822−09 are $(7.8\pm1.5)\times10^{-6}$ photons from 50–150 MeV/cm²s and $(4.7\pm1.3)\times10^{-7}$ photons > 150 MeV/cm²s for PSR 0740−28. The light curves for both pulsars are shown in FIGURE 9 and do not display a double maximum structure.

If one plots the gamma-ray luminosity of these pulsars against their loss of rotational energy under the assumption of a constant moment of inertia $I = 10^{45}$ g cm², one obtains the relation shown in FIGURE 10. This appears to indicate that, with decreasing $\dot{E}$, an ever larger proportion of $\dot{E}$ is radiated in gamma rays. Indeed, near $\dot{E} = 10^{34}$ erg/s, the entire energy loss appears to be generated in the form of gamma rays.

In order to investigate if this could be a selection effect, the rotational energy loss

Energy range	PSR 0531 + 21	PSR 0833 − 45
Pulsed emission*	photon cm^{-2}s^{-1}	photon cm^{-2}s^{-1}
>50 MeV	$(5.9\pm0.8)\ 10^{-6}$	$(15.6\pm0.8)\ 10^{-6}$
>200 MeV	$(1.3\pm0.3)\ 10^{-6}$	$(5.3\pm0.4)\ 10^{-6}$
Continuous emission, 2−σ upper limit	photon cm^{-2}s^{-1}	photon cm^{-2}s^{-1}
>50 MeV	1.7×10^{-6}	1.6×10^{-6}
>200 MeV	0.5×10^{-6}	0.7×10^{-6}
Pulsed fraction, 2−σ lower limit		
>50 MeV	78%	91%
>200 MeV	72%	89%
Contribution to total pulsed flux >50 MeV		
first pulse	(50 ± 9) %	(32 ± 3) %
interregion	<15 %	(30 ± 4) %
second pulse	(50 ± 9) %	(38 ± 4) %

*Only statistical errors included

FIGURE 5. Results on the Crab and Vela gamma-ray point sources, due to observations of COS-B.

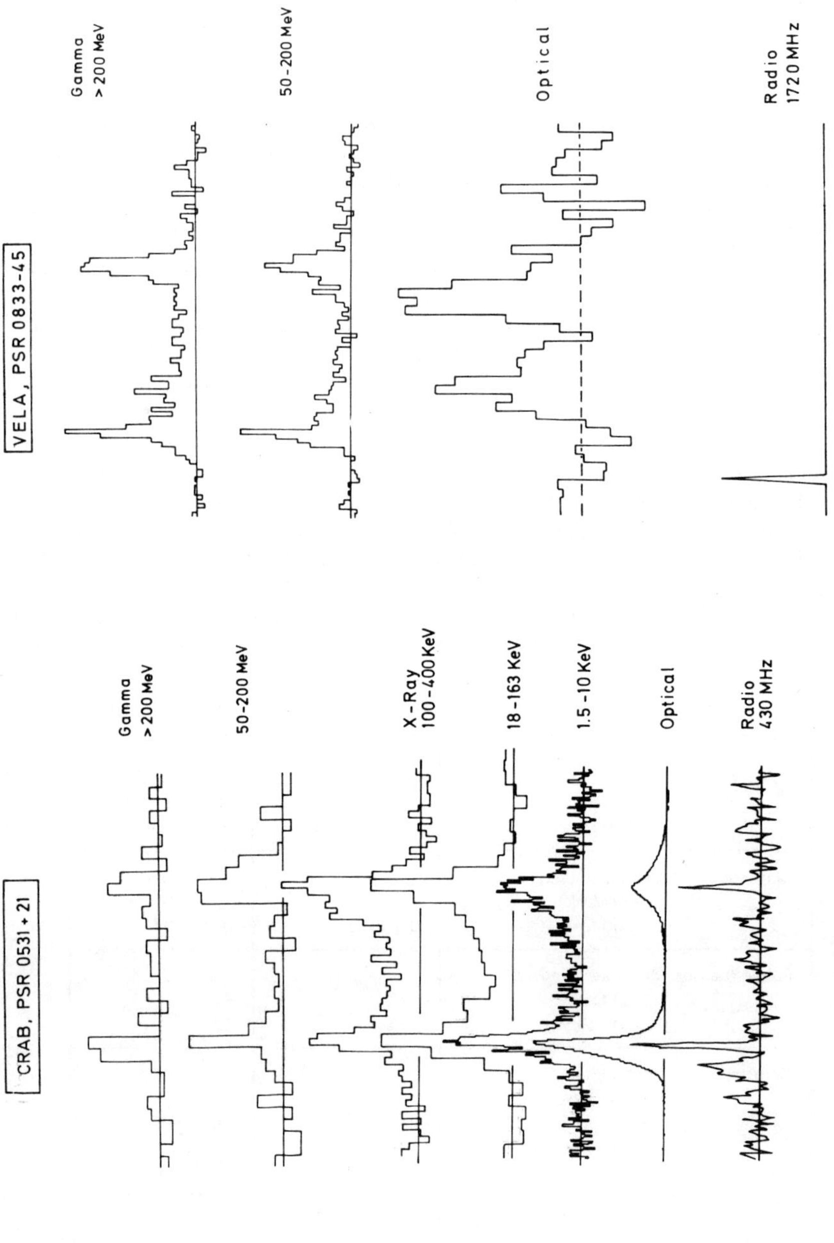

FIGURE 6. Light curves of the Crab and Vela pulsar in various parts of the electromagnetic spectrum. The gamma-ray data are from COS-B.

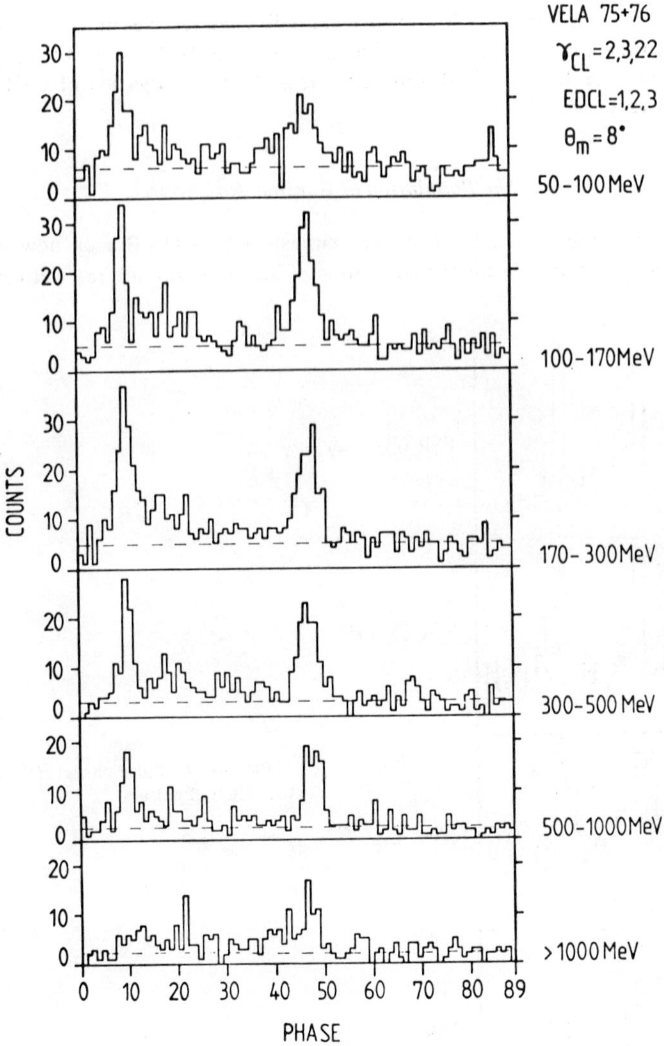

FIGURE 7. Energy-dependent pulse structure of Vela (PSR 0833−45).[18]

$\dot{E}$ of all pulsars with $\dot{E} > 10^{33}$ erg/s is also indicated in FIGURE 10, taken from Taylor and Manchester.[23] The entire sample available above $\dot{E} = 10^{35}$ erg/s consists of four pulsars of which three have been seen by COS-B. While COS-B has not yet looked at 1930+22, SAS-2 claims an upper limit,[15] which is indicated in FIGURE 10.

On the other hand, depending on distance, the minimum luminosity required for detection will be several 10^{33} erg/s in gamma rays. Thus, the pulsars below $\dot{E} = 10^{35}$ erg/s will either be discovered close to the $L_\gamma = \dot{E}$ line, or they will not be observable.

As a result, one will have to live with this evidence until much more sensitive instruments are available, unless the present indication receives much more support by the discovery that the remaining five pulsars with $10^{34} \leq \dot{E} \leq 10^{35}$ erg/s are all discovered close to $L_\gamma = \dot{E}$. In this context, a close investigation of PSR 1930+22 would be very valuable.

The Population of Gamma-Ray Sources

The catalogue of localized sources published by COS-B may now be used to discuss the question of what the population of localized gamma-ray sources (FIGURE

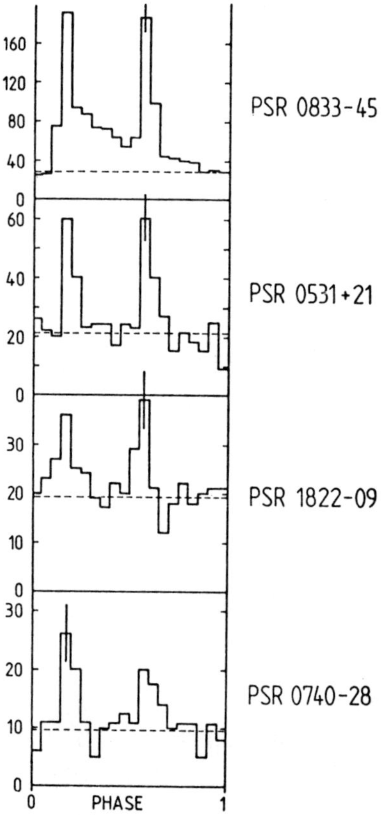

FIGURE 8. Pulse structure of the four COS-B pulsars.[14]

4) consists of. Since, however, the inner part of the galaxy shows very high emissivity and is of a complex nature, it is unclear how many localized gamma-ray sources one might have missed there, taking into account the limited angular resolution of gamma-ray instruments. The question of population is thus connected to the question of whether the large gamma-ray brightness of the central part of the galaxy is due to many unresolved localized sources being there, or due to more cosmic rays interacting

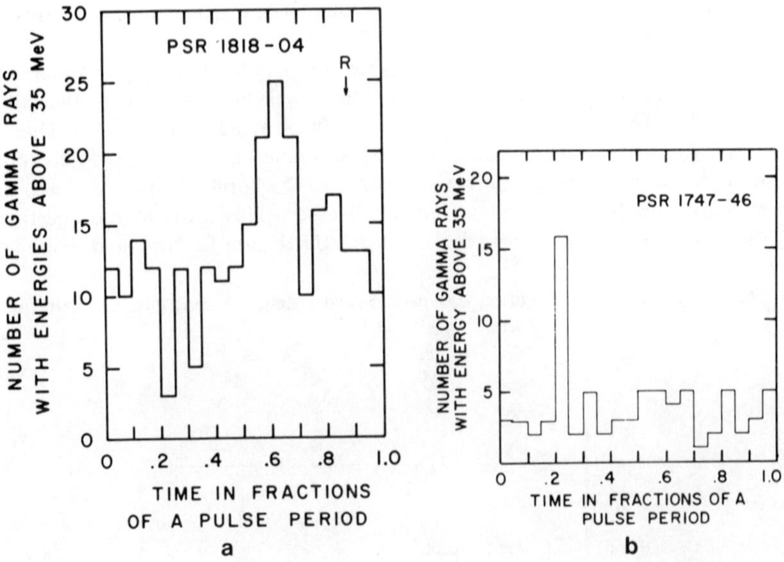

FIGURE 9. Pulsed gamma ray emission from pulsars. Distribution of gamma ray arrival times in fractions of a radio pulse period for gamma rays above 35 MeV from the directions of (a) PSR 1818−04 and (b) PSR 1747−46. In (a), the arrow marked R is the expected position of the pulse based on radio observations.[15]

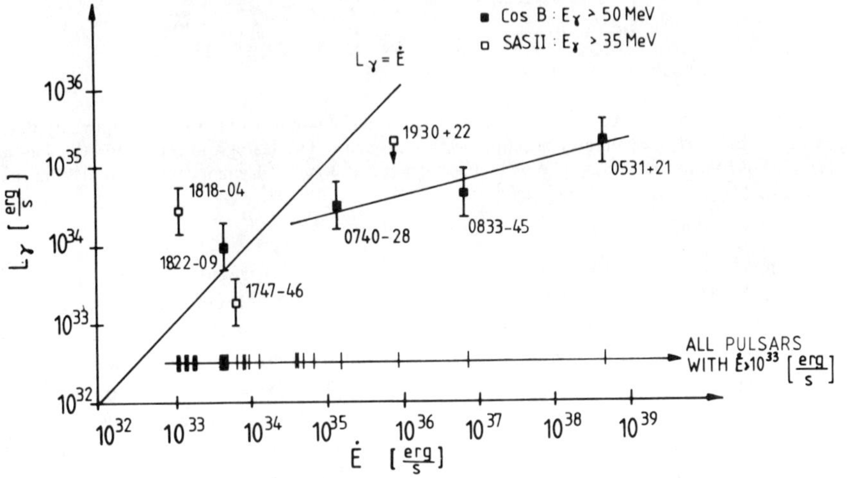

FIGURE 10. A plot of gamma-ray luminosity L_γ versus the rotational energy loss $\dot{E}$ of the pulsar. ($\dot{E}$ is taken from reference 24). Apparently, the fraction of $\dot{E}$ turning up in gamma rays increases with decreasing $\dot{E}$.

there with more matter than are available as interaction partners near the position of the sun in the galaxy.

This discussion is as old as the first positive results of gamma-ray astronomy.[25-27] As has been stressed above, the gamma-ray disc is very narrow, perhaps typically 2° full width. This corresponds to a scale height of 100 pc at a distance of 5 kpc. One may then ask if gamma-ray sources could be very young objects,[27-29] perhaps closely related in space where their progenitors were born in the form of stars. Indeed, such correlations have been claimed in individual cases. This leads to the question of whether a correlation exists between the stellar birth rate function and gamma-ray sources.[30]

The stellar birth rate function has been investigated, for example, by counting the

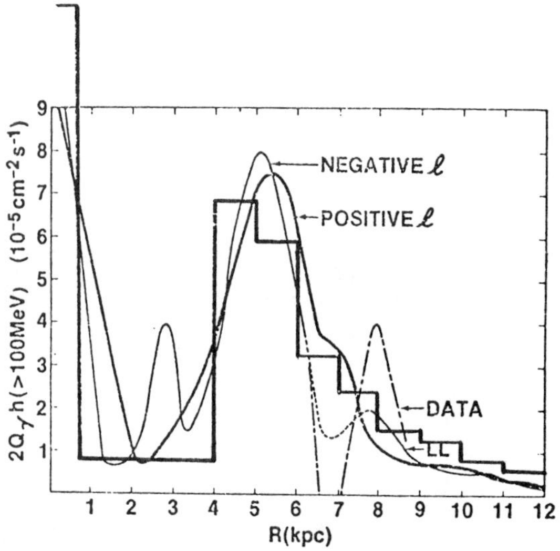

FIGURE 11. The smooth lines are the result of an unfolding of gamma-ray line intensity data assuming cylindrical symmetry of gamma-ray emissivity. The histogram represents the source function of Lyman continuum photons after Mezger,[31] if 3×10^{12} Lyman continuum photons are produced per gamma ray > 100 MeV.[27]

Lyman continuum photons (Lyc) emitted by the radio HII regions plus the extended low-density HII regions (see Mezger[31] for further references). One may now compare this stellar birth rate function with the number of gamma-ray sources observed in those regions of the galaxy where one finds favorable conditions of detection, namely low background and low space density of sources. If one now scales the luminosity of gamma-ray sources according to the stellar birth rate (i.e., the Lyc photons emitted), one obtains the average gamma-ray emissivity due to point sources as a function of galactocentric distance (FIGURE 11), in comparison with that deconvoluted from gamma-ray observations.[32] It is then possible to derive the apparent brightness distribution of gamma-ray sources in the galactic disc, and this is shown in FIGURE 12

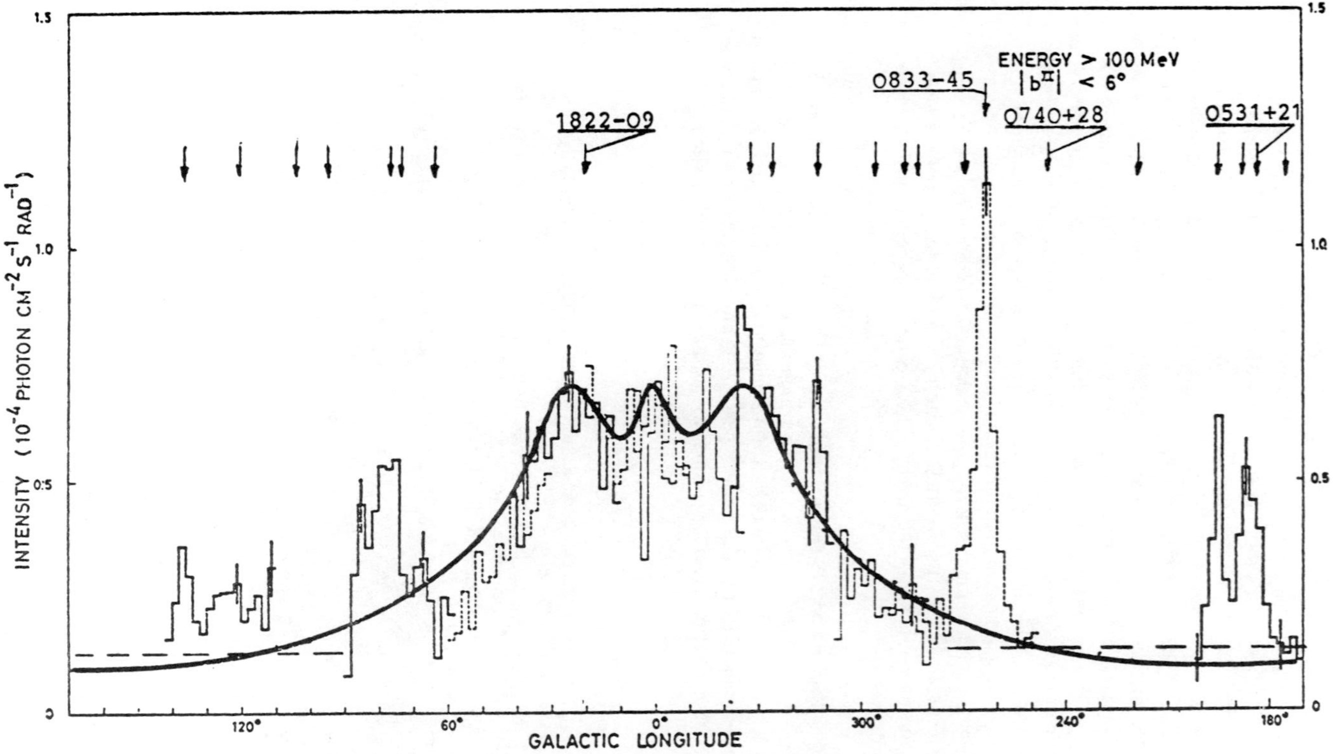

FIGURE 12. Longitude profile of the intensity of gamma rays of energy > 100 MeV over the latitude range $|b^{II}| < 6°$. The dashed line roughly indicates the line intensity expected from cosmic rays interacting with interstellar gas in the region $l^{II} = 180 \pm 80°$. The full line is the distribution in apparent brightness of O stars as given by their Lyman continuum flux after Mezger[31]; it has been normalized to the average line intensity of $< 10^{-5}$ photons > 100 MeV cm^{-2} s^{-1} rad^{-1} contributed by the COS-B point sources from the two viewing periods centered on 126° and 185°, respectively.[27]

in comparison with gamma-ray data. Thus, it is entirely possible that the bulk of the disc emission comes from point sources if they scale like the stellar birth rate.

Some interesting figures resulting from this comparison are that 3×10^{11} Lyc photons are emitted per single gamma ray > 100 MeV, or about 10^{49} gamma rays $>$ 100 MeV per birth of one solar mass. The galaxy has a luminosity of about 10^{42} photons > 100 MeV per second, and ~ 5000 sources are required to produce the bulk of this emission, since the COS-B collaboration has concluded that the typical luminosity of a gamma-ray source would be $\gtrsim 10^{35}$ erg s^{-1} (corresponding to about 2×10^{38} photons > 100 MeV per second). From such a comparison one also expects that typically $\gtrsim 15$ sources should be visible within any one "observation period" of COS-B directed towards the complex regions near the galactic center (each observation period being 30° wide in longitude), and this illuminates the problem of source resolution with an instrument of angular resolution 2°.

Interesting as such a comparison may be, it does of course not constitute a proof which could be obtained only by eventual discovery and identification of the gamma-ray sources and by establishing the astrophysical connection between birth rate and gamma-ray sources.

Note added (September 1979): Recently, radio measurements of pulsars PSR 1822 and PSR 0740 have become available which are contemporary to the COS-B gamma ray observations. The radio periods thus derived differ significantly by several hundred nanoseconds from the gamma ray periods. While this new situation is being investigated, these identifications as gamma ray pulsars are open to question.

REFERENCES

1. GRAML. F., F. P. PENNINGSFELD & V. SCHÖNFELDER. 1977. Proc. 12th ESLAB Symposium, Frascati. ESP SP **124:** 301.
2. SCARSI, L., K. BENNETT, G. F. BIGNAMI, G. BOELLA, R. BUCCHERI, W. HERMSEN, L. KOCH, H. A. MAYER-HASSELWANDER, J. A. PAUL, E. PFEFFERMANN, R. STIGLITZ, B. N. SWANENBURG, B. G. TAYLOR & R. D. WILLS. 1977. Proc. 12th ESLAB Symposium, Frascati. ESA SP **124:** 3.
3. GRAML, F., F. P. PENNINGSFELD & V. SCHÖNFELDER. 1978. Proc. Esrange Symposium, Ajaccio. ESA SP **135:** 287.
4. SWANENBURG, B. N., K. BENNETT, G. F. BIGNAMI, P. CARAVEO, W. HERMSEN, G. KANBACH, J. L. MASNOU, H. A. MAYER-HASSELWANDER, J. A. PAUL, B. SACCO, L. SCARSI & R. D. WILLS. 1978. Nature **275:** 298.
5. SCHÖNFELDER, V. 1979. This volume.
6. MUSHOTZKY, R. F., S. S. HOLT & P. J. SERLEMITSOS. 1978. Astrophys. J. **225:** L115.
7. BARR, P., N. E. WHITE, P. W. SANFORD & J. C. IVES. 1977. M.N.R.A.S. **181:** 43.
8. IVES, J. C., P. W. SANFORD & M. V. PENSTON. 1976. Astrophys. J. Lett. **207:** L159.
9. KATZ, J. 1976. Astrophys. J. **206:** 910; see there for further references.
10. POZDNYAKOV, L. A., I. M. SOBOL & R. A. SYUNYAEV. 1976. Soviet Astron. Lett. **2:** 55.
11. BLUMENTHAL, G. R. 1971. Phys. Rev. D **3:** 2308.
12. BLUMENTHAL, G. R. & R. J. GOULD. 1970. Rev. Mod. Phys. **42:** 237.
13. MAYER-HASSELWANDER, H. A., K. BENNETT, G. F. BIGNAMI, R. BUCCHERI, N. D'AMICO, W. HERMSEN, G. KANBACH, F. LEBRUN, G. G. LICHTI, J. L. MASNOU, J. A. PAUL, K. PINKAU, B. SACCO, L. SCARSI, B. N. SWANENBURG & R. D. WILLS. 1979. This volume.
14. BUCCHERI, R., K. BENNETT, G. F. BIGNAMI, N. D'AMICO, W. HERMSEN, D. J. H. HUIZING, G. KANBACH, G. G. LICHTI, J. L. MASNOU, H. A. MAYER-HASSELWANDER, J.

A. PAUL, B. SACCO, B. N. SWANENBURG & R. D. WILLS. 1979. Astron. Astrophys. In press.
15. ÖGELMAN, H., C. E. FICHTEL, D. A. KNIFFEN & D. J. THOMPSON. 1976. Astrophys. J. **209:** 584.
16. PINKAU, K. 1970. Phys. Rev. Lett. **25:** 603.
17. LINGENFELTER, R. E. & J. C. HIGDON. 1975. Proc. 14th Intern. Cosmic Ray Conference, München. **2:** 487.
18. KANBACH, G. 1978. 6th Europ. Cosmic Ray Symposium, Kiel.
19. STURROCK, P. A. 1971. Astrophys. J. **164:** 529.
20. RUDERMAN, M. A. & P. G. SUTHERLAND. 1975. Astrophys. J. **196:** 51.
21. ÖGELMAN, H., S. AYASLI & A. HACINLIYAN. 1976. *In* 'The Structure and Content of the Galaxy, and Galactic Gamma Rays'. NASA CP-002.
22. SALVATI, M. & E. MASSARO. 1978. Astron. Astrophys. **67:** 55.
23. HARDING, A. K., E. TADEMARU & L. W. ESPOSITO. 1978. Astrophys. J. **225:** 226.
24. TAYLOR, J. H. & R. N. MANCHESTER. 1975. Astrophys. J. **80:** 794.
25. CLARK, G. W., G. P. GARMIRE & W. L. KRAUSHAAR. 1970. IAU Symposium No. 37. L. Gratton, Ed. Reidel. Dordrecht.
26. KRAUSHAAR, W. L., G. W. CLARK, G. P. GARMIRE, R. BORKEN, P. HIGBIE, C. LEONG & T. THORSOS. 1972. Astrophys. J. **177:** 341.
27. PINKAU, K. 1979. Nature **277:** 17; see there for a more complete discussion.
28. MONTMERLE, T. 1978. Preprint SAp 78/27. Centre d'Etudes Nucléaires de Saclay, France.
29. PANAGIA, N. & G. ZAMORANI. 1979. Preprint. Laboratorio di Radioastronomia CNR. Bologna, Italy.
30. BURTON, W. B. 1976. Ann. Rev. Astron. Astrophys. **14:** 275.
31. MEZGER, P. G. 1978. Astron. Astrophys. **70:** 565.
32. STECKER, F. W. 1976. *In* 'The Structure and Content of the Galaxy, and Galactic Gamma Rays'. NASA CP-002.

GAMMA-RAY LINES FROM THE GALACTIC CENTER*

M. Leventhal

Bell Laboratories
Murray Hill, New Jersey 07974

C. J. MacCallum†

Sandia Laboratories
Albuquerque, New Mexico 87115

The history of the detection and identification of the first gamma-ray lines from outside the solar system is a fascinating scientific detective story that I would like to review and bring up to date. It is a story in several chapters, the first of which was written by the Rice University group led by R. C. Haymes. A second chapter was added this past year by the joint Bell Labs/Sandia Labs group of which I am a member. My principal scientific colleague in this work has been Dr. C. J. MacCallum of Sandia Labs. Several other chapters are currently in the press and I will tell you what I know about them. The gamma-ray lines with which we are concerned come from the Galactic Center (GC) direction and the principal line of interest is the 511 keV line from electron–positron annihilation.

The story begins in the early 1970s when the Rice University group went to Argentina and flew the telescope shown in FIGURE 1 on a balloon platform to search for gamma-ray lines from the GC. This instrument is the forerunner of all present gamma-ray line telescopes. It is built entirely from active NaI scintillator material and has a 4-inch-diameter crystal as the central element. The instrument has angular resolution ~24° FWHM and energy resolution ~100 keV. The NaI shield is operated in anticoincidence with the central detector and performs three separate functions: (1) it defines the field of view, (2) it ensures that only full-energy events are recorded, and (3) it shields the central detector from the large atmospheric gamma-ray background.

The Rice group originally reported the GC spectrum shown in FIGURE 2, based upon two balloon flights with this instrument.[1] There is clear evidence here for a continuum of gamma-ray emission falling off toward high energy. This is thought to arise from the many point sources of hard x-rays in the field of view. However, there is also strong evidence for a line feature at the puzzling energy of 476 ± 24 keV, close to but not quite in agreement with the electron–positron annihilation energy.

These observations stimulated several theoretical suggestions concerning the possible origin of the line feature. Fishman and Clayton[2] noticed that the lowest-lying nuclear energy level of the ^{7}Li isotope is at 478 keV and suggested that GC cosmic rays were generating the feature via the $^7\text{Li}(p,p')^7\text{Li}^*$ reaction. Kozlovsky and Ramaty[3] suggested that a more efficient way of getting to the excited state of ^{7}Li was through cosmic-ray interactions of the form $^4\text{He}(\alpha, p)^7\text{Li}^*$. However, both suggestions suffer from considerable difficulty in accounting for the large flux in the line, reported to be $(1.8 \pm 0.5) \times 10^{-3}$ photons cm^{-2}s^{-1}. A second class of models was proposed

*The balloon and helium for the Bell Labs/Sandia Labs experiment were provided by NASA.

†Research supported by USERDA.

0077–8923/80/0336–0248 $1.75/1 © 1980, NYAS

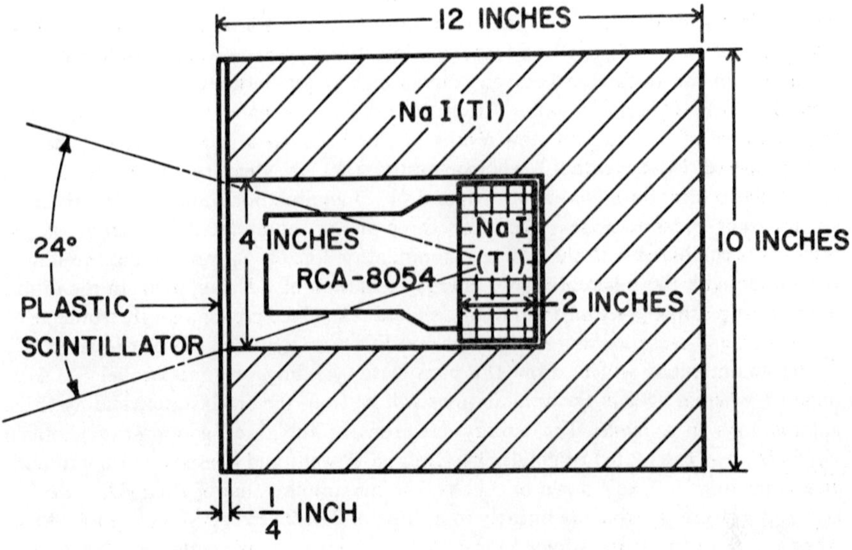

FIGURE 1. Schematic diagram of the original Rice University NaI scintillation gamma-ray telescope.

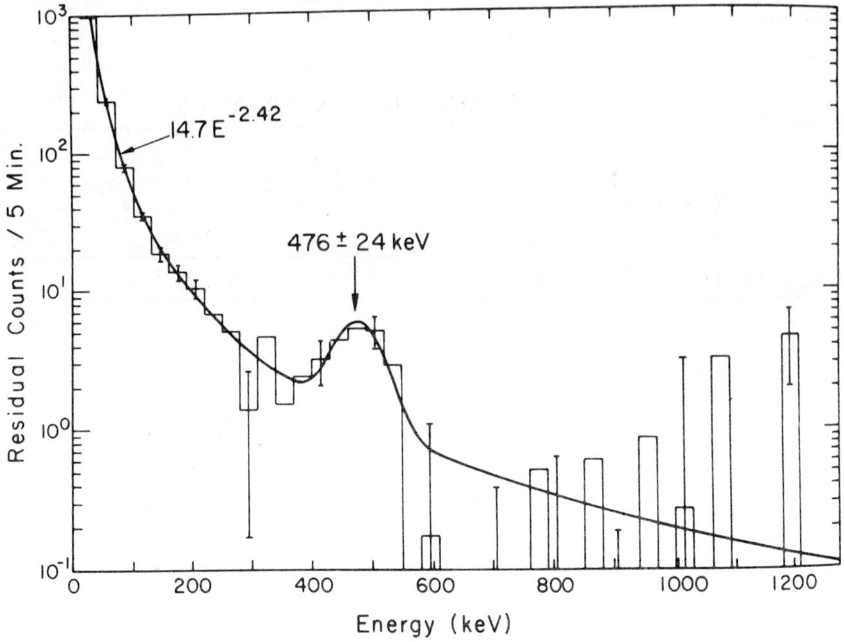

FIGURE 2. A histogram representation of the measured flux from the GC region based upon balloon flights in 1970 and 1971 by the Rice group.

independently by Guthrie and Tademaru[4] and Ramaty, Börner, and Cohen.[5] These workers suggested that the line feature arises from positron annihilation radiation occurring on the surface of neutron stars, which is gravitationally redshifted down from 511 keV. This model suffers from the need to populate the GC region with a large number of very small (~few tenths $M_\odot$) neutron stars. A third class of models was proposed by Leventhal,[6] who suggested that the feature might be due to interstellar positronium ($ps = e^+ + e^-$ bound state) annihilation radiation. In this case, an apparent redshift arises when the three-photon annihilation spectrum of $^3S_1$ ortho-*ps* is combined with the 511 keV annihilation line of $^1S_0$ para-*ps* and the sum is convoluted with the inherently poor energy resolution of a NaI system. In the dilute interstellar plasma, positron annihilation should take place predominantly from one of the two ground-state configurations of *ps* (see FIGURE 3). Should this be the case, 25% of the annihilations will be from the para state, yielding two antiparallel 511 keV photons, whereas 75% of the annihilations will be from the ortho state, yielding three gamma rays in a plane. The energy spectrum of the three gamma-ray photons, $P_T(E)dE$, was calculated originally by Ore and Powell[7] and consists of a continuum extending from 511 keV down to 0 keV. The maximum value of $P_T(E)dE$ is at 511 keV and it decreases roughly linearly to a value of 0 at 0 keV. As shown in FIGURE 4, when a pure *ps* spectrum is folded through an instrument with resolution ~100 keV, a result for the line center is obtained that is consistent with the Rice observations. What is going on here is that the instrument is including three-photon radiation in the line in an asymmetric fashion and shifting the apparent line center. A difficulty with the *ps* model is the need to populate the interstellar medium with a large number of positrons.

POSITRONIUM ANNIHILATION

	↓↑ $^1S_0$ PARA STATE	↑↑ $^3S_1$ ORTHO STATE
FRACTION	25 %	75 %
LIFETIME	1.3×10^{-10} sec	1.4×10^{-7} sec
MODE	2γ	3γ

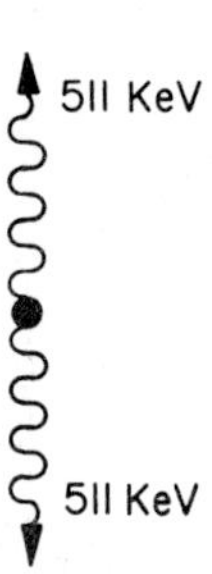

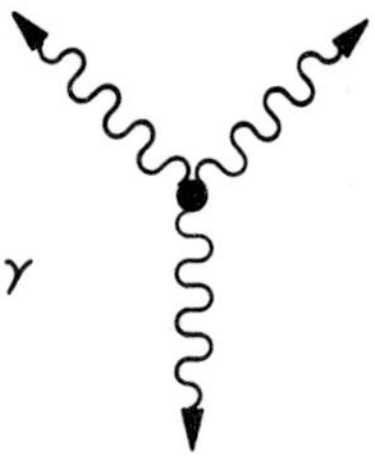

FIGURE 3. Characteristics of positronium annihilation from the ground electronic state.

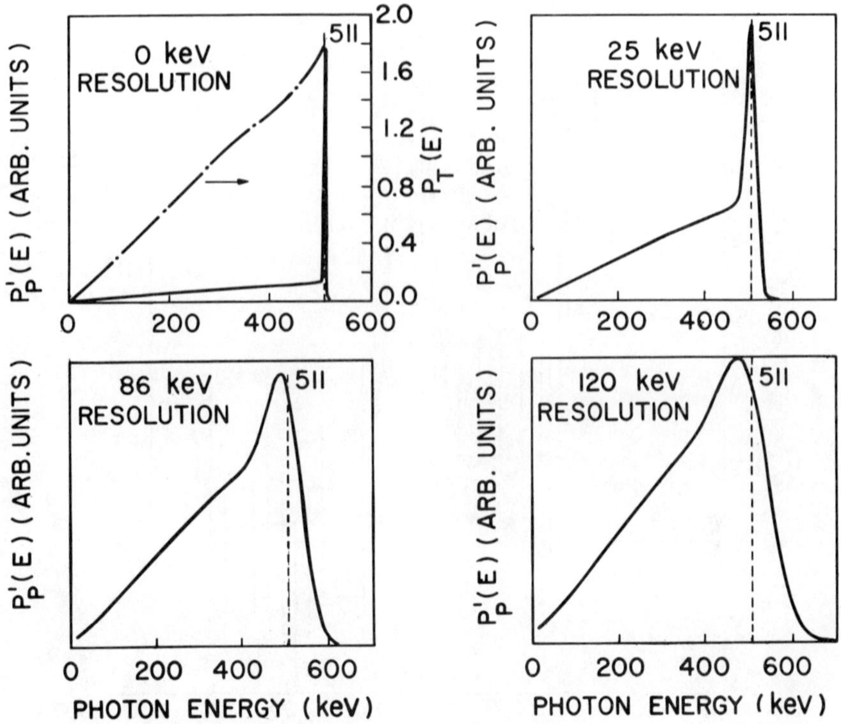

FIGURE 4. Plots of the detected positronium annihilation function $P'_p(E)$ for various energy resolutions (FWHM) of the γ-ray detector. A Gaussian width of 3.2 keV has been assumed for the singlet annihilations. The broken curve in the upper left-hand box is a plot of the pure, triplet annihilation function $P_T(E)$.

Since their earlier work, the Rice group has confounded us all by repeating their experiment with an improved NaI system and reporting the feature to be at 530 ± 11 keV with a reduced flux of $(8.0 \pm 2.3) \times 10^{-4}$ photons $cm^{-2}s^{-1}$. As shown in FIGURE 5, the feature is now to the high-energy side of 511 keV.[8] These workers suggest that the quantitative differences in flux and average energy between their earlier and later flights may be due to any of several factors, including a narrower beamwidth in the later experiment (13° FWHM versus 24° FWHM), time dependence, or different aiming directions. They also indicate their belief that the line is really the 511 keV free-positron-annihilation line shifted to higher energies by a statistical fluctuation in their data.

In addition, the Rice group now report additional broad line features at higher energies, as shown in FIGURE 6. Possible line activity is indicated in the regions from 1.2 to 2.0 MeV and 4.0 to 5.0 MeV. It has been suggested that these may be Doppler-broadened nuclear deexcitation lines produced by cosmic-ray interactions. In particular, all cosmic-ray interaction models[9,10] predict a prominent line at 4.4 MeV from the deexcitation of the first excited state of C^{12}. However, it should be noted that the reported detection is at a marginal level of statistical significance ($\sim 3.5\sigma$).

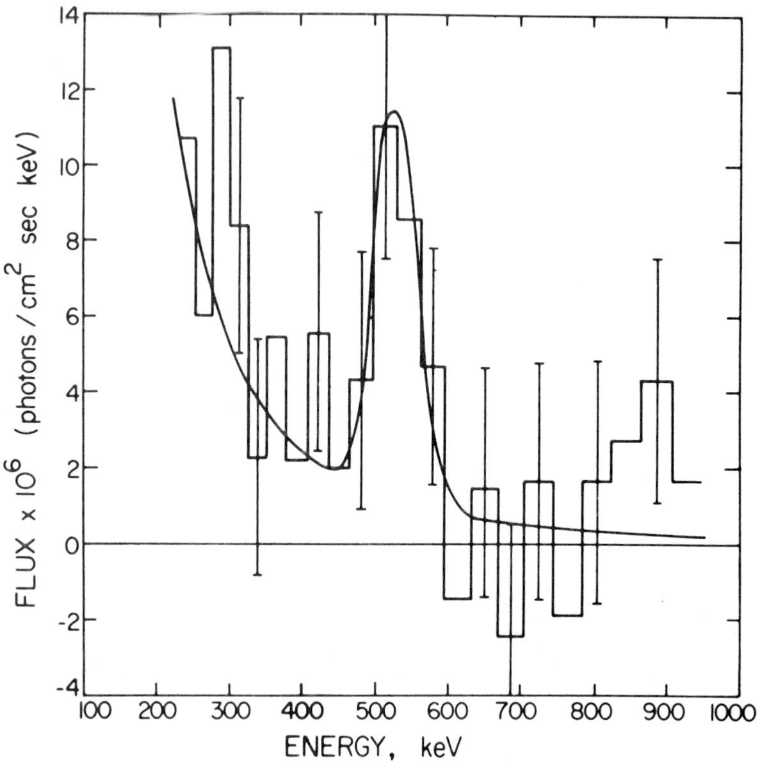

FIGURE 5. Measured differential energy spectrum of the GC region below 1 MeV, based upon one balloon flight in 1974 by the Rice group. The spectral line near 500 keV is detected at the 3.5σ level.

Clearly, the Rice group has done some remarkable pioneering work. However, the need for independent study of this problem with instruments of greater energy resolution is also apparent. The Bell Labs/Sandia Labs group has now completed the first phase of such a study. On 11–12 November 1977, a balloon-borne, high-resolution (3.2-keV FWHM at 511 keV) telescope was flown over Alice Springs, Australia, by our group to search for GC gamma-ray lines. The telescope is shown in FIGURE 7. The basic improvements made over earlier telescopes is the replacement of the central scintillator by a high-purity germanium diode, which must be operated at liquid nitrogen temperature, and the scaling up of the anticoincidence shield to massive proportions.

Observations of the GC direction ($\alpha = 17^h44^m$, $\delta = -28°0'$) were made at a mean atmospheric depth of 3.6 g cm^{-2}. A total of 17.3 hours of data was accumulated in alternate 20-min target-background segments, with the telescope maintained at the same zenith angle but reflected about the north–south direction in azimuth for the background measurements. Since the GC rose very nearly in the east and set in the west, the background may be said to have risen in the west and set in the east.

However, the background observations swept over the celestial sphere with fixed declination, $\delta = -28°0'$, and with right ascension α varying between 8^h49^m and 3^h09^m, paying no special attention to any particular point. This method of data acquisiton allows equal background measurement to the east and the west and eliminates the effect of any hypothetical asymmetry.

In FIGURE 8 we show energy spectra in the vicinity of 511 keV. Three clearly significant, cosmic-ray-induced telescope lines are evident in these data and have been seen by us in previous balloon flights. These are at 439 keV from ^{23}Na (n,n') ^{23}Na*, at 472 keV from ^{23}Na (n, γ) ^{24}Na* and/or ^{76}Ge (n, γ) ^{77}Ge, and at 511 keV from positron annihilation. The instrumental 511 keV line seen by all workers in the field arises from cosmic-ray generated positrons in the telescope and the atmosphere. A feeble unidentified line is also evident at 497 keV and has been seeen in previous flights. A striking characteristic of this data is the near doubling of the 511 keV line count rate above continuum when the GC is in view. One must now address the questions of (1) whether this effect may be induced by some systematic error in the experiment, and (2) whether or not the result is statistically significant.

Of the two dozen instrumental lines visible in our spectra, only the 511 keV line shows a statistically significant difference between target and background count rates. The data were also divided into six subgroups corresponding to GC ascending, near zenith, and descending, on each of two days. The background 511 keV line rate was essentially the same for each subgroup and a positive target–background excess was obtained for each subgroup. Hence the effect is independent of the direction in which

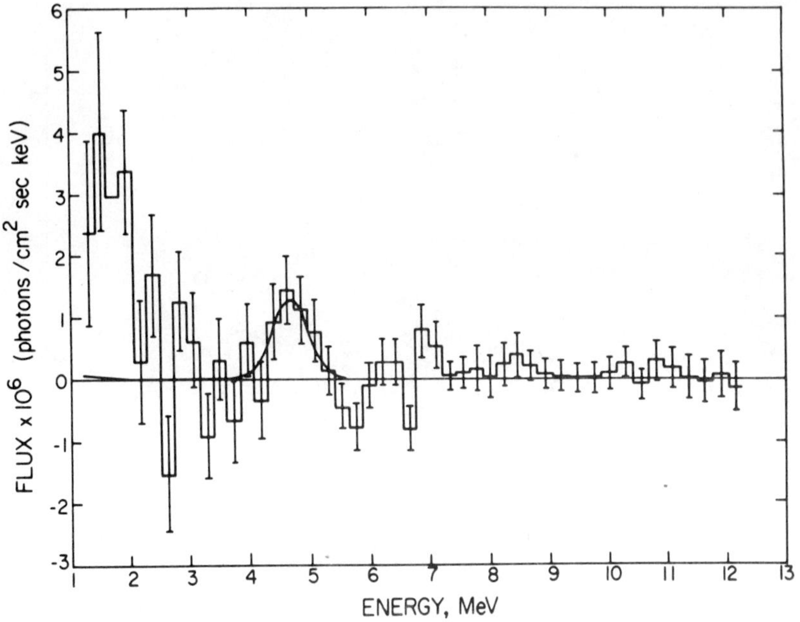

FIGURE 6. Measured differential energy spectrum of the GC region above 1 MeV based upon one balloon flight in 1974 by the Rice group.

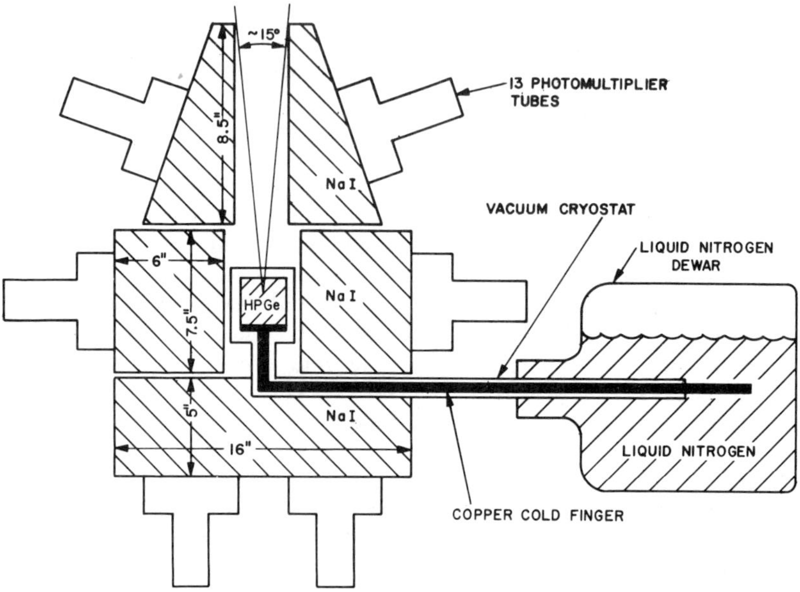

FIGURE 7. Schematic diagram of the Bell Labs/Sandia Labs germanium high-resolution telescope. The central detector is an ~ 130 cm^3 high-purity germanium diode.

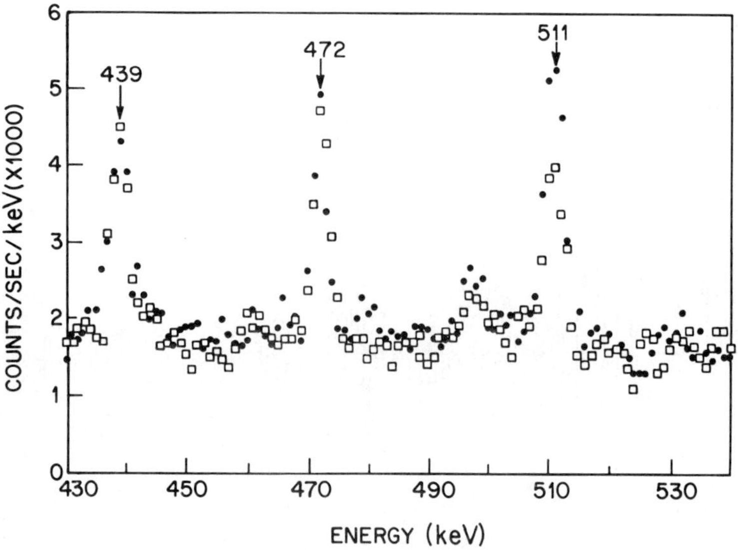

FIGURE 8. Energy spectra in the vicinity of 511 keV for the sum of all GC target (circles) and GC background (squares) measurements. Raw data corrected only for drifts in the energy calibration. Each point represents a 1 keV energy bin. Note that of the instrumental lines shown, only the 511 keV line shows an enhanced count rate in the target spectrum.

data is being taken. The azimuthal dependence of the instrumental 511 keV line intensity was studied in flight at an enhanced count rate with the anticoincidence shield off. FIGURE 9 shows the 511 keV line count rate as a function of azimuth angle for a fixed elevation angle of 60°. These data are consistent with there being no azimuthal dependence at all. A considerable effort has been made to think of a systematic effect that might be biasing the data. None has been found. Our conclusion is that we are not being fooled by some systematic error.

The raw target data of FIGURE 8 contain an excess of counts in the energy bins 509 through 512 keV that is significant at the 4.5σ level. However, each of the 20 min segments have been treated equally here. Two real physical effects have been ignored that should be included in the data analysis. As the telescope points down closer to the

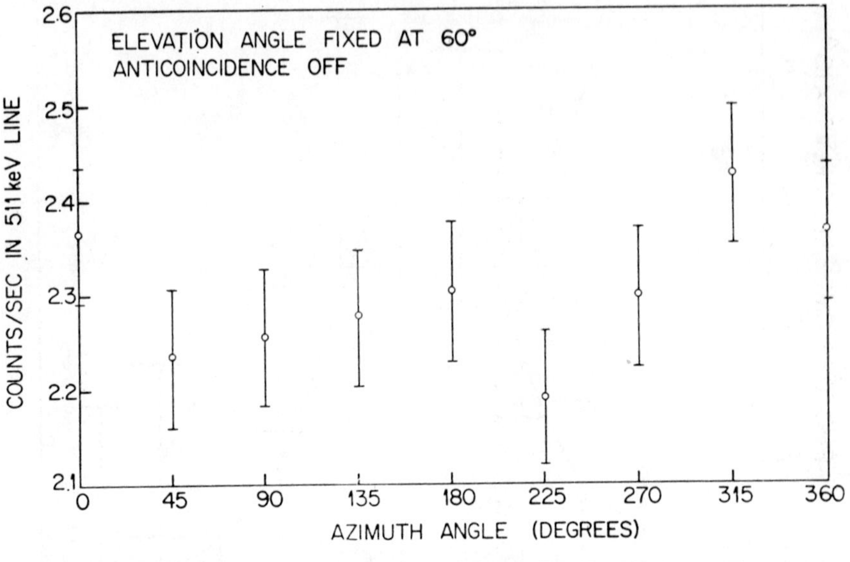

FIGURE 9. Count rate in the 511 keV line versus azimuth angle. North is 0°. Data were taken with the anticoincidence shield off, which increased the count rate in the line by a factor of about 100.

horizon, an increasing amount of atmospheric attenuation is obscuring any astrophysical 511 keV line. Also, the instrumental background at 511 keV is increasing. Data taken close to the horizon was inherently noisier than data taken near zenith. It is our belief that the question of statistical significance should be asked after a minimal amount of data massaging. When target–background pair difference rates are formed, corrected for atmospheric attenuation, weighted inversely as to their variances, and averaged, the excess in the 511 keV line becomes significant at the 5.6σ level. The increase in statistical significance that occurs when the data are weighted in a manner appropriate to an extraterrestial source is additional evidence for such an origin as opposed to an instrumental or atmospheric origin. When detector area, efficiency, and dead-time factors are included, the excess represents a 511 keV line flux of

$(1.22 \pm 0.22) \times 10^{-3}$ photons $cm^{-2}s^{-1}$ at the top of the atmosphere with a FWHM $\lesssim 3.2$ keV. This result is consistent with the line flux reported earlier by the Rice group. Our final result is shown in FIGURE 10. Taken in conjunction with the earlier work, the existence of a GC 511 keV line now seems unambigiously established. While it seems likely that the radiation is coming from an extended region of space, were it all to originate at the distance of the GC (d = 10 kpc), 8.1×10^{42} annihilations per second are required, corresponding to a luminosity of 1.4×10^{37} erg sec^{-1} in the 511 keV line. An additional 2.5×10^{43} annihilations per second are required if one accepts the existence of a *ps* continuum.

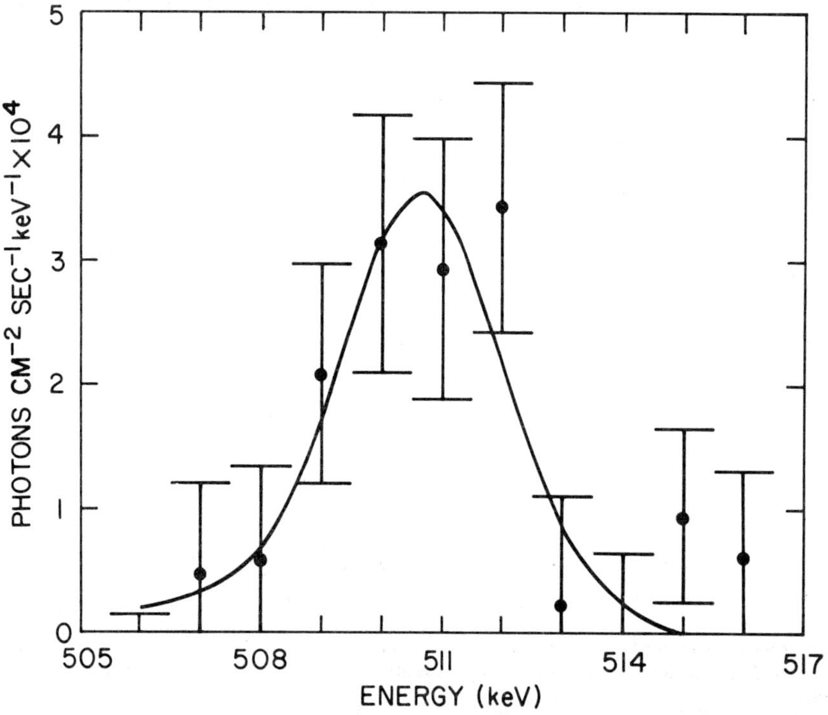

FIGURE 10. Differential photon spectrum from the GC direction for the energy range near 511 keV. The line center was found at 510.7 ± 0.5 keV with a width limited by system resolution at $\lesssim 3.2$ keV.

It is important to determine whether evidence for a three-photon continuum is to be found in our data. Because of the good energy resolution of the system, no apparent redshift of the 511 keV peak is expected. Rather, the three-photon continuum would give rise to a small excess of counts in the channels extending down from 511 keV to lower energy. However, a general x-ray–γ-ray inverse power law continuum emanating from the GC direction is seen very clearly in the data at lower energy. The presence of this additional continuum radiation greatly complicates the task of extricating any potential three-photon spectrum.

In FIGURE 11 the GC difference spectrum extrapolated to the top of the

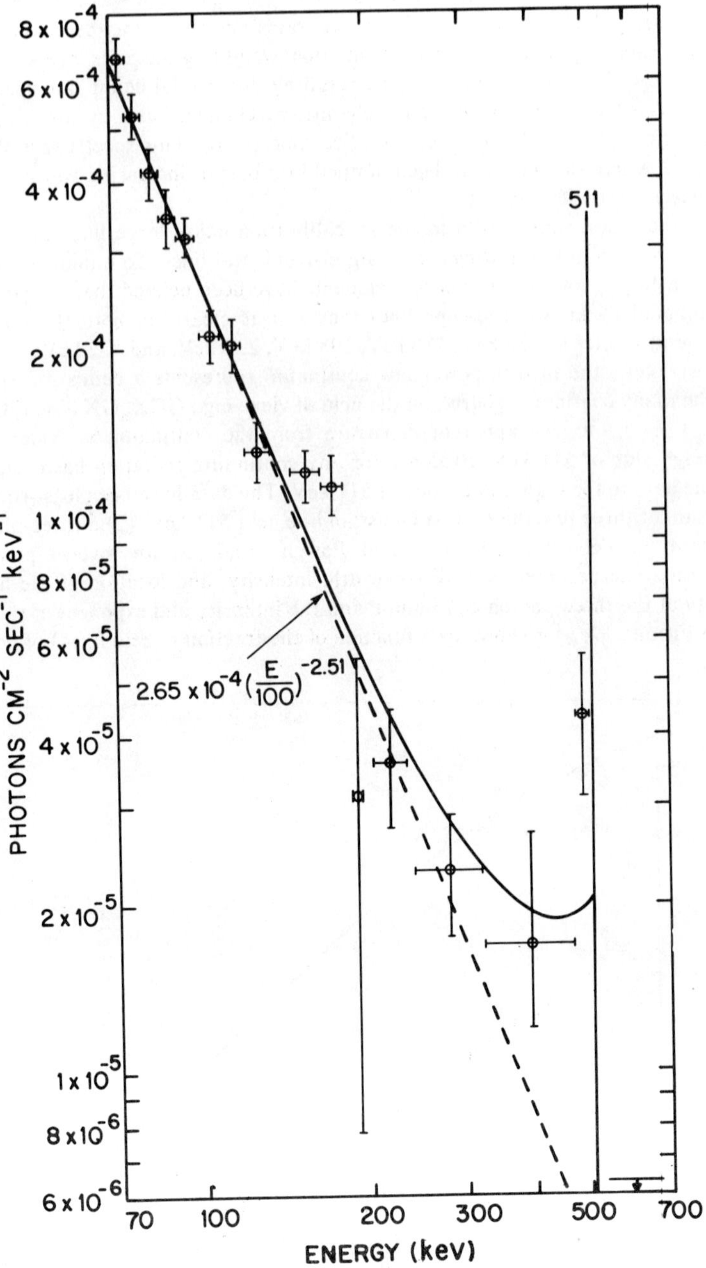

FIGURE 11. Differential photon spectrum from the GC direction for the energy range 70–700 keV. The solid line is a least-squares fit to the data as discussed in the text. The parameter f was found to be 0.92.

atmosphere is plotted between 70 and 700 keV. This spectrum was obtained by first forming target–background pair difference rates on a channel-by-channel basis, correcting each pair for atmospheric attenuation, weighting the pairs inversely as their variances, and forming an average. The resulting differential count rate spectrum is then converted to a photon flux on a channel-by-channel basis by dividing by the detector area, photopeak efficiency, and live-time factor. This spectrum is shown in FIGURE 11 where the data have been lumped into bins of increasing width to reduce the apparent statistical scatter.

In some cases, slight drifts in energy calibration cause large fluctuations in the difference spectrum in the wings of strong instrumental lines. To minimize confusion in curve fitting, a total of five energy segments have been deleted that are potentially contaminated by known telescope lines that appear clearly in both the target and background spectra at 139 keV, 185 keV, 198 keV, 238 keV, and 472 keV.

Most likely, the inverse power law continuum represents a composite spectrum from the many continuum sources in the field of view, e.g., GCX, GX 3 + 1, GX 5 − 1, GX 1 + 4, etc. An apparent departure from the continuum is evident to the low-energy side of 511 keV. Evidence for any continuum radiation has disappeared from the data to the high-energy side of 511 keV. The data have been least-squares fit to the sum of three functions: (1) a Gaussian line near 511 keV, (2) a three-photon *ps* continuum as determined by Ore and Powell,[7] and (3) an inverse power law continuum. The parameters were the width, intensity, and location of the line, the intensity of the three-photon continuum, and the intensity and exponent of the power law. In FIGURE 12, χ^2 is shown as a function of the fraction f (held fixed) of positrons

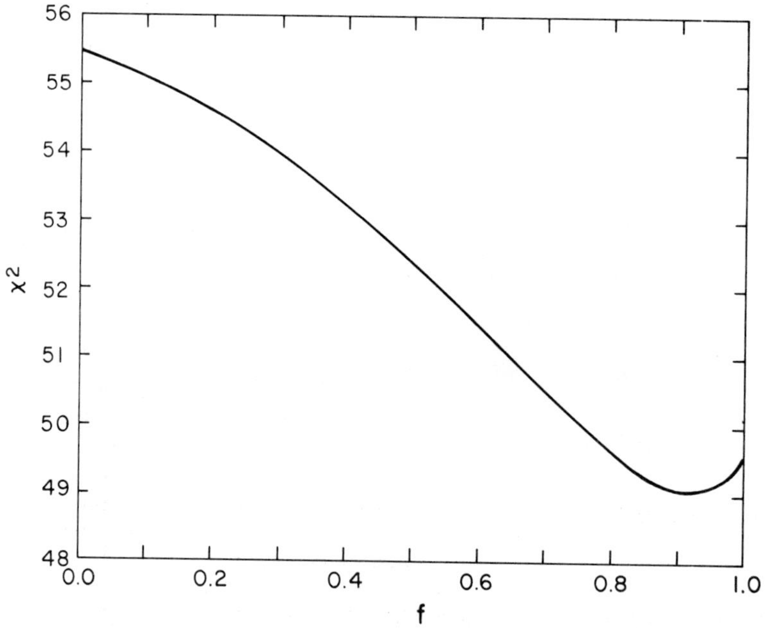

FIGURE 12. χ^2 versus f, the fraction of positrons annihilating via the positronium state, for 57 degrees of freedom and five parameters.

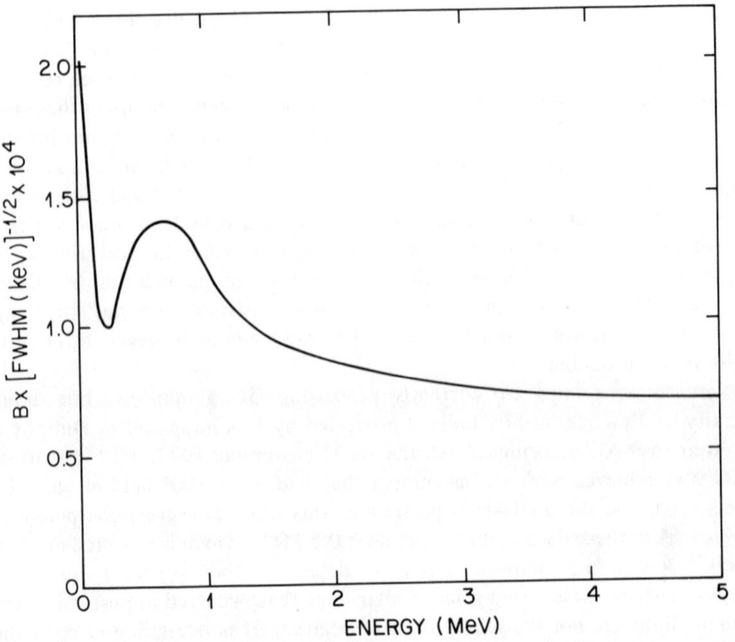

FIGURE 13. The 2σ GC line flux sensitivity B achieved by the Bell Labs/Sandia Labs group versus energy. This curve is invalid in the near vicinity of instrumental lines. See reference 21 for a tabulation of most of these lines.

allowed to annihilate via the *ps* state. For $f = 1$, corresponding to 100% *ps* annihilation, 75% of the intensity is in the three-photon continuum and 25% in the line at 511 keV. For $f = 0$, there is no three-photon continuum at all. This function has a minimum in the vicinity of 100% *ps* annihilation, which is very suggestive since this is what one expects on theoretical grounds for many astrophysical plasmas.[11] However, on the basis of the χ^2 test alone, even the fit for the case $f = 0$ is acceptable. Application of the F test for an additional term[12] indicates, at better than the 99% confidence level, though, that a statistically significant improvement in the fit has been made by inclusion of the three-photon continuum. We believe that strong but not conclusive evidence for the three-photon continuum has been presented. However, other explanations for the excess of counts to the low-energy side of 511 keV cannot be ruled out at this time.

The Bell Labs/Sandia Labs experiment did not achieve adequate sensitivity to test the reality of the broad GC line features reported[8] by the Rice group to be hundreds of keV in width. In FIGURE 13, the 2σ line sensitivity achieved by our group is shown versus energy. The reported broad line fluxes are two to three times below our sensitivity limit. It is interesting to note that the achieved sensitivity for sharp line features is $\sim 10^{-4}$ photons $cm^{-2}s^{-1}$ over most of the gamma-ray line region. This is just the sensitivity projected for the gamma-ray spectrometer to be launched aboard HEAO-C in 1979.

An automated computer search of our GC data was made to see if any other lines were apparent at any energy with any width. This search yielded three candidate lines at 170 keV, 1.611 MeV, and 3.70 MeV, which are shown in FIGURES 14–16. They have been detected in the final spectrum at the 4.1σ, 4.9σ, and 3.9σ level, respectively. However, it has not been possible to make a reasonable identification of these features and since an occasional statistical fluctuation is expected at this level, our judgment is that they should not yet be taken seriously. The Bell Labs/Sandia Labs group is planning on repeating the GC observation in April 1979. Should any of these features reappear in subsequent measurements or in the measurements of other workers, this judgment will be revoked. A similar search for *negative* (therefore certainly spurious) lines yielded several candidates at the 3.0–3.5σ level of significance, but none with higher significance. Such a chance result is quite consistent with statistical expectations. A spurious result at the 4.9σ level of significance is, however, three orders of magnitude less probable.

Several other groups are currently processing GC gamma-ray line data. The University of New Hampshire balloon group led by E. Chupp and P. Dunphy flew a NaI system over Alice Springs, Australia, on 21 November 1977. A 12-hr drift scan of the GC was achieved with an instrument that had an ~ 100° field of view. Only a progress report on the analysis is possible at this time. However, one potential line feature exists in these data at an energy near 0.5 MeV, which is detected at about the 6σ level.[13] But before claiming a positive detection, these workers must convince themselves that particles from a large solar flare that occurred almost coincidentally with their flight are not the source of this radiation. It is intriguing to note, though, that if their 0.5 MeV line turns out to have an astrophysical origin, it corresponds to a flux compatible with the Rice and Bell/Sandia results.

L. Peterson and J. Matteson from UCSD have employed the phoswich-type scintillation telescope aboard HEAO-A to search for GC gamma-ray lines. Only a fraction of their data has been processed to date. This preliminary analysis indicates that the broad features reported by the Rice group between 1 and 2 MeV and 4 and 5 MeV are not present. The sensitivity already achieved by these workers is a factor of 2–3 greater than that achieved in the earlier balloon experiment. The sensitivity required to detect the sharp line feature at 511 keV reported by the Bell Labs/Sandia Labs group has not been achieved in the preliminary analysis and will require processing of the full body of GC data.[14]

A joint CESR Toulouse/SEP Saclay group has flown a balloon-borne germanium telescope over Brazil in December 1976 and February 1977 to search for gamma-ray lines from the GC. While analysis of this data is still in progress, these workers have indicated that they definitely did not see the line at 476 keV originally reported by the Rice group, whereas evidence for a GC 511 keV line does exist in their data.[15]

In light of the above discussion, it would appear that the only GC gamma-ray line that has been conclusively detected is the 511 keV positron annihilation line. It is interesting to consider possible origins for this radiation. In FIGURE 17 we list the many possibilities that have been discussed in the literature.

A secondary positron component of the interstellar medium is expected from cosmic-ray interactions with the cold interstellar matter leading to positive pion decay and the production of beta-unstable nuclei in spallation reactions. Ramaty, Stecker, and Misra[16] have indicated that the lower-energy positrons from spallated CNO beta emitters should make the dominant contribution to any 511 keV annihilation line.

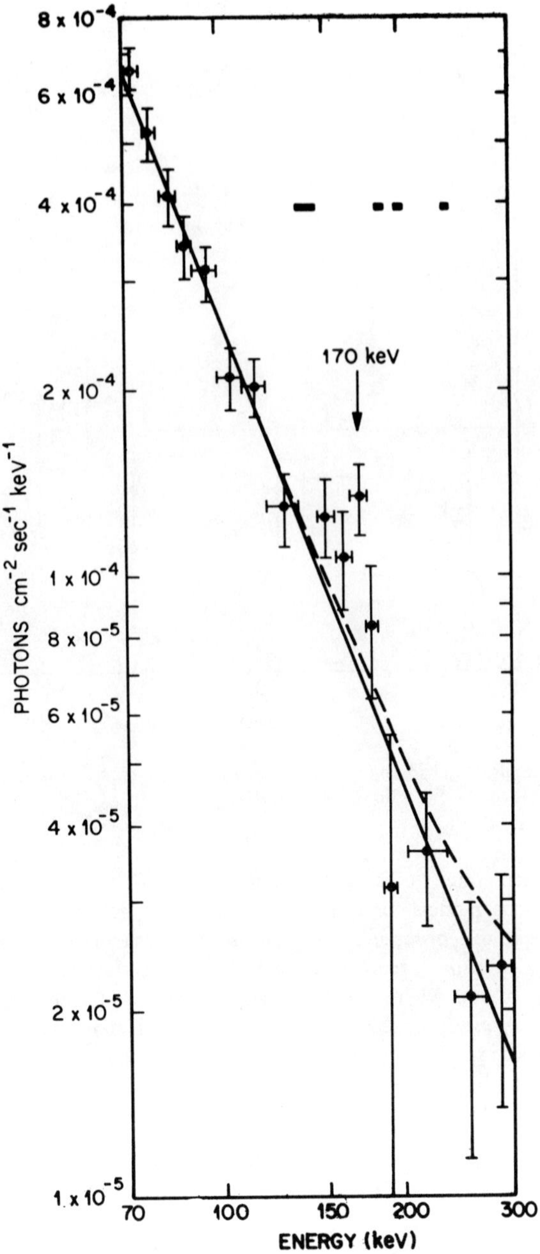

FIGURE 14. Candidate GC gamma-ray line at 170 keV corresponding to a flux of (7.4 ± 1.8) × 10^{-4} photons $cm^{-2}s^{-1}$ in a line of width ~ 12 keV. The bars indicate energy regions that have been excluded from curve fitting because of potential contamination with instrumental lines.

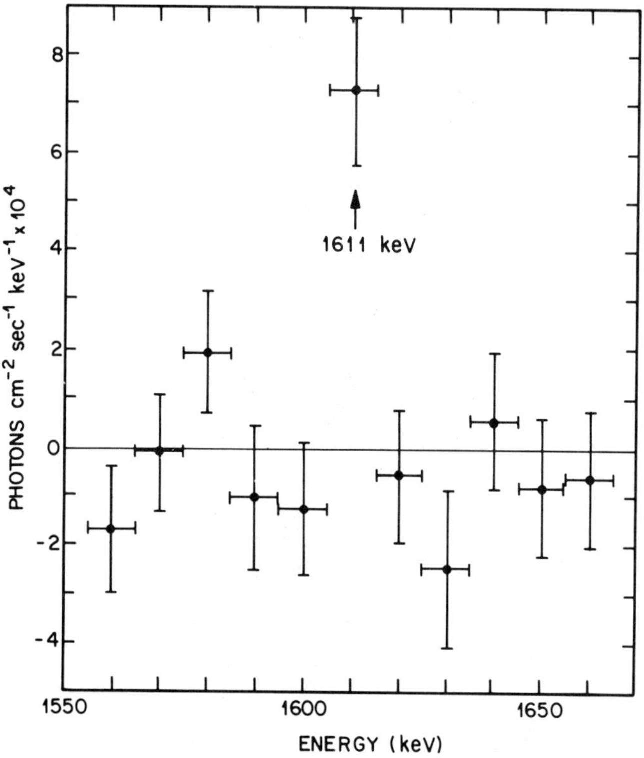

FIGURE 15. Candidate GC gamma-ray line at 1.611 MeV corresponding to a flux of $(7.3 \pm 1.5) \times 10^{-4}$ photons $cm^{-2}s^{-1}$ in a line of width ~ 10 keV.

These workers attempted to estimate a line flux from the galactic center direction based upon various models for the cosmic-ray spectrum. The largest of their flux estimates is about one order of magnitude smaller than the current observations. However, the calculations are very sensitive to the assumed intensity of low-energy cosmic rays ($E \lesssim 100$ MeV), about which very little is known because of solar modulation effects. Support for this hypothesis would come with the detection of Doppler broadened nuclear gamma-ray lines that must be excited in collision processes occurring in parallel with the production of CNO beta emitters. Rygg and Fishman[9] have calculated the expected relative intensity of the 511 keV line to various nuclear lines. They find six lines that should have an intensity in excess of the 511 keV intensity, e.g., ^{7}Li (478 keV), ^{24}Mg (1.37 MeV), ^{12}C (4.43 MeV), etc. Evidence for the existence of some of these broad lines is to be found in the galactic center data of Haymes *et al.*,[8] as discussed earlier. Hence a cosmic-ray interaction origin for the annihilation line must be taken seriously. It should be pointed out, however, that the low-energy cosmic-ray densities considered by Ramaty, Stecker, and Misra already lead to problems regarding the stability of the galaxy if they are distributed over the entire disc. If it is necessary to increase this density an order of magnitude to bring the calculations into agreement with observation, the hypothesis may become untenable.

During the past decade, elaborate explosive nucleosynthesis calculations have been performed that purport to follow the complex nuclear reaction matrix that takes place when stars explode and hydrodynamically expand. A remarkable result of these calculations has been their ability to reproduce large portions of the observed cosmic abundance table. The calculations indicate that many of the stable species should be produced first as beta-unstable radioactive elements that often decay with the emission of a positron. Colgate[17] and Clayton[18] have indicated that supernova-produced ^{56}Co or ^{44}Sc could yield enough positrons to account for the observed feature. Clayton and Hoyle[19] have pointed out that nova-produced ^{22}Na is an abundant source of positrons. In principle, this could also account for the feature. Support for either of these hypotheses would come with the detection of other nuclear gamma-ray lines predicted to accompany the decay of explosively-produced radioactive matter.[20] No believable evidence for these lines has yet been produced or exists in our data.[21,22]

Radio pulsars like the Crab are spinning neutron stars whose radiations are powered by a gradual loss of rotational kinetic energy. Although pulsar models that attempt to explain the energy transfer processes involved must still be considered

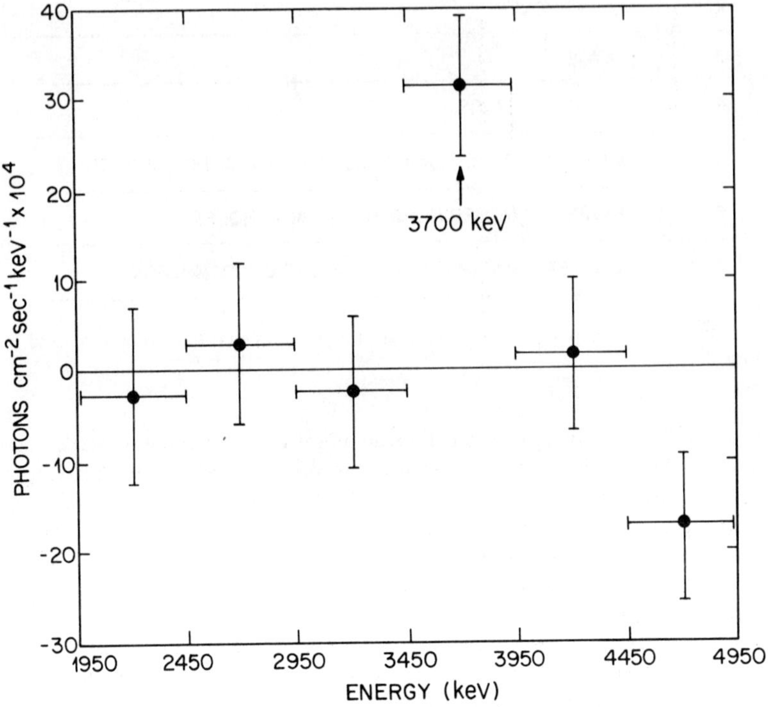

FIGURE 16. Candidate GC gamma-ray line at 3.70 MeV corresponding to a flux of $(3.1 \pm 0.8) \times 10^{-3}$ photons $cm^{-2}s^{-1}$ in a line of width ~ 500 keV.

uncertain, it is interesting to note that the more widely acccpted models generate large positron fluxes in the corotating magnetosphere above the star via various pair production processes.[23] For the Crab in particular, positron fluxes as large as 10^{41} per second have been envisioned.[24] The large number of radio pulsars within the field of view of our γ-ray telescope could presumably be injecting an adequate number of positrons into the interstellar medium to account for the observed feature.

Recently, Lingenfelter, Higdon, and Ramaty[25] have suggested that thermonuclear reactions occurring in the hot accretion disc around a massive black hole ($\sim 10^3$–$10^6 M_\odot$) could account for the 511 keV feature. In this model, a significant fraction of the positrons are produced with enough energy to escape into the interstellar medium and give rise to a sharp annihilation line. In addition, a whole series of gravitationally broadened nuclear deexcitation lines are expected from the accretion disc itself.[26]

PROPOSED SOURCES OF GALACTIC POSITRONS

1	COSMIC RAYS
2	SUPERNOVAE
3	NOVAE
4	RADIO PULSARS
5	MASSIVE BLACK HOLE AT THE GALACTIC CENTER
6	EVAPORATING PRIMORDIAL BLACK HOLES
7	MATTER – ANTIMATTER SYMMETRIC COSMOLOGIES

FIGURE 17. A list of possible sources of galactic positrons that have been discussed in the literature.

Hawking[27] originally pointed out that quantum effects will cause black holes to radiate like a blackbody whose temperature is inversely proportional to mass, leading possibly to a final explosion. However, this effect appears important only for small black holes presumably made in the early universe. In the present epoch, primordial black holes of initial mass near 10^{15} grams will be approaching their most luminous phase. Calculations indicate that for these primordial black holes, the energy will be radiated as 1% gravitons, 45% neutrinos, 45% electrons and positrons, and 9% photons.[28] By attributing the entire observed 511 keV line flux to positrons from the evaporation or explosion of primordial black holes, it is in principle possible to set an upper limit on the number of such objects within the galaxy. A greatly simplified calculation yields a limit of $\lesssim 10^{22}$ primordial black holes,[29] which is somewhat smaller than the present upper limit determined from the diffuse gamma-ray background.

It is not possible at this time to reach any firm conclusion on the question of positron genesis. Perhaps more than one mechanism is involved. Only cosmic-ray interaction processes can be identified as a contributor with some certainty. The key to the solution of this mystery may come with the detection of other weak, characteristic gamma-ray lines predicted to accompany the 511 keV line in many models and with an accurate determination of the spatial extent of the emitting region. Future experiments with germanium systems of greater sensitivity and/or spatial resolution such as those planned for the Gamma Ray Observatory should clarify the situation further.

Finally, we close with an intriguing note about the medium in which the annihilation is taking place. Crannell, Joyce, Ramaty, and Werntz[11] and recently Bussard, Ramaty, and Drachman[30] have considered the more general problem of positron annihilation in a partially ionized plasma at arbitrary temperature and density. They find that the relative amount of *ps* annihilation to free annihilation (which always yields two 511 keV gamma rays) and the width of the 511 keV line are sensitive functions of the plasma parameters. If the possible detection of the three-photon continuum is taken at face value, it and the observed linewidth imply that the positrons are annihilating in a medium of temperature less than 10^5 K and ionization fraction greater than 0.05. Possible sites for the annihilation would then be H II regions at the GC, the warm ionized component of the interstellar medium, or young supernova remnants.

References

1. Johnson, W. N. & R. C. Haymes. 1973. Astrophys. J. **184:** 103.
2. Fishman, G. J. & D. D. Clayton. 1972. Astrophys. J. **178:** 337.
3. Kozlovsky, B. & R. Ramaty. 1974. Astrophys. J. Lett. **191:** 43.
4. Guthrie, P. & E. Tademaru. 1973. Nature (London) Phys. Sci. **241:** 77.
5. Ramaty, R., G. Börner & J. M. Cohen. 1973. Astrophys. J. **181:** 891.
6. Leventhal, M. 1973. Astrophys. J. Lett. **183:** 147.
7. Ore, A. & J. L. Powell. 1949. Phys. Rev. **75:** 1696.
8. Haymes, R. C., G. D. Walraven, C. A. Meegan, R. D. Hall, F. T. Djuth & D. H. Shelton. 1975. Astrophys. J. **201:** 593.
9. Rygg, T. A. & G. J. Fishman. 1973. Proc. 13th Int. Cos. Ray Conf. (Denver) **1:** 472.
10. Lingenfelter, R. E. & R. Ramaty. 1978. Phys. Today **31:** 40.
11. Crannell, C. J., G. Joyce, R. Ramaty & C. Werntz. 1976. Astrophys. J. **210:** 582.
12. Bevington, P. R. 1969. Data Reduction and Error Analysis for the Physical Sciences. McGraw-Hill. New York, N.Y.
13. Chupp, E. L. Private communication.
14. Matteson, J. L. Private communication.
15. Albernhe, F., D. Boclet, P. Durouchoux, J. F. Le Borgne, R. Rocchia & G. Vedreene. COSPAR. Proc. 21st Plen. Meeting, Innsbruck, Austria, June 1978:329.
16. Ramaty, R., F. W. Stecker & D. Misra. 1970. J. Geophys. Res. **75:** 1141.
17. Colgate, S. A. 1970. Astrophys. Space Sci. **8:** 457.
18. Clayton, D. D. 1973. Nature (London) Phys. Sci. **244:** 137.
19. Clayton, D. D. & F. Hoyle. 1974. Astrophys. J. Lett. **187:** 101.
20. Clayton, D. D. *In* Explosive Nucleosynthesis. D. N. Schramm & W. D. Arnett, Eds.: 264. University of Texas Press. Austin, Tex.
21. Leventhal, M., C. J. MacCallum & A. C. Watts. 1977. Astrophys. J. **216:** 491.
22. Leventhal, M., C. J. MacCallum & P. D. Stang. 1978. Nature (London) **275:** 200.
23. Cheng, A. F. & M. A. Ruderman. 1977 Astrophys. J. **214:** 598.
24. Sturrock, P. A. 1971. Astrophys. J. **164:** 529.

25. LINGENFELTER, R. E., J. C. HIGDON & R. RAMATY. 1978. *In* Gamma-Ray Spectroscopy in Astrophysics. T. L. CLINE & R. RAMATY, Eds.: 252. NASA Tech. Memo. 79619.
26. HIGDON, J. C. & R. E. LINGENFELTER. 1977. Astrophys. J. Lett. **215:** 53.
27. HAWKING, S. W. 1974. Nature (London) **248:** 30.
28. PAGE, D. N. & S. W. HAWKING. 1976. Astrophys. J. **206:** 1.
29. LEVENTHAL, M., C. J. MACCALLUM & P. D. STANG. 1978. *In* Gamma-Ray Spectroscopy in Astrophysics. T. L. CLINE & R. RAMATY, Eds.: 169. NASA Tech. Memo. 79619.
30. BUSSARD, R. W., R. RAMATY & R. J. DRACHMAN. 1979. Astrophys. J. **228:** 928.

GAMMA-RAY OBSERVATIONS OF SEYFERT GALAXIES AND THE PROBLEM OF THE EXTRAGALACTIC GAMMA-RAY BACKGROUND

V. Schönfelder

Max-Planck-Institut für Physik und Astrophysik
Institut für Extraterrestrische Physik
8000 Munich 40
Federal Republic of Germany

This paper on the gamma ray emission from Seyfert galaxies and on the origin of the diffuse cosmic gamma-ray component is mainly based on results that we obtained from our balloon-borne Compton telescope, which allows gamma-ray astronomical observations in the energy range between 1–20 MeV. Therefore, in the first part of this paper, I shall briefly describe the observations and then, in the second part, discuss their consequences.

A schematic drawing of the Compton telescope is shown in FIGURE 1. It consists of two large-area, mosaic-like scintillation detectors, which are both entirely surrounded by thin anticoincidence plates of plastic scintillator in order to reject charged particles (not shown in this schematic drawing). A gamma ray is identified by a Compton collision in the upper detector and a subsequent interaction in the lower detector. By measuring the energy losses E_e' and E_e'' in both detectors and by measuring the locations in both interactions, the direction of the scattered gamma ray and the scattering angle φ at the position of the first interaction can be determined. If only events within a certain range of scattering angles φ are accepted from a celestial gamma-ray point source, the directions of the scattered gamma rays form an image circle in the plane of the lower detector around the direction of the source. The gamma-ray source thus can be identified.

The telescope was flown for the first time on a balloon in October 1977 from Palestine, Texas. FIGURE 2 shows those regions of the sky that were observed during the flight. The telescope axis always pointed towards the zenith. The balloon reached float altitude some hours after the transit of the Cygnus region. It stayed there for 20 hours, and during this time the anticenter region and the regions around the galactic pole passed through the field of view. The field of view inside which the detection efficiency drops to half its value, is indicated by the circle; it is typically 40° FWHM.

In a first kind of analysis the 1–20-MeV gamma-ray counting rate within the total field of view—normalized to a constant float altitude—was determined and is shown in FIGURE 3 as a function of time in universal time (UT). The transit times of some typical celestial objects are indicated; of course, they must not necessarily be identical with gamma-ray sources. This curve already indicates that several gamma-ray sources were in the field of view. First, there is a significant counting rate increase above the background level, which is indicated by the dashed line during the transit time of the anticenter region. It lasts for 6–7 hours and this already indicates that more than one source (namely the Crab, which we identified from the pulsar light curve) contributes to the counting rate. In the case of one single source the enhancement should last only 3 hours.

0077–8923/80/0336–0267 $1.75/1 © 1980, NYAS

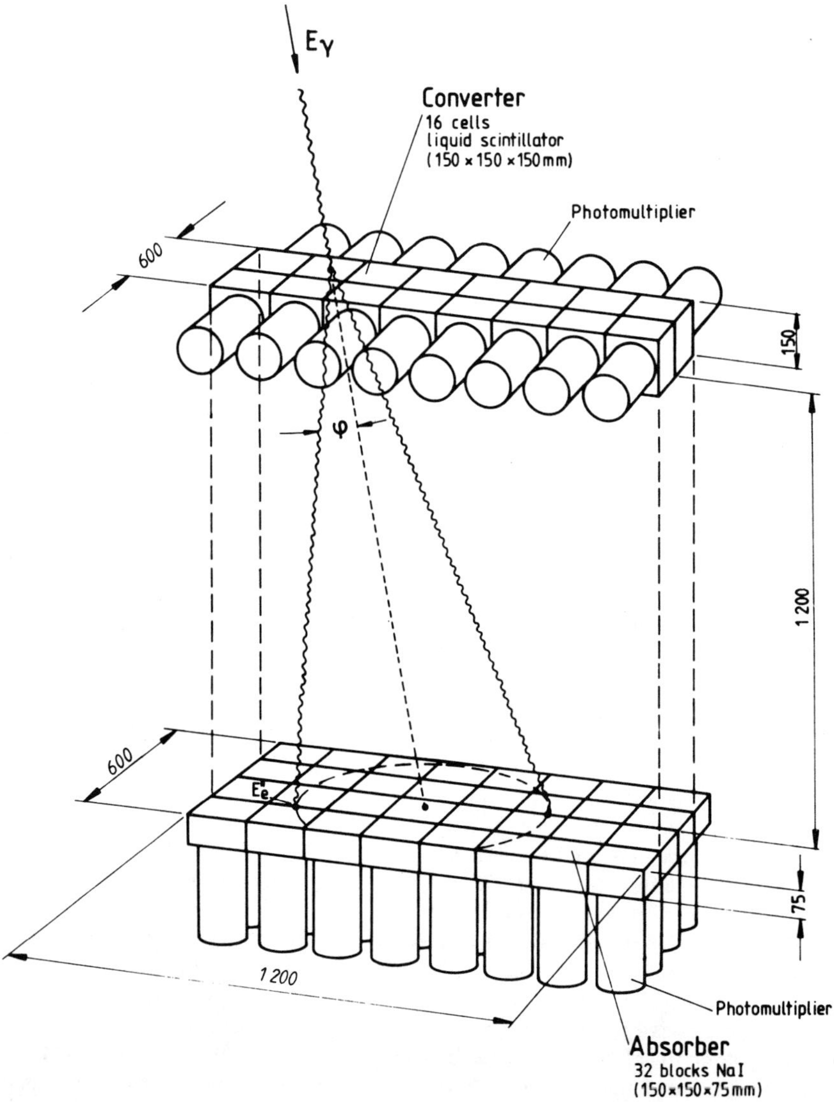

FIGURE 1. Schematic drawing and principle of the Compton telescope.

A second counting rate excess is seen during the transit time of the Seyfert galaxy NGC 4151, which just lasts for 3 hours. Considering the large field of view of 40°, an identification is, of course, difficult. During the last few months we therefore have concentrated on locating the position of the source more accurately.

FIGURE 4 shows a section of FIGURE 3 from 15 to 21 UT, however, with higher resolution. Into this counting rate curve the response function of the telescope has been

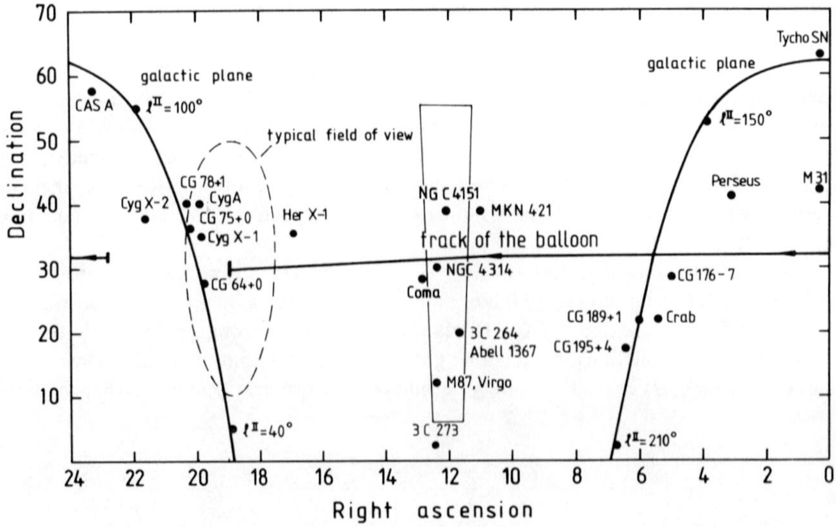

FIGURE 2. Sky map of those regions that were observed during the balloon flight.

drawn—after proper normalization—assuming that there is a celestial source at the position of NGC 4151. The agreement between the response function and the observed counting rate is quite good, especially at the right side of the curve and during the transit time, where a minimum is expected and indeed observed. At the left side of the curve there still seems to be a contribution from another source, as becomes clear from FIGURE 3.

Reasonable fits of the response function to the observed counting rate can no longer be obtained if the source position is outside ±10° of the right ascension and +15° or −35° of the declination of NGC 4151. In FIGURE 2 this error box is indicated together with the known x-ray sources inside or near to it. Besides NGC 4151, possible sources are only the Virgo cluster with M 87 and NGC 4151 and the Abell cluster

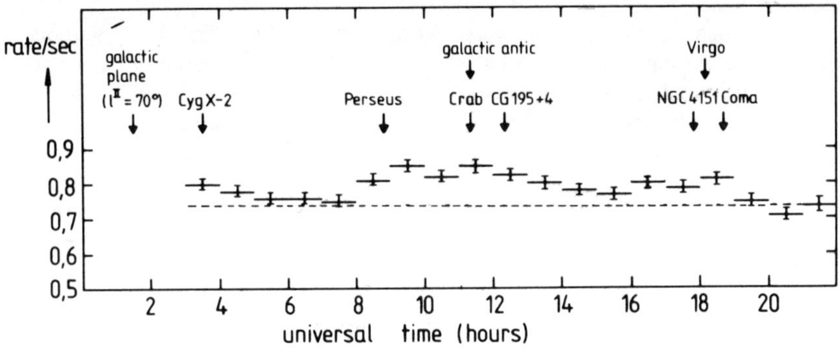

FIGURE 3. Total 1–20-MeV counting rate at float altitude as a function of universal time.

1367 with 3C 264. Already outside, but still close to the error box, are Coma and the quasar 3C 273.

Additional information about the identity of the source can be obtained from the energy spectrum. In FIGURE 5 observations of NGC 4151 at x-ray energies are compared with those at gamma-ray energies. Our flux values between 1–3 MeV nicely connect to the extrapolation from x-ray energies. Above 3 MeV, however, the spectrum steepens considerably.[1] The recent upper limits from the SAS-2 experiment (reference 2) between 35–200 MeV indicates that the spectrum above 3 MeV goes like E^{-3} or steeper.

For other possible sources in or near the error box there is not such a consistency between the derived gamma-ray fluxes and the x-ray data as far as they are available.

It therefore seems that NGC 4151 is indeed the most probable candidate for the source, though a contribution from a neighboring source within the field of view to the counting rate excess cannot definitely be ruled out at present. During further analysis

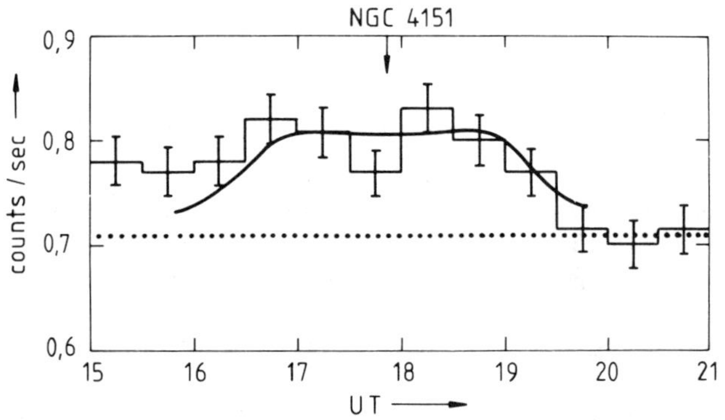

FIGURE 4. Section of FIGURE 3. The response function of the telescope for a source at the position of NGC 4151 is indicated.

of our balloon data it should be possible to locate the true position of the source more accurately.

If the identification of the source with NGC 4151 is correct, its gamma-ray luminosity is extremely high: it is 10^{45} erg/s below 3 MeV, compared to 7×10^{42} erg/s at x-ray energies between 2–10 keV. At least at x-ray energies this powerful energy source seems to be concentrated to a very small volume. An increase in the 2–10-keV flux by a factor of 1.7 over a time scale of $\approx$ 3 days was observed in 1976 by Elvis.[3] An increase in the 2–60-keV band by a factor of 2 within 1.5 days was recently observed by OSO-8,[4] and from Uhuru observations[5] flares of only typically 1000-s duration in the 2–10-keV band were reported very recently. From these variations an upper limit on the dimension of the x-ray source of 2×10^{13} to 2×10^{15} cm can be derived.

What may be the reason for the cutoff in the gamma-ray spectrum? On the one side, synchrotron self-Compton models have been proposed to explain the whole spectrum. In this case the existence of relativistic electrons is postulated, which produce mm, infrared, or optical radiation by the synchrotron process. The synchro-

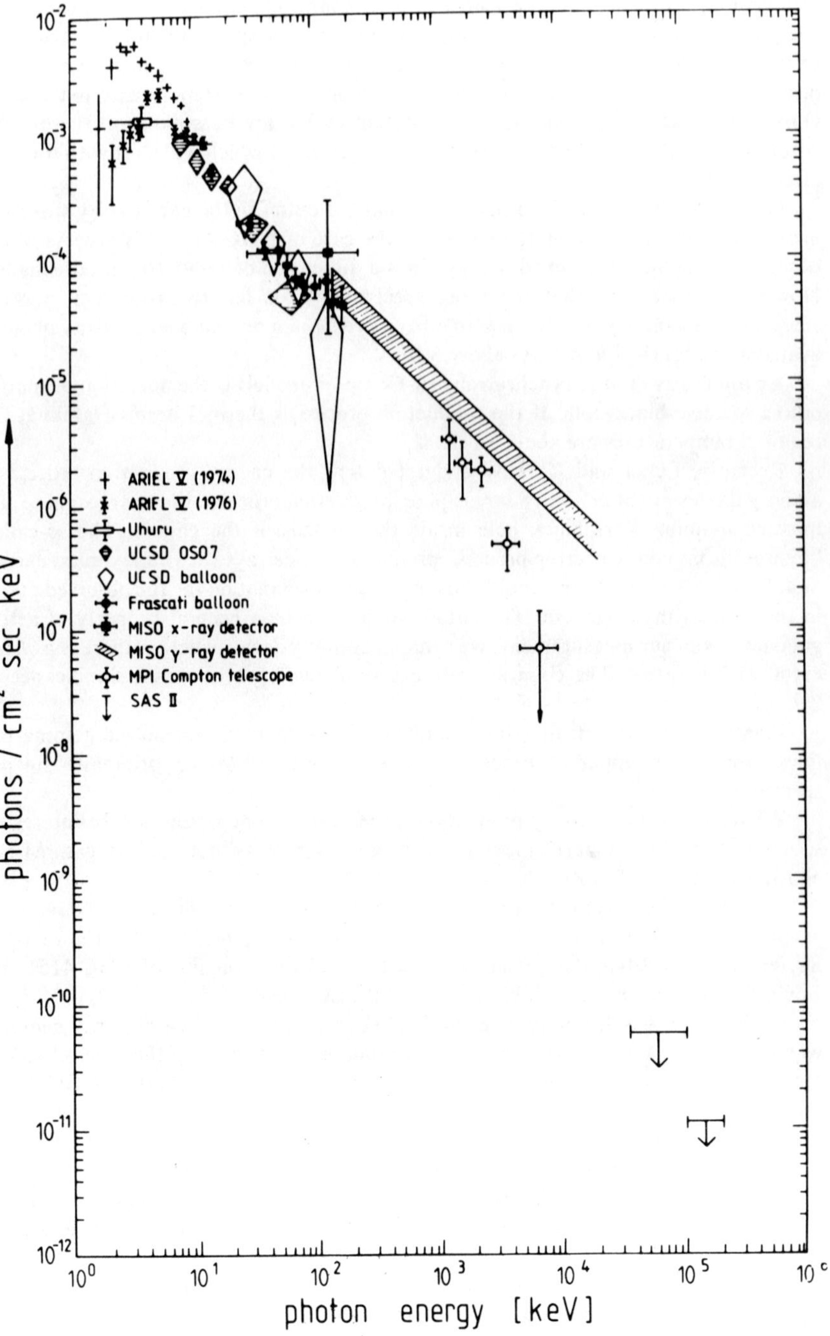

FIGURE 5. X- and gamma-ray spectrum of NGC 4151 (from references 1 and 2).

tron photons are then Compton scattered by the same electrons to x- and gamma-ray energies. In these models the difficulties in understanding the photon spectrum are shifted towards an understanding of the electron sources. Furthermore, one has to postulate that there is a cutoff in the electron spectrum at some maximum energy. Only if the electrons are not injected continuously but are burstlike, will there be an automatic cutoff in the electron spectrum, the position of which will shift with the time after injection.

One could also think of the possibility that the cutoff in the gamma-ray spectrum is caused by photon–photon absorption. In the case of NGC 4151 this process would become important if the emitting region has a dimension of about 10^{14} cm or smaller. However, it turns out that the x-ray spectrum is too flat to cause such a rapid steepening at gamma-ray energies. In other words: there are not enough x-ray photons available to absorb gamma rays above 3 MeV.

An alternative to the synchrotron self-Compton models is the accretion of matter onto a massive black hole. If the production process is thermal bremsstrahlung, the required temperatures are about 10^{10}K.

Recently, Leiter and Kafatos[6] suggested that the energy production process in active galaxies might be the "Penrose photoproduction process" in the ergosphere of a massive spinning Kerr black hole inside the nucleus of the galaxy. The so-called "Penrose Compton scattering process" produces gamma rays with a maximum energy at 4.3 m_ec^2 = 2.15 MeV and could thus provide an explanation for the observed cutoff in the gamma-ray spectrum. The cutoff would then be a general property of active galaxies. From our measurements we cannot definitely decide whether there is a *sharp* cutoff at 2.15 MeV. The counting rate excess above 2.5 MeV has a significance of 2 σ.

Finally, at the present time it cannot be excluded that still unresolved gamma-ray lines produce an apparent excess at MeV energies above an originally steeper spectrum of NGC 4151.

Whatever the gamma-ray production mechanism is, one would now be interested to know if further Seyfert galaxies can be expected to be detected as gamma-ray emitters in the near future.

If all x-ray Seyferts with x-ray luminosities equal to or higher than NGC 4151 would have the same gamma-ray luminosity as NGC 4151, then only 2 of the 16 x-ray Seyferts listed by Elvis *et al.*[7] would at least have 0.3 times the flux of NGC 4151 and would therefore be just detectable with present instruments.

In this case, the differential spectral shape between x- and gamma-ray energies would be similar and close to $\sim E^{-1}$ for both sources. Since most of the other observed x-ray Seyferts have x-ray luminosities considerably higher than that of NGC 4151, their spectra would be steeper.

If there would be proportionality between x- and gamma-ray luminosity, 6 of the 16 objects would be at least 0.3 times as strong as NGC 4151. However, later in this talk I shall show that proportionality between x- and gamma-ray luminosity is not possible for all x-ray Seyferts because the resulting integrated flux would be higher than the observed diffuse cosmic flux.

Indeed, if it should turn out that the high gamma-ray luminosity is typical for x-ray Seyferts or at least a subclass of x-ray Seyferts, then this would have important consequences for the understanding of the diffuse cosmic gamma radiation, which I now should like to outline in the rest of my talk.

Most of the existing measurements in the low-energy gamma-ray domain were performed on the diffuse cosmic gamma-ray component. This is the continuation of the diffuse x-radiation, which, to a high degree, is isotropic and therefore seems to be of extragalactic origin. At gamma-ray energies, no isotropy measurements of the same accuracy are available.

The energy spectrum of the diffuse cosmic gamma-ray component is shown in FIGURE 6 (from Schönfelder[1]). There has been a long debate in the literature whether there is a bump in the energy spectrum at MeV energies above the straight line extrapolation from x-ray energies, which follow an $\sim E^{-2.3}$ law.

The presently existing measurements at MeV energies tend to converge and to support the existence of a bump. Above about 5 MeV the spectrum is very steep, namely $\sim E^{-3}$. Our most recent, still unpublished energy spectrum that was derived from the same balloon flight, from which the NGC 4151 data were obtained, is also included (labeled "this experiment"). These new measurements between 1 and 20 MeV nicely connect the low-energy gamma range with the high-energy range, which was investigated by the SAS-2 experiment. The high flux values or upper limits reported between 10 and 40 MeV are no longer of any significance.

Various attempts were made to interpret the diffuse cosmic gamma-ray component. The general interest in its origin is due to the fact that it may contain signatures of cosmological significance. The transparency of the universe for gamma rays is so high that it should be possible to look backwards to epochs of the universe with red shifts around 100.

In most cases the diffuse cosmic gamma-ray component was interpreted as the result of interactions in intergalactic space. The following six processes have been suggested or investigated:

1. primordial cosmic-ray interactions with intergalactic matter producing π^0-decay gamma rays,[8]
2. inverse Compton collisions of cosmic-ray electrons leaking from galaxies with the 2.7 K radiation,[9,10]
3. bremsstrahlung production of nonthermal electrons with intergalactic matter,[11,12]
4. interaction of extragalactic cosmic-ray protons $> 10^{15}$ eV with the 2.7 K radiation producing electron–positron pairs and subsequent cascades,[13]
5. quantum-mechanical decay of small black holes,[14]
6. matter–antimatter annihilation in a baryon symmetric universe producing π^0-decay gamma rays.[15]

In the first four cases there are severe difficulties in matching the observed shape of the energy spectrum and in producing sufficiently high gamma-ray intensities.

Primordial gamma-ray emission was investigated in the last two models. Page and Hawking[14] derived an upper limit on the number of small black holes in the universe from the observed gamma-ray background flux. The expected spectrum from this process does not at all match the observed background spectrum. Stecker *et al.*[15] calculated the π^0-decay gamma-ray spectrum in a baryon symmetric universe from matter–antimatter annihilation. The derived gamma-ray spectrum gives indeed a nice fit to the observations for an open universe cosmology. The main questions to be solved here are whether we really live in a baryon symmetric universe and, if this is the case,

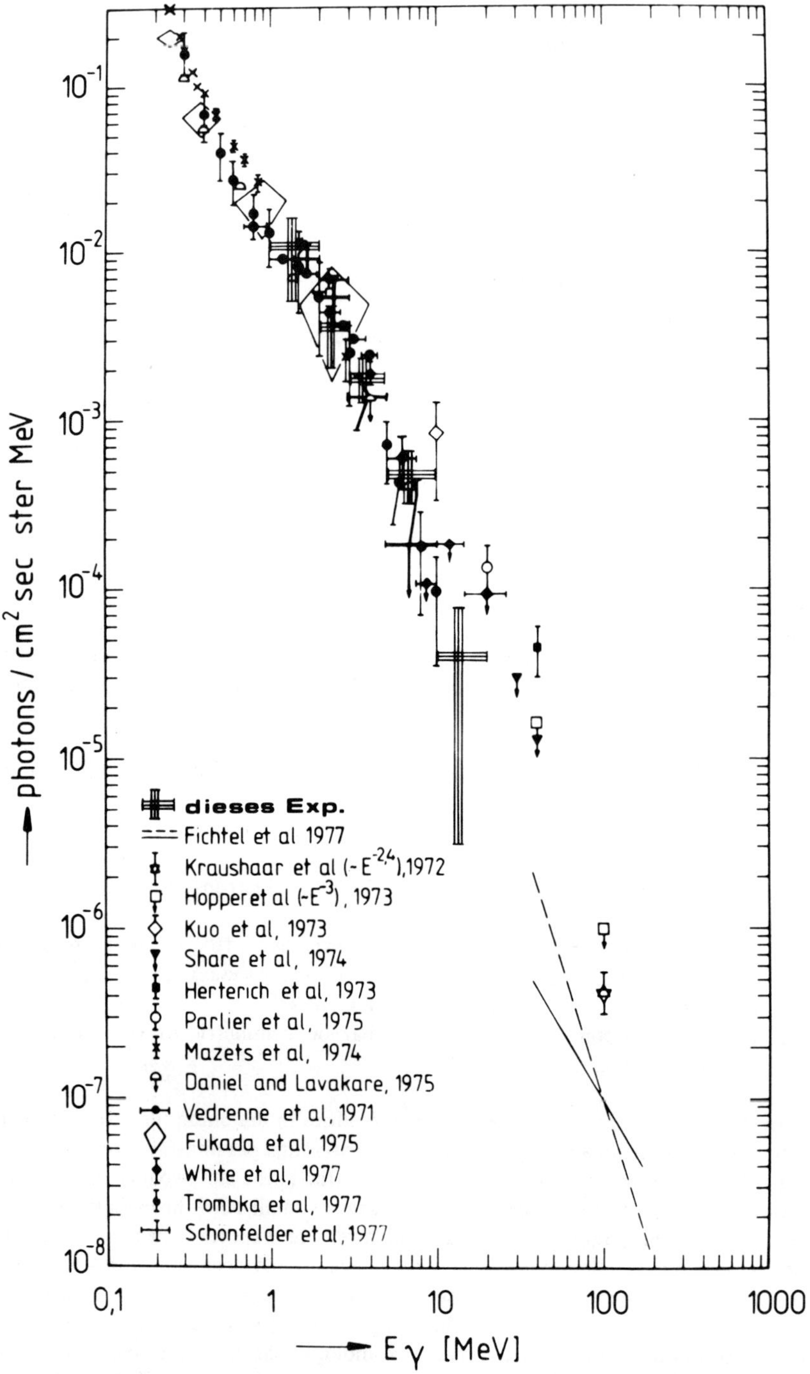

FIGURE 6. Energy spectrum of the diffuse cosmic gamma-ray component (from reference 1).

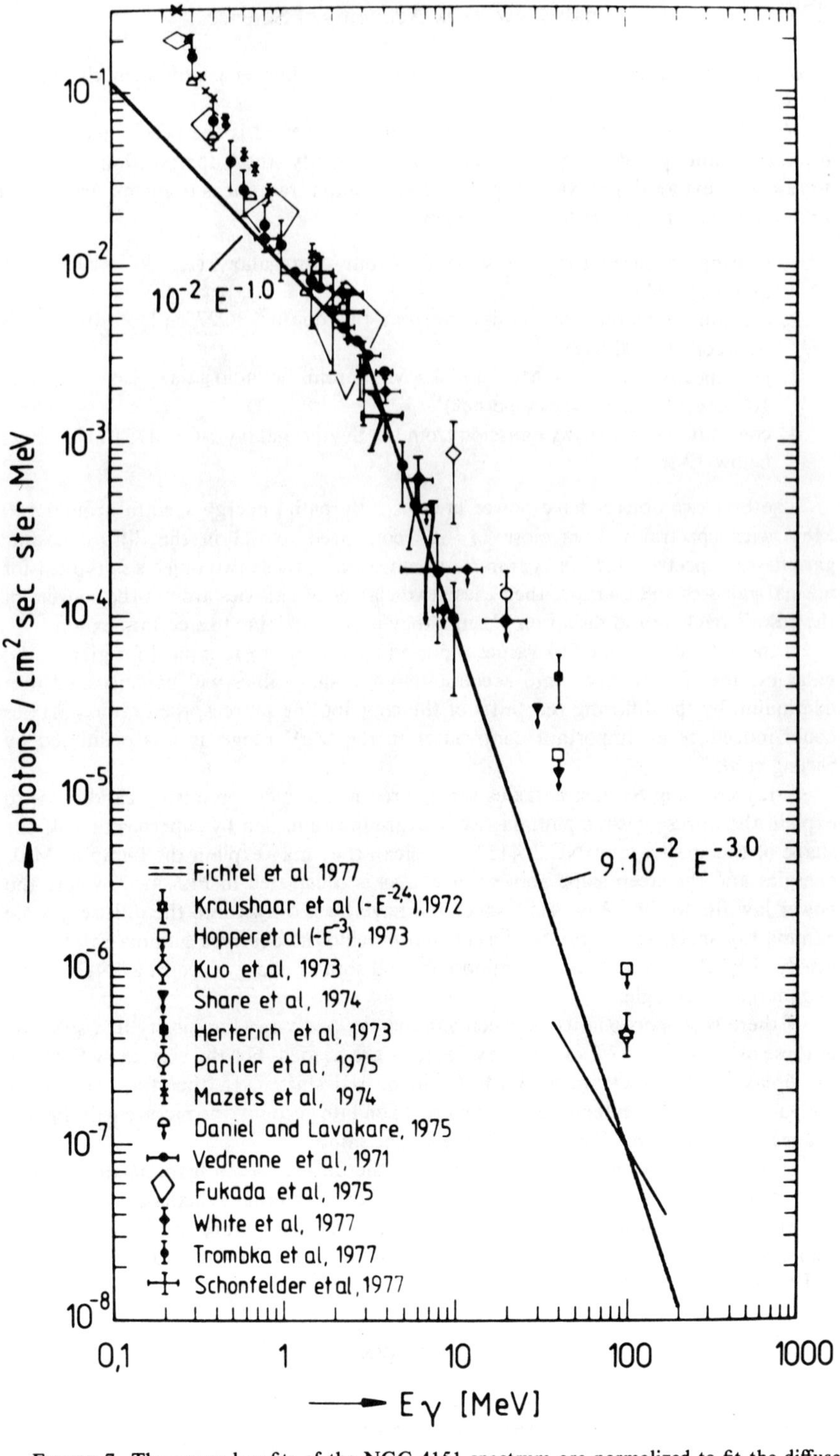

FIGURE 7. The power law fits of the NGC 4151 spectrum are normalized to fit the diffuse cosmic gamma-ray spectrum (from reference 1).

whether a separation into regions of matter and antimatter could occur during the early phases.

An alternative to these "diffuse" production processes is the possibility that the diffuse cosmic gamma-ray component is in reality the superposition of many unresolved external galaxies. Up to now, gamma ray measurements from four different kinds of galaxies have been reported:

1. continuum gamma-ray emission from our own galaxy ($L_\gamma \approx 4 \times 10^{38}$ erg/s above 100 MeV),
2. continuum gamma-ray emission from the quasar 3C273 (3×10^{46} erg/s between 50–500 MeV),[16]
3. gamma-ray lines at 1.6 MeV and 4.5 MeV from the radio galaxy Cent A (4.8×10^{43} erg/s for both lines together)[17]
4. continuum gamma-ray emission from the Seyfert galaxy NGC 4151 (10^{45} erg/s below 3 MeV).

The first two objects have power laws as differential energy spectra around 100 MeV with spectral indices close to -2 (compared to -3 in the diffuse cosmic gamma-ray spectrum). If the gamma-ray emission of these two objects is typical for normal galaxies and quasars, then these two classes of galaxies are not the sources of the cosmic background radiation, though they may contribute to a certain extent.[18]

If the gamma-ray line flux values reported from Cent A are typical for giant radio galaxies, and if it is taken into account that the sharp lines will be smoothed to a continuum by the differing red shifts of the contributing sources, then radio galaxies could indeed be an important contributor in the MeV range as was estimated by Strong *et al.*[19]

X-ray-emitting Seyfert galaxies are at present the most promising candidates to explain the diffuse cosmic gamma-ray background radiation by superposition. If the shape of the spectrum of NGC 4151 is typical, they may explain the bump at MeV energies and the steep slope above 5 MeV, as is illustrated in FIGURE 7, where the power law fits to the NGC 4151 spectrum have been drawn into the diffuse cosmic gamma-ray spectrum. X-ray Seyferts could provide the observed gamma-ray flux, if nearly all of them have similar luminosities and similar energy spectra as NGC 4151 at gamma-ray energies.

If there is proportionality between gamma-ray and x-ray luminosity in a subclass of these objects, only 13% of all x-ray Seyferts are needed. For the very early Seyferts the break in the spectrum at 3 MeV will be red shifted. If, therefore, red shifts and/or—to a certain extent—evolution are taken into account, the picture will slightly change; the conclusion, however, will still be the same.

To solve the question about the origin of the diffuse cosmic gamma-ray component, it will be necessary to observe a large number of external galaxies at gamma-ray energies. This will probably not be possible in the very near future, say in the next few years, but certainly in about six years, when the large Gamma Ray Observatory (GRO) will be in orbit.

REFERENCES

1. SCHÖNFELDER, V. 1978. Nature **274:** 344.
2. FICHTEL, C. E. 1978. NASA Tech. Memo. 79640.

3. ELVIS, M. 1976. M.N.R.A.S. **177:** 7.
4. MUSHOTZKY, R. F., S. S. HOLT & P. J. SERLEMITSOS. 1978. NASA Tech. Memo. 79551.
5. TANANBAUM, H., G. PETERS, W. FORMAN, R. GIACCONI & C. JONES. 1978. Astrophys. J. **223:** 74.
6. LEITER, D. & M. KAFATOS. 1978. Submitted to Nature.
7. ELVIS, M., T. MACCACARO, A. S. WILSON, M. J. WARD, M. V. PENSTON, R. A. E. FOSBURY & G. C. PEROLA. 1978. M.N.R.A.S. **183:** 129.
8. STECKER, F. W. 1969. Astrophys. J. **157:** 507.
9. FELTEN, J. E. & P. MORRISON. 1966. Astrophys. J. **146:** 686.
10. BRECHER, K. & P. MORRISON. 1969. Phys. Rev. Lett. **23:** 802.
11. ARONS, J. & R. MCCRAY. 1969. Astrophys. J. **158:** L91.
12. STECKER, F. W. & D. L. MORGAN. 1972. Astrophys. J. **171:** 201.
13. STRONG, A. W., J. WDOWCZYK & A. W. WOLFENDALE. 1973. Proceedings of the 13th International Cosmic Ray Conference, Denver. Vol. 1:27.
14. PAGE, D. N. & S. W. HAWKING. 1976. Astrophys. J. **206:** 1.
15. STECKER, F. W., D. L. MORGAN & J. BREDEKAMP. 1971. Phys. Rev. Lett. **27:** 1469.
16. SWANENBURG, B. N., K. BENNETT, G. F. BIGNAMI, P. CARAVEO, W. HERMSEN, G. KANBACH, J. L. MASNOU, H. A. MAYER-HASSELWANDER, J. A. PAUL, B. SACCO, L. SCARSI & R. D. WILLS. 1978. Nature (London) **275:** 298.
17. HALL, R. C., C. A. MEAGON, G. D. WALRAVEN, F. T. DJUTH, D. H. SHELTON & R. C. HAYMES. 1975. Proceedings of the 14th International Cosmic Ray Conference, Munich. Vol. 1:84.
18. LICHTI, G. G., G. F. BIGNAMI & J. A. PAUL. 1978. Astrophys. Space Sci. **56:** 403.
19. STRONG, A. W., A. W. WOLFENDALE & D. M. WORRAL. 1976. M.N.R.A.S. **175:** 23P.

ISOLATED SYSTEMS IN GENERAL RELATIVITY

Jürgen Ehlers*

Max-Planck-Institut für Physik und Astrophysik
8000 Munich 40
Federal Republic of Germany

INTRODUCTION

Almost all quantitative information about the gravitational interaction has been obtained from studies of nearly isolated systems composed of a few macroscopic bodies such as the solar system or double stars, and in the foreseeable future clean, precise tests of gravity-theories are expected to be based exclusively on observations of such systems, too, as can be seen, e.g., from the lectures of J.H. Taylor and C.M. Will at this conference.

At present Einstein's general theory of relativity (GR) still appears to be the most successful theory of gravity. Therefore it is important to have a reliable mathematical description of isolated systems based on that theory.

It appears to be widely believed that the implications of GR for systems of slowly moving, well-separated and therefore weakly-coupled bodies—even bodies with strong self-fields such as neutron stars—have been deduced satisfactorily with an accuracy sufficient to calculate not only quasistationary, first post-Newtonian effects of order $(v/c)^2 \sim Gm/rc^2$ like "anomalous" perihelion advances, but even the dominant, secular gravitational radiation reaction effects on the orbits, of order $(v/c)^5$. Such calculations have been applied, e.g., to the binary system containing the pulsar PSR 1913 + 16. A minority of relativity-theorists including the present author does not share this opinion, for reasons given below, and in references 1–4 and the papers and books quoted therein.

If one argues by analogy with Newtonian mechanics and Maxwellian electrodynamics one obtains "plausible" results concerning, e.g., gravitational radiation damping. One can derive from a linearized version of the GR-field equation Einstein's quadrupole formula[5] for the gravitational-radiation power loss, and equate that power loss to the rate of decrease of the Newtonian total energy expressed in terms of Keplerian elements, considered as secularly changing due to perturbing, gravitational-radiation reaction forces.[6,7] The main shortcoming of this and similar arguments and calculations is, in my opinion, not that they employ approximations which have not been rigorously mathematically justified—that they share with many approximations used in physics—but rather that they:

1) employ *notions* which are *not well defined* in terms of basic concepts of GR, such as "gravitational field energy," "total mass and linear momentum" of a gravitationally bound body interacting with other such bodies, "point particle," "gravitational radiation reaction force," "near zone," "radiation zone";

2) use *laws* which have *not* been *established* within GR, such as an "energy balance between radiation and material sources";

*I dedicate this paper to my son Martin Ehlers.

0077–8923/80/0336–0279 $1.75/1 © 1980, NYAS

3) depend essentially on *ad hoc assumptions* which not only are *without foundation* within GR itself, but for which there are indications that they may be *incompatible* with the fundamental assumptions of GR or with each other, such as global coordinate conditions, particular global splittings of the metric into a flat background and a "small" perturbation, noncovariant "outgoing radiation conditions," negligibility of various kinds of "small" terms, etc. (For discussions of some of these difficulties and proposals to overcome them see references 3, 4, and 8–12).

It seems to me to be an important challenge to find derivations of observable relativistic effects, particularly structure and radiation effects, of isolated systems which are free of such shortcomings, and which are not based on mere analogies, however plausible, with Newton's or Maxwell's theory. Needed are, in particular, approximation methods which have been rigorously justified at least in theories simpler than Einstein's, and which permit if not an error estimate, at least a reasonable guess about error bounds.

In the following pages I shall describe some problems of the relativistic theory of isolated systems, hoping to stimulate more research in this area. As will be seen, some progress has been made in this field since the last Texas symposium, but the main questions remain unanswered. Classical GR is not a closed subject; it poses difficult, accessible problems related to real observations.

ISOLATED SYSTEMS

A general-relativistic model of an isolated system is an asymptotically flat† spacetime (M, g_{ab}) with a physically reasonable stress-energy-momentum tensor T^{ab} satisfying the Einstein-Hilbert equation

$$G^{ab} = T^{ab}, \tag{1}$$

and hence also the local law of motion

$$T^{ab}{}_{;b} = 0. \tag{2}$$

(Here I take $c = 1$, $8\pi G = 1$, and choose the signature $(+---)$ for the metric.) If the system consists of N bodies separated by empty space, T^{ab} has to vanish outside of N timelike, connected, spatially compact world tubes.‡

Such an isolated system may contain incident as well as outgoing gravitational radiation. If, in some interval of advanced time, the system is not subject to incident radiation or, more generally, if such radiation is not in resonance with a mode of the system, then presumably the emission of radiation will be accompanied by a damping of the motion of its material sources. (Asymptotically for "$t \rightarrow +\infty$", damping may already be implied by spatial and null asymptotic flatness since the system then has a finite total ADM-energy momentum, part of which is radiated to future null infinity. For simple model theories such a behaviour has been deduced.[14] In GR a correspond-

†Instead of asymptotic flatness one can require (M, g_{ab}) to fit asymptotically into a cosmological model,[13] a possibility which appears not to have been investigated extensively.

‡In this mechanical model the bodies are idealized as having boundaries, and contributions of nongravitational radiation to T^{ab} are not taken into account. Such contributions could be added, but are of secondary importance to celestial mechanics.

ing theorem has not been rigorously formulated, let alone been proved.) The presence or absence of incident and emitted radiation can be expressed in terms of the asymptotic behavior of the gravitational field near past and future null infinity, respectively.[11,12,15]

Proposals concerning the precise meanings of the terms "asymptotically flat," "physically reasonable T^{ab}," "condition for absence of incident radiation" will be mentioned below. We cannot, however, be sure that any of such proposed definitions are adequate until we know that models of isolated systems with the stipulated properties do, in fact (in a mathematical sense) exist and are sufficiently general to describe those physical situations for which the concepts have been invented. It appears that we are still far from that kind of knowledge. (The degree of understanding in this field bears some similarity to that in quantum field theory: There are reasonable proposals for the ground rules ("axioms"), a few exact results, no existence theorems of sufficient generality; but in spite of this various ill-founded approximation methods, plagued by infinities and other difficulties and not related to the exact parts of the theory, give remarkably good agreement with observations.)

Before discussing later in this paper the problem of constructing models of isolated systems—a task which up to now can be tackled by formal approximation methods§ only, except for one-body problems—I shall consider two topics which have been treated within the exact theory: the description of bodies as a whole and the derivation of their overall laws of motion, and the asymptotics of gravitational fields near various parts of infinity. The concepts, results and methods of these investigations would seem to be needed, at least in principle, before the main, global problem of constructing isolated systems can be treated rigorously, and they may also help in improving approximation methods.

Bodies and Laws of Motion

In order to represent mathematically the bodies (planets, stars, black holes) of an isolated system in GR one has to specify either the stress-energy-momentum tensor T^{ab} in equations 1 and 2 or the parts of the event horizon which bound the black holes.

In the following it will be assumed, as part of the "physical reasonableness" of T^{ab}, that $T_b{}^a$ maps the interior of each half of the null cone into itself, a condition which implies that for all local observers the energy density T^{00} is positive, the magnitude of the density of linear momentum is less than T^{00}, and the stresses are also dominated by T^{00}.

The model of a point particle supported by a timelike world line $z(t)$, given formally by

$$T^{ab}(x) = m \int \dot{z}^a \dot{z}^b (-g g_{cd} \dot{z}^c \dot{z}^d)^{-1/2} \delta(x, z)\, dt, \tag{3}$$

is incompatible with the field equation 1. (There does not exist a solution of (1)

§By "formal approximation methods" I mean methods for which it is not known whether they do, in fact, provide approximate solutions to any well-defined problem of the full theory, due to lack of convergence proofs, error estimates, etc.

representing a spherically symmetric gravitational field generated by a distributional source (**3**). If (**3**) is used formally in approximation methods to solve (**1**), the metric not only diverges at the particle and thus makes equation 3 mathematically meaningless, but in addition has the wrong signature near the particle, even in linear approximation. Disregarding these trivial facts leads to spurious difficulties which may obscure the real problems. Using the word "renormalization" is no cure if such a procedure is not carried out beyond the linear approximation.¶ The theorem that equations 2 and 3, together with the assumption that g_{ab} is a C^1 Lorentz metric in a neighborhood of the particle's world line, imply the geodesic law, is by itself not relevant for GR since its assumptions contradict equation 1. A derivation of the geodesic law has to be based on assumptions which are compatible with (**1**), not only with (**2**). Essential steps towards such a derivation are contained in references 16–19, 32). Therefore ordinary bodies (as opposed to black holes) have to be represented in classical GR primarily as extended, deformable structures, that is, connected, spatially compact submanifolds with timelike boundary hypersurfaces, as has been emphasized by Fock,[24] Dirac,[20] Bondi,[21] Chandrasekhar[22] and others. Of course, one would like to describe bodies for most purposes of celestial mechanics by a world line and a few parameters, but the possibility and nature of such a simplified, approximate model has to be *deduced* from the primary model since ad hoc assumptions like (**3**) do not work.

I remark in passing that a Newtonian mass point with its gravitational field, represented in the spacetime formulation of Newtonian theory,[23] can be obtained as an (exact) Newtonian limit[25] of the Kerr or Schwarzschild spacetime for fixed mass and angular momentum parameter.[26] Combined with well-known black-hole theorems this indicates that the simplest "objects" of relativistic mechanics, corresponding to Newtonian mass points, are black holes. Some groundwork towards a theory of isolated systems of black holes has been done by Gibbons[27] and D'Eath.[19,28]

Due mainly to the work of Dixon[17] and some recent extensions of it by Ehlers and Rudolph[18] and by Schattner,[29] a framework for the description of (ordinary) extended bodies in GR is available. For any such body, existence and uniqueness of a center-of-mass world line $l{:}z^a(s)$ (contained in a convex hull of supp T^{ab}) have been established;[29] a timelike linear momentum 4-vector $p^a(s)$, an angular momentum bivector $S^{ab}(s)$ and a sequence of multipole moments $I_n^{\cdots}(s)$ have been defined covariantly[17] on l in terms of g_{ab} and T^{ab}; and it has been shown[17,18] that these dynamic *body variables* satisfy a finite system

$$\dot{z}^a = (p_b p^b)^{-1/2} \cdot p^a + V^a(g_{bc}, R^b{}_{cde}; p^b, S^{bc}, F^b, L^{bc})$$
$$\dot{p}^a + \frac{1}{2} R^a{}_{bcd} \dot{z}^b S^{cd} = F^a \qquad (\mathbf{4})$$
$$\dot{S}^{ab} + 2\, \dot{z}^{[a} p^{b]} = 2\, L^{ab}$$

of ordinary tensor differential equations. (The "force" F^a and "torque" L^{ab} depend on the multipole moments and the field g_{ab} · V^a measures the deviation of the CM-tangent vector $\dot{z}^a$ from the unit vector collinear with p^a and is small for nearly spherical bodies.[18] The explicit forms of F^a, L^{ab} and V^a are given in references 17, 18. One has $p_a V^a = 0$, $p_a S^{ab} = 0$.) Conversely, the field T^{ab} can be reconstructed from g_{ab}

¶In this connection, see also reference 68.

and the body variables (z^a, p^a, S^{ab}, $I_n^{\cdots}$), provided (**4**) holds; the reconstructed T^{ab} obeys (**2**). (See reference 17 and, for some mathematical extensions, see reference 30.) The last two statements are particularly important since they show that the body variables describe the body *completely,* and that the equations 4 contain the same information as the local equation 2. Even the existence of such "global" body variables is remarkable. In the terminology proposed by Havas and Goldberg[31] equations 4 are the "laws of motion" for extended bodies. They relate the evolution of the body variables to the gravitational potential g_{ab} due (at least partly) to these bodies themselves. These laws are to be regarded as intermediate results on the way towards "equations of motion," viz. evolution equations containing body variables and, perhaps, nondynamic fields like a background metric only. (It is very questionable whether exact "equations of motion" exist in GR, in contrast to "laws of motion." I suspect that equations of motion exist as approximations only.)

The expressions F^a and L^{ab} in equations 4 are bilinear functionals of the multipole moments and certain "gravitational field strengths" derived from the metric g_{ab} and its connection. It is expected that these field strengths depend, within a particular body, mainly on that body itself and vary rapidly within the body, and that they depend only weakly on the other bodies of the system if the distances between the bodies are much larger than their sizes.

As in Newtonian mechanics and Maxwell-Lorentzian electrodynamics one would like to subtract *self fields* from the laws of motion (**4**) in such a way that the remaining external fields (for each body) vary only slowly within the body so as to permit a multipole approximation of the external forces and torques. Some self-field terms would be expected to represent radiation reaction forces, others might provide (finitely) "renormalized" variables $\hat{p}^a$, $\hat{S}^{ab}$. Such a program has not been carried out, but in view of the mentioned linearity it seems accessible, at least in conjunction with approximation methods to solve (**1**).

The Dixon-framework for the description of bodies in GR also offers a way to introduce simple body models such as "dynamically rigid"[17,18] or "dynamically spherical" bodies, defined by suitable invariance properties of the system of multipole moments along the CM-world line l. Such body models might be useful relativistic analogs of Newtonian mass points or rigid bodies which avoid the inconsistencies inherent in combining (**1**) with (**2**).

Unfortunately, no possibility to combine the Dixon framework, which is based on (**2**) alone, with the field equation 1 has yet been suggested. Nevertheless, hitherto this approach is the only one in which dynamic body variables such as 4-momentum have been defined covariantly which obey various nontrivial, physically suggestive laws. (For further results see references 17, 18, 29, 30).

Asymptotics

Suppose a spacetime (M, g_{ab}, T^{ab}) obeys the field equation (**1**) and contains a finite number of bodies as considered in the previous section. One would interpret it as a model of an isolated system only if its gravitational potential g_{ab} could be viewed as due to the bodies of the system (and, perhaps, some incoming radiation), and not to any outside matter. One cannot express this isolatedness by requiring, e.g. that g_{ab} be the "retarded field of T^{ab}" since, 1) there are no Green functions associated with the *nonlinear* field equation 1, and 2) a "flat, nonMachian part" of the metric representing "pure inertia" would at any rate not be determined by T^{ab}. It is, however, possible to formulate conditions on (M, g_{ab}) which express that the metric approaches that of flat spacetime "at infinity." Due to the basic work of Bondi, Sachs *et al.*,[15] the geometrization and extension of their asymptotic description along future null cones to spacelike and past null infinity by Penrose,[15] and extensive subsequent work by Geroch, Schmidt, Sommers, Ashtekar and others, there exists now an eleborate, flexible framework for describing in detail various kinds of asymptotic flatness. For systematic accounts and references see Geroch,[11] Schmidt[34,35] and Ashtekar and Hansen.[36] The last paper contains what is perhaps the most comprehensive notion of asymptotic flatness which has been proposed so far.

The key idea, due to Penrose, is to compress all distances "near infinity" by a conformal rescaling $g_{ab} \rightarrow \hat{g}_{ab} = \Omega^2 g_{ab}$ of the physical metric g_{ab}, and to attach to the physical spacetime a boundary of "ideal events" at which both the conformal factor Ω and the unphysical metric $\hat{g}_{ab}$ remain well behaved, with $\Omega = 0$, although g_{ab} "becomes infinite" there. If this is possible (such that several technical conditions are satisfied[11,35,36]), the spacetime is said to be *asymptotically flat*.

The boundary of an asymptotically flat spacetime consists of future and past null infinity, called $\mathcal{I}^+$ and $\mathcal{I}^-$, respectively, and spacelike infinity, $\mathcal{I}^0$. $\mathcal{I}^+$ is the set of ideal end points of null geodesics (test signals) leaving the system; $\mathcal{I}^-$, the set of ideal initial points of null geodesics incident upon the system; and, at least in the Kerr spacetime and its specializations, $\mathcal{I}^0$ consists of a single point i^0, the common ideal end point of all spacelike geodesics, which also compactifies all spacelike, global Cauchy hypersurfaces of the physical spacetime. The union $\mathcal{I}^+ \cup \{i^0\} \cup \mathcal{I}^-$ forms the "null cone at infinity" with vertex i^0.

One merit of this extension of the physical spacetime by an ideal boundary is that the asymptotic behavior of fields at large distances from the material sources need not be described in terms of ill-definable limits, but can be described in terms of local properties at the boundary, like differentiability. A price one has to pay for this "finitization" is that one has to be very careful in the specification of the degree of differentiability of the conformal factor Ω and the unphysical metric $\hat{g}_{ab}$ at infinity, particularly at i^0, in order not to lose physically needed generality. (Differentiability at infinity reflects the rate at which fields in physical spacetime decrease. Simple analogy: A three times continuously differentiable function $f: \mathcal{R} \rightarrow \mathcal{R}$ has the property $f(x) = a + bx^{-1} + cx^{-2} + \mathcal{O}(x^{-3})$ if the function $g: \mathcal{R} - \{0\} \rightarrow \mathcal{R}$ defined by $g(x) = f(x^{-1})$ can be extended to $\mathcal{R}$ such that the extended function has continuous derivatives up to the third order at 0.)

An additional merit of the latest approach[36] is that within it the point i^0 representing spatial infinity has a tangent space T_{i^0} with a unique Lorentz metric g_{ab}. This tangent space provides a home for the Arnowitt-Deser-Misner 4-momentum $P_0{}^a$, the coordinate-independent, intrinsic nature of which had remained somewhat obscure so far. Now $P_0{}^a$ has gained the respectable status of a vector at i^0. What is more,‖ also the Bondi-Sachs-Penrose 4-momentum $P_+{}^a(\Sigma)$, defined on cross-sections Σ of future

‖The following remarks are based on a preprint[37] which became available only after the Texas symposium. They are included here since they partly answer an old question discussed in my talk, and permit me to simplify the subsequent remarks about total energy-momentum.

null infinity (or $p_-{}^a(\Sigma)$, at past null infinity), can be considered as a vector at i^0, so that $P_0{}^a$ and $P_+{}^a(\Sigma)$ can be meaningfully related to each other.[37]

$P_0{}^a$ is interpreted as the *total energy-momentum* of the system, including all contributions from the gravitational field and the material sources. It is a *time-independent* property of the state (= Cauchy data) of the system. On the other hand, the difference $P_+{}^a(\Sigma_1) - P_+{}^a(\Sigma_2) =: P_R{}^a(\Sigma_1, \Sigma_2)$ represents the energy-momentum carried out of the system by gravitational radiation between outgoing null hypersurfaces ending at Σ_1 and Σ_2, respectively, with Σ_2 lying in the future of Σ_1. Thus, *the function* $\Sigma \rightarrow P_+{}^a(\Sigma)$ *informs about the evolution* of the system.

Ashtekar and Magnon-Ashtekar have shown[37] that if the 4-momentum carried by the gravitational waves emitted between the infinite past and some retarded instant Σ is finite, then—as had been conjectured for some time—$\lim_{\Sigma \to i^0} P_+{}^a(\Sigma) = P_0{}^a$ holds, whence one can state the *asymptotic energy-momentum balance*

$$P_0{}^a = P_+{}^a(\Sigma) + P_R{}^a(\Sigma), \qquad (5)$$

valid for any cross section Σ of $\mathscr{I}^+$, i.e. for any retarded time. The 4-momentum radiated out before Σ, $P_R{}^a(\Sigma)$, is known[38,11] to be positive** or zero—gravitational waves carry positive energy. If $P_+{}^a(\Sigma)$, the 4-momentum "left over" at the instant Σ, were also positive, then, from (5), $P_0{}^a$ would be positive, too, and (5) would imply that the total amount of energy that a system could possibly emit in gravitational waves could never exceed the total energy, $P_0{}^0$.

Unfortunately, general conditions for $P_+{}^a(\Sigma)$ to be positive have not been found, although the representation[39]

$$P_+{}^a(\Sigma) = \int_H \tau^{ab} dV_b \qquad (6)$$

of $P_+{}^a$ as a flux of the Landau-Lifshitz energy momentum complex τ^{ab} through a spacelike, asymptotically null hypersurface H terminating at Σ, evaluated in special asymptotically orthonormal coordinates, as well as an inequality derived by Streubel[39] suggest at least physically reasonable *sufficient* conditions for this to be the case.

On the other hand, general conditions have been found under which the ADM-energy, $P_0{}^0$, is a strictly positive functional of asymptotically flat initial data supported by spacelike Cauchy hypersurfaces,[40,41] whence under those conditions $P_0{}^a$ is positive for an asymptotically empty and flat spacetime in the sense of reference 36. (This positivity is *not* related to an intrinsic arrow of time, for the 4-vector $P_0{}^a$ is *defined* relative to a conventional time-orientation of spacetime such that its positivity is preserved under a switch of time-orientation.) However, in contrast to a positivity statement about $P_+{}^a$, the positivity of the ADM-energy or even of $P_0{}^a$ does not, by itself, imply restrictions about the energy that can be emitted in the form of gravitational waves.

In order to draw physically useful conclusions from the constancy (hypersurface independence) and positivity of $P_0{}^a$ it appears to be necessary to decompose, for "late" Cauchy hypersurfaces when radiation has travelled away from its sources, the total $P_0{}^a$ into summands assignable to the source and the radiation, respectively, and to establish positivity separately for these parts. The possibility of such a procedure is

**We call a 4-vector *positive* if it is contained in the interior of the future null half cone. We assume spacetime to be time-oriented.

suggested by a formula[42] analogous to (**6**) for $P_0{}^a$, but I am unaware that it has been carried out.

Related to the problem just mentioned is the question of the connection between the Dixon 4-momenta p^a of the bodies considered in the previous section and the asymptotic quantities $P_0{}^a$, $P_+{}^a$. Expressions of $P_0{}^a$ and/or $P_+{}^a$ in terms of the p^a and field contributions would seem to be needed to relate the asymptotic energy-momentum balance (**5**) to the motion of the bodies and, thus, to observable properties of isolated systems.

The problem of defining asymptotic angular momenta, to establish for them a law analogous to (**5**) etc. has been considered, but will not be discussed here. (See references 12, 36, 39, 43.)

Whereas future null infinity is used to describe outgoing radiation, incident radiation can be described naturally in terms of asymptotic fields on past null infinity, $\mathcal{I}^-$. The analog of equation 5 for $\mathcal{I}^-$ shows: The condition

$$\underline{P}^a = \text{const} \tag{7}$$

which implies $\underline{P}^a = P_0{}^a$) is necessary and sufficient for absence of incoming gravitational radiation. This condition is equivalent to the vanishing of the advanced time-derivative of the asymptotic shear at $\mathcal{I}^-$. (For these and alternative conditions see references 39, 44.)

For asymptotically empty and flat spacetimes the metric, curvature, etc. can be represented, in a neighborhood of null infinity (far radiation zone), by asymptotic series in inverse powers of a radial coordinate, with coefficients depending on angles and retarded (or advanced) time; see, e.g. reference 45 and the references therein.

I conclude this section with some critical remarks taken partly from B.G. Schmidt.[35] The concepts of asymptotic flatness have been invented to describe the gravitational field of isolated systems far away from the sources. The main motivations came from linearized theory, stationary exact solutions and peeling properties of algebraically special vacuum fields. Since neither an exact solution of Einstein's field equation describing a bounded source emitting gravitational radiation nor even an existence proof for such a solution is known, the definitions of asymptotically flat spacetimes given so far have to be considered as *proposals*, the adequacy of which, for their intended purpose, is open to questions and needs to be investigated further. In particular, it is not known whether "typical" radiation fields fall off, and sources behave sufficiently quietly in the distant past, in order that a null- or space-like infinity exists and/or has the differentiability which has been assumed (conjectured) in currently used definitions of asymptotic flatness. Some perturbation calculations on scalar, massless waves in Schwarzschild spacetime by B.G. Schmidt and J.M. Stewart[46] and computations on the relativistic Kepler problem by M. Walker and C.M. Will[47] indicate that the usual differentiability requirements at $\mathcal{I}$ are, in fact, too strong even for simple examples which belong to the intended domain of application of the notion of asymptotic flatness. In this connection it may also be appropriate to remember that it is not known, in spite of extensive work on the initial value problem for Einstein's equation, whether the maximally developed spacetime[48,49] belonging to asymptotically flat initial data has even a piece of null infinity, and whether Cauchy data associated with boosted Cauchy hypersurfaces[36] are asymptotically flat.

It seems important to keep in mind that asymptotics is not a theory for its own sake, but a *tool* for analyzing and, I hope, constructing models of isolated systems and deriving properties of them. As a tool, it is important in so far as it serves these purposes. Since exact models of radiating systems are not in sight, it appears to be urgent to connect the concepts and constructions of asymptotics with approximation methods, even if that requires sacrificing some of its geometrical beauty or modifying its assumptions.

Approximation Methods and Equations of Motion

Attempts to construct relativistic models of isolated systems containing more than one body have been made since the beginning of general relativity by means of formal, analytical approximation methods and, only recently, by numerical methods. For the latter, see L. Smarr[50] and the contribution by D. Eardley to these proceedings.

In this section I shall outline a new, improved version of a weak-field, slow-motion approximation method; indicate results which have been obtained by means of it; discuss its drawbacks and suggestions to overcome them; and make some remarks about alternative methods. A general review of approximation methods will not be attempted here.

The approximation method to be outlined is a modification, proposed and sketched by myself[4] and elaborated by D. Kerlick[51] and A. Caporali,[52] of the method of J.L. Anderson and T.C. Decanio.[53] Theirs as well as the new method incorporate ab initio Fock's no-incoming-radiation-condition[24,44] into a systematic iteration scheme which uses the harmonic gauge throughout, in contrast to the method of Chandrasekhar et al.[54] The new method is, at least formally, an improvement in two respects: 1) All approximate metric components, Christoffel symbols etc. needed to set up the 2½ post-Newtonian equations of motion are finite; no divergent expressions occur during the calculations leading up to those equations. 2) The components of the metric and the connection at some coordinate time are obtained as functionals of the Newtonian Cauchy data (mass density and coordinate 3-velocity of matter) pertaining to that same instant. Hence the post-Newtonian forces in the equations of motion do not contain time derivatives. Thus, initial data for the matter variables determine, at least formally, a unique solution, and classical perturbation theory can be applied directly. On the other hand, the new method does give divergent terms in the third PNA. (This will be discussed below.)

The new method is based on the *reduced Einstein equation*

$$\mathfrak{g}^{cd}\partial_{cd}\mathfrak{g}^{ab} + Q^{ab}(\mathfrak{g}, \partial\mathfrak{g}) = 2(-g)T^{ab}, \qquad \mathbf{(8)}$$

obtained by writing equation (1) in terms of the contravariant metric density $\mathfrak{g}^{ab} := \sqrt{-g}\, g^{ab}$ ($g := \det g_{ab} = \det \mathfrak{g}^{ab}$) and omitting terms containing $\mathfrak{g}^{ab}{}_{,b}$. Clearly, a solution of (**8**) obeys (**1**) provided the *harmonic gauge condition*

$$\mathfrak{g}^{ab}{}_{,b} = 0 \qquad \mathbf{(10)}$$

is also satisfied. One now chooses an auxiliary flat Lorentz metric η^{ab} and defines the *metric deviation*

$$k^{ab} := \mathfrak{g}^{ab} - \eta^{ab}. \qquad \mathbf{(11)}$$

There is no invariant measure of the "smallness" of k^{ab} at a point (E. Eder[55] has shown that no natural distance function exists on the set of Lorentzian metrics on a vector space.) For the following approximation scheme in which k^{ab} is computed first and **(11)** is then used to find $\mathfrak{g}^{ab}$, g_{ab} etc., it is sufficient to assume that for $0 \leq x \leq 1$, det $(\eta^{ab} + xk^{ab}) < 0$; let the set of these k^{ab} be denoted as $\mathfrak{K}$. (This mild weak-field assumption holds even for neutron stars.) One verifies easily:

1) $\mathfrak{g}^{ab} = \eta^{ab} + k^{ab}$ is Lorentzian if k^{ab} is in $\mathfrak{K}$.

2) $(-g)$, $(-g)^{-1/2}$, g_{ab} and g^{ab} are analytic functions of k^{ab} on $\mathfrak{K}$.

3) The partial derivatives $(-g)_{,c}, \ldots, g^{ab}{}_{,c}$ are (in an obvious sense) analytic functions of k^{ab} and $k^{ab}{}_{,c}$ provided k^{ab} is in $\mathfrak{K}$. Analogous statements hold for higher derivatives and, therefore, for $\Gamma_{bc}{}^{a}$, R_{ab} etc.

One now works in orthonormal coordinates with respect to η^{ab} so that **(10)** can be replaced by

$$k^{ab}{}_{,b} = 0, \tag{12}$$

and rewrites **(8)** (modulo **(10)**) as

$$\frac{1}{2} \Box k^{ab} = g(T^{ab} + t^{ab}) + (k^{a[b}k^{d]c})_{,cd} \tag{13}$$

where $\Box := \vec{\nabla}^2 - \partial_{tt}$ is the flat-space wave operator. In $\mathfrak{K}$, the Landau-Lifshitz energy-momentum pseudo-tensor t^{ab} of the gravitational field depends analytically on k^{ab} and its first derivatives. The last term on the r.h.s. of **(13)** contains even second derivatives of k^{ab}; its divergence vanishes identically. Hence, **(12)** and **(13)** imply the (nontensorial) conservation law

$$\{-g(T^{ab} + t^{ab})\}_{,b} = 0. \tag{14}$$

Choose as a source model an isentropic perfect fluid,

$$T^{ab} = (\rho + p)U^{a}U^{b} - pg^{ab}, \qquad p = p(\rho), \qquad |p| < \rho, \tag{15}$$

and define as *matter variables* the density ρ and the coordinate 3-velocity $\vec{v} := (U^{\lambda}/U^{0})$. Because of $g_{ab}U^{a}U^{b} = 1$ and **(11)**, U^{0} and T^{ab} can be considered as functions of ρ, $\vec{v} = (v^{\lambda})$ and k^{ab};

$$T^{ab} = T^{ab}(\rho, \vec{v}, k^{cd}) \tag{16}$$

The last condition in **(15)** implies that $T_{b}{}^{a}$ satisfies the energy condition formulated in the section entitled "Bodies and Laws of Motion."

In view of the remarks made after equation 11, equation 13 is of the form

$$-\frac{1}{2}\Box k = (1 + f_1(k))T(\rho, \vec{v}, k) + f_2(k, \partial k) + f_2(k, \partial k, \partial^2 k) =: 2\pi\Delta. \tag{17}$$

Here tensor indices have been suppressed, and f_ν denotes a (numerical or matrix valued) analytic function of k and the indicated derivatives which starts with terms of order ν in k and its derivatives.

If k satisfies Fock's condition for absence of incoming radiation,[24,44]

$$k = o(1), \quad \left(\frac{\partial}{\partial r} + \frac{\partial}{\partial t}\right) k = o(r^{-1}) \tag{18}$$

uniformly in neighborhoods of past-directed, flat-space null cones, then (**17**) implies, via Kirchhoff's identity, the integro-differential equation

$$k(\vec{\mathbf{x}}, t) = \int_{\mathcal{R}^3} \Lambda(\vec{\mathbf{x}} + \vec{\mathbf{y}}, t - |\vec{y}|) \frac{d^3y}{|\vec{y}|}. \tag{19}$$

Postponing objections against (**18**), let us proceed with the algorithm. One now formally Taylor-expands the integrand with respect to its time argument at t, obtaining

$$k = \sum_{n=0}^{\infty} \frac{(-1)^n}{n!} I_{n-1}(\partial_t{}^n \Delta), \tag{20}$$

where

$$I_n(f)(\vec{x}, t) := \int f(\vec{x} + \vec{y}, t)\, |\vec{y}|^n d^3y. \tag{21}$$

The right hand side of equation 20 contains time derivatives of all orders due to the retardation expansion, just as the field expansion in Lorentz's electron theory. Convergence of the series in (**20**) is neither to be expected (except for stationary situations) nor relevant; it is part of a formal algorithm.

One can now carry out the following *iteration*:[4,51] Put $k = 0$ in Λ and in equation 14. Define, guided by (**20**), $\underset{(1)}{k}{}^{ab}$ *by*

$$\begin{aligned} \underset{(1)}{k}{}^{0a} &:= I_{-1}(\underset{(0)}{\Lambda^{0a}}) - I_0(\delta_t \underset{(0)}{\Lambda^{0a}}) \\ \underset{(1)}{k}{}^{\lambda\mu} &:= 0 \text{ for } \lambda, \mu = 1, 2, 3, \end{aligned} \tag{22}$$

and eliminate $\dot{\rho}, \dot{\vec{\nu}}$ by means of $(\mathbf{14})_{k=0}$, to obtain $\underset{(1)}{k}{}^{ab}$ as a functional of $\vec{\rho}$ and $\vec{\nu}$.

Then, insert $\underset{(1)}{k}{}^{ab}$ into Λ (equation 17) and (**14**) and compute $\underset{(2)}{k}{}^{ab}$ from (**20**), keeping for each component two more terms of the series than were used in (**22**), and eliminate time derivatives of ρ and $\vec{\nu}$ by means of the "improved" equation 14, to obtain $\underset{(2)}{k}{}^{ab}$ as a functional of ρ, $\vec{\nu}$ and their spatial derivatives. Continue in this way to obtain $\underset{(3)}{k}{}^{ab}$ etc.††

If the density ρ has compact support this procedure gives convergent integrals for $\underset{(1)}{k}{}^{ab}$ and $\underset{(2)}{k}{}^{ab}$. Most contributions to $\underset{(3)}{k}{}^{ab}$ are also well-defined, but some of its terms diverge.[51] To order the various terms in $\underset{(\nu)}{k}{}^{ab}$, $\underset{(\nu)}{T}{}^{ab} := T^{ab}(\rho, \vec{\text{v}}, \underset{(\nu)}{\text{k}}'')$ and the expressions derived from them, like $\underset{(\nu)}{\Gamma_{bc}{}^{a}}$ etc, one can either keep the speed of light c in

††This algorithm to obtain $\underset{(1)}{k}$, $\underset{(2)}{k}$ etc. is neither a formal power series expansion nor an iteration in the usual sense. It is similar to a *Hilbert expansion* used to obtain so-called "normal" asymptotic series solutions of Boltzmann's integro-differential equation.

all equations and order terms according to powers of c^{-1}, or use physical arguments to assign orders of magnitude to $\rho, \vec{v}, dp/d\rho$ etc. to obtain a dimensionless small parameter for this purpose.

If $\underset{(N)}{k}^{ab}$ is in $\mathfrak{R}$, it determines a Lorentz metric $\underset{(N)}{g}^{ab}$. There is no reason for the pair $(\underset{(N)}{g}_{ab}, \underset{(N)}{T}^{ab})$ to obey Einstein's equation approximately. But if one imposes on such a pair an *approximate equation of motion* by inserting $\underset{(N)}{g}_{ab}, \underset{(N)}{T}^{ab}$ into the local law of motion (**2**) and keeping consistently terms up to some assigned order of magnitude, or, alternatively, imposes the conservation law (**14**) or the gauge condition (**12**), then one might expect—from the way the algorithm has been set up—that those solutions of the approximate equations of motion which describe slowly moving bodies with small "compactness" (= Schwarzschild radius/geometrical radius), do indeed give approximate solutions of (**1**) in a near zone containing the bodies, small compared to a wavelength characteristic of the source motion. This expectation is supported, at least for the first post-Newtonian approximation, by some plausible estimates[52] of the "error tensor" $E_{ab} := R_{ab} - (T_{ab} - \frac{1}{2} g_{ab} T)$ associated with a pair $(\underset{(N)}{g}_{ab}, \underset{(N)}{T}^{ab})$, but no exact error estimates are known.

If the local, first post-Newtonian equations of motion (which are equivalent to those of Chandrasekhar[22] and Synge[56]) are integrated over the bodies, and if these bodies are assumed to be well-separated, nearly spherical, and nonrotating, then one recovers for the center-of-mass motions the Einstein-Infeld-Hoffmann equations, in accordance with older results, provided one identifies the positions and masses of EIH with suitable quantities defined within the 1PN-framework.[52,57] (In connection with this derivation, and in view of the remarks about the self field problem, the following may be of interest. If, within the 1PNA, one subtracts the self potential of a body from the total potential, then because of the nonlinear dependence of the potential on the source variables, the resulting difference potential still depends on the body under consideration, and the self potential depends on the other bodies too. What distinguishes the self potential of a body from the difference potential is that, within that body, the former varies on the scale of the size of the body, whereas the latter varies slowly, on the scale of the distance to the other bodies. This should be valid also in the exact theory, as emphasized by Dixon.[17]) If one assumes the bodies to be axially symmetric and rotating, one obtains in addition the spin-precession formulae (Caporali and Spyrou, to be published) derived previously by different, in my opinion, less satisfactory methods.[58–60]

As far as I am aware, the second post-Newtonian, local hydrodynamic equations of motion have not been used to obtain equations of motion for bodies as a whole.

Being in possession of the improved, infinity-free 2½ PN local equations of motion derived by Kerlick, one can repeat the Chandrasekhar-Esposito[54] analysis to identify radiation damping force densities, and set up a *near-zone energy-momentum balance* which can then, hopefully, be connected with the asymptotic balance (**5**) as indicated in FIGURE 1. This work has not yet been completed.

Let us now look at the approximation method *critically*. A first question mark should be attached to equation 10 or 12, respectively. It has not been shown that harmonic coordinates exist globally, as was assumed above. Although Y. Choquet-Bruhat, D. Christodoulou and M. Francaviglia[61] have recently shown that such coordinates do exist globally prior to some Cauchy-hypersurface if the metric is in a

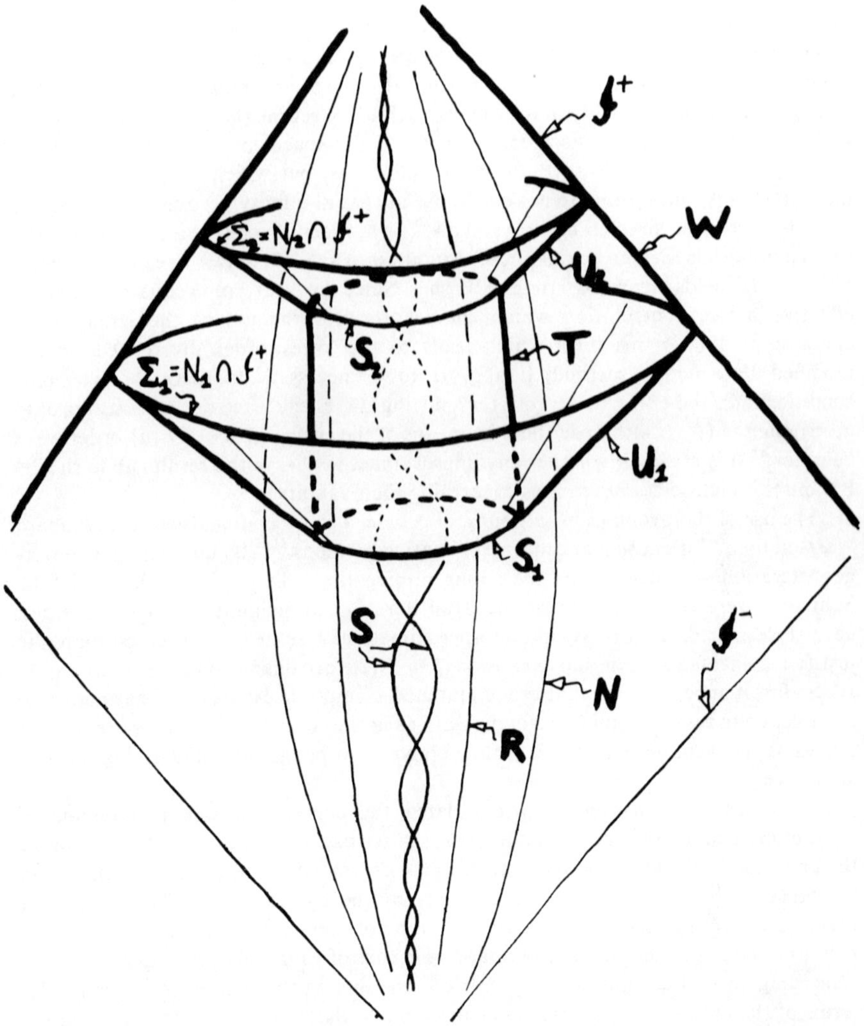

FIGURE 1. Connection between near-zone and asymptotic energy-momentum balances. Outside R is the radiation zone, inside N the near zone containing the source S. S_1 and S_2 are oriented, spacelike cross sections of the near zone; T is a timelike cylinder connecting the spherical boundaries of S_1, S_2, respectively; U_1, U_2 are outgoing null hypersurfaces; and W is the part of future null infinity which is between the cross sections Σ_1 and Σ_2. $P(S_1)$ etc. denotes the 4-momentum flux of the Landau-Lifshitz energy-momentum complex through S_1 etc. (as in equation 6 of the text), computed in a suitable, approximately orthonormal coordinate system. Then the *inner balance* is $P(S_2) - P(S_1) = -P(T)$; it can be approximately evaluated by means of the near-zone approximation, provided the S_i are chosen suitably. This relation is substantially equation 110 in reference 22, interpreted geometrically. The *outer balance* is $P(W) \approx P(T)$, since[62] $P(U_i) \approx 0$ for outgoing radiation. The *asymptotic balance* (text, equation 5) is $P(W) = P_R(\Sigma_1,\Sigma_2)$ = Bondi-flux. In summary, then, $(P(S_2) - P(S_1))_{\text{near zone}} \approx -(P_R(\Sigma_1,\Sigma_2))_{\text{asympt.}}$. This sketch indicates how one may be able to obtain an "overall balance," provided one carries out several estimates, defines suitable coordinates, etc. Such a reasoning should also provide inequalities to define a near zone and a radiation zone; it has not been carried out, as far as I am aware.

certain sense asymptotically stationary in the infinite past, this still does not seem to cover generic cases of physical interest. Secondly, and more importantly, condition (**18**) in harmonic coordinates may not express absence of incoming radiation correctly; for, as is well known, already in the Schwarzschild spacetime the obvious splitting (**11**) with respect to harmonic coordinates leads to flat-space, past-directed null cones which are *not* asymptotic to the correct null cones, but which deviate from them logarithmically and terminate not on $\mathscr{I}^-$ but at spatial infinity, i^0; and even along the curved-space, past-directed null rays, the $\mathbf{k}^{ab}$ do not seem to obey the conditions[9] (**18**). (This point needs further clarification.) Moreover, using (**13**) or (**19**) one propagates all higher-order fields along the wrong null cones. Since the last term in equation 13 is the effective source contribution which compensates for the use of the wrong wave operator, and it is this term which leads to the lowest-order divergences in the modified slow-motion method, it appears to be necessary to relax the harmonic condition (**10**), and either to improve the splitting (**11**) at each step of the iteration, or to propagate the $(N + 1)$th-order field by means of the curved-space, (N-th)-order wave operator.[63] It is not clear whether any improvement will leave the results up to the 2½ PN order intact, so doubt remains concerning their validity.

The use of the wrong propagation is, of course, a shortcoming also of the standard "fast-motion," Lorentz-invariant approximation methods.[31,64] By combining a convergent iteration with one using "the wrong propagation," D. Christodoulou and B.G. Schmidt[65] were able to prove that the latter iteration, although it in general diverges, nevertheless leads to an asymptotic approximation. Useful and exact estimates to justify a truncated retardation expansion (**19**) → (**20**) are unknown to me.

So far, it appears to me that use of matched asymptotic expansions[65] has not led to more concrete results or more exact derivations in the context of isolated systems than the more conventional method outlined above, although the underlying idea is attractive.

A. Rosenblum[64] has recently calculated the energy loss due to emission of gravitational radiation during small angle scattering, using a fast-motion analog of the first post-Newtonian approximation. Since he treated the bodies as points, he had to regularize infinite, linear self field-terms, and he had to assume the existence of higher-order regularization procedures, though not their form. His result differs by a factor of about 2.3 from the one obtained by means of Einstein's quadrupole formula. According to this calculation the contributions due to the nonlinear, second-order terms of the metric are numerically comparable to those from the linear approximation. This result may cast some light on the antidamping obtained long ago by S.F. Smith and P. Havas[67] on the basis of a linear, fast approximation calculation in which, perhaps, nonlinear terms would over-compensate the linear ones. However, the bound state problem has not been treated yet by a fast method beyond the linear stage. Rosenblum's result accentuates the need for further work to clarify the gravitational radiation damping problem.

Summary

The first post-Newtonian approximation that includes the EIH-equations has been obtained by different formal approximation methods. The results agree; there do not appear to be serious objections against its validity. Thus the theoretical basis for the four classical tests of GR and the spin precession effects, for bodies of arbitrary compactness and for black holes, seems to be firm. A corresponding statement cannot be made, in my opinion, for the higher approximations, particularly not for gravitational radiation reaction effects, for reasons which I have discussed.

Even at the 1PN level a connection of the approximation schemes with intrinsic, coordinate-free formulations would seem to be desirable for *understanding*, if not for computing effects. The basic problem of rigorously justifying the formal approximation is perhaps outside the reach of present mathematics.

Can a combination of approximation methods with exact approaches help to overcome the difficulties and unclarities discussed above, and to obtain answers to "concrete" questions? Can one *formulate questions* pertaining to, say, "secular effects of the emission of radiation on the motion of binary systems," *by means of intrinsic concepts* of Einstein's theory *without* reference to approximation methods, and compute such effects by methods which do not use assumptions contradicting the underlying precise formulation? Can the recent work on the dynamics of GR, i.e. the evolution of Cauchy date, help to improve approximations? Maybe at one of the future Texas symposia we shall get some answers.

References

1. Havas, P. 1973. *In* Ondes et Radiations Gravitationelles. Coll. Int. CNRS No. **220:** 383. Paris.
2. Papapetrou, A. 1974. Lectures on General Relativity. Reidel. Dordrecht, Holland.
3. Ehlers, J., A. Rosenblum, J. N. Goldberg & P. Havas. 1976. Ap. J. **208:** L77.
4. Ehlers, J. 1977. *In* Proceedings of the International School of General Relativistic Effects in Physics and Astrophysics. Institute-Report MPI-PAE/Astro **138:** 45. Munich.
5. Einstein, A. 1918. Sitzungsber. Preuss. Akadem. Wiss., Physik.-Math. **Kl.:** 154.
6. Peters, P. C. 1964. Phys. Rev. **136:** 1224.
7. Wagoner, R. V. 1975. Ap. J. Lett. **196:** L63.
8. Trautman, A. 1958. Lectures on General Relativity. (Mimeographed notes). King's College. London.
9. Isaacson, R. A. & J. Winicour. 1968. Phys. Rev. **168:** 1451; Madore, J. 1970. Ann. Inst. Henri Poincaré **12:** 285, 365.
10. Dixon, W. G. & J. M. Bird. 1975. Ann. Phys. (N.Y.) **94:** 320; Bird, J. M. 1976. Ann. Phys. **101:** 345.
11. Geroch, R. P. 1977. *In* Asymptotic Structure of Spacetime. F. P. Esposito & L. Witten, Eds.: 1. Plenum Press. New York, N.Y.
12. Ehlers, J., Ed. 1979. Isolated Gravitating Systems in General Relativity. North-Holland Publ. Company. Amsterdam.
13. Hawking, S. W. 1968. J. Math. Phys. **9:** 598.
14. Aichelburg, P. C. & R. Beig. 1976. Ann. Phys. **98:** 264; Beig, R. *In* ref. 12:307.
15. Bondi, H., M. G. J. van der Burg & A. W. K. Metzner. 1962. Proc. Roy. Soc. **A269:** 21; Sachs, R. K. 1962. Proc. Roy. Soc. **A270:** 103; Penrose, R. 1965. Proc. Roy. Soc. **A284:** 159.
16. Infeld, L. & A. Schild. 1949. Rev. Mod. Phys. **21:** 408.
17. Dixon, W. G. 1979. Ref. 12: 156. See also the papers cited therein.
18. Ehlers, J. & E. Rudolph. 1977. Gen. Rel. Grav. **8:** 197.
19. D'Eath, P. D. 1975. Phys. Rev. **D11:** 1387.
20. Dirac, P. A. M. 1961. Lecture at the Intern. Conference on General Relativity and Gravitation held in Jablonna, Warsaw.
21. Bondi, H. 1964. *In* Lectures on General Relativity. A. Trautman, F. A. E. Pirani, H. Bondi, Eds.: 375. Prentice-Hall. Englewood Cliffs, New Jersey.

22. CHANDRASEKHAR, S. 1970. *In* Relativity. M. Carmeli & S. I. Fickler, Eds.: 81. Plenum Press. New York, N.Y.
23. KÜNZLE, H. P. 1972. Ann. Inst. Henri Poincaré. **17:** 337.
24. FOCK, V. 1959. The Theory of Space, Time and Gravitation. Pergamon Press. London.
25. DAUTCOURT, G. 1964. Acta Physica Polonica **25:** 637; KÜNZLE, H. P. 1976. Gen. Rel. Grav. **7:** 445.
26. EHLERS, J. 1979. Festschrift für P. Mittelstaedt. To be published.
27. GIBBONS, G. W. 1974. Commun. Math. Phys. **35:** 13.
28. D'EATH, P. D. 1975. Phys. Rev. **D12:** 2183.
29. SCHATTNER, R. 1979. Gen. Rel. Grav. **10:** 377, 395.
30. SCHATTNER, R. 1979. Ph.D. Thesis. University of Munich. To be published.
31. HAVAS, P. & J. N. GOLDBERG. 1962. Phys. Rev. **128:** 398.
32. WEYL, H. 1923. Raum, Zeit, Materie (5th ed.). Springer Verlag. Berlin.
34. SCHMIDT, B. G. 1977. See ref. 11: 429.
35. SCHMIDT, B. G. 1979. *In* ref. 12: 11.
36. ASHTEKAR, A. & R. O. HANSEN. 1978. J. Math. Phys. **19:** 549.
37. ASHTEKAR, A. & A. MAGNON-ASHTEKAR. 1979. Preprint. Université de Clermont-Fd. 63170 Aubière, France.
38. SACHS, R. K. 1962. Phys. Rev. **128:** 2851.
39. STREUBEL, M. 1978. Dissertation. University of Hamburg. (Available as report MPI-PAE/Astro 165 from Max-Planck-Institut für Physik und Astrophysik, Munich.)
40. CHOQUET-BRUHAT, Y., A. E. FISCHER & J. E. MARSDEN. 1979. *In* ref. 12.
41. SCHOEN, R. & S. T. YAU. 1979. Commun. Math. Phys. **65:** 45.
42. MISNER, C. W., K. S. THORNE & J. A. WHEELER. 1973. Gravitation. Freeman and Comp. San Francisco.
43. WINICOUR, J. 1968. J. Math. Phys. **9:** 861.
44. LEIPOLD, G. & M. WALKER. 1977. Ann. Inst. Henri Poincarè **27:** 61.
45. EXTON, A., E. T. NEWMAN & R. PENROSE. 1969. J. Math. Phys. **10:** 1566; LUDWIG, G. 1976. Gen. Rel. Grav. **7:** 293.
46. SCHMIDT, B. G. & J. M. STEWART. 1978. Report MPI-PAE/Astro 161 of the MPI (see reference 39).
47. WALKER, M. & C. M. WILL. 1979. Phys. Rev. **D19:** 3483, 3495.
48. CHOQUET-BRUHAT, Y. & R. P. GEROCH. 1969. Comm. Math. Phys. **14:** 329.
49. HAWKING, S. W. & G. F. R. ELLIS. 1973. The Large-Scale Structure of Spacetime. University Press. Cambridge.
50. SMARR, L. 1977. Ann. N.Y. Acad. Sci. **302:** 569.
51. KERLICK, D. To be published in Gen. Rel. Grav.
52. CAPORALI, A. 1979. Ph.D. Thesis. University of Munich. To be published.
53. ANDERSON, J. L. & T. C. DECANIO. 1975. Gen. Rel. Grav. **6:** 197.
54. CHANDRASEKHAR, S. & F. P. ESPOSITO. 1970. Ap. J. **160:** 153.
55. EDER, E. 1979. Report MPI-PAE/Astro 186 of the MPI (see reference 39).
56. SYNGE, J. L. 1970. Proc. Roy. Irish Acad. **69A:** 11; MCCREA, J. D. & G. O'BRIEN. preprint DIAS-TP-78-23. Dublin.
57. HOGAN, P. A. & J. D. MCCREA. 1974. Gen. Rel. Grav. **5:** 79; SPYROU, N. 1978. Gen. Rel. Grav. **9:** 519.
58. BARKER, B. M. & R. F. O'CONNELL. 1975. Phys. Rev. **D12:** 329.
59. BÖRNER, G., J. EHLERS & E. RUDOLPH. 1975. Astron. Astrophys. **44:** 417.
60. MCCREA, J. D. & G. O'BRIEN. 1978. Gen. Rel. Grav. **9:** 1101.
61. CHOQUET-BRUHAT, Y., D. CHRISTODOULOU & M. FRANCAVIGLIA. In preparation.
62. ISAACSON, R. A. 1968. Phys. Rev. **166:** 1263, 1272.
63. THORNE, K. S. & S. J. KOVÁCZ. 1975. Ap. J. **200:** 245.
64. ROSENBLUM, A. 1978. Phys. Rev. Lett. **41:** 1003.
65. CHRISTODOULOU, D. & B. G. SCHMIDT. 1979. Commun. Math. Phys. **68:** 275.
66. BURKE, W. L. 1979. In ref. 12: 220; also 1971, J. Math. Phys. **12:** 401.
67. SMITH, S. F. & P. HAVAS. 1965. Phys. Rev. **138:** 495.
68. MÅNSSON, B. 1978. Gen. Rel. Grav. **9:** 1049.

SUPERGRAVITY

J. Wess

Karlsruhe University
7500 Karlsruhe
Federal Republic of Germany

Supergravity is the gauge theory based on the following algebra:

$$[P_m, Q_\alpha]_- = 0, \qquad [P_m, P_n]_- = 0, \qquad \{Q_\alpha, \overline{Q}_\beta\}_+ = 2i\gamma_{\alpha\beta}{}^m P_m. \tag{1}$$

P_m is the energy-momentum four-vector and Q_α is a Majoranor spinor. This specifies the transformation properties of P_m and Q_α under Lorentz transformations.

A relativistic field theory is a field theory based on the Poincaré algebra. In the same way, a supersymmetric theory is a field theory based on the supersymmetry algebra 1. The surprising thing is that such theories exist and that they meet all the requirements of a quantum field theory.[1]

In supersymmetric theories, there is a spin 3/2 current $y_{\alpha m}$ in addition to the energy-momentum tensor θ_{mn}. This current is conserved $\partial^m y_{\alpha m} = 0$. Its charge $Q_\alpha = \int d^3x y_{\alpha 0}$ satisfies the algebra 1 due to the commutation relations of the field operators.

Theories which possess such currents have remarkable properties:

They have a structure in which the Poincaré group is tied into a larger algebra—the algebra 1—in a nontrivial way. This is only possible because of the anticommutator in 1. This would not be possible for a Lie algebra with commutators only, due to a well-known no-go theorem.[2]

These theories are usually less divergent than one would expect from power counting without taking into account the relations which follow from the conservation law of $y_{\alpha m}$.

Because the translations are part of the algebra which has to be gauged, supergravity will be a theory invariant under general coordinate transformations. We expect a theory where the gravitational field couples to matter fields in a well-defined way.

We expect that supergravity will be less divergent than gravity. This is of great interest because up to now all gravitational theories are meaningless when quantized. The gravitational interaction alone is one-loop renormalizable on-mass shell, while all couplings of the gravitational field to matter fields of spin 0, 1/2, or 1 are not even one-loop renormalizable. It has been shown that supergravity, where the graviton couples to a spin 3/2 field in a consistant way, is even two-loop renormalizable on mass shell.[3] It is just the presence of the spin 3/2 field that improves the situation in comparison to pure gravity.

Secondly, the fact that the Poincaré group is a nontrivial part of the symmetry allows us to construct a model for a nontrivial unification of gravity with other interactions. Generalized no-go theorems[4] tell us that supergravity and extended supergravity, which I will explain in a moment, are essentially all possible models in which such a unification can be studied in a serious way, i.e., in a way which is not from the very beginning in conflict with the axioms of quantum field theory.

0077–8923/80/0336–0295 $1.75/1 © 1980, NYAS

Extended supergravity, as mentioned above, is based on the algebra:

$$\{Q_\alpha{}^i, \overline{Q}_\beta{}^j\} = 2i\gamma_{\alpha\beta}{}^m P_m \delta^{i\gamma}$$

$$[P_m, Q_\alpha{}^i]_- = 0$$

$$i = 1 \ldots N.$$

It is easy to see that this algebra has, for $P^2 = 0$, a mass shell representation on helicity states where the helicity ranges from j to $j + (N/2)$. For a field theoretical description we also have to add the states with opposite helicity. If we do not want to have a state with spin larger than two we are restricted to $N = 4$, if we start from $j = 0$. It is however possible to start from $j = -2$ and to arrange the theory in such a way that this representation contains already the states with opposite helicity. In this case we can go up to $N = 8$. This seems to be the most interesting model for the study of a unification scheme. In terms of "physical" fields it contains a field with spin 2 (graviton), eight spin 3/2 fields, 28 vector fields, 56 spin 1/2 Majorana fields, and 70 fields with spin zero. This is quite a playground and I am sure that this model will keep us busy before we understand all its features.

However, already the simpler models are very interesting. The $N = 2$ model contains one spin 2, two spin 3/2, and one vector field. This is a theory where the graviton and the Maxwell field interact in a fully covariant way. Its field-theoretical properties are remarkable, divergencies cancel in the one- and two-loop approximation, and we arrive at a theory which is at least two-loop renormalizable.

I would like to emphasize that this cancellation only occurs because there are spin 3/2 fields in the theory. They seem to be able to compensate for the short-range singularities of a gravitational theory.

At present we have good reason to believe that all extended supergravitation theories are, at least, two-loop renormalizable. This property disappears, however, if we try to couple other multiplets to the gravitational one in a supergauge invariant way.

Supergravity ($N = 1$) will be described in terms of a physical spin 2 and spin 3/2 field. A supersymmetric extension of the pure gravity Lagrangian leads to a Lagrangian that is invariant under supergauge transformations if one is willing to use the equations of motion. Such Lagrangians were constructed by Deser and Zumino[5] and Ferrara, Freedman, and van Nieuwenhuizen.[6] These Lagrangians give rise to a consistent coupling of the graviton to a spin 3/2 field—a nontrivial achievement. In addition, they yield a theory which is at least two-loop renormalizable.

A more systematic approach, which allows the construction of Lagrangians that are invariant also off-mass shell, is based on the concept of differential geometry in superspace.[7]

The concept of superspace has been introduced by Salam and Strathdee.[8] An element of this space Z^M will be parametrized by the four commuting variables X^m, space and time, and by four anticommuting, spinorial variables θ^μ. The generators of the supersymmetric algebra 1 can be represented in terms of differential operators:

$$P_m = i\frac{\partial}{\partial X^m}, \qquad Q_\alpha = \frac{\partial}{\partial\theta^\alpha} - i(\gamma^m\theta)_\alpha \frac{\partial}{\partial X^m}. \tag{2}$$

They, in turn, generate a motion in superspace:

$$\delta X^m = e^m - i\bar{\xi}\gamma^m\theta$$
$$\delta\theta^\alpha = \bar{\xi}^\alpha. \tag{3}$$

e^m and ξ^α are infinitesimal parameters of the curve generated. The tangent vectors to this curve can be represented as follows:

$$D_m = i\frac{\partial}{\partial X^m}, \qquad D_\alpha = \frac{\partial}{\partial\theta^\alpha} + i(\gamma^m\theta)_\alpha \frac{\partial}{\partial X^m}, \tag{4}$$

and we compute

$$\{D_\alpha, \bar{D}_\beta\}_+ = -2i\gamma_{\alpha\beta}{}^m D_m. \tag{5}$$

A flat superspace is a space in which the coordinate lines can be chosen to be the curves generated by the supersymmetric transformation—the curves from above. If we want to allow more general spaces we have to generalize the concept. We introduce general coordinate lines with tangent vectors

$$E_A{}^M \frac{\partial}{\partial Z^M}.$$

$E_A{}^M$ is a generalized vierbein. It is a function of the variables X^m and θ^μ. We call such a function a superfield and it should be understood as a power series in θ^μ which ends after the fourth power due to the anticommuting nature of the variables.

The index M of $E_A{}^M$ transforms like a contravariant vector index under general coordinate transformations in superspace. This gives a coordinate independent meaning to the expression $E_A{}^M(\partial/\partial Z^M)$. The index A can transform under a local group which is at our disposal. This is the so-called structure group and we shall choose the Lorentz group. In order to define the concept of covariant differentiations with respect to this structure group we have to introduce the connection $\phi_{MA}{}^B$. Again, the index M transforms under general coordinate transformations covariantly, whereas A and B transform under the structure group inhomogeneously in such a way that

$$E_A{}^M\left\{\frac{\partial}{\partial Z^M} - \phi_{MA}{}^B\right\}V_B = \mathcal{D}_A V_B \tag{6}$$

transforms like a tensor under the structure group if V_B transforms like a vector.

The commutation relation

$$\{\mathcal{D}_A, \mathcal{D}_B\}_\pm = -R_{AB} - T_{AB}{}^C\mathcal{D}_c \tag{7}$$

defines the tensor quantities $R_{AB,C}{}^D$ as the curvature tensor and $T_{AB}{}^C$ as the torsion. The sign of the commutation relation above depends on the spinorial or tensorial nature of the indices A and B.

The curvature tensor and the torsion are covariant tensor quantities. They are functions of the vierbein and the connection and they can be used to impose covariant equations for these quantities. These equations are of two different types. One set of equations reduces the number of independent component fields and does not have as a consequence any differential equations in the four-dimensional x space. These equations are called superspace constraints. The second set of equations leads to field equations for the remaining independent component fields. Curvature tensor and torsion have to satisfy the Bianchi identities; these are the Jacobi identities of the above commutator relation:

$$\sum_{\text{Cyclic } ABC} \{\mathcal{D}_A T_{BC}{}^D + T_{AB}{}^F T_{FC}{}^D - R_{AB,C}{}^D\} = 0$$

These identities allow us to express $R_{AB,C}{}^D$ in terms of the torsion.[9] Therefore, all equations can be formulated in terms of the torsion only. We found that the right constraint equations are[10]

$$T_{\alpha\dot\beta}{}^c = 2i\gamma_{\alpha\dot\beta}{}^c, \qquad T_{\alpha\beta}{}^\gamma = 0, \qquad T_{ab}{}^c = 0, \qquad T_{\alpha b}{}^c = 0. \tag{8}$$

It is now possible to use the constraints together with the Bianchi identities to express the curvature tensor and the torsion in terms of the superfields[11]

$$R, \qquad \mathcal{G}_{\alpha\dot\beta}, \qquad W_{\alpha\beta\gamma}. \tag{9}$$

The field equations can be formulated in terms of these superfields. They are

$$R = 0, \qquad \mathcal{G}_{\alpha\beta} = 0, \qquad \mathcal{D}^\alpha W_{\alpha\beta\gamma} = 0. \tag{10}$$

These equations follow from the superspace Lagrangian[12]

$$L = \int d^4X\, d^4\theta \det E_M{}^A. \tag{11}$$

REFERENCES

1. WESS, J. & B. ZUMINO. 1974. Nucl. Phys. B **70:** 39; 1974. Phys. Lett. B **49:** 52.
   FAYET, P. & S. FERRARA. 1977. Phys. Rep. C **32.**
2. COLEMAN, S. & J. MANDULA. 1967. Phys. Rev. **159:** 251.
3. GRISARU, M. T., P. VAN NIEUWENHUIZEN & J. A. M. VERMASEREN. 1976. Phys. Rev. Lett. **37:** 1662;
   P. VAN NIEUWENHUIZEN & J. A. M. VERMASEREN. 1977. Phys. Lett. B **65:** 263; 1977. Phys. Rev. D **16:** 798;
   S. DESER, J. KAY & K. STELLE. Phys. Rev. Lett. 1977. **38:** 527;
   S. DESER & J. KAY. 1978. Phys. Lett. B **76:** 400;
   J. G. TAYLOR. 1978 Proc. R. Soc. London A **362:** 493.
4. R. HAAG, J. T. ŁOPUSZAŃSKI & M. SOHNIUS. 1975. Nucl. Phys. B **88:** 257.
5. DESER S., & B. ZUMINO. 1976. Phys. Lett. B **62:** 335.
6. FREEDMAN, D. Z., P. VAN NIEUWENHUIZEN & S. FERRARA. 1976. Phys. Rev. D **13:** 3214.
7. ARNOWITT, R., P. NATH & B. ZUMINO. 1975. Phys. Let. B **56:** 81;
   AKULOV, V. P., D. V. VOLKOV & V. A. SORAKA. 1975. JETP Lett. **22:** 396 (English 187);
   WESS, J. & B. ZUMINO. 1977. Phys. Lett. B **66:** 361;
   BRINK, L., M. GELL-MANN, P. RAMONAT & J. H. SCHWARZ. 1978. Phys. Lett. B **74:** 336;
   BREITENLOHNER, P. 1977. Phys. Lett. B *67;*
   1977. Nucl. Phys. B **124:** 500.
   SIEGEL, W. & S. JAMES GATES, JR. 1977. Nucl. Phys. B **147:** 77;
   WESS, J. Supersymmetry–Supergravity. Lectures given at the VIII GIFT Seminar (Salamanca 1977). Springer Tracts.
8. SALAM, A. & J. STRATHDEE. 1974. Nucl. Phys. **76B:** 477.
9. DRAGON, N. 1979. Z. Physik. **C2:** 29.
10. WESS, J. & B. ZUMINO. 1977. Phys. Lett. B **66:** 361.
11. GRIMM, R., J. WESS & B. ZUMINO. 1979. Nucl. Phys. B **152:** 255.
12. WESS J. & B. ZUMINO. 1978. Phys. Lett. **74B:** 51.

SUMMARY OF WORKSHOP ON GAUGE THEORIES OF GRAVITATION

Friedrich W. Hehl

University of Cologne
D-5000 Köln 41
Federal Republic of Germany

This workshop was considered to be a forum for discussions of gauge approaches to gravity: in particular, those discussions connected with the Poincaré, the Lorentz, the translation, and the deSitter group. Supergravity, the gauge theory of the supersymmetry group, was excluded from the workshop since it was covered in a plenary lecture of J. Wess.

There were two half-hour review lectures. The rest were ten-minute communications, each followed by five minutes of discussion. Because of current interest at that time, Y. Ne'eman did not speak on "Poincaré Gauge Theory" as originally planned, but on his new unified gauge model of the Weinberg-Salam theory.

The program was as follows:

Review Lectures

W. Thirring (Vienna): A general introduction on gauge and Yang-Mills theories.

Y. Ne'eman (Tel Aviv & Austin): Irreducible gauge theory of a consolidated Salam-Weinberg model.

Communications

Quantization

E. E. Fairchild (St. Louis): Renormalizable quantum gravity.

Frameworks for Gravitational Gauge Theory

C. L. Tsou (Peking): Some research on the gauge theories of gravitation.

P. von der Heyde (Cologne): Poincaré gauge theory with quadratic Lagrangian.

J. Nitsch (Cologne): Translational gauge theory and its weak field approximation.

M. Camenzind (Zürich): Currents in the SO (3, 1) gauge theory of gravity.

J. G. Taylor (London): On gauge theories on arbitrary non-metric manifolds.

Specific Consequences

M. Soffel (Frankfurt): Vacuum with spin in the ECSK-Theory of gravitation.

Dj. Sijacki (Belgrade): Multilicity-free unitary representations of $\overline{GL}(4, R)$ and $\overline{\text{Diff}}\ R^4$.

R. Güven (Istanbul): Solution for gravity coupled to non-Abelian plane waves.

J. W. Moffat (Toronto): U(3, 1) gauge theory of gravity.

0077–8923/80/0336–0299 $1.75/1 © 1980, NYAS

In my judgement, the state of the art in gravitational gauge theories can be described as follows: the possible limitations of Einstein's theory of gravity become visible and the supposedly more fundamental microphysical gravitational theories of a gauge type, underlying the macrophysical Einstein theory, become more and more established.

Since the lectures and the communications given at the workshop will not be published, I have collected a list of references of related work, enriched by some references to the work of B. N. Frolov (Moscow), J. P. Hsu (North Dartmouth), and G. Sardanashvily (Moscow) who were unable to present their communications in Munich. Finally, three recent review articles on gauge theories of gravity are mentioned that, together with the literature cited in them, should give a fairly complete picture of the investigations undertaken so far.

References

1. Thirring, W. 1978. Gauge theories of gravitation. Acta Physica Austriaca, Suppl. XIX: 439. See also Wallner, R. P. 1979. Notes on recent U_4-Theories of gravitation. GRG Journal.
2. Ne'eman, Y. 1979. Irreducible gauge theory of a consolidated Salam-Weinberg model. Phys. Lett. **81B:** 190.
3. Fairchild, E. E. 1978. Renormalizable, unitary, and possibly finite quantum gravity. Preprint. Washington University. St. Louis, Mo. See also Fairchild, E. E. 1977. Yang-Mills formulation of gravitational dynamics. Phys. Rev. **D16:** 2438.
4. Zou Zhenlong (C. L. Tsou) *et al.* 1979. Some researches on gauge theories of gravitation. Scientia Sinica **22:** 628. Gou Hanying (H. Y. Kuo) 1979. Einstein principle, gauge theory of gravitation and quantum geometrodynamics. Preprint. Academia Sinica. Peking.
5. von der Heyde, P. 1976. The field equations of the Poincaré gauge theory of gravitation. Phys. Lett. **58A:** 141.
   von der Heyde, P. 1976. Is gravitation mediated by the torsion of spacetime? Z. Naturf. **31a:** 1725.
6. Nitsch, J. & F. W. Hehl. 1979. Translational gauge theory of gravity: post-Newtonian approximation and spin precession. Preprint. University of Cologne.
7. Camenzind, M. 1978. Weak and strong sources of gravity: an SO (1, 3)-gauge theory of gravity. Phys. Rev. **D18:** 1068; and Camenzind, M. 1978. Interactions, currents and gravity. Preprint. University of Hamburg.
8. Taylor, J. G. 1978. Towards a geometrization of nature. Phys. Rev. **D18:** 3544.
9. Soffel, M., B. Müller & W. Greiner. 1979. Vacuum with spin in the ECSK-Theory of gravitation. Phys. Lett. **70A:** 167.
10. Ne'eman, Y. & Dj. Sijacki. 1979. Unified affine gauge theory of gravity and strong interactions with finite and infinite GL (4, R) spinor fields. Ann. Phys. (N.Y.)
11. Güven, R. 1979. Solution for gravity coupled to nonAbelian plane wave. Phys. Rev. **D19:** 471.
12. Moffat, J. W. 1979. New theory of gravitation. Phys. Rev. **D19:** 3554.
13. Frolov, B. N. 1978. Tetrad palatini formalism and quadratic Lagrangians in the gravitational field theory. Acta Phys. Polonica **B9:** 823.
14. Hsu, J. P. 1979. Implications of general covariance and maximum four-dimensional Yang-Mills gauge symmetry. Phys. Rev. Lett. **42:** 934.
15. Sardanashvily, G. 1978. Izv. VUZ USSR (Physics) **1:** 10; **7:** 137.
16. Ne'eman, Y. 1979. Gravity, groups and gauges; Trautman, A. 1979. Fibre bundles, gauge fields, and gravitation; Hehl, F. W., J. Nitsch & P. von der Heyde 1979. Gravitation and poincaré gauge field theory with quadratic Langrangian. *In:* Einstein Commemorative Volume. Alan Held *et al.*, Eds. Plenum Press, New York. To be published.

SUMMARY OF WORKSHOPS ON QUANTUM GRAVITY

C. J. Isham

Blackett Laboratory
Imperial College
South Kensington
London SW7, England

Two quantum gravity workshops were held at the symposium and lasted for a total of seven hours. The subjects covered reflected to a fair extent the topics of dominant interest in recent years. The main areas of discussion were quantum field theory in a curved space-time and supergravity plus three talks on more general approaches to quantum gravity. Since most of the contributions will have already appeared in print by the time this article is published, I will be very brief in my summary of their contents and leave interested readers to contact the individual authors for further details.

The study of quantum field theory in a curved space-time is important as a precursor to quantum gravity proper and also because of its potential applications in "big bang" cosmology. There have been (and still are) numerous arguments concerning the correct mathematical structure and physical interpretation that should be adopted. Typically the problem revolves around what is meant by a "vacuum state" $|0>$ and a "particle" in such a quantum field theory. This is often linked with the construction of a regularized and/or renormalized vacuum expectation value $<0|\hat{T}_{\mu\nu}|0>$ of the quantized energy-momentum tensor.

Bernard Kay (University of Berne and Imperial College) discussed the vacuum self-energy of a quantized field propagating on a locally flat space-time M (such as a cylinder). He showed that this "Casimir effect" could be obtained by constructing a vacuum state s for M as a state on the normal Minkowski space field theory with Fock vacuum ω. The energy-momentum tensor is normal ordered with respect to ω so that $\omega(:T_{00}:) = 0$ and then the "Casimir" energy of the cylinder is defined as $s(:T_{00}:)$. This approach emphasizes that there is basically a renormalization being performed (unlike the pure regularization philosophy of ζ function methods) and that it is unnecessary to make any explicit reference to zero-point energies. The technique is especially appropriate to interacting field theories where some form of renormalization is essential anyway.

The problem of defining vacuum states was discussed by Helmut Rumpf (Vienna) who presented the latest developments in the covariant in/out formalism that he has been developing for some time. The basic idea is to define a natural generalization of the Minkowski-space Feynman function as the resolvent of the self-adjoint differential operator obtained from the linear field equations. This works even for *non*asymptotically flat space-times or, more generally, non-cutoff-external potentials. It is a global construction and gives a different result from the more local Hademard–de Witt expression. Rather surprisingly (in a de Sitter space example) even the local singularity structure of the two-point functions are different! Rumpf's propagator predicts particle production in this particular case and he has also applied it to expanding isotropic cosmologies leading to infinite production by the initial singularity. He feels

0077–8923/80/0336–0301 $1.75/1 © 1980, NYAS

that the implicit definition of an "in" vacuum provided by his method is the correct one to use in back-reaction calculations of the $G_{\mu\nu} = <0|\hat{T}_{\mu\nu}|0>$ type.

This back-reaction problem has been investigated by Jim Hartle (University of Santa Barbara) who believes that its correct inclusion will remove the infinite nature of the number of particles produced in a big bang cosmology as well as leading to isotropization of an initially anisotropic metric. His technique is to evaluate the net production by computing the imaginary part of the effective action up to the one-loop level. He demonstrated his method for conformally coupled scalar particles with a metric of the form $ds^2 = a^2(\eta)\,[-d\eta^2 + (e^{2\beta})_{ij}dx^i dx^j]$ where the anisotropy parameter β_{ij} is a traceless 3×3 matrix. β is regarded as a small perturbation about the conformally flat metric $ds^2 = a^2(\eta)\,[-d\eta^2 + i dx^i dx^i]$ enabling an unambiguous "bare" vacuum to be defined in terms of conventional Minkowski-space quantum field theory. The resulting effective action Γ is a function of a and β and, in this simple model, the back-reaction effects are incorporated by minimizing Γ with respect to these parameters. This results in a more isotropic metric and only finite particle production.

The cosmological theme was continued by Jürgen Audretsch (University of Konstanz) who explored the analogy between quantum field theory in a curved space-time and both pair production by time-dependent electric fields and barrier reflection in one-dimensional Schrödinger wave mechanics. He exploited this analogy to carry across exact results from the latter to the curved space-time case. He neglected the back-reaction effects and defined the particle states by various adaptations of the WKB methods employed for example, by Parker and Fulling. These different possibilities were compared and contrasted and he illustrated the application of his techniques to a number of specific cosmological metrics.

The problem of defining a vacuum state in the presence of an initial singularity has also been studied by Doug Eardley (Yale University) in a rather novel example. He considered the special case of a marginally naked singularity which forms inside and just on, but not outside, the absolute event horizen associated with the collapse of a cloud of matter. The initial and final quantum states can be defined with respect to the event-horizon structure in analogy with the method pioneered by Stephen Hawking in his famous black-hole quantum-radiation calculations. Eardley and his collaborator Williams employ a Vaidya (spherically symmetric imploding null fluid) space-time and arrive at the rather startling conclusion that the outgoing "thermal" radiation contains components of both positive and negative temperature. The exact origin of the negative temperature is not yet clear but the potential implications in, for example, the removal of the singularity by a back-reaction effect are quite considerable.

One interesting recent development has been the study of self-interacting quantum fields in a curved space-time. Peter Hajicek (University of Berne) has developed a formalism proposed originally by Schwinger for computing directly expectation values of the form $<0, \text{in}\,|\,T\phi(x_1)\cdots\phi(x_n)\,|\,0, \text{in}>$ rather than the more usual $<0, \text{out}\,|\,T\phi(x_1)\cdots\phi(x_n)\,|\,0, \text{in}>$. These expressions differ if there is particle production (as frequently occurs in a curved background space-time) and the former makes sense in space-times in which there is only one asymptotic temporal region. This involves the use of two external sources and leads to a modified form of perturbation theory.

The effects of self-interactions in cosmological particle creation was discussed by

Larry Ford (King's College, London). In a $\lambda\phi^4$ interaction, Feynman graphs occur with the form

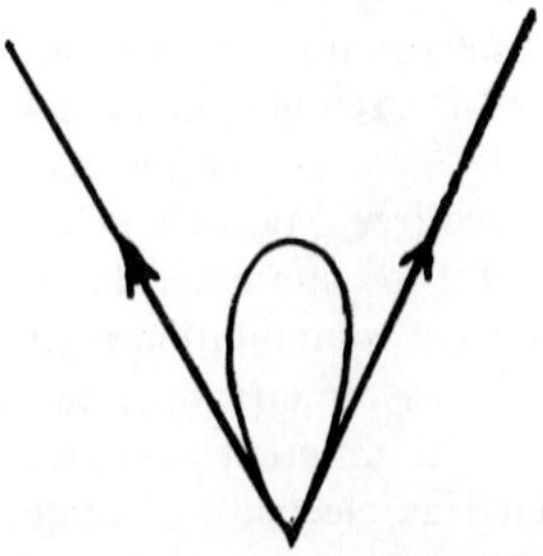

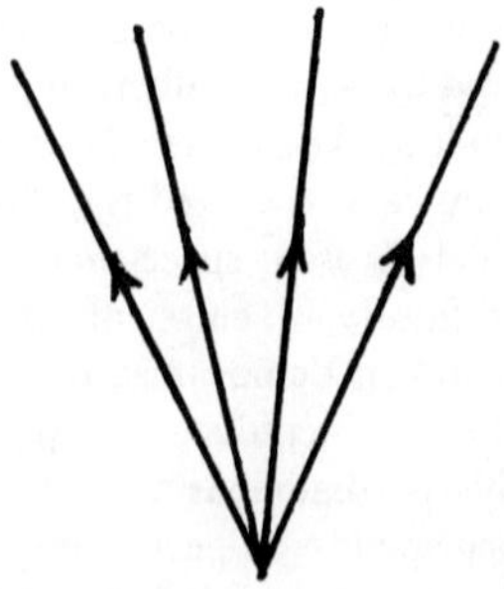

The average number of particles produced is of order λ and a number of other interesting results are obtained. These include the "Casimir" energy on a cylinder $\mathcal{S}^1 \times \mathcal{R}^3$ and self-mass generation via the Feynman graph

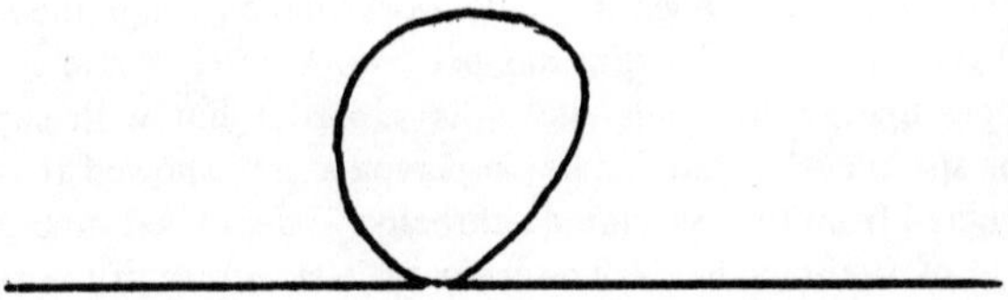

A rather startling result is the apparent violation of electromagnetic gauge invariance when quantum electrodynamics is studied on this model $\mathcal{S}^1 \times \mathcal{R}^3$ space. This manifests itself in the nonvanishing of $q^\mu \pi_{\mu\nu}(q)$ where $\pi_{\mu\nu}(q)$ is the vacuum polarization tensor. It is not absolutely clear at the moment whether this "anomaly" really does signal a breakdown of gauge invariance or whether it is to be associated more with the definition of particle states and the LSZ formalism on this topologically nontrivial space-time. Evidently this is a phenomenon which deserves further investigation.

One difficulty which invariably arises in any use of an interacting field theory is the development of algorithms for a perturbation theory. In the case of Minkowski space this is resolved by the use of momentum-space propagators and the Feynmann–Dyson expansion. Calculations performed directly in configuration space are rather difficult and therefore Nick Birrell's (King's College, London) description of his development, with John G. Taylor, of curved space-time momentum-space techniques, was of considerable interest. These methods enable all quantities of interest (in a topologically trivial space-time) to be calculated in terms of T matrices satisfying Lippmann–Schwinger-type equations. Techniques for solving these equations can be borrowed from low-energy nuclear physics and have already been successfully applied to compute particle production in a cosmological background.

Another area which has received considerable study in recent years is the role played by the topology of space-time in quantum gravity. A lot of work in this direction has been done by Hawking's group, none of whom could unfortunately attend the symposium due to a collective shortage of travel funds, and a small group at Imperial College (regretfully I felt that, in my capacity as chairman, it would be inappropriate to invite myself to speak as this would lead to the type of situation much loved by W. G. Gilbert and A. Sullivan). However, this sad situation was remedied by a talk by Abhay Ashtekar (University of Clermont-Ferrand) on his work with A. Sen. They have analyzed the Hilbert-space quantization of a Maxwell field in a Schwarzschlid–Kruskal space-time M. The nonvanishing of $H^2(M; \mathcal{R})$ leads to a two parameter family of classical, everywhere regular, static source-free Maxwell fields and Ashtekar demonstrated the existence of an associated two parameter family of unitarily inequivalent representations of the usual canonical commutation relations. This is analogous to the situation arising in spontaneous symmetry breaking and corresponds to performing quantum fluctations around the different backgrounds. The parameters labelling the family can be interpreted as electric and magnetic charge with naturally occurring superselection rules.

One of the major developments in quantum gravity has been the invention of supergravity theories. Attention at the moment is primarily focussed on the development of superspace formalisms for both the usual and extended theories and on finding a theory that is both renormalizable and free of ghosts. The resolution of the latter problem is also crucial in the successful implementation of the $R + R^2$ type of gravity theories that were much discussed in Hehl's workshop on gauge theories.

John G. Taylor (King's College) has been working for some time on superspace formulations of supergravity. His theory is superflat but with supertorsion in the fermion sector specified by that for flat superspace. He showed that the Lagrangian det V_M^A constructed from the associated achtbein V_M^A described supergravity but with a nonminimal set of auxiliary fields. The invariance structure of the tangent space and world space indices are respectively a seven-parameter Lie subgroup of $GL(4, \mathcal{R})$ and the general supercoordinate transformation group. He discussed the use of both the Wess–Zumino gauge and a superfield generalization of the harmonic gauge and displayed the associated ghost propagators and vertices. A complete superfield perturbation theory can be obtained in his theory but a powercounting argument suggests a lack of renormalizability. The talk concluded however with the statement and proof of a theorem that the $N = 2$ extended supergravity theory is renormalizable.

John Moffet (University of Toronto) has also been studying superspace methods. He has formulated a theory based on a non-Riemannian supergeometry. His action principle leads to equations of motion in a curved superspace that contains a global supersymmetric space as a special solution. This space again has zero curvature but constant torsion and exists as a continuous limit of the curved non-Riemannian superspace. The local gauge group is a complex extension of $Sp(4)$ which, in the Bose sector, contains $U(3, 1)$ as a subgroup. Moffet had already previously investigated this Bose sector theory and shown that although it is not equivalent to conventional general relativity, it is nevertheless consistent with all known experimental facts. A preliminary investigation has shown the supergravity version to be free of ultraviolet divergences but, regretfully, it contains the almost inevitable ghost problems. Work on possible ghost-killing mechanisms is currently being actively pursued. If successful it would lead to a truly finite and unified theory of all fundamental interactions.

The final speaker on supergravity was Kelly Stelle (Imperial College) who described his work with Peter West on the construction of supergravity models in which new terms can be added that are individually invariant and do not require changes in the supersymmetric transformations of the gravitational multiplet. He has developed techniques which generalize to local supersymmetry the Wess–Zumino scalar and vector multiplets and combination rules of rigid supersymmetry. A significant step was the discovery of differential operators that supercovariantly relate the multiplets. This has been applied to the discussion of gauge invariance for the Maxwell multiplet in locally supersymmetric models and has been connected to the superconformal transformations in the locally supersymmetric version of the Fayet–Iliopoulos term.

There is, throughout the world, significant activity in many other aspects of quantum gravity than those discussed above. Indeed the first contribution at the first workshop was from Peter Bergmann. He discussed the canonical quantization of gravity and emphasized the role played by the various gauge groups present in the theory. Of particular importance are the diffeomorphism group and a new group associated with the Cauchy data on a spacelike hypersurface. There is a subtle interplay between these gauge groups, the constraint equations of canonical general relativity and the associated operator-ordering problem. This whole subject is of considerable interest not only to people working in canonical quantum gravity but also to those involved in supergravity. The Lie algebra of local supersymmetry typically has field dependent "structure constants" and only closes after using the equations of motion (unless auxiliary fields are added). This situation is closely related to the ideas about which Bergmann spoke and it is no coincidence that many workers in supergravity are currently deeply interested in the canonical structure of their theories.

Most people involved in quantum research would probably admit to a belief that something pretty dramatic happens at the Planck length $(G\hbar/c^3)^{1/2} \simeq 10^{-33}$ cm. Indeed this lies behind much of the topological work referred to earlier and it has always been anticipated that twistor theory, when completely developed, would incorporate some such phenomenon. Roger Penrose (Mathematical Institute, University of Oxford) reviewed some of the recent developments in this area with special reference to the problem of space-time singularities and time asymmetry. In particular he restated his belief that the CPT theorem may break down in these extreme conditions—a possibility which arises naturally in the twistor formalism. Particle physicists have tended to be unwilling to see one of their most cherished principles thus cast aside. However, recent developments in instanton theory make it seem more feasible. In the Yang Mills case it is quite easy to get CP violation and nonconservation of fermion number and in the gravitational case the use of asymptotic locally Euclidean solutions leads to a violation of a number of discrete space-time symmetries.

The final talk at the workshop was by Herbert Jehle (George Washington University) who expounded his idea that the equations of motion for a system of gravitationally interacting particles can be usefully formulated in terms of a Schrödinger-type equation in which the $\hbar^3$ characterizing the quantum uncertainty principle is replaced by a macroscopic phase-space volume characterizing the coarse grainedness of the phase-space mass distribution. Applications of these ideas to the solar system lead to remarkable agreement with experimental data concerning the orbital motion of the planets.

Papers Presented at the Workshops on Quantum Gravity

Ashtekar, A.: On the role of space-time topology in quantum phenomena: Superselection of charge and emergence of nontrivial vacua.
Audretsch, J.: Cosmological particle creation as above-barrier reflection.
Bergmann, P. G.: Canonical quantization of gravity.
Birrell, N. D.: Momentum space calculations in curved space-time quantum field theory.
Eardley, D. M.: Particle creation by some naked singularities.
Ford, L.: Interacting quantum fields in non-Minkowskian space-times.
Hajicek, P.: Generating functional for expectation values.
Hartle, J.: Pair creation and the effective action.
Jehle, H.: New approaches to statistical dynamics of gravitational systems.
Kay, B.: The Casimir effect without magic.
Moffet, J. W.: Supergravity theory in superspace.
Penrose, R.: Twistor theory.
Rumpf, H.: Distinguished quantum states in generic external gravitational fields.
Stelle, K.: Supergravity in the component formalism.
Taylor, J. G.: Quantized superfield supergravity.

THE CONFRONTATION BETWEEN GENERAL RELATIVITY AND EXPERIMENT*

Clifford M. Will†

Institute of Theoretical Physics
Department of Physics
Stanford University
Stanford, California 94305

INTRODUCTION

The principal cornerstone of all scientific theory is observation and experiment. One of the strengths of this Texas symposium is that it brings together astrophysical theorists and observers and forces them to confront theories with observations (and vice versa). However, the symposium title also contains the word "relativistic," so at this point it is appropriate to focus on the confrontation between relativity—specifically general relativity—and experiment. This is especially timely in view of the near-coincident occurrence of two momentous events in this field: the beginning of the year of the centenary of Einstein's birth, and the announcement of the probable first discovery of the existence of gravitational radiation.

The confrontation between gravitation theory and experiment can be discussed on three levels. At a very fundamental level, a variety of experiments test the foundations of the gravitational interaction. These experiments verify, for example, that classical gravity is a geometric, curved-space phenomenon, but do not test general relativity itself. The first tests of general relativity occur at the "post-Newtonian" level, where experiments in the solar system play an important role. Finally, new arenas for testing general relativity are becoming crucial, in particular, the binary pulsar, where recent measurements by J. H. Taylor and his colleagues may provide make-or-break tests of gravitational theory.

TESTS OF THE FOUNDATIONS OF GRAVITATION THEORY

There is a class of experiments that probe the nature of gravity at a very fundamental level, such as the Eötvös experiment, the gravitational redshift experiment, and the Hughes–Drever experiment. These experiments test what has come to be called the "Einstein Equivalence Principle" (EEP), which states that "test" bodies follow geodesics of a space-time metric, and that in local freely falling frames of that metric, the nongravitational laws of physics take on universal forms, independent of the location and velocity of the frame. If EEP is valid, it is then possible to show that the correct classical theory of gravity must be a "metric" theory, one in which all the physical effects of gravitation are produced by space-time geometry.

A simple way to discuss and classify experiments that test EEP has arisen from the

*Supported in part by the National Science Foundation [PHY 76–21454] and by the National Aeronautics and Space Administration [NSG 7204 S3].

†Alfred P. Sloan Foundation Research Fellow.

0077–8923/80/0336–0307 $1.75/1 © 1980, NYAS

work of Dicke, Nordtvedt, and Haugan.[1] One separates EEP into three subprinciples: the "Weak Equivalence Principle" (WEP), which states that "test" bodies fall with the same acceleration; the "absence of preferred-location effects" in local experiments involving nongravitational forces; and the "absence of preferred-frame effects" in such experiments. Although these subprinciples may involve very different kinds of experiments, theoretically they are intimately connected, in the sense that if one of them is violated, then one of the others must be violated. This bears out a 1960s conjecture by Schiff that the validity of WEP alone is sufficient for the validity of EEP.

FIGURE 1 summarizes these principles and their experimental tests.[1,2] Any of the nongravitational interactions, strong (S), electromagnetic (E), or weak (W), could violate EEP by means of a "nonmetric" coupling to gravitation of strength $\alpha_A g_A$ (A = S, E, or W), where g_A is the nominal coupling constant of the interaction. Experiments that check EEP thereby set upper limits on possible anomalous values of the α's, shown in FIGURE 1. These include the Eötvös experiment (Moscow version 1971, equality of acceleration for Al and Pt to parts in 10^{12}), the gravitational redshift experiment (Vessot–Levine hydrogen-maser rocket experiment 1976, accuracy one part in 10^4), and the Hughes–Drever experiment (isotropy of interial mass, verified to one part in 10^{23}). Another possible preferred-location effect that would violate EEP is a cosmological time dependence of the nongravitational constants. Major improve-

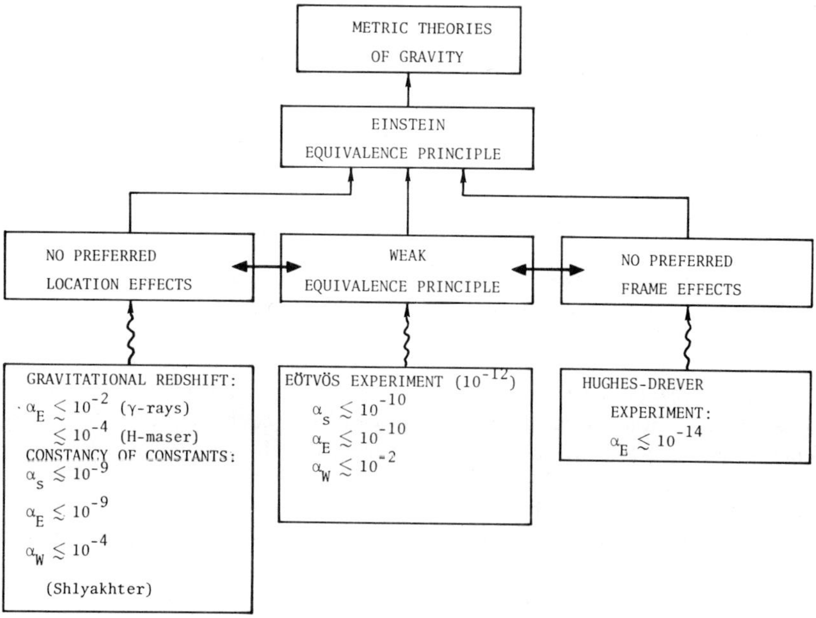

FIGURE 1. The Einstein equivalence principle. (For theoretical discussion, see reference 1; for summary of the experiments, see reference 2.) The horizontal arrows represent the intimate connection between the three subprinciples discussed by Dicke, Nordtvedt, and Haugan.

TABLE 1

THE PPN PARAMETERS AND THEIR SIGNIFICANCE

PPN Parameter	Significance	Value in General Relativity
γ	How much spatial curvature does mass produce?	1
β	How "nonlinear" is gravity?	1
ξ	Are there gravitational preferred-location effects?	0
α_1	Are there gravitational preferred-frame	0
α_2	effects?	0

ments in the upper limits on such variations have been made by Shlyakhter,[3] using analyses of fission yields from the Oklo Natural Reactor, a natural, sustained fission reactor believed to have occurred in West Africa around 2×10^9 years ago. The results shown in FIGURE 1, represent upper limits on the amount of variation of the constants over one Hubble time (2×10^{10} yr).

SOLAR-SYSTEM TESTS OF POST-NEWTONIAN GRAVITY

The Parametrized Post-Newtonian Formalism

The experimental evidence shown in FIGURE 1 that supports EEP gives confidence that the correct theory of gravity must be a metric theory. In the weak-field, slow-motion, or "post-Newtonian" limit appropriate to the study of experiments in the solar system, most metric theories of gravity can be analysed in terms of a "theory of theories of gravity" that classifies them in terms of a set of arbitrary dimensionless parameters whose values vary from theory to theory. This theory of theories is known as the Parametrized Post-Newtonian (PPN) formalism.[4] One version of the PPN formalism restricts attention to theories of gravity that possess momentum and energy conservation laws, called "semiconservative" theories. In this version there are five PPN parameters, γ, β, ξ, α_1 and α_2, whose significance and values in general relativity are shown in TABLE 1. TABLE 2 lists a number of recent semiconservative metric theories of gravitation and shows their PPN parameter values.

One can now regard solar-system tests of post-Newtonian effects as measurements of the "correct" values of these parameters. It is convenient to separate solar-system experiments into two classes: the "classical tests," and tests of the "strong equivalence principle".

The Classical Tests

Three solar-system experiments, light deflection, time delay, and perihelion shift, can be called the three "classical tests." This terminology differs from popular usage in which the term "classical tests" refers to gravitational redshift, light deflection, and perihelion shift. But the gravitational redshift is a test of EEP, not of general relativity

TABLE 2

METRIC THEORIES OF GRAVITY*

Theory (Gravitational Fields Present)	PPN Parameters γ	β	ξ	α_1	α_2	Obeys SEP?	Can Be Made Identical to General Relativity? Post-Newtonian	Strong Field, Grav. Waves
(a) Purely Dynamical Theories								
(i) General relativity (g)	1	1	0	0	0	Yes	Yes	Yes
(ii) Scalar-tensor (g, ϕ)								
BWN-Bekenstein	$\frac{1+\omega}{2+\omega}$	$1+\Lambda$	0	0	0	No	Yes ($\omega \to \infty$)	Yes ($\omega \to \infty$)
Brans–Dicke	$\frac{1+\omega}{2+\omega}$	1	0	0	0	No	Yes ($\omega \to \infty$)	Yes ($\omega \to \infty$)
(iii) Vector-tensor (g, K)	1	1	0	0	$K_0^2/(1+\frac{1}{2}K_0^2)$	No	Yes ($K_0 \to 0$)	Yes ($K_0 \to 0$)
(b) Theories with prior geometry								
(iv) Bimetric theories								
Rosen (g, η)	1	1	0	0	$(C_0/C_1) - 1$	No	Yes	No
Lightman–Lee (g, η, h)	γ'	β'	0	α_1'	α_2'	No	Yes	No
Rastall (g, η, K)	1	1	0	0	α_2'	No	Yes	No
(v) Stratified theories								
Ni (g, η, t, ϕ, K)	a	b	0	$-4\,(a+1+\frac{1}{2}e) - (d+1)$		No	Yes	No

*For detailed discussion and references, see reference 2.

itself, and so should not be included in this class. Furthermore, the "time-delay" test, discovered by Shapiro in 1964, is, in its theoretical interpretation, on an equal footing with the light-deflection test. Actually, this latter test has yielded the most precise results of the three.

The light-deflection and time-delay tests are related in the sense that they measure effects on the propagation of photons in the curved space-time around the Sun; in fact, they depend on the PPN parameter γ in the same way. A ray of light that passes the Sun at a distance d (in units of solar radii) is deflected by an angle

$$\delta\theta = \tfrac{1}{2}(1 + \gamma)\ 1''.75/d, \tag{1}$$

and a ray of light that similarly passes the Sun on a round trip, say from the Earth to Mars, suffers a delay that leads to an excess round trip travel time

$$\delta t = \tfrac{1}{2}(1 + \gamma)\ 250[1 - 0.16 \ln d]\ \mu s. \tag{2}$$

The most precise light-deflection experiments are those that employ radio interferometry to measure the deflection of radio waves from quasars; since 1969 a series of such measurements has yielded values of the coefficient $\tfrac{1}{2}(1 + \gamma)$ in agreement with general relativity at levels of accuracy approaching 1.5 per cent (see FIGURE 2).[5]

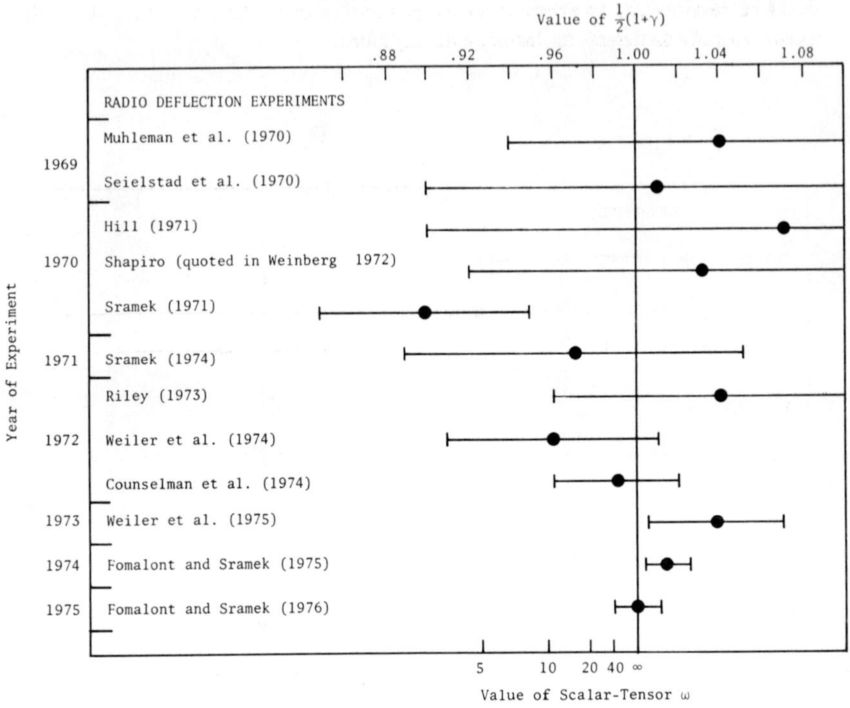

FIGURE 2. Light-deflection measurements using radio interferometry.[5] Bottom scale shows the value of the coupling constant ω of scalar-tensor theories that would give the corresponding value of $\tfrac{1}{2}(1+\gamma)$. (After reference 2.)

Measurements of the time delay using radar ranging to planets and spacecraft have been carried out since 1967; recent data from ranging to the Viking orbiters and landers on Mars have yielded results in agreement with general relativity with published errors between 0.5 and 0.2 percent (FIGURE 3).[6,7]

The third of the classical tests is the perihelion shift, but here the situation is more complicated. In terms of the PPN formalism and Newtonian gravitation, the predicted advance of Mercury's perihelion due to the Sun is, in arcseconds per century,

$$\omega = 42.98\,[\tfrac{1}{3}\,(2 + 2\gamma - \beta)] + 0.013\,(J_2/1 \times 10^{-7}), \qquad (3)$$

where J_2 is the quadrupole moment of the Sun. From radar ranging to the inner planets, the measured perihelion shift agrees with 42″.98 per century within about half of one percent. However, measurements of the visual solar oblateness by Dicke and Goldenberg in 1967 were interpreted as corresponding to a value $J_2 = (2.5 \pm 0.2) \times 10^{-5}$, which would contribute an anomalous 3″ per century to the shift. Later measurements by Hill and colleagues yielded an upper limit $J_2 < 0.5 \times 10^{-5}$. Thus there remains some uncertainty in the interpretation of perihelion shift measurements as tests of relativistic gravity.[2] A direct, unambiguous measurement of J_2 would be provided by a mission under study by NASA, known as the Solar Probe, a spacecraft that would approach the Sun to within four solar radii. Feasibility studies indicate that J_2 could be measured to an accuracy of ten per cent of its nominal value 1×10^{-7}, that of a centrifugally flattened, uniformly rotating Sun.

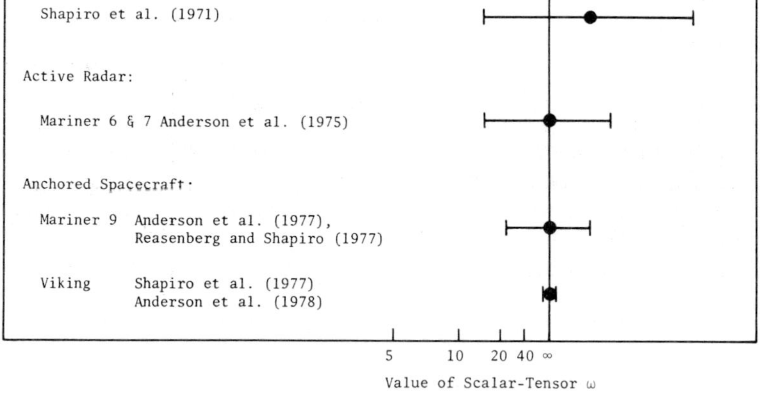

FIGURE 3. Radar time-delay measurements. (After reference 2.)

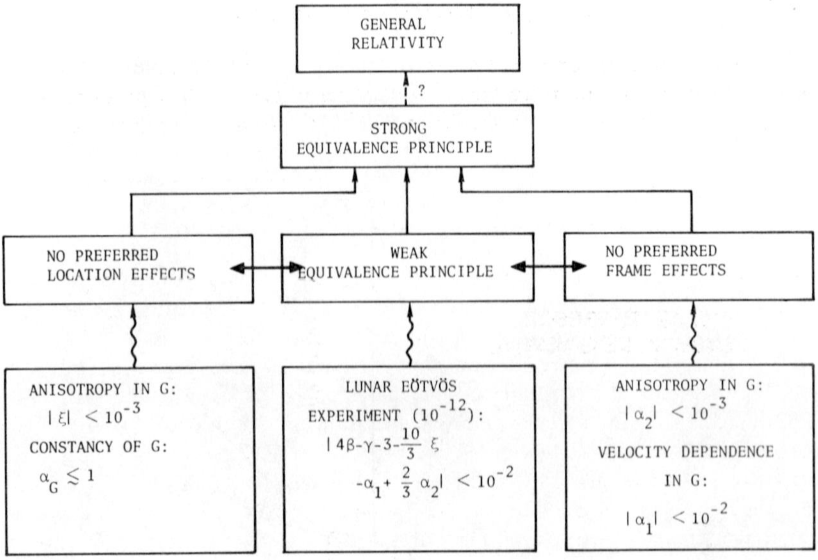

FIGURE 4. The strong equivalence principle (as in FIGURE 1.)

Tests of the Strong Equivalence Principle

A number of tests of post-Newtonian gravity can be viewed as tests of the "Strong Equivalence Principle" (SEP). This principle has many of the essential features of EEP, except that it also incorporates the effects of local gravitational interactions. For example, for SEP to be valid, "test" bodies, bodies that are small compared to inhomogeneities in external gravitational fields, yet that themselves contain significant self-gravitational binding energy, must fall with the same acceleration (WEP). In addition, local gravitational experiments (such as Cavendish experiments), should show no preferred-location or preferred-frame effects[1] (see FIGURE 4 and compare FIGURE 1). Although it is impossible to go beyond this to formulate a precise statement of SEP in parallel with that of EEP, it can be argued, at least heuristically, that SEP implies the presence of *only one* gravitational field, namely the physical metric. Some authors have gone further to argue, using primarily the techniques of field theory, that this implies general relativity uniquely.[8]

The "gravitational WEP" has been verified to one part in 10^{12} by the lunar laser ranging experiment (the possibility of a violation of WEP here is called the "Nordtvedt effect"), setting a limit on a combination of PPN parameters shown in FIGURE 4. Geophysical measurements have set limits on preferred-frame and preferred-location effects in the local gravitational constant, and a variety of measurements have limited cosmic variations in the gravitational constant to a factor around unity in one Hubble time.

Putting the Squeeze on Theory Space

These experiments serve to limit the "theory space" available to alternative theories of gravity. To illustrate this, we take three of the five PPN parameters, γ-1, β-1, and α_2, and define a three-dimensional part of "theory space"; FIGURE 5 then

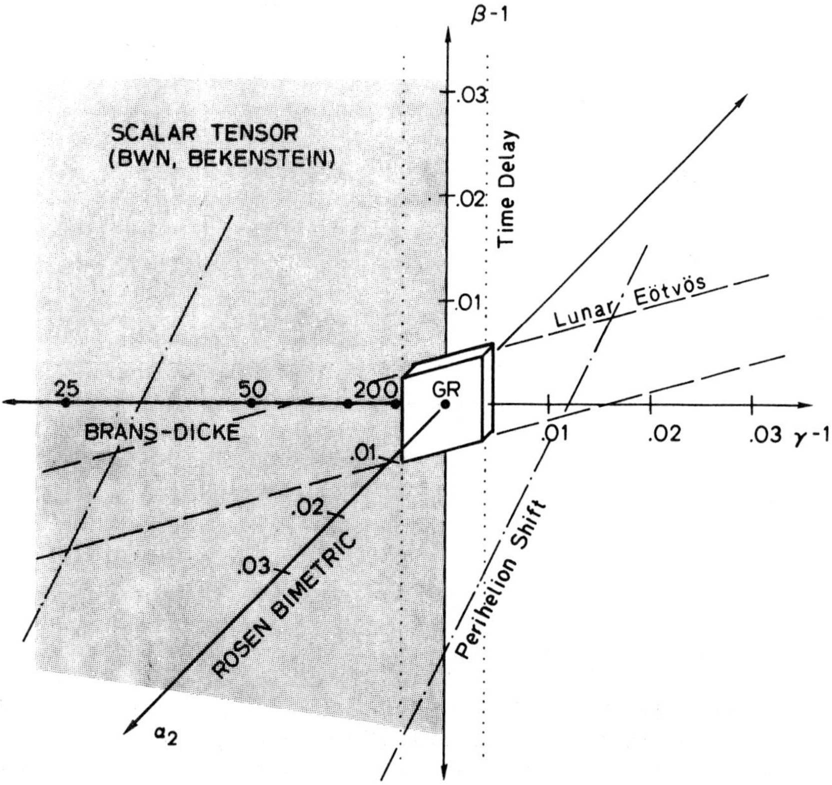

FIGURE 5. The (β-1)–(γ-1)–α_2 space. The Brans–Dicke theory occupies the negative (γ-1) axis (β=1), while the generalized scalar-tensor theories of Bergmann, Wagoner, Nordtvedt, and Bekenstein occupy the half-plane (γ–1) $<$ 0. The numbers on the negative (γ-1) axis are the corresponding values of ω. The Rosen bimetric theory occupies the negative α_2 axis, and general relativity resides at the origin. Shown are limits on the PPN parameters placed by the Viking time-delay (dotted lines), lunar laser-ranging (dashed lines) and perihelion-shift (dot-dashed lines) measurements. The remaining available PPN "theory space" is the box shown, of thickness 2×10^{-3} in the α_2 direction.

demonstrates the available region in this space as limited by the results of time-delay, perihelion-shift, laser-ranging, and geophysical experiments. One conclusion that is apparent from FIGURE 5 is that the Brans–Dicke theory must be constrained to have its coupling constant $\omega \gtrsim 200$. In the limit $\omega \rightarrow \infty$, this theory becomes indistinguishable from general relativity in all its predictions. On the other hand, the Rosen bimetric theory can have $\alpha_2 \sim 0$, thereby satisfying the experimental constraints, yet remain a completely distinct theory from general relativity (as we shall see below). Many other theories have this property (see TABLE 2).

Because of this, one must look for new arenas for testing relativistic gravity, arenas that may involve astrophysical systems far outside the solar system. At this point, it becomes useful to adopt a slightly different attitude toward the significance of observational data from such systems. In the solar system, the physics that underlies most experiments is extremely well understood, so the observational data can be used to perform "clean" tests of gravitational effects, whose results can be viewed as squeezing available theory space, as in FIGURE 5. However, when complex astrophysical systems such as the binary pulsar are used as gravitational testing grounds, one can no longer be so certain about the underlying physics. In such cases, a better approach to the confrontation between theory and observation may be to assume that an individual theory is correct, then use the observations to make statements about the possible compatible physics underlying the system. The viability of a gravitation theory would be called into question if the resulting "available physics space" were squeezed into untenable, unreasonable, or *ad hoc* positions. Such a method will be most powerful for theories, primarily those with "prior geometry" (see TABLE 2), that make qualitatively different predictions in such systems. We shall see this procedure in operation when general relativity and the bimetric theory are confronted with the new data from the binary pulsar.

New Arenas for Testing General Relativity

Gravitational Radiation

One new arena for testing relativistic gravity that involves gravitational effects in sources outside the solar system is gravitational radiation.[9] Once gravitational waves (GW) have been detected and studied by laboratory GW telescopes, one can imagine a number of possible tests of gravitation theory. General relativity predicts that an arbitrary GW possesses only two states of polarization, while essentially all other metric theories predict *more* than two (up to six). Measurements of the number of polarization states in observed radiation could test some theories. A further test could be provided by comparisons of the arrival times of GW and electromagnetic pulses from an event such as a supernova. The two waves propagate with the same speed, according to general relativity, and so should arrive simultaneously. In many other theories, the speeds are different. Another aspect of gravitational radiation, the multipolarity of the waves and the amount of radiation damping, will be discussed when we deal with the binary pulsar.

Cosmology

Cosmological observations may provide tests of some gravitation theories, but because of uncertainties in our quantitative knowledge of the detailed physics of the early universe and its subsequent evolution, we must adopt the philosophy of "squeezing physics space" in order to make progress. For instance, some cosmological models in the Rosen bimetric theory predict a low-density "bounce" of the universe in the past, and so must invent new mechanisms to explain the cosmic helium abundance and the existence and isotropy of the 3-K microwave background radiation.[10]

The Binary Pulsar

The most exciting new extra-solar-system gravitational testing ground is the binary pulsar. Discovered in the summer of 1974 by Hulse and Taylor,[11] it is a pulsar of nominal pulse period 59 ms in a close binary system with an as yet unseen companion. From detailed analyses of the arrival times of pulses (see the paper by Taylor, Fowler, and McCulloch in this volume), extremely accurate orbital parameters have been obtained (see TABLE 3). Because the orbit is so close ($\sim 1\ R_\odot$) and because there is no evidence of an eclipse of the pulsar signal or of mass transfer from the companion, it is generally believed that the companion is compact: a white dwarf, a neutron star, or a black hole. Thus the orbital motion is thought to be relatively "clean." Furthermore, the data acquisition is also "clean" in the sense that the observers can keep track of the pulsar phase with an accuracy of 50 μs, despite gaps of up to six months between observing sessions. The pulsar has shown no evidence of "glitches" in its pulse period.

Three factors make this system a promising testing ground for relativity: the relatively large size of relativistic effects ($v_{\rm orbit}/c \sim 10^{-3}$); the short orbital period (8 hours), allowing secular effects to build up rapidly; and the cleanliness of the system (see reference 2 for review and references). However, there are still some uncertainties in the underlying physics, primarily involving the nature of the companion, that force us to adopt the philosophy of "squeezing physics space." We shall assume in turn that particular gravitation theories are correct, then shall ask whether the available physical configurations consistent with the data are physically reasonable or natural, or whether *ad hoc* assumptions or models must be made. In this sense, the relativistic theory of gravity is being used *as a tool for measuring astrophysical parameters,* a new role for general relativity (or its competitors).

There are four basic pieces of data on the binary pulsar that we shall use for this purpose: the mass function f_1, the periastron shift $\dot{\omega}$, the redshift-Doppler parameter $\gamma_{\rm RD}$, and the rate of change of the orbit period $\dot{P}_{\rm b}$. Given a relativistic theory of gravity, these can be expressed in terms of physical parameters of the system, such as the masses of the pulsar m_1 and companion m_2, and the inclination angle i of the orbit relative to the plane of the sky.

Now, in theories of gravity that violate SEP (e.g., the Rosen bimetric theory), the strong self-gravitational fields of the neutron star (that presumably is the pulsar) and possibly of the companion, if it is a neutron star or a black hole, cause the masses that determine the dynamics of the system to be renormalized.[12] For example, if m_1 and m_2 are the masses of the bodies as measured by Keplerian *test*-body orbits far from them, then the Newtonian limit of their *mutual* orbital equations of motion is given by‡

$$d\mathbf{v}/dt = -\mathcal{G} m_1 m_2 \mathbf{r}/r^3, \tag{4}$$

‡We use units in which the local gravitational constant far from gravitating matter and the speed of light are unity.

where $\mathcal{G}$ is a function of the self-gravitational binding energies per unit mass, denoted s_1 and s_2, of the bodies. In general relatively, $\mathcal{G} \equiv 1$, while in the Rosen bimetric theory, $\mathcal{G} = 1 - \frac{4}{3} s_1 s_2$. Such an effect could be regarded as a renormalization of the effective gravitational constant by the presence of the highly relativistic bodies (violation of SEP). In such a case, the measured orbital period P_b is related to the total mass $m(=m_1 + m_2)$ and the semimajor axis a by $(P_b/2\pi) = (a^3/\mathcal{G}m)^{1/2}$. A consequence of this is that the mass function f_1 is related to m_1, m_2, and sin i by

$$f_1 = \mathcal{G}\,(m_2 \sin i)^3/m^2. \tag{5}$$

A similar effect occurs in the periastron shift. Even though a given theory may predict the same periastron shift as general relativity in the strict weak-field post-Newtonian limit, (as does the bimetric theory) the presence of strong fields in the pulsar can cause a renormalization of the predicted shift. The periastron shift per orbit, $\Delta\omega$, can be written most conveniently in the form $\Delta\omega = \mathcal{P}\mathcal{G}^{-1}[6\pi m/a(1-e^2)]$, where $\mathcal{P}$ is also a function of s_1 and s_2. In general relativity $\mathcal{P} \equiv 1$, while in the Rosen bimetric theory, $\mathcal{P}$ can be as small as 0.3, depending upon the mass and structure of the pulsar and its companion. When the known eccentricity and orbital period of the system are used, the rate of periastron advance becomes

$$\dot{\omega} = 2.10\mathcal{P}\mathcal{G}^{-4/3}\,(m/\mathrm{m}_\odot)^{2/3} \text{ deg yr}^{-1}. \tag{6}$$

Unfortunately, here, as in the case of Mercury's perihelion shift, there may be complications. If the companion is a rapidly rotating white dwarf, centrifugal flattening could produce a quadrupole moment (J_2) and make an additional contribution to the periastron shift. If the companion is a neutron star or a black hole, no such effects occur.

The parameter γ_{RD}, which measures the effect on the pulse arrival times of the gravitational redshift due to the companion's gravitational potential and of the second-order doppler shift due to the pulsar's orbital motion, is also affected by renormalizations, namely

$$\gamma_{\mathrm{RD}} = 2.93\,(m_2/\mathrm{m}_\odot)^{2/3}\,(m_2/\mathcal{G}m)^{1/3}\,(1 + \mathcal{G}m_2/m) \text{ ms}. \tag{7}$$

Finally, the possibility of a secular change in the orbit period P_b has an immediate interpretation in terms of gravitational radiation damping. According to the "quadrupole" formula of general relativity, radiation damping should cause the orbit energy, and thereby the period of the orbit, to decrease according to[13]

$$\dot{P}_b/P_b = -\,1.9 \times 10^{-9}\,(m_1 m_2/\mathrm{m}_\odot^2)\,(m/\mathrm{m}_\odot)^{-1/3}. \tag{8}$$

However, in most (if not all) alternative theories, dipole gravitational radiation can also exist, generated by the varying dipole moment of the gravitational binding energy of the two bodies. The result, in the bimetric theory, is an orbit period increase, given by[12]

$$(\dot{P}_b/P_b)_{\mathrm{dipole}} = 2.1 \times 10^{-4}\,(m_1 m_2/m\mathrm{m}_\odot)\,(s_1 - s_2)^2 \text{ yr}^{-1}. \tag{9}$$

The measured values for these quantities as announced by Taylor at this Texas symposium are shown in TABLE 3.[14] The constraints that these values place (via Equations 5–9) on the system are most conveniently displayed on a plot of m_1 versus m_2. In general relativity, the results are shown in FIGURE 6. One constraint is provided

TABLE 3

PARAMETERS OF THE BINARY PULSAR*

Pulse period	$P = 0.059029995269 \pm 2$ s
Derivative of period	$\dot{P} = (8.64 \pm 0.02) \times 10^{-18}$ s s^{-1}
Projected semimajor axis	$a_1 \sin i = 2.3424 \pm 0.0007$ light s
Orbital eccentricity	$e = 0.617155 \pm 0.000007$
Binary orbit period	$P_b = 27906.98172 \pm 0.00005$ s
Mass function	$f_1 = 0.13227 \pm .00004$ m$_\odot$
Periastron shift	$\dot{\omega} = 4.226 \pm 0.002$ deg yr^{-1}
Redshift-Doppler	$\gamma_{RD} = 0.0047 \pm 0.0007$ s
Derivative of orbit period	$\dot{P}_b = (-3.2 \pm 0.6) \times 10^{-12}$ s s^{-1}

*After reference 14.

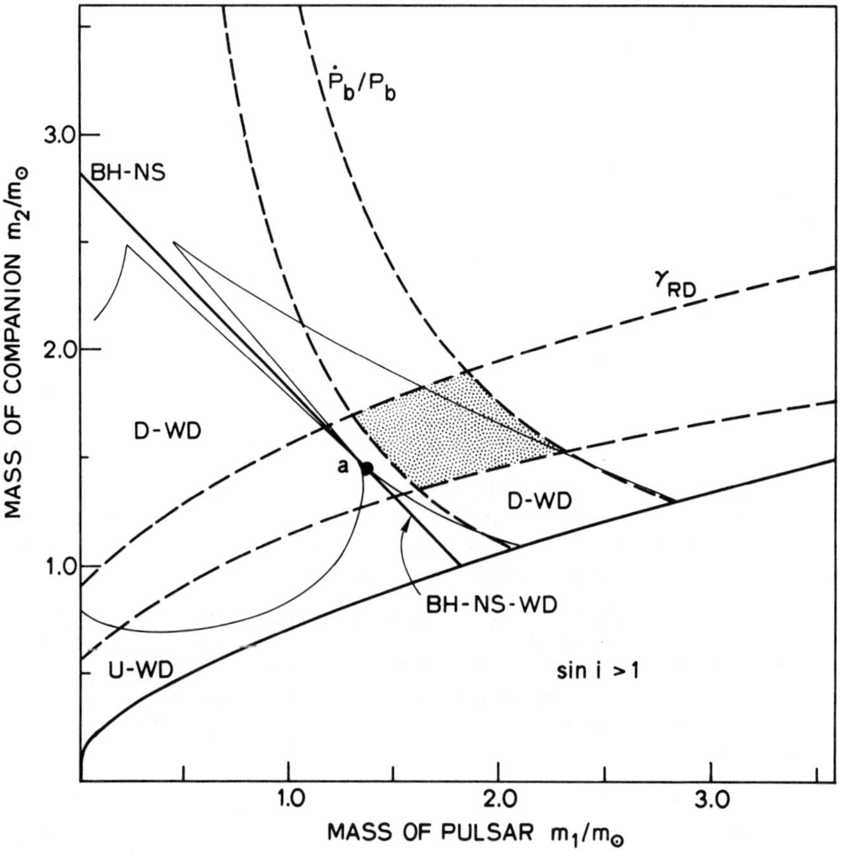

FIGURE 6. The m_1–m_2 plane in general relativity. The shaded region fits all the formal observational constraints, however, the point marked "a" is the most likely configuration, with $m_1 = 1.39 \pm 0.15$ m$_\odot$, $m_2 = 1.44 \pm 0.15$ m$_\odot$. (After reference 14.)

by the mass function f_1 and by the fact that sin $i \leqslant 1$. The periastron shift constrains the system to lie along the straight line BH–NS–WD, *if* the companion is a black hole, neutron star, or nonrotating white dwarf. However, if the companion is a rapidly [uniformly (U) or differentially (D)] rotating white dwarf, the system could lie in the regions denoted U and D. The constraints set by γ_{RD} and by $\dot{P}_b/P_b$ then leave the shaded region available. The deviation ($\sim$1.2 σ of the $\dot{P}_b$ data) between the shaded region and the BH–NS–WD line cannot be called significant at present, and so a natural interpretation of the system is that of a pulsar of mass $m_1 = 1.39 \pm 0.15$ m$_\odot$ and a companion of mass $m_2 = 1.44 \pm 0.15$ m$_\odot$ (point "a" of FIGURE 6). There are two possible (though unlikely) additional contributions to $\dot{P}_b/P_b$. If the companion is a rotating white dwarf, and *if* a strong source of viscosity such as magnetism or turbulence is present in its atmosphere, then tidal dissipation of orbital energy could account for some $\dot{P}_b$. However, if the white dwarf rotates too rapidly, the effect can cause $\dot{P}_b$ to increase. Another possible source of $\dot{P}_b$ is a distant third body that gives the binary system an acceleration. Such an acceleration will lead to both $\dot{P}$ and $\ddot{P}$, for both the orbital and pulsar periods. A measurement of an upper limit on the second time derivative of the pulse period, $\ddot{P}$, could rule out such a possibility by setting a lower limit on the mass of the third body given by $m_3 \gtrsim 60$ m$_\odot(\delta/10^{-9}\,\mathrm{yr}^{-1})^7(\eta/10^{-13}\,\mathrm{yr}^{-2})^{-4}$, where δ is the amount of $\dot{P}_b/P_b$ that the third body is assumed to produce, and η is the upper limit on $\ddot{P}/P$ for the pulsar.

Thus, barring these remote possibilities, general relativity leads to a natural physical configuration for the system, and the results support the conclusion that the measurement of $\dot{P}_b$ represents the first observation of the effects of gravitational radiation.

In the Rosen bimetric theory, on the other hand, the situation is very different. The self-gravitational renormalizations of masses (factors of $\sim$ 2–3) lead to qualitative differences in the m_1–m_2 plane, FIGURE 7. For example, the companion cannot be a nonrotating white dwarf, since such a configuration would violate the condition sin $i \leqslant 1$. If the companion is a neutron star, the system lies along the curve "NS", with total mass $\sim$7 m$_\odot$. Black holes do not exist in this theory. When the γ_{RD} constraint§ is folded in, the theory is left with a major problem. Dipole gravitational radiation causes the system to gain energy and the period to increase at rates shown in FIGURE 7. The theory then must produce a mechanism (tidal dissipation or a third body) to account for the observed period decrease. The contrived and *ad hoc* nature of such mechanisms casts serious doubt upon the viability of this theory.

This result may, in fact, apply to many other theories, particularly those with "prior geometry" (see TABLE 2). In such theories, SEP is violated, and the differences between the theories and general relativity become larger the stronger the gravitational fields. Thus one can expect qualitative mass renormalizations similar to those in the bimetric theory. Furthermore, all such theories are expected to predict dipole gravitational radiation of magnitude comparable to that in the bimetric theory. So it is

§There is an additional SEP-violating effect that modifies the dependence of γ_{RD} slightly. If the local gravitational constant felt by the pulsar depends on the proximity of the companion, the pulsar's structure, moment of inertia, and rotation rate may vary with the orbital motion.[15] This produces an effect of identical signature to the redshift-Doppler effects. In the Rosen bimetric theory, the result is a 20% modification of the effective value of γ_{RD}; this does not alter our general conclusions.

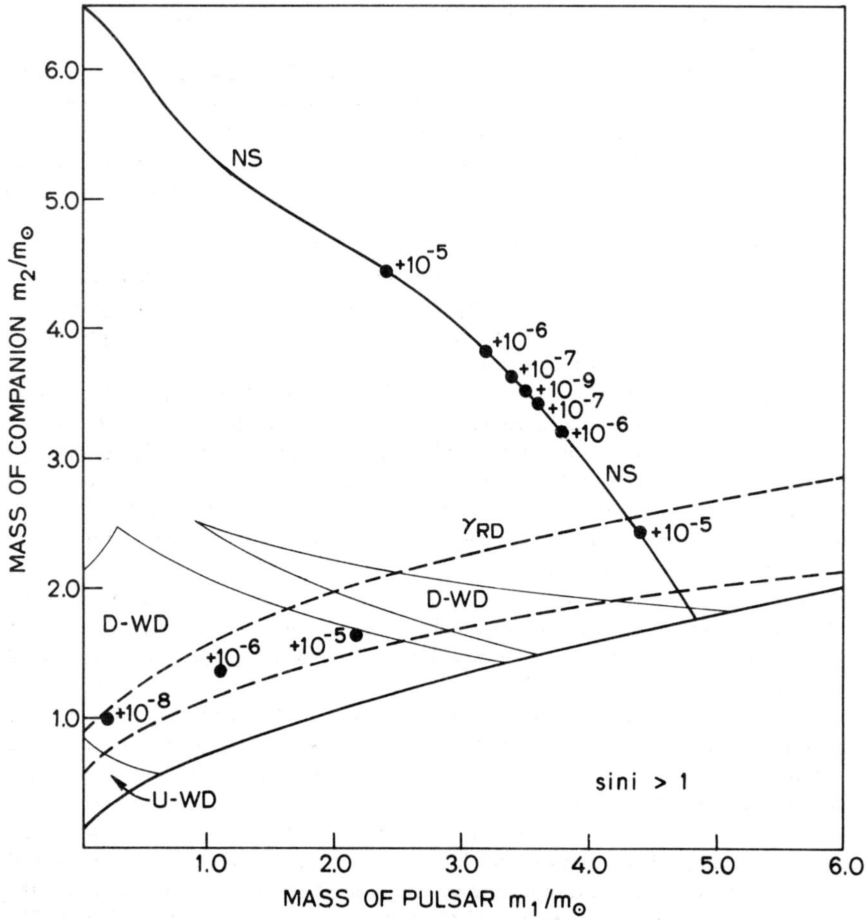

FIGURE 7. The m_1–m_2 plane in the Rosen bimetric theory. Note the scale of masses is almost double that of FIGURE 6. The numbers shown are the predicted values of $\dot{P}_b/P_b$ due to gravitational radiation, including dipole gravitational radiation.

very likely that the binary pulsar data will be able to rule out a broad class of alternative gravitation theories.

On the other hand, the class of "purely dynamical" theories has the property that the effects of the additional gravitational fields can be made as small as one chooses, both in weak field *and* in strong field or gravitational radiation situations, by choosing sufficiently weak coupling constants ($\omega^{-1} \rightarrow 0$ in Brans–Dicke, for instance). Thus, Brans–Dicke theory, with $\omega \gtrsim 200$, is consistent with the present binary pulsar data, even though it, too, predicts dipole gravitational radiation. Thus the binary pulsar can yield do-or-die tests of a wide range of gravitation theories, and can help put severe constraints on others.

CONCLUSIONS

Shortly after Eddington reported the successful measurement of the deflection of starlight by the Sun, Einstein was asked what would have happened if Eddington had gotten a different value for the deflection than that predicted by general relativity. Einstein is reputed to have replied, "Well, it would be too bad for God." As we begin 1979, the year of the centenary of Einstein's birth, it is clear that both God and Einstein are "hanging in." But with new discoveries such as the binary pulsar, it is also clear that the confrontation between gravitation theory and experiment will continue to be an exciting challenge for some time.

ACKNOWLEDGMENTS

I am grateful to Joe Taylor for providing me with the binary pulsar results in advance of publication, and to Douglas Eardley for helpful discussions.

REFERENCES

1. HAUGAN, M. P. 1979. Energy conservation and the principle of equivalence. Ann. Phys. (N.Y.) **118:** 156–186.
2. WILL, C. M. 1979. *In* General Relativity: An Einstein Centenary Survey. S. W. Hawking & W. Israel, Eds. Cambridge University Press. Cambridge.
3. SHLYAKHTER, A. I. 1976. Nature (London) **264:** 340.
4. WILL, C. M. 1974. *In* Experimental Gravitation. B. Bertotti, Ed.: 1–110. Academic Press. New York.
5. FOMALONT, E. B. & R. A. SRAMEK. 1977. Comm. Astrophys. **7:** 19–33.
6. SHAPIRO, I. I., R. D. REASENBERG, P. E. MACNEIL, R. B. GOLDSTEIN, J. P. BRENKLE, D. L. CAIN, T. KOMAREK, A. I. ZYGIELBAUM, W. F. CUDDIHY & W. H. MICHAEL, JR. 1977. J. Geophys. Res. **82:** 4329–4334.
7. CAIN, D. L., J. D. ANDERSON, M. S. W. KEESEY, T. KOMAREK, P. A. LAING & E. L. LAU. 1978. Bull. Am. Astron. Soc. **10:** 396.
8. WEINBERG, S. 1972. General Relativity and Cosmology: 171. Wiley. New York.
9. EARDLEY, D. M., D. L. LEE & A. P. LIGHTMAN. 1973. Phys. Rev. D **10:** 3308–3321.
10. CAVES, C. M. 1977. *In* Proceedings of the 8th International Conference on General Relativity and Gravitation: 104. Preprint.
11. HULSE, R. A. & J. H. TAYLOR. 1975. Astrophys. J. Lett. **195:** 51–53.
12. WILL, C. M. & D. M. EARDLEY. 1977. Astrophys. J. Lett. **212:** 91–94.
13. WAGONER, R. V. 1975. Astrophys. J. Lett. **196:** 63–65.
14. TAYLOR, J. H., L. A. FOWLER & P. M. MCCULLOCH. 1979. Nature **227:** 437–440.
15. EARDLEY, D. M. 1975. Astrophys. J. Lett. **196:** 59–62.

SUMMARY OF WORKSHOP ON GRAVITATIONAL RADIATION

E. Amaldi

Instituto di Fisica
dell' Università, Roma
Rome, Italy

Workshop number eight was devoted to problems directly connected to the detection of gravitational radiation. Therefore, a few valuable papers proposed for the workshop but dealing with different subjects were assigned to a different section which, regrettably, at the last moment did not take place because of unexpected organizational difficulties.

In opening the discussion the chairman pointed out that, today, practically all attempts to detect gravitational radiation emitted from extraterrestrial sources are made in two separate frequency regions: (1) medium frequencies (mf) ($50 \leqslant \nu_g \leqslant 10^4$ Hz) and (2) very low frequencies (vlf) ($10^{-5} \leqslant \nu_g \leqslant 10^{-2}$ Hz). The corresponding sources are illustrated in FIGURE 1,[1] which summarizes the theoretical results obtained by various authors. In the figure the abscissa indicates the frequency ν_g of the emitted gravitational waves and the duration τ_g of the event, with the only exception being the Crab pulsar, which is a periodic source (of period τ_g). The ordinates give the spectral energy density $F(\nu)$ in GPU (1 Gravitational Pulse Unit = 10^5 erg/cm^2 Hz)^2 as well as the corresponding *amplitude h* of the wave, i.e., the amplitude of the deviation $h = h$ ($\mathbf{r}$, t) of the metric tensor g_{ik} from the Minkowski metric. This is connected to the variation δl of the distance l between pairs of points belonging to planes perpendicular to the direction of propagation by the well-known relation

$$\frac{\delta l}{l} = \frac{1}{2} h. \tag{1}$$

The numbers attached to the various points in the figure indicate the number of solar masses converted into gw emitted isotropically in all directions.

The majority of the detectors of mf gravitational radiation under development today belong to two well-known classes: resonant (or Weber-type) bars and broadband laser interferometers. The setups designed for revealing vlf gravitational radiation are based on Doppler tracking of interplanetary spacecraft.[3]

These devices were discussed in part 1 of the workshop, while part 2 consisted of two contributions regarding other possible methods for detecting gravitational radiation or at least recognizing its presence in distant regions.

The workshop ended with the presentation by J.R. Wilson of the Lawrence Livermore Laboratory of a film on neutron star collapse.

PART 1: GRAVITATIONAL RADIATION DETECTORS

Detection of Medium Frequency Gravitational Radiation by Resonant Antennas

The discussion was devoted to the detection of bursts of gravitational radiation emitted in catastrophic events because of their greater intensity with respect to that of

0077–8923/80/0336–0322 $01.75/1 © 1980, NYAS

periodic sources (FIGURE 1). This is also the reason why all experimental groups search for bursts of gravitational radiation, with the exception of the Tokyo group,[5] which tries to detect a steady emission of gw correlated to the radio emission from the Crab pulsar.

Resonant detectors,[4] usually high-Q cylindrical bars, start to vibrate with an

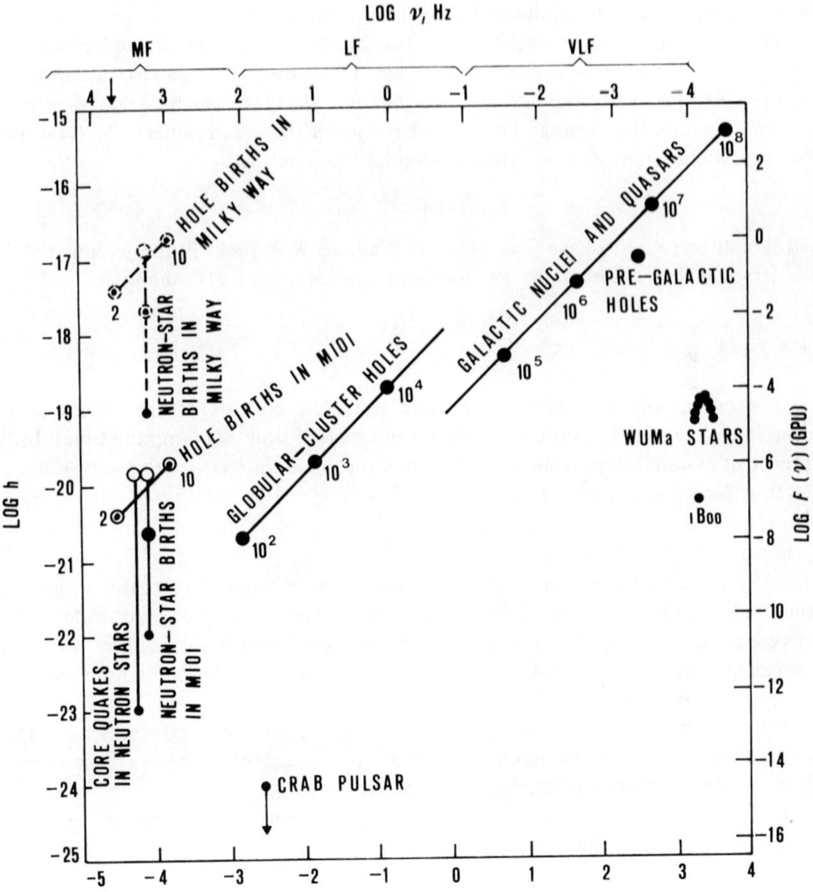

FIGURE 1. Intensities and frequencies of gravitational radiation sources.[1]

appreciable amplitude in one of their longitudinal modes (usually the fundamental longitudinal mode of angular frequency ω_0) whenever the incoming gravitational radiation burst contains a Fourier component of the same angular frequency with an appreciable amplitude. Once a mode has been excited by the passage of the burst, it continues to vibrate with its own angular frequency ω_0 and with its intrinsic damping time $\tau_0 = Q/\omega_0$. This, clearly, should be as long as possible (high Q) for allowing an adequate determination of the bar's initial excitation.

In the case of broadband laser interferometer the variation of the length l of the interferometer arms follows the time variation of $h(t)$ as shown by the relationship (**1**).

One of the more delicate aspects of a resonant antenna is the transducer, i.e., the device that converts the relative variation $\delta l/l$ undergone by the length l of the bar into an electromagnetic signal that can be amplified and recorded.

The signal due to a gravitational radiation burst that we can indicate for brevity as a voltage $V_s(t)$* should be separated from the "total noise" $V(t)$ due to composition of the Brownian motion of the bar, the noise of the transducer, the back reaction on the bar, and possibly the (voltage and current) noise of the preamplifier. The total noise $V(t)$ can be described as a vibration of angular frequency ω_0

$$V(t) = re^{i(\omega_0 t + \varphi)},$$

with amplitude r and phase φ varying (at random) with time. By projecting $V(t)$ on two orthogonal axis, we obtain the *stochastic variables* $x(t), y(t)$ such that

$$r^2(t) = x^2(t) + y^2(t)$$
$$tg\varphi(t) = y(t)/x(t) \tag{2}$$

The output of most present resonant antennas consists of the values of the variables $x(t)$ and $y(t)$ recorded at regular intervals of time Δt (sampling time). In the absence of external signals the distribution of the values of $r^2(t)$ (with $t = t_0 + n \cdot \Delta t$, $n = 0, 1, 2, \ldots$) is given by the exponential law

$$F(r^2) = 2\sigma_0^{-2} \exp(-r^2/2\sigma_0^2), \tag{3}$$

where σ_0^2 includes the various noises mentioned above. Signals $V_s(t)$ due to gravitational radiation bursts (or disturbances of various types) are always embedded in the background (**3**) from which they should be disentagled. This is obtained by employing convenient algorithms, the simplest of which is the variables (**2**) that can be indicated as a *direct algorithm.*

Different from "noises," disturbances can *not* be treated mathematically. They can be eliminated only by observing signals in coincidence between two or more detectors placed at great distances from one another.

E. Amaldi recalled that a wide category of algorithms is that of *predictive algorithms* defined by the general relationship[6]

$$\rho_i^2 = [x(t) - x_{pi}(t)]^2 + [y(t) - y_{pi}(t)]^2, \tag{4}$$

where $x_{pi}(t), y_{pi}(t)$ are the values of the stochastic variables x and y predicted for time t.

The various algorithms of this category differ from each other for the complexity of the operations involved for computing the predicted values.[7]

We call *zero-order predictive algorithm* (ZOP) ρ_z^2 that obtained by introducing into (**4**)

$$x_{p0}(t) = x(t - \Delta t), \qquad y_{p0}(t) = (t - \Delta t). \tag{5}$$

*Voltage mode. In general the output will consist of a voltage $V_s(t)$ and a current $I_s(t)$.

It has been used in the past by many authors and is still advantageously used today because it reduces to a minimum the manipulation of the data.

In the *first-order predictive algorithm* (FOP) the best statistical use is made only of the last measured values of x and y:

$$x_{p1}(t) = \frac{R(\Delta t)}{R(0)} x(t - \Delta t),$$

$$y_{p1}(t) = \frac{R(\Delta t)}{R(0)} y(t - \Delta t).$$

Here $R(\Delta t) = R_{xx}(\Delta t) = R_{yy}(\Delta t)$ is the autocorrelation function of the recorded data.

Other predictive algorithms use explicitly two or more preceding measurements. The most refined algorithm of this category is that of Wiener and Kolmogorov,[7] which uses all the past and all the future data.† It has been conveniently adapted to the case of discrete data.[6]

The distribution functions for all these algorithm are similar to (**3**), i.e.,

$$F(\rho_i^2) = 2\sigma_i^{-2} \exp(-\rho_i^2/2\sigma_i^2), \tag{6}$$

where, however,

$$\sigma_i^2 < \sigma_0^2.$$

FIGURE 2 shows the experimental distributions obtained by the Rome group[6] for three of the above defined algorithms applied to the same set of data collected with an Al test bar of $M = 25$ kg, operated at $T = 4.2$ K.[8]

The instrumental sensitivity of a resonant antenna is expressed by the signal-to-noise ratio, which is conveniently defined as[9]

$$\text{SNR} = \rho_{ig}^2/2\sigma_i^2, \tag{7}$$

where ρ_{ig}^2 is the signal due to gravitational radiation. The ratio (**7**) always depends on (a) the wideband noise $S_0[(nV)^2/\text{Hz}]$ of the transducer + electronics, (b) the mass M of the bar, (c) the composite parameter

$$\beta Q/T, \tag{8}$$

and (d) the adopted algorithm. As customary, β is the ratio of the energy in the transducer to the energy in the bar. A higher sensitivity is obtained by decreasing S_0, increasing M and the composite parameter (**8**), and choosing the algorithm that maximizes the SNR (**7**).

The minimum detectable spectral energy density is obtained by equating the SNR to 1. The result can be expressed in the simple form[8]

$$F(\nu) \geqslant 782\, T_{\text{eff}}^i/M_{\text{kg}}\ \text{GPU} \tag{9}$$

The numerical value of the constant corresponds to Al at $T = 4.2$ K. It contains universal constants (G and c) and is inversely proportional to v_S^2 (v_S = velocity of sound in the bar material). The effective temperature T_{eff}^i expresses in a temperature scale the value of $\bar{\sigma}_i^2$.

The goal of the Stanford–Louisiana–Rome collaboration is to operate at each station in coincidence an Al bar of $M = 5000$ kg at a temperature $T \ll 4.2$ K. The approach of these groups usually is described as "large mass–low temperature." It

†Past and future are limited by the damping time $\tau_0 = Q/\omega_0$ of the bar. Therefore the start of the manipulation of the data should be retarded a time interval of the order of τ_0.

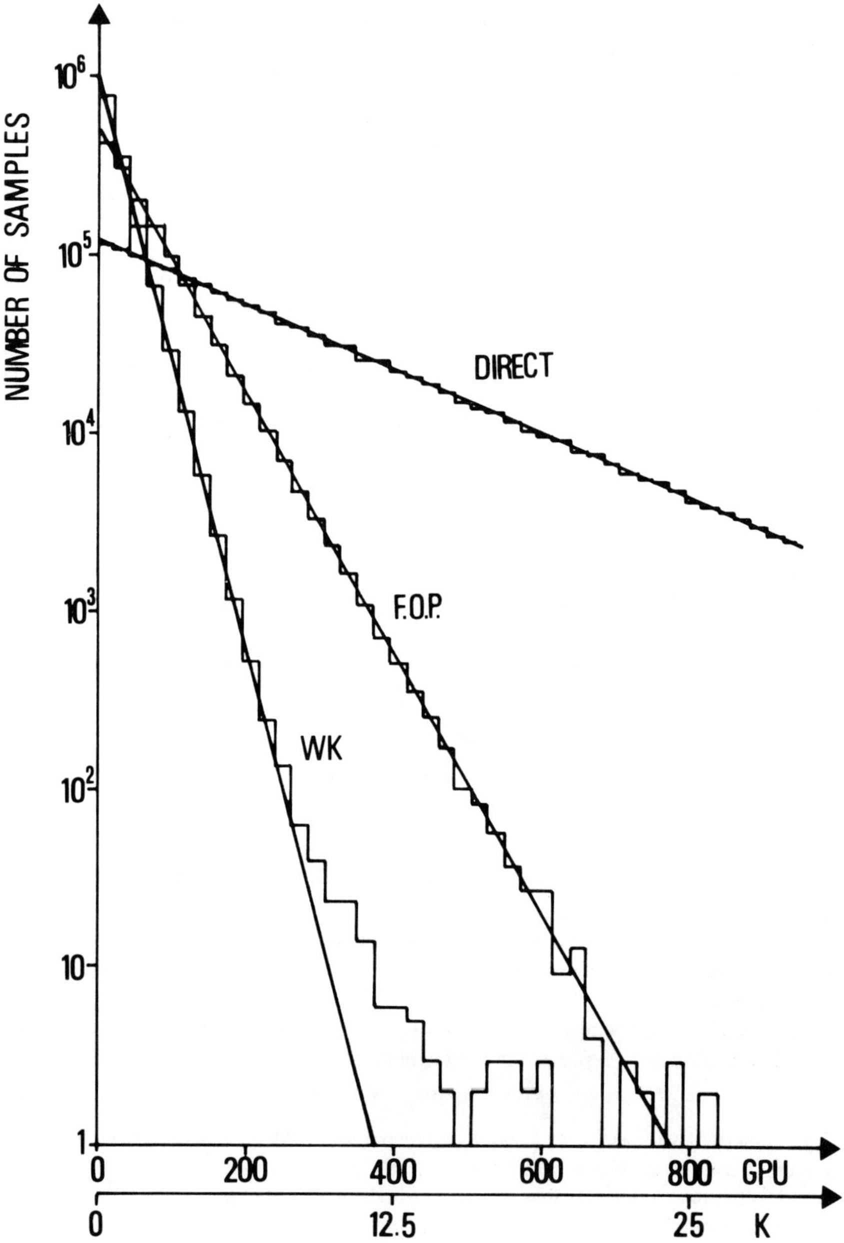

FIGURE 2. Analysis of data with different algorithms.[6]

differs from the approach "high Q–low temperature" developed by other groups (Moscow, Rochester, Regina; see Douglass' contribution).

Intermediate bars have been operated at Stanford and Rome.

The Rome intermediate bar ($M = 390$ kg, $\nu_0 = 1795$ Hz) was kept in operation at 4.2 K for 39 days (from the 10th of June to the 19th of July 1978).[10]

From the data collected in 19 days during July, $T_{\text{eff}}{}^{\text{K}} = 1.53$ K was obtained corresponding to[11]

$$F(\nu) \simeq 2 \text{ GPU or } h \sim 2 \times 10^{-17}.$$

The effective temperatures of the two previous more sensitive experiments, at room temperature, are:

$T_{\text{eff}} = 18.3$ K for the Bell Telephone experiment[12] (3.7 tons)

$T_{\text{eff}} = 7.3$ K for the Munich experiment[13] (1.7 tons)

The Rome group has started to work on the $M = 5000$ kg bar, which, according to (**9**), will allow an increase in sensitivity of at least a factor 12 due to the higher mass of the bar. At present one of the main limitations on the value of T_{eff} derives from resonant noise (or back reaction) of the bar induced by the current noise of the preamplifier FET.

D.H. Douglass of the University of Rochester discussed the high Q–low temperature gravitational wave detectors for the amplitude–phase type of detection process used by all groups until now. He expresses qualitatively the detection criterion by the relation

$$M\omega^2 \Delta x_g{}^2 > kT\frac{\omega\tau}{Q} + n\hbar\omega\left(\frac{1}{\beta\omega\tau} + \beta\omega\tau\right), \qquad \textbf{(10)}$$

where Δx_g $(= \delta l)$‡ $= hl$ is the displacement of the end of the bar, τ $(= \Delta t)$ the sampling time, and

$$n = E_t/\hbar\omega\,, \qquad \textbf{(11)}$$

where E_t is the characteristic noise energy associated with the transducer. Equation 10 shows that the signal, i.e., the energy deposited in the bar by a gw burst, should be greater than the sum of three terms: the Brownian motion of the bar, the wideband noise of the transducer, and the back action of the transducer on the bar.[14]

By requiring that the signal should be greater than each of the three terms separately, the three following conditions are derived:

(I) $\beta Q > \dfrac{kT}{n\hbar\omega} = 3.3 \times 10^7 \dfrac{1}{n}\left(\dfrac{T}{1.4\text{ K}}\right)\left(\dfrac{10^3\text{ Hz}}{\nu}\right);$

(II) $\beta\omega\tau = 1;$

(III) n as small as possible.

Condition (II) can always be satisfied by a convenient choice of the sampling time τ.

Thus all groups pursuing the resonant antenna approach *must* have large Q and small n. If we accept that $n \sim 1$ is about the limit of the amplitude–phase detection scheme, then one has $M\omega^2\Delta x_g{}^2 > 2\hbar\omega$ or $\Delta x_g = (2\hbar/M\omega)^{1/2}$. Expressed in terms of the strain, this becomes

‡The notations used by the previous speaker are given in parentheses.

$$\left(\frac{\delta l}{l}\right) = \frac{\Delta x_g}{l} > \left(\frac{2\hbar\omega}{\pi^2 v_s^2 M}\right)^{1/2}$$

$$\sim 2.3 \times 10^{-21}\left(\frac{10^6\text{ g}}{M}\right)^{1/2}\left(\frac{\nu}{10^3\text{ Hz}}\right)^{1/2}\left(\frac{5\times 10^5\text{ cm/s}}{v_s}\right). \tag{12}$$

Strains of 10^{-21} are close to the values expected from supernovae events in the Virgo cluster (FIGURE 1).

Condition (I) shows the importance of high Q. Since $\beta \lesssim 1$, Qs greater than 10^7 are needed for a temperature greater than 1 K. Or, stated another way, if the Q is less than 10^7, then to reach the limit (**3**) it is necessary to cool the bar below 1 K.

Another important technical comment is that if $Q \sim 10^{10}$ (values close to this have been measured in silicon and sapphire[15]) then β need only be as large as 10^{-3} at $T \sim 1$ K.

The method of achieving small n is not yet a solved problem. Each of the various groups is pursuing what to them is a promising approach.

These approaches all involve the invention of new technology and include such things as superconducting electromagnetic resonators, ultrastable oscillators, superconducting quantum interference devices (SQUID), maser amplifiers, etc. As one example of the state of the art the Rochester group is developing a scheme based upon the DC SQUID for which

$$n \sim S_\Phi/2\hbar L_S, \tag{13}$$

where S_Φ is the spectral density of the flux noise and L_S is the SQUID loop inductance. Using the best values of S_Φ and L_S reported in the literature, one calculates $n \sim 10^3$. The prospects for considerable improvement are excellent. Tesche and Clarke[16] have estimated that the limit for the SQUID is $n \sim 1$.

The strain limit of the amplitude-phase detection scheme given by (**2**) may not be sufficient to detect the Virgo cluster events mentioned above, This limit, which is sometimes called the "quantum limit," caused some concern to the workers in this field several years ago. Fortunately, study of this problem has shown that there is no such limit.[17] Schemes which might accomplish this are now called back-action evading measurements.[18]

Since there are no confirmed claims of detection of gravitational radiation it is informative to describe the improvement of the sensitivity of the detectors that has occurred in recent years. In 1973 the room temperature detectors had a strain noise of $(\delta l/l)^2_{\text{noise}} \sim 10^{-32}$ and in 1976 this had been reduced to $(\delta l/l)^2_{\text{noise}} \sim 10^{-34}$. In 1979, one can expect to have cooled resonators with $(\delta l/l)^2_{\text{noise}} \sim 10^{-36}$. The improvement during this time interval is at the rate $10^6/10$ years. THis, according to Douglass, is quite good. For comparison, in the field of radio astronomy the rate of improvement of the equivalent figure of merit has been a factor of 20 every 10 years over a 45 year period beginning with Jansky.

Broadband Laser Interferometers

W. Winkler of the Max–Planck Institute für Physik und Astrophysik in Munich, reported on the work carried on in Munich by H. Billing, K. Maischberger, A. Rüdiger, R. Schilling, L. Schnupp, and himself.

This kind of detector may be able to see a strain in space of several 10^{-21} if the limiting noise is the shot noise of the photocurrent of a 100-W laser, and if the effective baseline is half the wavelength of the gravitational radiation. A short description may be found in the Proceedings of the Meeting of Experimental Gravitation, Pavia, Italy, September 1976.[19]

To test the feasibility of such an experiment, they built first a small and rigid interferometer and then a prototype interferometer with 3-m arm length; the light source is a 1-W single-mode Ar laser. The central block containing the beamsplitter and the two near mirrors of the optical delay lines, and the two far mirrors are suspended separately on thin steel wires in a vacuum of 10^{-6} torr. The big amplitudes of the pendula at their resonance frequency are avoided by active damping. The interferometer is kept at the desired point of operation by a servo loop including a Pockels cell to adjust the phase properly. Several noise sources have already been investigated; they reported here on two new ones.

(1) The laser beam shows lateral displacements in the kHz region of several $\text{Å}/\sqrt{\text{Hz}}$ and a jitter in angle of several 10^{-10} radian/$\sqrt{\text{Hz}}$. This noise in position and orientation leads to an interferometer signal if the setup is not totally symmetric. Allowed tolerances for a 1-W laser are: angle between the beamsplitter and the plane of symmetry $<10^{-6}$ radian; difference in optical thickness of the components in the two arms ~ several 10μ; difference in wedges of the components ~ several minute of arc.

(2) Scattered light is disturbing in several respects. The most critical one occurs in connection with the frequency jitter in the case of long effective baselines: if scattered light with an amplitude s is added to one of the two main beams running back to the beamsplitter with amplitude 1, an interferometer signal of $\delta L \approx sL/\delta\nu/\nu$ is produced, where $\delta\nu/\nu$ is the relative amplitude of the frequency jitter and L the path difference between the main beam and the scattered light. Their laser shows a $\delta\nu/\nu/\sqrt{\text{Hz}} \approx 10^{-13}/\sqrt{\text{Hz}}$ in the kHz region; L can be as long as 100 km. In order to remain below the shot noise limit, s has to be smaller than 10^{-8}.

The sensitivity of their rigid interferometer came close to the shot noise limit at 100-mW laser power; the remaining signal was due to thermal noise of the components of the temporarily designed setup and partly to the noise in the position of the laser beam.

R.W.P. Drever reported on the work on laser interferometry done at Glasgow (R.W.P. Drever, J. Hough, J.R. Pugh, W.A. Edelstein, H. Ward, N.A. Roberston, M. Ford).

After the early work at Glasgow on a divided-bar gravity wave detector with piezoelectric sensing they decided a few years ago to switch to "free-mass" gravity wave detectors using laser interferometry displacement sensing. In order to develop some of the new optical and other techniques required they have begun work using the main parts from an old divided-bar detector, modified for optical sensing.[20] A special multireflection system, based on a two-dimensional extension of the White cell, is used to monitor the relative motions of the inside faces of the two bars and lead to a fringe shift in a Michelson inteferometer. The test bars are suspended by pendulum suspensions having high Q (low damping) and a period of the order 1 Hz. High-frequency phase modulation of one arm of the interferometer is used to reduce laser intensity noise, and an electrostatic force feedback system holds the optical paths in the two arms of the interferometer nearly equal, thereby minimizing frequency fluctuation noise. At present, with relatively poor mirrors, and about 200 mw of output light to the detecting photodiode, the apparent noise level of the system corresponds to about 10^{-14} m/Hz$^{1/2}$ in the region of 1 kHz. It seems possible that much of this present noise is related to translational motions of the laser beam, and they are now preparing to monitor and perhaps control this. While this development of optical techniques continues, they have begun construction of a rather large interferometric detector, using four test masses placed at the corners of a square of side 10 m. It is planned to monitor all four sides of the square using interferometric techniques, giving some redundancy and possible discrimination against local disturbances. They plan to use this apparatus for development of further optical and other techniques for future detectors, but its sensitivity may be at least comparable with that of other present detectors.

Background Gravity Wave Detection with Doppler Tracking of Spacecraft

R.W. Hellings of the Jet Propulsion Laboratory, Pasadena, California, reported on work done by J.D. Anderson, P.S. Callahan, F.B. Estabrook, and H. Wahlquist.

In round-trip Doppler tracking of spacecraft, a microwave signal is sent from Earth to the spacecraft, coherently transponded at the spacecraft, and received again on Earth where the Doppler shift is measured. The passage of a gravity wave through the solar system will affect the signal, imparting three pulses to the Doppler record for every pulse in the wave, the first and last pulses being inverse to each other and separated by a round-trip light time ($2l$).

The sensitivity of the tracking network to an isotropic background of gravity waves was considered—gravity waves coming from a stochastic combination of known sources (Rosi and Zimmermann, 1976), from early cosmological black hole formation (Bertotti and Carr, 1979), or from a cosmic closure density of waves of primordial origin. The amplitude spectral densities (units h^2/Hz) of these sources are given in FIGURE 3.

The spacecraft tracking network itself has noise in it due to its various elements—the hydrogen maser clock/frequency standard, the receiver electronics, the Metric Data Assembly sampling software, interplanetary plasma, and the earth's troposphere. The spectral densities of these noise sources are given in FIGURE 4, where three separate lines represent plasma noise when the microwave tracking signal is S band, X band, or S/X band.

Comparison of FIGURES 3 and 4 shows that the system noise is stronger than any background signals. However, the unique signature of the gravity wave in the Doppler record may be used to dig into the noise. In so-called "three-way" tracking, where the receiving station is different from the transmitting station, only the gravity waves will be autocorrelated in the Doppler record at the round-trip light time. Thus, the autocorrelation function of the Doppler record in the three way mode provides a tool that will dig into the noise by a factor of $1/\sqrt{n}$, where n is the number of data points taken. This increase in sensitivity can make a closure density below a 10^{-3} Hz cutoff accessible now with an X band system.

PART 2: OTHER DETECTORS

F. Pegoraro reported on an electromagnetic detector for gravitational waves proposed by F. Pegoraro and L.A. Radicati of the University of Pisa and Ph. Bernard and E. Picasso of CERN.[21]

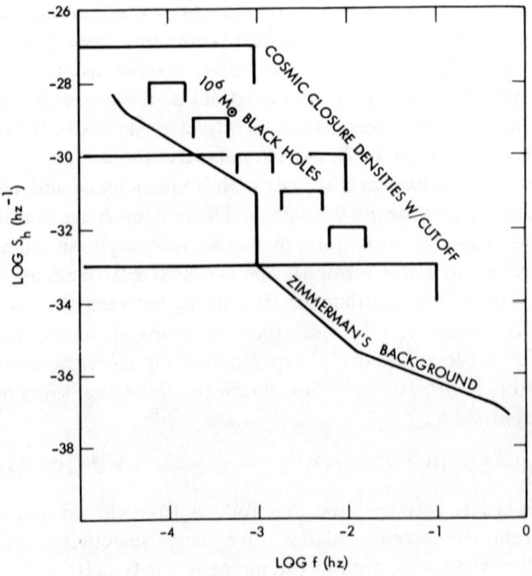

FIGURE 3. Background spectral density.

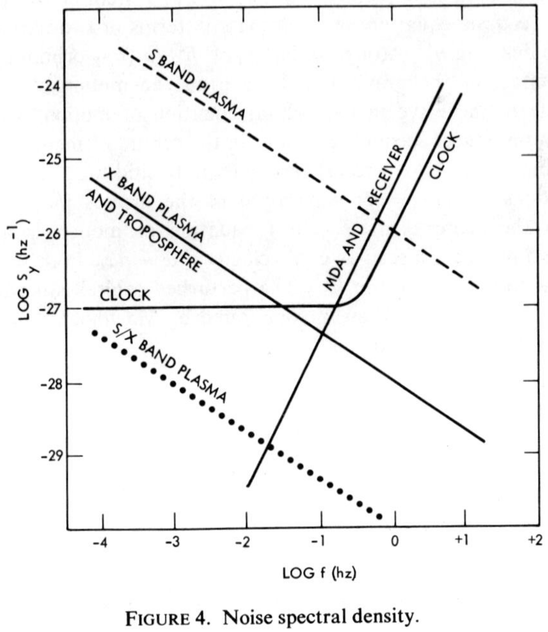

FIGURE 4. Noise spectral density.

A gravitational wave induces a time-dependent anisotropic dielectric tensor that can be monitored by using electromagnetic resonators. These detectors possess the following advantages: they are easily tunable over a wide range of frequencies, they can be sensitive only to quadrupole perturbations and can reach extremely high Q values. On the other hand, their small effective mass is disadvantageous in the absorption cross section and in the Brownian noise threshold.

The authors consider a two-level system with frequencies ω_1 and ω_2 with $\omega_1 - \omega_2 = \Omega$ and $\omega_1, \omega_2 \gg \Omega$, where Ω is the gw frequency. The two levels are obtained by coupling two identical high-frequency resonators, placed at right angles in the plane perpendicular to the gw. This geometrical configuration is sensitive to quadrupole perturbations of the dielectric tensor. The gw induces transitions between the two levels and the behavior of the resonator can be described in terms of an equivalent-frequency converter. By referring to a cylindrical superconducting niobium microwave cavity in the GHz range, stored with $10^{31}/\omega^3$, they obtain the following limits on the minimum detectable gw amplitude $h_{\min}$:

$$h_{\min} \geqslant 3 \times 10^{-23} (10^{12}/Q)^{1/2}, \quad h_{\min} \gtrsim 4 \times 10^{-18} (10^{12}/Q)^2$$

The first limit is a sensitivity bond that follows from the minimum power that is detectable at present; the second, which is a frequency selectivity requirement, is due to the finite width of the levels, and, at the moment, for $\Omega = 10^3$ rad s^{-1} Ω is the most restrictive condition.

M.S. Turner of the Enrico Fermi Institute, at the University of Chicago, Illinois, reported on the influence of a weak gravitational wave on a bound system of two point masses.

He considers the effect of a weak gravitational wave (dimensionless amplitude $h^{TT} \ll 1$) on a nonrelativistic (orbital velocity $v \ll c$) system of two point masses by expanding the geodesic equation for each mass in terms of two parameters, v/c and h^{TT}. The terms linear in h^{TT} through order $(v/c)^3\, h^{TT}/a$ ($a \sim$ orbital size) are used to calculate the effect of the gravitational wave on the motion of the masses. The pertubation due to the wave in the orbital equation of motion is at lowest order $(v/c)^2\, h^{TT}/a$ and produces secular variations in the orbital elements for $\omega = n\omega_0$ ($\omega =$ frequency of the wave, $\omega_0 =$ orbital frequency). In addition, for any ω there are periodic variations in the proper separation of the masses $[\delta a \sim a\, h^{TT}(t)]$. The pertubation in the center-of-mass (c.m.) equation of motion is at lowest order $(v/c)^3\, \mathrm{h}^{TT}/a$ and produces a secular c.m. velocity if $\omega = n\omega_0$. Both secular effects are gauge invariant to linear order in h^{TT}. The perturbed orbital equation of motion at order $(v/c)^2\, h^{TT}/a$ is different than the one found by Mashhoon[22] but is related by a simple coordinate transformation.

References and Notes

1. Whalquist, H. D., J. D. Anderson, F. B. Estabrook & K. S. Thorne. 1977. Proc. Symp. Experimental Gravitation, Pavia, Italy, September 17–20, 1976. B. Bertotti, Ed. Atti Convegni Lincei **34:** 335.
2. The use of this unit, introduced by C. W. Misner, is justified by the following remark: 1 $M_\odot$ converted into gravitational radiation, emitted isotropically over 4π and in a frequency band of 1 kHz, gives on the Earth a spectral energy density of 1.5 GPU if the event takes place at the center of our galaxy.

3. ESTABROOK, F. B. & H. D. WAHLQUIST. 1975. Gen. Rel. Grav. **6:** 439.
4. Resonant Gravitational Wave detectors are being developed at Louisiana State University, University of Maryland, The University of Moscow, The Peoples Republic of China (Peking and Canton), The University of Regina, The University of Rochester, The University of Rome, Stanford University, The University of Tokyo, and the University of Western Australia.
5. HIRAKAWA, H., K. TZUBONO & M.-K. FUJIMOTO. 1978. Phys. Rev. D **17:** 1919.
6. BONIFAZI, P., V. FERRARI, S. FRASCA, G. V. PALLOTTINO & G. PIZZELLA. 1978. Data analysis algorithms for gravitational wave experiments. Nuovo Cimento. **1c:** 465.
7. PAPOULIS, A. 1965. Probability, Random Variables, and Stochastic Processes. McGraw-Hill. New York.
8. AMALDI, E., P. BONIFAZI, F. BORDONI, C. COSMELLI, V. FERRARI, U. GIOVANARDI, I. MODENA, G. V. PALLOTTINO, G. PIZZELLA & G. VANNARONI. 1978. Measurements with a small ($M = 24.4$ kg) cryogenic gravitational wave antenna at $T = 1.5$ K. Nuovo Cimento. **1c:** 341.
9. AMALDI, E. & G. PIZZELLA. 1979. Search for Gravitation Waves. *In* Albert Einstein, 1879–1979. F. De Finis, Ed. Academic Press. New York.
10. AMALDI, E., C. COSMELLI, S. FRASCA, I. MODENA, G. V. PALLOTTINO, G. PIZZELLA, F. RICCI, P. BONIFAZI, F. BORDONI, V. FERRARI, U. GIOVANARDI, B. IAFOLLA, B. PAVAN, S. UGAZIO & G. VANNARONI. 1978. Initial operation of the $M = 390$ kg cryogenic gravitational wave antenna. Nuovo Cimento. **1c:** 497.
11. AMALDI, E., P. BONIFAZI, V. FERRARI, G. V. PALLOTTINO & G. PIZZELLA. 1978. Main Features and Performance of the $M = 390$ kg Cryogenic Gravitational Wave Antenna. Internal Report of the Instituto di Fisica of the University of Rome, January 1979.
12. DOUGLASS, D. H., R. Q. GRAM, J. A. TYSON & R. W. LEE. 1975. Phys. Rev. Lett. **35:** 480.
13. BILLING, H., P. KAFKA, K. MAISCHBERGER, F. MEYER & W. WINKLER 1975. Lett. Nuovo Cimento **12:** 11.
14. The transducer noise and the back reaction noise have been assumed to be statistically independent, which is not true in all measurement schemes. Furthermore the noise of the amplifier is ignored in this presentation.
15. MCGUIGAN, D. F. *et al.* 1978. J. Low Temp. Phys. **30:** 621; BRAGINSKY, V. B. *In* Proc. Accad. Naz. Lincei, International Symp. on Exp. Gravitation, Pavia, (September 1976).
16. TESCHE C. D. & J. CLARKE. 1977. J. Low Temp. Phys. **29:** 301.
17. BRAGINSKY, V. B. *et al.* 1978. Pis'ma Zh. Eksp Teor. Fiz. **27:** 296; THORNE, K. S. *et al.* 1978. Phys Rev. Lett. **40:** 667; UNRUH, W. G. 1978. Phys. Rev. D **18:** 1764.
18. THORNE, K. S., *et al.* Proceedings of "Sources of Gravitational Radiation: A Summer Workshop." Seattle, Washington, July 24, 1978.
19. WINKLER W. 1977. Proc. Symp. Experimental Gravitation, Pavia, Italy, September 17–20, 1976. B. Bertotti, Ed. Atti Convegni Lincei **34:** 351.
20. DREVER, R. W. P., J. HOUGH, W. A. EDELSTEIN, J. R. PUGH & W. MARTIN. 1977. Proceedings of the International Symposium on Experimental Gravitation, Pavia, 1976. Accademia dei Lincei. Roma. p. 365.
21. PEGORARO, F., L. A. RADICATI, PH. BERNARD & E. PICASSO. 1978. Phys. Lett. **A68:** 165.
22. MASHOON, B. 1978. Astrophysic. J. **223:** 285.

EVOLUTION AND EXPLOSION OF MASSIVE STARS*

Thomas A. Weaver

University of California
Lawrence Livermore Laboratory
Livermore, California 94550

S. E. Woosley†

Board of Studies in Astronomy and Astrophysics
University of California at Santa Cruz
Santa Cruz, California 95064
and
University of California
Lawrence Livermore Laboratory
Livermore, California 94550

INTRODUCTION

The evolution and final fate of massive stars ($10\ M_\odot \lesssim M \lesssim 100\ M_\odot$) has long been postulated to play a central role in such important astrophysical phenomena as the origin of the chemical elements,[1,2] supernova explosions,[3,4] black holes,[5] neutron stars,[6] and cosmic rays[7]—many of which have already been the subject of vigorous discussion during this conference. In the past, these phenomena have tended to be treated separately and related to each other and to observations by idealized parameterizations that, while offering a good deal of qualitative insight into what might happen, can be criticized as isolated "fits" rather than self-consistent, quantitative explanations of such events.

This talk is mainly concerned with our attempt[8,9]—within the context of current uncertainties involving supernova core physics, 2d-effects, convection, mass loss, and nuclear reaction rates—to build self-consistent evolutionary models of complete massive stars, starting from their observable zero-age main sequence configurations, and evolving through their various hydrostatic nuclear burning stages, iron core collapse, bounce, outward-going shock formation, and finally, explosive nucleosynthesis and supernova light curve formation. Previous attempts in this direction, notably by Arnett,[10–15] Iben,[16,17] and others,[18–22] while pioneering many of the required tools and concepts, have been limited either to the early nuclear burning stages or to the treatment of simplified stellar cores. Our model[8,9] incorporates implicit hydrodynamics with a new treatment of time-dependent convection and semiconvection, and a careful treatment of the complexities of the advanced stages of nuclear burning. This model has been used to completely evolve Population I stars of 15 and 25 $M_\odot$, with calculations for other masses and compositions in progress. We find that these models, even though they neglect mass loss, rotation, magnetic fields, and other two-dimensional effects, yield excellent agreement with the observed properties of type-II

*Work performed under the auspices of the U.S. Department of Energy by the UCLLL under contract number W-7405-ENG-48.

†Work performed in part under NSF contract No. AST-76-10933. Present address is University of California at Santa Cruz.

0077–8923/80/0336–0335 $01.75/1 © 1980, NYAS

TABLE 1
NUCLEAR BURNING STAGES OF 25 $M_\odot$ STAR

Burning Stage	Temperature	Density	Timescale
Hydrogen	5 keV	5 g/cc	7×10^6 years
Helium	20 keV	700 g/cc	5×10^5 years
Carbon	80 keV	2×10^5 g/cc	600 years
Neon	150 keV	4×10^6 g/cc	1 year
Oxygen	200 keV	10^7 g/cc	6 months
Silicon	350 keV	3×10^7 g/cc	1 day
Collapse	600 keV	3×10^9 g/cc	seconds
Bounce	3 MeV	10^{14} g/cc	milliseconds
Explosive	100–600 keV	varies	.1–10 seconds

supernovae while predicting the ejection of newly created elements in the mass range from oxygen to iron (and quite possibly beyond) in ratios close to those observed in the solar system,[23] but with greatly enhanced abundances relative to hydrogen. Further, these models allow us to make a number of detailed predictions about supernova remnant abundances, γ-ray lines, supernova fluid velocities, etc., that hopefully will motivate more comprehensive observations of massive stars and supernovae.

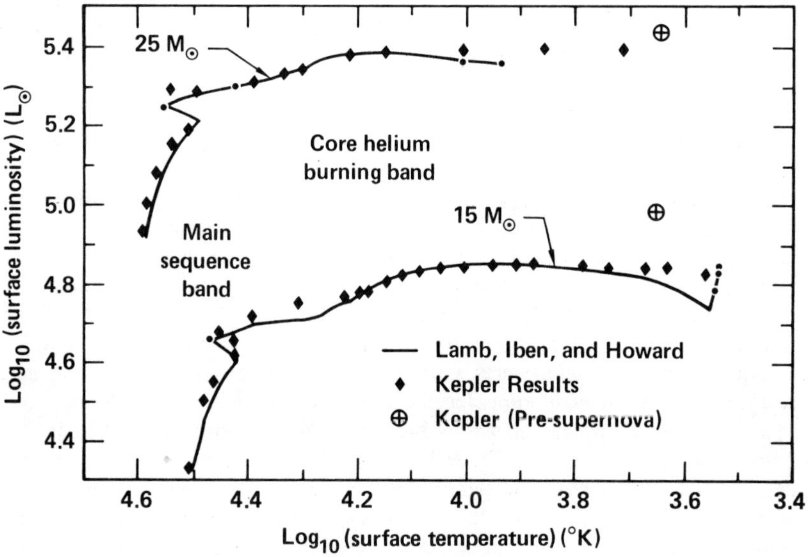

FIGURE 1. Evolutionary paths of the present 15 $M_\odot$ and 25 $M_\odot$ model stars in the H–R diagram (diamonds) compared with the results of Lamb, Iben, and Howard[17] (solid lines). Virtually all of the evolutionary motion of these stars in the H–R diagram occurs before the onset of core carbon burning. The final post carbon burning (i.e., presupernova) location is shown as a circled cross for each present stellar model.

PRESUPERNOVA EVOLUTION

TABLE 1 presents the temperatures, densities, and overall timescales that characterize the various presupernova nuclear burning stages for a 25 $M_{\odot}$ Population I star. It is evident that the timescales of the successive burning phases shorten dramatically, largely due to the overwhelming neutrino losses that occur at the temperatures and densities necessary to burn the higher-Z elements. As shown in FIGURE 1, massive stars typically begin their main sequence lives as blue giants that slowly increase in radius and luminosity as they burn hydrogen and then expand dramatically at nearly constant luminosity into blue, yellow, and finally red supergiants as they burn and exhaust helium in their cores. At this stage, about ⅔ of the star's mass resides in an envelope of nearly constant 10^{-8} g/cc density, and 10^5 K temperature. This envelope surrounds a $\sim 3 \times 10^{10}$ cm radius mantle containing a carbon–oxygen core which rapidly contracts to a central density above 10^5 g/cc and a central temperature $\sim 7 \times 10^8$ K, at which point carbon burning can commence. By that time neutrino losses directly from the star's core have displaced optical radiation loss from its surface as the dominant source of energy loss that nuclear reactions must make up to keep the star in thermal equilibrium. Indeed, the subsequent evolution takes place so rapidly compared with the thermal relaxation time of the envelope that the outward appearance of the two stars remains fixed at the positions in the H–R diagram (shown by the circled crosses in FIGURE 1) that they have reached during carbon burning. Thus, unless one has a neutrino-sensitive detector, the first indication one has of the complex series of nuclear reactions that take place during the last few thousand years of the life of massive stars [and which are quite probably responsible for the production of most of the heavy ($Z \geq 8$) chemical elements now present in the universe] is when the shock wave from the collapsing, bouncing core strikes the surface resulting in the characteristic type-II supernova explosions discussed below.

The subsequent evolution of the core is the story of a complex balance between neutrino losses, degeneracy pressure, nuclear reactions, and time-dependent convection, with the core losing its battle against gravitational collapse at an ever more rapid rate, as shown in FIGURE 2. The temporary drops in core temperatures occur when a given fuel has been exhausted in the star's center while a nuclear burning shell still remains active within the inner Chandrasekhar mass (i.e., the maximum mass of material which can support itself on degeneracy pressure alone). When the current nuclear fuel has been exhausted in this entire inner region, as well as in any exterior region to which it is joined by convective mixing, gravity overwhelms the core's residual degeneracy pressure, and it contracts and heats until the next fuel is ignited. Complex off-center burning shells also develop during neon, oxygen, and silicon burning due to composition gradients and the density dependence of the neutrino loss rate which often favors ignition at lower densities than those which occur at the innermost part of a fuel shell.

PRESUPERNOVA STRUCTURE

FIGURES 3 and 4 show the structure of our 25 solar mass model at a point early in iron core collapse when peak collapse velocities have nearly reached ~1000 km/sec, and are rapidly accelerating. Additional detail and characteristics of a 15 $M_{\odot}$

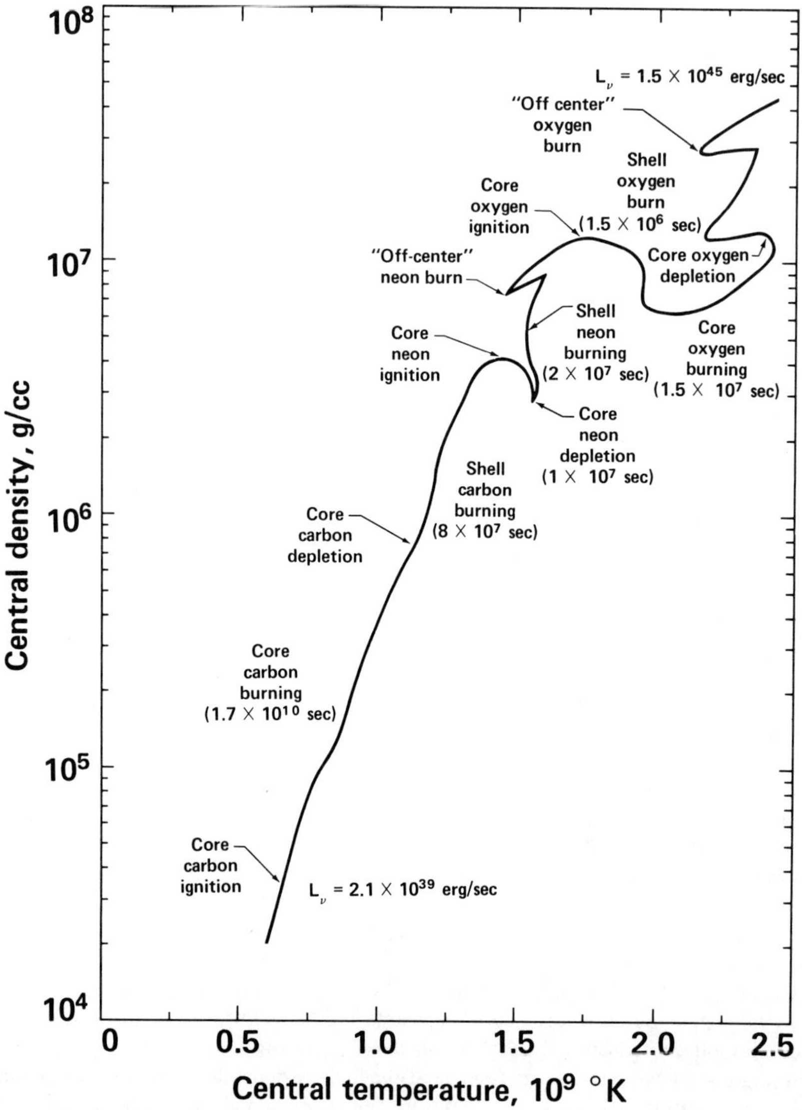

FIGURE 2. Thermodynamic history of the center of a 25 $M_\odot$ Population I star during carbon, neon, and oxygen burning. Burning characteristics and elapsed time spent during each phase are given along the evolutionary path and the dramatic increase in the star's total neutrino luminosity (L_ν) during these phases is noted. "Off-center" shell burning, in which the contracting shell of nuclear fuel first ignites at a location outward of its inner boundary due to the large density dependence of the neutrino losses, is a common occurrence as shown. This results in temperature and composition inversions that are typically transient.

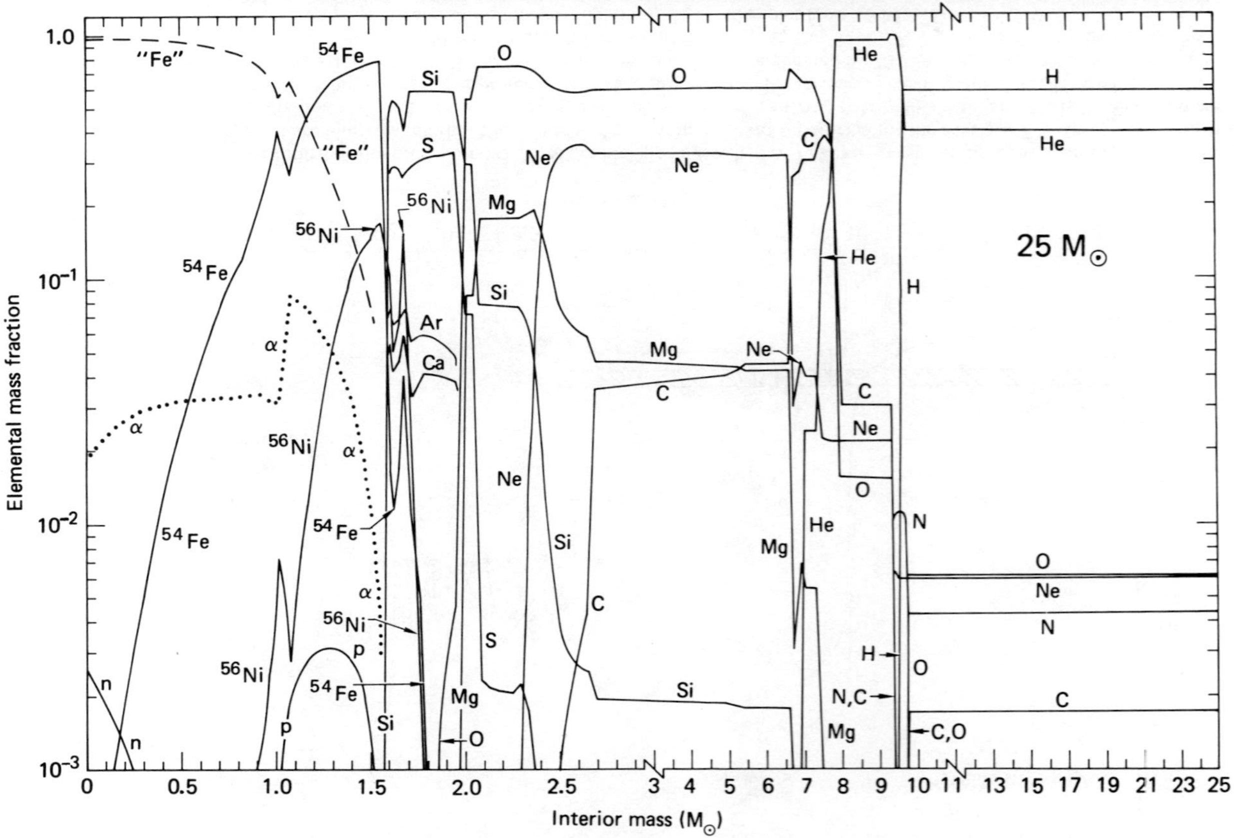

FIGURE 3. Presupernova composition profile of a 25 $M_\odot$ Population I star as a function of the interior mass coordinate as given by WZW.[8] The structure is shown at the point at which collapse velocities in the iron core have nearly reached 1000 km/sec and are rapidly increasing. In the region interior to 1.61 $M_\odot$ where a 131-isotope quasiequilibrium nuclear-burning network was used to treat quasistatic silicon burning and neutronization, the curve labeled ^{56}Ni includes minor contributions from other $A \approx 2Z$ iron peak nuclei; the curve labeled ^{54}Fe includes those iron peak isotopes with $A = 2(Z + 1)$, while that labeled "Fe" includes all other isotopes with $Z \geq 22$ (e.g., ^{56}Fe, ^{58}Fe, and major contributions from highly neutronized iron peak species such as ^{48}Ca, ^{66}Ni, ^{50}Ti, etc.). For additional details and a similar plot for a 15 $M_\odot$ star, see WZW[8].

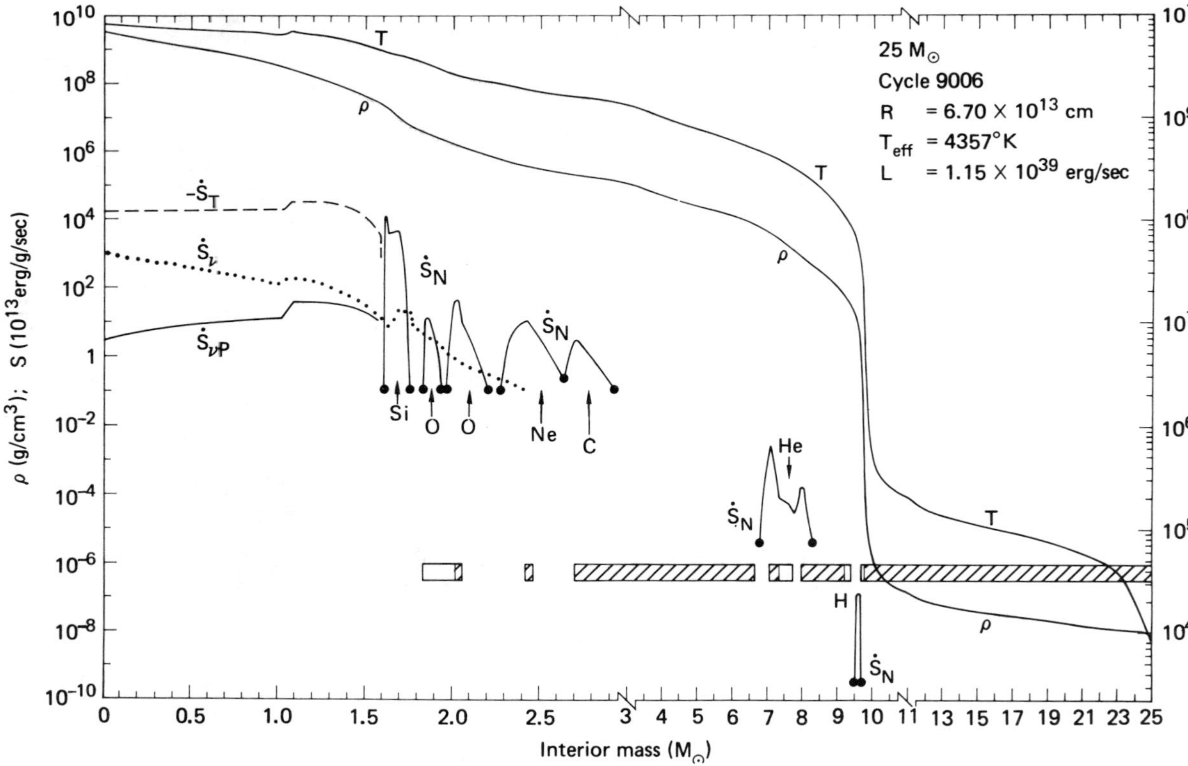

FIGURE 4. The thermodynamic structure of the presupernova 25 $M_\odot$ model star described in FIGURE 3 is shown as a function of mass fraction. Note that the density and temperature profiles are plotted such that the curves will maintain constant separation for the case $\rho \propto T^3$. Here $-\dot{S}_T$ is the total local energy loss rate due to both neutrino emission and nuclear photodisintegration, $\dot{S}_\nu$ is the total neutrino energy loss rate, and $\dot{S}_{\nu P}$ is the neutrino energy loss rate due to the thermal plasma processes. The nuclear energy generation rate profiles for the various nuclear burning shells are labeled $\dot{S}_N$, and the principal nuclear fuel is indicated. All energy generation and loss rates share the common scale denoted by "S." Active convective regions are indicated by striped bars, while semiconvective and convectively neutral regions are shown as outlined bars. In this figure, R, T_{eff}, and L denote the photospheric radius, effective temperature, and optical luminosity, respectively. For additional details and a similar plot for a 15 $M_\odot$ star, see WZW.[8]

presupernova star are given by Weaver, Zimmerman, and Woosley[8] (henceforth WZW). It is evident that the mantle of this star has evolved into an "onionskin" structure with distinct H, He, C, Ne, O, and S burning shells. These shells are typically convective, but are separated by density gradients sufficiently large to prevent convective mixing between them. Such a presupernova structure recapitulates the various evolutionary stages experienced by the star during its hydrostatic evolution, as one proceeds from the core towards the surface.

Hydrostatic silicon burning occurs over a sufficiently long timescale that the iron peak elements that result are substantially neutronized by electron capture during the burning itself. The resulting "iron" cores have masses of 1.56 $M_\odot$ and 1.61 $M_\odot$, and central electron abundances of 0.427 and 0.439 moles/g, respectively, for 15 and 25 $M_\odot$ stars. The composition is dominated by very neutron-rich species such as ^{48}Ca, ^{66}Ni, ^{54}Cr, and ^{50}Ti near the center and gradually shifts through progressively less neutron-rich species, dominated successively by ^{58}Fe, ^{56}Fe, and ^{54}Fe, as one proceeds outward. The region around $Y_e = 0.46$ where ^{56}Fe is the dominant iron peak constituent (as it is observed to be in the solar system) covers at most a few tenths of a solar mass, while non-neutronized iron ($Y_e = 0.50$, i.e., ^{56}Ni or $^{54}Fe + 2p$), which beta decays naturally to the observed iron abundances, is present only in the narrow flashing region of the 25 $M_\odot$ star's silicon shell located at an interior mass coordinate of 1.65 $M_\odot$ (see FIGURE 3). There exist discontinuities in composition and Y_e at an interior mass of approximately 1 $M_\odot$ which are fossil remnants of the outermost extent of the central convective region during core silicon burning. Similarly, there is an abrupt fall in Y_e at the edge of the iron core corresponding to the outermost extent of the convective region associated with shell silicon burning. It is important to note that this convective mixing has allowed the iron cores to grow substantially beyond the Chandrasekhar mass ($\approx$1.2–1.3 $M_\odot$ for the current Y_e values) before collapse. The general character of these results appears relatively insensitive to the choice of electron-capture rates.

Core Collapse and Outward-Going Shock Wave Formation

It is apparent from FIGURE 4 that photodisintegration of the iron-peak elements to alpha particles provides the dominant source of core energy loss during the early stages of collapse. Neutrino losses due to neutronization and thermal plasma processes[24] are about a factor of 30 and 1000, respectively, less important, and the Y_e profile has not yet changed significantly from its hydrostatic value. The collapse of the iron core by these processes is a natural result of the preceding evolution and is in no way artificially induced.

These collapsing presupernova models were used as input to Wilson's detailed supernova core code[25–28] and resulted in explosions or not, depending on the details of his assumptions about the physical processes taking place in the core. He and Arnett will discuss the current status of our understanding of these processes later this morning. In our present work we have sought to avoid entanglement in the very complex physics and attendant considerable uncertainties involved in current core calculations (including the probable importance of two-dimensional rotational effects), and have concentrated our attention on *observable* phenomena such as light

TABLE 2
EXPLOSION ENERGIES FOR SEVERAL SUPERNOVA MODELS

Model	Mass ($M_\odot$)	Explosion Energy (10^{51} erg)	Model	Mass ($M_\odot$)	Explosion Energy (10^{51} erg)
15A	15	1.3	25A	25	1.4
15B	15	3.3	25D	25	4.0

curves, velocities, and nucleosynthesis which we find to be relatively insensitive to the exact manner in which the core behaves, as long as it generates a sufficiently strong outgoing shock wave to eject the mantle and envelope ($\sim 10^{51}$ erg). Because of the widely differing response times of the core and the rest of the star, such a decoupling can be made by replacing the core with a time-dependent inner boundary condition (including an outward neutrino flux). This inner boundary condition acts like a piston which first falls in, then accelerates sharply outward to drive a shock wave into the mantle, and finally generates a rarefaction wave behind the shock as it falls back to the surface of the newly formed neutron star. Such inner boundary conditions are either taken from various (exploding) calculations of Wilson and Bowers[28] based on our initial models, or represent parameterizations of their results to give differing shock energies. Wilson will discuss examples of such calculations in the following presentation. The total explosion energies resulting for the four supernova models we will consider in this talk are given in TABLE 2. In each case, the explosion energies refer to the sum of all kinetic and electromagnetic (but not neutrino) energies released by the supernova. Sensitivity tests in which approximately the same net energy was deposited into the mantle by differing neutrino fluxes and piston trajectories (as might occur, for example, due to different bounce densities) showed no substantial differences in nucleosynthetic yield, final velocity profiles, or photometric properties, suggesting that net explosion energy alone is a sufficiently accurate characterization of core processes for our present purposes. A possible exception to this involves the position of the mass cut, which is somewhat obscured in the present calculation by the necessity of taking piston positions near the edge of the collapsing core where neutrino flux attenuation is small, and densities are within the range of our model. It will be argued below, however, that on the basis of nucleosynthetic, supernova light curve, and neutron star observations, the mass cut very probably occurs in the sharp density gradient near the edge of the neutronized core, as we have assumed.

EXPLOSIVE NUCLEOSYNTHESIS

The shock wave generated by the bounce of the core propagates out through the rest of the star, temporarily heating the material to temperatures sufficient to cause substantial nuclear processing in the silicon, oxygen, and neon layers just above the core. The extent of the nuclear processing which occurs is limited by the rapid, nearly adiabatic expansion of the shock-heated material, which, together with the momentum initially deposited by the shock, serves to accelerate the matter beyond its escape velocity. For the range of core explosion energies studied ($1\text{–}5 \times 10^{51}$ erg), this shock is not particularly strong with respect to the energy density of the presupernova mantle. Indeed, shock Mach numbers of about 2 and compressions of 3–4 are typical—in contrast to the sevenfold compression which occurs in the high shock-energy limit. This circumstance implies that most of the energy in the immediately post-shock material resides as internal instead of kinetic energy. Another interesting feature of the explosion is that the density and temperature in the mantle at any given instant is roughly constant behind the shock. These characteristics allow the shock to be thought of with reasonable accuracy as an expanding bubble of radiation containing most of the energy of the core explosion. The temperature (T_s) that a given zone in the mantle reaches is then just determined by its *presupernova radius* (r_0) and the total internal energy (E_0) by the relation:

$$T_s(r_0) \approx \left[\frac{3E_0}{4\pi r_0^3 a}\right]^{1/4} \quad (1)$$

where a is the radiation energy density constant. Peak shock temperatures are shown in FIGURE 5 for the weakest and strongest 25 $M_\odot$ explosions considered, and the energy/radius scaling is found to be approximately as predicted by Equation 1. During the few tenths of a second for which these temperatures typically persist, material heated above ~4 × 10^9 K (generally initially silicon or oxygen) will be

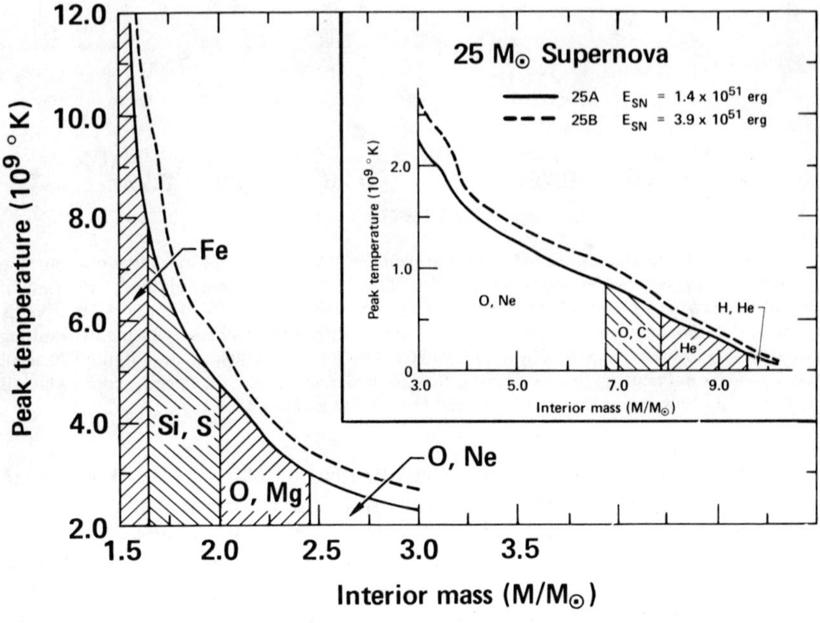

FIGURE 5. Peak explosion temperatures as a function of mass fraction and pre-explosion composition for the 25 $M_\odot$ model supernovae given in TABLE 2. The peak temperature experienced in each zone scales roughly as the fourth root of the explosion energy as is expected from the simple radiation-dominated model described in the text. Substantial explosive burning of neon, oxygen, and silicon are expected to occur but a significant amount of hydrogen, helium, or carbon burning appears unlikely.

processed into non-neutronized iron peak elements[29] (predominantly radioactive ^{56}Ni), while peak temperatures between 3 and 4 × 10^9 K will produce elements from silicon to calcium,[29] typically from material originally rich in oxygen or neon. Peak shock temperatures between 2 and 3 × 10^9 K, which mainly induce production of elements between oxygen and phosphorous, have traditionally been associated with the explosive burning of carbon.[30,31] We find, however, that the zones that reach such temperatures in our massive star models are instead composed mainly of neon and oxygen with only a trace of carbon. The zones experiencing "explosive neon burning" are particularly interesting not only because they have gone essentially unstudied to date, but also because they appear to offer an ideal site for a pre-explosive hydrostatic

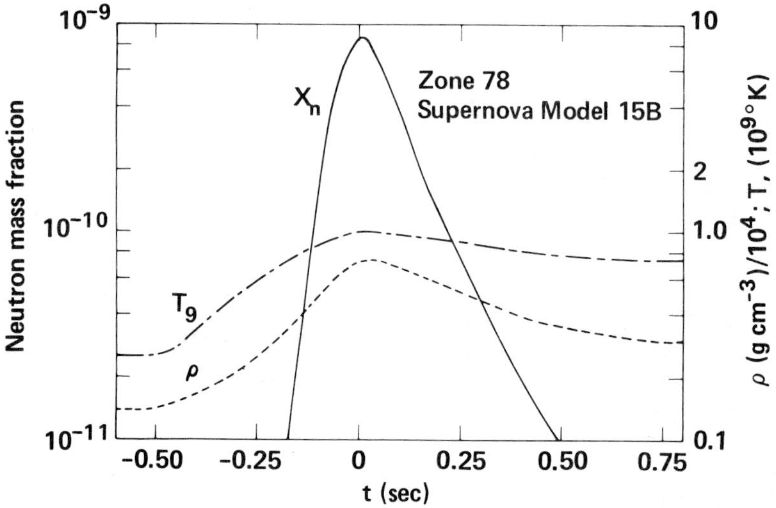

FIGURE 6. Thermodynamic conditions and neutron flux experienced by the helium-burning shell during the supernova explosion of a 15 $M_\odot$ star. Time is measured relative to the time of shock wave passage. The initial composition of this zone contained 4.5% ^4He and 1.2% ^{22}Ne by mass with the remainder in the form of ^{12}C, ^{16}O, and a distribution of heavier elements resulting from a pre-explosive *s*-process. These explosion characteristics will lead to a limited form of *r*-process. The total neutron fluence is insensitive to the ρ and T_9 risetime during shock (which is here artificially slow due to the use of artificial viscosity to mediate the shock).

s-process,[32] as well as for the photodisintegration-based "*p*"-process of Woosley and Howard,[33] which produces proton-rich nuclei from *s*-process seed at the 2–3 billion degree temperatures characteristic of explosive neon burning. In addition, sufficient neutrons are released in the explosive burning of some neon zones to induce a limited *r*-process.[33,34] The mix of isotopic products from these various competing processes is a sensitive function of the peak temperature. It is thus tempting indeed to tentatively identify explosive neon-burning zones as the source of many *r*- and *p*-process isotopic anomalies that have recently been observed by Wasserberg and his collaborators[35–37] in inclusions in the Allende meteorite. Further, our calculations show these "neon" zones to contain about 1% aluminum after shock wave passage. About one part in 10^3 of this aluminum is in the form of radioactive ^{26}Al ($\tau_{1/2} = 7.3 \times 10^5$ yr) whose fossil remains as an anomalous component of ^{26}Mg have also been observed in many Allende samples. While ^{26}Al is also produced in substantial quantities relative to ^{27}Al in the late phases of hydrostatic carbon burning and at the end of hydrogen burning (via the $^{25}Mg(p, \gamma)^{26}Al$ reaction),[9,38] the total *mass* of ^{26}Al synthesized in these latter two cases is small compared to the production in the "neon" zones.

For peak temperatures below about 2×10^9 K, the shocked material is largely ejected without substantial explosive nuclear processing. An exception occurs at the inner edge of the helium shell where peak temperatures in the vicinity of 10^9 K and densities near 10^4 g cm^{-3} are sufficient to generate an intense neutron pulse that lasts for about half a second (see FIGURE 6). This flux is sufficient to drive a "mini-r" process[39,40] that shifts iron-peak isotopes upwards by several mass units and produces even heavier r-process elements from the enhanced s-process seed already present in the region. While the details of this process are currently being worked out in collaboration with Blake and Schramm,[41] it appears possible to produce substantial amounts of r-process isotopes having masses between 56 and 80 by this process. Such isotopes tend to be underproduced in the higher neutron-flux conditions required for the classical (as yet siteless) r-process that produces heavier isotopes.[42]

Judgment of any nucleosynthetic study must ultimately rest on how well the overall ratios of elements and isotopes it produces reproduce the observed solar abundances when averaged over a suitable distribution of objects. The final composition profiles of the ejected ashes of 15 and 25 $M_\odot$ stars are shown in FIGURE 7, and are to be compared to the presupernova composition profiles given in FIGURE 3 and WZW.[8] It is apparent that explosive processing is primarily responsible for the elements between silicon and iron, while those from oxygen to magnesium are formed hydrostatically and have abundances largely unmodified by explosive ejection. It is also evident that 25 $M_\odot$ stars produce about a factor of 5 greater mass per star of $Z > 2$ ashes than do 15 $M_\odot$ stars. Indeed, the mass *fraction*, Z_{ej}, of the high-Z ashes in stars we have studied in the 15–50 $M_\odot$ range can be well-fitted by the function

$$Z_{ej} \approx 0.5 - \frac{6.3}{M} \tag{2}$$

where M is the mass of the star in solar masses. When this fraction of high-Z products is multiplied by the stellar birth rate as a function of mass as compiled by Scalo and Miller,[43] the resulting curve (shown in FIGURE 8) indicates the relative average contribution to nucleosynthesis of stars in a given mass range (see also references 44 – 46). While this curve peaks at 18.5 $M_\odot$, its *median* value is 29 $M_\odot$, a number fairly insensitive to the choice of upper-mass cutoff. Roughly speaking then, one might expect a star of about 25 $M_\odot$ to be a "typical nucleosynthetic object" (at least among massive stars which undergo iron-core-collapse-induced supernova explosions). Thus it is very pleasing that the elements in the mass range between oxygen* and iron that are produced and ejected by the 25 $M_\odot$ supernova are enhanced by a large and almost uniform factor of 25 relative to hydrogen, within the factor-of-two errors in the solar abundance determinations[23] [as shown in FIGURE 9 (top)]. Indeed, the solar abun-

*Note that carbon and nitrogen are expected to be produced copiously in lower-mass stars,[49] so that their underproduction here does not pose a problem.

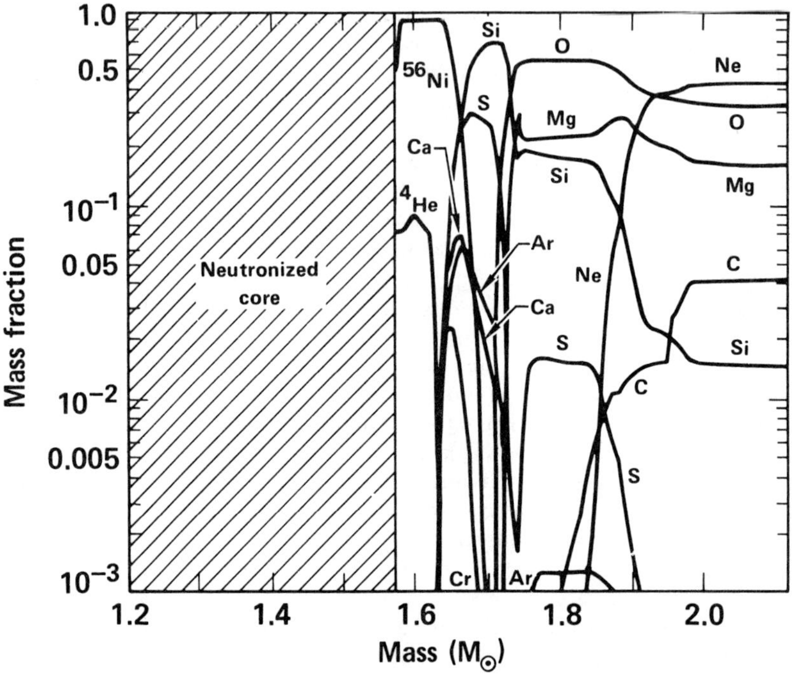

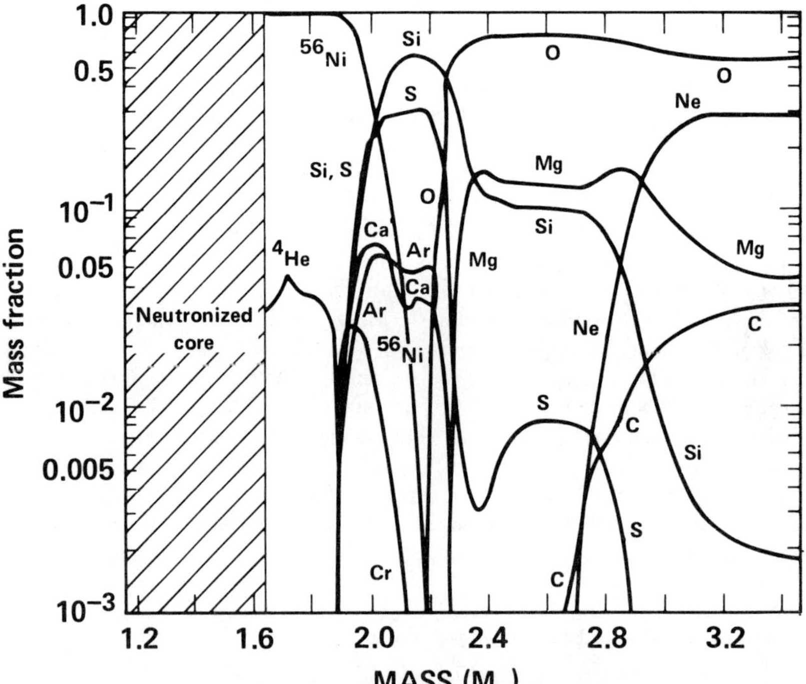

See caption on following page.

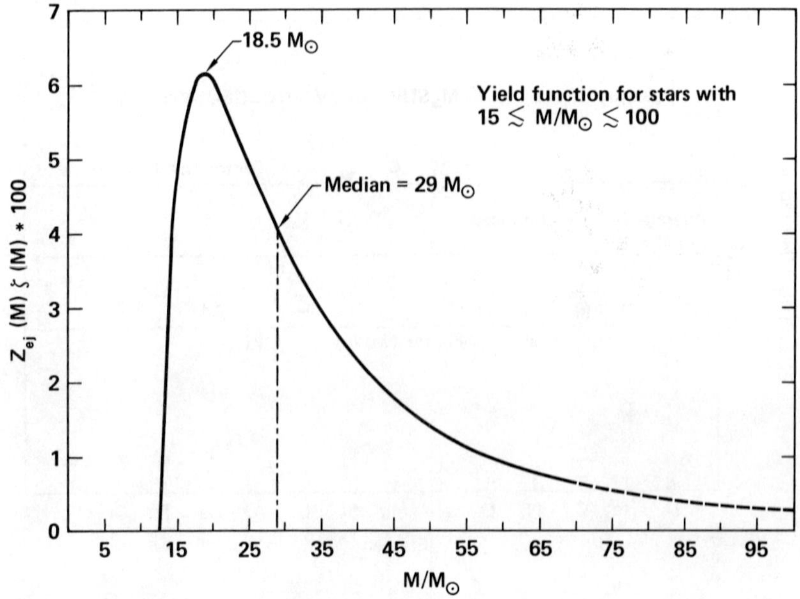

FIGURE 8. The product of the Miller–Scalo[43] stellar birth function and the Weaver–Woosley[9] supernova yield function as a function of stellar mass.

dance of Ne, the most discrepant element, is difficult to determine experimentally. Recent planetary nebulae observations by Kaler[47] suggest a value 50% higher than Cameron[23] gives for the solar system. This would have the effect of lowering the points shown for Ne in FIGURE 9 (top) by 50%, bringing it into excellent agreement with the other elements. As expected, the overall enhancement factors for the 15 $M_\odot$ star shown in FIGURE 9 (bottom) bear considerably less resemblance to the solar abundances, and that star is found to be dominantly enhanced in neon, magnesium, and silicon. Note, however, that many elemental ratios in the 15 $M_\odot$ star are still quite close to their solar values.

These figures also serve to demonstrate that the energy of a supernova explosion has relatively little effect on its net nucleosynthetic yield. The main effect of increased energy is to shift the regions experiencing explosive neon and oxygen burning outward in the star while leaving their total masses relatively unchanged, resulting in more neon burnup, increased iron production, and other small changes. This nucleosyn-

FIGURE 7. The final elemental composition of the 15 $M_\odot$ (top) and 25 $M_\odot$ (bottom) post-supernova models having explosion energies of 1.3×10^{51} erg (Model 15A) and 1.4×10^{51} erg (Model 25A), respectively, is shown as a function of the interior mass coordinate. The term "neutronized core" refers to material which contains such large concentrations of rare neutron-rich isotopes (e.g., ^{58}Fe, ^{48}Ca) that only a small fraction could, on the average, be ejected into the interstellar medium. For mass zones exterior to those shown in the figure, the pre-explosive composition (ashes of hydrostatic-burning phases) is ejected into the interstellar medium virtually unmodified.

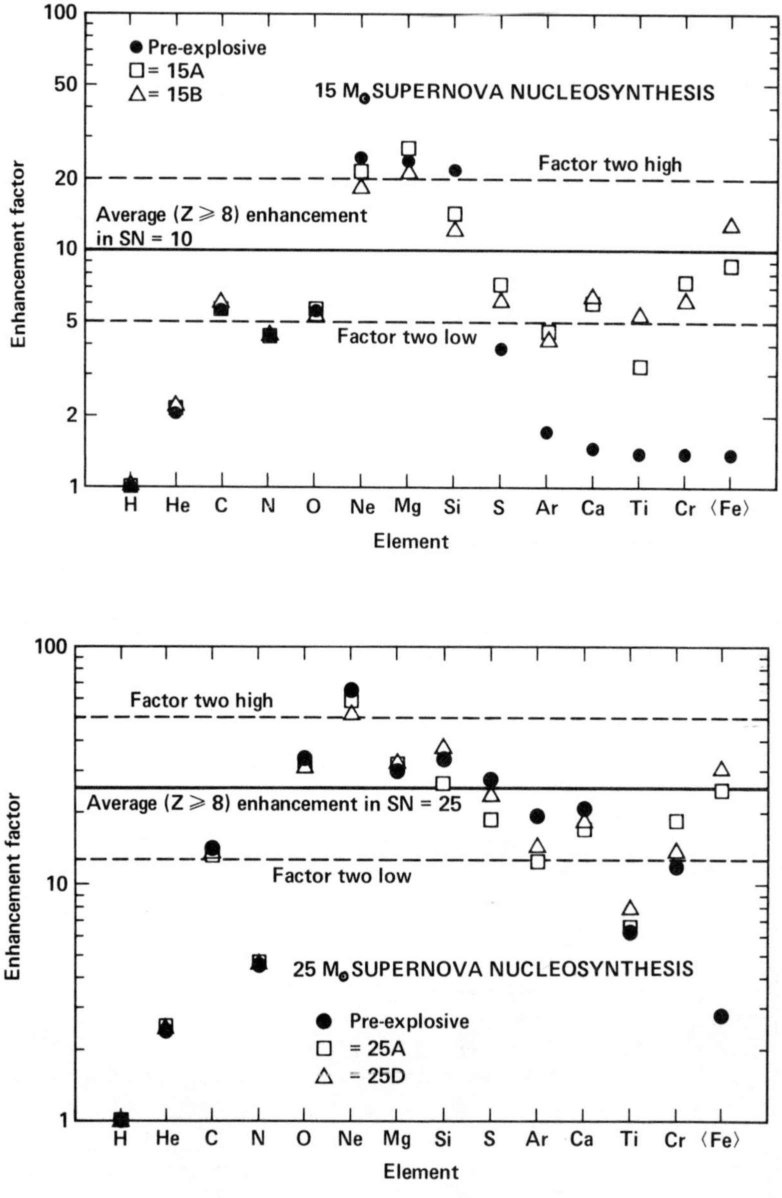

FIGURE 9. The enhancement factor, defined as the mass of an element in all material external to the neutronized core divided by the mass of hydrogen and normalized to a similar ratio in the solar system,[23] is given as a function of mass for 15 $M_\odot$ (top) and 25 $M_\odot$ (bottom) Population I supernovae. Black circles represent enhancement factors in the pre-supernova star at the time of core collapse. Open squares and triangles represent the nucleosynthesis of relatively low-energy (15A, 25A) and high-energy (15B, 25D) explosions, respectively.

thetic insensitivity to explosion energy is quite reassuring in the face of our uncertain understanding of the complex supernova core physics.

TYPE-II SUPERNOVA LIGHT CURVES

Representative electromagnetic displays produced by our two 15 $M_\odot$ supernova models are shown in FIGURE 10, compared to photometric data for Supernova 1969ℓ in NGC 1058, perhaps the best observed type-II supernova.[49,50] SN 1969ℓ is characteristic of a common subclass of type-II supernovae which shows a 2–3 month initial plateau in its visual emission ($M_v \approx -17$) followed by a rapid decline of about 2 stellar magnitudes and then a slower decline at a rate of ~3 magnitudes/year. (SN 1970g in NGC 5457 = M101 is another recent example of this subclass).[51] It is apparent that the theoretical and observational results are in excellent agreement. As FIGURE 10 shows, an increase in the energy of the core explosion produces a roughly linear increase in the optical brightness together with a shorter "plateau." Although we have chosen to focus our comparison on 15 $M_\odot$ supernovae which are expected to occur more frequently,[43] the results for our 25 $M_\odot$ models are generally quite similar (see Weaver and Woosley[9]). Note in particular that this agreement has *not* been achieved by normalizing the observational absolute magnitude so as to provide the best fit.

FIGURES 11 and 12 compare derived observational SN 1969ℓ results[49,50,52] for the temperature and radius of the supernova photosphere with the corresponding theoretical results. FIGURE 13 shows a comparison between observed absorption line velocities and photospheric velocities for the low-energy 15 $M_\odot$ explosion. The agreement between theory and observation is uniformly within the observational errors, and allows a confident description of the general physical processes which are occurring (see also references 14, 15, and 53 for conclusions based on parameterized models).

The initial sharp spike in temperature and luminosity at small photospheric radius corresponds to the breakout of the supernova shock through the surface of the star's supergiant envelope. This is accompanied in theory[9,54] by a ~2000 sec long, soft x-ray pulse with an equivalent blackbody temperature $\sim 1.5 \times 10^5$ K and a peak luminosity $\sim 10^{45}$ erg/sec. The surface is then rapidly cooled by radiative emission and hydrodynamic expansion, balanced in part by radiative diffusion from below. The photosphere expands rapidly in physical size, but slowly recedes with respect to the envelope material as the optical thickness of that material decreases due to rarefaction. After about 20 days, the photosphere has cooled to roughly 6000 K, the temperature at which hydrogen recombines at the extremely low densities ($\sim 10^{-13}$ g/cc) prevalent in the envelope. Over the next 2 months or so, a transparency wave eats its way into the envelope with the photospheric temperature remaining near 6000 K (see FIGURES 11 and 12). In doing so, material traveling at lower and lower velocities is uncovered until the speed of the cooling wave exceeds the fluid velocity, and the photosphere physically recedes until it encounters the slowly moving (<1000 km/sec), relatively dense, and very optically thick mantle. At this point a transient increase in the photospheric temperature occurs and persists for ~2–3 days due to the uncovering and rapid cooling of the hot surface of the mantle. This phenomenon is potentially observable, although non-LTE radiation effects in the overlying envelope may tend to mask it. The sharpness of this final recession and the resulting fall in luminosity is

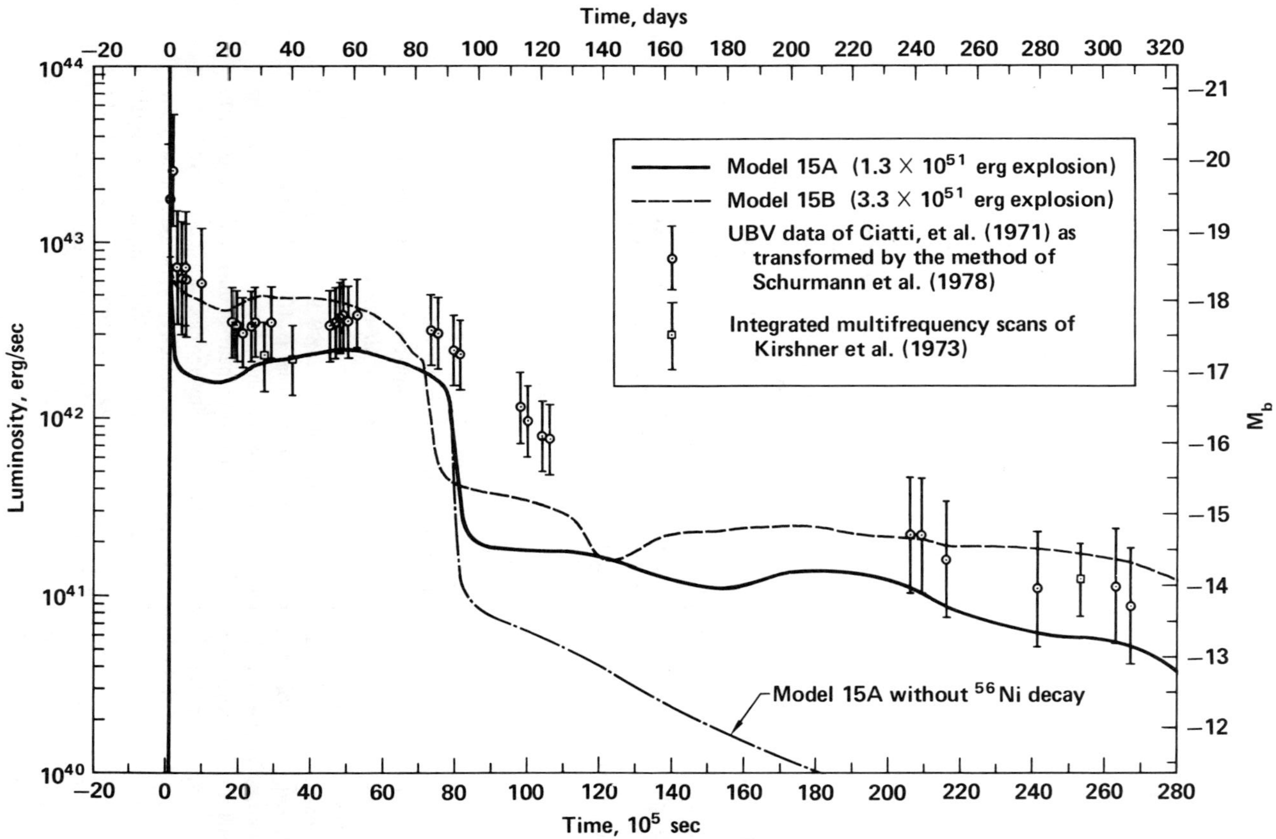

FIGURE 10. Light curve of type-II supernova 1969ℓ compared to 15 M⊙ theoretical models. Here, M_b is the absolute bolometric magnitude of the supernova and time is measured with respect to core bounce for the theoretical models and Julian day 2,440,560 for the observations. The solid and dotted lines give the bolometric luminosity for models 15A and 15B, while the data points for 1969ℓ are either derived from the UBV data of Ciatti *et al.*[49] as transformed by the method of Schurmann *et al.*[15] (open circles) or obtained by integrating the multifrequency scans of Kirshner *et al.*[50] (open squares). The dot-dashed curve shows the result when ^{56}Ni decay is artificially suppressed in model 15A.

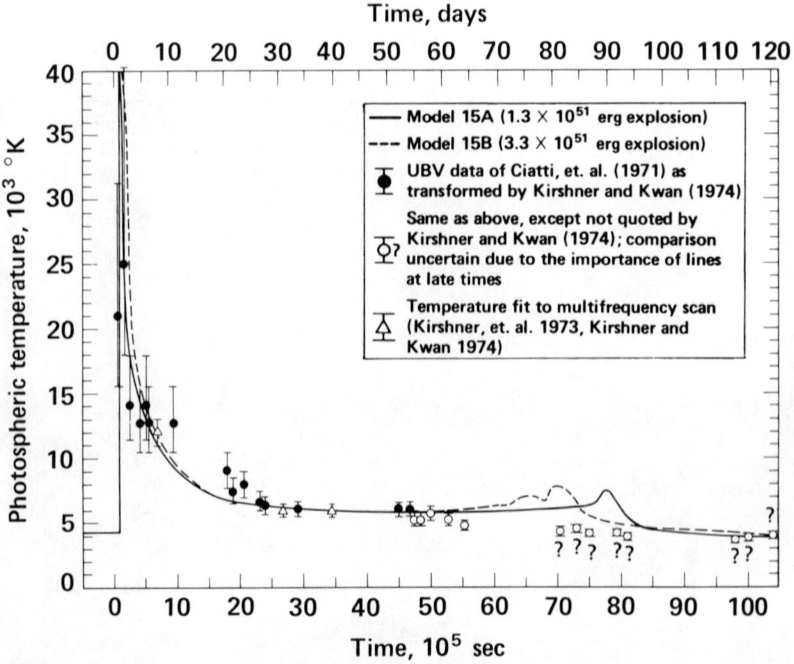

FIGURE 11. Photospheric temperature of type-II supernova 1969ℓ compared with 15 $M_\odot$ theoretical models. The solid and dotted lines give the results for models 15A and 15B, while the solid circle data points represent the data of Ciatti *et al.*[49] as transformed by Kirshner and Kwan.[52] The open circle data points are obtained by extrapolating the transformation methods of Kirshner and Kwan[52] to the late time data of Ciatti *et al.*,[49] and are associated with "?" marks where lines dominate the spectrum and such color/temperature transformations become dubious. The open triangle data points are derived from fits to multifrequency scans.[50]

probably artificially abrupt due to our relatively simplified treatment of recombination and non-LTE radiation effects.

At late times the luminosity is due to thermalized radiation from the decay of the explosively generated† ^{56}Ni ($\tau_{1/2}$ = 6.1 days) and its daughter ^{56}Co ($\tau_{1/2}$ = 78.5 days) diffusing out of the mantle. As the figures show, models with ^{56}Ni decay turned off display a much more sharply falling luminosity tail as the residual thermal energy in the mantle diffuses out over a characteristic time of only 1–2 months. In models containing energy input from radioactivity the convolution of radiation escape and ^{56}Co decay times produces a luminosity decline slower than the 78-day ^{56}Co half-life, particularly in the 25 $M_\odot$ case. [Indeed it is tempting to speculate that all supernova tails are due at late times to ^{56}Co decay with the apparent decay time either shortened by incomplete absorption of the decay γ-rays (e.g., type-I supernova) or lengthened by the temporary trapping of the thermalized decay energy by an optically-thick

†The mass of ^{56}Ni produced during explosive silicon burning is 0.08, 0.12, 0.36 and 0.43 $M_\odot$ in models 15A, 15B, 25A, and 25D, respectively.

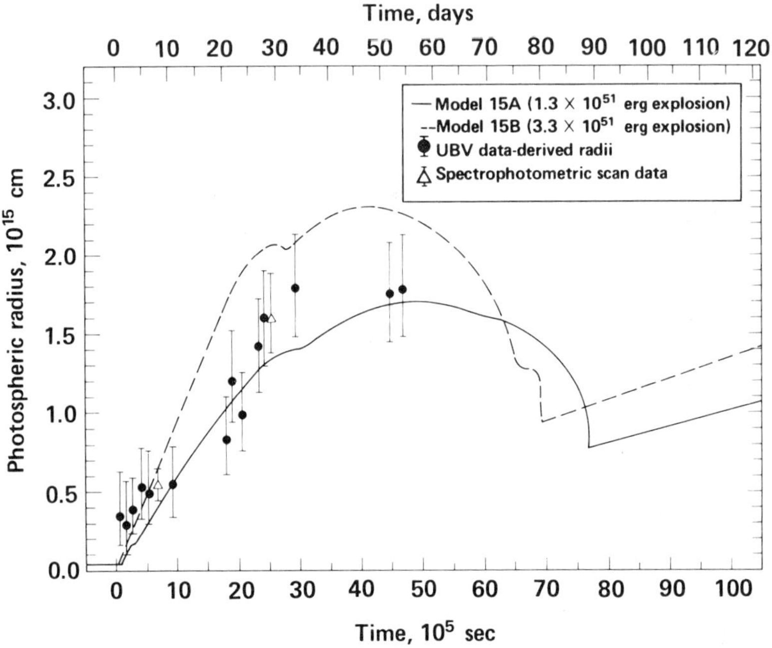

FIGURE 12. Photospheric radius of type-II supernova 1969ℓ compared to 15 $M_\odot$ theoretical models. Solid and dotted lines give the results for models 15A and 15B, while the 1969ℓ data points are either derived from UBV data[49] or multifrequency scans.[50]

envelope. (See reference 55 for related models.)] The ripples in the theoretical curves at late times are due to the (possibly numerical) formation of density clumps (see FIGURE 14 and reference 56) in the mantle (which appear to offer excellent sites for grain formation). It is likely two-dimensional instabilities will also occur[57] (as is suggested by observations of supernova remnants), which should have the effect of increasing the noise, but damping the size of transient excursions in the luminosity.

Useful information about supernova models can be obtained from their light curves. At early times ($\lesssim$3 months) the light curve principally conveys information about the structure of the envelope of the presupernova star. More massive, extended envelopes produce prolonged luminosity plateaus and slower photospheric velocities for fixed shock energies. The behavior of the tail of the light curve, on the other hand, yields information about the size and density of the mantle and, potentially, its composition. The deduced amount of ^{56}Ni may serve to indicate the position of the mass cut between the core and the mantle for a star of given mass. Additional information comes from γ-ray lines resulting from the decay of ^{56}Co and perhaps ^{60}Co in young supernovae placed in nearby galaxies. The observed flux is critically dependent on the time at which the mantle becomes transparent to γ-rays and thus on its density profile and fluid velocities. Detailed predictions of γ-ray line strengths and profiles are currently being calculated by Axelrod (at LLL)[58] based on our supernova models, and should be complete before the launch of HEAO-C.

SUPERNOVA REMNANT MASS

The above results have direct implications for the masses of the neutron star or black hole remnants likely to result from massive star evolution. In particular, as was pointed out by WZW,[8] the neutronized iron cores that form in these stars contain such large quantities of ^{54}Fe and rare highly-neutronized iron peak species that at most ~0.03 $M_\odot$ of the core's mass can be ejected on the average if massive overproduction of these isotopes is to be avoided. Further, our explanation for type-II supernova light curves suggests that the 0.1 – 0.4 $M_\odot$ of explosively generated ^{56}Ni which lies directly above the neutronized core must be at least partly ejected in order to provide the energy to produce late-time light curve tails. While additional stellar models and light curve observations are obviously needed to clarify the statistics and generality of these conclusions, the direct implication is that the initial mass of a condensed supernova remnant should fall within 0.1 – 0.2 $M_\odot$ of the mass of the presupernova iron core. (1.56 and 1.61 $M_\odot$, respectively, for our 15 and 25 $M_\odot$ models.) The mass cut thus falls quite plausibly in the steep density gradient occurring at the edge of the core, and indeed most of the calculations of Wilson[25–28] that lead to ~10^{51} erg explosions exhibit such behavior. Allowing for the 10% mass decrease due to neutrino emission, the corresponding observable neutron star remnant masses would be 1.40 and 1.45 $M_\odot$, respectively. These masses are in excellent agreement with the precisely measured masses of the components of the binary pulsar 1913 + 16 as reported by Taylor[59] to be

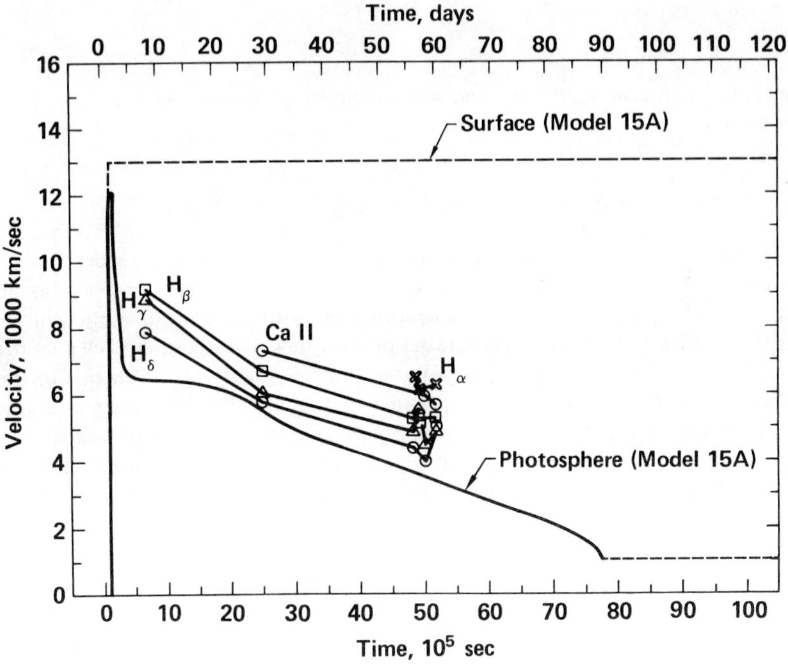

FIGURE 13. Surface and photospheric velocities as a function of time from model 15A compared with absorption lines velocities observed for supernova 1969ℓ.[50]

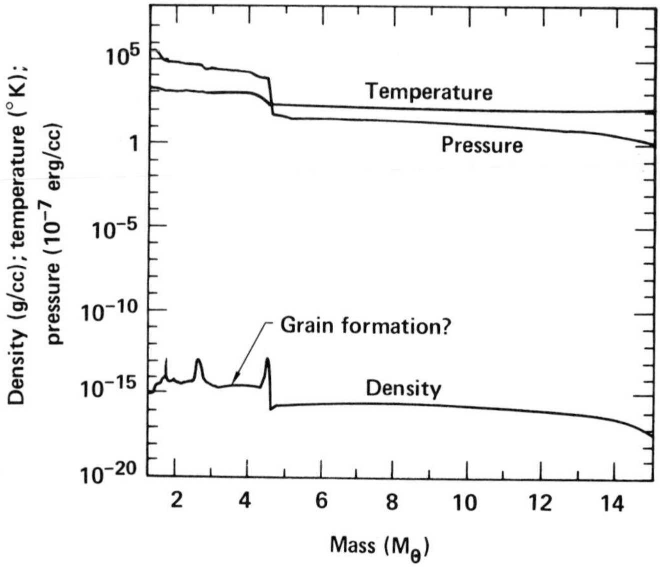

FIGURE 14. Density, temperature, and pressure plotted as a function of interior mass for model 15A at a time since core bounce (7.12×10^7 sec) when conditions for grain formation in the mantle appear most favorable.

1.39 ± 0.15 and 1.44 ± 0.15 $M_\odot$, and with other less accurate observations of neutron star masses.[60]

Conclusions and Need for Future Work

Despite the remarkable agreement of our theoretical supernova models with the relatively sparse available data for type-II explosions, it would be erroneous to assume that we now understand such phenomena in their entirety. Instead, we would hope that the detailed predictions that come out of these models concerning such quantities as photospheric temperature, velocity, radius, and luminosity as a function of time; abundances in supernova remnants and in the galaxy; and γ-line fluxes will aid in justifying (and help provide a basis for interpreting) further extensive observations of supernovae and highly-evolved massive stars. The excellent supernova search and observation program at Asiago[61] provides a promising start in this direction and deserves support. Frequent multifrequency scans of known supernovae when used in conjunction with our temperature and density predictions and taking non-LTE effects into account should yield good abundance data not only for the surfaces of these stars, but also for the underlying layers. Another exciting application of such extensive observations is that when combined with further refinements in the theory of supernova atmospheres they should allow the accurate use of supernova as both standard[61] and "nonstandard"[15,62] candles for determining distances to distant galaxies. Computerized sky searches of the kind pioneered by Colgate[63] would also be quite valuable in discovering supernovae in an early enough phase to study shock break-out and to uniformly calibrate light curves at early times. The importance of detecting and analyzing the predicted γ-lines from young (~1-year old) supernova remnants cannot be overstated. Data from the ^{56}Co decays will yield valuable information about the interior of supernovae and the "mass cut" as well as directly confirming in the strongest possible way the existence of explosive nucleosynthesis. We look forward to HEAO-C (or a galactic supernova).

On the theoretical front, we cannot claim a true understanding of supernovae until such (real) complications as mass loss, Raleigh–Taylor instabilities, rotation, magnetic fields, and uncertainties in convection theory have been properly addressed. This may not occur in our lifetimes! In the more immediate future we see as requisite extensions of our own (benchmark) work the consideration of stars of other masses and populations (only Population I 15 $M_\odot$ and 25 $M_\odot$ stars have been treated here) and extensive non-LTE studies of the emission/absorption features of the photospheres in our models (hopefully to be performed by others). The evolution and possibly explosive demise of stars with $M < 10\ M_\odot$ and $M > 100\ M_\odot$ deserves careful attention and may ultimately prove to be important nucleosynthetic sites. The collapsing cores of stars in the range $10 \lesssim M/M_\odot \lesssim 100$ obviously need further study. By treating the core bounce as a simple inner-boundary condition we have temporarily averted a confrontation with the uncertainties that accompany the study of neutrino physics and the equation of state for matter at (super-) nuclear density. While we have not addressed the core-bounce problem it is at least reassuring that the bulk of observed (and observable) supernova properties do not appear to be especially sensitive to the precise manner in which energy is coupled into the mantle and envelope providing, of course, as we apparently have been repeatedly assured by nature, that the cores of these stars do in fact blow up!

Acknowledgments

We gratefully acknowledge the use of unpublished supernova core models of J. R. Wilson and R. L. Bowers, as well as the encouragement and support of Lowell Wood and Edward Teller. We have profited by numerous private conversations with many researchers in the fields of supernova dynamics, nucleosynthesis, and observations. Chief among this group (whose names must be omitted for lack of considerable space) are W. A. Fowler and W. D. Arnett who, by both suggestion and example, have served to motivate much of this work.

References

1. Hoyle, F. 1946. MNRAS. **106:** 343.
2. Burbidge, E. M. G. R. Burbidge, W. A. Fowler & F. Hoyle. 1957. Rev. Mod. Phys. **29:** 547.
3. Colgate, S. A. & R. H. White. 1966. Ap. J. **146:** 626.
4. Fowler, W. A. & F. Hoyle. 1964. Ap. J. Suppl. Ser. **9:** 201.
5. Oppenheimer, J. R. & G. M. Volkoff. 1939. Phys. Rev. **55:** 374.
   Zeldovitch, Ya. B. 1964. Sov. Phys. Dokl. **9:** 195.

6. Oppenheimer, J. R. & H. Snyder 1939. Phys. Rev. **56:** 455.
Penrose, R. 1969. Riv. Nuovo Cimento Numero Speciale **1:** 252.
7. Colgate, S. A. 1969. *In* Proc. 11th Intern. Conf. Cosmic Rays. Akademia Kiado, Budapest.
Colgate, S. A. & M. H. Johnson 1960. Phys. Rev. Lett. **5:** 235.
8. Weaver, T. A., G. B. Zimmerman & S. E. Woosley. 1978. Ap. J. **225:** 1021.
9. Weaver, T. A. & S. E. Woosley. 1979. In preparation.
10. Arnett, W. D. 1969. Astrophys. Space Sci. **5:** 180.
11. Arnett, W. D. 1972. Ap. J. **176:** 681; **176:** 699; **173:** 393.
Arnett, W. D. 1973. Ap. J. **179:** 249.
12. Arnett, W. D. 1974. Ap. J. **193:** 169; **194:** 373.
Arnett, W. D. 1977. Ap. J. Suppl. Ser. **35:** 145.
13. Arnett, W. D. 1978. Ap. J. **219:** 1008.
14. Falk, S. W. & W. D. Arnett. 1973. Ap. J. Lett. **180:** L65.
Falk, S. W. & W. D. Arnett. Ap. J. Suppl. Ser. **33:** 515.
Arnett, W. D. & S. W. Falk. 1976. Ap. J. **210:** 733.
15. Shurmann, S. R., W. D. Arnett & S. W. Falk. 1978. Preprint.
16. Iben, I. 1975. Ap. J. **196:** 525.
17. Lamb, S. A., I. Iben & W. M. Howard. 1976. Ap. J. **207:** 209.
18. Barkat, Z. 1971. Ap. J. **163:** 433.
19. Ikeuchi, S., K. Nakazawa, T. Murai, R. Hoshi & C. Hayashi 1971. Prog. Theor. Phys. **46:** 1713; and 1972. *ibid.* **48:** 1870.
20. Endal, A. S. 1975. Ap. J. **195:** 187; **197:** 405.
21. Sugimoto, D. 1970. Prog. Theor. Phys. **44:** 599; see also 1971. *ibid.* **45:** 461.
22. Rakavy, G., G. Shaviv & Z. Zinamon. 1967. Ap. J. **150:** 131.
23. Cameron, A. G. W. 1973. *In* Explosive Nucleosynthesis. D. N. Schramm, Ed. University of Texas Press. Austin, Texas.
24. Beaudet, G., V. Petrosian & E. E. Salpeter. 1967. Ap. J. **150:** 979.
25. Wilson, J. R. 1971. Ap. J. **163:** 209.
26. Wilson, J. R. 1974. Phys. Rev. Lett. **32:** 849.
27. Wilson, J. R. 1978. *In* Physics and Astrophysics of Neutron Stars and Black Holes, LXV Corso. R. Ruffini and R. Giaconni, Eds. Societe Italiana di Fisica. Bologna, Italy.
28. Wilson, J. R. & R. L. Bowers. 1978. Private communication; Wilson, J. R. 1978. This conference.
29. Woosley, S. E., W. D. Arnett & D. D. Clayton. 1973. Ap. J. Suppl. Ser. **26:** 231.
30. Arnett, W. D. & J. W. Truran. 1969. Ap. J. **157:** 339.
31. Pardo, R. C., R. G. Couch & W. D. Arnett. 1974. Ap. J. **191:** 711.
32. Lamb, S. A., W. M. Howard, J. W. Truran & I. Iben. 1977. Ap. J. **217:** 213.
33. Woosley, S. E. & W. M. Howard. 1978 Ap. J. Suppl. Ser. **36:** 285.
34. Howard, W. M., W. D. Arnett, D. D. Clayton & S. E. Woosley. 1972. Ap. J. **175:** 201.
35. Lee, T., D. A. Papanastassiou & G. J. Wasserburg. 1977. Ap. J. Lett. **211:** L107.
36. Lee, T., D. A. Papanastassiou & G. J. Wasserburg. 1978. Ap. J. Lett. **220:** L15.
37. McCulloch, M. T. & G. J. Wasserburg. 1978. Ap. J. Lett. **220:** L15; see also Geophys. Res. Lett. **5:** 599.
38. Arnett, W. D. & J. P. Wefel. 1978. Ap. J. Lett. **224:** L139.
39. Truran, J. W., J. J. Cowan & A. G. W. Cameron. 1978. Ap. J. Lett. **222:** L63.
40. Thielemann, F. K. M. Arnoùld & W. Hillebrandt. 1979. Astron. Astrophys. **74:** 175.
41. Blake, B. T. & D. N. Schramm. 1978. Private communication.
42. Seeger, P. A. W. A. Fowler & D. D. Clayton. 1965. Ap. J. Suppl. Ser. **11:** 121.
43. Miller, G. E. & J. M. Scalo. 1979. Ap. J. Suppl. Ser. In press.
44. Arnett, W. D. & D. N. Schramm. 1973. Ap. J. **184:** L47.
45. Talbot, R. J. & W. D. Arnett. 1973. Ap. J. **186:** 51.
46. Ostriker, J. P., D. O. Richstone & T. X. Thuan. 1974. Ap. J. **188:** L87.
47. Kaler, J. B. 1978. Ap. J. **225:** 527.
48. Shields, G. A. 1978. Ap. J. **219:** 599 and 565.
Farland, G. J. & G. A. Shields. Ap. J. **226:** 172.

49. Ciatti, F., L. Rosino & F. Bertola. 1971. Mem. Soc. Astron. Ital. **42:** 163.
50. Kirshner, R. P., J. B. Oke, M. V. Penston & L. Searle. 1973. Ap. J. **185:** 303; and reference cited therein.
51. Barbon, R., F. Ciatti & L. Rosino. 1973. Astron. Astrophys. **29:** 57.
52. Kirshner, R. P. & J. Kwan. 1974. Ap. J. **193:** 27.
53. Chevalier, R. A. 1976. Ap. J. **207:** 872.
54. Lasher, G. J. & K. L. Chan. 1978. Preprint.
See also Klein, R. I. & R. A. Chevalier. 1978. Ap. J. Lett. **223:** L109.
55. Colgate, S. A. & C. McKee. 1969. Ap. J. **157:** 623.
56. Isenberg, P. A. 1977. Ap. J. **217:** 597.
57. Chevalier, R. A. & R. I. Klein. 1978. Ap. J. **219:** 931.
58. Axelrod, T. S. 1979. Ph.D. Thesis. University of California at Santa Cruz. In preparation.
59. Taylor, J. H. 1978. This conference.
60. Lewin, W. H. G. 1978. This conference; Bahcall, J. N. 1978. Annu. Rev. Astron. Astrophys. **16:** 241.
61. Rosino, L. 1977. *In* Supernovae. D. N. Schramm, Ed. Reidel. Dordrecht, Holland.
Rosino, L. & G. Di Tullio. 1974. In Supernovae and Supernova Remnants. C. B. Cosmovici, Ed. Reidel. Dordrecht, Holland.
62. Branch, D. 1977. *In* Supernovae. D. N. Schramm Ed. Reidel. Dordrecht, Holland.
63. Colgate, S. A., E. P. Moore & R. Carlson. 1975. Publ. Astron. Soc. Pac. **87:** 565.
Colgate, S. A., E. P. Moore & J. Colburn. 1975. Appl. Opt. **14:** 1429.

NEUTRINO FLOW AND STELLAR CORE COLLAPSE*

James R. Wilson

Lawrence Livermore Laboratory
Livermore, California 94550

The present paper presents the results of a detailed numerical calculation of the collapse of the iron core of a fifteen solar mass star at the end of its normal evolution. The initial model for these calculations is taken from a fifteen solar mass stellar model at the start of core instability. This model was described by T. Weaver in the previous talk. The present talk should be considered as a continuation of the papers given at the 1974 Texas symposium,[1] the 1975 Varrena lectures,[2] and the 1978 Battelle lectures.[3] The principle changes since reference 2 are in the equation of state, the electron capture model, and the complete proper treatment of neutrino degeneracy. For details of the latest changes see reference 3. The new equation of state incorporates ideas given in Bethe *et al.*[4] The major change in the equation of state is the use of a much larger statistical weight for heavy nuclei than was previously used by the author. Since the Battelle lectures a convective model has been added to the calculations. Epstein[5] has pointed out that, since neutronization is strongest at the outside of the core when the central density rises above about 10^{12} g/cm^3 and neutrinos in the core are effectively trapped, there will be a positive gradient of mean particle mass. A region of convective instability may then arise due to the positive mean atomic weight gradient. To model this effect a convective mass flux term

$$m = 4\pi r^2 c_s \frac{\partial r}{\partial p} P \frac{\partial \rho}{\partial r}$$

is introduced when the system is unstable. Here r is radius, c_s is sound speed, P is pressure, and $\partial\rho/\partial r$ is the rate of change of density with radius above the rate that would hold for stability. All intensive material quantities are transported with this mass flux when convection occurs.

The effect of the new equation of state is to allow the collapse to proceed to about twice nuclear density before bounce occurs. Much higher material infall velocities are realized than in the calculations of references 1 and 2. In the calculations presented in the talk at Munich corresponding to this paper, when the standard cross sections appropriate to the Weinberg–Salam theory with $\sin^2\theta_w = 0.25$ were used, the outward shock wave formed from core bounce was sufficiently dissipated by neutrino emission that no explosion was produced. Following the talk it was suggested by Schramm and others that, since the neutrino emission is so critical to the propagation of the shockwave, I recheck the neutrino pair emission rate. The fit to the neutrino emission rate was found to be two to four times too large in the density and temperature range of most importance for shock propagation. A new fit to the neutrino pair emission rate

*Work performed under the auspices of the U.S. Department of Energy under contract W-7405-Eng-48.

0077–8923/80/0336–0358 $01.75/1 © 1980, NYAS

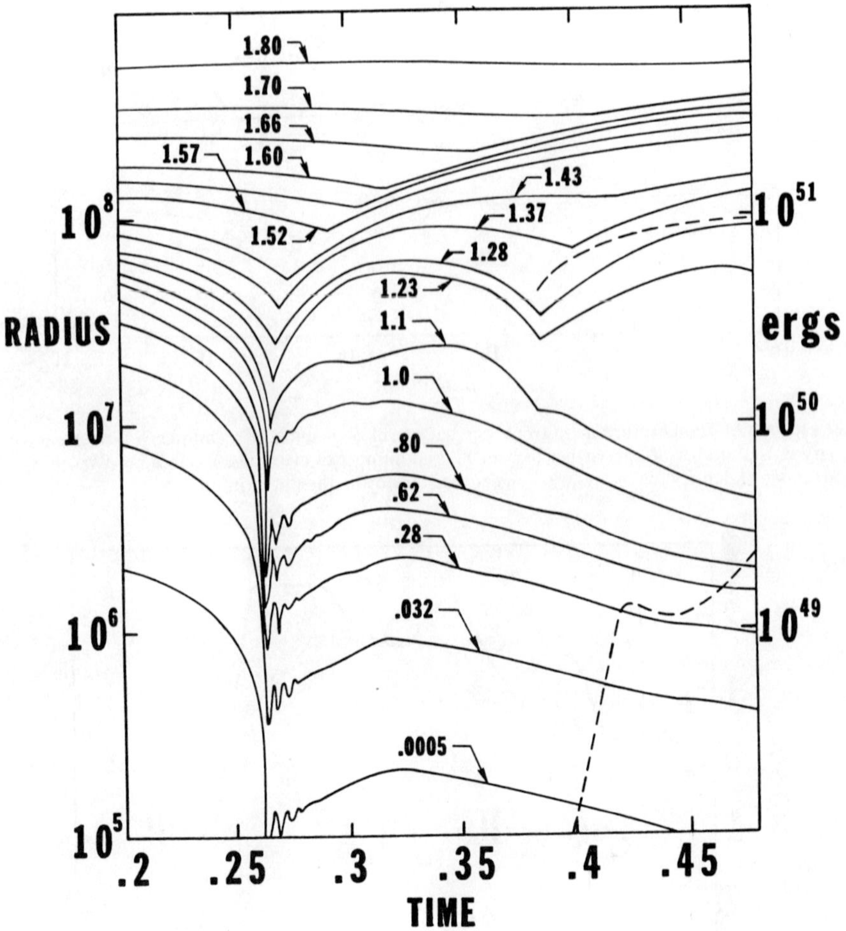

FIGURE 1. Radius versus time for selected mass points. The labels above each of the lines are the mass inside the line in units of solar mass. The top dashed curve is the excess of kinetic plus internal energy over gravitational energy of outward moving material. The lower dashed curve is the excess of kinetic over gravitational energy of outward moving material. Radius is on left scale, energy is on right scale.

has now been made. With the improved emission rate the shock is able to produce an explosion in a calculation using the standard neutrino cross section.

The new calculation will now be described in detail. As before, the core collapses until the central density reaches about 4.0×10^{14} g/cm^3 and a temperature of 16 MeV (see FIGURE 1). The central core of about mass 0.8 $M_\odot$ rings several times before settling down to a slow quasistatic evolution. The inner 0.8 $M_\odot$ is never shocked. In FIGURE 1 we see that the shock wave from core bounce propagates on out. At the end of the calculation 0.05 $M_\odot$ has acquired an excess kinetic energy compared to its

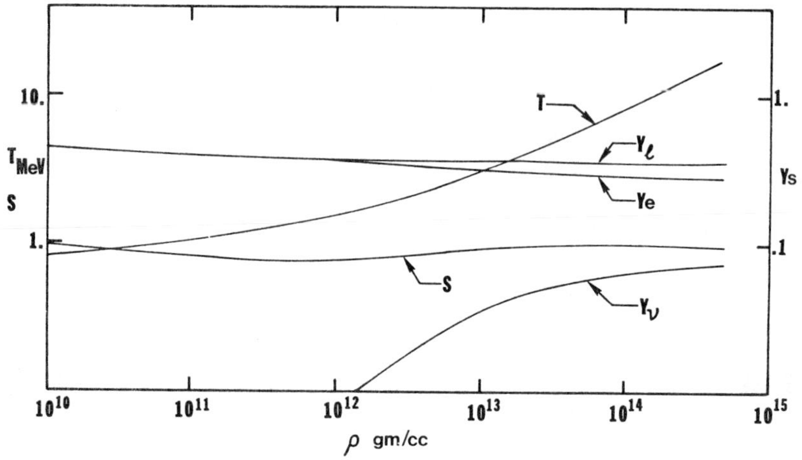

FIGURE 2. Temperature T, entropy per baryon in S in units of k, number of electrons per baryon Y_e, number of leptons per baryon Y_l, and number of electron neutrinos per baryon Y_ν as functions of density for the central zone of the star during the initial infall.

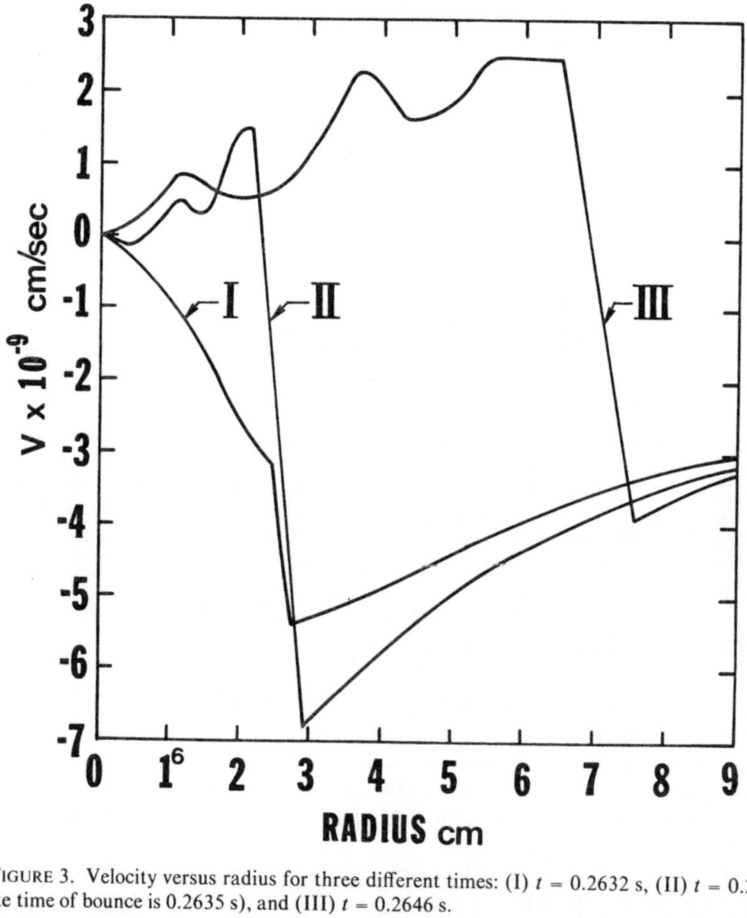

FIGURE 3. Velocity versus radius for three different times: (I) $t = 0.2632$ s, (II) $t = 0.2637$ s (note time of bounce is 0.2635 s), and (III) $t = 0.2646$ s.

gravitational energy of 2.0×10^{49} ergs. The total of kinetic plus internal minus gravitational energy for outward moving material is 9.5×10^{50} ergs. The final explosion energy should be some substantial fraction of the latter energy. From FIGURE 1 we see that the total excess energy has leveled off in time but the excess kinetic energy is still rising rapidly. At time 0.42 s we see a sharp change in kinetic energy. At that time only one zone was contributing to the excess kinetic energy. At the end four zones are above escape velocity and several more inside have almost reached escape velocity.

In FIGURE 2 the temperature and entropy of the central zone are given as functions

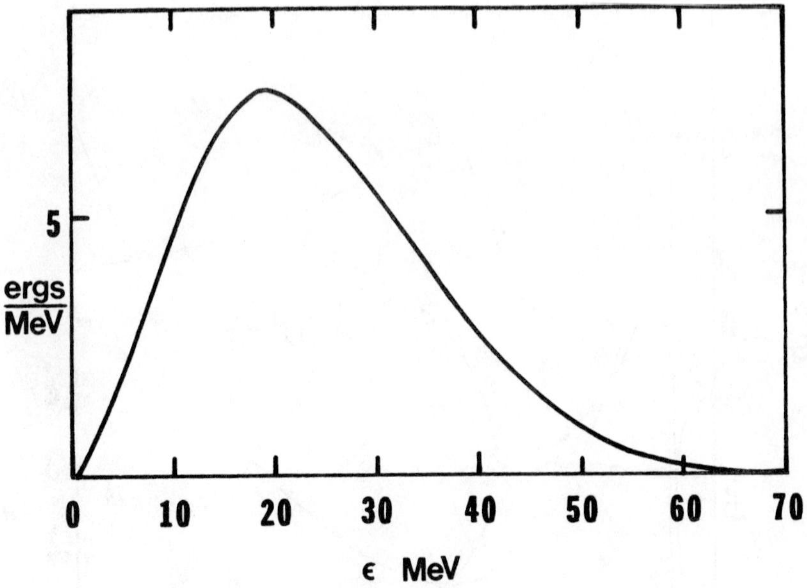

FIGURE 4. Spectrum of mu and tau neutrinos for the first burst of neutrinos from the shock wave. No scale for intensity. The integrated energy in the mu, tau burst is 6×10^{50} ergs. The electron neutrinos should have a similar burst energy distribution. However, the burst electron neutrinos are hard to get from the edit because of the neutronization background. Seven percent of the burst is antielectron neutrinos so probably there is a fast burst of hot electron and antielectron neutrinos of about 0.8×10^{51} ergs each with the above spectrum.

of the central density during the initial infall. Note that the dimensionless entropy per nucleon first falls until about neutrino trapping time and then rises back to about its initial value. Later when the central zone neutronizes, the entropy rises to about 1.5 but then falls back to 1.0, a value it seems to like.

FIGURE 3 shows several velocity profiles before and after bounce. Note that a shock wave has formed on the ingoing motion before bounce. This shock heats the material between 0.8 and 1.0 $M_{\odot}$ enough to about double its entropy. The main shock wave forms at about the 1 $M_{\odot}$ after the core reverses its infall. Between times II and III the luminosity from the shock rises to about 2×10^{54} erg/s. In that interval the shock wave power, $4\pi r^2 \rho v^3$, is also about 2×10^{54} erg/s. For a very short time neutrino emission damps the shock but the temperature soon falls to values insufficient to radiate the shock energy. The width of the 2×10^{54} erg/s burst is only about a half of a millisecond. The peak infall kinetic energy is 6.5×10^{51} ergs, the kinetic energy at time II is 4×10^{51} ergs, and at time III 2.6×10^{51} ergs. Half the energy of the early

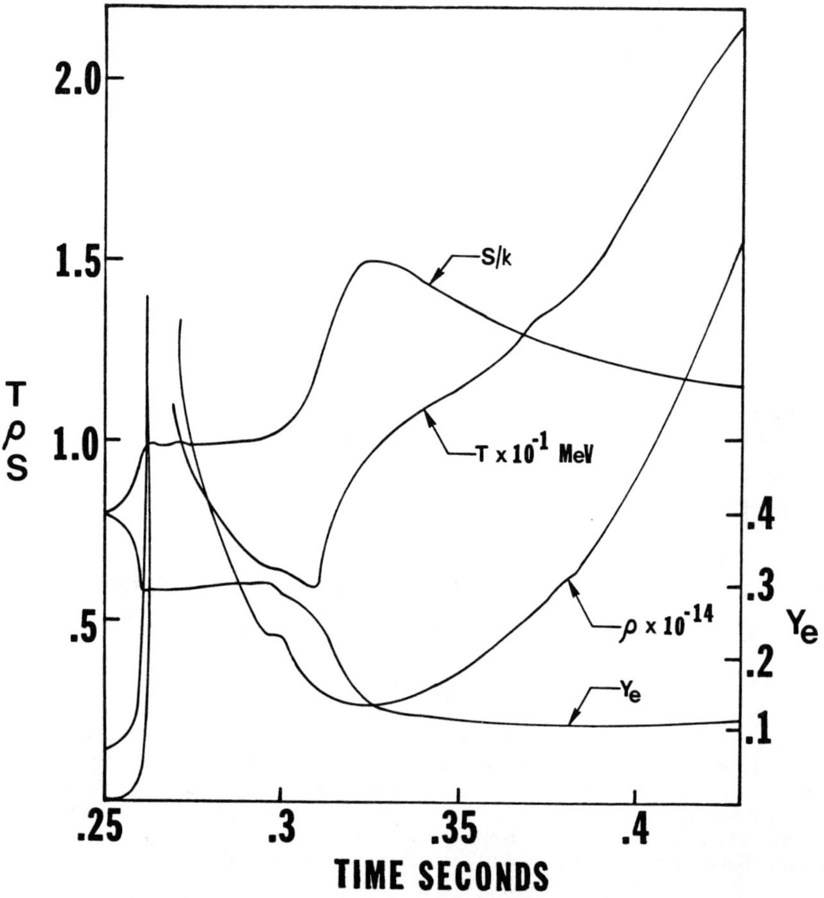

FIGURE 5. Temperature T, density ρ, entropy per baryon S, and number of electrons per baryon Y_e as functions of time for the central zones. The time during which the core was strongly oscillating has not been shown.

burst of neutrinos is electron neutrino, half is mu, tau, $\overline{\text{mu}}$, $\overline{\text{tau}}$, and about 7% antielectron neutrinos. After the short burst the luminosity falls to 1.0×10^{53} erg/s, almost all of which is electron capture neutrinos. FIGURE 4 shows the burst spectrum.

After the above burst from the shock wave the luminosity falls back to 1×10^{53} erg/s and is composed mostly of electron neutrinos with mean energy about 10 MeV. Then the luminosity slowly rises during the latter part of the calculation due to the compressional heating of the core. At the end of the calculation the total luminosity is about 3×10^{53} erg/s. The mu and tau neutrino luminosities are about 70% of the total emission, and the electron and antielectron luminosities are each about 15% of the total luminosity. The mean electron neutrino energy is about 11 MeV and mean mu, tau neutrino energy is about 18 MeV.

In the Munich talk results were presented for collapse models in which centrifugal rotational forces were added (averaged over angle). In those calculations the principle result was that if enough initial rotational energy was added to produce a bounce at a central density less than about 10^{14} g/cm^3 then neutronization at high density was just enough to expand the core and add support to the main outward shockwave. Turning back to FIGURE 1 we see that after core bounce the core fluffs up, slowly expanding until a time of about 0.32 s. What appears to be happening in this nonrotating model, as in the rotating model, is that as neutronization works its way into the core the pressure rises, because initially most of the pressure is from electrons ($p = 2\,\epsilon\rho/3$) and after neutronization the pressure is from free baryons ($p = 2\,\epsilon\rho/3$). If no neutrino energy was lost the pressure would almost double. However, the neutrinos do take an appreciable fraction of the energy so the rise is rather modest. Also some energy of capture goes into nuclear decomposition. In the central zone it is found that $\partial \log P/\partial \log Y_e = 0.25$ during neutronization. In order to get pressure enhancement the density must be large enough (10^{13} g/cm^3) that the electron degeneracy energy is much larger than the nuclear binding energy and the mean energy of lost neutrinos. If the density is too large then the neutrinos take too long to leak out. It should be noted that this effect could have appeared in earlier nonrotating calculations, but the conditions were never quite right before. In FIGURE 5 the central values of density, temperature, entropy, and electron number per baryon are shown versus time. The neutronization proceeds in through the core as a wave since after neutronization the density falls, the temperature rises, decomposing the nuclei, and the electron number per baryon falls. All this greatly increases the neutrino mean free path behind the neutronization wave. At time 0.30 s, just before neutronization, the central zone has a neutrino mean free path of 6 m; at time 0.32, just after neutronization, the central mean free path is 660 m. Of course part of the increase of the mean free path is also due to a change in spectrum of the neutrinos. They are now cooler. This slow expansion of the core is probably important because it reduces the degree of rarefaction behind the outward going shock wave. In FIGURE 1 at time equal 0.38 s we see a secondary infall and bounce. At the end of the calculation the shock from the secondary bounce is going more slowly than the primary shock and may not catch up. It will affect the mass cut that appears to be around 1.5 $M_\odot$

In FIGURES 6a, 6b, and 6c, profiles through the star at three different times are given. In FIGURE 6c most of the variables are bumpy out in the region 6×10^8 to 6×10^9cm. This is probably due to combination of the initial compositional layering and the second shock just going through this region. Note also the bump in T and the drop in Y_e for both FIGURES 6a and 6b. This is a mark of the neutronization wave. In FIGURE 6a the infalling unshocked material just outside the shock is moving in at a sizeable fraction of the outward velocity behind the shock. For the times of FIGURES 6b and 6c the inward unshocked velocities are small compared to the outward velocity. The point where the mean free path equals the radius is a rough indicator of the neutrino photosphere. At the time FIGURE 6a this photospheric radius is at 7.1×10^6 cm for electron neutrinos and at 3.0×10^6 cm for mu neutrinos. At the time of FIGURE

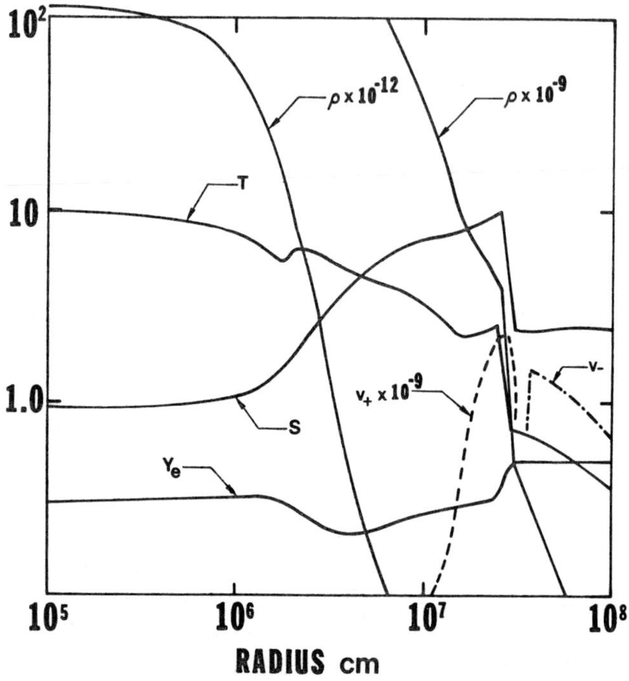

10²
10
1.0
ρ x 10⁻¹²
ρ x 10⁻⁹
T
S
Yₑ
v₊ x 10⁻⁹
v₋
10⁵
10⁶
10⁷
10⁸
RADIUS cm

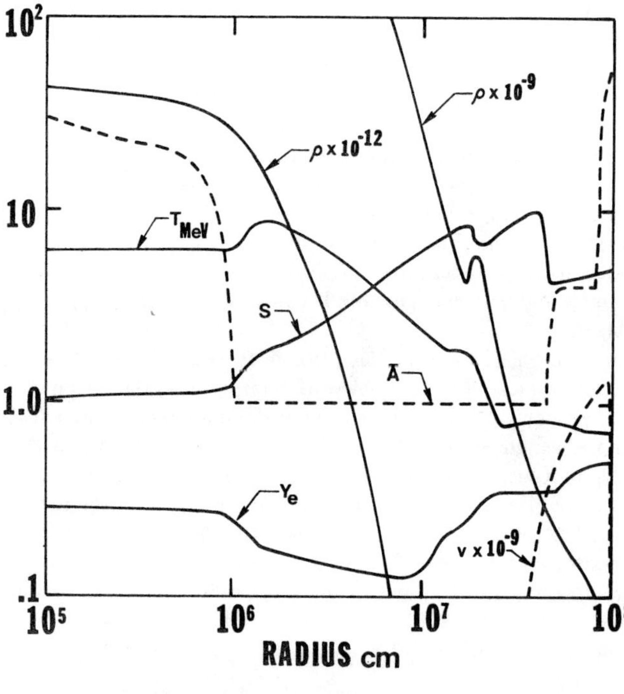

10²
10
1.0
.1
ρ x 10⁻⁹
ρ x 10⁻¹²
T MeV
S
Ā
Yₑ
v x 10⁻⁹
10⁵
10⁶
10⁷
10⁸
RADIUS cm

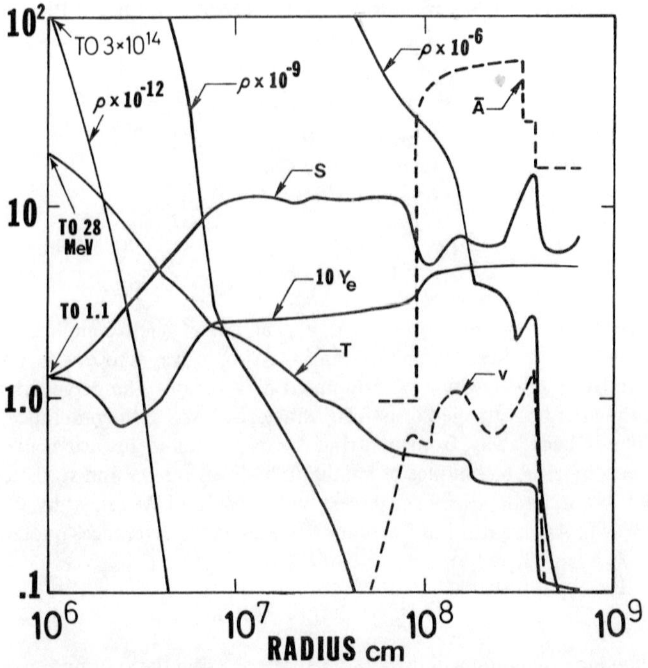

FIGURE 6. Temperature T, density ρ, entropy per baryon S, electron number per baryon Y_e, mean atomic weight A, and velocity are given as functions of radial position for three different times. The top left graph (a) is at $t = 0.270$ s, a time shortly after bounce when the core oscillations have almost damped out. The bottom left graph (b) is at $t = 0.30$ s, when the core is almost at maximum extension before recollapse. The graph above (c) is at $t = 0.46$ s, the end of the calculation.

6b the corresponding radii are 6.0×10^6 cm for electron neutrinos and 3.5×10^6 cm for mu neutrinos. At the time of FIGURE 6c the radii are 4.2×10^6 cm for electron neutrinos and 3.3×10^6 cm for mu neutrinos. The photosphere thus defined is always close to a density of 10^{11} g/cm^3 for the electron neutrinos.

REFERENCES

1. WILSON, J. R., R. COUCH, S. COCHRAN, J. LEBLANC & Z. BARKOT. Ann. N.Y. Acad. Sci. **262:** 54.
2. WILSON, J. R. 1978. Neutrino flow and gravitational collapse. Proc. Int. Sch. Phys. "Enrico Fermi" Vol. LXV, p. 644. North-Holland, Amsterdam.
3. WILSON, J. R. 1979. *In* Sources of Gravitational Radiation. L. Smarr, Ed. Cambridge University Press, Cambridge.
4. BETHE, H. A., G. E. BROWN, J. APPLEGATE & J. M. LATTIMER. 1978. Equation of state in gravitational collapse of stars. Nordita preprint. Copenhagen.
5. EPSTEIN, R. I. 1978. Lepton driven convection in supernovae. Nordita preprint. Copenhagen.

THE PHYSICS OF GRAVITATIONAL COLLAPSE: AN OVERVIEW*

W. David Arnett

Enrico Fermi Institute
University of Chicago
Chicago, Illinois 60637

INTRODUCTION

The gravitational collapse of stars is a problem of such complexity that it is valuable to review our ideas, emphasizing in a simple way those which seem most fundamental. Such a review will be attempted now because the perspective of many workers in the area has changed drastically since the last Texas conference. The shift in emphasis has been away from neutrino "transport" and toward neutrino "trapping," so that powerful techniques of adiabatic hydrodynamics and statistical physics may be used. For more detail see the papers by Weaver and Woosley,[1] by Wilson,[2] and by Bethe, Brown, Applegate, and Lattimer,[3] as well as the references quoted here.

THE STARTING POINT

The following discussion will involve calculations using the assumption of spherical symmetry. The validity of this assumption is not well-tested. As a star evolves and its inner core region contracts, angular momentum might be conserved locally, so that the core could spin up to break-up velocity. On the other hand, angular momentum might be transferred outward (turbulent convection and magnetic fields might be the culprits), so that the core would remain slowly rotating and spherical. Ideally, both cases should be examined. There are several reasons for seriously examining the spherical case: (1) It is solvable and we probably would not believe more complex results until this simpler case is worked out; (2) no white dwarf or neutron stars have been detected rotating anywhere near break-up velocity; and (3) present observational evidence on masses of neutron stars suggests that they lie near the Chandrasekhar limit, which is a natural unit of mass only for slowly rotating (nonmixed) systems.

It is now widely accepted that single massive stars undergo extensive mass loss. If a massive star is a member of a close binary system, it may lose mass by transfer to its companion (or out of the system). These phenomena modify our interpretation of the stages of hydrogen and perhaps helium burning, but later stages are thought to be too swift to be affected.

Thermonuclear energy generation in massive stars is more sensitive to temperature than is energy loss by photon diffusion or neutrino emission. Consequently, nuclear burning regions are cooled by convection. Thus an evolved star undergoes a sequence of generally well-separated stages of nuclear processing, with the ashes of one stage becoming the fuel for the next.[4] FIGURE 1 gives a simplified visualization of the process. Because the Coulomb barrier increases with increasing charge number Z,

*This work was supported in part by NSF grant AST 78-20392 at the University of Chicago.

0077–8923/80/0336–0366 $01.75/1 © 1980, NYAS

heavier nuclei are fused at higher temperatures. The process ends with fusion to the iron-group nuclei, which have the maximum binding energy per nucleon.

Convection does not mix the whole star, and unburned fuel survives in colder, outer regions. As one burning stage ends in the core, it may begin in a shell further from the center. Thus the star tends to develop a layered structure, a quantization of burning in space as well as time. This proceeds until a core is formed[5] with a mass above the Chandrasekhar limit, and composed of silicon-burning ashes (iron-group nuclei). Then the core begins to contract hydrostatically. The effective adiabatic exponent Γ is close to critical, so the restoring forces are small. Central temperature and density are $T_c \simeq 6 \times 10^9$ K and $\rho_c \simeq (3 \text{ to } 6) \times 10^9$ g cm^{-3}.

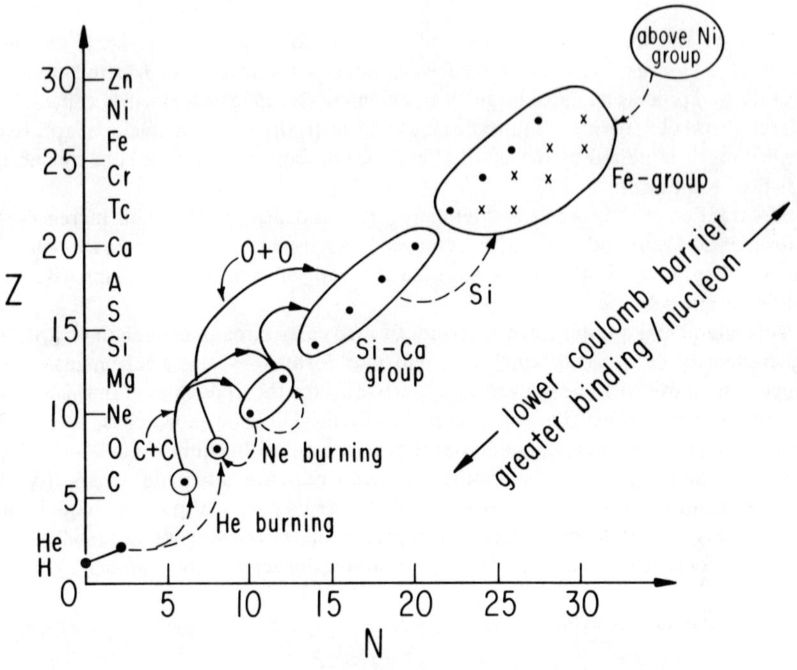

FIGURE 1. A schematic diagram of nuclear-burning stages. See reference 4 for details and references.

This general behavior is found for both helium stars and hydrogen stars of a variety of masses. Note that stellar cores must burn past the $O^{16} + O^{16}$ stage in order to have a low enough energy release to make collapse possible.

MICROPHYSICS OF COLLAPSE

For an assembly of nucleons, weak interactions have the important ability to change the net charge state; that is, the total neutron/proton ratio. By charge conservation this corresponds to changing the mole number of electrons Y_e. With the experimental discovery of neutral currents one feels more secure in the prediction of neutrino–antineutrino pair emission processes (real pair, plasmon, photoneutrino, neutrino-bremsstrahlung, etc.) which dominate stellar cooling after helium burning. These do not change Y_e however; the processes that do so are lumped together and called "neutronization." Increasing density gives higher electron Fermi energy, allowing electron capture to occur, which would not happen at lower densities. The dominant processes are capture on nuclei,

$$e^- + (Z, A) \rightarrow \nu + (Z - 1, A),$$

and on free protons,

$$e^- + p \rightarrow \nu + n.$$

The rate of the former process depends upon which nuclei are present, and how abundant they are. The capture on free protons is the inverse of neutron decay, a superallowed process having a large matrix element. Nuclei resist electron capture due to large threshold energies required to make terrestrially-unstable nuclei. Capture on free protons is inhibited by the small abundance of free protons boiled out by nuclear statistical equilibrium.

Contraction of the core causes temperature and density to rise. This increases the neutronization rate and robs the core of electron pressure. It also increases nuclear photodisintegration. Both effects increase the rate of contraction, giving rise to a hydrodynamic collapse.[6]

A fundamental question: are neutrinos formed early through enough to escape? At higher density ($\rho \gtrsim 10^{12}$g cm^3) the neutrinos formed by this neutronization are trapped and move with the matter: $\tau_{\text{diffusion}} >> \tau_{\text{infall}}$ for the timescales τ. In this regime the neutrino distribution function approaches Fermi–Dirac form, which is specified by a temperature and a chemical potential for the neutrinos. In equilibrium $T_\nu \rightarrow T_{\text{matter}}$, so that T and Y_ν, the mole number of neutrinos, are adequate to specify the thermodynamic properties ($Y_\nu = 0.5\ Y_e$ if the neutrino chemical potential equals that for electrons). Thus to calculate the properties of the system we need only three quantities, for example, density ρ, entropy S, and Y_e (or lepton mole number $Y_l = Y_e + Y_\nu$).

Let us explicitly list the regimes.

$$3 \times 10^9 \text{g cm}^{-3} \lesssim \rho \lesssim 10^{11} \text{g cm}^{-3} \quad \textbf{(1)}$$

This is the regime at which nuclear statistical equilibrium (NSE) is obtained, due to the speed of reactions involving strong and electromagnetic (but not weak) interactions. By specifying ρ, T and Y_e one can solve the coupled (very nonlinear) "Saha" equations (i.e., equate chemical potentials) to obtain pressure, entropy internal energy, and nuclear abundances. The solution is simple in principle but not trivial in practice.

$$10^{11} \text{g cm}^{-3} \lesssim \rho \lesssim 10^{12} \text{g cm}^{-3} \quad \textbf{(2)}$$

In this regime neutronization and neutrino trapping occur. Thus specifying Y_e for a NSE calculation requires computation of neutrino production, absorption, and transfer, coupled to the hydrodynamic behavior, making this perhaps the most computationally-demanding regime.

$$10^{12}\text{g cm}^{-3} \lesssim \rho \lesssim 3 \times 10^{13}\text{g cm}^{-3} \quad (3)$$

Neutrinos are trapped so that lepton number (i.e., $Y_l = Y_e + Y_\nu$) is a good state variable. This regime is technically no more difficult than (1) above.

$$3 \times 10^{13}\text{g cm}^{-3} \lesssim \rho \lesssim 4 \times 10^{14}\text{g cm}^{-3} \quad (4)$$

The density is so high that free nucleon interaction and nuclear compressibility need to be included.

$$\rho \gtrsim 4 \times 10^{14}\text{g cm}^{-3} \quad (5)$$

At greater than nuclear density the nuclei and nucleons merge into a giant, warm nucleus. All the problems associated with neutron star equations of state arise.

In all of these regimes, low entropies are expected. Thus the nuclei do not dissociate much and the leptons are at least moderately degenerate. This simplifies the theoretical analysis. Note that knowledge of the two quantities, entropy S and electron number Y_e, allows us to specify the equation of state for a density in any of these regimes.

Neutrino Astrophysics of Collapse

As collapse proceeds, the denser core evolves faster than the outer regions (the "mantle," which we define as the processed matter above the silicon-burning shell). This is illustrated in FIGURE 2, which shows "snapshots" of structure at several moments, from the initial instability until well after neutrino trapping.[6] An 8 $M_\odot$ helium star is illustrated. The entire evolution shown takes one second, mostly spent in the early evolution away from hydrostatic equilibrium. The various burning shells that are active are shown.

The structure of the core is illustrated schematically in FIGURE 3. As we move inward we pass a neutrino "photosphere" (sometimes called a "nusphere" or "neutrinosphere"), which is about one mean free path in. Then there is a neutronization shell where most of the electron capture occurs, and a neutrino trapping region inside which the neutrino diffusion time exceeds the infall time. These regions overlap each other, allowing some loss of lepton number. The degree of this overlap determines the value of one of the state variables (lepton mole fraction, Y_l), which we need to specify the equation of state at higher density. The inner core may have a core shock (see below).

The neutrino distribution function (energy weighted) is shown for various shells in FIGURE 4. In the inner zones it closely approximates a Fermi–Dirac distribution at the local temperature, density, and mole number of neutrinos Y_ν. As we move outward the distribution function peaks at lower neutrino energy and decouples from the matter with a broad maximum at a neutrino energy around 8 MeV. The slight excess over the degeneracy limit represents an improvement over previous numerical results, but more recent calculations approach the limit from below. (These errors are negligible for our purposes).

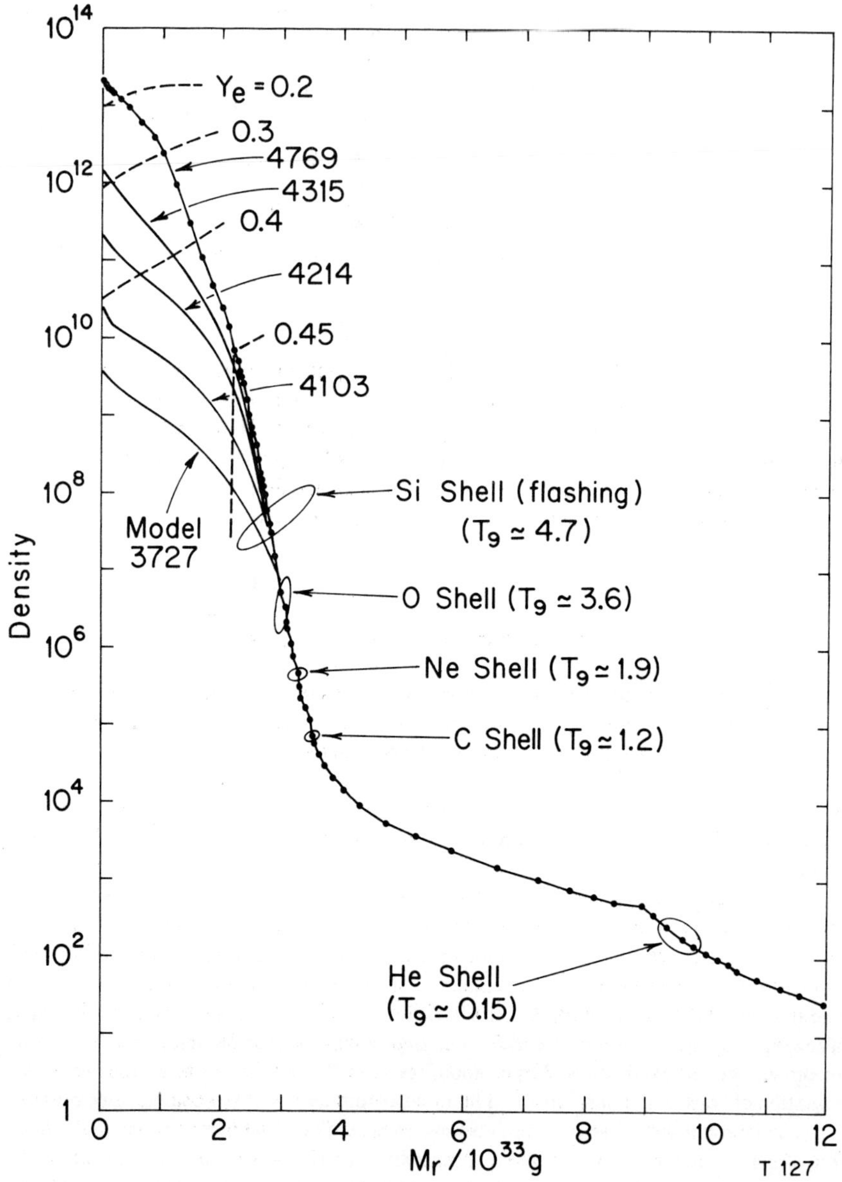

FIGURE 2. Density structure at several times during the collapse. Density is plotted versus Lagrangian mass coordinate M_r; the dots represent individual zone centers. For details see reference 6.

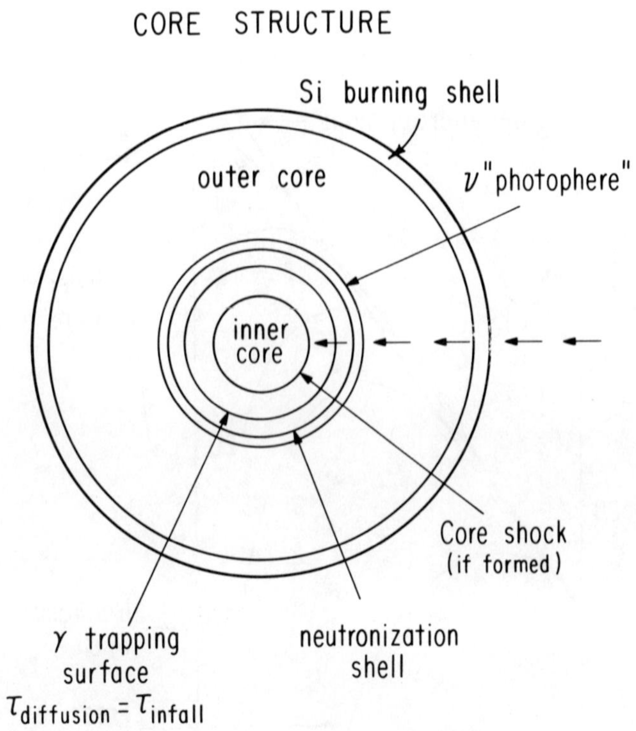

FIGURE 3. Core structure, showing neutrino "photosphere," neutrino trapping surface, neutronization shell, core shock, and silicon-burning shell.

How does the other state variable, the entropy, change? The first law of thermodynamics is

$$\dot{E} + P\dot{V} = \dot{Q}$$
$$= T\dot{S} + \sum_i \dot{N}_i\mu_i,$$

where μ_i is the chemical potential of the species i. Let $N_i = \rho\mathcal{A}Y_i$ where $\mathcal{A}$ is Avogadro's number, $\sigma = S/k$ is the dimensionless entropy per nucleon, and q is the added energy per nucleon due to heat flow. Now[7]

$$\dot{q} = kT\dot{\sigma} + \sum_i \mu_i\dot{Y}_i$$
$$\simeq \langle\epsilon_\nu\rangle_{\text{escape}} Y_\nu/\tau_{\text{escape}},$$

where $\langle\epsilon_\nu\rangle_{\text{escape}}$ is the mean energy of escaping (nontrapped) neutrinos and τ_{escape} is the average escape time. The loss of leptons by neutrino diffusion is

$$\dot{Y}_l = \dot{Y}_e + \dot{Y}_\nu$$
$$\simeq -Y_\nu/\tau_{\text{escape}}.$$

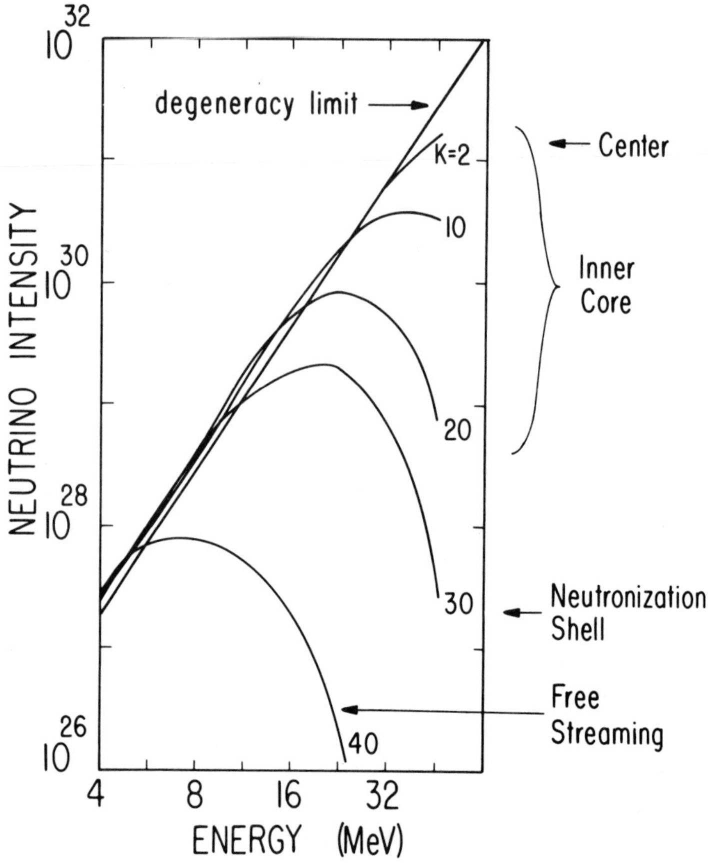

FIGURE 4. Energy-weighted distribution function for neutrinos, shown at several points in the collapsing core. Zone numbers ($K = 2$ is the center) are given. At low energy the degenerate limit is slightly exceeded; see text however.

Now the neutronization reactions have the form

$$e^- + p \rightleftharpoons n + \nu,$$

so $\dot{Y}_e = \dot{Y}_p = -\dot{Y}_n$ for these trapped species, and

$$\sum_i \mu_i \dot{Y}_i = \dot{Y}_e(\mu_e + \mu_p - \mu_n) + \mu_\nu(-\dot{Y}_e - Y_\nu/\tau_{\text{escape}}),$$

which can be written as

$$kT\dot{\sigma} = (-\dot{Y}_e)(\mu_\nu - <\epsilon_\nu>_{\text{escape}}) + (-\dot{Y}_e)(\mu_e + \mu_p - \mu_n - \mu_\nu).$$

The first term on the RHS is a "downscatter" term due to the fact that the escaping neutrinos have been degraded in energy. This may be the largest effect. The second term, of perhaps comparable size, is due to the lack of weak interaction (neutronization) equilibrium. C. Pethick calls this the "out-of-whackness" term. Both of these have a maximum in the neutronization region. Typical values for the total change in entropy are $\Delta\sigma \lesssim 1$, which are small.

To examine the effects of neutrino transport on explosion (?), consider an extreme case. We arbitrarily set the free proton abundance Y_p so large ($Y_p \simeq 10^{-3}$, not 10^{-6}) that the neutronization shell moves out toward the neutrinosphere. The results are shown in FIGURE 5. The electron mole fraction (in the core) goes from $Y_e \simeq 0.45$ to 0.08; most leptons escape! There is a large amount of neutrino transport; in fact, a standing accretion shock is set up with neutrino stress stopping the incoming matter. However, there is not the slightest hint of explosion. The neutrinos *dissipate* the infall kinetic energy, not *reflect* it. The failure of this highly favorable and extreme case illustrates the difficulty with the hypothesis that neutrino transport gives rise to an explosion.

ADIABATIC HYDRODYNAMICS OF COLLAPSE

Let us consider what is probably a more realistic limit: strongly trapped neutrinos. In this limit we can neglect the effects of neutrino escape, to first order, and fudge our equation of state a bit to correct this neglect. The approximation is very good for

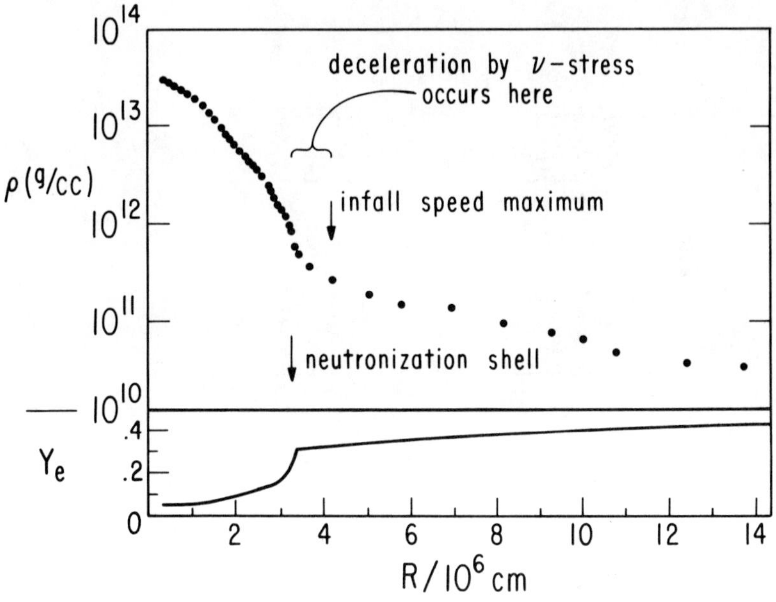

FIGURE 5. Density and electron mole number versus radius during collapse for an (arbitrarily) extreme neutrino transport case (see text). Notice the standing shock at the neutronization shell; the infalling matter is smoothly decelerated by neutrino stress.

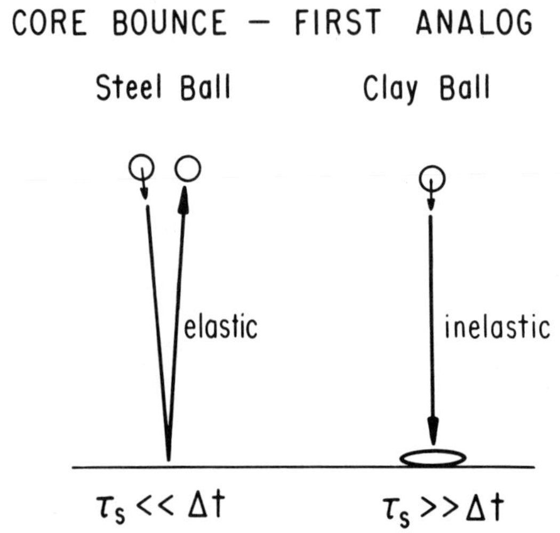

FIGURE 6. First analog for core bounce. The analogy is not adequate; see text.

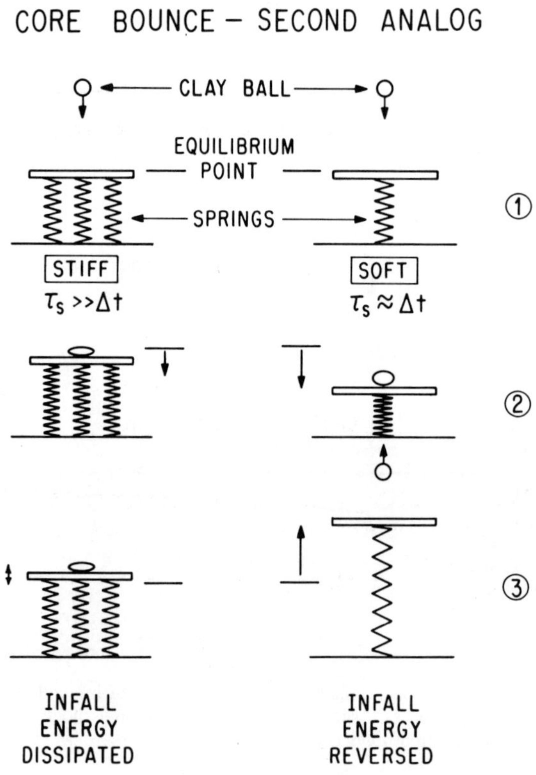

FIGURE 7. Second analog for core bounce.

$\rho \gtrsim 10^{12}$g cm^{-3}. Ken van Riper[8] has examined the hydrodynamics of collapse in this approximation for spherical symmetry. In the *Newtonian* case he finds that the results are sensitive to the effective adiabatic exponent Γ, i.e., to the equation of state. For $\Gamma \simeq 4/3$, a vigorous "reflected" shock ejects matter. For $\Gamma \gg 4/3$, at high density (nucleon–nucleon repulsion) he found a standing accretion shock, with no mass ejection or explosion.

To highlight the physics underlying these important results, let us consider some analogs whose behavior is more intuitively obvious. These analogs are not a complete representation of our problem, so we must not take them too literally. Consider first

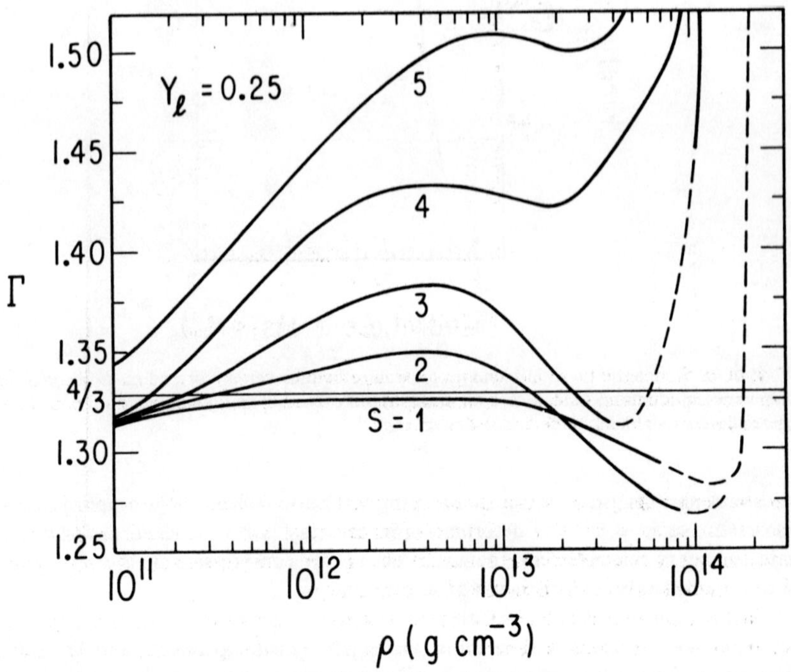

FIGURE 8. Adiabatic exponent Γ versus density for various values of S, the dimensionless entropy per nucleon. For $S \simeq 1$ to 2, $\Gamma \simeq 4/3$. See reference 9 for details.

the bounce of a ball dropped to the floor, as shown in FIGURE 6. A steel ball bounces while a clay ball splats. The difference is that for a steel ball the communication time across the ball, $\tau_s \simeq D/\mathcal{C}$, where D is the diameter and $\mathcal{C}$ the sound speed, is much smaller than the time available for communication $\Delta t \simeq D/v_{\text{fall}}$, where v_{fall} is the velocity before impact. Clay has less resistance to deformation, hence a slower sound speed; for $\tau_s \gg \Delta t$ the impact is dissipative. Ordered kinetic energy is degraded into heat rather than being preserved by rebound. In the stellar situation we have $v_{\text{fall}} \simeq \mathcal{C}$ and $\tau_s \simeq \Delta t$, so this analog is not quite appropriate.

A better analog is shown in FIGURE 7. A clay ball is dropped onto a platform supported by springs. If the springs are stiff we recover the $\tau_s \gg \Delta t$ case. However, if

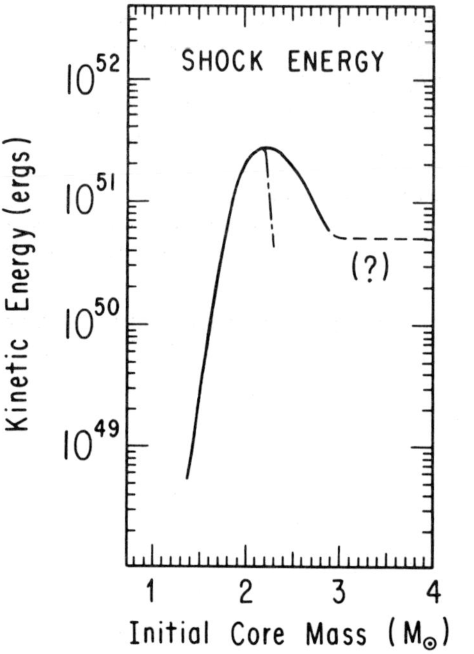

FIGURE 9. Schematic of kinetic energy from core bounce versus core mass, from reference 10. For some equations of state, a core mass slightly in excess of that giving maximum explosion energy collapses to a black hole (dot-dashed curve).

we use weak springs, there is a small restoring force and hence a large amplitude. This increases Δt, so $\tau_s \simeq \Delta t$. The deformation of the clay ball is reduced and the infall kinetic energy is converted into potential energy by compression of the spring, and then returned as outwardly directed kinetic energy.

Note that the "ejection" shock that Van Riper found was formed as the pressure wave from core rebound steepened in the star's density gradient, and should be distinguished from a "standing accretion" shock which does not give an explosion. Also note that in the Newtonian case $\Gamma \simeq 4/3$ corresponds in general relativity to a mass M close to the critical (Oppenheimer–Volkoff) mass M_{ov}. This suggests that to get a velocity *reversal* at bounce we want little dissipation, so we want:

(1) little neutrino transport (strong trapping);

(2) no accretion shock (weak restoring force).

The second condition also gives a large amplitude for core oscillation.

The Γ to be expected from a realistic equation of state, according to Lamb, Lattimer, Pethick and Ravenhall,[9] is shown in FIGURE 8 as a function of density for several values of entropy. Initially the entropy is $\sigma \simeq 1$ ("S" in the figure) so that if $\Delta\sigma \simeq 1$ we would look at the curves for $S = 1$ to 2. For the lower densities, Newtonian results apply and Γ is close to the value of $4/3$ which allows explosion. At $\rho \gtrsim 10^{14}$g cm^{-3}, Γ increases, increasing the pressure gradient force, but general relativistic effects tend to cancel this by increasing gravity. FIGURE 9 shows schematically the results of hydrodynamic calculations[10] including these effects.

The explosion energy is a maximum for $M_{core} \lesssim M_{ov}$ and of the order of several times 10^{51}erg. Note that lower masses can give modest explosion energies. For many choices of "neutron star" equation of state, a black hole results from a slightly larger M_{core} (see dot-dashed curve). FIGURE 10 shows such a case, illustrating the suddenness of the transition with increasing mass. It is interesting that for plausible equations of state and core masses determined by evolutionary sequences, explosions do result.

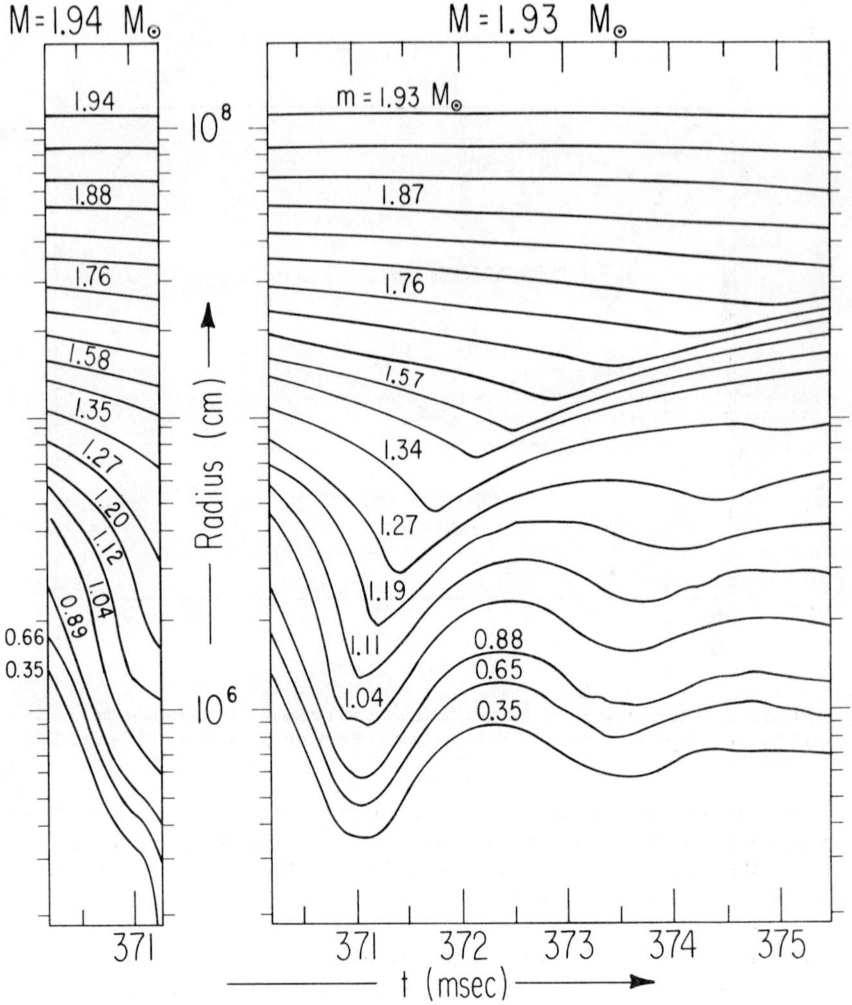

FIGURE 10. Various Lagrangian radii versus time for a collapse to a black hole and for an explosion. From reference 10.

These results suggest:

(1) For such an explosion, we have a prediction of the mass range for the dense remnant $M_{\text{Chandrasekhar}} \lesssim M$ (collapsed remnant) $\lesssim M_{ov}$. Observations of x-ray binaries give $M_{rem} \simeq 1$ to 2 $M_{\odot}$ and the binary pulsar[11] PSR 1913 + 16 gives masses of 1.39 ± 0.15 and 1.44 ± 0.15, which are consistent with this relation.

(2) The bounce occurs at greater than nuclear density. Either the effects of general relativity[12] or the attractive part of the nucleon–nucleon interaction seems to be adequate to insure this result, which may be important for gravitational radiation.

A number of new and interesting questions have arisen.

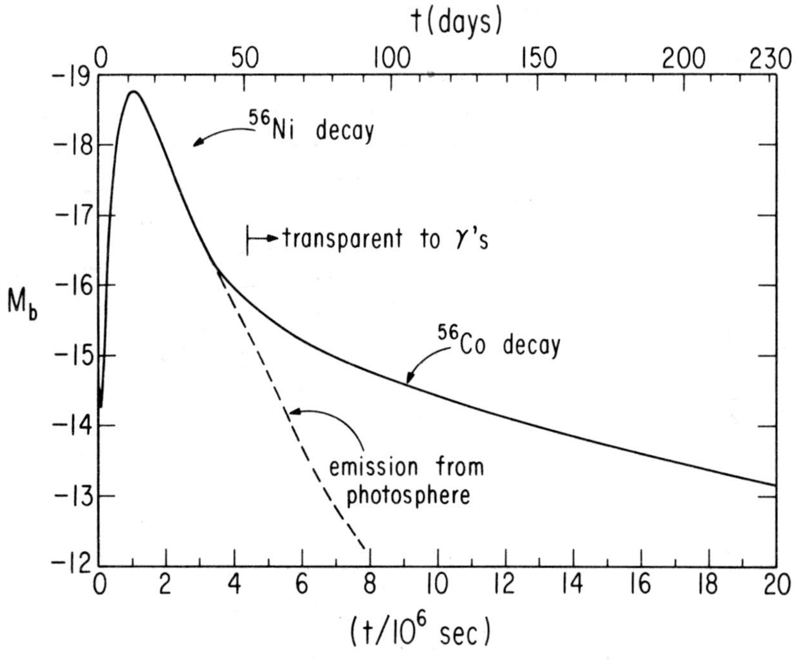

FIGURE 11. Light curve from the core collapse and explosion of a low-mass helium star (about 2.5 $M_{\odot}$ at collapse). Note similarity to a type-I supernova; see reference 18 for details.

(1) Does the Saenz–Shapiro effect,[13] a growth of asymmetry after several core bounces, insure a large quadrapole moment and profuse gravitational radiation?

(2) What is the neutrino shock structure[14,15] for a reflected shock propagating to $\rho \lesssim 10^{12}$g cm^{-3} (which might be observable)?

(3) What effects will Epstein's neutrino convection[16] have?

(4) What is the best equation of state, especially for $\rho \gtrsim 10^{14}$g cm^{-3}?

(5) At the very high bounce densities, even a slowly rotating precollapse model may become nonspherical. What are the characteristics of the 2- and 3-dimensional hydrodynamics of core collapse?

This new viewpoint leads to insight into other problems. FIGURE 11 shows the calculated light curve for a 4 $M_\odot$ helium star that evolved to core collapse (losing about 1.5 $M_\odot$ of helium envelope) and underwent core bounce, making ^{56}Ni and ejecting it.[17] The similarity with a type-I supernova is striking, and I suggest[18] that this is in fact what a type-I supernova is. Shortly after peak light the ejected matter becomes transparent to gamma rays; instruments in operation at present, or soon to be so, should be able to detect line radiation from ^{56}Co decay in a modestly bright type-I supernova.[18] Amusingly enough, using this model to infer distance gives a Hubble constant of $H_0 \simeq 65$ km s^{-1} Mpc^{-1} (independent of the calibration of the standard distance scale).

References

1. Weaver, T. & S. E. Woosley. 1979. Evolution and explosion of massive stars. This volume and references therein.
2. Wilson, J. 1979. Stellar collapse and supernovae. This volume and references therein.
3. Bethe, H. A., G. E. Brown, J. Applegate & J. Lattimer. 1979. Equation of state in the gravitational collapse of stars. Preprint.
4. Arnett, W. D. 1974. Nuclear Reactions and nurtinos in stellar evolution. *In* Late stages of stellar evolution, IAU Symposium No. 66. R. Tayler, Ed. D. Reidel. Dordrecht, Holland.
5. Arnett, W. D. 1977. Advanced evolution of massive stars. VII. Silicon burning. Astrophys. J. Suppl. **35:** 145.
6. Arnett, W. D. 1977. Neutrino trapping during gravitational collapse of stars. Astrophys. J. **218:** 815.
7. Lamb, D. Q. & W. D. Arnett. Entropy generation in stellar collapse. In preparation.
8. Van Riper, K. A. 1978. The hydrodynamics of stellar collapse. Astrophys. J. **221:** 304.
9. Lamb, D. Q., J. M. Lattimer, C. J. Pethick & D. G. Ravenhall. 1979. Hot dense matter and stellar collapse. Preprint.
10. Van Riper, K. A. & W. D. Arnett. 1978. Stellar collapse and explosion: Hydrodynamics of the core. Astrophys. J. Lett. **225:** L129.
11. Taylor, J. 1979. This volume.
12. Van Riper, K. A. 1979. General relativistic hydrodynamics and the adiabatic collapse of stellar cores. Astrophys. J. **225:** L129.
13. Saenz, R. A. & S. L. Shapiro. 1978. Gravitational radiation from stellar collapse: Ellipsoidal models. Astrophys. J. **221:** 286; and S. L. Shapiro. 1979. *In* Sources of gravitational radiation. L. Smarr, Ed. Cambridge University Press. Cambridge, England.
14. Bruenn, S. W., R. Buchler & W. Yueh. 1978. Shock structure and neutrino radiation in stellar collapse. Astrophys. J. Lett. **221:** L86.
15. Kazanas, D. 1979. Neutrino mediated shocks in supernovae. Preprint.
16. Epstein, R. E. 1979. Lepton driven convection in supernovae. Mon. Not. R. Astron. Soc. In press.
17. This is qualitatively similar to some of the ideas proposed in Colgate, S. A. & C. McKee. 1969. Early supernova luminosity. Astrophys. J. **157:** 623.
18. Arnett, W. D. 1979. On the theory of type-I supernovae. Astrophys. J. Lett. **230:** L37.

NEUTRINO ASTRONOMY*

David N. Schramm

Department of Astronomy, Astrophysics and Physics and the Enrico Fermi Institute The University of Chicago Chicago, Illinois 60637

INTRODUCTION

It is a well known phenomena in astronomy that whenever a new "window" is opened on the universe, totally unexpected discoveries come through it. One merely has to remember QSO's, pulsars, and x-ray and γ-ray bursters, to name but a few. A brand-new window is beginning to be opened which does not involve any part of the electromagnetic spectrum but instead involves neutrinos. As with the electromagnetic spectrum, one can certainly attempt to predict what phenomena one should expect to see with this new window, but also, as with the electromagnetic spectra, some of the most exciting discoveries may be those which are not predicted.

This review will summarize the three basic energy windows of the neutrino sky:

1. The low-energy ($E_\nu \sim 1$ MeV) solar neutrino situation will be briefly commented on but not reviewed in detail since there have already been many such reviews.
2. The intermediate-energy ($10 \lesssim E_\nu \lesssim 50$ MeV) neutrinos from gravitational collapse will be discussed. Here, the basic collapse scenario will be briefly reviewed. In addition, some discussion will be presented of the relative advantages and disadvantages of ν versus gravitational wave astronomy for the detection of collapse. It will be shown that ν astronomy provides an effective way to resolve some of the long-standing problems in collapse calculations.
3. The ultrahigh-energy ($E_\nu \gtrsim 10$ GeV) neutrinos will also be discussed. In particular it should be noted that whenever high-energy proton collisions occur anywhere in the universe, π's and κ's are produced which eventually decay to yield ultrahigh-energy neutrinos. Such collisions occur both when cosmic rays hit the Earth's atmosphere and when cosmic rays hit interstellar gas. Maximal interest surrounds the fact that such collisions will occur in the vicinity of astronomic objects which accelerate particles. In particular, some of the most exciting objects in the sky—pulsars, Seyferts, quasars, etc.—are thought to accelerate particles. Thus, these objects may be visible as "point" neutrino sources. Since proposed ν detectors such as the *D*eep *U*nderwater *M*uon *a*nd *N*eutrino *D*etector (DUMAND) should have angular resolution of better than 1 sq. deg., this means that the exciting objects may indeed be pulled out of the ν background as point sources.

After discussing the expected neutrinos in each of these energy regions, the review will then discuss the existing and proposed detectors for the intermediate and ultrahigh-energy neutrinos. It will be shown that in addition to observing neutrinos of

*This work was supported in part by NSF grant AST 78-20402 at The University of Chicago.

0077–8923/80/0336–0380 $01.75/1 © 1980, NYAS

astrophysical origin, these detectors can be used for other interesting experiments as well. In particular, they can set limits upon, or measure, *the lifetime of the proton.* It will be shown that such a measurement is extremely important to the new unification theories and may be directly related to the origin of a net baryon number for the universe.

LOW-ENERGY NEUTRINOS ($E_\nu \sim 1$ MeV)

It is well known (see reference 1) that standard main sequence stars should produce ν_e's as a result of burning 4H $\rightarrow$ ^{4}He. Because of the $1/r^2$ effect it is obvious (just as in the case of optical light) that the neutrino flux hitting earth in this energy range should be dominated by our sun. Ray Davis' solar neutrino experiment[2] has been well described in many places and will not be reviewed here other than to note that Davis' average counting rate is now 2.2 ± 0.4 SNU (where 1 SNU = 10^{-36} captures/^{37}Cl/s). The standard theoretical model for the sun predicts a counting rate for Davis of 5 SNU.[3] However merely taking standard stated allowed uncertainties in the various stellar model input parameters one can construct solar models yielding from ~3 to ~10 SNU. To go below 3 SNU does require some nonstandard assumption such as inhomogeneous composition, repeated mixing, a black hole in the center, or perhaps even ν decay or ν mixing. One should note that the minimum standard theory of 3 SNU and the observed flux of 2.2 ± 0.4 SNU are only $\sim 2\sigma$ apart. Thus perhaps all the worry about the solar neutrino problem is only a very bad case of statistics working against us. (It is also interesting that the overall average has, for no obvious reason, been rising in recent years.) However, regardless of the cause for the discrepancy between theory and observation, the best way to resolve the issue is to build a new solar neutrino experiment using ^{71}Ga which can see the basic *pp* generated ν's and not merely the "fringe" neutrinos from ^{8}B and ^{7}Be which are observed in the ^{37}Cl experiment of Davis.

INTERMEDIATE-ENERGY NEUTRINOS ($10 \lesssim E_\nu \lesssim 50$ MeV)

The dominant neutrinos in this energy range come from gravitational collapse. The details of various collapse calculations have been presented in other papers in this volume[4,5] and the basic neutrino physics is reviewed by Freedman *et al.*[6] Although various details such as the equation of state, the strength of the bounce, the role of neutrinos in shock wave mediation and mass ejection, and the possible role of magnetic fields and rotation are clearly matters which are subject to much debate, there are some general considerations which are universally agreed upon.

The first of these is the fact that, in order to form a neutron star remnant, the $\sim 10^{53}$ erg binding energy must be radiated away. The only two ways to rapidly carry away large amounts of energy from a collapsing core are via either gravitational radiation or neutrinos. Obviously, for a spherical symmetric collapse, the gravitational radiation loss is zero. Even for the optimal cases of nonspherical rotating collapse with multiple bounces, Shapiro[7] has found that the bulk of the binding energy still comes off in neutrinos and at most a few percent is in gravitational waves. The underlying reasons for this can be seen in the calculations of Kazanas and Schramm[8] which showed that for all but supernuclear densities, even nonspherical perturbations were more rapidly damped by neutrinos than by gravitational waves.

From the above we see that one expects most of the 10^{53} erg to be radiated in neutrinos. Even if some new and totally unexpected radiation mechanism were found, there should still be a minimum of 10^{52} erg radiated in electron neutrinos from the neutronization of the core to make a neutron star. The rest of the 10^{53} erg emitted neutrinos will be radiated in pairs via thermal processes such as $e^+e^- \rightarrow \nu\bar{\nu}$. Since these thermal processes go via neutral as well as charged currents (see reference 6 and references therein), the collapsing core will not only radiate $\nu_e\bar{\nu}_e$ but also $\nu_\mu\bar{\nu}_\mu$ and $\nu_\tau\bar{\nu}_\tau$ and perhaps $\nu_?\bar{\nu}_?$. However because of the cosmological arguments of Yang *et al.*[9] we can be fairly confident that there is at most one additional two-component neutrino species beyond the τ. One further point here is that because of the neutronization ν_e's filling phase space, the $\nu_e\bar{\nu}_e$ pairs will initially be inhibited by a factor ~10. The time scale on the emission of the neutrinos is dependent on the equation of state of the collapsing matter, with neutrinos coming from densities $\gtrsim 10^{11}$g/cm^3 (the ν photosphere) having to diffuse out. For some models there is an initial burst from the neutronization electron neutrinos followed slightly later by the pair neutrinos smoothly diffusing out on a time scale of 100's of ms. In other models the initial neutronization neutrinos blend smoothly into the pair neutrinos and in still other models there are multiple bounces of the core with each bounce producing a burst of neutrinos. Because of difficulties in determining the equation of state of hot, dense collapsing matter, it may be that the only convincing way to resolve these differences is by detecting the neutrinos from a collapse and determining their time spectrum.

In the past year the understanding of the equation of state has increased significantly, thanks to the efforts of people such as Jim Lattimer, Gerry Brown, Hans Bethe, Chris Pethick, Geoff Ravenhall, and Don Lamb (see reference 10). They have shown that during the collapse from $\rho \sim 10^{10}$ g/cm^3 to nuclear densities at $\rho \sim 2 \times 10^{14}$ g/cm^3, the dominant pressure term comes from the $\gamma = 4/3$ degenerate electron gas. Although it used to be thought that a transition to a $\gamma \sim 5/3$ thermal nucleon gas would occur before nuclear densities, these workers showed that instead of photodisintegration occurring, the thermal energy will merely cause the complex nuclei to go into excited states, thus resulting in fewer free nucleons. Only when the nuclei come into physical contact with each other at $\rho \gtrsim 2 \times 10^{14}$ will the nucleon interactions beat out the electron pressure. With this new result, it means that the strength of the resultant hydrodynamic bounce will depend on the still mysterious nucleon–nucleon interaction. Thus, in an indirect way, neutrino astronomy could shed some light on the nature of the nucleon–nucleon interaction.

It should be noted that since the nature of the core bounce is not resolved, then the role of the neutrinos in any subsequent mass ejection is still unresolved as well. However it is now apparent that as a result of the recent accelerator experiments confirming $SU(2) \times U(1)$ the basic neutrino interaction physics is much better understood and thus effects the hydrodynamic less than the equation of state uncertainties. Neutrinos can transport both energy and momentum to the eventually ejected matter (see reference 6 and references therein), but they can also damp the shock as well (see reference 4), thus the final role they play in the mass ejection will depend on the underlying hydrodynamics, which are primarily determined by the equation of state.

Regardless of how these questions of the supernova mechanism are resolved, it seems clear that the bulk of the 10^{53} erg binding energy does come off in neutrinos including not only ν_e, but also $\nu_\mu\bar{\nu}_\mu$, and $\nu_\tau\bar{\nu}_\tau$ and $\bar{\nu}_e$. The exact value of the Weinberg angle will determine the admixture of ν_μ's and μ_τ's (see reference 6) with the current best value of $\sin^2\theta_\omega \sim 0.25$ resulting in the minimum admixture. (For $\sin^2\theta_\omega = 1/4$ the neutral vector current for ν_μ and ν_τ production vanishes and only the axial vector current contributes.)

ULTRAHIGH-ENERGY NEUTRINOS ($E_\nu \gtrsim 10$ Gev)

This subject has been rather thoroughly reviewed in the DUMAND summer workshop proceedings[12,13] and in papers by Berenzinsky and Zatsepin,[14] Margolis *et al.*[15] and Eichler and Schramm[16] and will be only briefly described here. The basic point is that whenever high-energy proton collisions occur in the universe π's and κ's are produced. These mesons then decay to produce neutrinos. Places where such high-energy collisions occur can be classified as follows.

A. Diffuse sources
   1. *Atmospheric*
      High-energy cosmic rays hitting the Earth's atmosphere.
   2. *Interstellar*
      Cosmic rays hitting the interstellar gas and dust.
   3. *Extragalactic*
      Integral of distant sources possibly enhanced by higher number of high-energy particles early but degraded by redshift.

B. Point sources
   1. *The Sun*
      a. High-energy solar flares, particularly those on the far side of the sun, will result in downward moving particles which intercept the sun and yield neutrinos.
      b. The galactic cosmic-ray flux will also hit the sun and produce neutrinos.
   2. *Pulsars and other galactic cosmic-ray sources*
      If cosmic-ray sources have any surrounding material or a nearby stellar companion then the newly produced cosmic rays will collide and produce ν's. Cyg X-3 may be just such a source as is indicated by the high-energy γ-ray data.
   3. *The galactic center*
      The high density of cosmic rays and interstellar gas in the dirction of the galactic center may enable it to stand out above the background.
   4. *Active galactic nuclei, Seyferts, and quasars*
      Many standard models for these objects involve large fluxes of high-energy particles. If true, then these objects might be detectable as point neutrino sources. The high-energy γ flux from 3C 273 lends support to these ideas.

One point that should be remembered is that if the high-energy proton collisions occur in an optically thin region (less than ~ 100 g/cm^2 column thickness between us and the site), then the π^0 γ rays which are also produced in the *pp* collisions will escape and should also be detectable. If the observed γ-ray fluxes from 3C 273 and Cyg X-3 are due to π^0's then these objects should also be detectable in neutrinos (see detailed discussions in reference 13). It should also be remembered that because of the lack of observable high-energy γ rays from some objects, these objects either do not have high-energy proton collisions or the high-energy collisions are occurring in an optically thick region. Since particle acceleration presumable requires a low-density region, the observation of neutrinos from these objects would imply that the acceleration region was surrounded by some sort of cocoon of denser material of thickness $\gtrsim 100\ \mathrm{g/cm^2}$.

The diffuse neutrino sources mentioned above may provide a powerful source of neutrinos for particle physics investigations at energies far above that achievable in terrestrial accelerators. They may also be used (as emphasized by Eichler) to set limits on post big bang deuterium production. However the primary excitement for astronomers comes from the point sources. Here one is closely examining the inner workings of some of the most exciting objects in the sky. As we will discuss later, the proposed DUMAND detector should be able to resolve the direction of incident neutrinos to better than a square degree. This may enable point sources as weak as ~3 counts/yr for $E \gtrsim 1$ TeV to be pulled out of the background of ~0.3 counts/yr/sq. deg.

Neutrino Detectors and the Decay of the Proton

Since the ^{37}Cl solar neutrino detector has been discussed in detail many places, I will not describe it here. Nor will I go into the proposed ^{71}Ga–^{71}Ge detector which would see the main-line neutrinos from the *pp* chain.

The detectors of the collapse neutrinos and of the ultrahigh-energy neutrinos are, at the present time, basically of the Cherenkov type and have been described in some detail by Chudakov[17] and Lande.[18]

For collapse neutrinos, the reaction is either

$$\bar{\nu}_e + p \rightarrow n + e^+ \{[\sigma \sim 7.5 \times 10^{-44} E_\nu (\mathrm{MeV}) \mathrm{cm}^2]$$

or

$$\nu_e + e \rightarrow \nu_e + e \{[\tau \sim 1.5 \times 10^{-44} E_\nu (\mathrm{MeV}) \mathrm{cm}^2]$$

each producing a rapid lepton and thus Cherenkov radiation.

It should be noted that for water, as is used for example in the Mount Blanc Tunnel and the Homestake Gold Mine detectors, there are ~5 electrons per proton, thus the relative detection efficiency of $\bar{\nu}_e$'s to ν_e's is $\sim 1/E_\nu$(MeV). For the typical 10 MeV ν's from collapse this yields $\sigma(\nu_e)/\sigma(\bar{\nu}_e) \sim 1/10$. But remember that in collapse, there are initially ~10 times more ν_e's emitted than $\bar{\nu}_e$'s because of neutronization. Thus one might expect two equal signals, one from the ν_e neutronization burst and the second from the subsequent pair cooling.

At the present time a gravitational collapse in the center of the galaxy could be detected by operating detectors at the Homestake Gold Mine, Barberton (Ohio), the Mt. Blanc Tunnel, and the Soviet Caucasus. The detectors at Homestake and in the Caucasus each have a mass sufficiently large (several hundred tons) that they could detect $\gtrsim 10^{53}$ erg of neutrinos from *anywhere* in our glaxay. In principle, these should provide an early warning system for any collapse event since the neutrino burst should be coincident with (or slightly precede by ~sec) any gravity wave burst and precede the optical light peak from a supernova by the order of a week. So far the only "collapselike" signal recorded was the "Lande event"[19] of several rapid millisecond pulses. This event could conceivably have been the multiple bounces of a collapsing stellar core, however, no optical burst was ever seen; thus, either there was no optical burst, it was behind some dense obscuration, or the event was not caused by collapse.

In discussing the detection of gravitational collapse, there is frequently some debate between neutrino advocates and gravitational wave advocates. (For a review of this topic see reference 20.) In actual fact the two methods complement each other. For our galaxy it is clear that neutrino detectors offer the surest way to see a collapse. With 10^{53} erg emitted in neutrinos, the existing detectors mentioned above should have no difficulty in seeing a collapse. Since the fraction of the 10^{53} erg binding energy coming off in gravitational waves could be as small $\sim 10^{-12}$ (see Turner and Wagoner[21]), one cannot guarantee any gravity wave detection. Of course Gr–ν coincidences would obviously show GR generated above the minimum. Unfortunately the supernova rate in the galaxy is less than one every ~15 years (see Tammann[22]). If one could see all the collapses in the Virgo cluster then the rate would be better than a supernova a year. However neutrino astronomy runs into difficulties in looking at extragalactic supernovae because of the $1/r^2$ effect. If the flux drops by the $1/r^2$ then the detector size must increase by the same factor to yield a detectable signal. Thus if it takes a ~1000 ton detector to see our entire galaxy out to ~10 kpc then it would take a 10^7 ton detector to see the Virgo cluster at 10 Mpc. In order to see ~10 MeV neutrinos one must outfit a detector with photomultipliers at the density of $\gtrsim 1$/ton. Thus it is clear that a 10^7 ton detector with that density of photomultipliers is prohibitively expensive. Later, when we discuss DUMAND, we will discuss 10^9 ton detectors but then we will concentrate on high-energy neutrinos so that the photomultiplier spacing can be $\gtrsim 20$ meters. By pushing present technology to its limits and putting the maximum-mass detector in the mine and by utilizing clever geometry, Lande may marginally be able to see extragalactic supernovae. It is clear that even with the bulk of the binding energy coming off in ν's, extragalactic collapse neutrino astronomy is not easy.

TABLE 1

NEUTRINO DETECTION	GRAVITY WAVE DETECTION
Fermions	Bosons
$\sigma \propto E^2$	$\sigma_{max} \propto \frac{\lambda^2 \Gamma_g}{\Gamma_{TOT}} \propto \frac{1}{E^2} \frac{\Gamma_g}{\Gamma_{TOT}}$ yields $\sigma \propto 1/E$
Incoherent	Coherent
$\lambda \ll$ detector size	$\lambda >$ detector size
(SN puts out at ~10 MeV-limited number at low E because of Pauli exclusion.)	(If gravitons produced they are at low energies because they are bosons.)
Optimal for SN in our galaxy because bulk of binding energy in ν's.	If fraction of binding energy in GR is not zero, then GR may be optimal for SN in Virgo Cluster.

Fortunately, gravity waves don't have the same problems. The reasons are straightforward and are listed in TABLE 1. Basically, gravity waves are detected coherently with a cross section increasing at low energies, whereas neutrinos are detected incoherently with a cross section which increases at higher energies. (Only extremely low-energy ν's could have a large enough wave length to coherently interact via neutral currents with a macroscopic detector. This may eventually be of some use in detecting anisotropies in the 2 K cosmic background neutrino flux but is of no importance for $E_\nu \gtrsim 10$ MeV collapse neutrinos.)

To detect ultrahigh-energy neutrinos one again can utilize Cherenkov detectors, where now the interaction produces a hadron shower and a fast muon which produce optical flashes. An alternative detector posibility is to *hear* the acoustic pulse generated by the hadron shower (see reference 12 and references therein). Unfortunately, the acoustic method only works for $E_\nu \gtrsim 10^4$ TeV where the fluxes are probably extremely low.

Current detectors, such as Lande's and Reines' can already set interesting limits on the fluxes from extraterrestrial objects. However, unless we are extremely lucky, a definitive detection of an extraterrestrial point neutrino source will have to await the construction of the *D*eep *U*nderwater *M*uon *a*nd *N*eutrino *D*etector (DUMAND). This is a proposed cubic kilometer array of photomultipliers (and hydrophones) which would be set up for shielding purposes ~5 kilometers underwater (see references 12 and 13). Depending on the geometry of the array, such a detector should be able to resolve incident neutrino angles to better than a square degree.

In addition to neutrino astronomy, big neutrino detectors also have the ability to do other things of interest. For example, DUMAND will be able to examine ultrahigh-energy neutrino interaction which should be able to produce intermediate vector bosons and perhaps even Higgs particles. In the rest of this section I thought I would discuss another interesting bit of physics which big neutrino detectors can do, namely, measuring the lifetime of the proton. Recent new developments in grand unification theories have indicated that proton decay is indeed expected at some level and that this may be directly related to the origin of the net baryon number of the universe. Since proton decay would be a background for neutrino astronomy experiments, they are able to set lifetime limits.

Although there are many different versions of the detailed argument, (see preprints by Weinberg, Susskin, Gaillard, and Nanopoulis) the basic idea is as follows. There exists some grand unification scheme which unites the strong, weak, and electromagnetic forces through some superweak interaction which involves a massive boson. This massive unification boson interacts in such a way as to not necessarily conserve baryon number (since it unifies both weak and strong forces) and it must contain some CP violation so as to generate the observed properties of kaons. If, following Weinberg,[23] we call the unification boson X, then because of CP violation and baryon nonconservation, the mean net baryon number following X and $\overline{X}$ decay may be nonzero. If in the early universe $kT \gtrsim m_x c^2$, then X and $\overline{X}$'s will be produced. As the universe expands, kT drops below $m_x c^2$ and X creation ceases. The X's and $\overline{X}$'s that are left will then decay and if the superweak interaction is frozen out at that time, the reverse reactions will not occur and the universe will end up with a net baryon number related to the CP violation component of this interaction. Thus the microscopic physics yields the ratio n_B/n_γ which was previously thought to be an arbitrary initial condition of the universe.

It is obvious that if such a unification interaction exists then at some level the proton will be unstable with a lifetime $\tau_p \sim m_x^4/\alpha^2 m_p^5$ where α is the fine-structure constant and m_p is the proton mass. Reines[25] has been able to set a limit of $\geq 10^{29}$ yr on the proton lifetime from his neutrino astronomy experiments. Lande's experiment should be able to set a limit of $\sim 10^{31}$ yr within a year. Current predictions from $SU(5)$ and $SO(10)$ theories[26] imply that $M_x \sim 10^{15}$ GeV and $\tau_p \sim 10^{33\pm1}$ yr, thus explicit testing of these ideas may not be too far off. One interesting side light here is that to push τ_p limits beyond $\sim 10^{31}$ yr requires identifying and eliminating the astronomical neutrino background. Considering how hard it has been to get successful neutrino telescopes going it's curious that the next generation of experiments may need to be so sensitive that the neutrino counts are the background.

Summary

Neutrino astronomy is beginning to be a vibrant and active field with experiments concentrating on three energy regions. The discoveries from these three regions may be some of the most exciting in astronomy in that they probe in a unique manner some of the most interesting objects in the sky. It may also be that a by-product of these experiments may be the proof that the proton is unstable and this may tell us about the origin of the net baryon number of the universe.

As a final thought, it is fun to contemplate the nature of an open universe 10^{33} yr old when all the protons are decaying.

Acknowledgments

I would like to especially thank Drs. Lande, Steinberg, and Galeotti for their help in bringing me up to date on the current status of their detectors. I would also like to thank any previous collaborators for their help on various aspects of this work, particularly Dave Eichler on the ultrahigh-energy neutrinos.

References

1. Burbidge, M., G. Burbidge, W. A. Fowler & F. Hoyle. 1957. Rev. Mod. Phys. **29:** 547.
2. Davis, R. 1978. Proc. Brookhaven Solar Neutrino Conf. **1:** 1.
3. Bahcall, J. N. 1979. Space Sci. Rev. In press.
4. Wilson J. R. This volume.
5. Arnett, W. D. This volume.
6. Freedman, D. Z., D. N. Schramm & D. Tubbs. 1977. Ann. Rev. Nucl. Sci. **27:** 37.
7. Shapiro, S. 1978. Proc. Workshop on Gravitational Radiation. Seattle, Wash.
8. Kazanas, D. & D. N. Schramm. 1977. Ap. J. **214:** 819.
9. Yang, J. D., N. Schramm, G. Steigman & R. Rood. 1979. Ap. J. **277:** 697.
10. Bethe, H., G. Brown, J. Applegate & J. Lattimer. 1979. NORDITA preprint.
11. Weinberg, S. 1978. Proc. Tokyo Conference on High Energy Physics.
12. Roberts, A. Ed. 1977. 1976 DUMAND Summer Workshop, Fermilab.
13. Roberts, A. Ed. 1979. 1978 DUMAND Summer Workshop, Scripps.
14. Berenzinsky, V. & G. Zatsepin. 1975. Astrophys. Space Sci. **36:** 17.
15. Margolis, S., D. N. Schramm & R. Silberberg. 1978. Ap. J. **221:** 990.
16. Eichler, D. & D. N. Schramm. 1978. Nature **275:** 904.

17. RYZAHSKAYA, G. G. & A. Z. CHUDAKOV. 1978. Proc. Neutrino '77, Soviet Acad. Science.
18. DEAKYN, K., W. FRATI, K. LANDE, C. LEE, R. STEINBERG & E. FENYVES. 1978. Proc. Neutrino '78, Purdue University., p. 887.
19. FRATI, W. K. LANDE, J. SUPPLEE, G. BOZOKI, E. FENYVES & C. K. LEE. 1975. Ann. N.Y. Acad. Sci. **262:** 219.
20. KAZANAS, D. & D. N. SCHRAMM. 1978. Proc. Workshop on Gravitational Radiation. L. Smarr Ed., Seattle, Wash.
21. TURNER, M. S. & R. V. WAGONER. 1978. Proc. Workshop on Gravitational Radiation. Seattle, Washington.
22. TAMMANN, G. *In* Supernovae. 1977. D. N. Schramm, Ed. Reidel, Dordrecht.
23. WEINBERG, S. 1979. Harvard University preprint.
24. TURNER, M. S. & D. N. SCHRAMM. 1979. Nature. In press.
25. REINES, F. 1979. Proc. Neutrino '79. Bergen, Norway. In press.
26. NANAPOLOUS, G. 1979. Proc. Madison Conf. on Proton Stability. D. Cline, Ed. University of Wisconsin.

MAGNETOROTATIONAL MODEL OF SUPERNOVAE EXPLOSIONS

G. S. Bisnovatyi-Kogan

Space Research Institute
Moscow 117810, USSR

INTRODUCTION

Many efforts in the construction of the physical quantitative model of supernovae (SN) explosions are based on spherically symmetrical (SS) calculations. There have been many achievements with this approach, but one essential problem is not yet solved; one must explain the formation of the neutron star–pulsar as a result of the strong SN explosion.

Two types of SN models in the SS approximation are usually considered. In the first type the explosion follows the development of a hydrodynamical instability at the endpoint of the evolution of massive stars $M > (8\text{–}10)M_\odot$, after formation of an iron core. The reflection of the infalling matter from the surface of the stable neutron core (bounce), the nuclear explosion in the outer parts of the collapsing star, and neutrino deposition may give in principle an SN explosion in this model. All these effects were taken into account in the elaborate calculations made by Nadyozhin,[1-3] but the explosion was not obtained. The main reason of this is the action of a strong neutrino flux from the neutrino-opaque core, which leads to a rather slow contraction and slow nuclear burning without bounce and an explosion.

The second type of SS model is based on a consideration of thermal instability and nuclear explosion in the degenerate C-O core, which may form in the evolutionary process of stars with $(4\text{–}6) \leqslant M \leqslant (8\text{–}10)\ M_\odot$. Extensive calculations in this model have been made[4] for different initial conditions that take into account the important neutrino heating of matter in the mantle of the star, which leads to ignition of the nuclear fuel. For initial central densities $2 \times 10^9 < \rho_c < 9 \times 10^9$ g/cm^3 a total disruption of the star with energy output 10^{50}–10^{51} erg was obtained, and in the case $9 \times 10^9 < \rho_c < 3 \times 10^{10}$ g/cm^3, a neutron star was formed, but the kinetic energy output was less than 6×10^{49} erg, which is not sufficient to explain SN explosions (see also 5 and 6).

MAGNETOROTATIONAL (MGR) MODEL: QUALITATIVE PICTURE

Strong explosion of SN energetics may occur after neutron star formation with rotation as the energy source. The mechanism of the transformation of the rotational energy into the kinetic energy of the outburst is based on the magnetic connection between rotating stellar layers. The MGR mechanism of SN explosion was first considered in reference 7. In reference 8 an analogous mechanism was proposed for the energy support of the Crab nebula by a central rotating magnetized body, which later was discovered as the pulsar.

If a presupernova star rotates sufficiently rapidly, then during contraction the

0077–8923/80/0336–0389 $01.75/1

outer layers will be supported by centrifugal forces and will form a differentially rotating flattened configuration ("disk") around a stable neutron star. Such a system may be formed in both types of SN models discussed above, if the presupernova mass does not exceed (2–3) $M_{\odot}$.

The magnetic field that connects the "disk" and the star linearly increases in time during differential rotation. The magnetic pressure grows and the angular momentum flows into the outer parts of the system. The compression wave that is formed near the surface of the neutron star propagates outward, is amplified, and leads to an outburst of matter with large velocities. So, the rotational energy of the system is transforming into the kinetic energy of the outbursting matter (see also reference 9). In order to evaluate the effectiveness of this transformation it is necessary to make numerical calculations.

MGR Model: Quantitative Consideration

The simplified model of MGR explosion was calculated in reference 10. The two-body mechanical model—neutron star + envelope—was considered with magnetic connection between them. For some initial conditions the strong outburst was obtained and the importance of magnetic pressure was emphasized.

Magnetohydrodynamical calculations in the one-dimensional cylindrical model have been made.[11,12] A rigidly rotating core and a differentially rotating envelope of equal mass were considered as initial conditions representing the neutron star and the "disk." The initial magnetic field was taken to be radial outside the core, whose radius was supposed to be constant. The following set of equations was solved numerically in the envelope:[11]

$$\frac{\partial}{\partial t}\left(\frac{1}{\rho}\right) = \frac{\partial (r v_r)}{\partial \vartheta}, \qquad \frac{\partial r}{\partial t} = v_r, \qquad r\frac{\partial \vartheta}{\partial t} = v_\vartheta,$$

$$\frac{\partial v_r}{\partial t} - \frac{{v_\vartheta}^2}{r} = -r\frac{\partial P}{\partial \vartheta} - \frac{1}{8\pi r}\frac{\partial}{\partial \vartheta}(rH_\vartheta)^2 + g,$$

$$g = -2G\frac{2\pi\,(M_0 + \vartheta)}{r},$$

$$\frac{\partial (r v_\vartheta)}{\partial t} = \frac{rH_r}{4\pi}\frac{\partial (rH_\vartheta)}{\partial \vartheta}, \qquad (1)$$

$$\frac{\partial}{\partial t}\left(\frac{H_\vartheta}{\rho r}\right) = rH_r\frac{\partial}{\partial \vartheta}\left(\frac{v_\vartheta}{r}\right),$$

$$rH_r = \mathrm{A} = \text{const},$$

$$\frac{\partial \epsilon}{\partial t} = -P\frac{\partial}{\partial \vartheta}(r v_r) - f_\nu(\rho, T),$$

where t is time, r the Eulerian radius, φ the azimuthal angle, ρ the density, $\varphi(d\varphi = \rho r dr)$ the Lagrangian mass coordinate, V_r and V_φ the radial and azimuthal velocities, P the pressure, ϵ the internal energy, T the temperature, H_r and H_φ the radial and azimuthal components of magnetic field strength, g the gravitational force, M_0 the mass of the core with the units of cylinder length and one radian, f_ν the neutrino energy losses, and the time derivatives are taken in the Lagrangian sense.

System (**1**) was solved in the region $t > 0$, $0 < \varphi < M$, $R_0 < r < R(t)$, where $M = M_0$ is the mass of the envelope of the cylinder of unit length and one radian; R_0 and $R(t)$ are the internal and external radii of the envelope. The total angular momentum of the system was supposed to be constant and the following boundary conditions were considered:

$$P(M,t) = 0, \qquad H_\vartheta(M,t) = 0,$$
$$v_r(0,t) = 0, \qquad v_\vartheta(0,t) = R_0\omega(t), \tag{2}$$

where $\omega(t)$ is the angular velocity of the core. The functions $P(\rho, T)$, $\epsilon(\rho, T)$, and

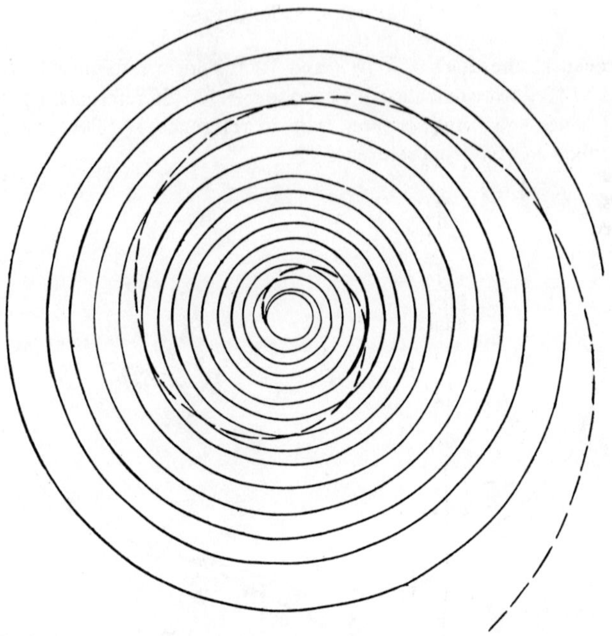

FIGURE 1. The form of the field lines near the core at $t_\alpha = 7$ for $\alpha = 10^{-2}$ (dashed line) and $\alpha = 10^{-4}$ (solid line).

$f_\nu(\rho, T)$ have been represented by the interpolation formula, taking into account the major properties of the matter.[11]

The transformation of the rotational energy into the energy of the outburst, formation of the shock wave, neutrino emission, and other features of the process were obtained in the numerical calculations of the system (**1**), (**2**) made in references 11 and 12. The important physical parameter of the problem is the ratio of the initial magnetic energy to the rotational energy:

$$\alpha = \frac{A^2}{4\pi M V_0^2}. \tag{3}$$

If the core is near the limit of its rotational instability $R_0 = 10^6$ cm, $2\pi\ M_0 = 0.5 \times 10^{-6}$, $V_\varphi(0, 0) = V_0 - \sqrt{10^{-6} GM_\odot}$, which corresponds to a neutron star with $M = M_\odot$, then

$$H = \sqrt{\alpha}\sqrt{\frac{8\pi G M_0^2}{R_0^2}} \simeq 5 \times 10^{17}\ \sqrt{\alpha}\ G. \quad (4)$$

For $H = 5 \times 10^{13}$ G, as on the surface of the pulsar with the strongest field,[13] we have $\alpha = 10^{-8}$. Inside the neutron star the magnetic field may be stronger. If $H_{in}/H_{out} \simeq 10^3$ is the ratio of the field in a sunspot to the average solar field, then we may consider $\alpha = 10^{-2}$.

Numerical Results

In reference 11 the cases $\alpha = 10^{-1}$ and 10^{-2}, and in reference 12 the cases $\alpha = 10^{-2}$, 10^{-4}, and 10^{-8} were calculated. The case $\alpha = 10^{-2}$ in reference 12 was computed for physical times ~40 times greater than in reference 11. The description of the results will follow mainly from reference 12.

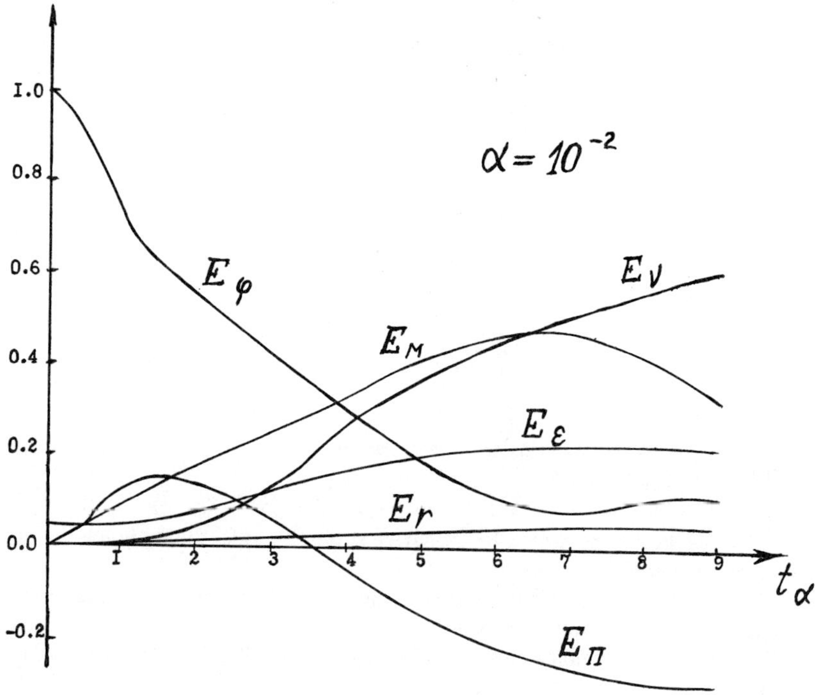

FIGURE 2. Dependence on time t_α of energies E_φ, rotational; E_r, radial kinetic; E_ε, internal; E_M, magnetic; E_Π, potential; and E_ν, total neutrino losses; for $\alpha = 10^{-2}$. All values are normalized to $(E_\varphi)_{max} = 10^{47}$ erg/cm.

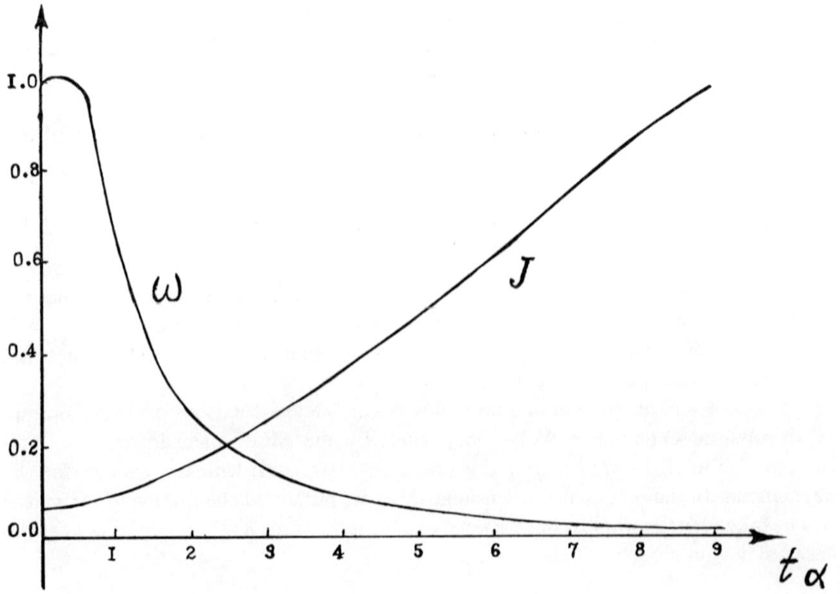

FIGURE 3. Dependence on time t_α of specific angular momentum J and angular velocity ω in one of the last mass intervals. The values J and ω are normalized to their maximal values.

It was found that the transformation coefficient for the change from rotational to radial kinetic energy η changes very slowly. For $\alpha = 10^{-1}$–10^{-8} we have $\eta = 8$–3.5%, the outbursting mass changes very little, and $\mu \approx 0.13M$ in all cases.

It is necessary to note that in the cylindrical model the finite amount of matter cannot flow to infinity because of the divergence of the cylindrical gravitational potential. So the outflowing part of the matter is stored far from the core $(10^3$–$10^4)$ R_0 and performs the magnetorotational oscillations. A comparison of the cylindrical and spherical potential energies permits an evaluation of the mass fraction μ and coefficient η. It is suitable to introduce t_α instead of the physical time t: $T_\alpha = t\alpha^{1/2}$. All normalized physical functions, for example, $V_{r\alpha} = V_r\alpha^{-2}$ and $H_{\vartheta\alpha} = H_\vartheta\alpha^{1/2}$, depend on t_α in a very similar manner for all $\alpha \lesssim 10^{-2}$. One may state that the solution tends to the limiting one at $\alpha \to 0$. From this point of view it is easy to understand, that η and μ do not depend on α at $\alpha \to 0$.

The magnetic pressure is important for the MGR explosion. The maximum of magnetic energy E_M is reached at $t_\alpha \approx 6.5$, for all α, and $(E_M)_{max} \approx 0.5\ E_{rot,init.}$. At the end of the process the system core + envelope separates into the rigidly rotating core + 0.85 envelope, which merges together, and the ~0.15 envelope, which oscillates at a large radius in the cylindrical model and which would be burst out with a large velocity in the real, spherical case. A significant part of the energy, $\sim 0.6\ E_{rot,init.}$, is emitted as neutrinos and $\sim 0.2\ E_{rot,init.}$ is provided by gravitational contraction to the rigidly rotating central system. The residual angular velocity ω of the central part at $t_\alpha = 10$ is $\sim 1/3$ of the initial ω_0 for all α. The sign of the ω of the central object changes during the oscillations in our calculations, which indicates that in the real spherical case, where continual oscillations do not take place because of the outflow of the matter to infinity, the sign of ω may change several times and the sign of residual ω in the remnant neutron star may be opposite to the initial one. Some results of the calculations are presented on FIGURES 1–3 from references 11 and 12.

The time of the energy production τ increases with decreasing α. For evaluation in the cylindrical model we may adopt

$$\tau \approx 10^2 \frac{R_0}{\sqrt{2\pi G M_0}} \alpha^{-1/2} \approx 0.02\, \alpha^{-1/2}\ \text{s}.$$

The magnetic field drops more rapidly for the real spherical case, which may increase the time τ by one or two orders of magnitude, so for $\alpha \simeq 10^{-8}$ we may have $\tau \sim 10$ hours. Such slow energy production gives a better light curve for SNI,[14] where the presupernova star must be a compact object.[15]

In conclusion, one may note that cylindrical model calculations give very promising results for explanation of SN explosions by the MGR mechanism. $\eta = 3.5\%$ corresponds to $E_{SN} = 3.5 \times 10^{51}$ erg ($E_{rot,max} \simeq 10^{53}$ erg), which is enough for this phenomena. In the real three-dimensional case the picture of the motion of the matter will be much more complex, but we hope to obtain the same η. The calculations for this case are now in progress.

REFERENCES

1. NADYOZHIN, D. K. 1977. Astrophys. Space Sci. **49:** 399.
2. NADYOZHIN, D. K. 1977. Astrophys. Space Sci. **51:** 283.
3. NADYOZHIN, D. K. 1978. Astrophys. Space Sci. **53:** 131.
4. GERSHTEIN, S. S., L. N. IVANOVA, V. S. IMSHENNIK, M. YU. KHLOPOV & V. M. CHECHETKIN. 1978. Supernovae of types I and II and the neutrino mechanism of the explosions of degenerate carbon–oxygen stellar cores. Preprint. Inst. Appl. Math. Acad. Sci. USSR N 104.
5. NOMOTO, K., D. SUGIMOTO & S. NEO. 1976. Astrophys. Space Sci. **39:** L37.
6. MAZUREK, T., D. MEIER & J. WHEELER. 1977. Astrophys. J. **215:** 518.
7. BISNOVATYI-KOGAN, G. S. 1970. Astron. Zh. **47:** 813.
8. KARDASHEV, N. S. 1964. Astron. Zh. **41:** 807.
9. KUNDT, W. 1976. Nature (London) **261:** 673.
10. AMNUEL, P. R., O. Ch. GUSSEINOV & F. K. KASUMOV. 1972. Astron. Zh. **49:** 1139.
11. BISNOVATYI-KOGAN, G. S., YU. P. POPOV & A. A. SAMOCHIN. 1976. Astrophys. Space Sci. **41:** 287.
12. ARDELYAN, N. V., G. S. BISNOVATYI-KOGAN & YU. P. POPOV. Investigation of the Magnetorotational supernova explosion in the cylindrical model. Astron. Zh. In press.
13. TAYLOR, J. H. & R. N. MANCHESTER. 1975. Astron. J. **80:** 794.
14. NADYOZHIN, D. K. & V. P. UTROBIN. 1977. Astron. Zh. **54:** 996.
15. SHKLOVSKYI, I. S. 1978. Astron. Zh. **55:** 726.

SUMMARY OF WORKSHOP ON STELLAR COLLAPSE AND SUPERNOVAE

Wolfgang Hillebrandt

Max-Planck-Institute für Physik und Astrophysik
8000 Munich 40
Federal Republic of Germany

Because detailed descriptions of the collapse of massive stars to neutron stars, followed presumably by supernova explosions, have been given elsewhere in this volume (see the articles by Arnett,[1] Schramm,[2] Weaver and Woosley,[3] and Wilson[4]) we can concentrate on some still controversial aspects of the problem which have been discussed in the workshop. Hydrodynamical studies (K.A. van Riper, N. Sack) showed that the quantity of key importance in all computer simulations of stellar collapse is the equation of state at high densities ($\rho \gtrsim 10^{12}$ g cm^{-3}) and temperatures ($T \gtrsim 10^{10}$ K). In van Riper's computations the collapse of the iron core of a massive star was simulated assuming spherical symmetry and adiabatic collapse in general relativistic hydrodynamics. An ideal gas equation of state was used, the adiabatic index (a function of density only) being essentially a free parameter. He found that supernovalike explosions can result from purely hydrodynamic processes (a core bounce initiating a reflected shock wave) if the bounce has a large amplitude while the velocity of the infalling matter is not too large. Such conditions are met if the adiabatic index does not exceed the critical value ($\gamma = 4/3$ in Newtonian hydrodynamics) by much. His models showed that both neutron star formation and explosion of the envelope are possible if the hot neutron star is formed near the maximum stable mass. These results suggest that neutron star masses should lie within a narrow range just below the maximum mass of a cold neutron star. In addition it was pointed out that general relativistic effects may become very important if the adiabatic index should turn out to be slightly larger than $4/3$ for densities $10^{11} \lesssim \rho$ g $\lesssim 10^{13}$ g cm^{-3}. (Such an equation of state was presented by Lattimer and will be discussed later.) While in Newtonian hydrodynamics the bounce can then be expected at around $\rho \simeq 10^{12} - 10^{13}$ g cm^{-3}, in general relativity the collapse will continue (because of $\gamma_{crit} > 4/3$) essentially to nuclear matter density.

In a similar Newtonian study Lichtenstadt and Sack (presented by N. Sack) discussed the effects of shock heating on the equation of state near bounce. They assumed that the pressure can be decomposed into a "cold" (degenerate) and a "thermal" (nondegenerate) part. While the cold pressure was a function of density only, the thermal pressure depended on both density and internal energy. Since shock heating increases the internal energy, the thermal pressure increases as well, in some cases giving rise to mass ejection. Energy dissipation in the shock was treated by the standard artificial viscosity technique. They found that the thermal stiffness (i.e., the adiabatic constant of the thermal part) determines whether or not there is mass ejection and also the kinetic energy of the ejected material. This result could be expected since this quantity determines what fraction of the kinetic energy of the infalling matter is transformed in an irreversable way into (thermal) pressure.

0077–8923/80/0336–0395 $01.75/1 © 1980, NYAS

One should bear in mind, however, that the artificial viscosity pressure is only a numerical method to make physical quantities continuous across the shock front. Therefore the magnitude of the artificial viscosity heating is somewhat arbitrary. Since, on the other hand, the results of van Riper and Lichtenstadt and Sack are strongly dependent on the effectiveness of artificial viscosity heating, we feel that their studies cannot answer the question completely of whether hydrodynamic effects alone can cause supernova explosions. Nevertheless, their papers clearly show the need for reliable equations of state.

Consequently, several talks dealt with the problem of calculating the properties of matter at high temperatures and densities. J.M. Lattimer presented the results obtained in collaboration with Bethe *et al.*[5] and Lamb *et al.*[6]. First of all they could show that during core collapse the entropy per baryon due to neutrino trapping will always stay rather small ($S/k \simeq 1$–2, where k is Boltzmann's constant). This in turn means that nuclei will not disintegrate before they touch at roughly nuclear matter density, because the entropy of free nucleons is too large. Therefore the adiabatic index will be close to $4/3$ (pressure of degenerate leptons and of photons) up to approximately $\rho = 1$–2×10^{14} g cm^{-3} and then increase very rapidly to values of around 2.5. Numerical calculations were performed for a system of electrons, electron neutrinos, neutrons, protrons, α particles, and heavy nuclei in nuclear statistical equilibrium for both fixed electron concentration and fixed lepton concentration in β equilibrium. Heavy nuclei were described by a compressible-liquid-drop model. For low values of the entropy ($S \lesssim 3k$) their computations were in very good agreement with the qualitative considerations mentioned earlier. As a very interesting result they obtained an adiabatic index slightly larger $4/3$ for $S \gtrsim 1k$ and $4\times10^{11} \lesssim \rho \lesssim 10^{13}$ g cm^{-3}, which may influence computer simulations of stellar collapse significantly.

Because the equation of state is very sensitive to changes in the specific entropy and lepton number (and in nuclear statistical equilibrium both quantities are only changed by nonequilibrium weak interactions and neutrino losses) electron captures by free protons and heavy nuclei have to be considered. Methods of how to calculate nuclear β transitions under stellar conditions of high temperatures and densities were presented by K. Takahashi (in collaboration with K. Yokoi) and by M.J. Newman (in collaboration with G.M. Fuller and W.A. Fowler). The approach of Takahashi and Yokoi was based on the idea of the so-called β-strength function defined as an energetical average of squared nuclear matrix elements times the final nuclear level density. They applied the gross theory of β decay to calculate the transition rates, including transitions from thermally excited states. Their method, which was also applied to lower temperatures and densities, where matter is only partially ionized, can be used to predict transition rates even in cases where no or only few experimental data are available. It was pointed out, however, that the ordinary β decay expansion formalism breaks down when electron momentum times nuclear radius (in natural units) becomes larger than one. So at very high densities ($\rho \gtrsim 10^{12}$ g cm^{-3}) and for heavy nuclei the standard theory cannot be applied. This need not yield severe restrictions for supernova computations because at such high densities matter is close to β equilibrium anyway and then the composition is determined not by the individual reaction rates but by the chemical potentials of the constituents.

The approach of Newman *et al.*, on the other hand, utilizes as much experimental information as possible in order to compute astrophysical β-transition rates. Consequently their method is more applicable to the precollapse stage and the neutronization occurring there than to the collapse phase. They argued that very often several log *ft* values or nuclear matrix elements have been measured (from the ground state and perhaps from a low-lying isomeric state if one exists) and that this number can often be extended by microscopic reversibility and mirror-nucleus considerations. At low temperatures and moderate densities these log *ft* values will often dominate the astrophysical rate, which can then be computed quite accurately. These arguments, which hold for intermediate mass nuclei and (fast) Fermi transitions, fail in cases where unmeasured Gamow–Teller transitions become important. Matrix elements from these transitions have to be estimated. They can in principle be obtained from large-scale detailed shell model calculations, but this process, while yielding rates of high reliability, is extremely time consuming. Therefore, for large numbers of nuclei (and beyond the *sd* shell) such calculations are hopeless. Newman *et al.* proposed instead to assume an average log *ft* value for all unmeasured Gamow–Teller transitions and to vary it in order to determine the sensitivity of the rates to this assignment. They found that often a particular transition dominates the stellar rate at a given temperature and density. To ignore the role of discrete states would then make order of magnitude errors in the astrophysical rate.

R. I. Epstein and C. J. Pethick (presented by R. I. Epstein) investigated with a simple, parametrized model of a stellar core the dependence of specific entropy and lepton number on electron-capture rates from the onset of dynamic infall until the collapse becomes almost adiabatic due to neutrino trapping. In their model the electron capture rates were taken to be proportional to the rates obtained with the naive shell model. For both the proportionality constant and the average excitation energy of the daughter nucleus, a range of values were considered. They found that for a rather wide range of initial conditions and collapse time scales, the changes in entropy and lepton number are insensitive to large changes in the electron-capture rates. This gives some hope to those working on very detailed supernova models because, as it was stated earlier, those rates are presently not very well known. Moreover, they found that the increase in entropy varies directly with the nuclear excitation due to an electron capture, that the decrease in lepton number is greater if the initial entropy is higher, and that neutrino blocking is important even before neutrino trapping is complete.

As was mentioned earlier, the matter in the cores of massive collapsing stars is a mixture of leptons, free nucleons, and highly excited and neutron rich nuclei. In order to compute the equation of state the properties of such nuclei have to be understood. While in most computations nuclei are treated as an ideal Boltzmann gas and their ground-state and excitation properties are calculated from mass laws and level density formulas, J. R. Buchler and R. I. Epstein (presented by R. I. Epstein) used a finite temperature Thomas–Fermi model instead. In their approach shell and pairing effects were neglected, which however may only be a reasonable approximation at high-excitation energies. At low temperatures the free energy of the nucleus was approximately given by $F \simeq F_0 - aT^2$, where a is the level density parameter and describes the lowest-order temperature corrections. The level density parameter was computed and was found to agree well with the general trend of the measured values for nuclei with mass numbers in the range 50–250.

Hydrodynamical simulations of stellar collapse and supernova explosions usually neglect effects from rotation and magnetic fields and it is often argued that those effects will always be too small to modify the dynamics significantly. E. Müller (in collaboration with W. Hillebrandt) demonstrated, however, that this need not be true. They proposed a supernova model in which mass ejection was caused by a two-step process. In the first stage, the rebounce of the core at nuclear density generates a hydrodynamical shock wave. If this shock wave is insufficient for mass ejection, reimplosion of the stellar mantle results. A typical time scale for reimplosion was found to be of the order of 0.5 to 1 s. Since during collapse angular momentum is essentially conserved, the core spins much more rapidly than the mantle, thus amplifying the magnetic field by winding the magnetic field lines. For sufficiently large values of the initial magnetic field strength, the magnetic pressure becomes comparable to the thermal pressure in a small region near the collapsed core, pushing a second magnetohydrodynamical pressure wave. Results of numerical computations (in a one-dimensional model) were presented which indeed showed mass ejection with energies of around several times 10^{50} erg for plausible initial stellar models, leaving behind a neutron star of roughly 1 to 1.5 $M_{\odot}$.

All papers discussed so far were based on detailed models of very massive stars which presumably give rise to type-II supernova explosions. Type-I supernovae, on the other hand, may be related to less massive stars since they also occur in ordinary old red elliptical galaxies. The question remains then why some low mass stars explode as supernovae, whereas most of them do not. An alternative, proposed by J. C. Wheeler, would be that some small fraction of massive stars, due to any of the several mixing mechanisms catalyzed by rotation and/or magnetic fields, live longer than normal on the main sequence. He claimed that this hypothesis may also provide a viable explanation of the blue straggler phenomenon. Mixing would stir in fresh fuel, lower the burning temperature and tend to cause evolution parallel to the main sequence, thus enhancing the apparent degree of straggle. Mixing would result in a higher than normal helium abundance and hence may play a role in the production of helium stars which also may have lifetimes which are excessive for their kinematic population. Following Wheeler, certain helium stars having relatively high mass, low hydrogen abundance, and belonging to old disk or halo populations are empirically viable candidates for the pregenitors of type-I supernovae.

The special case of the supernova in the Crab, which was probably also of type I, was discussed by R. K. Kochhar. He proposed that the Crab nebula is the result of an explosion of a helium star in a close binary already containing a neutron star. In this model the explosion was caused by the complete spiralling in of the neutron star into the helium star which was totally disrupted. The neutron star was spun up in this process and is now identified with the Crab-pulsar. Arguments in favor of that idea are the relatively small mass and the large helium abundance (not accompanied by large overabundances of other heavy elements) of the Crab nebula. According to Kochhar the energy released in an explosion caused by the spiralling in would be much less than that in a normal supernova explosion. Consequently a Crab-like nebula around a fast spinning pulsar would disperse in times much shorter than those for normal supernova remnants and this may explain the absence of supernova remnants around some pulsars.

Also from their radio properties, Crab-like supernova remnants are different from most others. They have a characteristic filled-center radio form and show high degrees of linear polarization. K. W. Weiler reported that by now five such objects (including the Crab) have been found and that two of them are associated with historical supernovae. Because of their radio characteristic he named them "plerions." He suggested that there are at least two types of supernovae: a) those which form a strong shock wave and quickly develop a shell of radio emission, and b) those where the shock wave is weak or nonexistent and where the appearance of the remnant is not controlled by the properties of the original explosion. In the latter case a pulsar with very short rotation period and large slow down rate builds and maintains a plerion-type remnant through conversion of rotational energy into other forms — in particular, relativistic electrons. He concluded that because they are dependent on the rotational energy lost by a rapidly rotating pulsar, plerions should be short-lived objects compared to the normal, shell-type supernova remnants (around 10^4 yr compared to 10^5 yr). This conclusion gets support from the observational fact that only 5% of all supernova remnants are of the plerion-type but 2 out of 6 historical supernovae have led to the formation of plerions.

Supernova explosions in close binary systems were discussed by J.P. De Cuyper. In his model he considered two stars of given mass which revolve around their common center of mass in circular orbits with constant absolute velocities. One of them was assumed to explode instantaneously as a supernova. The supernova remnant was assumed to have a certain mass and due to asymmetric mass loss to have received an extra kick velocity. The mass loss of the companion due to stripping and ablation, as well as the radial velocity received from the impact of the supernova shell, were also considered. Analytic expressions for the probability of having a bound system after the supernova explosion were derived by assuming a random distribution of the direction of the kick velocity. He found that if more than half of the total mass leaves the system, the survival probability is small ($< \frac{1}{2}$) for all values of impact and kick velocity. If, on the other hand, less than half of the mass is lost, for a variety of parameters (mass loss, kick velocity, and impact velocity) the system always remains bound.

K. J. Fricke, W. W. Ober, and S. E. Woosley (presented by W. W. Ober) investigated explosions of primordial supermassive stars with zero metal content. Such objects have been discussed earlier in the literature in connection with quasars and active galactic nuclei and as an efficient source for the enrichment of the pregalactic matter with heavy elements. The new result of Fricke *et al.* is that the CNO leak $^{15}O(\alpha, \gamma)^{19}Ne$ in "hot hydrogen burning" allows for a higher energy generation and thus also stars with $Z = 0$ show explosions. In particular, the collapse and nuclear explosion of a nonrotating supermassive star of 10^5 $M_\odot$ was investigated. The relativistic equation of motion, the (nonrelativistic) continuity equation, and the energy equation coupled to a simple nuclear reaction network (consisting of 29 nuclei and linked by 48 reactions) were solved by a finite differencing scheme. After around 4×10^4 s from the onset of the dynamical instability the collapse was stopped and the star exploded with a kinetic energy of $\simeq 10^{56}$ erg. Temperatures up to 10^9 K and densities of around 200 g cm^{-3} were achieved for a time scale of the order of 10^3 s. The final chemical composition was CNO $= 3 \times 10^{-7}$ and Fe $= 2 \times 10^{-5}$. It was argued that the rather large ratio of Fe to CNO is due to shortcomings in the adopted nucleosynthesis network and to uncertainties in some theoretical reaction rates. Nevertheless, primordial supermassive stars (if they existed) could produce an early burst of metal enrichment to explain the chemical abundances of old stars.

F. R. Tangherlini derived classical and quantum restrictions for a test particle travelling in the gravitational field of a Schwarzschild black hole. Because in general relativity the gravitational mass of the particle increases (in contrast with Newtonian theory) the condition that the test body exerts negligible influence on the central body becomes invalid for sufficiently small distances from the Schwarzschild radius. The effects of complementarity and the uncertainty principle were also discussed. His analysis lead to an extremely small quantum distance below which the Schwarzschild geodesic motion is expected to break down. Both the classical and quantum restrictions were smaller than the Planck length.

In a final paper, J. Pachner presented a new method for the numerical integration of spherically symmetric Einstein equations. He claimed that several of the results obtained in earlier studies of relativistic spherically symmetric collapse, like bounce, shock waves, etc., are due to numerical instabilities rather than physical effects. According to him the main source of the instability is in the three-point Lagrange differentiation formula used in the finite difference method for computations of derivatives with respect to the radial coordinate. In his method the two constraint equations are replaced by three differential equations which can be integrated by the standard Runge–Kutta method with very high accuracy. As a numerical example, a zero-pressure collapse was computed and very high accuracy was found in comparison to the analytic solution. In addition, the collapse of a star consisting of an ideal fluid with an equation of state $p = \frac{2}{3} \rho\epsilon$ was computed. In that example the extremely high pressure gradient in the small central region was able to reverse the collapse to an explosion.

Acknowledgments

My thanks go to all colleagues who contributed to the workshop by presenting papers with their newest results or by making lively discussions possible. This summary could not have been written without the help of the speakers, who provided me with abstracts of their talks, which in many cases I could incorporate into this article essentially unchanged.

List of Papers Presented at the Workshop on Stellar Collapse and Supernovae*

Buchler, J. R. & R. I. Epstein:* Thomas-Fermi models of warm nuclei.

De Cuyper, J.-P.: Asymmetric mass ejection and post-supernova orbits of close binaries.

Epstein*, R. I. & C. J. Pethick: Lepton loss and entropy generation in collapsing stellar cores.

Fricke, K. J., W. W. Ober* & S. E. Woosley: Exploding primordial supermassive stars with $Z=0$.

Kochhar, R. K.: On the origin of the Crab nebula.

Lattimer, J. M.: Hot dense matter and stellar collapse.

Müller*, E. & W. Hillebrandt: A magnetohydrodynamic supernova model.

Newman*, M. J., G. M. Fuller & W. A. Fowler: The importance of discrete nuclear states in stellar beta decay.

Pachner, J.: A new method for the numerical integration of spherically symmetric Einstein equations.

SACK*, N. & I. LICHTENSTADT: The dependence of stellar collapse on the equation of state.

TANGHERLINI, F. R.: Classical and quantum restrictions on geodesic motion near a Schwarzschild black hole.

VAN RIPER, K. A.: General relativistic stellar collapse and neutron star formation.

WEILER, K. W.: Crab-like supernova remnants (plerions).

WHEELER, J. C.: Helium stars and supernovae.

YOKOI, K. & K. TAKAHASHI:* Astrophysical beta transitions.

*Asterisks denote speakers.

REFERENCES

1. ARNETT, W. D. 1979. Physical aspects of collapse. Ann. N.Y. Acad. Sci. This volume.
2. SCHRAMM, D. N. 1979. Neutrinos from the formation of neutron stars and black holes. Ann. N.Y. Acad. Sci. This volume.
3. WEAVER, T. A. & S. E. WOOSLEY. 1979. Evolution and explosion of massive stars. Ann. N.Y. Acad. Sci. This volume.
4. WILSON, J. R. 1979. Collapse and supernovae. Ann. N.Y. Acad. Sci. This volume.
5. BETHE, H. A., G. E. BROWN, J. APPELGATE & J. M. LATTIMER. 1979. Equation of state in the gravitational collapse of stars. Nucl. Phys. In press.
6. LAMB, D. Q., J. M. LATTIMER, C. J. PETHICK & D. G. RAVENHALL. 1979. Phys. Rev. Lett. **41:** 1623.

CONDENSED MATTER—GENESIS OF A CONCEPT

Walter Sullivan

The New York Times
New York, New York 10036

Having attended most of the Texas Symposiums on Relativistic Astrophysics since that memorable one in Dallas during December, 1963, I have been asked to offer a layman's view of those meetings.

I doubt that I had ever heard of a black hole before the first Texas symposium, held only a few weeks after the Kennedy assassination a few blocks away. Indeed probably few astrophysicists had paid the idea much heed until the ferment consequent to the discovery of quasars[1] that led to that meeting.

The invitation to the symposium had gone out the previous June, signed by two of those at the present one—Peter G. Bergmann and Ivor Robinson—as well as E.L. Schücking and the late Alfred Schild. It said, in part:

> For more than ten years, the nature of the strong extragalactic radio sources has been one of the most fascinating problems of modern astronomy. For a time, it was believed that such radio sources were due to collision of galaxies. But it has emerged in the course of the last few years that this explanation is untenable in most of the cases. The spectacular nature of strong radio sources becomes clear if one considers the enormous amount of energy involved.

This, the invitation added, "has so far ruled out nearly all of the explanations and theories put forward to explain such extraordinary events." In February, it continued, Fred Hoyle and William Fowler had suggested "that energies which lead to the formation of radio sources could be supplied through the gravitational collapse of a superstar. Such an object, with a mass between one hundred thousand and one hundred million solar masses, would be located in the center of the galaxy. The gravitational collapse of this superstar could supply the necessary energy if it were to shrink down close to the Schwarzschild radius."[2]

The opening session of that first meeting was presided over by J. Robert Oppenheimer, the man who, with his colleagues at Berkeley, had set forth the first modern definition of the black hole. Present, as well, were John Archibald Wheeler, who coined the term, and Martin Schwarzschild, whose father had defined the Schwarzschild radius. At the close of the symposium, Philip Morrison pointed out that some participants, notably Hoyle and Wheeler, "have been bold enough to speculate on completely new physics at the basis of these events. I remain interested but unpersuaded." Nevertheless he spoke for many at the conference when he said, "The sense of wonder and excitement that has been generated is amazing."

The intent of this presentation is to look back further, to the origins of the idea that matter can occur in extremely condensed states. Newton regarded the property of solidity as deceptive. Matter is "much more rare and porous than is commonly believed," he wrote in the 1704 edition of his *Optiks*. While the modern concept of the atom was still more than two centuries off, he argued (in the 1717 edition of that

0077–8923/80/0336–0402 $01.75/1 © 1980, NYAS

classic work) that matter is formed of particles with empty spaces in between. The particles themselves, he said, are formed of even tinier particles, likewise distributed through space, and so on "perpetually" to smaller and smaller levels of matter.

Henry Pemberton, a popularizer of Newton's ideas, wrote that the entire universe "may be compounded of no greater a portion of solid matter, than might be reduced into a globe of one inch only in diameter, or even less." Joseph Priestley, the discoverer of oxygen, who was born seven years after the death of Newton, went even further. Not only might all the material of the solar system "be contained within a nutshell," he said, but the concept of solidity might be entirely erroneous. "There might be no such thing in nature," he said. It is almost as though he glimpsed the concept of a singularity.[3]

The first indication that great condensations of matter might occur was Friedrich W. Bessel's 1844 observation of the perturbed motion of Sirius, followed nineteen years later by Alvan Clark's detection of the companion responsible for such motion. Bessel was at the observatory in Königsberg, Prussia (now Kaliningrad, in the Soviet Union). Alvan Clark was testing an eighteen-inch lens he had made for the Dearborn Observatory in Illinois. The companion, unlike other small stars, had an extremely white spectrum, like Sirius itself. Being of roughly one solar mass it clearly was extremely dense. As Sir Arthur Eddington put it in 1927, its message said: "I am composed of material 3,000 times denser than anything you have come across; a ton of my material would be a little nugget that you could put into a matchbox."

How can one reply to such a message? Eddington asked rhetorically. "The reply which most of us made in 1914 was—'Shut up. Don't talk nonsense.' "

The stage was now set for Subrahmanyan Chandrasekhar to define what prevented such "white drawfs" from collapsing to even greater density. In 1926 Ralph H. Fowler in Britain had proposed that such objects are formed when a star, having exhausted its nuclear fuel, collapses to a state like "a gigantic molecule in its lowest quantum state."

Chandra—as he is now widely known—was a youth of 19 when he traveled to Britain to work under Fowler. He had sprung from a tradition of excellence in his native India, where his uncle, Sir Chandrasekhara Venkata Raman, had recently identified the Raman effect in light scattering that won him a Nobel prize. A year or two earlier Fermi and Dirac had worked out the statistics that helped Chandra apply the exclusion principle to the white dwarf problem. The resulting resistance of electrons to compression would not withstand the pressure of a collapsing body greater (in his revised version) than 1.4 solar masses. "A star of large mass," he wrote, "cannot pass into the white drawf stage, and one is left speculating on other possibilities."[4]

His bold suggestion was not greeted warmly. The *Astrophysical Journal* (which, for almost two decades, he later edited), rejected his original paper. One of his mentors, Edward A. Milne, the cosmologist, wrote him: "To me it is clear that matter cannot behave the way you predict." Eddington recognized the possibility of a black hole, but treated it as ridiculous. He wrote:

> Chandrasekhar shows that a star of mass greater than a certain limit . . . has to go on radiating and radiating and contracting, and contracting until, I suppose, it gets down to a few kilometers' radius when gravity becomes strong enough to hold the radiation and the star can at last find peace.

This, he said, was almost a *reductio ad absurdum:* "I think that there should be a law of Nature to prevent the star from behaving in this absurd way."[5]

That there might be a counterpart to the "electron pressure" that prevents further collapse of a white dwarf involving neutrons, and leading to the concept of a neutron star, evolved from two somewhat separate lines of reasoning. One sought to explain supernova explosions and their consequences, such as the Crab nebula. The other concerned what happens to matter within the core of a very massive star. In the latter case George Gamow proposed what amounted to inverse beta decay—electrons and protons combining to form neutrons. He envisioned a "gas of neutrons" so dense as to be "analogous to the conditions inside an atomic nucleus."[6]

In 1931, the year Chandra published his first white dwarf analysis, Lev Landau, then in Zürich, noted that for the cores of more massive stars "there exists in the quantum theory no cause preventing the system from collapsing to a point." Since stars much more massive than the Chandrasekhar limit are observed and do not show "such ridiculous tendencies," he assumed some form of "pathological regions" exist in the cores to prevent further collapse.[7] His calculation, however, disregarded the role of temperature in a star that is still generating nuclear energy. He later proposed that stars derive their energy from progressive collapse of their cores into closely packed neutrons. To generate the energy radiated by the sun in two billions years, he said, only two percent of the sun need have collapsed in this manner.

Following up on these lines of reasoning Fritz Zwicky and Walter Baade in California proposed in 1933 "that supernovae represent the transitions from ordinary stars into neutron stars, which in their final stages consist of extremely closely packed neutrons."[8] Baade cited a star near the center of the Crab nebula with an unusual spectrum as a possible example of such a star.

As described so beautifully at the last symposium by Jocelyn Bell Burnell, the discovery of pulsars provided dramatic confirmation of such speculation.

The story of the final step toward the ultimate condensation—the black hole—usually begins with Oppenheimer and his students. But this ignores the contribution of Karl Schwarzschild as he lay dying of pemphigus, then an invariably fatal disease, during the winter of 1915–16. Although exempt because of his age (in his 40's) and his distinguished post as head of the Potsdam Observatory, he had volunteered for service in the Kaiser's army and had been stricken with this insidious disease on the Russian front.

A few months earlier Einstein had published the elements of his general theory, and Schwarzschild had written three classic papers, one of which defined the Schwarzschild radius in terms of that theory.[9] He spelled out, in terms of a point source of gravity, the distance at which relativity effects would generate such elements of a black hole's event horizon as an infinite red shift. He sent the paper to Einstein—then living in nearby Berlin—who replied on January 9, 1916:

My esteemed Colleague!

I have read your paper with the greatest interest. I had not expected that the exact solution of the problem could be formulated so simply. The analytical treatment of the problem appears to me splendid. Next Thursday I will present the work, with several explanatory words, to the Academy.

After an extensive discourse on another theoretical question he ended:

> Wishing you the best
> Your devoted
> A. Einstein[10]

Schwarzschild was thus the first to awaken interest in the relativistic effects of superpowerful gravitational fields and Eddington was well aware of them in 1929 when he wrote his classic work, *The Internal Constitution of the Stars*. He noted that a star like Betelgeuse, with ten to one hundred solar masses, exerts very potent gravity, and would do so even more strongly were it as dense as the sun, whereas it is fifty million times more voluminous than our parent star, its diameter being comparable to that of the earth's orbit. He speculated, however, on the effects if so large a star were of solar density:

> Firstly, the force of gravitation would be so great that light would be unable to escape from it, the rays falling back to the star like a stone on the earth. Secondly, the red shift of the spectral lines would be so great that the spectrum would be shifted out of existence. Thirdly, the mass could produce so much curvature of the space-time metric that space would close up around the star, leaving us outside (i.e., nowhere).[11]

This was five years before Chandrasekhar wrote his initial paper on the lower limit for white dwarf formation, provoking Eddington's comment that there should be a law of nature "to prevent the star from behaving in this absurd way." Eddington also drew attention to the early speculation of Laplace that stars might be invisible if their gravity was strong enough.

Laplace, born twenty-two years after Newton's death in 1727, thought in terms of the corpuscular theory of light, favored by Newton's followers. In the previous century Ole Rømer had obtained a surprisingly accurate figure for the speed of light by observing the lag in occultations of Jovian moons when Jupiter was on the far side of the sun (225,000 kilometers per second, compared to the modern figure of about 299,780 kms^{-1}.)

From this Laplace calculated the strength of a gravitational field sufficient for its escape velocity to exceed the speed of light. The discussion of this question appears in the 1798 edition of Laplace's *Système du Monde,* but not in later editions, as though he had misgivings about it. Because so much that Laplace had to say anticipates modern astronomy, it is worth quoting his discussion in full:

> Several stars display in their color and their brightness very remarkable periodic variations: there are others that have appeared all of a sudden, and that have disappeared, after having, for some time, displayed a bright light. What prodigious changes must have taken place on the surface of these great bodies, to be so evident at the distance that separates us? How much they must surpass those that we observe on the surface of the sun, and convince us that nature is far from being always and everywhere the same? All these bodies that have become invisible, are in the position where they were observed, since they did not at all change it during their appearance; there exist in celestial space, invisible bodies as large, and perhaps in as great numbers, as the stars. A luminous star of the same density as the earth, and whose diameter was two hundred and fifty times greater than that of the sun, would not, because of its attraction, allow any of its rays to arrive at us; it is thus possible that the largest luminous bodies of the universe are even in this way invisible.[12]

Little did Laplace realize the extent of the "prodigious changes" that are taking place in distant parts of the universe. It has been noted that his estimate of the gravitational field necessary to make an object invisible is remarkably close to that derived by Schwarzschild in a very different way.

It was not, however, the speculations of Laplace, Schwarzschild, or Eddington that finally led Oppenheimer and his students to define the nature of a black hole in relativistic terms. It was the work of Landau and others on the cores of stars and the ability of neutrons to resist collapse.

One of Oppenheimer's co-workers was George M. Volkoff, a graduate student who had been born in Moscow, brought to Canada, and then educated both in Harbin, Manchuria, and the University of British Columbia (where he is now). Regarding the ability of neutrons to resist collapse, they wrote in the *Physical Review* of February 15, 1939, either their formulation of the problem "fails to describe the behavior of highly condensed matter," or, they said, "the star will continue to contract indefinitely, never reaching equilibrium."[13]

The final step was then taken by Oppenheimer and Hartland S. Snyder, another graduate student who, having grown up in Utah in an environment of rough ways and manual labor, stood out in Oppenheimer's more elitist entourage. He was, however, a mathematical wizard. Their paper, in 1939, carried the argument to its conclusion: nothing could stop the collapse.

"The star thus tends to close itself off from any communication with a distant observer; only its gravitational field persists," they wrote. To an external observer the process would appear to take an infinite time, because of relativistic time dilation, but internally the isolation would occur within about a day, if one ignored such effects as rotation and residual radiation. "Of course," they added, "actual stars would collapse more slowly. . . ."[14]

Thus the theoretical stage was set for what has become one of the most exciting periods in the history of astronomy—a period marked by the discovery of x-ray binaries and catastrophic and perplexing activity in such objects as radio galaxies, quasars, Seyfert galaxies, BL Lacertae objects, and x-ray bursters. It has led many to believe black holes may be a reality, and others to argue, with Eddington, that "there should be a law of Nature to prevent the star from behaving in this absurd way."

References and Notes

1. Those objects were then, of course, still referred to as "quasi-stellar radio sources" since Hong-yee Chiu had not invented the shorter term. It was, in fact, several years before Chandrasekhar, as editor of the *Astrophysical Journal,* permitted the use of the shorter term with a note saying "the journal regrets that it must now concede." The longer term is now less appropriate because many quasars are silent at radio wavelengths.
2. Robinson, I., A. Schild & E. L. Schücking, Eds. 1965. Quasi-stellar Sources and Gravitational Collapse—Including the Proceedings of the First Texas Symposium on Relativistic Astrophysics, Chicago, Illinois: XI–XVII. University of Chicago Press.
3. The quotations from Newton, Pemberton, and Priestley are taken from Arnold Thackray. 1969. *Ambix* **15:** 29–43.
4. Chandrasekhar, S. 1931. The London Edinburgh and Dublin Philosophical Magazine and Journal of Science. **11:** 592–96; 1931. Mon. Not. R. Astron. Soc. **91:** 456–66; 1932. Astrophys. J. **74:** 81–82.

5. CHANDRASEKHAR, S. 1932. Observatory **92:** 160–71, contains citations from A. S. Eddington and E. A. Milne.
6. GAMOW, G. 1937. Structure of Atomic Nuclei and Nuclear Transformations: 234–38. Oxford University Press. New York.
7. LANDAU, L. 1932. Phys. Z. Sowjetunion **1:** 285–88.
8. BAADE, W. & F. ZWICKY. 1934. Phys. Rev. **45:** 138; 1934. Proc. Natl. Acad. Sci. **20:** 254–59 and 259–63. ZWICKY, F. 1938. Astrophys. J. **88:** 522–25; 1939. Phys. Rev. **55:** 726–43.
9. SCHWARZSCHILD, K. 1916. Sitzungsber: 189–96. K. Preuss. Akad. Wiss.
10. A photocopy of this letter was generously provided by S. Chandrasekhar and the translation is largely his.
11. EDDINGTON, A. S. 1926. The Internal Constitution of the Stars: 6. Cambridge University Press. London.
12. LAPLACE, P. S. Exposition du Système du Monde, 2nd ed. Paris: VII, 546–49.
13. OPPENHEIMER, J. R. & G. M. VOLKOFF. 1939. Phys. Rev. 374–81; OPPENHEIMER, J. R. & R. SERBER. 1938. Phys. Rev. **54:** 540.
14. OPPENHEIMER, R. & H. SNYDER. 1939. Phys. Rev. **56:** 455–59.

PULSAR RADIATION MECHANISMS*

M. Ruderman

Department of Physics
Columbia University
New York, New York 10027

INTRODUCTION

The first pulsar was discovered eleven years ago. Several hundred are now known. Active and extinct pulsars probably constitute about one hundreth of all stars in our Galaxy. Observers have presented a rich treasure of data from which to deduce the behavior of these rotating magnetized neutron stars. Theoreticians have apparently found it easy to understand them for they have produced not only a theory of pulsars but dozens of theories of pulsars. (The signal-to-noise ratio seems poor; perhaps there is no signal.) There are many families of "models" characterized by different presumed radiation mechanisms and locations from which the radiation is emitted.[1-4] Nor is there unanimity about how the needed coherence is brought about to give the enormous radio brightness temperature (up to 10^{28} K), about how the appropriate plasma configurations are developed, or even about how rotational kinetic energy of the spinning neutron star is conveyed to the radiating region.

In most models, emitted radiation is in a roughly conical beam that corotates with the star. In Polar Cap Models[43,5] (PCM), the radiation cone originates within the star's corotating magnetosphere with a cone axis near that of the star's dipole magnetic field. In Light Cylinder Models (LCM),[6-8] the cone originates at the "light cylinder" where a corotating magnetosphere would achieve the speed of light. In LCM's the cone axis is tangential to the light cylinder and perpendicular to the rotation axis of the star. Still other models put the source in suburban regions near, but beyond, the light cylinder or even in rural areas, very much further away.[9]

Observational data and the direct interpretations of them have not yet been decisive in establishing even these geometrical properties of the emitted radiation.[10-12] But the data are at least suggestive of an emission cone whose properties, radio luminosity, and spectrum are not sensitive to the period of the pulsar,[2] except for the different rates with which the beam is rotated by the spinning neutron star. (There is, however, a great scatter in parameters among pulsars that may obscure any but rather strong period differences.)

A great number of suggestions have been made to explain pulsar radio emission. Some attribute it to charge acceleration perpendicular to the local ambient magnetic field (synchotron, curvature, or synchro-Compton radiation); others assume it originates from charge acceleration along B (plasmon–plasmon collisions, steady electric fields) or that it is Cerenkov radiation.[13] The needed coherence is assumed by some to come from maser action,[14,3] by others from two-stream plasma instability bunching,[5,15] from radiation reaction induced plasma bunching,[16] or from a variety of other plasma instabilities. There is not yet any unanimity. Polar cap models in which the radiation

*Research supported in part by a grant from the National Science Foundation

0077–8923/80/0336–0409 $01.75/1 © 1980, NYAS

results from an outward-streaming highly relativistic quasi-turbulent plasma are, however, probably the most popular class among model builders.

In the following I shall outline a polar cap model that is significantly different in detail from previously published PCM's. It seems to lead naturally to a cone of coherent radio emission whose properties resemble those observed.

Current Flow in a Pulsar Magnetosphere

There is not yet an entirely convincing, much less rigorous, solution for a spinning neutron star even with an axisymmetric magnetic field whose axis is oriented along the spin (Ω) direction. For $r < R$ (the stellar radius),

$$\mathbf{E} = \frac{\mathbf{V}}{c} \times \mathbf{B} \tag{1}$$

$$\mathbf{v} = \Omega \times \mathbf{r} \tag{2}$$

$$\mathbf{E} \cdot \mathbf{B} = 0. \tag{3}$$

For $r > R$, $\mathbf{E} \cdot \mathbf{B} = 0$ everywhere only if (a) plasma flow along $\mathbf{B}$ is ignorable, (b) inertial effects from finite particle mass are negligible; this clearly must fail in the region $r \rightarrow r_{\mathrm{lc}} \equiv 2\pi cP^{-1}$, where $v \rightarrow c$, and (c) there are no regions along a field line where there is no plasma at all. When these conditions are satisfied, Equation 3 should hold, and Equations 1 and 2, the corotating magnetosphere, follow.[17] Where contributions to $\mathbf{B}$ from magnetosphere currents are negligible ($r \ll r_{\mathrm{lc}}$),

$$\rho \equiv \frac{\nabla \cdot \mathbf{E}}{4\pi} = \frac{\Omega \cdot \mathbf{B}}{2\pi c}. \tag{4}$$

There are very large potential drops between points on the different field lines that thread the rotating conductor. It is presumed that there is a current flow through the rotating star caused by this EMF. Otherwise the star with this geometry cannot slow down. If it does the circuit consisting of the spin-generated stellar EMF and conducting "wires" along $\mathbf{B}$ outside the star cannot be closed within the light cylinder $r < r_{\mathrm{lc}}$ and still conserve both total energy and total angular momentum.[18] This is consistent with expectations based upon the properties of the magnetosphere plasma since, where plasma collisions are unimportant, conductivity is high along $\mathbf{B}$ lines, but negligible perpendicular to them within the magnetosphere. Therefore current flowing through the star should flow out and in only along "open" field lines that penetrate the light cylinder. If the magnetospheric $\mathbf{B}$ is approximated by that of a dipole in the neutron star center, the EMF at the stellar surface between open field lines is

$$\mathrm{EMF} \sim \frac{10^{13}}{P^2} B_{12} \text{ Volts}, \tag{5}$$

where B_{12} is the surface dipole field in units of 10^{12} G. The current density j along open field lines driven by the EMF is estimated to be of order ρc.[17] The resulting $(\mathbf{J} \times \mathbf{B}) \times \mathbf{r}$ torque within the star causes the stellar spin-down and leads to the required $B_{12} \sim 1$ in this type of model for a spinning neutron star. (The parameters and assumed magnetosphere structure have been shown to be independent of the assumption of alignment between Ω and the stellar dipole moment.[19]) That the spinning star actually acts as a homopolar generator in this fashion has not been rigorously proved and clearly depends upon the behavior of currents at and beyond the light cylinder. Model details depend on the unknown structure of the stellar magnetic field near the star as well as on trans-light-cylinder processes.

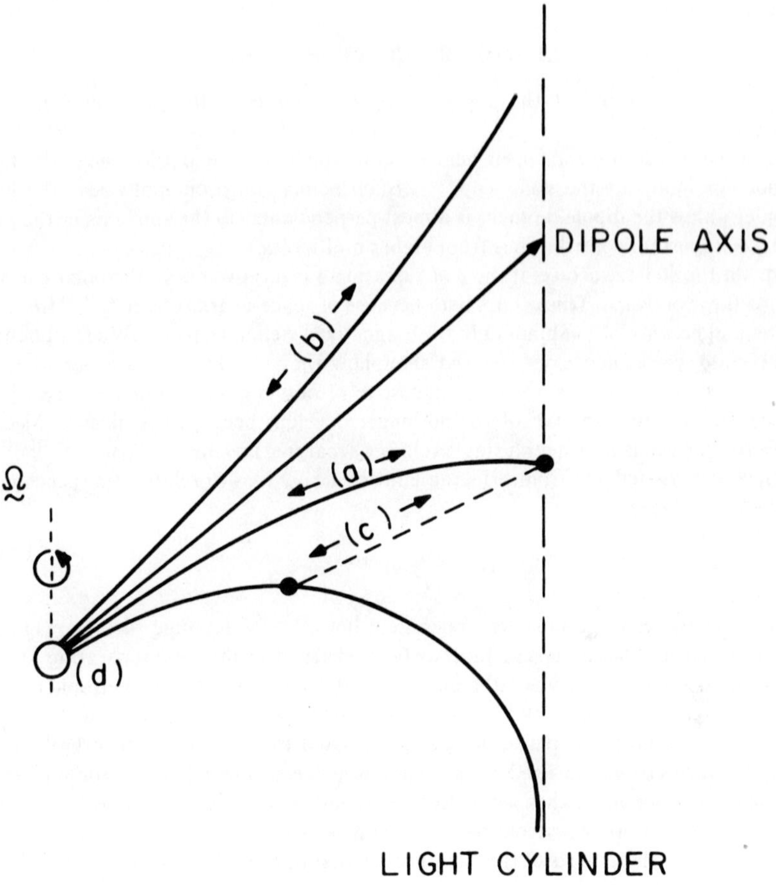

FIGURE 1. Regions along open field lines that have been considered for possible potential drops along **B**. The dotted straight line (c) is the locus of $\mathbf{\Omega} \cdot \mathbf{B} = 0$.

Wherever the presumed current flow of this homopolar generator meets resistance, potential drops will develop along the flow. Four regions within the magnetosphere shown in FIGURE 1 have been investigated for possible significant $\mathbf{E} \cdot \mathbf{B} \neq 0$ when a (charge separated) one-component plasma flows out of (or into) the polar cap region of a neutron star:

a. Along an open field line with decreasing $\hat{\Omega} \cdot \hat{\mathbf{B}}$ (all the open field lines for the completely aligned model with a dipole field). This flow can have $\mathbf{E} \cdot \mathbf{B} = 0$ all the way to the light cylinder.[20,21]

b. Along an open field line with increasing $\hat{\Omega} \cdot \hat{\mathbf{B}}$ (open field lines curving toward the spin axis when the dipole moment and spin axis are not aligned.[3,24,25,34] As long as only a single charge component flows, this $\mathbf{E} \cdot \mathbf{B}$ would arise from the inconsistency between $\nabla \cdot \mathbf{j} = 0$, $\nabla \cdot \mathbf{B} = 0$, $\mathbf{j} = \rho c\hat{\mathbf{B}}$, and surface boundary conditions. Potential drops as large as

$$\Delta V = (\hat{\Omega} \cdot \hat{\mathbf{B}} - \hat{\mathbf{B}}_s)\, 10^{13}\, P^{-2} \text{ Volts,} \tag{6}$$

with $\hat{\Omega} \cdot \hat{\mathbf{B}}_s$ the cosine of the angle between Ω and $\mathbf{B}$ at the polar cap, may be achieved.

c. Where $\rho = 0$ on an open field line on which there is outflow near the light cylinder or inflow at the polar cap.[26,27] (Such points can occur only near the light cylinder unless the dipole moment is almost perpendicular to the spin axis or the near field has important contributions from higher multipoles.)

d. At the stellar surface, if the ρ of Equation 4 is positive above the polar cap and plasma flows outward. This occurs both because of space-charge effects[23,28–30] from ion inertia and because of ion binding in the magnetized stellar surface.[31] We shall outline below many consequences of (d) and show how they should lead to coherent radio emission whose properties are very suggestive of observed ones. Further, one of the consequences is that the ΔV of (b) no longer develops because the plasma injected above the surface is no longer a single (charge separated) component one.

Space-charge effects from bringing emitted heavy ions to relativistic speeds near the stellar surface cause

$$\Delta V \sim \frac{10^{12}}{P} B_{12}{}^{1/2} \text{ Volts,} \tag{7}$$

there along the involved polar cap open field lines.[23,28–30] The huge surface magnetic field also causes the stellar surface to be a sharp discontinuity separating stellar matter with a mass density $> 10^4$ g cm^{-3} from the almost empty magnetosphere with rest mass density of order 10^{-10} g cm^{-3}. The binding energy (E_b) of atoms (assumed iron) into the condensed phase has been estimated to vary from 1 to 8 keV as the surface B varies from 1 to 5×10^{12} G.[31] The energy needed to pull atoms (or ions) from the stellar surface must then be at least as great as this binding energy. To supply freely currents of order ρc from parts of the polar cap despite the condensed matter binding energy, those parts of the cap must be maintained at a temperature

$$kT \sim E_b/30. \tag{8}$$

If the polar cap temperature falls below this value ions will not be liberated by the surface sufficiently rapidly to maintain the homopolar generator. The entire stellar EMF between polar cap open field lines could develop along $\mathbf{B}$ there. This would give a potential drop,

$$\Delta V \sim 10^{13}/P^2 \text{ Volts,} \tag{9}$$

across a 10^4 cm gap at the surface.[15] Except for the possible measurement of cyclotron resonance on the surface of Her X-1[33] giving $B_{12} \sim 5$, estimates of the surface field are all based upon inward extrapolations of the field from at least hundreds of stellar radii away. These assume a pure dipole configuration. Usually this gives values of B_{12} very considerably less than 5, which would imply that the actual surface field contains many multipoles and is complicated. Therefore B_{12} and thus E_b and kT of Equation 8 are not well known and may vary greatly among pulsars. In typical pulsars, whose "ages" are much greater than 10^5 yr, the average surface temperature is expected to be considerably less than that of Equation 8, unless there is special heating of the polar cap. But such heating should indeed occur because of electron–positron discharges at the polar cap[5,15,32] whenever ΔV exceeds about 10^{12} V. These can occur because of several hundred MeV curvature radiation from 10^{12}-eV electrons or positrons that have traversed the gap moving along curved **B** lines. Subsequent pair creation by these γ rays in the same huge **B** field ("Sturrock discharge") is a source for additional pairs. The resulting e^-–e^+ avalanche discharge, and consequent limiting of ΔV, occurs as soon as

$$\Delta V \sim 10^{12}\, B_{12}^{-1/7}\, P^{-1/7}\, s_6^{4/7}\ \text{Volts} \qquad \textbf{(10)}$$

is reached,[15] where s_6 is the radius of curvature of polar cap **B** field lines in units of 10^6 cm (the stellar radius). There are a variety of other processes for initiating and sustaining such e^-–e^+ discharges but all that have been examined also limit the ΔV that can grow above a polar cap to of order 10^{12} V.[30] The discharge that initiates a flow of 10^{12} eV positrons moving out along polar cap field lines also sends a similar current of electrons back onto the cap, thus keeping the cap hotter than it would otherwise be. A patch on the polar cap can thus be kept at a temperature $\sim 10^6$ K, just enough to permit almost the demanded ion flow. A stronger e^+–e^- discharge could heat the patch more but the increased ion flow would (unless Equation 7 is relevant) bring ΔV below the threshold of Equation 10.[23,30,34] A decreased e^-–e^+ would give less polar cap heating, result in more effective ion binding, and thus push ΔV up toward the limit of Equation 9, where the e^-–e^+ discharge would run away. Therefore the symbiotic relationship between polar cap ion flow and an e^-–e^+ discharge is, on the average, a stable one. Because of field line curvature and almost tangential emission of the curvature radiation γ rays that make the e^-–e^+ pairs, the discharge always moves outward on the convex side of field lines as shown in FIGURE 2. It must continually be restarted on a time scale of order 10–10^2 μs, the time it would take to reach either the boundary or an inner field line with no curvature. Reignition is most likely to start on just those field lines that thread an already heated patch of the polar cap: ions are emitted mainly from that region and these achieve a Lorentz $\gamma \sim 10^3$ when accelerated across the 10^{12}-V gap. The high energy tail of keV photons from the 10^6 K x-ray hot patch are seen as MeV x-rays in the ion rest frame and these can make e^-–e^+ pairs by the conventional pair production process in the ion coulomb field.[30] Thus there is a tendency for a discharge to renew itself in the same place, even if it is extinguished for a while (nulling). However, the coupled-discharge hot patch can have a slow $\mathbf{E} \times \mathbf{B}$ drift around the polar cap since $\mathbf{E} \cdot \mathbf{B} \neq 0$ must be accompanied by a component of **E** perpendicular to **B** as long as $\nabla \times \mathbf{E} \sim 0$. This leads[15,39] to a ratio of drift period P_3 to pulsar period P,

$$\frac{P_3}{P} \sim \frac{6B_{12}}{P^2}. \qquad \textbf{(11)}$$

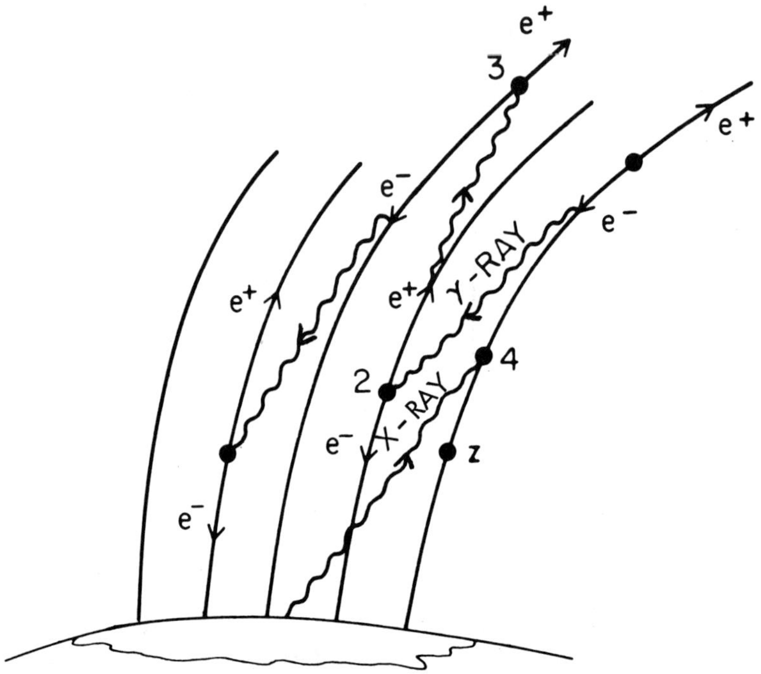

FIGURE 2. A pair is created at point 1 on a curved **B** field line along which there is a 10^{12} V potential drop. An electron accelerated downward along **B** radiates a several 10^2 MeV γ ray directed along **B.** This creates another $e^{\pm}$ pair at 2 on a neighboring field line. The process repeats and the discharge moves to the left. It is restarted continually at 4 by a keV thermal x ray from the patch (heated by the incoming electrons) that makes an $e^{\pm}$ pair in the Coulomb field of the relativistic positive ion also released from the patch.

The main problem in associating this coupled 10^{12}-V drifting-ion–e^+ injection into the region above the polar cap with observed pulsar behavior is to understand how these currents lead to strong coherent radio emission. We turn next to a consideration of how that may come about.

RELATIVISTIC ION, ELECTRON, AND POSITRON FLOWS: COHERENT RADIO EMISSION

As long as the stellar EMF of Equation 9 exceeds about 10^{12} V, the "gap" $\Delta V \sim 10^{12}$ V. The primary positive current densities consist of ions with $\gamma \sim 500$ and

$$j_z \sim \frac{2 \times 10^{21} B_{12}}{P} \text{ cm}^{-2}\,\text{s}^{-1}, \qquad \textbf{(12)}$$

where $j_z = n_z Zc$, and positrons with $\gamma_+ \sim 2 \times 10^6$ and

$$j_+ \sim 10^{18} \left(\frac{E_b}{\text{keV}}\right)^4 \text{ cm}^{-2}\,\text{s}^{-1}. \qquad \textbf{(13)}$$

The positron current density approximation assumes Equation 8 and an energy input into the heated polar cap patch from back electron flow equal to the blackbody radiation from it. (This input differs in two important ways from a previous model that assumed that ion binding was always sufficient to prevent their liberation from the stellar surface[15]: j_+ is greatly reduced by as much as several orders of magnitude if $E_b \sim 1$ keV, and the reduced positron flow is always accompanied by a very much larger monoenergetic[22] ion flow.) In the faster pulsars even the space-charge potential drop of Equation 7 is large enough to exceed the ΔV of Equation 10 (or that for other $e^- - e^+$ discharge breakdowns). For these a j_+ outflow would then accompany j_z even if $E_b = 0$, such that[30]

$$j_+ \sim j_z/500. \quad \textbf{(14)}$$

Each "primary" e^+ or e^- traversing the 10^4-cm 10^{12}-V surface gap is a source for the curvature radiation photons of 10^2–10^3 MeV each. At least one of these photons materializes into an $e^\pm$ pair within the gap to maintain the discharge. These $e^\pm$ produced in the gap in turn are accelerated through 10^{12} V and repeat the process that formed them. But a very much larger number of $e^\pm$ "secondaries" is made by curvature gamma rays above the surface gap because of more volume for pair materialization, larger angles between photon trajectories and **B**, and continuing curvature radiation from the 10^{12}-eV "primaries." These secondaries are created and flow in regions where their very presence insures a plasma conductivity high enough to maintain $\mathbf{E} \cdot \mathbf{B} \sim 0$. This relativistic secondary plasma has, initially, a Lorentz factor

$$\gamma^\pm \sim 10^2\text{–}10^3, \quad \textbf{(15)}$$

a current flow

$$j^\pm \equiv n^\pm c \sim (10^3\text{–}10^4) j_+, \quad \textbf{(16)}$$

and an energy density proportional to

$$\gamma^\pm n^\pm \lesssim \gamma_+ n_+. \quad \textbf{(17)}$$

(We note that for $E_b = 1$ keV, $P = 1$ s, and $B_{12} = 1$, $j_z \sim 2 \times 10^3\, j_+$; then, to within a factor ten, $\gamma_z n_z \sim \gamma^\pm n^\pm$ and $Z n_z \sim n^\pm$.) As the secondary $e^\pm$ plasma flows out along the guiding field lines its density drops with **B**; the velocities of the secondary e^+ and e^- begin to differ slightly as the net charge density $e|n^+ - n^-|$ adjusts to maintain Equation 4[25] and $\mathbf{E} \cdot \mathbf{B} \sim 0$ along a curving field line all the way out to near the light cylinder. This brings the minimum Lorentz factor ($\gamma_<$) down to ~ 10 for one of the charge components and leads to a spread of several orders of magnitude among the Lorentz factors of the secondary $e^\pm$ plasma. Threading this secondary plasma is the ion beam, which at any point is almost exactly monoenergetic. After an initial negligible thermal energy spread of only about a keV associated with their injection, all ions go through the same potential drop of $\sim 10^{12}$ V. Between the monoenergetic ion beam and the secondary $e^\pm$ plasma formed around it there is always a two stream instability and charge clumping no matter what the $e^\pm$ energy distribution. Plasma oscillations are then pumped by the energy in the relative streaming. These growing charge-clumping modes satisfy:

$$\omega \sim kc \sim 2(4\pi\gamma^\pm n^\pm e^2/m)^{1/2}, \quad \textbf{(18)}$$

and a growth rate

$$\operatorname{Im}\omega \sim \frac{\omega}{\gamma^{\pm}\gamma_z}\left(\frac{n_z m Z^2}{n^{\pm} M_z}\right)^{1/3}, \tag{19}$$

where M_z is the positive ion mass, and Ze is its charge. Because of field line bending we should replace $\gamma^{\pm}$ by that of the "slow" component that makes the dominant contribution to the plasma frequency in the aboratory frame. For a laboratory $\omega = 10^9$ s^{-1}, $\gamma^{\pm} \rightarrow \gamma_< \sim 10$, $\gamma_z = 500$, and $Zm_Z \sim n^{\pm}$ (cf. remarks after Equation 17), the e folding distance $c\,(\operatorname{Im}\omega)^{-1}$ is

$$l \sim 2 \times 10^6 \text{ cm}. \tag{20}$$

(This is very much smaller than that from the primary e^+ plasma both because $\gamma_z \sim 10^{-3}\,\gamma_+$ and because $Zm_Z \sim 10^3\,n_+$ for 1-keV surface ion binding. These charges dominate the increase from the larger ion rest mass so Equation 19 greatly surpasses the weaker growth rate from the e^+ beam.[36])

The rapidly growing two stream instability gives strong charge clumping and turbulence. The clumps are accelerated in directions parallel to **B** because of the acceleration associated with the growth of the plasma instability itself, and from plasma–plasma collisions and from possible residual ordered electric fields.[35] They are accelerated perpendicular to **B** because of the curvature of the guiding field lines. Charge acceleration parallel to **B** and perpendicular to it will be a source of electromagnetic radiation, mainly into orthogonal modes, whose relative intensity depends upon the relative magnitudes of ω and $\omega_{\text{curv}} = \gamma_<^{\,3}\, c\, s^{-1}$ (s is the radius of curvature of the field line). But quantitative calculations of the polarization and intensity of the coherent radiation ultimately emitted from the plasma are difficult and have not been attempted. Radiation reacts back on the clumping and can accelerate it.[16] The coupling among plasma and "radiation" modes is sufficiently strong that they cannot be considered separately.[37] The true and appropriate growing modes are an intimate combination of plasma motions along curved field lines and electromagnetic fields. A few general features are expected:

a. The emitted frequency of coherent radio waves should be about that of Equation 18.[38]
b. Radio frequency $\nu \sim 10^2$–10^3 MHz will then usually be emitted at a distance $r \sim 10^7$–10^8 cm from the neutron star in typical cases.
c. The observed radiation pulse width depends upon the size of the bundle of open field lines, which in turn is defined by the unknown behavior of **B** near the light cylinder. If **B** is approximated as a pure dipole for all r, $\theta \sim 2(\Omega r/c)^{1/2} \sin^{-1} \alpha$, with α the angle between the spin and dipole axes. For $\sin\alpha \sim 1$ and r between 10^7 and 10^8 cm, $\theta \sim 5°$–$15°$.
d. For emission in the intermediate radius region (10^2–10^3 MHz), where **B** should be dipolar, the separation between pulse envelope peaks should vary like $\nu^{-2/7}$ when account is taken of the decrease of both $n^{\pm}$ and $\gamma_<$ with diverging **B** in Equation (18).
e. Curvature radiation dominated emission will disappear near the dipole axes resulting in a hollow emission cone for it.
f. Strong radio emission features should drift within pulse envelopes with a

repetition period given by P_3 of Equation 11 and a separation related to the expected interval between hot e^-–e^+ discharge patches.[15]

g. The plasma properties are insensitive to the pulsar period P and depend mainly on the surface magnetic field through the surface ion binding energy.

h. The maximum available power that can be converted to coherent radio emission is $10^{30}\,p^{-2}$ erg s^{-1}.

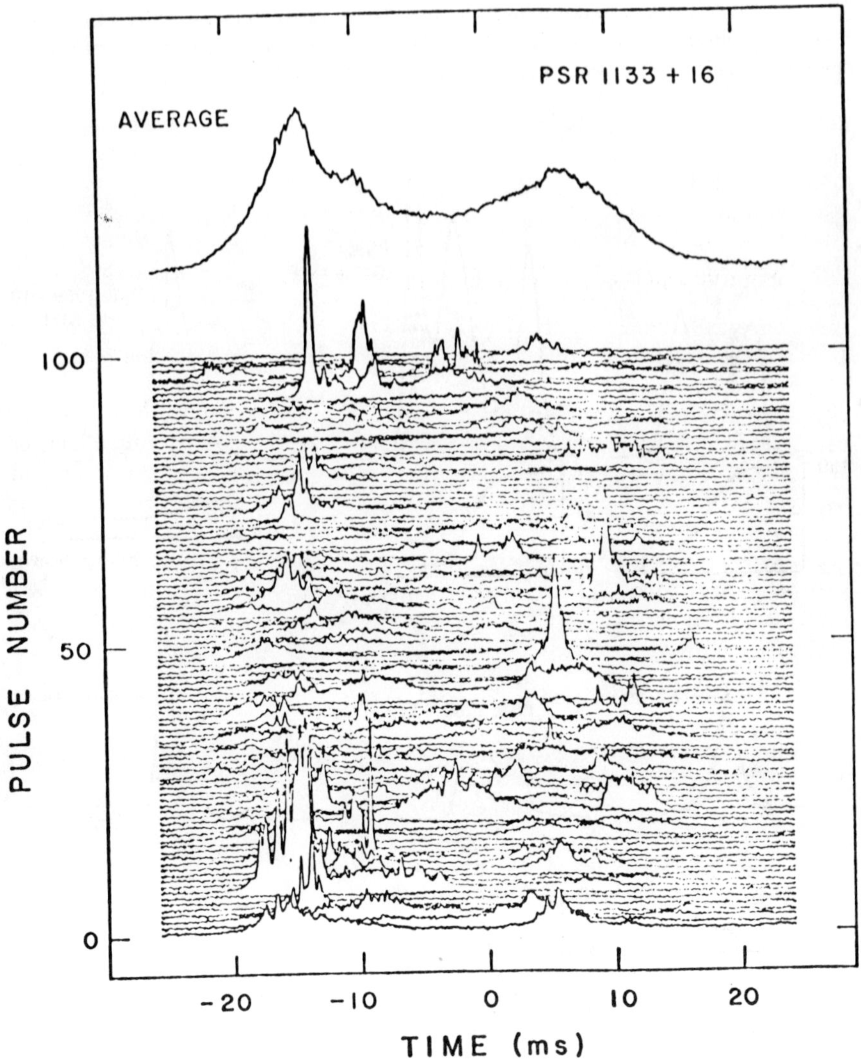

FIGURE 3. Successive individual radio pulses of PSR 1133+16 (Cordes[40]). The pulse envelope is at the top.

i. Emission into the two orthogonal polarization modes can be present simultaneously.

We turn next to a consideration of some of the vast amount of observational data with which to confront some of the expected theoretical features.

OBSERVATIONS

FIGURE 3[40] shows the radio intensity in 100 successive pulses of PSR 1133+16, a "typical" pulsar with a period $P = 1.2$ s. The envelope obtained by summing the 100 pulses is stable (over a scale of years except for the small increase in pulsar period).

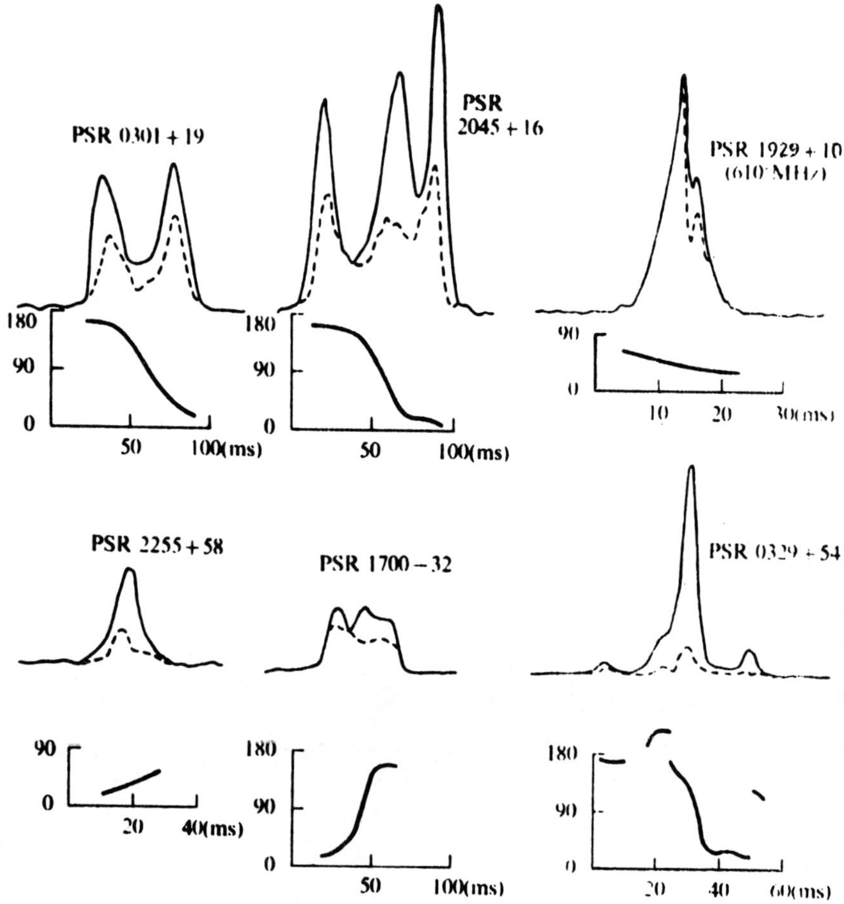

FIGURE 4. Polarization and pulse envelopes of six pulsars (Smith[2]). The graphs below show the position angle of the linear polarization. The broken lines above within the envelopes show the linearly polarized component.

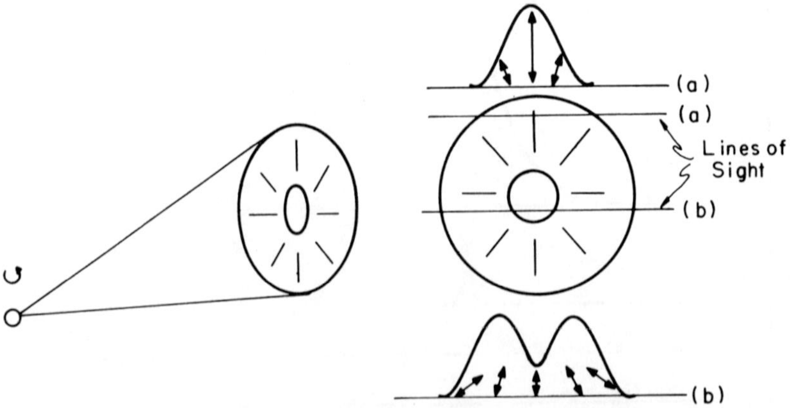

FIGURE 5. A schematic hollow radiation cone centered on a magnetic dipole axis rotating with a spinning neutron star. The polarization plane is assumed to be determined by the plane of the curving magnetic field. (a) If the line of sight cuts the edge only, the envelope is simple and the polarization swing small. (b) If it cuts through the center, the envelope is double peaked and the polarization swing almost 180°.

But there are large variations between and within pulses. The radio energy is characteristically emitted in subpulses several milliseconds wide. There appears to be a tendency for the position of a subpulse to drift irregularly within the overall envelope: on a time scale of order ten seconds it tends to shift between preferred positions near the leading and trailing edges of the pulse envelope. Within the subpulse there is quasi-periodic microstructure on a scale of less than 10^{-3} s. The microstructure correlation is strongly correlated over the radio frequency interval 320–2700 MHz.[11]

In PSR 1133+16 and pulsars with similar envelopes there is a small dependence of the separation between $\Delta\theta$ (the two envelope peaks of FIGURE 3 and the frequency of radio detection: $\Delta\theta \sim \nu^{-0.26}$ for $\nu < 10^3$ MHz and $\Delta\theta$ is not sensitive to ν between 10^3 and 10^4 MHz.[41] Where the magnetosphere **B** is accurately dipolar a theoretical estimate based on the above PCM gives $\nu^{-2/7}$. (Higher frequencies correspond to $r < 10^7$ cm, and there is probably no longer a reason to expect a pure dipole field to give an accurate (~2°) description of the direction of the local **B**.)

Cordes[42] has noted that the absence of aberration-type time displacements as a function of $\nu(r)$, despite the ν-dependent $\Delta\theta$, restricts the radii at which the emission takes place in models of the sort we are considering. He finds, for PSR 1311, r(40 MHz) $\lesssim 10^8$ cm and r(1400 MHz) $\lesssim 10^7$ cm, and for PSR 0950, r(40 MHz) $\lesssim 10^8$ cm and r(430 MHz) $\lesssim 10^7$ cm. These values are not inconsistent with those of the model presented above, but would have been embarrassingly small for a PCM in which $j_+ \gg j_z$.

FIGURE 4 shows some typical averaged pulsar linear polarization patterns.[2] Pulsars with simple single-peak envelopes tend to have polarization position angles that do not vary greatly through the pulse. In more complicated envelopes, mostly doubly peaked, the polarization position angle is likely to swing much more, even up to 180°. In

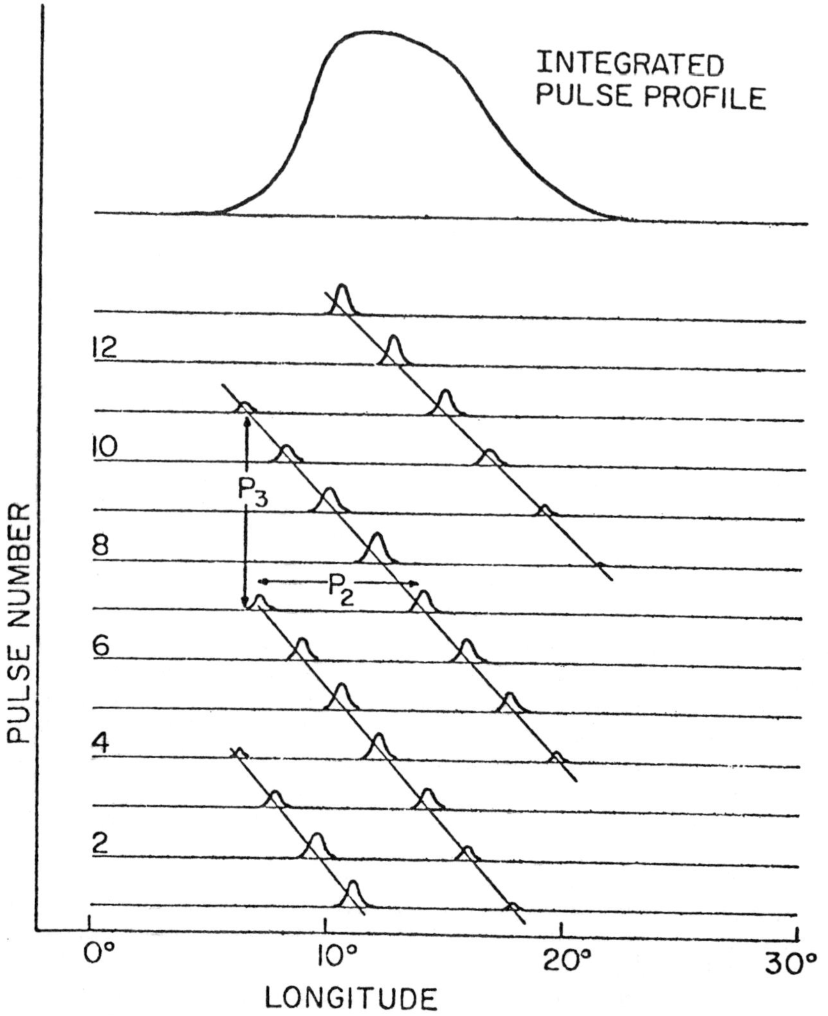

FIGURE 6. Definition of repetition period P_3 and separation period P_2 for subpulses moving within a pulse envelope.

PCM's this correlation between envelope shape and position-angle swing has been interpreted in terms of the varying aspect of the pulsar magnetic field presented as the emission cone swings across the radio telescope.[43] If the emission cone axis coincides with that of the pulsar's magnetic dipole, then, as long as the plane of the arriving dipole **B** lines determines the polarization position angle, the latter can vary greatly over the cone. Pulse envelopes suggest a hollow cone for the radii emission beam. When the observer's line of sight cuts the cone and passes through the central hollow two humps are seen with a polarization plane swinging greatly between both cone edges (cf. FIGURE 5). But if the line of sight cuts only an edge of the beam the envelope is "simple" and the swing in the plane of **B** is limited.

The long period drifting subpulse movement within the pulse envelope of PSR 1133 in FIGURE 3 may be present in almost all pulsars. The drift is characterized by a pattern repetition quasi-period P_3 and an interval between subpulses P_2, sketched in FIGURE 6. In typical pulsars P_2 is constant but P_3 fluctuates.[44,45] However, in at least one pulsar, PSR 0809+74 ($P = 1.3$ s), $P_3 = 11$ s with a $Q \equiv P_3/\Delta P_3 > 10^3$. This pulsar "turns off," and the subpulse reappears exactly where it disappeared[46]: the pulsar "remembers" a subpulse position that does not drift when radiation stops for brief periods. It is tempting to associate the observed drifting with that described by Equation 11. The smallest reported drifting subpulse period ratio, P_3/P, is for PSR 2303+30 with $P = 1.6$ s and $P_3/P \sim 2$. The largest is for PSR 1944+17 with $P = 0.4$ s and $P_3/P \sim 20$. Both are then consistent with Equation 11 and $B_{12} \sim 1$. Memory of where to turn on after nulling does not seem inconsistent with the probable reignition mechanism suggested above.

In some pulsars a number of the above properties are simultaneously observed.

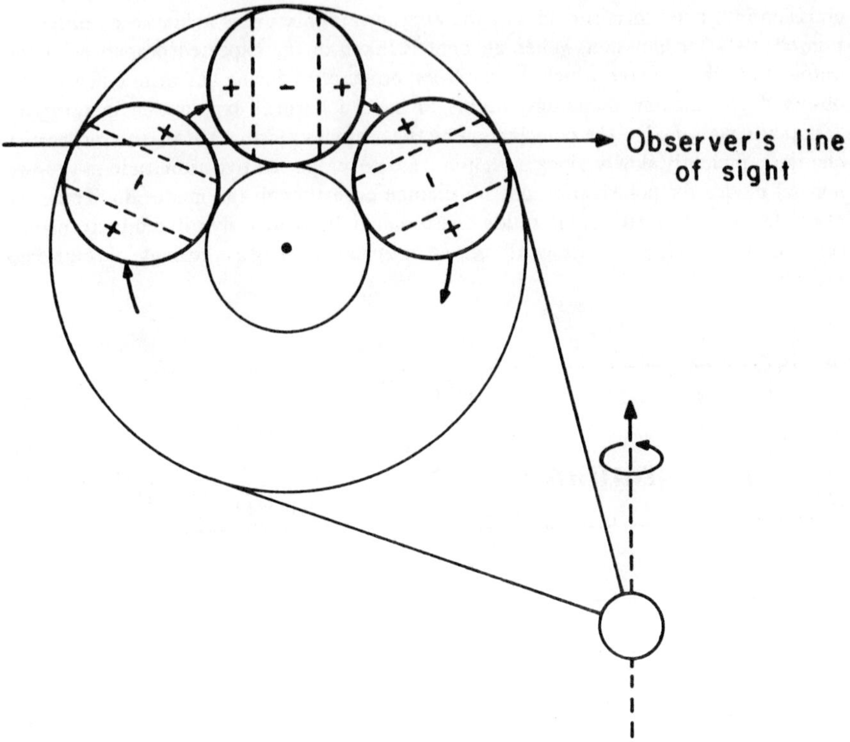

FIGURE 7. Motion of subpulses and their polarization within PSR 2016+28 as described by Backer, Rankin, and Campbell.[47] $P = 0.6$ s, $P_3/P = 3$, $P_2 = 7°$. The subpulses have a trailing and leading edge whose polarization mode is orthogonal to that of the central part of the subpulse. This subpulse polarization pattern keeps a fixed aspect as seen from the cone axis.

The time-dependent radiation pattern in PSR 2016+28 (P = 0.6 s) has been interpreted[47] in terms of subpulses (P_2 = 7°) drifting around a hollow cone (P_3 – 1.8 s) (cf. FIGURE 7). Most significant for implications about the radiation mechanism is the fact that in some (weak) subpulses (PSR 0329) the polarization seems to reach 100%. The very relativistic electrons (positrons) that are presumed to be the source of the coherent radio emission each give very narrow emission cones, so that they must be moving towards an observer if their radiation is seen. If charged particles are accelerated toward the observer, the observed polarization at any time (effectively the average over the very narrow beam) is zero. If the acceleration is perpendicular (e.g., curvature acceleration), the average for radiation into a vacuum does not exceed 80%. Therefore either the ambient medium during emission may not be approximated as a vacuum or polarization is determined after emission rather than by the emission mechanism.

FIGURES 8 and 9 show the observed position angles and circular polarization of PSR 2020+28 pulses.[49]. There is an obvious dichotomy in the polarization pattern. The polarization is in either of two orthogonal elliptical normal modes; the principal axes differ by 90° and the associated circular component has an opposite sign. The discontinuous jumping between the two modes implies that there is either incoherent emission into both normal modes or the appropriate mixture is achieved by propagation effects after emission. When account is taken of the expected properties of the ambient plasma through which the radiation propagates during and after emission the observed polarization dichotomy seems to have a natural explanation in terms of "adiabatic walking" of the polarization by the ambient plasma.[50] If plasma properties change sufficiently slowly along a photon trajectory, the relative amplitude in the two normal modes for polarization doesn't change even though the modes do. Thus, for example, in FIGURE 10 a polarization direction can be arbitrarily rotated by transmissions through a series of plates with rotated optic axes as long as the angle of rotation ϕ satisfies

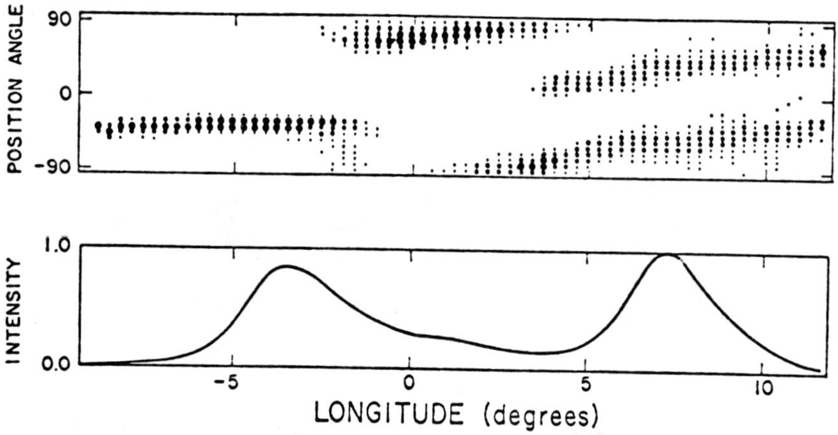

FIGURE 8. Polarization position angle of individual pulses and pulse envelope of PSR 2020+28 (Cordes, Rankin, and Backer[49]).

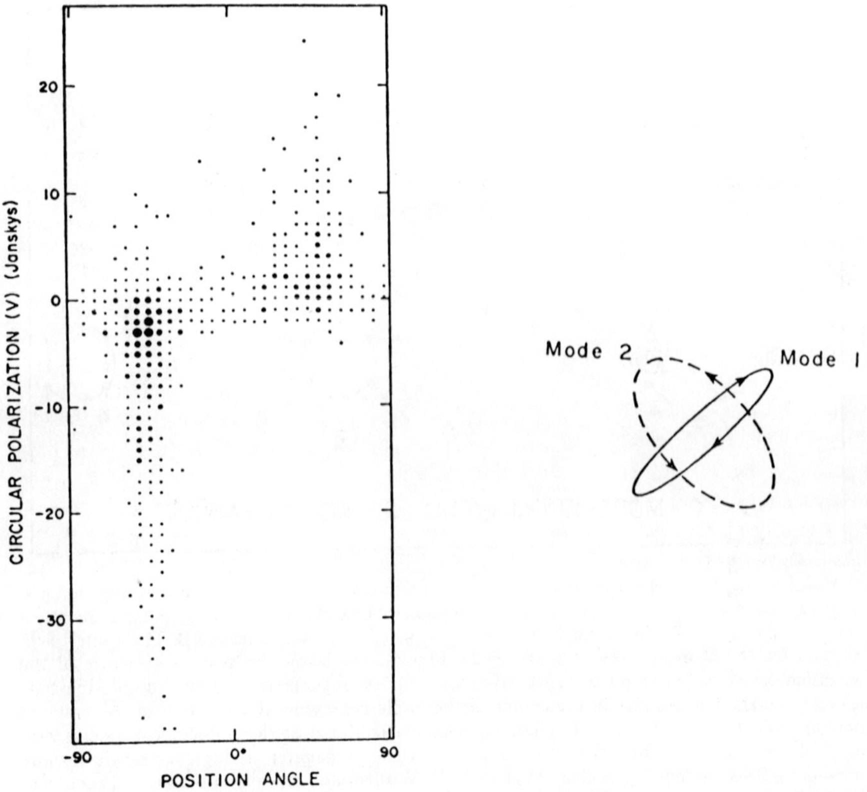

FIGURE 9. Circular polarization versus position angles of individual pulses of PSR 2020+28. The two favored modes are also shown. (Cordes, Rankin, and Backer[49]).

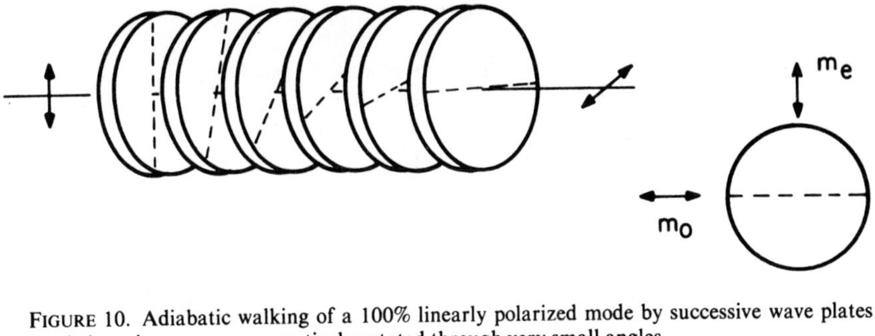

FIGURE 10. Adiabatic walking of a 100% linearly polarized mode by successive wave plates when their optic axes are consecutively rotated through very small angles.

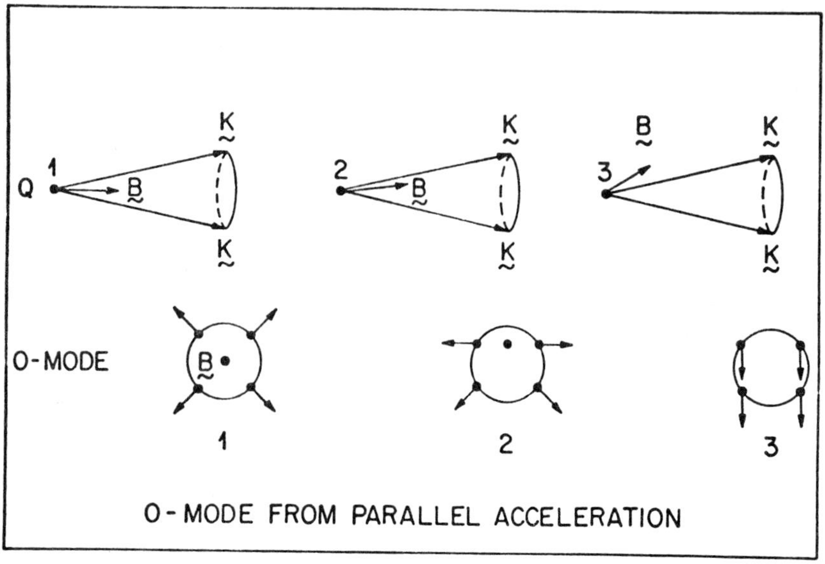

FIGURE 11. Ordering the polarization angles of a narrow cone of rays by adiabatic walking—at point 1, a charge Q is accelerated along the magnetic field **B** and gives O-mode radiation into a narrow cone of rays **k**, called the "emission cone," which is centered around **B**. The electric field vectors for the O-mode rays at 1 are given immediately below the side view picture of the emission cone, for four representative rays on the emission cone. We are looking into the cone along the axis, and the dot in the center of the circle represents the endpoint of the ambient magnetic field vector. At point 1, averaging the polarization over the emission cone yields zero net polarization. By point 2 the same cone of rays has propagated through the slowly varying medium until **B** no longer lies along the cone axis. With adiabatic walking the rays all remain in the O mode and the polarization vectors change appropriately. By point 3 the direction of **B** is well outside the emission cone, and the polarization angles are nearly aligned. Averaging the polarization over the emission cone now yields 100% polarization. For this diagram the O mode has been approximated as linearly polarized, an excellent approximation when e^+ and e^- distributions are nearly the same or when the ambient magnetic field is very large.

$$\left| \lambda \frac{\partial}{\partial s} \phi \right| \ll |\Delta n|, \tag{21}$$

where Δn is the difference in index of refraction between the two modes, ordinary (O) and extraordinary (E). For acceleration of very relativistic particles parallel to **B** the polarization averaged over the very narrow emission cone is zero. But as the radiated beam propagates in FIGURE 11 the angle between its wave vectors **k** and **B** becomes much larger than the beam opening angle γ^{-1} because **B** lines curve away from the beam axis. Then the average polarization becomes 100% since the beam is still 100% O-mode (**E** vector in the plane defined by **k** and **B** if there is either symmetry between e^- and e^+ in the plasma or $\gamma^{\pm} eB/mc \gg \omega$). A similar effect is achieved for the mixture of E-mode (**E** vector parallel to **k** × **B**) and O-mode radiation from acceleration perpendicular to **B** as indicated in FIGURE 12. We note from the directions and relative phases of the radiation **E** vectors that after adiabatic walking all interference terms between E and O mode for perpendicular acceleration vanish as well as those between O modes from parallel and perpendicular acceleration. Thus if adiabatic walking is an appropriate description for mode propagation away from the emission point, the "sudden" switching between the two modes expected for incoherent mixing (measured polarization exactly that of the dominant mode) is expected. Pulsar emitting plasma parameters do indeed indicate such behavior and the richness of observations is reflected by a richness of model possibilities.

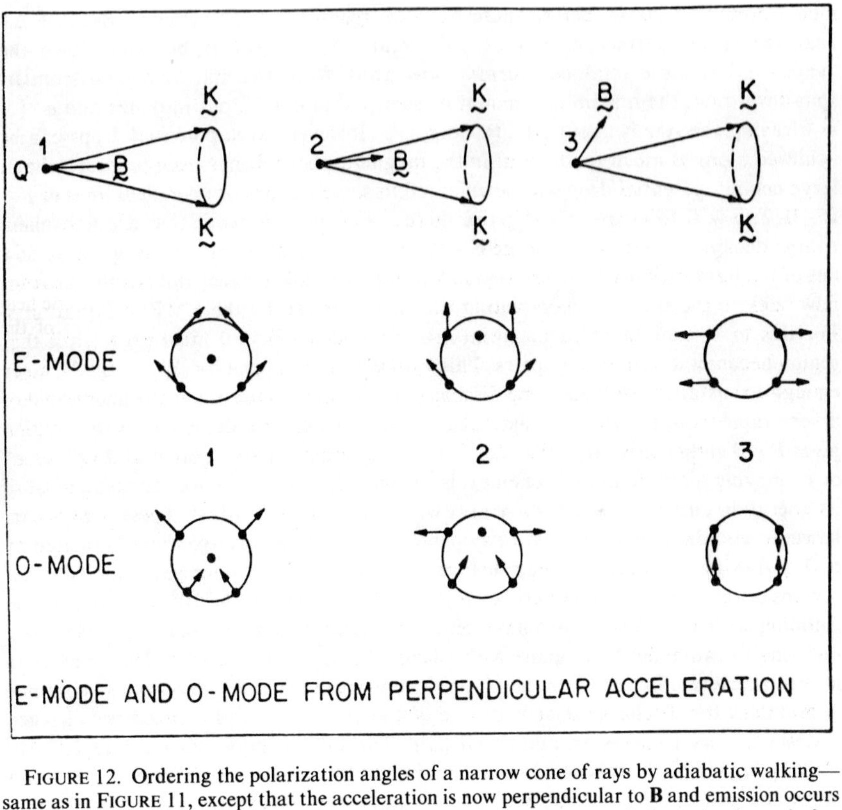

FIGURE 12. Ordering the polarization angles of a narrow cone of rays by adiabatic walking—same as in FIGURE 11, except that the acceleration is now perpendicular to **B** and emission occurs into both E mode and O mode. The E-mode and O-mode rays propagate independently, and after adiabatic walking (point 3) the polarization angles of the O-mode ray bundle and those of the E-mode bundle have been aligned separately into orthogonal directions. Comparison of FIGURES 11 and 12 shows that it will not be possible to observe interference between any of the three types of radiation (E mode and O mode from perpendicular acceleration and O mode from parallel acceleration) after adiabatic walking and an averaging over the emission cone. The electric vectors for any pair of the three types of radiation will combine with opposite phases for various rays, so that the interference terms sum to zero over the emission cone.

Gamma-Ray Emission

At least four pulsars[51] PSR 0531+21, 0833−45, 0740−28, and 1822−09, all characterized by a combination of large B and small P to give a relatively large total energy loss rate, give a double γ-ray pulse (10^2–10^3 MeV) similar to that of the Vela pulsar.[52–54] The radiated energy, 10^{34}–10^{35} erg s^{-1}, greatly exceeds that in the pulsar radio beams. In the slowest of the identified γ-ray pulsars, PSR 1822−09 with P = 0.8 s, the estimated γ-ray emmisivity in the range 50–500 MeV (10^{34} erg s^{-1}) approximates the measured total pulsar energy loss rate (based upon a reasonable 10^{45} g cm^2 neutron star moment of inertia). All four identified γ-ray pulsars give a double pulse separated by about 145°.

Energetic considerations preclude this energy coming from the e^+, $e^{\pm}$, or Z of the model presented above, before these particles reach the light cylinder. Because ΔV near the stellar surface cannot exceed about 10^{12} V without breaking down the "vacuum," and the total net current flow away from the star is known from its spin-down rate, the maximum amount of energy that can be put into ions and e^+ (or e^-) leaving the star is about $10^{30}\,P^{-2}$ erg s^{-1}. However, large potential drops can be achieved along $\mathbf{B}$ much further out in the magnetosphere. It has been proposed that a large enough potential drop will be achieved in some pulsars or open field lines at $\rho = \boldsymbol{\Omega} \cdot \mathbf{B}/2\pi c = 0$ to create e^-–e^+ pairs there.[27] Beyond this point the magnetosphere charge density changes sign if large $\mathbf{E} \cdot \mathbf{B}$ is to be avoided. Negative charges may flow out of the light cylinder from the region beyond that null surface, and positive charges flow back to the star, thus completing the circuit driven by the EMF of Equation 8. For this to be maintained a potential drop around $\boldsymbol{\Omega} \cdot \mathbf{B} = 0$ must grow until that region becomes a source of $e^{\pm}$ pairs. This will certainly happen for $\Delta V \gtrsim 10^{15}$ V, large enough to sustain a Sturrock-type discharge in the weak outer magnetosphere field of a very rapid pulsar.[55] But this may also occur when charge depletion in that region gives $\mathbf{E} \cdot \mathbf{B}$ sufficiently large that $\Delta V \sim 10^{13}$ V, a much smaller potential drop. An e^- or e^+ moving along $\mathbf{B}$ and traversing this region will achieve $\gamma \sim 10^7$ and give most of its energy to curvature radiation γ rays with $\hbar\omega \sim 10^2$–10^3 MeV. These γ rays can, however, collide with multi-keV x-ray photons from the polar cap, which is heated by both inflowing e^- from the near surface discharge of the "inner gap" and in other regions of the cap by the energetic e^+ from this "outer gap." The 10^2–10^3 MeV γ rays colliding with 1–10 keV x-rays have center of mass energies in excess of 1 MeV = 2 mc^2, and so can make e^+–e^- pairs with a near Thomson cross section. These pairs can in turn accelerate in the "outer gap," produce more curvature γ rays, and so continue to maintain the discharge if at least one of the 10^4–10^5 γ rays emitted per charged particle traversal makes an additional pair. The total energy emitted into 10^2–10^3 MeV γ rays for all pulsars that can maintain this process (EMF $\gtrsim 10^{13}$ V is a necessary condition) is $\Delta V J$, with J the total net curvature flow. The roughly symmetric double pulse follows naturally from the symmetry of the outer gap discharge with one beam directed toward the star's spin axis and one away from it. Aberration and travel time effects, which are important when the corotation velocity of the source region is not negligible relative to c and its position is not negligible relative to r_{lc}, can reduce the angular separation of the beams from 180° to 145°. An essentially similar model for γ-ray emission will result if the charge deficiency which gives rise to $\Delta V \sim 10^{13}$ volts comes from mechanism b) and equation 6, or if large centrifugal forces arise from relativistic corotation near the light cylinder. In such cases the two observed γ-ray beams would, however, come from opposite ends of the magnetosphere.

Conclusion

In summary, there does seem to exist a description of expected behavior of a spinning magnetized neutron star that is not (yet?) embarrassed by confrontation with observations of radio and γ-ray emission. Because of a present lack of understanding of magnetosphere currents at and beyond the magnetosphere, it is not yet possible to give a compelling derivation even though only current flows well within the light cylinder are relevant. Also, the complications in describing the growth and behavior of unstable modes of a relativistic multistream curving plasma preclude very quantitative predictions of coherent radio emission, although it must be present for the particular model discussed. But there seems to me to be no reason to be pessimistic about models of this sort as a reasonable representation of pulsars.

Acknowledgment

I am pleased to acknowledge stimulating conversations with Professors A. Cheng and P. Sutherland with whom I have collaborated on much of the theoretical work described above, and with Professor E. T. Scharlemann.

References

1. Manchester, R. & J. Taylor. 1977. Pulsars. W. H. Freeman. San Francisco. See references therein.
2. Smith, G. Pulsars. Cambridge University Press. Cambridge. See references therein.
3. Ginzburg, V. & V. Zholeznyakov. 1975. Annu. Rev. Astron. Astrophys. **13:** 511.
4. Sutherland, P. *In* Fundamentals of Cosmic Physics. V. Canuto, Ed. To be published.
5. Sturrock, P. 1971. Astrophys. J. **164:** 529.
6. Gold, T. 1968. Nature (London) **218:** 731.
7. Smith, G. 1970. M.N.R.A.S. **149:** 1.
8. Zheleznyskov. 1971. Astrophys. Space Sci. **13:** 87.
9. See, for example, W. Kundt. This volume.
10. Manchester, R., E. Tademaru, J. Taylor & G. Huguenin. 1973. Astrophys. J. **180:** 207.
11. Bartel, N. & W. Sieber. 1978. Astron. Astrophys. **70:** 307.
12. Bartel, N. 1978. Astron. Astrophys. **62:** 393.
13. Kolbenstredt, H. 1976. Preprint.
14. Ter Haar, D. 1972. Phys. Rep. C **3:** 58. See references therein.
15. Ruderman, M. & P. Sutherland. 1975. Astrophys. J. **196:** 51.
16. Goldreich, P. & D. Keeley. 1971. Astrophys. J. **170:** 463.
17. Goldreich, P. & W. Julian. 1969. Astrophys. J. **157:** 869.
18. Gold, T. Ann. N.Y. Acad. Sci. This volume.
19. Mestel, L. 1971. Nature (London) Phys. Sci. **223:** 149.
20. Mestel, L., P. Phillips & Y. M. Wang. 1979. M.N.R.A.S. In press.
21. Scharlemann, E., J. Arons, & W. Fawley. 1978. Astrophys. J. **222:** 297.
22. Buschauer, & G. Benford. 1977, M.N.R.A.S. **179:** 99.

23. Fawley, W., J. Arons & E. Scharlemann. 1977. Astrophys. J. **217:** 227.
24. Arons J. & E. Scharlemann. 1978. Astrophys. J. To be published.
25. Cheng, A. & M. Ruderman. 1977. Astrophys. J. **212:** 800.
26. Holloway, N. 1973. Nature (London) Phys. Sci. **246:** 6.
27. Cheng, A., M. Ruderman, & P. Sutherland. 1976. Astrophys. J. **203:** 209.
28. Michel, R. 1974. Phys. Rev. Lett. **33:** 1521.
29. Michel, R. 1974. Astrophys. J. **192:** 713.
30. Cheng, A., & M. Ruderman. 1977. Astrophys. J. **214:** 598.
31. Flowers, E., J.-F. Lee, M. Ruderman, P. Sutherland, W. Hillebrandt & E. Müller. 1977. Astrophys. J. **215:** 291.
32. Jones, P. 1977. Preprint.
33. Trumper, J., W. Pietsch, C. Reppin, W. Goves, R. Staubert & E. Kendzlorra. 1978. Astrophys. J. Lett. **219:** 105.
34. Arons, J. 1979. Space Sci. Rev. To be published.
35. Melrose, D. 1978. Astrophys. J. **225:** 557.
36. Benford, G. & R. Buschauer. 1977. M.N.R.A.S. **179:** 189.
37. Pellat, R. 1979. Space Sci. Rev. To be published.
38. Benford, G. 1979. Space Sci. Rev. To be published.
39. Ruderman, M. 1976. Astrophys. J. **203:** 206.
40. Cordes, J. 1979. Space Sci. Rev. To be published.
41. Sieber, W. 1973. Astron. Astrophys. **28:** 237.
42. Cordes, J. 1978. Astrophys. J. **222:** 1006.
43. Komesaroff, M. 1970. Nature (London) **225:** 612.
44. Oster, L. & W. Sieber. 1977. Astron. Astrophys. **58:** 303.
45. Oster, L. & W. Sieber. 1978. Astron. Astrophys. **65:** 179.
46. Unwin, S., A. Readhead, P. Wilkenson & M. Ewing. 1978. M.N.R.A.S. **182:** 711.
47. Backer, D., J. Rankin & D. Campbell. 1976. Astrophys. J. **197:** 481.
48. Proszynski, M. 1978. Preprint.
49. Cordes, J., J. Rankin & D. Backer, 1978. Astrophys. J. **223:** 961.
50. Cheng, A. & M. Ruderman. 1979. Astrophys. J. In press.
51. Buccheri, B., K. Bennett, G. Bignami, N. D'Amico, W. Hermsen, D. Huizing, G. Kanbach, G. Lichti, J. Paul, B. Sacco, B. Swanenburg & R. Wills. 1979. Astron. Astrophys. In press.
52. Manchester, R., P. Hamilton, W. Gross & L. Newton. 1976. Proc. Astron. Soc. Aust. **3:** 81.
53. Wallace, P., B. Peterson, P. Murdin, J. Danziger, R. Manchester, A. Lyne, W. Gross, F. Smith, M. Disney, K. Hartley, D. Jones & G. Wellgate. 1977. Nature (London) **266:** 692.
54. Buccheri, R. 1976. The Structure and Content of the Galaxy and Galactic Gamma Rays. C. Fichtel and F. Stecker, Eds.: 52. Goddard Space Flight Center. New York, N.Y.
55. Cheng, A. & M. Ruderman. 1977. Astrophys. J. **216:** 865.

WHAT CAN WE LEARN FROM THE CRAB NEBULA?

Wolfgang Kundt

Institut für Astrophysik der Universität Bonn
D-5300 Bonn
Federal Republic of Germany

INTRODUCTION

The Crab nebula is one of the best studied objects in astrophysics. It has been brilliantly described by Rees and Gunn,[1] and by many others. Yet there is no unique answer to be found, e.g., to the question: what exactly is the pulsar input into the nebula? When I tried to answer this question, I convinced myself of a number of inconsistencies in all existing models, and ended up making an innovative model with E. Krotscheck.[2]

The Crab nebula does not only tell us what a young pulsar can do. It also holds its secrets about the fate and mass of its progenitor star, about the mechanism that drives a supernova explosion, and about cosmic-ray production. Under the conservative hypothesis that the Crab is typical rather than exceptional, a complete understanding of its structure can give us keys to a better understanding of supernova remnants, (γ-ray) pulsars, binary evolution, nucleosynthesis, and perhaps even of the central engines in galaxies.[3] In this talk I shall concentrate on those properties which have a chance of being of general importance; justifications, if missing, can be found in reference 2.

ARE SUPERNOVA REMNANTS TURBULENT?

Almost every paper on supernova remnants mentions turbulence somewhere along the line. Yet wherever the motion of filaments has been measured, in the Crab, in Cas A, or elsewhere, there is evidence for ordered radial motion, at the 1% level. The filaments in the Crab form a thin network of luminous matter. Most of the space between them is filled by a relativistic medium, probably electrons and positrons, more or less in pressure equilibrium. But pressure equilibrium implies that the respective mass densities ρ_j behave inversely as the specific kinetic particle energies: $\rho_{\text{therm}}/\rho_{e^\pm} = m_p c^2/3kT_{\text{therm}} \approx 4 \times 10^8 T_4^{-1}$. In other words, the filaments (of a young remnant) are infinitely heavy. They move radially; is there a reason for the relativistic component to be turbulent?

The relativistic component has a sufficient (collision-free) electrical conductivity to freeze[2] its magnetic flux. When it overtakes a filament, it has to drag its flux across it, as sketched in FIGURE 1. This mechanism implies friction between the two components. The friction is expected to damp turbulence on scales large compared with filamentary spacing. Also, turbulence would raise the magnetic energy of the frozen-in flux. I therefore cannot see a reason for large deviations of the relativistic component from ordered radial motion. Krotscheck and I have made it an assumption of our model for the Crab, and we get reasonable results.

An often-quoted counter example to nonturbulent motion is the 300-yr-old

0077-8923/80/0336-0429 $01.75/1 © 1980, NYAS

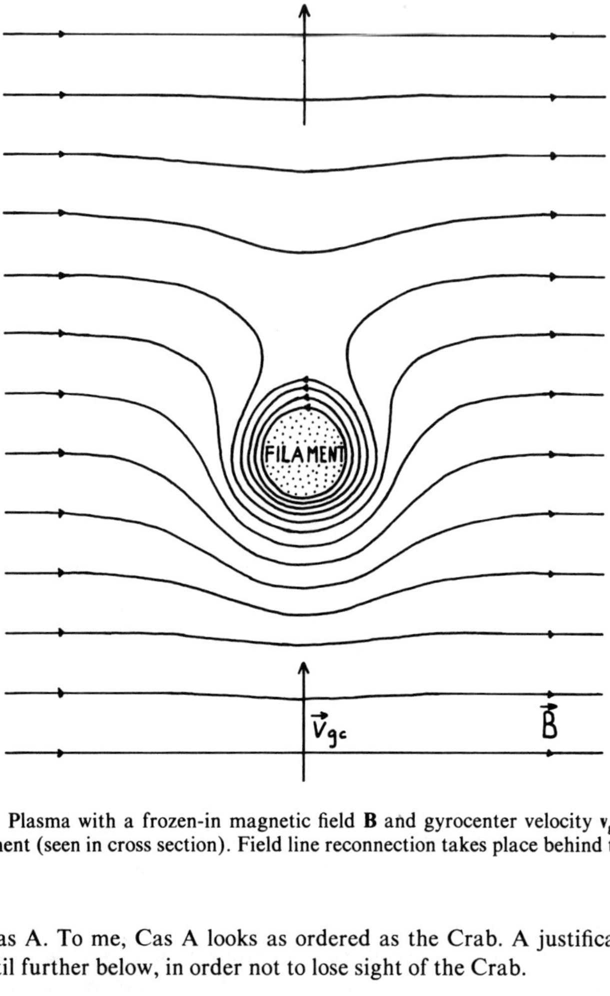

FIGURE 1. Plasma with a frozen-in magnetic field **B** and gyrocenter velocity $\mathbf{v}_{gc}$ overtakes a massive filament (seen in cross section). Field line reconnection takes place behind the filament.

remnant Cas A. To me, Cas A looks as ordered as the Crab. A justification will be delayed until further below, in order not to lose sight of the Crab.

Some Inferences from a Filamentary Structure

In the literature, supernova remnants are often treated as continuous detonation waves governed by hydrodynamic equations. The solutions show the formation of a reverse shock. Such a reverse shock seems to conform with an observed hot interior of Cas A.

But if the heavy component has the geometry of a thin network, its motion obeys the pressure-free equations, and no reverse shock can form. Instead, the interior of a young supernova remnant is heated by an impulsive implosion wave once its internal pressure drops below the ambient pressure. For the Crab, this stage lies still in the future.

A similar remark concerns the suggested hot-tunnel system connecting supernova remnants: I do not see the origin of the material that supposedly makes the skins of such tunnels.

If supernova shells have filamentary geometry, their calculated x-ray production rate has to involve a filling factor. Analogous remarks may apply to the line-forming material in galactic center sources.

On the Kinematics of the Crab Nebula

Figure 2 shows how I would like you to see the Crab nebula: It has an approximate cigar shape around an axis Ω_{SNR}, which may be the spin axis of the progenitor star's core before the supernova explosion. At its center is the pulsar, bracketed by a more or less symmetrical array of (thin optical) wisps. The relativistic wind from the pulsar moves supersonically up to an inner shock surface, the "injection sphere," which is almost spherical, of radius 1 ly. The injection sphere becomes apparent in Figure 3 (which has been produced in Bonn from Woltjer's analysis of Baade's continuum measurements, with a suitable color wedge). One quarter of the sphere is missing: an extraneous cloud inside the remnant obscures it. The outer

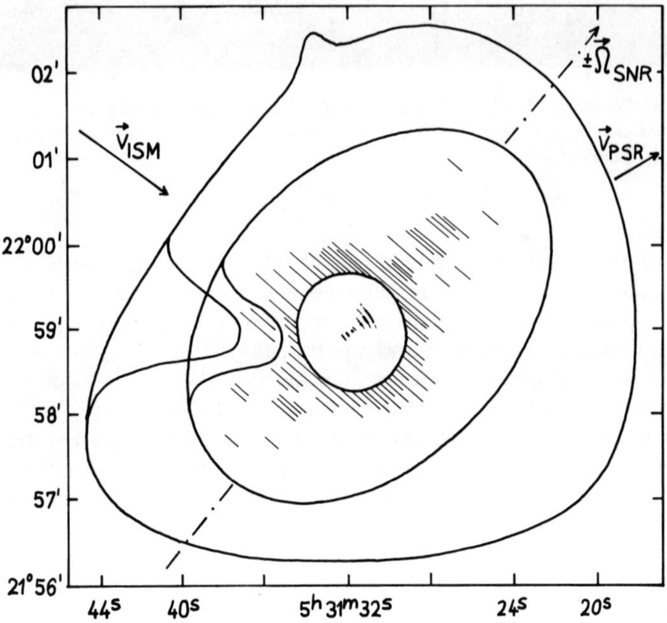

Figure 2. The Crab nebula: schematic isophotes. $\mathbf{v}_{PSR}$ is the pulsar's peculiar velocity. Ω_{SNR} may be the spin axis of the progenitor star's core, and $\mathbf{v}_{ISM}$ may be the relative velocity of the interstellar medium. The parallel, straight lines outside the injection sphere symbolize fine-structured regions of enhanced nonthermal emission.

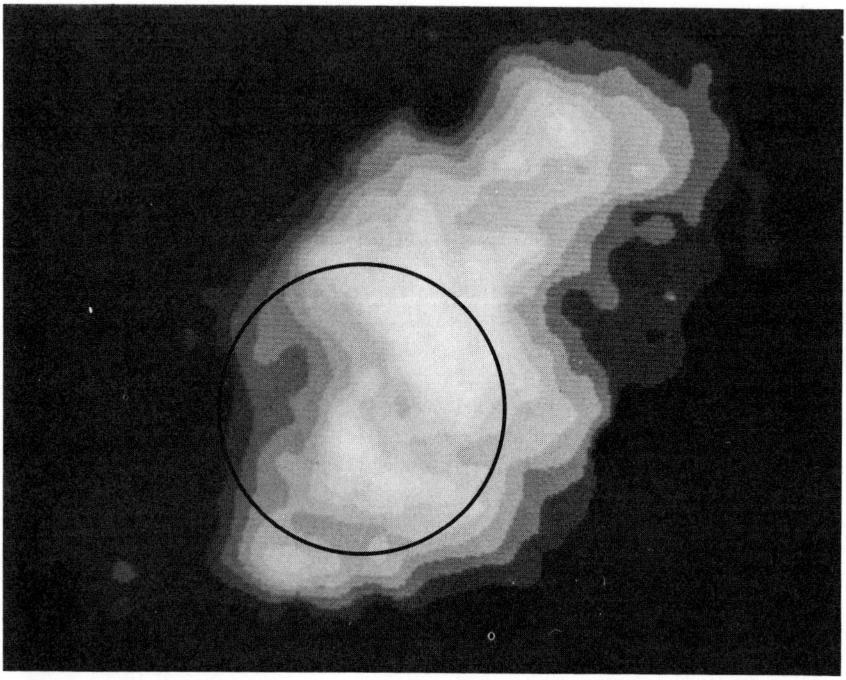

FIGURE 3. Computer-simulated picture of the Crab nebula, at wavelengths $\lambda = 52-64 \times 10^{-6}$ cm. All stars except PSR 0531+21 have been removed. Intensities are taken from Woltjer's evaluation of Baade's plates.

contours of the nebula deviate increasingly from cigar shape, perhaps due to a beginning interaction with the interstellar medium. The normal to the wisp array may reveal the spin axis of the pulsar; it is somewhat inclined w.r.t. Ω_{SNR}. Still more inclined is the projected peculiar velocity of the pulsar.[10]

FIGURE 4 explains how we have approximated the dynamics of the nebula: Periodic electron–positron sheets, at separations $\lambda/2 = \pi c/\Omega = 5 \times 10^8$ cm, move radially from the pulsar to the outer edge of the nebula. On their journey, they radiate at decreasing frequencies, from γ-rays through x-rays, optical, infrared to radio. They encounter an "inner shock" after about one year, where their motion changes into gyration. An "outer shock" separates them from the interstellar medium. Instabilities at the outer edge are ignored in the figure. Individual particle orbits are drawn to the left: they change from straight lines to gyrations when crossing the injection sphere, with gyrocenter velocities ranging from c through 0.5 c to some 0.6×10^{-2} c at the outer edge.

A check on this model is given by the fact that the present number of radio-emitting electrons in the nebula equals the time-integrated rate of newly injected electrons, as determined from the optical luminosity. Here a shifted power-law extrapolation for the past pulsar luminosity has been assumed.

A Coarse Description of the Model

The observed spectrum of the Crab is sketched in Figure 5, dotted when pulsed. It has the shape of an elephant with a radio trunk, an optical forehead, x-ray back, γ-ray tail, and some ultrahard leftovers.

I interpret the hard part down to optical frequencies as due to the newly injected electrons, and the infrared and radio part as due to the electrons accumulated from earlier times. As an implication, there is no distinction between an x-ray nebula and an x-ray pulsar (at hard x-rays): The γ-ray and x-ray pulses are emitted after the electrons and positrons have left the speed-of-light cylinder of the pulsar and have been post-accelerated by the strong 30-Hz wave; c.f. Figures 6 and 7 which document my understanding of a pulsar at work. The pulses are increasingly smeared out by detouring of the radiating charges. This interpretation is suggested by the facts that (i) the high-energy pulse shape is energy-independent, (ii) there is a smooth transition from pulsed to unpulsed, near 3 MeV, (iii) the apparent size of the nebula shrinks with increasing photon energy; and (4) highly relativistic electrons must radiate a substantial fraction of their energy when encountering the (even partially) reflected 30-Hz waves (see below).

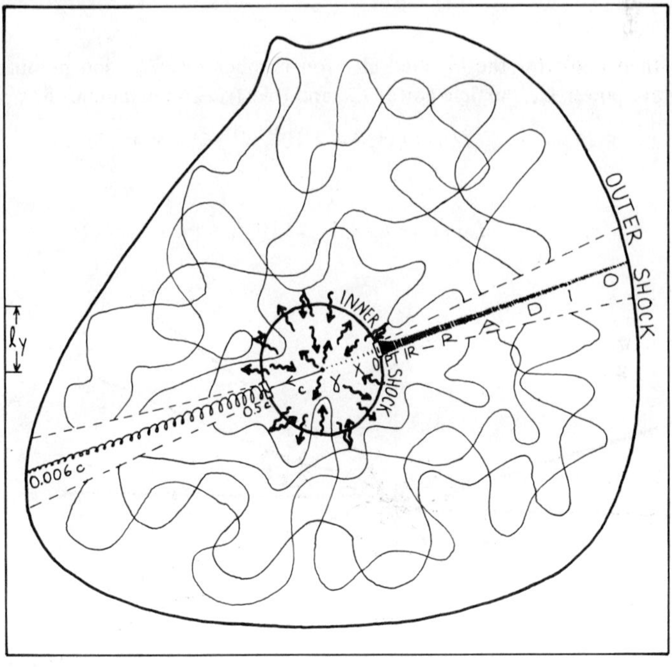

Figure 4. Schematic drawing of how I interpret the Crab nebula: The injection sphere, of radius 1 ly, is a bath of multiply reflected 30-Hz waves. Periodically emitted electron–positron sheets move radially outward, and radiate in sequence γ-rays, x-rays, optical, infrared, and radio waves. Individual particle orbits change from straight lines to gyrations at the inner shock surface.

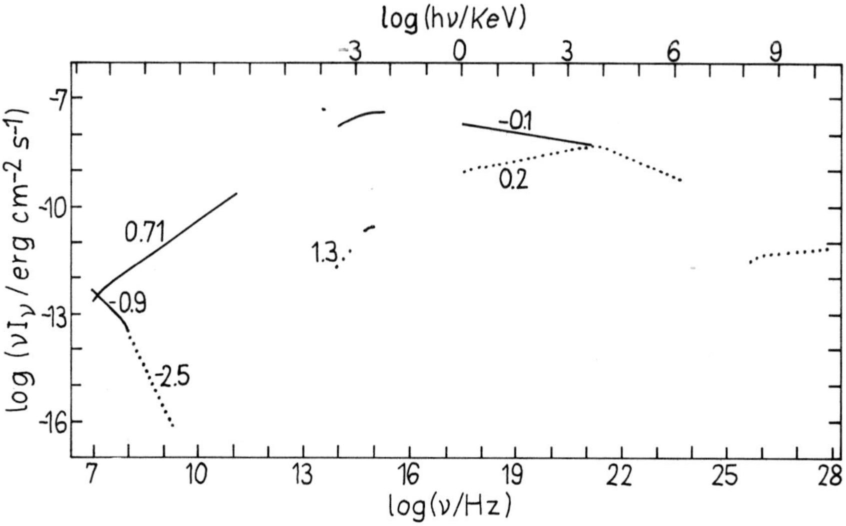

FIGURE 5. Spectral density of the Crab nebula and its pulsar (dotted).

One then finds[2] for the injected electron number rate $\dot{N}_{e^\pm}$, ion number rate $\dot{N}_i$, 30-Hz wave power L_W, particle power $L_{e^\pm}$, and (electric and magnetic) flux power L_F:

$$\dot{N}_{e^\pm} = L_{\text{nebula}}/\langle\gamma\rangle m_e c^2 = 4 \times 10^{38}\ \text{s}^{-1} = 1.2 \times 10^4 \dot{N}_{\text{GJ}}, \tag{1}$$

$$\dot{N}_i/\dot{N}_{e^\pm} < 10^{-2}, \tag{2}$$

$$L_W : L_{e^\pm} : L_F \approx 1 : 1 : 10^{-3}, \tag{3}$$

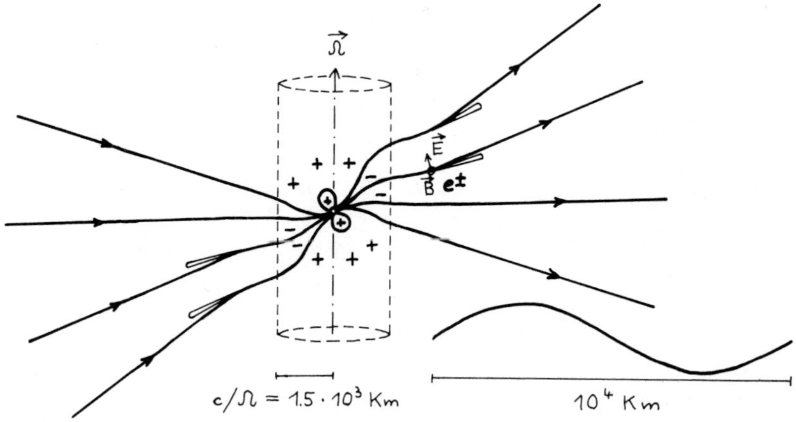

FIGURE 6. Pulsar at work, in profile (tentative): An inclined magnet spins around **Ω** and sets up a charged magnetosphere. Escaping relativistic electron–positron pairs radiate predominantly near the speed-of-light cylinder, of radius c/Ω, where they are seized by the strong (forming) low-frequency wave.

where $\dot{N}_{GJ}$ denotes the electron rate obtained from a (stationary) Goldreich–Julian magnetosphere by extracting the charges at the speed of light. The only uncertain factor in Equation 1 is the average Lorentz factor $\langle\gamma\rangle$ of the electrons radiating UV photons; it follows from synchrotron theory for a plausible (inferred) magnetic field strength of $\lesssim 10^{-3}$G. Equations 2 and 3 are the results of fitting our numerical solution to the observed boundary conditions.

Moreover one finds for the neutron star's moment of inertia I and polar surface magnetic field strength B:

$$3 \times 10^{38}\text{erg/s} \approx 2L_{\text{nebula}} \approx L_{\text{PSR}} = \begin{Bmatrix} I\Omega\dot{\Omega} & \Longrightarrow I = 6 \times 10^{44}\text{g cm}^2 \\ (2/3c^3)(BR^3\Omega^2)^2 & \Longrightarrow B = 3 \times 10^{12}\text{G} \end{Bmatrix}. \quad (4)$$

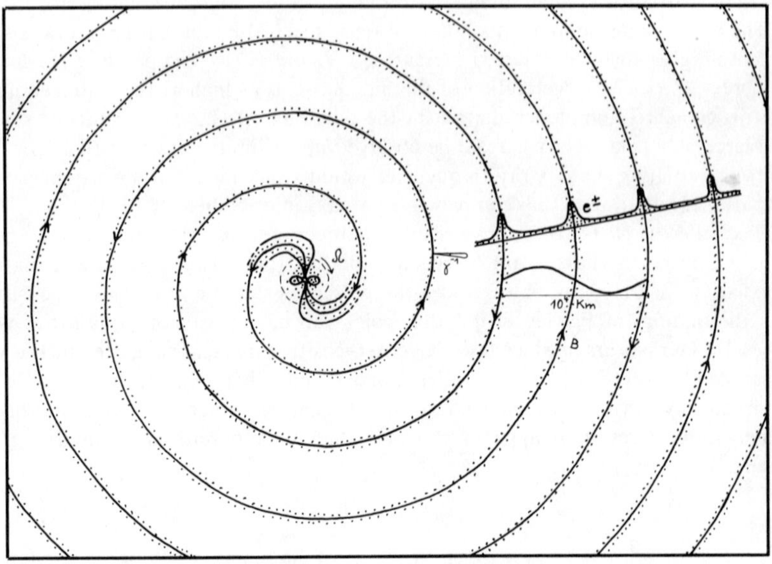

FIGURE 7. As in FIGURE 6, but seen from the top. The densities of the radially moving electron–positron sheets may correspond to the observed γ-ray light curve.

Finally, the hard part of the spectrum corresponds to an effective injected power-law distribution of electrons and positrons (by the pulsar) given by

$$N_\gamma\,\mathrm{d}\gamma \sim \gamma^{-2.2}\mathrm{d}\gamma \qquad \text{for } 3 \times 10^5 \lesssim \gamma \lesssim 10^9. \quad (5)$$

I shall now list some reasons for the claim that the relativistic component of the Crab nebula consists essentially of electrons and positrons: (i) The fact that $\dot{N}_e$ exceeds $\dot{N}_{GJ}$ by more than a factor 10^4 suggests, in reasonable pulsar models, that electrons are produced midway between the neutron star surface and the inner edge of the nebula, probably near the polar caps. Charge conservation implies that positrons are produced at an equal rate. (ii) Ions would conserve their stiffness (γ-factor) throughout the nebula, but there are no stationary solutions coming close to satisfying the boundary conditions for this case. (iii) The Gunn–Ostriker mechanism (for post-accelerating the charges) gets suffocated if too high a density of charges with unequal e/m ratios is injected; the charges would acquire equal velocities, hence all the power would go into the heavier component. In the case of the Crab, the critical rate is $\dot{N}_e^{\text{crit}} = 10^{35}\gamma^2 s^{-1}$ where γ is the average Lorentz factor of the slower component at injection. (iv) There are no solutions for near-corotating pulsar magnetospheres with one-component currents (along the open field lines).

Fate of the Low-Frequency Waves—γ-Ray Pulsars

A crucial point of our model is that the 30-Hz pulsar waves are repeatedly reflected near the inner edge of the nebula, c.f. FIGURE 4. This result is not only consistent with direct calculations, but is demanded by the following observed facts: (i) The pressure inside filaments, inferred from spectral-line intensity ratios, is some 30 times higher than the injected pressure $p = L/4\pi r^2 c$. (ii) The pressure needed for the observed 8% post-acceleration of the filaments[10] is as high as that inferred under (i). (iii) Synchro-Compton radiation in the outgoing 30-Hz wave would be several 10^{-3} circularly polarized, whereas the observed upper limit is of order 10^{-4}. (iv) The electrons radiating up to x-ray frequencies would reach much larger distances from the pulsar if radiating in the (comparatively weak) injected field.

The reflected 30-Hz waves give rise to a number of important inferences. To begin with, the outgoing charges are forced to radiate strongly, and one gains a natural explanation for the γ-ray pulses and x-ray pulses, as stated above. This is consistent with the findings of Fawley *et al.*[4] that polar cap models cannot explain the γ-ray pulses. If older pulsars produce fewer electron–positron pairs, as suggested by the Vela pulsar and, theoretically, by a smaller unipolar induction voltage, the power of the low-frequency wave is transferred to fewer charges which then acquire higher γ-factors. As a result, it appears plausible that older pulsars are primarily γ-ray pulsars.

A Lunar Dust Atmosphere?

If there is a continuous transition from pulsed to unpulsed x-rays, whereby the pulses are smeared out by detouring, the shape of the source should shrink homologously from a radius of some 40″ at soft x-rays (KeV) to $\lesssim 10^{-3}$″ at MeV energies. This prediction is at variance with the lunar occultation experiments,[5,6] which show only mild shrinking, in proportion to $E^{-0.15}$, up to 20 KeV. Moreover, these lunar occultation experiments find an elliptical emission region whose major axis is at right angles to that of the nebula.

Even though some post-acceleration is expected to occur in the magnetic flux tubes around filaments, I cannot believe that these occultation experiments reveal the true shape of the hard x-ray Crab. Rather, both size and shape may be mocked up by scattering in a lunar dust atmosphere, of variable height between 1 and 30 Km, particle masses $\lesssim 10^{-15}$g, and number density some 10^4 cm^{-3}. Such an atmosphere would be electrostatically supported, raised on the night hemisphere through negative charging by solar wind electrons, and on the day hemisphere through positive charging by the photoelectric effect. A transverse ellipse, as observed, corresponds to a lunar atmosphere which was twice as high on the night side as on the day side.

Filamentary Structure and Mass of the Thermal Component

The fact that the thermal component of the Crab consists of more than 10^2 filaments is not only observed optically, but is equally obvious from temporal changes in the dispersion measure of the pulsar, and in the scintillation decorrelation time, on timescales of days and greater.

A much larger number of thinner filaments is implied by our model through the fact that the 30-Hz waves are some 30 times reflected. For this to be true, the optical depth τ for the 30-Hz waves must be ≥ 6. Approximating typical filaments by cylindrical threads of diameter d, length l, mass density $\gtrsim 10^{-21}$ g cm^{-3}, one finds $d \leq 4 \times 10^{15}$ cm $(M/3\mathrm{M}_\odot)$ for their average diameter, and $N \geq 0.6 \times 10^4(3\mathrm{M}_\odot/M)(r/l)$ for their typical number.

These estimates contain the mass M of the nebula in units of 3 $\mathrm{M}_\odot$. This factor is probably not smaller than unity: Apart from direct estimates of the amount of luminous matter in filaments, made recently by Davidson,[7] there are the following two dynamical arguments. (i) If the Crab pulsar was one of two neutron stars liberated during the 1054 supernova explosion, as is suggested by evolutionary reasonings,[8,9] the mass of the ejected shell must exceed two neutron star masses $\approx 2.8\ \mathrm{M}_\odot$ (to guarantee fission). (ii) Wyckoff and Murray[10] have measured the projected peculiar motion of the Crab pulsar, between 1899 and 1976. When extrapolated backwards to the year 1054, the pulsar position almost coincides with the present center of mass of the nebula, defined as the divergent point of the proper motions of the filaments. Accounting for the error ellipses of the two measurements, I find a projected nebular velocity $\lesssim 30$ Km/s if pointing northwest, and much smaller otherwise. When this is compared with the pulsar's 1.2×10^2Km/s towards northwest, and account is taken of a joint presupernova velocity of the binary system, the conclusion can be drawn that the nebula must be at least as massive as two neutron stars.

Cigar Shapes of Plerions

The Crab nebula has an elliptical shape. Can we conclude that its 3-d shape is prolate? From the directions of linear polarization one finds that magnetic field lines are on average perpendicular to $\mathbf{\Omega}_{\mathrm{SNR}}$, corroborating the idea that a dipole field is drawn out into a coil around the rotation axis. But the next best filled-center supernova remnant (=plerion), 3C 58, looks different in linear polarization. We then realize that the directions of linear polarization tend to be aligned with the strongest filaments, and can therefore be deceptive.

Two alternative methods are better suited to determine the 3-d shape: (i) Trimble[11] measured the two peculiar velocity fields $\mathbf{v}_i(x, z)$ and $\mathbf{v}_i(y, z)$ of filaments in the two coordinate planes, with the z axis pointing toward us. When fitted by ellipses, the curves $|\mathbf{v}_i| =$ const reveal a prolate shape of the nebula if at a distance >1.3 Kpc. (ii) The 3727 Å [OII] line measured by Mayall and discussed by Shklovski shows a linear increase in redshift toward northwest, indicating a prolate shape inclined at some 25°.

The observed cigar shape of the Crab cannot be the result of an asymmetric post-acceleration. This is already clear from the fact that post-acceleration amounts to only 8% whereas the ratio of axes measures 1.6. It must therefore be due to the explosion event, and is a strong indication of a magnetically driven supernova.[13] As plerions seem to correspond to the second explosion in a massive binary, which has[9] the characteristics of a supernova of type I, I am tempted to conclude that a magnetic spring is responsible for the momentum transfer to the supernova shell in at least all type-I events. In type-II events, a generally larger mass in the envelope may homogenize the originally anisotropic magnetic pressure distribution, or else the remnant may be seen on-axis (Cas A?).

Is Cas A Turbulent?

Cas A contains three different expanding components: (i) A system of ($\gtrsim 10^2$) "fast-moving optical knots," with velocities 4–9 × 10^3 Km/s; the majority of them have been decelerated by 6% and form a luminous arc; (ii) a system of ($\gtrsim 30$) "quasistationary flocculi," with velocities of order 2 × 10^2 Km/s; and (iii) a somewhat irregular system of ($\gtrsim 30$) "compact radio peaks," whose expansion velocities are $\lesssim 30\%$ of those of the knots.[17] The system (i) shows ordered radial motion,[14], c.f.

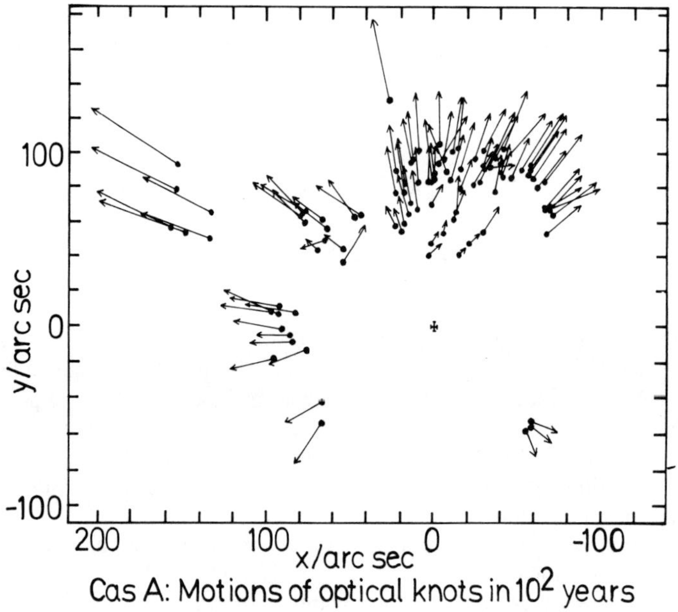

FIGURE 8. Observed motions of the (fast) optical knots in Cas A, copied from Kamper and Van den Bergh.[14] The divergent point (at the center) and its uncertainties are shown.

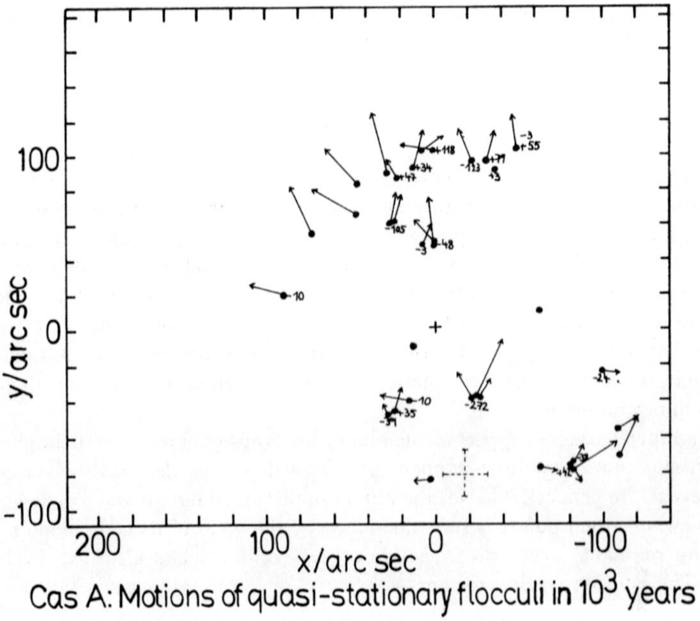

FIGURE 9. Observed motions of the quasistationary flocculi, after Kamper and Van den Bergh.[14] Their apparent divergent center is offset to the south.

FIGURE 8, like the filamentary system in the Crab. System (ii) is a novelty: despite its small velocity, it is almost coincident with the system of fast-moving knots, but has a southward displaced divergent center, c.f. FIGURE 9 (less than half of this offset is due to a systematic proper motion of the faint reference stars). Moreover, it is exclusively blueshifted.

The only reasonable interpretation[9] I can think of is that the flocculi are luminous interstellar material dragged along by the fast-expanding filaments (identified as fast knots). The offsets in blue shift and divergent center are then a measure of a large space velocity, of order (165 ± 15)Km/s, towards north-northeast and towards us (after correction for galactic rotation). But the binary system that must have absorbed the recoil momentum has not even been found yet, hence cannot be very massive. I conclude that Cas A cannot be more massive than some 3 $M_{\odot}$. This estimate is at variance with estimates[16] based on stellar evolution theory.

Once the quasistationary flocculi have been interpreted as a wake phenomenon of the fast moving filaments, there remains the question of whether the velocities of the radio peaks[17] in Cas A are a manifestation of turbulent motion, or simply irregular phase velocities. I favor the latter interpretation. In any case, I mistrust Scott and Chevalier's estimate of the cloud collision efficiency k used by Cowsik[18] on page 859: instead of their $k = 10^{-2}$ yr^{-1}, my estimate reads $k \ll 10^{-6}$ yr^{-1}. In other words, I do not think that any Fermi acceleration is presently going on in Cas A. Apart from the reasons given earlier,[15] there is the "soft radio flare" (factor 1.3 within ≤ 7 yr, at 38 MHz) reported by Read,[19] which is difficult to explain as a stochastic phenomenon inside the expanding shell. The source of cosmic rays[20] inside Cas A may rather be the unseen binary neutron star near its center.

A GUESS AT THE CRAB'S PAST

FIGURE 10 suggests an evolutionary scheme[9] that ties up the two types of supernovae, the two types of young supernova remnants (viz, the plerions and the kelifons), and the two types of pulsars. It is based on two hypotheses: (i) Sufficiently massive stars ($\gtrsim 5\ M_\odot$ at formation) make neutron stars; and (ii) massive stars form in binaries (or multiples). The latter hypothesis is plausible in view of the fact that collapsing clouds would have to reduce their angular momentum by a factor $10^{6\pm1}(\rho g^{-1} cm^3)^{1/6}$ in order to form isolated stars (of density ρ); as gravitational lever arms can be much longer than magnetic ones, excess angular momentum is likely to be dumped in orbital motion.

Yet a much stronger support of the binary hypothesis comes from its implications: The first supernova explosion happens at the birth site of the massive system, like type-II events. In general, it leaves the binary undisrupted but gives it a runaway kick. As the newly formed pulsar is soon quenched by the wind of its companion star, the expanding remnant turns into a shell source (=kelifon), like CAS A, Tycho, and Kepler. The neutron star's rotational energy is converted into cosmic rays until it is

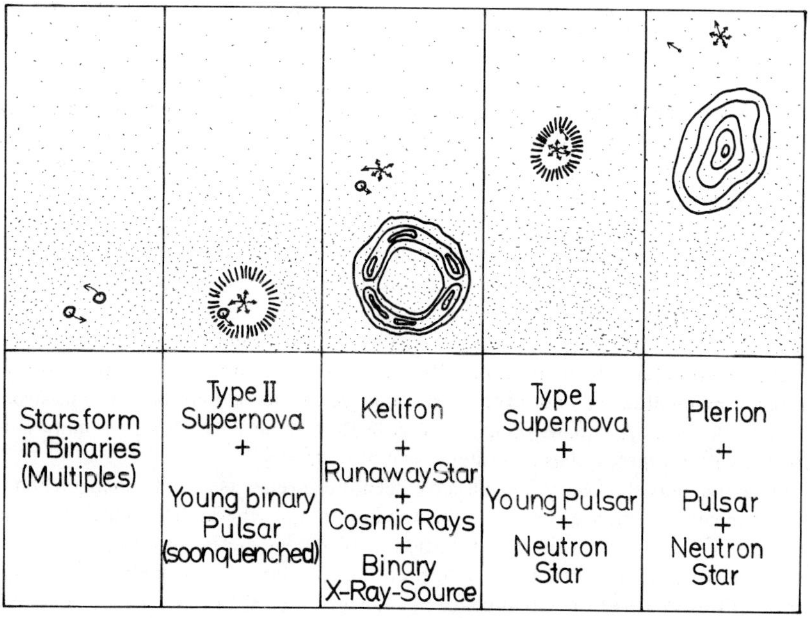

FIGURE 10. Tentative evolution of massive stars, which are assumed to form in binaries. Their remnants tend to move away from the galactic plane. In columns 3 and 5, the stellar remnants are drawn at a later epoch than the blown-off shells (when they are preferentially observed); they would otherwise be near the center of their shell.

sufficiently reduced for disk formation and accretion: now the system becomes a strong intermittent x-ray source.

Losses through stellar wind and ram pressure during the companion's explosion have reduced its mass when the second star explodes. This supernova occurs at some distance from the birth site; it is interpreted as of type I. The remaining two neutron stars thereby tend to be decoupled (by the mass loss; exception: the binary pulsar). At least the newly formed neutron star is a pulsar, like the Crab. It illuminates the expanding young remnant, and turns it into a plerion. More than ten plerions are presently known[21] in our galaxy.

Acknowledgments

Apart from E. Krotscheck, I should like to thank K. Weiler, M. Rees, C. Salter, and H. Sato for several informative discussions. Particular thanks are due to R. Beck for providing me with Figure 3.

References

1. Rees, M. J. & J. E. Gunn. 1974. Mon. Not. R. Astr. Soc. **167:** 1.
2. Kundt, W. & E. Krotscheck. 1979. Astron. Astrophys. In press.
3. Kundt, W. 1979. Astrophys. Space Sci. **62:** 335.
4. Fawley, W. M., J. Arons & E. T. Scharlemann. 1977. Astrophys. J. **217:** 227.
5. Ricker, G. R., A. Scheepmaker, S. G., Ryckman, J. E. Ballintine, J. P. Doty, P. M. Downey & W. H. G. Lewin. 1975. Astrophys. J. **197:** L83.
6. Kestenbaum, H. L., W. Ku, R. Novick & R. S. Wolff. 1975. Astrophys. J. **202:** L21.
7. Davidson, K. 1978. Astrophys. J. **220:** 177.
8. Kundt, W. 1977. Naturwissenschaften **64:** 493.
9. Kundt, W. 1978. The Two Types of Supernovae, Bonn, October. Preprint.
10. Wyckoff, S. & C. A. Murray. 1977. Mon. Not. R. Astr. Soc. **180:** 717.
11. Trimble, V. 1968. Astron. J. **73:** 535.
12. Shklovsky, I. S. 1968. Supernovae. : 215. John Wiley & Sons, London.
13. Kundt, W. 1976. Nature **26:** 673.
14. Kamper, K. & S. van den Bergh. 1976. Astrophys. J. **32:** 351.
15. Kundt, W. 1977. Nature **266:** 334.
16. Lamb, S. A. 1978. Astrophys. J. **220:** 186.
17. Bell, A. R. 1977. Mon. Not. R. Astr. Soc. **179:** 573.
18. Cowsik, R. 1979. Astrophys. J. **227:** 856.
19. Read, P. L. 1977. Mon. Not. R. Astr. Soc. **178:** 259.
20. Holloway, N., W. Kundt & Y.-M. Wang. 1978. Astron. Astrophys. **70:** L23.
21. Weiler, K. W. & N. Panagia. 1979. Vela X and The Evolution of Plerions. Bonn, August. Preprint.

EVIDENCE FOR THE EXISTENCE OF GRAVITATIONAL RADIATION FROM MEASUREMENTS OF THE BINARY PULSAR PSR 1913+16*

J. H. Taylor and P. M. McCulloch†

Department of Physics and Astronomy
University of Massachusetts
Amherst, Massachusetts 01003

Recent surveys[1,2] have increased the number of known pulsars to more than 320, yet there still exists only one known radio-frequency pulsar in a binary system. When PSR 1913+16 was discovered[3] in 1974, it quickly generated much interest among astrophysicists because it seemed destined to be the first pulsar to yield an accurate estimate of its mass. The pulsar also created excitement among relativists, because it appeared to be in a "clean" system with rather large expected post-Newtonian effects (references 4–7, and additional references therein). There even appeared to be a potential for using the system to probe the general theory of relativity beyond its first post-Newtonian approximation—a task that cannot be accomplished by tests carried out in the weak gravitational fields in the solar system—by observing the loss of orbital kinetic energy and angular momentum due to gravitational quadrupole radiation. Such a test is made possible by the unusually precise clockwork represented by the pulsar's natural periodicity of emission. (In fact, if one were given the task of designing an ideal machine for testing gravitation theories, the result might by a system rather similar to PSR 1913+16: an accurate clock of large mass and small size, moving at high speed in an eccentric orbit around a similar object located in otherwise empty space. To be sure, one would place the system somewhat closer to the earth than ~5 kpc, or would arrange for a more powerful transmitter to convey the clock pulses to terrestrial telescopes; but we cannot expect Nature to be concerned with the inadequacies of our instrumentation!)

This paper summarizes the results of four years of timing observations of PSR 1913+16, made with the 305-m telescope at the Arecibo Observatory, at frequencies near 430 and 1410 MHz. As is generally done in pulsar arrival-time measurements, received intensity data were typically accumulated for intervals of 5 minutes or so—in this case, long enough to produce a mean pulse profile with usable signal-to-noise ratio, yet still short compared to the orbital period of nearly 8 hours—and the resulting profile was then fitted to a template, or "standard profile," to determine its precise arrival time. Data on PSR 1913+16 have been acquired using a number of different receivers, dispersion-removing systems, and recording methods, in efforts to improve the measurement accuracy. Random errors in the measured pulse arrival times have been reduced[8,9] from ~300 μs in 1974 to ~50 μs in 1978.

Analysis of the timing data has been carried out by a method first described in detail by Blandford and Teukolsky[4] and further developed by Epstein.[6] The model

*Based on observations carried out in collaboration with L. A. Fowler, R. A. Hulse, and J. M. Weisberg.

†Permanent address: Physics Department, University of Tasmania, Hobart, Australia.

0077–8923/80/0336–0442 $01.75/1 © 1980, NYAS

presently in use describes the system in terms of 13 parameters of physical interest, listed in TABLE 1. The quoted values of the parameters were determined from a multiparameter fit to the arrival-time data, using a linearized least-square method. The observed times are first corrected from the location of the observatory to the barycenter of the solar system, including a relativistic correction that accounts for annual changes in gravitational potential at the earth. Adjustments are then made for the dispersive delay imposed by the interstellar medium and for the light travel time from the reference frame of the pulsar to the barycenter of its orbit. The latter correction includes Doppler shift terms of first and second order, gravitational red shift, and gravitational propagation delay; the description of the orbit path is carried out through post-Newtonian terms of order (v/c).[3] The resulting times in the pulsar's reference frame are then compared to times at which the pulsar "should" emit pulses, given its measured period and slowdown rate, and the differences are used to solve for improved parameter estimates.

TABLE 1
PARAMETERS DERIVED FROM TIMING DATA*

Right ascension (1950.0)	$\alpha = 19^h\,13^m\,12^s.474 \pm 0^s.004$
Declination (1950.0)	$\delta = 16°\,01'\,08''.02 \pm 0''.06$
Period	$P = 0.059029995269 \pm 2$ s
Derivative of period	$\dot{P} = (8.64 \pm 0.02) \times 10^{-18}$ s s^{-1}
Projected semimajor axis	$a_1 \sin i = 2.3424 \pm 0.0007$ light s
Orbital eccentricity	$e = 0.617155 \pm 0.000007$
Binary orbit period	$P_b = 27906.98172 \pm 0.00005$ s
Longitude of periastron	$\omega_0 = 178°.864 \pm 0°.002$
Time of periastron passage	$T_0 =$ JD 2442321.433206 ± 0.000001
Rate of advance of periastron	$\dot{\omega} = 4.226 \pm 0.002$ deg yr^{-1}
Transverse Doppler and gravitational redshift	$\gamma = 0.0047 \pm 0.0007$ s
Sine of inclination angle	$\sin i = 0.81 \pm 0.16$
Derivative of orbit period	$\dot{P}_b = (-3.2 \pm 0.6) \times 10^{-12}$ s s^{-1}

*Quoted uncertainties are twice the formal standard errors from the least-squares fit.

As indicated in TABLE 1, the set of 13 parameters includes the celestial coordinates of the system, α and δ; the period and period derivative of the pulsar "clock," P and $\dot{P}$; five "Keplerian" orbit parameters; and four parameters that measure effects of special and general relativistic origin. The first ten parameters are now known to accuracies of many significant digits, while the last three have been determined to within approximately 20 percent.

The measured values of $\dot{\omega}$, γ, $\sin i$, and $\dot{P}_b$, together with the well-determined Keplerian parameters, may be used to place constraints on the unknown masses of the pulsar and its companion (m_p and m_c, respectively). By inserting numerical values for the well-determined parameters, the relevant equations may be reduced to expressions involving only the four "relativistic" orbit parameters and the two unknown masses:

$$\dot{\omega}_{GR} = 2.11 \left(\frac{m_p + m_c}{M_\odot}\right)^{2/3} \text{deg yr}^{-1} \qquad (1)$$

$$\gamma = 0.002951 \left(\frac{m_c}{M_\odot}\right)\left(\frac{m_p + 2m_c}{M_\odot}\right)\left(\frac{m_p + m_c}{M_\odot}\right)^{-4/3} \text{ s} \quad (2)$$

$$\sin i = 0.5083 \left(\frac{m_c}{M_\odot}\right)^{-1}\left(\frac{m_p + m_c}{M_\odot}\right)^{2/3} \quad (3)$$

$$\dot{P}_b = -1.70 \times 10^{-12} \frac{m_p m_c}{M_\odot^2}\left(\frac{m_p + m_c}{M_\odot}\right)^{-1/3} \text{ s s}^{-1} \quad (4)$$

These equations follow directly from relations given in reference 8, and obviously contain more than enough information to determine the two masses. The redundancy is useful, however, because it permits a test of (a) whether the measured rate of advance of periastron $\dot{\omega}$ arises solely from the general relativistic effect quantified in Equation 1, and (b) whether the rate of loss of orbital energy measured by $\dot{P}_b$ agrees with a calculation based on quadrupole radiation within general relativity.[10,11]

The constraints placed upon the masses by the presently available data are summarized in FIGURE 1. Here we have plotted, on a plane of pulsar mass versus companion mass, curves bounding regions of plus and minus one standard deviation in each of the parameters $\dot{\omega}$, γ, $\sin i$, and $\dot{P}_b$. (The uncertainty in $\dot{\omega}$ is so small that just a single line is drawn for this parameter.) If one makes no assumptions about possible nonrelativistic contributions to $\dot{\omega}$, and assumes that Equation 4 correctly expresses the

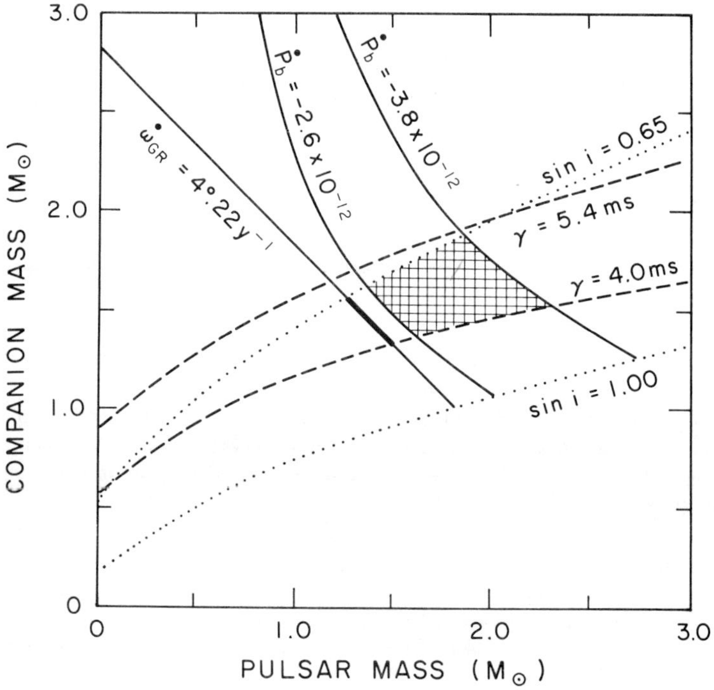

FIGURE 1. Curves delimiting the possible masses of PSR 1913+16 and its companion, based on the observed values of $\dot{\omega}$, γ, $\sin i$, and $\dot{P}_b$. The most probable masses lie on the thickened portion of the sloping straight line.

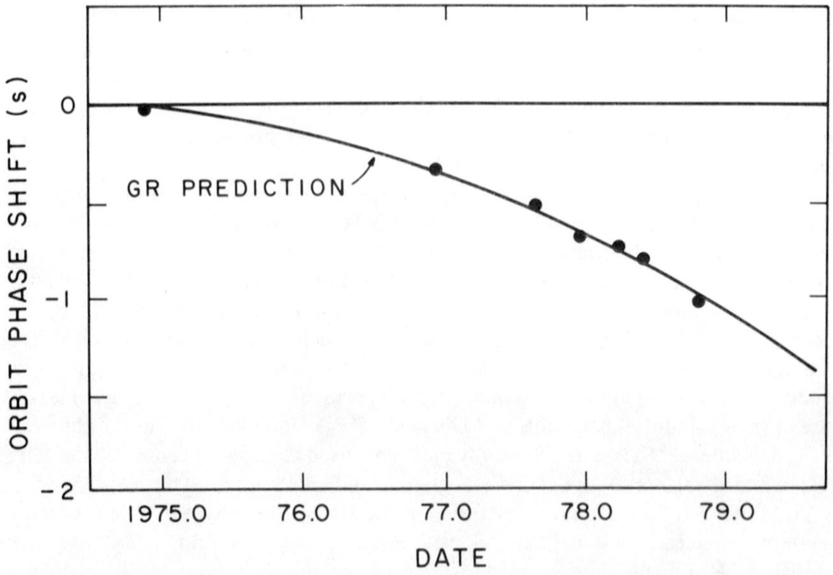

FIGURE 2. The points represent measured excess orbital phase, assuming a fixed value of P_b = 27906.981726 s. The plotted curve corresponds to the predicted rate of period change caused by gravitational radiation if $m_p = m_c = 1.4\ M_\odot$.

rate of orbital period change caused by gravitational radiation damping, then the most probable solution for the masses lies in the cross-hatched region in FIGURE 1, i.e., $m_p \approx m_c \approx 1.6\ M_\odot$. There is at present no evidence for nonrelativistic contributions to $\dot{\omega}$, so under the further assumption that $\dot{\omega} = \dot{\omega}_{GR} = 4.22$ deg yr^{-1}, Equation 1 constrains the masses to lie along the sloping straight line in the figure, on which $m_p + m_c = 2.83\ M_\odot$. The limits imposed by γ and sin i further reduce the region of interest to the thickened portion of this line, $m_p \approx m_c = (1.4 \pm 0.2)\ M_\odot$. This solution is a quite plausible one, because theoretical work suggests that neutron stars probably form most readily with masses close to the Chandrasekhar limit. Equation 4 then yields $\dot{P}_b = (-2.404 \pm 0.003) \times 10^{-12}$ for the prediction of orbital damping by gravitational radiation. This value may be compared to the measured quantity quoted in TABLE 1.

Because of the important implications of our measurement of $\dot{P}_b$, and its interpretation as evidence for the existence of gravitational waves, we provide further evidence of the validity of the measurement in FIGURE 2. This illustration is a plot of accumulating "orbit phase shift" versus time over the duration of our observations. If the orbital period P_b had remained fixed at its 1974.9 value, no accumulating phase errors would be expected. However, if orbital energy is lost via gravitational waves, causing a period derivative as calculated above, then excess orbit phase should accumulate as indicated by the parabolic curve. The points in FIGURE 2 represent the average orbit phase residuals determined during our seven best observing periods‡ in

‡Additional data were acquired during 1975–76, but systematic errors in the ephemeris then being used for pulsar period predictions caused the pulse profiles obtained near periastron to be of poor quality. Orbit phase measurements depend critically on just these data, so the resulting orbit phase measurements were not useful for the purposes of FIGURE 2.

1974–78. Obviously the data fit the predicted period variations quite well. It is well to point out that the prediction specifies only the amount of curvature of the quantity plotted in FIGURE 2, not the first-order slope. (If the measured points fell along a sloping straight line, it would indicate simply an error in the assumed orbital period.) The points as plotted are best fit by a parabola of ~30% greater curvature than the predicted curve, with an uncertainty of ~25%.

What mechanisms other than gravitational radiation damping might be invoked to explain the observed change of orbital period? Several possibilities have been considered and discussed in the literature,[5] but all of them now appear to be either implausible, *ad hoc,* or of negligible magnitude. For example, differential galactic rotation and mass loss from the system contribute at most about one percent[5,12] of the observed value of $\dot{P}_b$. Tidal interactions are utterly negligible if the companion object is another neutron star or a black hole. Dynamic considerations cannot yet rule out a rapidly rotating white dwarf companion,[5] but all permissible white dwarfs have masses close to the Chandrasekhar limit, and hence radii $R \lesssim 3000$ km. Because of the very strong dependence[5] ($\Delta\dot{P}_b \propto R^9$) of tidally induced period changes on companion radius, these white dwarfs also would have negligible influence on $\dot{P}_b$. The presence of a massive third body causing an acceleration of the system cannot yet be ruled out with certainty, but seems most unlikely on evolutionary grounds. Within a few years the observational limit on the second derivative of the pulsar period $\ddot{P}$ will provide a definitive test of this possibility.

We conclude that the most straightforward interpretation of our measurement of $\dot{P}_b$ is that gravitational waves exist, and carry energy away from this orbiting system at a rate consistent with the predictions of general relativity. We are aware that some disagreement exists about whether the quadrupole formula used[10] to derive Equation 4 has neglected important higher-order terms; Professor Ehlers has reviewed some of the arguments earlier in this symposium. Obviously it is important for the theorists to reach an agreement on this point. At present, we can remark that the experimental evidence suggests that any inaccuracies in the calculations are not very large.

REFERENCES

1. MANCHESTER, R. N., A. G. LYNE, J. H. TAYLOR, J. M. DURDIN, M. I. LARGE & A. G. LITTLE. 1978. Mon. Not. R. Astron. Soc. **185:** 409–421.
2. DAMASHEK, M., J. H. TAYLOR & R. A. HULSE. 1978. Astrophys. J. Lett. **225:** 31–33.
3. HULSE, R. A. & J. H. TAYLOR. 1975. Astrophys. J. Lett. **195:** 51–53.
4. BLANDFORD, R. & S. A. TEUKOLSKY. 1976. Astrophys. J. **205:** 580–591.
5. SMARR, L. L. & R. BLANDFORD. 1976 Astrophys. J. **207:** 574–588.
6. EPSTEIN, R. 1977. Astrophys. J. **216:** 92–100.
7. WILL, C. M. 1979. *In* Einstein Centenary Volume. Cambridge University Press.
8. TAYLOR, J. H., R. A. HULSE, L. A. FOWLER, G. E. GULLAHORN, J. M. RANKIN. 1976. Astrophys. J. Lett. **206:** 53–58.
9. MCCULLOCH, P. M., J. H. TAYLOR, & J. M. WEISBERG. 1979. Astrophys. J. Lett. In press.
10. PETERS, P. C. & J. MATHEWS. 1963. Phys. Rev. **131:** 435–440.
11. WAGONER, R. V. 1975. Astrophys. J. Lett. **196:** 63–65.
12. SHAPIRO, S. L. & Y. TERZIAN. 1976. Astron. Astrophys. **52:** 115–118.

SUMMARY OF WORKSHOP ON NEUTRON STARS AND PULSARS

N. J. Holloway

Astronomy Centre
University of Sussex
Falmer, Brighton BN1 9QH, England

This report is intended to be a report on the principal themes of the workshop, as seen from the chair, and not an exhaustive or detailed account of all the papers. The workshop was divided into two sections:

SECTION I. THE PULSAR MAGNETOSPHERE AND RADIATION

L. Mestel (Sussex) described some continuing work on the axisymmetric magnetosphere, involving a model with return flow of charges to the polar caps. Taking the Goldreich–Julian space-charge distribution as a starting point, with a negatively charged polar region ($\mathbf{\Omega} \cdot \mathbf{B}_0 > 0$), the model involves a slow outflow of electrons from the inner part of the polar cap and return of the same electrons at slightly lower latitudes around the pole. The motion of these electrons follows magnetic field lines out to the light cylinder, where the continuation of field line following would produce infinite energies. At this stage, the electrons cease to follow the magnetic field lines and dissipate energy in γ radiation, before returning to the star.

Poloidal motion of the electrons is at all points $\ll c$, and is controlled by the component $E_\parallel$ of the electric field parallel to $\mathbf{B}$, which is calculated to act in the opposite sense to that of the centrifugal forces at the light cylinder. Specific predictions of the model are that:

(i) Axisymmetric or near-axisymmetric rotators are primarily γ-ray emitters; and that

(ii) Particles do not attain high energies near the poles of the star.

(The above represents work done in conjunction with Y-M. Wang and P. Phillips of Sussex Astronomy Centre.)

P. G. Sutherland (Ontario) concentrated on the radio emission of pulsars and constructed his argument around three relationships:

(i) The pulse width is determined by the geometry of the open field lines that emanate from the polar caps.

(ii) The frequencies of microwave radiation indicate the magnetic field strength and Lorentz factors in the radiating region, on the assumption of a curvature radiation mechanism.

(iii) The coherence of the radio emission is a result of plasma instabilities above the polar caps.

Using the above, the pulse width implies a height for the emitting region of about 10 star radii, with field curvature radii of about 10^6m and Lorentz factors for the electrons or positrons of 350.

Dr. Sutherland suggested that plasma instability was caused by the passage of a cold ion beam (the primary emission from the stellar surface) through the e^+e^- plasma generated in a pair production cascade at the polar caps, and invoked the bunching mechanism of Goldreich and Keeley[1] to produce coherent bunches. (The above work was carried out in conjunction with A. Cheng and M. A. Ruderman)

Other papers and discussion centered around the basic question of the location of the active region in pulsars.

E. Asseo (Palaiseau) supported the idea of emission near the polar caps with some further calculations on instabilities above Ruderman and Sutherland polar gaps.[2] Growth of instabilities seemed to be possible over a wide range of the plasma and cyclotron frequencies.

Several speakers supported the ideas of Mestel *et al.* on γ-ray emission:

A. Treves (Milan) pointed out that the γ-ray luminosity of pulsars did not depend on period as strongly as the overall energy loss rate (as calculated from P and $\dot{P}$) but only as P^{-1} so that the slow (and possibly nearly aligned) rotators produced a large fraction of their energies in γ-rays, as Mestel's model would predict.

T. Gold (Cornell) emphasized the different nature of the Crab radio and optical emission—the former being very spiky and the latter relatively smooth. He also supported the model of Mestel *et al.* on the grounds of its conformity to the angular momentum and energy loss ratio ($\dot{E}/\dot{L} = \Omega$)—a conformity that is automatically incorporated into light cylinder emission models of this type.

M. Ruderman (Columbia) pointed out that the radio emission mechanisms associated with polar caps could exist without violating the angular momentum requirements because of their very low ($<10^{-2}$ of the total) contribution to the overall energy and angular momentum losses. He pointed out that closure of the currents (to avoid charge accumulation on the star) was still an unsolved problem in this scenario. Gamma-ray production in the outer gaps, between the oppositely charged regions of the corotating magnetosphere, still seemed to be a possibility, but only in those pulsars with strong magnetic fields or rapid rotation.

G. Lichti (COS-B team, Garching) presented some γ-ray data from the COS-B satellite, particularly for PSR 1822−09 and PSR 0740−28. The data showed an asymmetrical double-peaked structure, and a tendency (such as can be seen with so few data points) for slower pulsars to radiate a greater fraction of their energy in γ-rays.

In summarizing the papers of the workshop, the chairman suggested that there now seemed to be considerable evidence for the idea that the emission of pulsar γ-rays occurred at or near the light cylinder, where their appreciable contribution to the overall energy losses could be made to accord with the angular momentum loss requirement. However, the radio emission never amounted to a significant fraction of the energy loss and so could escape this argument. Thus the location of the radio emission could not yet be specified from data currently available.

Section II. Other Topics in Neutron Star Astronomy

Unlike the first part of the workshop, this part did not revolve around a particular theme. Thus most of the papers are unrelated to each other.

G. Greenstein (Amherst) presented a detailed theory of pulsar glitches, based on the idea of a temperature-dependent frictional coupling between the neutron star's crust and core. The injection of some 10^{33} J of energy (possibly by a small starquake, $\delta r/r \sim 10^{-9}$) can result in a 50% rise in the superfluid core temperature (it has a very small specific heat) and a sudden increase in the frictional accelerating torque on the crust (since the core is overtaking the crust). The model predicts that the glitch is not a discontinuous change in pulsar rotation, but a smooth and fairly rapid rise, from a few minutes (for a 50% change in T_{core}) to a few days (for a 10% change) in duration.

E. van den Heuvel (Amsterdam) and P. McCullough (Amherst) commented on some properties of the binary pulsar. McCullough described the measurement of pulsar-shape changes—an indication of the pulsar's beam shape if the changes can be interpreted on the basis of the precession of the spin axis. Van den Heuvel pointed out that the value of $P\dot{P}$ for this pulsar is very small, indicating a weak magnetic field of order $<10^7$ tesla at the star's surface. This pulsar might be a very old neutron star, which has already been through the life cycle of an x-ray source, and has been spun up to the limit of $P = 0.05$ s in an accretion phase.

There is a problem with this interpretation: there are no short-period single pulsars with very low $\dot{P}$, although binary systems are not usually expected to survive two supernova events, as the binary pulsar system seems to have done, but are expected to produce two single pulsars, one old and one new.

Also on the subject of the binary pulsar, P. Crane (E.S.O.) described the detection of a star of 20.8 magnitude at about 0.26″ from the binary pulsar position. On the basis of the field density of such stars, he reckoned the probability of such a chance coincidence at 18%—not low enough to claim a definite identification of the companion, but low enough to be interesting.

W. Kundt (Bonn) described a new nonblackhole model of Cygnus X-1. The mass of the x-ray source deduced from the standard theory of eclipsing binaries, and the assumption of a $25M_\odot$ companion, is about 6–7 $M_\odot$. This, he suggested, comprises a normal neutron star of $\sim 1\ M_\odot$ with a massive disc of 5–6 $M_\odot$ accreting at a rate near the Eddington limit. The model seemed a bit too new for the audience to respond to it at that time.

G. Auriemma (Frascati) presented evidence for a spectral absorption line in the x-ray transient of 30 July 1976. There was some doubt as to whether the feature could be explained as emission at ~100 keV or absorption at 70 keV, but the profile seemed to favor an absorption.

The workshop closed with a significant development in the history of the Texas Symposium—the first address by an astronomer from the Republic of China: Dr. L. Z. Fang (Peking) presented some theoretical work on nucleon states at very high (neutron star) densities, and in particular on the "abnormal neutron" state. He suggested that the existence of this abnormal state could raise neutron star mass limits to about 3 $M_\odot$. The chairman expressed the wish that Dr. Fang would be but the forerunner of many astronomers from a country with several thousand years of astronomical heritage.

References

1. Goldreich, P. & D. Keeley. 1971. Astrophys. J. **170:** 463.
2. Ruderman, M. A. & P. G. Sutherland. 1975. Astrophys. J. **196:** 51.

GALACTIC BULGE SOURCES, WHAT ARE THEY? (X-RAY SOURCES IN GLOBULAR CLUSTERS AND BURST SOURCES)

Walter H. G. Lewin and George W. Clark

Massachusetts Institute of Technology
Center for Space Research
Cambridge, Massachusetts 02139

INTRODUCTION

Most of the 50 x-ray sources that lie within 30 degrees of the direction of the galactic center have the following characteristics:

- Soft x-ray spectra (in comparison to the massive x-ray binaries of Population I)
- Variability on time scales of minutes to days
- Luminosities greater than 10^{36} erg/sec
- No periodic pulsations
- No eclipses

About two dozen produce x-ray bursts. As a class they are often referred to as the galactic bulge sources, a somewhat misleading name since most of them do not lie within the strictly defined galactic bulge.[80] Nevertheless, their location outside the probable regions of current or recent star formation identify them as being members of an older stellar population. At least seven high-luminosity, variable, nonpulsing, noneclipsing sources with soft spectra lie in globular clusters, and of these at least four (probably six) emit bursts. The similarities of the bulge and cluster sources leads one to consider them together as a class of high-luminosity old Population II x-ray sources.

What could these objects be?

In this paper we will argue on the basis of the observed properties that the majority of these sources are neutron stars in low-mass binary systems.

PROPERTIES OF GALACTIC BULGE SOURCES

Soft Spectra

Considering the uncertain state of knowledge about the physics of accretion, the softness of the spectra does not tell us much about the nature of the compact objects (i.e., whether they are white dwarfs, neutron stars, black holes of a few solar masses, or massive black holes of $\sim 10^2\ M_\odot$). However, it does suggest that the area of the x-ray emitting regions of Population II sources is larger than for the highly magnetized pulsating sources of Population I.

Variability

The observed variability in the persistent x-ray emission down to time scales of minutes does not tell us much either about the nature of the compact objects.

0077–8923/80/0336–0451 $01.75/1 © 1980, NYAS

High Luminosity

On the basis of the high luminosity ($>10^{36}$ erg/sec) one can rule out white dwarfs for the bulge sources as a class (see Kylafis and Lamb[46] and references therein).

Absence of Periodic Pulsations

More than twenty of these objects have been observed intensively in search of periodic pulsations. None were found. Cominsky *et al.*[16] recently analyzed the persistent flux of seven burst sources. They found upper limits in the order of 1–3% to pulsations with periods from ~1.6 to 10^3 s (TABLE 1). Spada, Rappaport, and Li (private communication) searched for periodicities in seven of the brightest (nonburst-

TABLE 1
UPPER LIMITS TO PERIODICITIES IN THE PERSISTENT COMPONENTS OF TYPE I BURST SOURCES*

	Upper Limits (90% Confidence) to Pulsed Fraction† in %, for Periods in the Ranges Listed		
Source	2 ms–2 s	1.6 s–66 s	66 s–10^3 s
1636−53		1.9	3.5
1659−29		3.5	9.5
1728−34		3.6	9.2
1735−44	8	1.9	13.0
1820−30	1.5¶	1.5	1.5‡
1837+05		1.1	6.1
1916−05		1.3	15

*This table is from Cominsky *et al.*[16]
†Fraction of total flux
‡Period range 66–200 s.
¶F.K. Li (private communication). Upper limits of ~3% were found for seven bright sources (not bursters) near the galactic center in the range ~2 ms–2 s (see text).

ing) sources near the galactic center. They didn't find any either. Their upper limits are typically 3% in the range ~2 ms–2 s.

The absence of detected pulsations does not tell us much about the nature of the compact objects. On the basis of this information alone they could be black holes as well as neutron stars.

Pulsations are expected from accreting neutron stars only if they have strong magnetic fields ($B \sim 10^{12}$ G), as is presumably the case among the binary x-ray pulsars of Population I. Old neutron stars may have lost their magnetic fields or their rotation and magnetic dipole axes may have become coaligned (references 22 and 30 and references therein) or differences in their formation may have prevented the development of strong magnetic fields.[41] In fact, if the accretion onto a neutron star is funneled by a strong magnetic field, thermonuclear flashes may be suppressed.[40,43,87] This may explain why Type-I x-ray burst sources never pulse and why x-ray pulsars never burst.[41] Thus the absence of pulsations is not a strong argument against neutron stars.

Absence of Eclipses

No x-ray eclipses have been found among these objects. However, this does not disprove the binary nature of these systems. Joss and Rappaport[42] have shown that an x-ray source powered by Roche lobe overflow from a low-mass companion (less than or about 0.3 solar masses) has a low probability (about 25%) of being eclipsed by its companion. The eclipse probability varies as the cube root of the mass of the companion and would therefore be less than 10% for a companion of 0.01 solar masses. Milgrom[75] proposed a model in which the accretion disk casts an x-ray shadow in the orbital plane, thereby excluding from our detection those systems that would otherwise show eclipses.

Globular Cluster X-Ray Sources

The similarities between the bulge and cluster sources favor the assumption that they are basically the same kind of systems.[9] Therefore an examination of the circumstances of their occurrences in globular clusters, based on knowledge of the clusters themselves, may throw light on the nature of the Population II sources in general.

The structure of a globular cluster can be broadly characterized by three observables, which are the core radius R_c, the tidal radius R_t, and the central brightness B_c (reference 44). Peterson and King[77] summarized the data as of 1975, and several studies updating and extending their work have been carried out since then with special attention to the clusters containing x-ray sources (e.g., references 4, 5, 10, 45, 78). The data show a striking correlation between the presence of an x-ray source and the quantity B_c/R_c, which is a measure of the central density of luminous stars (Figure 1). We found that the mean value of this measure for seven clusters taken at random from among the 72 for which data are available did not exceed the mean value for the seven x-ray clusters (NGC 1851, 6440, 6441, 6624, 6712, 7078, Liller I) in 3000 trials, in spite of the presence among the x-ray clusters of NGC 6712, which stands out as peculiarly low in central density.

The expectation that the cluster sources would be found to lie close to the cluster cores followed naturally from the recognition of the correlation with high central density. This expectation was confirmed by position measurements, carried out with the RMC detectors on SAS-3, which showed that in five cases the positional error circles with radii of 20 to 30 arc sec (90% confidence) included the optical centers of the clusters.[37]

The evidence clearly points away from any explanation of the cluster sources based on the evolution of primordial binary systems that might be found in all clusters and toward an explanation based on the formation of the x-ray sources under the circumstances peculiar to the dense cores of the most centrally condensed clusters. One likely circumstance is a high concentration of compact remnants of the short-lived massive stars that were presumably present at the beginning, i.e., neutron stars

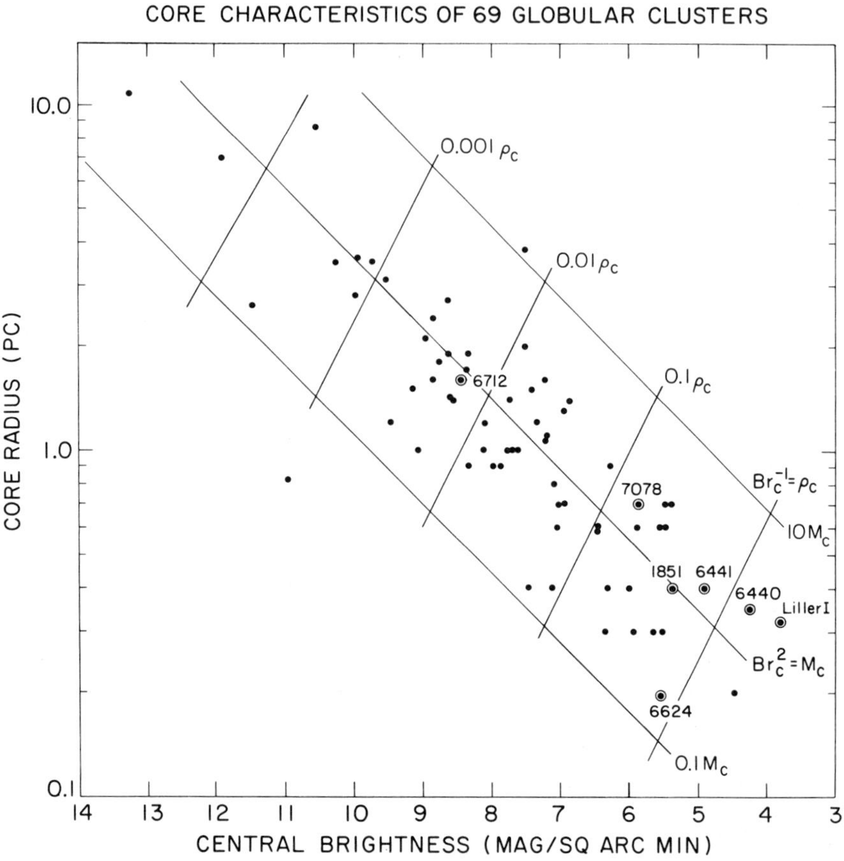

FIGURE 1. A log–log plot of the core radius R_c versus the central core brightness B_c (in V mag/sq arc min) for 69 globular clusters. The grids are rough loci of constant core masses M_c and core densities ρ_c. The core masses have a range of $\sim 10^5$. The globular clusters with x-ray sources have relatively high core densities except NGC 6712. [From F. K. Li (thesis).]

and black holes that were not ejected from the clusters by the processes of their formation and subsequently settled into the cores. The frequency of close encounters between such remnants and nuclear burning stars is certainly greater in a dense core. Clark[11] suggested that the cluster sources are neutron stars that have acquired close nuclear burning companions by capture. Fabian, Pringle, and Rees[20] estimated that captures could occur with sufficient frequency in typical dense cores through tidal dissipation of orbital energy in two-body encounters, and Press and Teukolsky[79] confirmed the viability of the mechanism by detailed calculations. Hills[32] examined binary formation through star exchange interactions between neutron stars and primordial low-mass binaries, though the apparently low occurrence frequency of binaries in clusters makes the importance of this process less certain. Capture through three body encounters in a dense subcluster may also be significant.[21]

It has been proposed that the deep potential well of a centrally concentrated cluster might retain gas lost by stars and that a black hole of sufficient mass, formed perhaps in a collapse of the dense core, could accrete this gas at the rate required to produce the observed x-ray luminosities.[2,82] Grindlay and Liller[100] found optical evidence for gas in the cores of four x-ray clusters in amounts ($<0.1\ M_\odot$) sufficient to supply a black hole if its mass exceeds $10^2\ M_\odot$. However, no satisfactory mechanism to produce bursts by a black hole of mass $>10\ M_\odot$ has been put forward.

On the basis of only (i) the average luminosity, (ii) the soft spectra, and (iii) the preferential occurrence in dense cluster cores, there is insufficient basis for choosing between the neutron star, the low-mass black hole and the massive black hole hypotheses. However, a new approach has become available using precise position measurements. Bahcall and Wolff[3] showed that the expectation distance of an object of mass M from the center of a dynamically relaxed cluster of stars of mass m is $0.9R_c$ $(m/M)^{1/2}$ arc seconds. In typical x-ray clusters R_c is about 7 arc seconds (see TABLE 2). Thus an x-ray binary with four times the average mass ($\sim 0.5\ M_\odot$) of cluster stars (reference 98) would have an expectation distance of about 3.5 arc seconds from the center, while a black hole with more than a hundred times the average mass would rarely be found outside of 1 arc second. The position measurements from SAS-3 permitted the conclusion that the cluster sources are more massive than the visible stars.[37]

The Einstein Observatory can determine x-ray source positions with error radii that are typically about 2 arc seconds. Preliminary results from that observatory, reported by J. Grindlay[29] in April 1979, show that some cluster sources lie more than 1 arc second away from their cluster centers. Thus, if the burst sources constitute a uniform population as we suspect (see below), then they are not massive ($M \gtrsim 100\ M_\odot$) black holes.

Another possibility to keep in mind about the low-mass binary hypothesis is that it implies a fair chance of finding two x-ray sources within the same cluster. Of the ~20 known centrally condensed clusters, 6 contain an x-ray source. Therefore the chance of finding two or more in a single centrally condensed cluster is $\sim(6/20)^2 = 0.09$. Obviously, the discovery of two in one would independently exclude the supermassive black hole hypothesis.

TABLE 2

GLOBULAR CLUSTER X-RAY SOURCES*

		Burst Source	References	R_c (arc sec)
NGC 1851	4U1513−40	Yes	14,23,37	7
Terzan 2	4U1722−30	Yes	26,29,85	7
NGC 6441	4U1746−37	Probably	37,63	9
Liller I	MXB 1730−335	Yes	17,50,68	7
NGC 6624	4U1820−30	Yes	12,13,27,37	6
NGC 6712	4U1850−08	Almost certainly	37,66,84	49
NGC 7078	4U2131+11	No?	37	10
†Kron 3	4U0026−73	No?	26	
†NGC 6440	MX1746−20	No?	24	

*For more detailed references and for accurate source locations, see references 8, 60 and 61.
†These two associations are uncertain.

The anomaly in central density presented by the case of NGC 6712 is especially intriguing in the light of calculations by Shapiro[81] which suggest that core collapse may be followed by puffing up and dissolution of the remaining cluster through injection of kinetic energy from interactions with a massive central body. Perhaps x-ray binaries and a massive black hole are produced in the processes of core collapse, and in NGC 6712 we are witnessing the aftermath as the cluster proceeds to oblivion. Some consequences of this theme of evolution and death in globular clusters have been explored by Lightman, Press and Oldenwald.[67]

X-Ray Burst Sources

So far we have only been able to reject white dwarfs and massive black holes as a class for the bulge sources. We will now turn to the last characteristic, namely that many of the Population II sources produce x-ray bursts. We believe that the wealth of information on bursts harvested over the last few years has given us evidence that these objects are neutron stars and not low-mass black holes.

X-ray bursts were discovered in 1975 by Grindlay et al.[27] Since then bursts from over 25 sources have been detected, and an important distinction between two types of bursts was recognized by Hoffman, Marshall, and Lewin.[35] They suggested that two very different mechanisms were involved, and introduced the classification of Type-I and Type-II bursts. We summarize in Table 3 the characteristics of these two types of bursts and we list the most promising mechanisms that produce these bursts.

The Rapid Burster (MXB 1730−335)

The rapid burster became a Rosetta stone when it was recognized that it produced two different types of bursts[35] (Figures 2 and 3). Details of this unique object are

Table 3

Classification of X-Ray Bursts*

Type I (~25 Sources)
• Burst intervals of hours–days (longer?)
• Distinct spectral "softening" during burst decay
o Cooling of a "blackbody"†
o Thermonuclear flashes? (Helium is the most probable fuel.)
o Nuclear energy
Type II (Rapid Burster + Others?)
• Burst intervals of seconds–minutes
• No distinct spectral softening during burst decay
o "Blackbody" radiation (no significant cooling)
o Instabilities in accretion flow
o Gravitational potential energy

*This classification was introduced by Hoffman, Marshall, and Lewin.[35] It was based on the two phenomenological differences as listed under the solid dots. The characteristics listed as open circles are interpretations (e.g., see references 35, 40, 41, 71).

†It is not clear at this time how "black" the blackbodies are (references 33, 34, 85, 86, 93, 94).

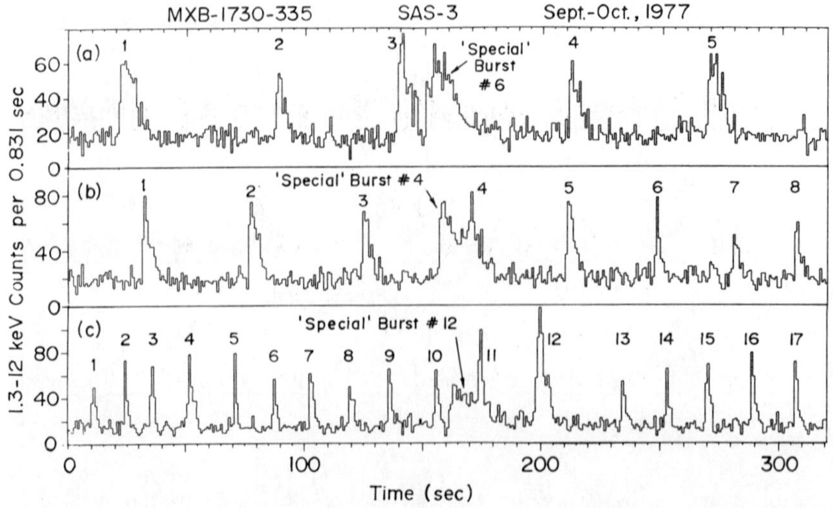

FIGURE 2. Discovery of Type-I and Type-II bursts from the rapid burster. The Type-I bursts ("special") occur independently of the sequence of the rapidly repetitive Type-II bursts (numbered independently). (From Hoffman, Marshall, and Lewin.[35])

described in two recent review articles on burst sources.[60,61] We only summarize here the conclusions from these papers.

- The rapid burster is a recurrent transient with a period of about 6 months [the last turn-on came on schedule in March 1979 (reference 70)].
- The "persistent" emission from this object is released in the form of Type-II bursts,[35] the energy of which is derived from the gravitational potential energy of accreting material.[71]
- The rapid burster at times also produces Type-I bursts at intervals of ~3–4 hours.[35] These are probably the result of thermonuclear flashes of the freshly accreted material on the surface of a neutron star (reference 41 and references therein).
- The spectral data of both Type-I and Type-II bursts from the rapid burster can be fit reasonably well by blackbody radiation.[71] For the Type-I bursts a radius of ~8 km is found for the emitting region and this remains approximately constant throughout the bursts. During the burst decay the temperature decreases rapidly (maximum temperature $\sim 30 \times 10^6$ K). For Type-II bursts a constant temperature of $\sim 18 \times 10^6$ K is found and the radius of the emitting region decreases from ~16 km to less than a few km. For short Type-II bursts this can happen in only a few seconds.[71]
- The average power in Type-II bursts is about 120 times that in Type-I bursts (when present).[35] This is consistent with the Type-I bursts being due to the release of nuclear energy (about 1 MeV per nucleon for helium-burning) and the Type-II burts being due to the release of gravitational potential energy (about 100 MeV per nucleon when it falls into a ~1 solar mass neutron star) (references 39–41, 47, 69, 96 and references therein).

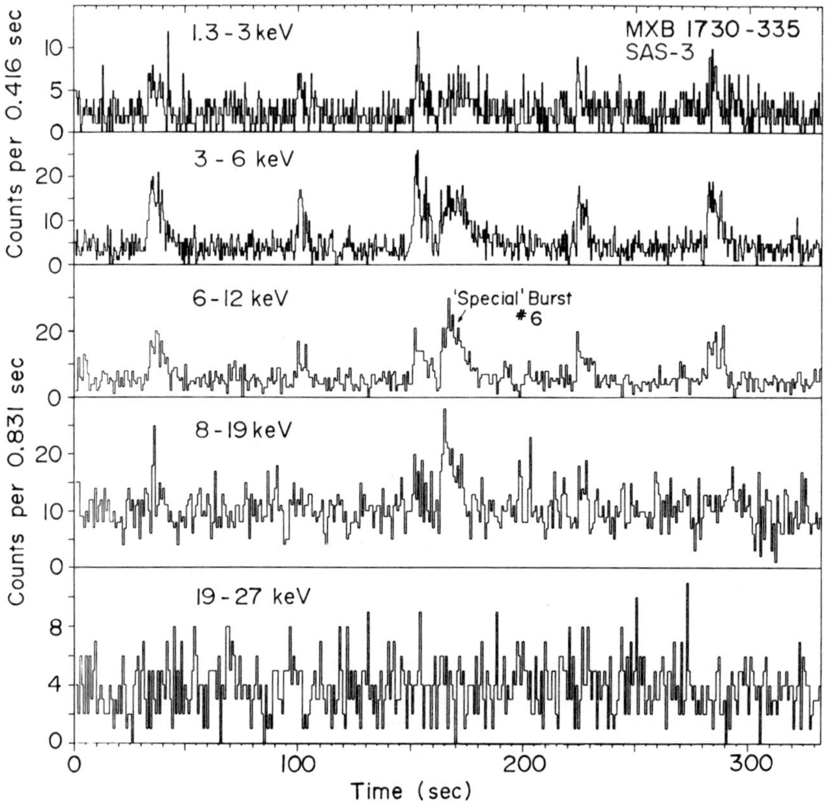

FIGURE 3. Spectral information for the data shown in FIGURE 2a. The Type-I burst spectrum softens during burst decay; this softening is absent in the rapid (Type-II) bursts. (From Hoffman, Marshall, and Lewis.[35])

Recently, in April 1979, puzzling infrared and radio bursts have been reported from the direction of the rapid burster (IAU Circulars 3344 and 3347).

It is of interest to note that Type-II bursts may have been observed from several known binary x-ray sources.[35,60,61]

Type-I Burst Sources

A sky map of Type-I burst sources is given in FIGURE 4. Five of them are located in globular clusters (including the rapid burster). FIGURE 5 shows five bursts from different burst sources. The characteristic spectral softening during burst decay is clearly visible. FIGURES 6–8 show a series of bursts from various sources. They illustrate that bursts from one and the same source generally look similar although there are exceptions (e.g., see FIGURE 9 and the caption of FIGURE 10).

In FIGURE 10 we show a burst from MXB 1850 − 08 in the globular cluster NGC 6712. The burst profile is somewhat unusual. The double-peaked structure at high-energy x rays is reminiscent of bursts from MXB 1743 − 29 (see FIGURE 8 and reference 51).

Burst intervals can be regular or irregular on time scales of hours to days. A correlation between these intervals and the persistent x-ray emission was found in two burst sources: MXB 1659 − 29 and the source MXB 1820 − 30 in the globular cluster NGC 6624.

In May 1976, the persistent flux of NGC 6624 increased by a factor of about 5 while the burst intervals gradually decreased by 50%. The persistent flux continued to increase and the bursts stopped completely.[13] Bursts have only been detected from this source when the persistent flux was in a "low" state.[12,13,27]

In October 1976, MXB 1659 − 29 produced bursts at fairly regular intervals of ~2.5 h; no persistent flux could be detected.[49] In March 1978, the persistent flux was high (~8% of the Crab) and no bursts were observed.[58,59] In several other cases (MXB 1837 + 05, MXB 1735 − 44, and MXB 1636 − 53) large fluctuations observed for burst intervals were not correlated with the persistent x-ray flux.[15,34,64]

In a thermonuclear flash model (reference 41 and references therein) in its simplest form one would expect that the burst intervals would decrease for an increasing persistent x-ray flux; the higher the supply of accreted material the sooner the flash goes off (except for very high[40,41] and very low[47] values of the accretion rate

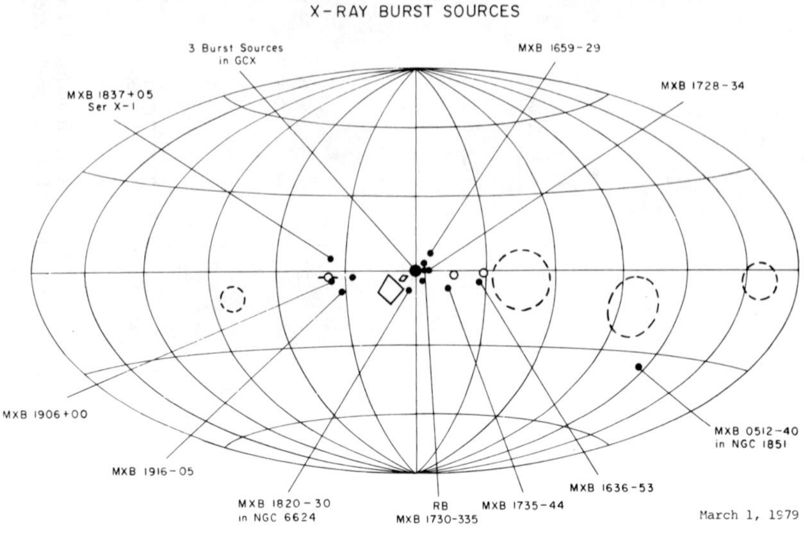

FIGURE 4. Sky map (galactic coordinates) of 25 x-ray burst sources as of March 1, 1979. Only one burst was observed from each of the four sources indicated with a dashed circle or a horizontal bar. There are at least six more burst sources probably all within 30° of the galactic center. Their positions are poorly known and they are not shown here (see also Evans, Belian, and Conner[19] and comments in Lewin.[54] At least 20, but possibly all sources shown, are of Type I.

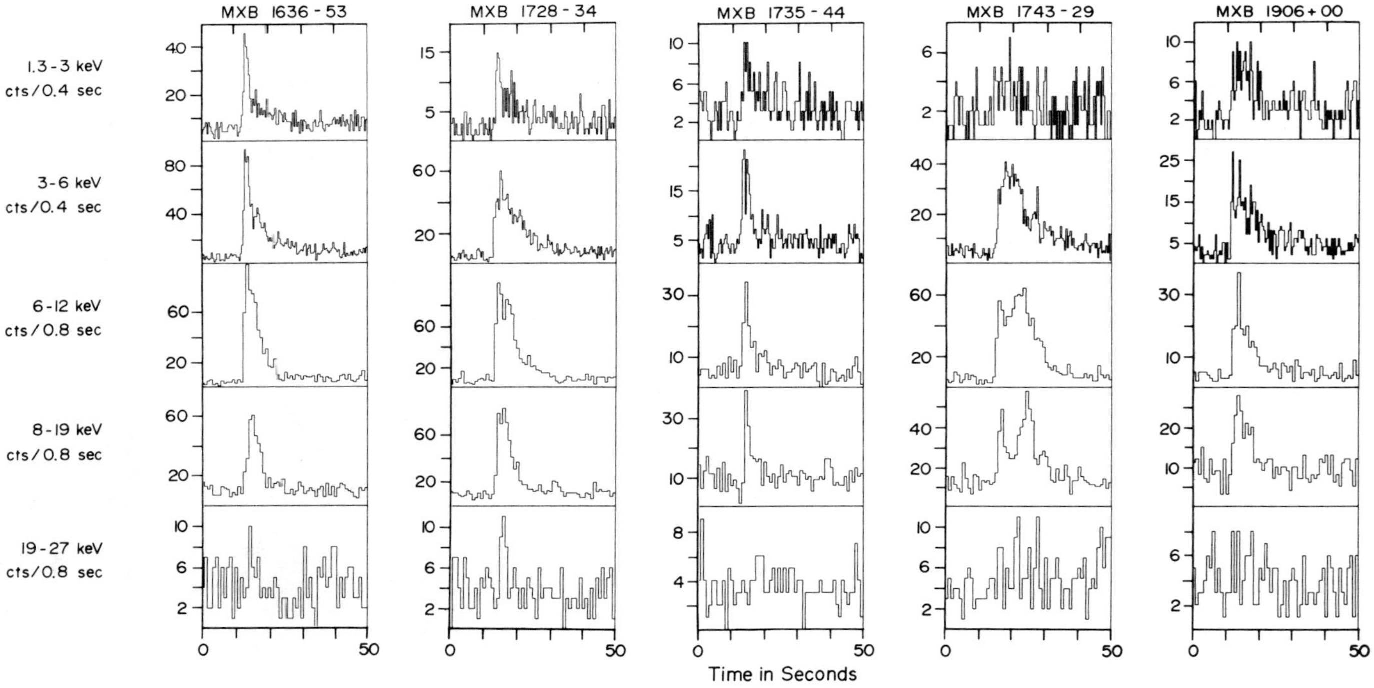

FIGURE 5. Profiles (five energy channels) of Type-I bursts from five different sources (SAS-3 data). Note that in all cases the gradual decay (tail) persists longer at lower energies than at higher energies. The burst profiles are reasonably distinctive for each particular source. (From Lewin and Joss.[55])

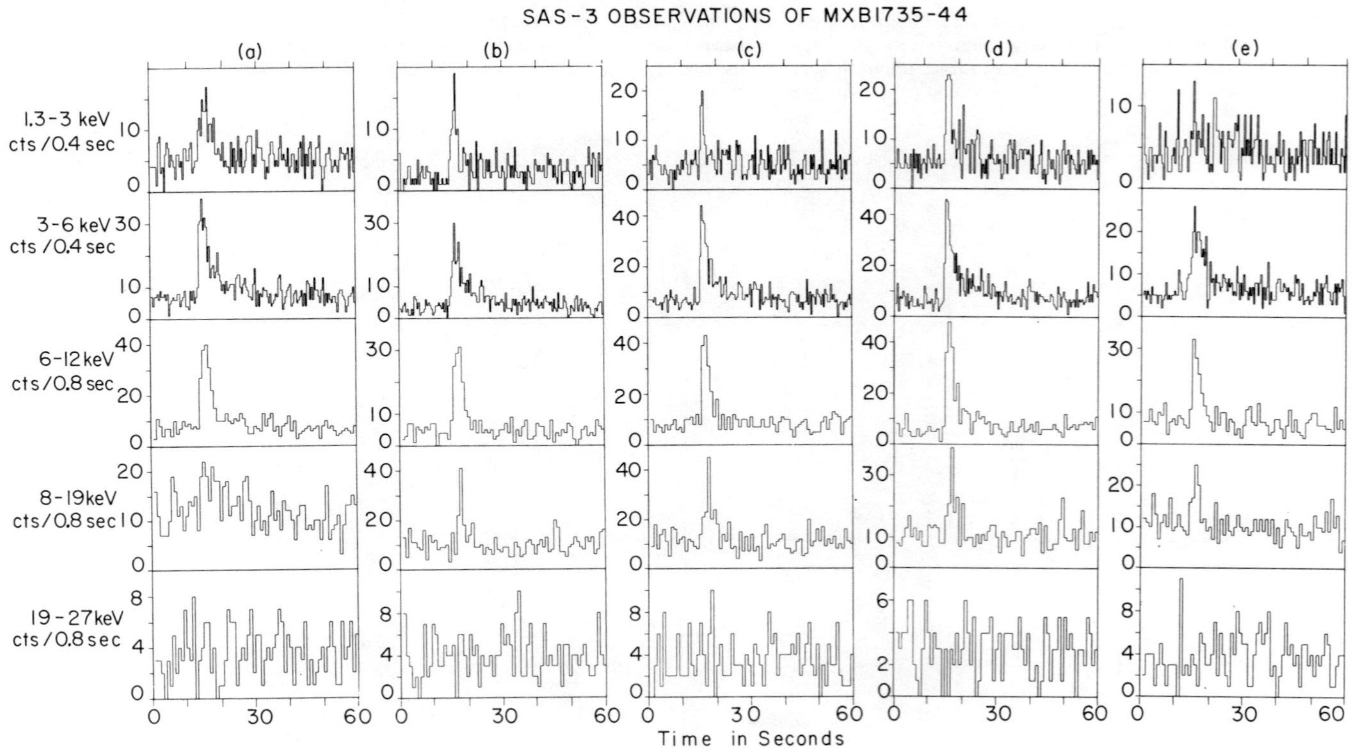

FIGURE 6. Raw counting rate plots of five bursts from MXB 1735 − 44. The burst profiles are similar. Notice the cooling during burst decay. Burst intervals from this source are highly irregular. (From Cominsky *et al.*[15]).

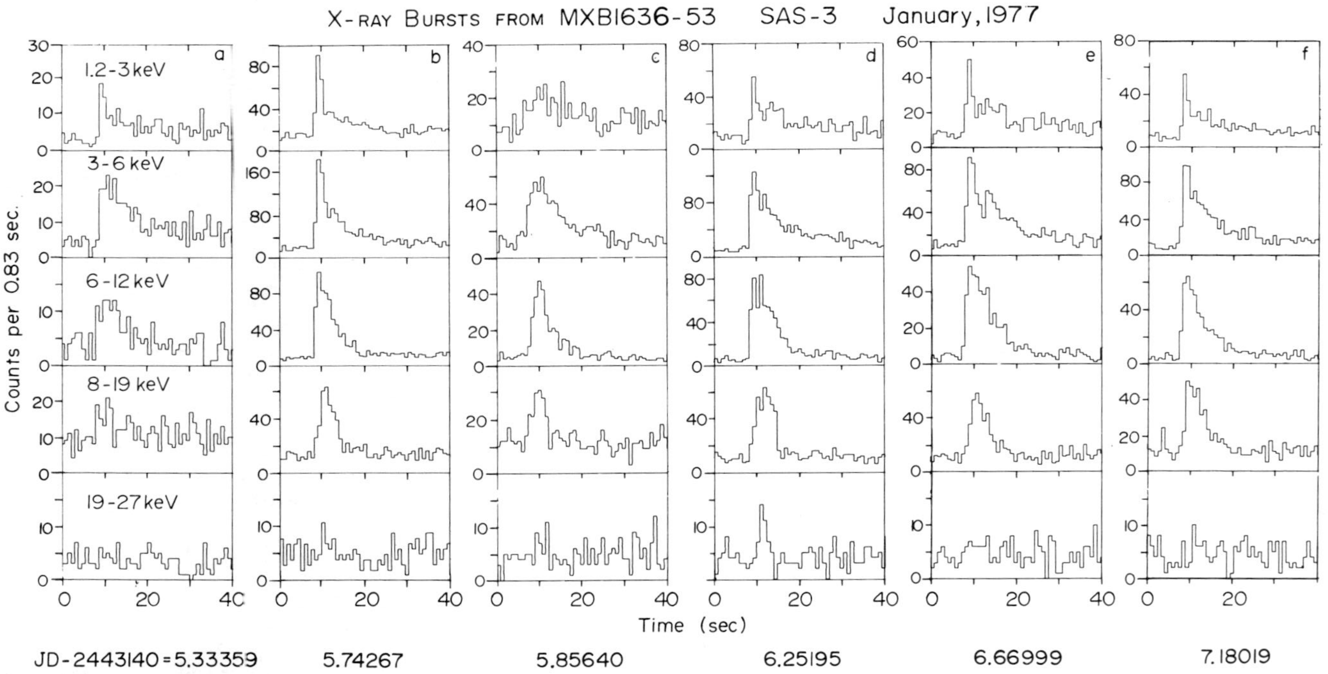

FIGURE 7. Raw counting rate plots of five energy channels from the SAS-3 horizontal tube (HT) detector system for six x-ray bursts from MXB 1636 – 53. The times at the bottom, in reduced Julian days = JD-2,443,140, correspond to the beginning of the time bin that contained the maximum. The bursts have similar shapes except the third one, which doesn't show the narrow feature in the 1.2–3-keV channel. This burst occurred unusually soon (2.7 h) after the preceding burst. Intervals between the other bursts range from 9.5 to 12.2 h. (From Hoffman, Lewin, and Doty.[34]).

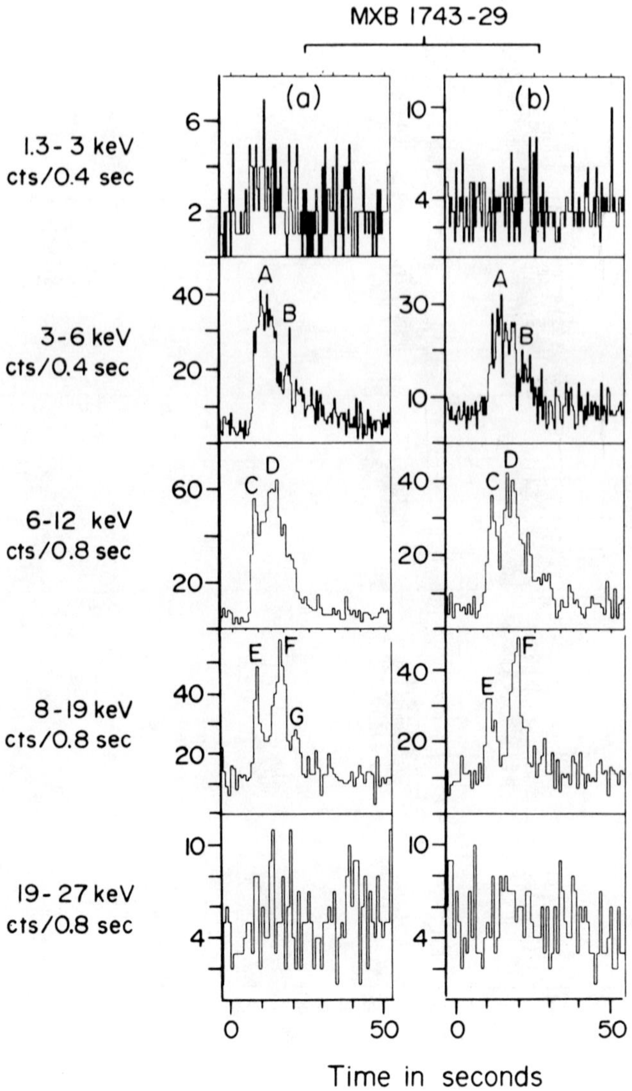

FIGURE 8. Two bursts (raw counting rate plots) from MXB 1743 − 29, which is one of at least three burst sources located within 1° of the galactic center. This source produced bursts at intervals of ~35 h. They all showed the double (sometimes triple) peaks at high energies. (Modified from Lewin *et al.*[51])

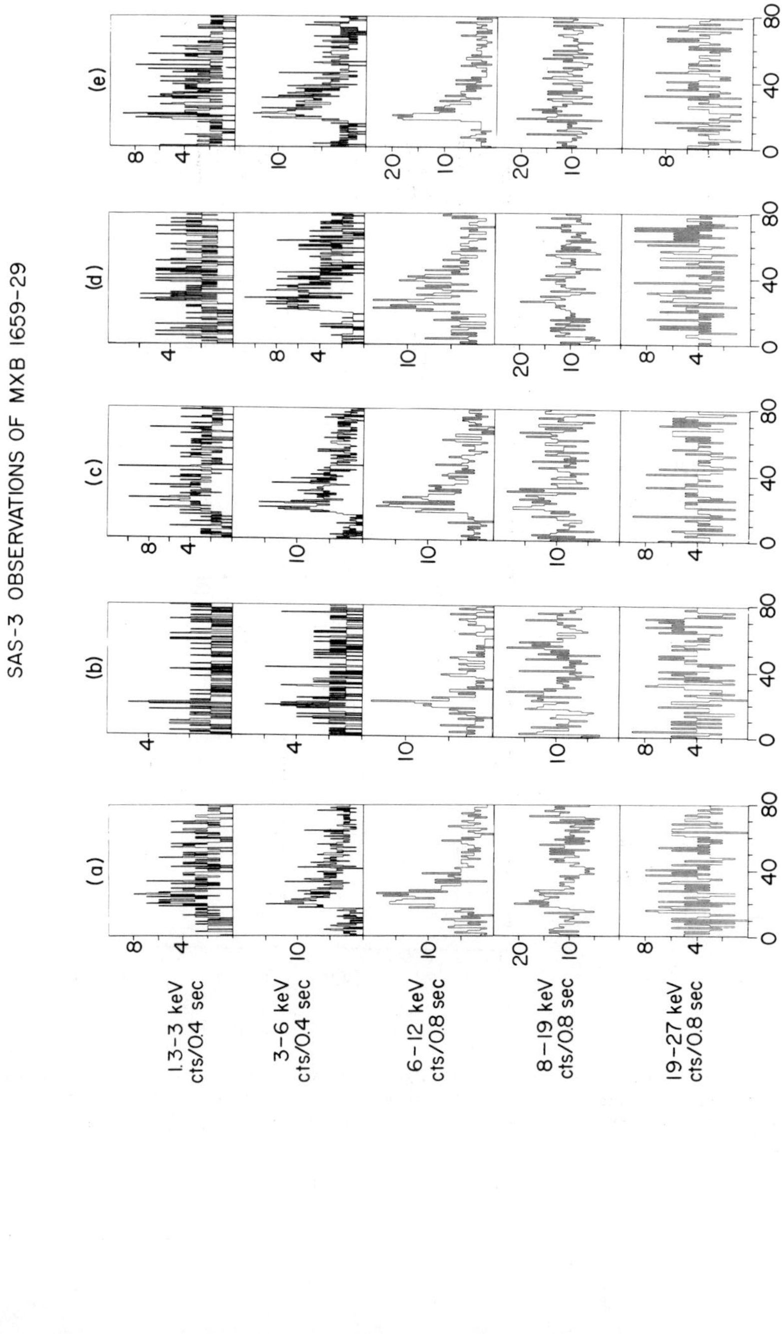

FIGURE 9. Five bursts (raw counting rate plots) from MXB 1659 − 29. The bursts profiles show quite a bit of variation. The burst intervals were the most regular (~2.5 h) of all burst sources. (From Lewin *et al.*[62]).

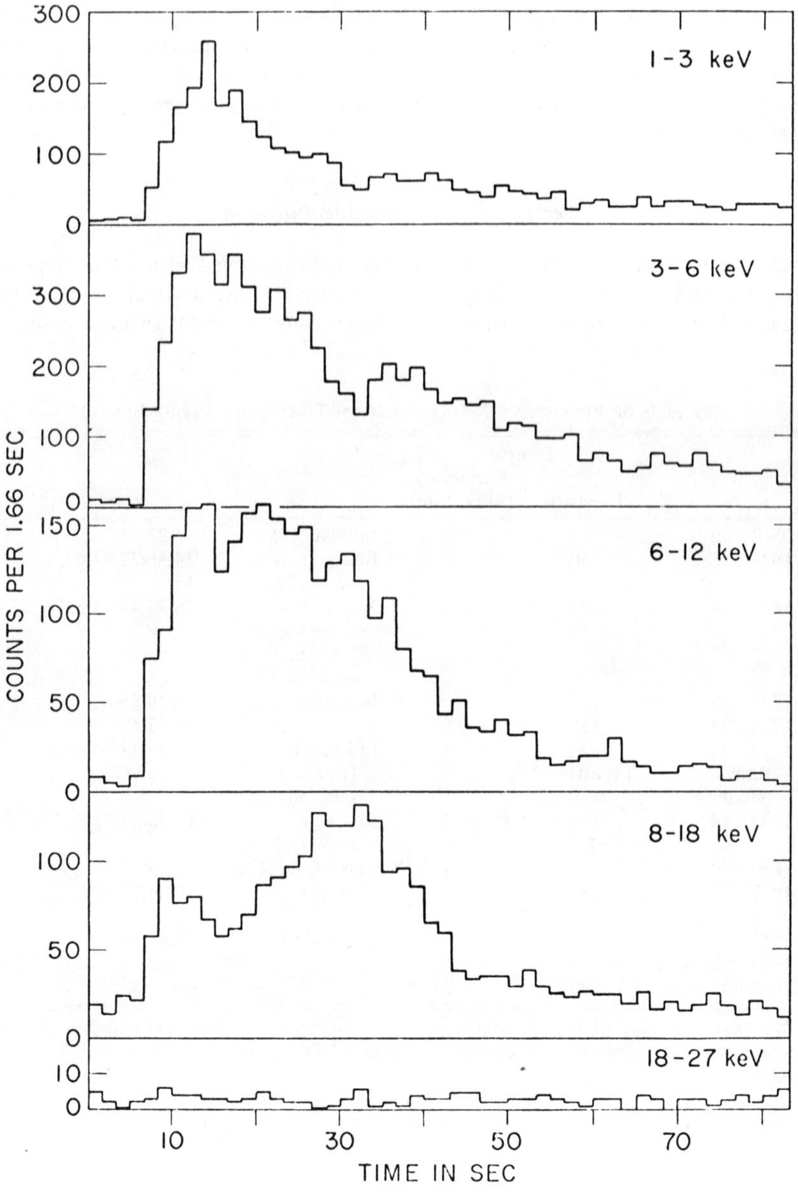

FIGURE 10. An x-ray burst from MXB 1850 – 08 (in the globular cluster NGC 6712). It is quite different in shape and total luminosity from two other bursts (not shown here) also observed from this source. The double-peaked structure in the higher energy x-ray data is also observed in several other burst sources (e.g., see FIGURE 8). (From Li and Clark.[66])

when bursting is expected to stop). This is observed to some extent (see above); however, more detailed calculations are required to explain why the burst intervals can change by an order of magnitude (and more) without an appreciable change in the luminosities of the persistent x-ray sources. If the bursts are due to thermonuclear flashes, apparently the helium can burn steadily for extended periods without producing bursts.

Persistent Emission—Identifications

In TABLE 4 we list burst sources that have been identified with a high degree of certainty with a source of persistent x-ray emission. Other associations have been suggested but they are less certain (see references 60 and 61). In most cases, the

TABLE 4

SOURCES OF PERSISTENT X-RAY EMISSION THAT EMIT TYPE-I BURSTS*

	Uhuru† Intensity	Uhuru† Variability (Max/Min)		References‡
0513−40	18	3	In NGC 1851	14,23,37
1636−53	250	2	Blue star $m_v \sim 17.6$	34,38,72,84,90
1659−29	~80	>15	Transient x-ray source UV star, $m_v \sim 18$	18,25,49,55,58,59,62
1722−30	7		In Terzan 2	26,29,85
1728−34	150	5	?	33,38,48
1730−335 Rapid burster	~50 (Type-II bursts)	>10	In Liller I (transient x-ray source)	17,35,48,50,68,71
1735−44	210	1.7	Blue star $m_v \sim 17.5$	15,28,38,57,72,73
1820−30	320	3	In NGC 6624	12,13,27,37
1837+05 Ser X-1	280	2	Blue star $m_B \sim 19.2$	31,64,84,91
1850−08	9		In NGC 6712	37,66,84
1906+00	4			17,52
1916−05	20	2		6,17,56

*It is expected that all Type-I burst sources have an associated persistent emission of x rays. Those listed here are virtually certain associations. Other associations suggested are less certain.[60,61]

†From 4U Catalogue[24] (except 1659−29 and 1730−335).

‡For more detailed references and a list of source locations, see references 8,60, and 61.

energy emitted in persistent emission (at high burst activity) is $\gtrsim 10^2$ times that in Type-I bursts.[53,95] This would be expected if Type-I bursts were produced by thermonuclear flashes (see above and reference 41).

In the case of MXB 1659 − 29 the energy in persistent emission was less than 25 times that in Type-I bursts in October 1976.[55] This value is perhaps uncomfortably low for the nuclear flash model in its present form and it suggests that either the persistent emission comes partly from the disk and is not isotropic[75] and/or that the available energy in the thermonuclear flash is more than 1 MeV per nucleon. If the latter is true, it might indicate that some of the hydrogen is ignited when the helium flash occurs (hydrogen fusion releases ~7 MeV per nucleon; helium fusion releases only ~1 MeV per nucleon). The role of hydrogen in the flash is discussed by Taam and Picklum.[88]

In TABLE 4 we also list the identifications of the burst sources with faint blue stars (in four cases).

Burst Spectra

Why do we believe that burst sources are neutron stars? Of course, the fact that the thermonuclear flash model, worked out by Joss,[40] is capable of explaining such a large number of the burst properties is very persuasive. However, historically the belief in neutron stars preceded the paper by Joss.[40]

It was first pointed out by Swank *et al.*[85] that the burst spectra fit that of a blackbody reasonably well. For one particular burst they derived radii of the emitting region of the burst and found ~100 km during the first 10 s and ~15 km later on in the burst (they assumed a distance of 10 kpc). Hoffman, Lewin, and Doty[33,34] later showed that radii of burst regions for two other sources (MXB 1728 − 34 and MXB 1636 − 53) were ~10 km. They observed a distinct cooling in the bursts (see FIGURES 5 and 7) but found that the emitting regions remained constant throughout the bursts. The above dimensions suggest that neutron stars are involved and certainly not black holes in excess of a few hundred solar mass as proposed by Grindlay and Gursky.[99]

Van Paradijs[93] carried this a step further in a study of ten burst sources for which good broadband spectral data are available from SAS-3. He showed that all ten, interpreted as blackbodies emitting isotropically, have nearly the same ratio of radiating surface area to peak burst luminosity. If the peak luminosities are assumed to be the Eddington limit for a 1.4-$M_{\odot}$ star with a hydrogen envelope (1.8×10^{38} erg/sec), the radii are found to have an average value of 7 km with a scatter of only about 20%. In a later paper Van Paradijs[94] makes corrections for the gravitational red shift and adjusts the maximum burst luminosity such that ten sources are distributed evenly around the galactic center (at an assumed distance of 9 kpc). He again concluded that burst sources are neutron star and standard candles (at burst maximum). Recently, Swank, Eardley, and Serlemitsos,[86] assuming modified black-body spectra and correcting for the gravitational redshift, have come to similar conclusions.

On the basis of the spectral information discussed above and the success of the thermonuclear flash model, we conclude that it is very likely that x-ray burst sources are neutron stars.

Radio/Optical/X-Ray Bursts

It was our hope and expectation that the simultaneous detection of radio, optical, infrared, and x-ray bursts could make a significant contribution to explaining the mechanism of the bursts and to defining the geometry (e.g., their binary character) of the systems.

A worldwide coordinated burst watch was organized in 1977 by the SAS-3 group. Forty-four observatories participated from 14 different countries. A total of 120 bursts were observed by SAS-3 in 35 days from ten different Type-I burst sources. No simultaneous positive detection was made in either the radio, optical, or infrared. Upper limits were reported for MXB 1636 − 53, MXB 1837 + 05, and MXB 1916 − 05 (references 1, 7, 89, 90, 92).

On June 2, 1978, an optical burst was observed simultaneously with an x-ray burst from MXB 1735 − 44.[28] The fast rise and the small (~2.8 s) delay between the x-ray and optical bursts[73] indicate that the optical region must be within ~3 lightseconds of the x-ray emitting region (FIGURE 11). This result, in combination with the absence of optical variability of this star (other than bursts) makes it likely that the optical burst comes from an accretion disk heated by the x-ray burst rather than from a possible stellar companion. (This, of course, does not exclude the existence of a companion star.)

Comparing the steady flux with the flux at burst maximum, the x-ray signal increased by a factor of 6.3 ± 1 and the optical signal by a factor of 1.5 ± 0.3. Calculations of the effects of x-ray heating predict that the factor by which the optical flux increases (at burst maximum) should equal that of the x-ray increase raised to the power β. The theoretical value of β for x-ray heating of a hot blackbody is ~0.25–1.0 (references 42, 74, 97), which is consistent with the observed value of 0.22 ± 0.1. Since the optical spectrum shows the characteristic features expected from x-ray heating, in combination with the observed value for β, it seems possible that most (perhaps all) of the optical emission (thus not only the optical burst but also the steady optical emission) comes from the inner part of an accretion disk rather than a companion star.

The fact that no periodic variability is observed in the optical emission is perhaps not surprising. If the x-ray source were in a binary system (we favor this idea), the companion star may be so small[42,75] that its contribution to the visible light is negligibly small. (The light flux due to its own nuclear burning would be small and the light due to x-ray heating could be negligible if the star is located in the x-ray shadow of the disk.[75])

Recently a possible optical burst was detected simultaneously with an x-ray burst from MXB 1837 + 05.[31] The optical burst is delayed by 1.4 ± 0.5 s relative to the x-ray burst; this delay was ~2.8 s for MXB 1735 − 44.[73]

TABLE 5 lists the ratios of optical (B range; $\Delta\lambda$ = 1000 Å) and x-ray energies (~2–15 keV) observed for the persistent emission and for the bursts. Notice that for the persistent source the value for MXB 1837 + 05 is 5.9 ± 0.6 times lower than for MXB 1735 − 44; for the bursts, the value for MXB 1837 + 05 is 7 ± 3 times lower. These data are therefore consistent with the idea that the two objects are very similar but that the optical extinction for MXB 1837 + 05 is a factor of ~6 times higher than for MXB 1735 − 44.

In the spring and summer of 1979 further simultaneous observations will be made. SAS-3 reentered the Earth's atmosphere on April 9, 1979 and will therefore not participate. However, the Japanese satellite Hakucho (Japanese for Swan) and UK-6

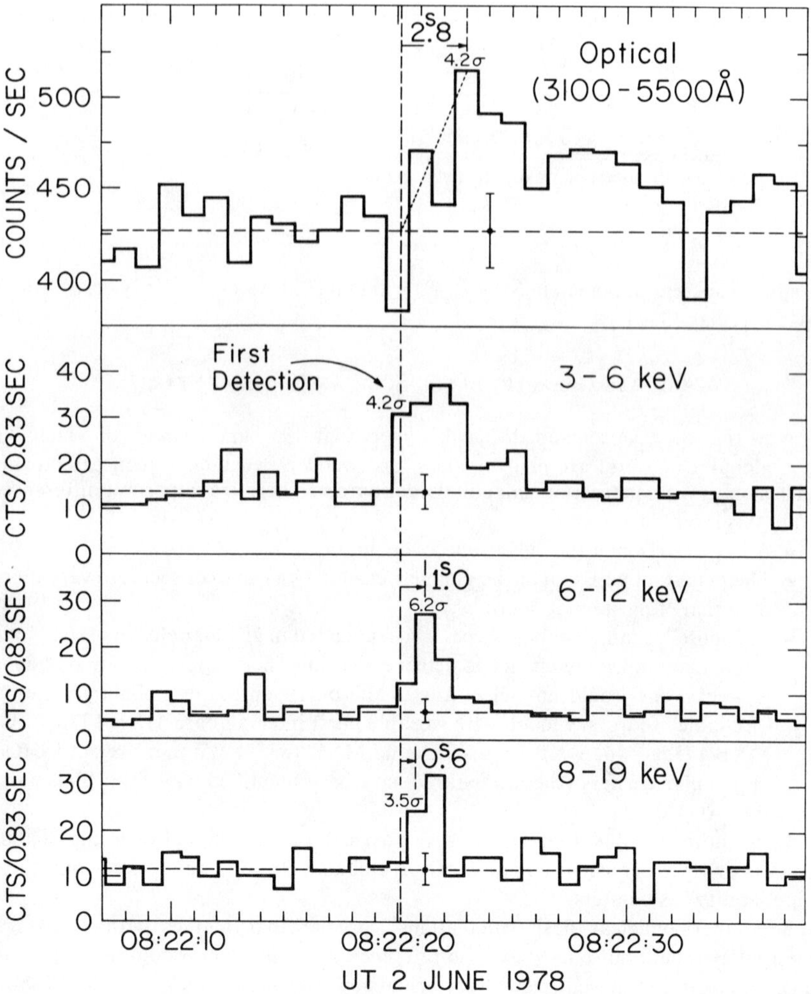

FIGURE 11. X-ray/optical burst from the x-ray burst source MXB 1735 − 44 as detected on June 2, 1978. The event was detected first in 3–6-keV x rays, as marked by the vertical dashed line, and ~2.8 s later in blue light. The background counting rates and their approximate ($\sqrt{N}$) uncertainties are indicated by horizontal dashed lines and error bars, respectively. The exact Poisson statistical significance of the first bin to exceed a 3σ detection threshold is also given. The dotted line shows that the optical data are equally consistent with a 3-s linear rise to maximum from the time the burst was first detected or with a 3-s delay followed by an abrupt rise to maximum. A tail is apparent in the optical data. (From McClintock *et al.*[73])

TABLE 5

OBSERVED RATIOS OF ENERGY DENSITIES*

(OPTICAL/X-RAY FOR PERSISTENT RADIATION AND BURSTS)

	$\left(\frac{E_{opt}}{E_x}\right)_{persistent}$	$\left(\frac{E_{opt}†}{E_x}\right)_{burst}$	References
MXB 1735−44	$(1.3 \pm 0.1 \times 10^{-4}$	$(2.0 \pm 0.6) \times 10^{-5}$	28,72,73
MXB 1837+05 (Ser X-1)	$(2.2 \pm 0.2) \times 10^{-5}$	$(2.8 \pm 0.9) \times 10^{-6}$	31

*No correction for interstellar extinction was made.

†Integrated for ~10 s.

(pending a successful launch in May 1979) will make the necessary x-ray observations (IAU Circular 3334).

ARE ALL GALACTIC BULGE SOURCES NEUTRON STARS?

Now that we have concluded that it is likely that ~25 burst sources (of which five lie in globular clusters) are neutron stars, one wonders whether we can extend this conclusion to the nonbursting galactic bulge sources. There are important differences within this large class:

- A large fraction of the bulge sources do not burst.
- The persistent emission of the globular cluster x-ray sources seems to vary more than that of most bulge sources.[65]
- The most "reliable" burst sources are not located in globular clusters (i.e., if one wants to optimize the chance to detect bursts in a short observing period of days to weeks, one would not select any of the burst sources in globular clusters). "Reliable" sources (though with variable burst intervals) are MXB 1728 − 34, MXB 1735 − 44, MXB 1636 − 53, and MXB 1837 + 05; they are not located in globular clusters (the latter three have been identified with faint blue stars; see TABLE 4).

It is quite possible that there is a physical connection between the higher variability of the persistent emission of the globular cluster x-ray sources and their "unreliability" as bursters.

Joss' thermonuclear flash calculations[40] predict that sources with very high luminosities should not burst (see also reference 95). Thus, one would not expect to detect bursts from the well-known very bright sources such as GX5 − 1 (4U 1758 − 25), GX3 + 1 (4U 1744 − 26), GX9 + 1 (4U 1758 − 20), GX13 + 1 (4U 1811 − 17), and GX17 + 2 (4U 1813 − 14).

It seems likely that the evolution of the bulge and cluster systems are very different (primordial binaries versus capture). There may be a large range in the magnetic fields, which could influence the burst behavior.[43] The difference in evolution may well account for the observed differences.

In spite of the above differences it seems likely that the majority of the bulge sources are neutron star systems similar to burst sources. Of course, it is quite possible that several of the nonbursting (Type-I) x-ray sources are black holes.

FAST TRANSIENTS (BURSTS?) WITH PRECURSORS

Before we leave the reader with the feeling that "all is well," we want to discuss some intriguing results that have not received much attention so far. Several exceptional transient phenomena (or very long x-ray bursts?) have been observed with SAS-3. Two were discussed by Hoffman *et al.*[36] In FIGURES 12–14 we show the profiles at different time scales; they show a great resemblance to x-ray bursts. The February 1977 event lasted ~1500 s. Its burst spectrum can be reasonably well represented by that of a cooling blackbody. The maximum temperature is 29×10^6 K (as in the case of Type-I bursts) (FIGURE 15). At a distance of 10 kpc the object has a

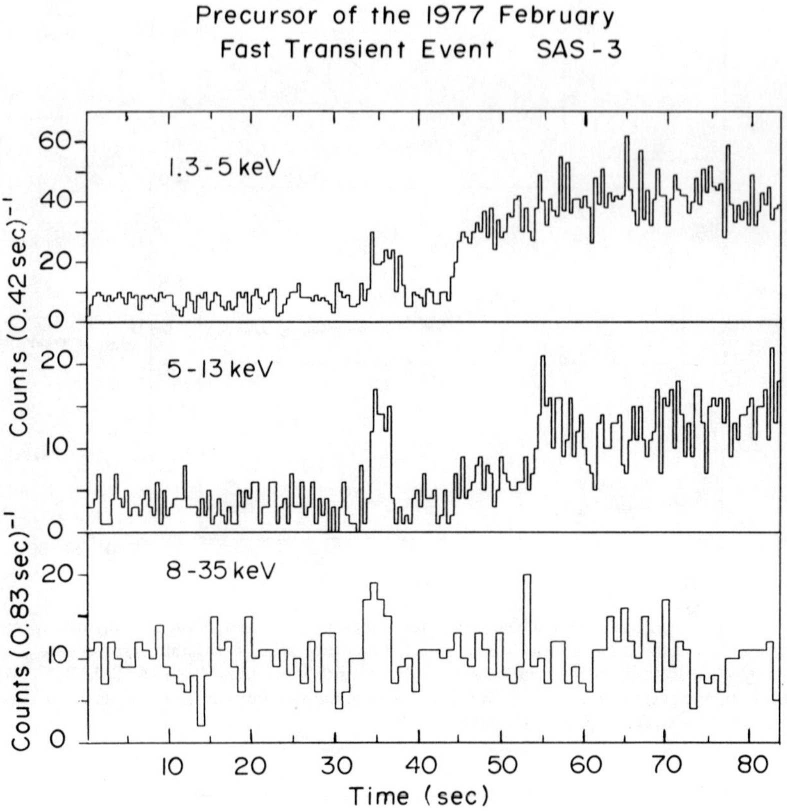

FIGURE 12. Light curves in three energy channels of the 1977 February 7 event as observed by the SAS-3 center slat detectors. The precursor is clearly visible and shows no obvious spectral changes in its evolution. After the precursor the signal drops to the preprecursor level before the "main" event begins. The delay between the precursor and the rise of the "main" event increases with energy and is ~80 s for the 8–35-keV channel. Thus in this energy channel it rises near t ~ 115 s (off scale). The "main" event shows significant spectral softening during its decay (see FIGURES 13 and 15). (Modified from Hoffman *et al.*[36])

radius of ~9 km (assuming that the source was near the center of the SAS-3 center slat collimator in which the event was observed).

Both events are different from Type-I bursts in that they have a precursor and their duration is long (1500 s in one case). It would be interesting to see whether these phenomena could also be interpreted in terms of a thermonuclear flash model. Extended SAS-3 observations (lasting for weeks), covering the large regions in the sky from which these events were seen, have only produced these two. Thus it seems that if the "bursts" are repetitive, the intervals are quite long. In this may lie the answer to the question why these "bursts" are so different.

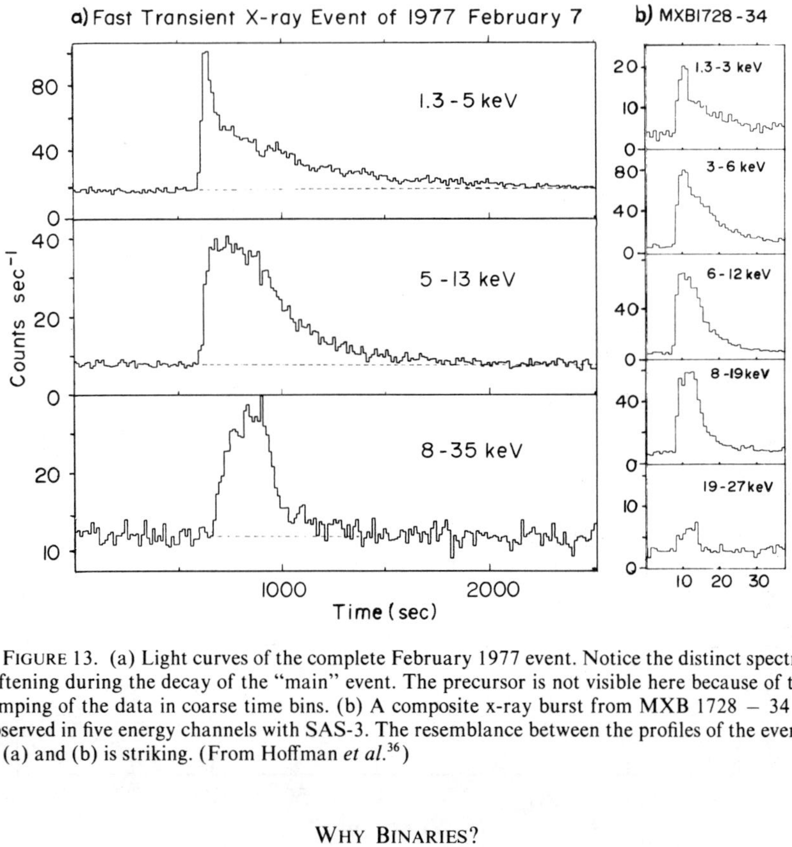

FIGURE 13. (a) Light curves of the complete February 1977 event. Notice the distinct spectral softening during the decay of the "main" event. The precursor is not visible here because of the lumping of the data in coarse time bins. (b) A composite x-ray burst from MXB 1728 − 34 as observed in five energy channels with SAS-3. The resemblance between the profiles of the events in (a) and (b) is striking. (From Hoffman *et al.*[36])

WHY BINARIES?

Above we have shown why we believe that most of the bulge sources, but in any case the Type-I burst sources, are neutron stars. In the absence of eclipses, are there any reasons to believe that these neutron stars are in binary systems?

Neutron stars can not be adequately powered by accretion from the interstellar medium to produce the persistent luminosity from the burst sources[83] unless they are in dense ($n > 10^6$ cm^{-3}) clouds. This seems unlikely since no low-energy absorption is

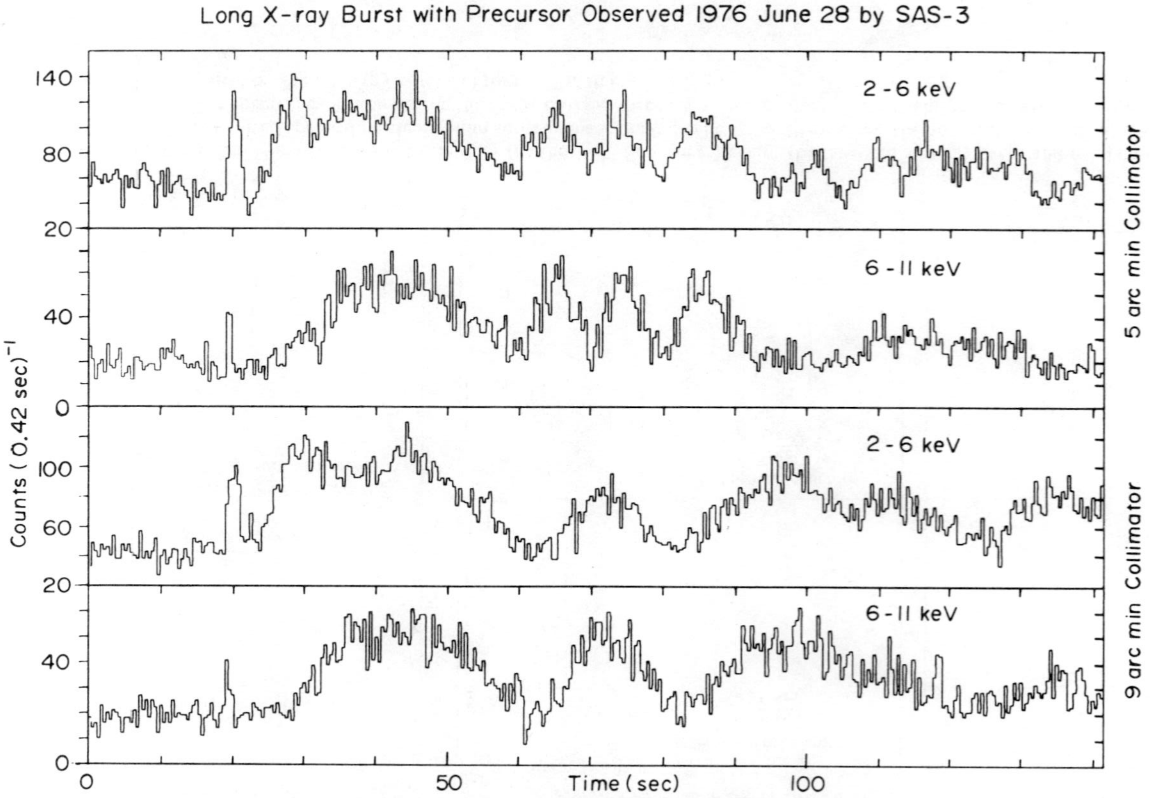

FIGURE 14. Light curves of the June 28, 1976 event observed by the SAS-3 modulation collimators (5 arc min and 9 arc min systems). A precursor is clearly visible (no spectral evolution). The "main" event rises later at high energies just as in the February 1977 event (see FIGURES 12 and 13). The intensity variations observed in the "main" event are not real; they are due to the rotating modulation collimator response. This event is described in detail by Hoffman *et al.*[36]

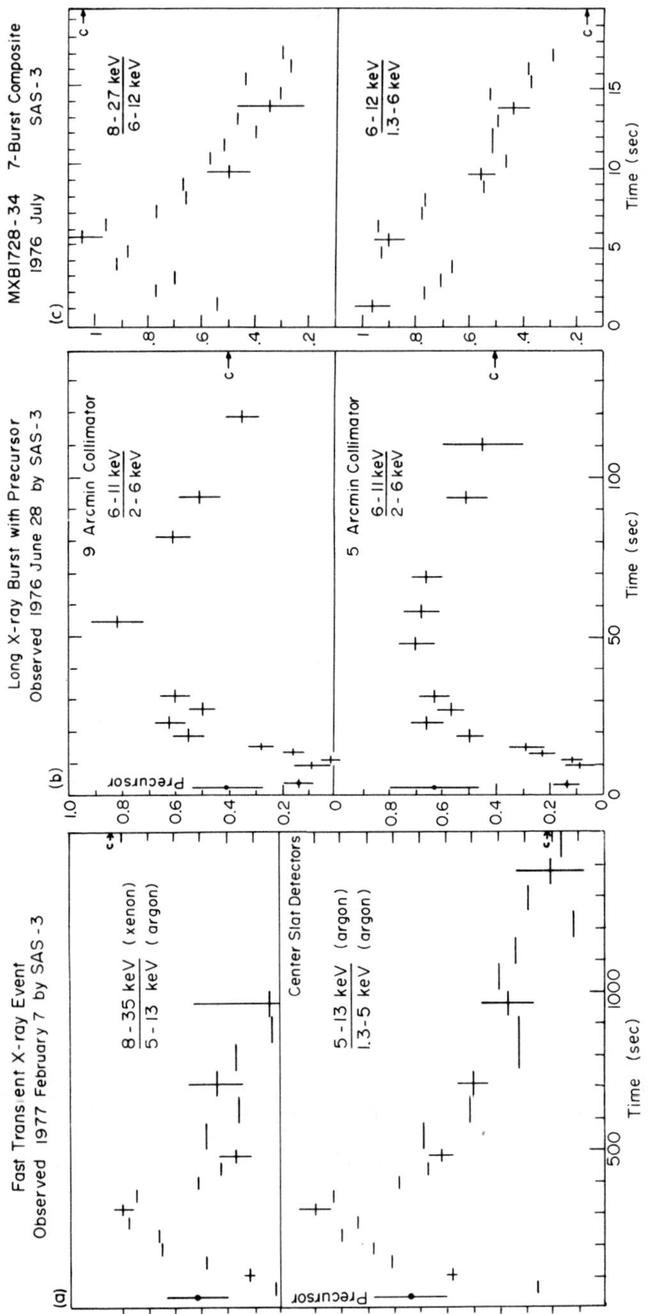

FIGURE 15. The evolution of spectral channel ratios with time for (a) the 1977 February 7 event, (b) the 1976 June 28 event and (c) a composite x-ray burst from MXB 1728 – 34. All show initial spectral hardening and subsequent softening. The "c" with an arrow (bottom right) in each case represents the spectral ratio of the Crab nebula. The spectral evolution during the decay of these three events is strikingly similar. This suggests that the peculiar events of June 1976 and February 1977 may also be Type-I x-ray bursts. (From Hoffman *et al.*[36])

observed in the x-ray spectra of the persistent emission from these sources. Thus it seems likely that a gas supply has to be nearby. The objects could be single neutron stars with a fossil accretion disk (see reference 76). If this were so it is difficult to see why they occur preferentially in centrally condensed globular clusters. We therefore believe it is more likely that the neutron stars are in binary systems.

Note Added in Proof

This article was written in April 1979. Since then several important results were reported.

(i) Canizares, McClintock, and Grindlay (preprint) found more evidence that the optical emission from burst sources is produced by an accretion disk (x-ray heating).

(ii) Lewin, Van Paradijs, Cominsky, and Holzner (preprint) have shown that, when examined in detail, Type-I bursts are not standard candles at burst maximum. They showed that the standard deviation in the maximum flux of 56 bursts from MXB 1735 − 44 was 37 percent (normalized to the mean value).

(iii) Infrared bursts first observed in April 1979 by Apparao *et al.* (IAU Circular 3344) from the Rapid Burster have been confirmed in September 1979 by A. W. Jones (private communication).

The infrared bursts are unusually bright. At $\sim 2.2\ \mu$ the average luminosity is $\sim 10^{37}$ erg/sec (band width $\sim 0.4\ \mu$; assumed isotropic emission). The total integrated burst energy is $\sim 10^{38}$ erg. To date it is still unclear whether the infrared bursts are associated with the x-ray bursts, but that seems likely. If this is so, it is expected, in view of the low infrared burst frequency, that they are associated with Type-I rather than the Type-II bursts which repeat on time scales of minutes. In that case, the bright infrared bursts are also expected from other Type-I burst sources and they would be much easier to detect than optical bursts.[101,102]

(iv) Long Type-II bursts of a few hundred seconds duration first detected by F. Marshall[70] in March 1979 were observed in large quantities about half a year later by Hakucho (Inoue *et al.* preprint; see also IAU Circular No. 3392).

(v) Fifteen optical bursts were detected from MXB 1636 − 53. Five of these occurred simultaneously with x-ray bursts detected by Hakucho (IAU Circular No. 3399).

(vi) Preliminary results from the Einstein Observatory (Grindlay, talks in the summer of 1979 and private communication) show that if globular cluster x-ray sources constitute a uniform population, their total mass (including a possible companion star) are larger than ~ 2 solar masses (2σ level of confidence).

Acknowledgments

The authors are grateful to L. Cominsky, P. C. Joss, D. Q. Lamb, and J. Van Paradijs for their valuable comments on this manuscript.

References

1. Abramenko, A. N., R. E. Gershberg, E. P. Pavlenko, & V. V. Prokof'eva, W. H. G. Lewin, J. van Paradijs, J. A. Hoffman & F. K. Li. 1978. M.N.R.A.S. **184:** 27P.
2. Bahcall, J. N. & J. P. Ostriker. 1975. Nature (London) **256:** 23.
3. Bahcall, J. N. & R. A. Wolf. 1976. Ap. J. **109:** 214.
4. Bahcall, N. E. & M. A. Hausman. 1977. Ap. J. **213:** 93.
5. Bahcall, N. E., B. M. Lasker & W. Wamsteker. 1977. Ap. J. Lett. **213:** L105.
6. Becker, R. H., B. W. Smith & J. H. Swank. 1977. Ap. J. Lett. **216:** L101.
7. Bernacca, P. L., A. Bianchini, A. Walker, D. Backman, C. Canizares, J. van Paradijs, J. A. Hoffman, J. Doty, H. Marshall, W. Wheaton, J. G. Jernigan & W. H. G. Lewin. 1979. M.N.R.A.S. **186:** 287.
8. Bradt, H. V., R. E. Doxsey & J. G. Jernigan. To be published in Advances in Space Exploration.
9. Canizares, C. R. 1975. Ap. J. **201:** 589.
10. Canizares, C. R., J. E. Grindlay, W. A. Hiltner, W. Liller & J. E. McClintock. 1978. Ap. J. **224:** 39.
11. Clark, G. W. 1975. Ap. J. Lett. **199:** L143.
12. Clark, G. W., G. Jernigan, H. Bradt, C. Canizares, W. H. G. Lewin, F. K. Li, W. Mayer & J. E. McClintock. 1976. Ap. J. Lett. **207:**L105.
13. Clark, G. W., F. K. Li, C. R. Canizares, S. Hayakawa, G. Jernigan & W. H. G. Lewin. 1977. M.N.R.A.S. **179:** 651.
14. Clark, G. W. & F. K. Li. 1977. IAU Circular No. 3092.
15. Cominsky, L., S. Holzner, J. van Paradijs & W. H. G. Lewin. In preparation.
16. Cominsky, L. W. Ossman, J. Doty, J. G. Jernigan, J. van Paradijs & W. H. G. Lewin. In preparation.
17. Doxsey, R., H. Bradt, H. Gursky, M. Johnston, D. A. Schwartz & J. Schwarz. 1978. Ap. J. Lett. **221:** L53.
18. Doxsey, R. E., J. E. Grindlay, R. E. Griffiths, H. Bradt, M. Johnston, R. Leach, D. Schwartz & J. Schwarz. 1979. Ap. J. Lett. **228:** L67.
19. Evans, W. D., R. D. Belian & J. P. Conner. 1976. Ap. J. Lett. **207:** L91.
20. Fabian, A. C., J. E. Pringle & M. J. Rees. 1975. M.N.R.A.S. **172:** 15P.
21. Fall, S. M. & M. A. Malkan. 1978. M.N.R.A.S. **185:** 899.
22. Flowers, E. & M. Ruderman. 1977. Ap. J. **215:** 302.
23. Forman, W. and C. Jones. 1976. Ap. J. Lett. **207:** L177.
24. Forman, W., C. Jones, L. Cominsky, P. Julien, S. Murray, G. Peters, H. Tananbaum & R. Giacconi. 1978. Ap. J. Suppl. **38:** No. 4 (4U Catalogue).
25. Grindlay, J. E. 1978. IAU Circular 3229.
26. Grindlay, J. E. 1978. Ap. J. Lett. **224:** L107.
27. Grindlay, J., H. Gursky, H. Schnopper, D. R. Parsignault, J. Heise, A. C. Brinkman & J. Schrijver. 1976. Ap. J. Lett. **205:** L127.
28. Grindlay, J. E., J. E. McClintock, C. R. Canizares, J. van Paradijs, L. Cominsky, F. K. Li, & W. H. G. Lewin. 1978. Nature (London) **274:** 567.
29. Grindlay, J. 1979. Spring APS Meeting, Washington, D.C. (results from the Einstein observatory).
30. Gunn, J. E. & J. P. Ostriker. 1970. Ap. J. **160:** 979.
31. Hackwell, J. A., R. D. Gehrz, G. L. Grasdalen, J. van Paradijs, L. Cominsky & W. H. G. Lewin. 1979. IAU Circular No. 3331. In preparation.
32. Hills, J. G. 1975. Astr. J. **80:** 1075.
33. Hoffman, J. A., W. H. G. Lewin & J. Doty. 1977. M.N.R.A.S. **179:** 57P.
34. Hoffman, J. A., W. H. G. Lewin & J. Doty. 1977. Ap. J. Lett. **217:** L23.
35. Hoffman, J. A., H. L. Marshall & W. H. G. Lewin. 1978. Nature (London) **271:** 630.
36. Hoffman, J. A., W. H. G. Lewin, J. Doty, J. G. Jernigan, M. Haney & J. A. Richardson. 1978. Ap. J. Lett. **221:** L57.
37. Jernigan, J. G. & G. W. Clark. Ap. J. In press.
38. Jernigan, J. G., K. M. V. Apparao, H. V. Bradt & R. E. Doxsey. 1977. Nature (London) **270:** 321.

39. Joss, P. C. 1977. Nature **270:** 310.
40. Joss, P. C. 1978. Ap. J. Lett. **225:** L123.
41. Joss, P. C. Ann. N.Y. Acad. Sci. This volume.
42. Joss, P. C. & S. Rappaport. 1979. Astron. Astrophys. **71:** 217.
43. Joss, P. C. & F. K. Li. Submitted to Ap. J.
44. King, I. R. 1966. Astron. J. **71:** 64.
45. Kleinmann, D. E., S. G. Kleinmann & E. L. Wright. 1976. Ap. J. Lett. **210:** L83.
46. Kylafis, N. D. & D. Q. Lamb. 1979. Ap. J. Lett. **218:** L105.
47. Lamb, D. Q. & F. K. Lamb. 1978. Ap. J. **220:** 291.
48. Lewin, W. H. G. 1976. IAU Circular No. 2922.
49. Lewin, W. H. G., J. A. Hoffman & J. Doty. 1976. IAU Circular No. 2994.
50. Lewin, W. H. G., J. Doty, G. W. Clark, S. A. Rappaport, H. Bradt, R. Doxsey, D. R. Hearn, J. A. Hoffman, J. G. Jernigan, F. K. Li, W. Mayer, J. McClintock, F. Primini & J. Richardson. 1976. Ap. J. Lett. **207:** L95.
51. Lewin, W. H. G., J. A. Hoffman, J. Doty, D. R. Hearn, G. W. Clark, J. G. Jernigan, F. K. Li, J. E. McClintock & J. Richardson. 1976. M.N.R.A.S. **177:** 83P.
52. Lewin, W. H. G., F. K. Li, J. A. Hoffman, J. Doty, J. Buff, G. W. Clark & S. Rappaport. 1976. M.N.R.A.S. **177:** 93P.
53. Lewin, W. H. G. 1977. M.N.R.A.S. **179:** 43.
54. Lewin, W. H. G. 1977. Ann. N.Y. Acad. Sci. **302:** 210.
55. Lewin, W. H. G. & P. C. Joss. 1977. Nature (London) **270:** 211.
56. Lewin, W. H. G., J. A. Hoffman & J. Doty. 1977. IAU Circular No. 3087.
57. Lewin, W. H. G., J. A. Hoffman, J. Doty, F. K. Li & J. E. McClintock. 1977. IAU Circular No. 3075.
58. Lewin, W. H. G., J. A. Hoffman, H. Marshall, F. Primini, W. Wheaton, L. Cominsky, J. G. Jernigan & W. Ossman. 1978. IAU Circular No. 3190.
59. Lewin, W. H. G. 1978. IAU Circular No. 3193.
60. Lewin, W. H. G. Proceedings of IAU/COSPAR Meeting held in June, 1978 in Innsbruck, Austria. To be published in Advances in Space Exploration.
61. Lewin, W. H. G. 1979. *In* Globular Clusters. D. Hanes and B. Madore, Eds. Cambridge University Press.
62. Lewin, W. H. G., L. Cominsky, W. Ossman & J. van Paradijs. In preparation.
63. Li, F. K. & G. W. Clark. 1977. IAU Circular No. 3095.
64. Li, F. K., W. H. G. Lewin, G. W. Clark, J. Doty, J. A. Hoffman & S. A. Rappaport. 1977. M.N.R.A.S. **179:** 21P.
65. Li, F. K., G. W. Clark & T. Markert. 1979. Preprint CSR-P-78-44.
66. Li, F. K. & G. W. Clark. 1979. Preprint.
67. Lightman, L. P., W. H. Press & S. F. Oldenwald. 1978. Ap. J. **219:** 629.
68. Liller, W. 1977. Ap. J. Lett. **213:** L21.
69. Maraschi, L. & A. Cavaliere. 1977. *In* Highlights of Astronomy **4,** Part I.
70. Marshall, F. J. 1979. IAU Circular No. 3336.
71. Marshall, H. L., M. P. Ulmer, J. A. Hoffman, J. Doty & W. H. G. Lewin. 1979. Ap. J. **227:** 555.
72. McClintock, J. E., C. R. Canizares, H. V. Bradt & R. E. Doxsey. 1977. Nature (London) **270:** 320.
73. McClintock, J. E., J. E. Grindlay, C. R. Canizares, J. van Paradijs, L. Cominsky, F. K. Li & W. H. G. Lewin. Nature (London). In press.
74. Milgrom, M. 1976. Astron. Astrophys. **50:** 273.
75. Milgrom, M. 1978. Astron Astrophys. Lett. **67:** L25.
76. Paczynski, B. & M. Jaroszynski. 1978. Acta Astron. **28:** 111.
77. Peterson, C. J. & I. R. King. 1975. Astron. J. **80:** 427.
78. Peterson, C. J. 1976. Astron. J. **81:** 617.
79. Press, W. H. & S. A. Tcukolsky. 1977. Ap. J. **213:** 183.
80. Salpeter, E. E. 1973. IAU Symposium #55 in X- and Gamma-Ray Astronomy. H. V. Bradt and R. Giacconi, Eds. Reidel Publishing Company. Dordrecht, Holland.
81. Shapiro, S. L. 1977. Ap. J. **217:** 281.
82. Silk, J. & J. Arons. 1975. Ap. J. Lett. **200:** L131.

83. Ostriker, J. P., M. J. Rees & J. Silk. 1970. Ap. J. Lett. **6:** 179.
84. Swank, J. H., R. H. Becker, S. H. Pravdo, J. R. Saba & P. J. Serlemitsos. 1976. IAU Circular No. 3010.
85. Swank, J. H., R. H. Becker, E. A. Boldt, E. A. Holt, S. H. Pravdo & P. J. Serlemitsos. 1977. Ap. J. Lett. **212:** L73.
86. Swank, J. H., D. M. Eardley & P. J. Serlemitsos. 1979. Preprint.
87. Taam, R. E. & R. E. Picklum. 1978. Ap. J. **224:** 210.
88. Taam, R. E. & R. E. Picklum. 1979. Preprint.
89. Takagishi, K., K. Nagareda, M. Matsuoka, M. Fujii, S. Sato, J. van Paradijs, J. A. Hoffman, J. G. Jernigan, W. Wheaton, F. Primini & W. H. G. Lewin. 1978. ISAS Res. Note. ISAS RN 60 CRS-P-78-31.
90. Thomas, R. M., M. L. Duldig, R. F. Haynes, L. W. Simons, P. Murdin, J. A. Hoffman, W. H. G. Lewin & W. A. Wheaton. M.N.R.A.S. In press.
91. Thorstensen, J., P. Charles & S. Bowyer. 1978. IAU Circular No. 3253.
92. Ulmer, M. P., R. M. Hjellming & W. H. G. Lewin. 1978. Private communication.
93. Van Paradijs, J. 1978. Nature (London) **274:**650.
94. Van Paradijs, J. Ap. J. Lett. In press.
95. Van Paradijs, J., P. C. Joss, L. Cominsky & W. H. G. Lewin. In preparation.
96. Woosley, S. E. & R. E. Taam. 1976. Nature (London) **263:** 101.
97. Milgrom, M. & E. E. Salpeter. 1975. Ap. J. **196:** 583.
98. Ostriker, J. P., L. Spitzer & R. A. Chevalier. 1972. Ap. J. Lett. **176:** L51.
99. Grindlay, J. & H. Gursky. 1976. Ap. J. Lett. **205:** L131.
100. Grindlay, J. E. & W. Liller. 1977. Ap. J. Lett. **216:** L105.
101. Kulkarni *et al.* 1979. Nature **280:** 819.
102. Apparao & Chitre. 1979. Preprint.

THERMONUCLEAR-FLASH MODELS FOR X-RAY BURST SOURCES*

P. C. Joss†

Center for Space Research and Department of Physics
Massachusetts Institute of Technology
Cambridge, Massachusetts 02139

INTRODUCTION

The observed properties of x-ray burst sources have been extensively reviewed by Lewin,[1] Lewin and Joss,[2] and Lewin and Clark.[42] The salient features of these sources include burst rise times of $\lesssim 1$ s, decay time scales of ~3–100 s, peak luminosities of $\sim 10^{38}$ ergs s^{-1}, total emitted energies of $\sim 10^{39}$ ergs per burst, spectra that can generally be well fitted by blackbody emission from a surface with a constant effective radius of ~10 km and a peak temperature of $\sim 3 \times 10^7$ K, and "tails" of softer x-ray emission that may persist for several minutes after the burst maximum. The intervals between bursts from a given source may be regular or erratic and are typically in the range $\sim 10^4$–10^5 s; many sources undergo burst-inactive phases that can last for weeks or months. Most burst sources are also sources of persistent x-ray emission, and the ratio of average persistent luminosity to time-averaged burst luminosity is typically $\sim 10^2$ during burst-active phases. (The properties of the "rapid burster," MXB 1730−335, are different from those of all other known burst sources and will be discussed separately below.) There are few correlations among the burst flux, burst intervals, and persistent x-ray flux from any given source, and the detailed burst shapes vary from one source to another and often vary with time in a given source.

A theoretical model for x-ray burst sources ideally should account for all these observational properties. Virtually all models that have been proposed so far[3–13] assume that the bursting phenomenon involves accretion of matter onto a collapsed object (degenerate dwarf, neutron star, or black hole). In all cases, the collapsed object serves at least one of two functions: its deep gravitational potential well allows the release of large amounts of gravitational energy in the form of x radiation by the accreting matter, and its small dimensions permit the x-ray emission to be released on the short time scale of an x-ray burst. The generic relations among the various proposed models are indicated schematically in FIGURE 1.

The proposed models can be broken down into two broad classes: (1) those invoking instabilities in the accretion flow onto a collapsed object; and (2) those that invoke thermonuclear flashes in the surface layers of an accreting neutron star. Models in class (1) were thoroughly reviewed by Lamb and Lamb[14] at the Eighth Texas Symposium on Relativistic Astrophysics. Although a few new models[10,11] in this class have been proposed in the past two years, relatively little progress has been achieved in understanding the complex hydrodynamics and radiative transfer that all

*This work was supported in part by the National Science Foundation under grant AST78-21993 and in part by NASA under contract NAS5-24441.

†Alfred P. Sloan Research Fellow.

0077–8923/80/0336–0479 $01.75/1 © 1980, NYAS

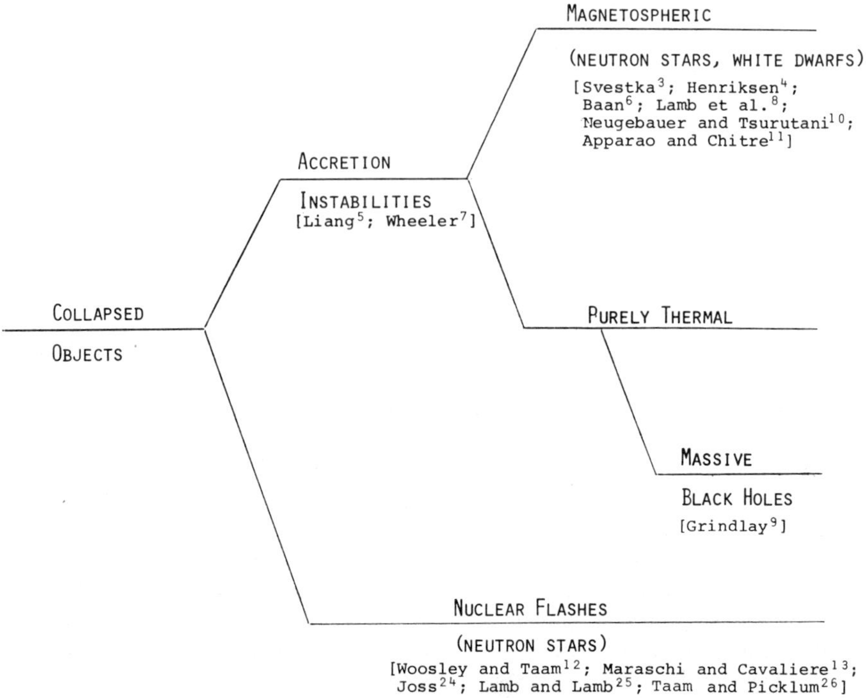

FIGURE 1. Generic relations among the various proposed mechanisms for x-ray bursts. "Magnetospheric" instabilities[3,4,6,8,10,11] rely in part on the intense magnetic field of the collapsed object, which must therefore be a degenerate dwarf or neutron star. "Purely thermal" instabilities[9] depend only upon the character of the heating and cooling rates of the accreting matter, and hence could conceivably be associated with accretion onto a massive black hole. Other proposed mechanisms[5,7] invoke instabilities in an accretion disk surrounding a collapsed object. This review focuses on thermonuclear-flash mechanisms[12,13,24–28]; for further discussions of accretion instabilities, see Lamb and Lamb[14] and Cowie *et al.*[15]

such models entail (see, however, Cowie, Ostriker, and Stark[15] for a recent discussion of physical constraints on many of these models). During the past year or two, the greatest amount of progress has instead been achieved in developing the thermonuclear-flash model, which has proved to be amenable to detailed numerical computations. In this review, I shall describe this progress and discuss the prospects for further developments in this field. An earlier version of some portions of this review was previously published in Comments on Astrophysics.[16]

HISTORICAL DEVELOPMENT

Our understanding of the character of nuclear processes on accreting neutron stars has had a rather tortuous development. When x-ray pulsars were discovered in 1972,[17–19] it was immediately recognized that these objects were probably neutron stars that were undergoing accretion from binary stellar companions. The following year, Rosenbluth *et al.*[20] pointed out that nuclear fusion in the surface layers would be an independent source of energy that might be radiated from the neutron-star photosphere. However, it was not until 1975 that Hansen and Van Horn[21] demonstrated that under a wide variety of conditions, the nuclear burning ought to be unstable and should lead to thermonuclear flashes.

Hansen and Van Horn[21] noted that the energy released in such flashes might produce variable x-ray emission from the neutron star. However, they also discovered that the characteristic time scale for thermonuclear runaway was usually $\lesssim 1$ s. This was considerably shorter than most of the time scales of variability from x-ray sources that were then known (with the exception of the periodic pulses from x-ray pulsars). Van Horn and Hansen[22] attempted to construct a hydrogen-flash model for the transient x-ray sources (which typically have rise times of a few days and decay time scales of weeks to months), but they were forced to resort to extremely low-mass ($\lesssim 0.15\ M_\odot$) neutron stars to get sufficiently thick hydrogen-rich envelopes and sufficiently long runaway time scales.

Following the discovery of x-ray bursts by Grindlay *et al.*[23] in 1975 September, a number of possible explanations for this phenomenon were soon advanced.[3-13] Among the early proposals was the suggestion by Woosley and Taam[12] and Maraschi and Cavaliere[13] that x-ray bursts result from thermonuclear flashes on accreting neutron stars. This suggestion spurred more detailed investigations by Joss,[24] Lamb and Lamb,[25] and Taam and Picklum[26] into the physics of nuclear flashes on accreting neutron stars and their possible relation to x-ray bursts and other phenomena. Most recently, Joss[27] and Joss and Li[28] have presented the results of detailed numerical computations of flashes of this type.

The Overall Physical Picture

Consider a neutron star undergoing accretion from a binary stellar companion. The freshly accreted matter will be rich in hydrogen and/or helium. However, at depths $\gtrsim 10^4$ cm beneath the surface of the neutron star, the density is sufficiently high that nuclear statistical equilibrium will be swiftly achieved; the predominant nuclei will have maximal binding energies, with atomic weights of ~60. (Still deeper in the star, these nuclei dissolve into a fluid in which neutrons are the primary constituent.) Hence, the accreting matter must pass through a series of nuclear burning shells as it is gradually compressed by the addition of new material. If the core of the neutron star is sufficiently hot or the accretion rate is sufficiently high, the temperature in the surface layers will be high enough that the burning will proceed via thermonuclear reactions, rather than electron capture or pycnonuclear reactions (which are driven by high densities rather than high temperatures). A sketch of the resultant structure of the neutron-star surface layers is given in FIGURE 2.

It was first realized by Hansen and Van Horn[21] that these burning shells will tend to be unstable to thermal runaway. The instability, known as the "thin-shell instability," was first discovered in a different context by Schwarzschild and Härm.[29] The existence and strength of the instability are a direct result of the strong temperature dependence of the thermonuclear reaction rates; in the case of neutron-star envelopes, the instability is further enhanced by the partial degeneracy of the burning material.

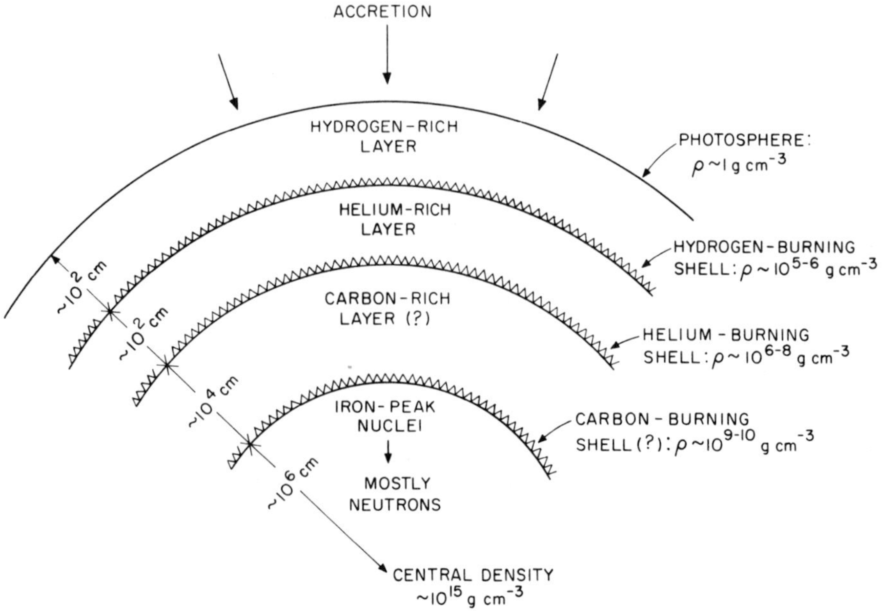

FIGURE 2. Schematic sketch of the surface layers of an accreting neutron star (from Joss[16]).

A cogent and thorough technical discussion of this type of instability has been given by Giannone and Wiegert.[30]

The *p–p* chains are insufficiently temperature sensitive to produce a thermal runaway in the hydrogen-burning shell of a neutron star. The instability of this shell is thus quenched by the saturation of the CNO cycle at very high reaction rates.[24,25] The saturation results from the appreciable lifetimes ($\sim 10^{2-3}$ s) of the beta-unstable nuclei N^{13}, O^{14}, O^{15}, and F^{17} that participate in the cycle. The next shell inward is the helium-burning shell, which should be unstable over a wide range of conditions. It is uncertain whether there will be any other significant burning shells, as the matter might already burn to quite heavy elements in the helium shell.[26,27] However, if a carbon shell exists, it is very likely to be unstable also.[12,26]

Dimensional analysis[24,25] indicates that the helium-burning flashes should have the following properties: (1) They should occur after the accumulation of $\lesssim 10^{21}$ g of fuel and release total energies of $\lesssim 10^{39}$ ergs per flash. (2) For accretion rates comparable with those observed in x-ray pulsars ($\lesssim 10^{17}$ g s^{-1}), the time interval between flashes would be $\sim 10^4$ s, very roughly. (3) The transport of energy through the surface layers should result in the emission of bursts of electromagnetic radiation from the neutron-star photosphere with rise times of ~ 0.1 s, peak luminosities of $\sim 10^{38}$ ergs s^{-1}, decay time scales of ~ 10 s, and peak blackbody temperatures of $\sim 3 \times 10^7$ K (if a full 10^{39} ergs of energy is indeed released in a single flash).

Carbon-burning flashes, if they exist, would involve a thicker layer of nuclear fuel and result in the release of substantially more energy. A carbon shell would be buried much deeper beneath the neutron-star surface ($\sim 10^4$ cm, compared to $\sim 10^2$ cm for the helium shell), so that the time scales for surface luminosity variations resulting from a carbon flash should be much longer than those for a helium flash.[24]

NUMERICAL MODELS

The above estimates, though very crude, suggest that thermonuclear flashes on accreting neutron stars could account for the observed properties of x-ray burst sources. With this encouragement, the first detailed numerical computations of the evolution of the surface layers of an accreting neutron star have recently been carried out.

Joss[27] has explored the evolution of the helium-burning shell. In these calculations, the neutron star was chosen to have a mass of $M = 1.4\ M_{\odot}$ and a radius of $R = 6.6$ km. A simplified nuclear reaction network was used, incorporating all of the energetically important reactions linking the nuclei from He^4 to Si^{28} and allowing the release of most of the available nuclear energy (see FIGURE 3). The accretion was assumed to be

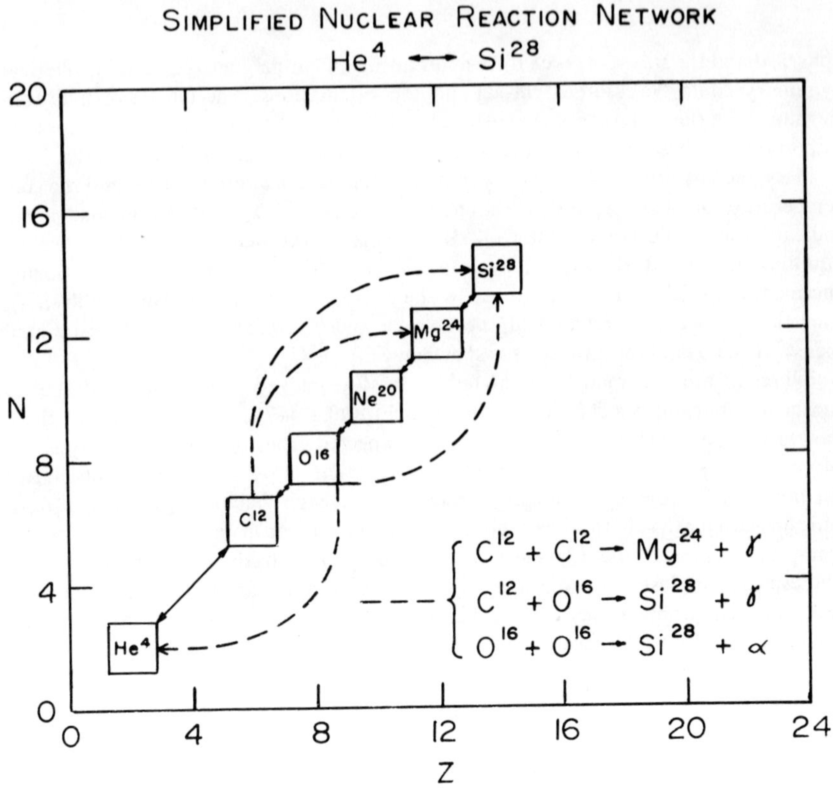

FIGURE 3. Simplified nuclear reaction network employed by Joss[27] and Joss and Li[28] for the helium-burning shell. Z is the atomic number and N the neutron number of each nucleus in the network. The network includes (i) the triple-α reaction; (ii) the (α, γ) reactions linking the alpha-particle nuclei from C^{12} to Si^{28}; (iii) the inverse (photodisintegration) reactions; and (iv) the $C^{12} + C^{12}$, $C^{12} + O^{16}$, and $O^{16} + O^{16}$ reactions (dashed lines).

TABLE 1

MODELS OF THE HELIUM-BURNING SHELL IN AN ACCRETING NEUTRON STAR*

M ($M_\odot$)	R (km)	B (10^{12} G)	$\dot{m}$ (10^{17} g s^{-1})	T_c (10^8 K)	τ_R (hr)	L_{max} (10^{38} ergs s^{-1})	τ_D (s)	E_{nuc} (10^{39} ergs)
1.41	6.57	0	0.3	2.52	15	2.6	3	2.2
1.41	6.57	0	1	2.52	4.4	2.7	3	2.1
1.41	6.57	0	1	3.87	1.4	2.2	1	0.7
1.41	6.57	0	3	5.74		No Flashes		
1.41	13.15	0	0.3	2.52	135	2.8	30	19
0.705	6.57	0	0.3	2.52	22	1.5	8	3.2
1.41	6.57	3	1	3.87	7.4	2.8	3	3.6
1.41	6.57	10	1	3.87		No Flashes		

*Results taken from Joss[27] and Joss and Li.[28] M = neutron-star mass; R = neutron-star radius; B = surface magnetic field strength (assumed uniform); $\dot{m}$ = mass accretion rate; T_c = core temperature; τ_R = recurrence intervals between helium-burning flashes; L_{max} = peak surface x-ray luminosity following a flash; τ_D = decay time scale of x-ray emission following a flash; and E_{nuc} = total nuclear energy liberated in a flash.

spherical and the star was taken to be nonrotating and unmagnetized, so that spherical symmetry could be assumed throughout the calculations. The effects of hydrogen burning upon the structure of the surface layers were also neglected. I shall discuss the importance of these assumptions and approximations in my concluding remarks.

Joss' models[27] contain two free parameters: the mass accretion rate $\dot{m}$ and the core temperature of the neutron star T_c. The values of $\dot{m}$ and T_c used in four models are indicated in TABLE 1 and FIGURE 4. If the core of the neutron star is in thermal equilibrium (i.e., if the heat flow into the core from the surface layers during thermonuclear flashes is just balanced by the heat lost from the core between flashes), then there is a unique relationship between $\dot{m}$ and T_c (reference 25); the estimated locus[27] of these equilibrium values is shown in FIGURE 4.

Three of the four models calculated by Joss[27] displayed thermonuclear flashes in the helium-burning shell (see TABLE 1 and FIGURES 4–7). The properties of these flashes were in good agreement with those expected from dimensional analysis.[24,25] More importantly, these calculations indicated that (1) a full $\sim 10^{21}$ g of matter accumulates on the neutron-star surface before each helium flash, (2) a flash consumes virtually all the available nuclear fuel and probably synthesizes mostly iron-peak elements, and (3) most of the energy of a flash is transported to the photosphere and lost as x radiation, rather than carried inward to heat the interior of the star. These properties of the flashes had not been discerned prior to the performance of detailed evolutionary computations, because they depend upon the highly nonlinear characteristics of the flash growth and decay.

The behavior of the helium-burning shell was further explored by Joss and Li,[28] who investigated the sensitivity of the flash properties to the assumed mass and radius of the neutron star. Joss and Li argued that since each helium-burning flash apparently consumes virtually all the available nuclear fuel, the nuclear physics essentially factors out of the problem and most of the mass and radius dependence of the flash properties follow from a few basic physical considerations. Thus, the recurrence interval τ_R between flashes is just the amount of time required for the base

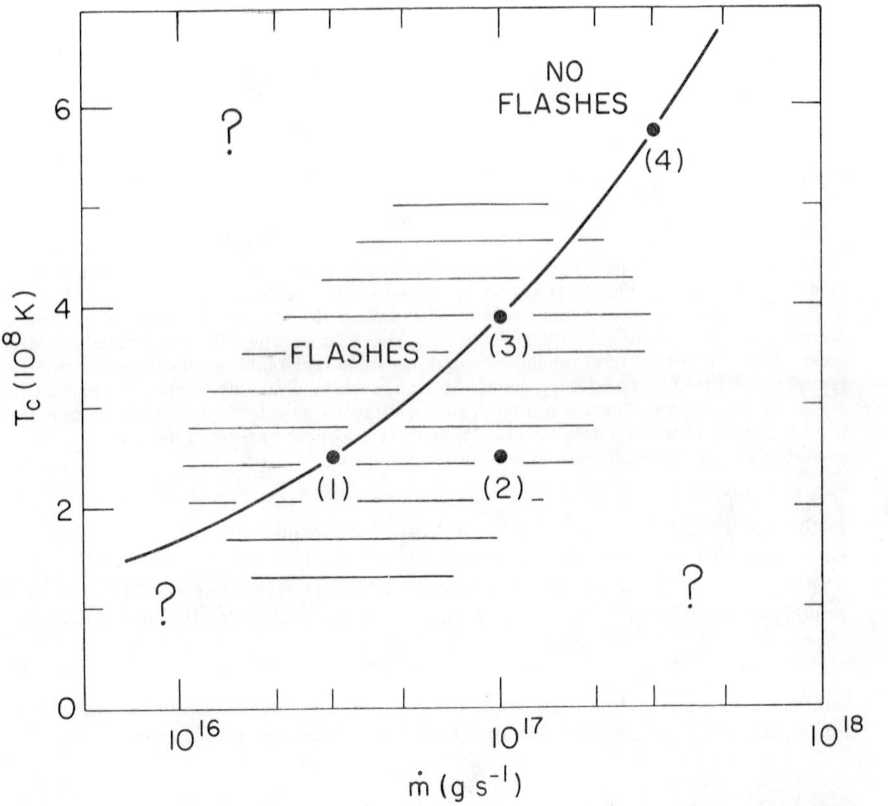

FIGURE 4. Mass accretion rates $\dot{m}$ and core temperatures T_c for the evolutionary models of the helium-burning shell by Joss.[27] Points (1) through (4) denote the parameter values used in four models. All models assumed a neutron-star mass of 1.4 $M_\odot$, a radius of 6.6 km, and no magnetic field. The solid curve passing through models (1), (3), and (4) is the estimated locus of parameter values for which the core of the neutron star is in thermal equilibrium (see text). Models (1), (2), and (3) all displayed thermonuclear flashes in the helium-burning shell; the hatched region denotes the range of parameter values for which flashes may be expected. The helium-burning shell in model (4) did not display flashes; the shell evidently becomes thermally stable at high values of T_c and $\dot{m}$, for reasons explained by Joss.[27] The behavior of the helium-burning shell at low values of T_c and $\dot{m}$ remains unexplored, but it is anticipated that flashing behavior will disappear at very low values of T_c and $\dot{m}$ ($T_c < 10^8$ K and $\dot{m} < 10^{15}$ g s^{-1}; reference 25). Also unexplored is the behavior of the helium-burning shell for parameter values far from the equilibrium curve (the upper left-hand and lower right-hand corners of the figure). A strong neutron-star magnetic field ($\gtrsim 10^{12}$ G) and the thermal effects of an outermost hydrogen-burning shell should both tend to suppress the region of flashing behavior to smaller values of $\dot{m}$ and T_c than indicated by these models[26-28] (see text). (From Joss.[16])

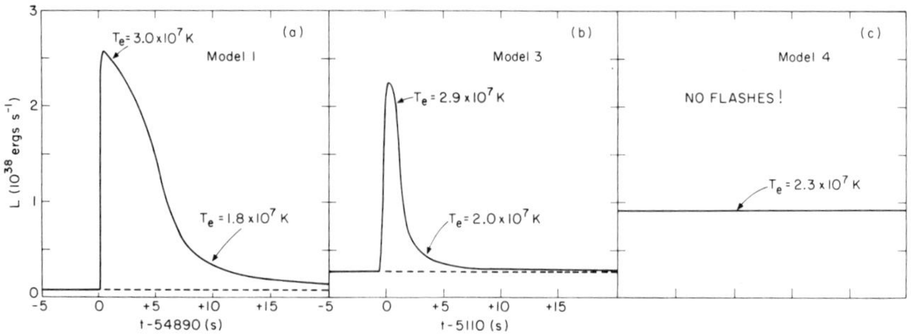

FIGURE 5. (a) The behavior of the surface luminosity L following a thermonuclear flash in model (1) by Joss[27] (see FIGURE 4). (b) Same for model (3). In each case, time $t = 0$ is at the start of accretion onto the neutron-star surface, the dashed line denotes the level of persistent accretion-driven luminosity, and the effective blackbody temperature (T_e) is indicated at a few points. The properties of these luminosity variations are in remarkably good agreement with the typical properties of observed x-ray bursts.[1,2] (c) The surface luminosity behavior for model (4). No flashes occur at the high core temperature and accretion rate of this model, so that the nuclear energy generation rate is roughly constant and never produces more than a small perturbation on the accretion-driven luminosity. (From Joss.[16])

of the helium-rich layer to reach the critical density and temperature for a flash to commence, which is roughly proportional to the surface area ($A = 4\pi R^2$) of the neutron star and the scale height in its surface layers. The scale height, in turn, is roughly inversely proportional to the surface gravity $g = GM/R^2$. It then follows that

$$\tau_R \sim Ag^{-1} \sim R^4 M^{-1}. \tag{1}$$

Similarly, the peak surface x-ray luminosity (L_{max}) following a flash just scales as the Eddington limit,[24] which is proportional to M but is independent of R; thus,

$$L_{max} \sim M. \tag{2}$$

Finally, the time scale (τ_D) of decline of surface x-ray emission following a flash is roughly directly proportional to the energy released in the flash, which is in turn directly proportional to τ_R and inversely proportional to the rate ($\sim L_{max}$) at which that energy escapes the star; thus,

$$\tau_D \sim \tau_R / L_{max} \sim R^4 M^{-2}. \tag{3}$$

Joss and Li[28] carried out numerical computations of the evolution of the helium-burning shell for neutron stars with a few different masses and radii (see TABLE 1 and FIGURE 8) and fitted their results for τ_R, L_{max}, and τ_B to power-law expressions in M and R. They obtained

$$\begin{aligned} \tau_R^{fit} &\sim R^{3.1} M^{-0.5}; \\ L_{max}^{fit} &\sim R^{0.1} M^{0.8}; \\ \tau_D^{fit} &\sim R^{3.1} M^{-1.4}. \end{aligned} \tag{4}$$

These expressions are in reasonably good agreement with Relations 1–3. The existence of these simple scaling laws suggests the intriguing possibility that once the physics of

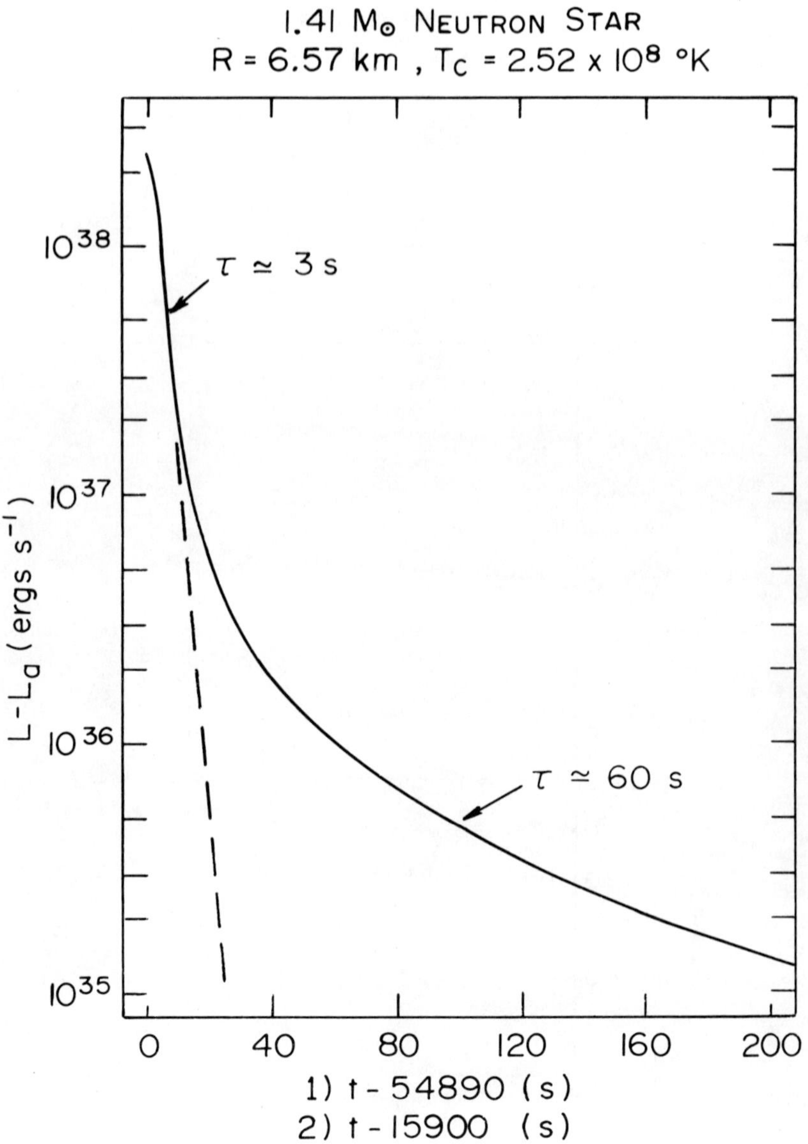

FIGURE 6. The decline from maximum x-ray luminosity in model (1) by Joss,[27] on time scales longer than those shown in FIGURE 5. L_a is the level of persistent accretion-driven luminosity from the neutron-star surface, so that $(L - L_a)$ is the excess luminosity due to the thermonuclear flash. The initial decline is well fitted by an exponential decay (dashed line) with a time constant of $\tau \simeq$ 3 s. However, for times greater than ~20 s after the burst peak, the decay time scale becomes much longer; 100 s after the peak, the local best-fit time constant is $\tau \simeq 60$ s. This "tail" of relatively soft x rays (blackbody temperature $\simeq 1.3 \times 10^7$ K) contains ~10% of the total burst emission. These properties are in good agreement with those of soft x-ray "tails" in many observed x-ray burst sources.[1,2] (After Joss.[27])

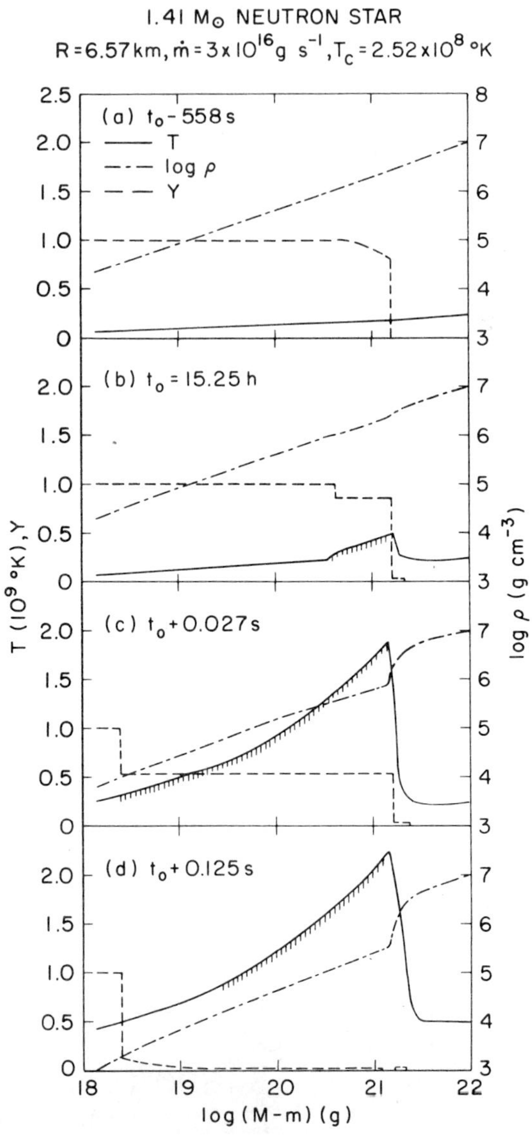

FIGURE 7. Structure of the surface layers of model (1) by Joss[27] prior to and during the first helium-burning flash, which begins near time t_0. M is the total mass of the neutron star and m (r) is the mass enclosed within a sphere of radius r, with $r = 0$ at the stellar center; thus $(M-m)$ is the total mass of the surface layers above level r. T is the temperature (left-hand scale), p the density (right-hand scale), and Y the fractional abundance of helium by mass (left-hand scale). The hatched regions indicate the extent of the convection zone generated by the flash. (a) Just prior to the flash; (b) near the start of the flash; (c) at the time when ~50% of the available fuel has been consumed; and (d) near the time of peak shell-burning temperature and peak surface luminosity. (From Joss.[27])

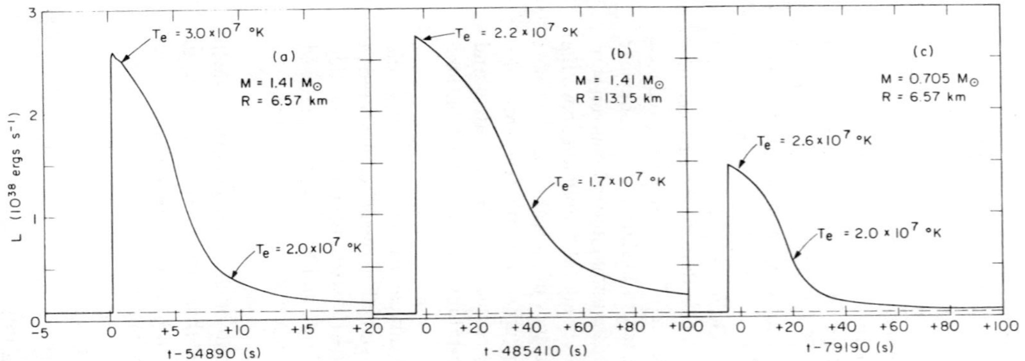

FIGURE 8. (a) Temporal evolution of the first x-ray burst from model (1) by Joss,[27] with neutron-star mass $M = 1.41$ $M_{\odot}$, radius $R =$ 6.57 km, core temperature $T_c = 2.52 \times 10^8$ K, accretion rate $\dot{m} = 3 \times 10^{16}$ g s^{-1}, and surface magnetic field $B = 0$. The notation is the same as in FIGURE 5. (b) Same as (a), but for a model by Joss and Li[28] that has the same parameters as model (1) except that $R = 13.15$ km. (c) Same as (a), but for a model[28] that has the same parameters as model (1) except that $M = 0.705$ $M_{\odot}$. Note the difference in the scale of time for (b) and (c) compared to that of (a). The differences among these three models can be largely explained by simple scaling arguments[28] (see text). (From Joss and Li[28].)

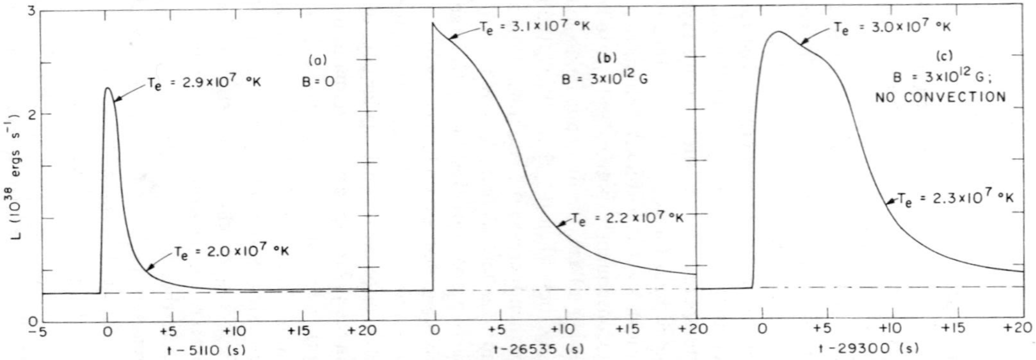

FIGURE 9. (a) Temporal evolution of the first x-ray burst from model (3) by Joss,[27] with $M = 1.41$ $M_{\odot}$, $R = 6.57$ km, $T_c = 3.87 \times 10^8$ K, $\dot{m} = 1 \times 10^{17}$ g s^{-1}, and $B = 0$. The notation is the same as in FIGURES 5 and 8. For $B \lesssim 1 \times 10^{12}$ G, the effect of the magnetic field upon the heat transport properties of the surface layers is unimportant, and the properties of the x-ray bursts are not significantly altered.[28] (b) Same as (a), but for a model by Joss and Li[28] that has the same parameters as model (3) except that $B = 3 \times 10^{12}$ G. The differences in the properties of the emitted x radiation from those of the $B = 0$ case can be understood in terms of the increased efficiency of radiative and conductive heat transport in the presence of the intense field (see text). The suppression of convection by the magnetic field was not taken into account for this model. (c) Same as (b) except that, in addition, it is (realistically) assumed that the strong magnetic field completely inhibits convection (see text). (From Joss and Li.[28])

neutron-star thermonuclear flashes is sufficiently well understood, it may be possible to deduce information on the masses and radii of neutron stars from the observed properties of x-ray bursts that result from such flashes.

Joss and Li[28] also computed the evolution of the helium-burning shell in models wherein the effects of an intense magnetic field B upon radiative, conductive, and convective heat transport were taken into account; the results of these calculations are summarized in TABLE 1 and FIGURE 9. The magnetic field was assumed to be vertical and uniform in strength across the entire surface of the neutron star; this is an adequate approximation for studying the effect of the field upon the heat transport properties of the surface layers. The radiative[31] and conductive[32] opacities are both appreciably suppressed by the field when $B \gtrsim 3 \times 10^{12}$ G. For fixed accretion rate and core temperature, the outer surface layers are then cooler than in the absence of a magnetic field (since the temperature gradient is flatter in response to the reduced opacities), so that a thicker helium-rich layer is needed to initiate a thermonuclear flash. Consequently, both τ_R and τ_D are larger than for an unmagnetized neutron star. Following a flash, the magnetic field should also strongly inhibit convection above the burning shell when $B \gtrsim 3 \times 10^{12}$ G; the reduced efficiency of outward heat transport would then increase the rise time of the surface x-ray emission from a few tenths of a second to a few seconds. For $B \gtrsim 1 \times 10^{13}$ G, the reduction of the radiative and conductive opacities is sufficiently large that the instability of the helium-burning shell is quenched and no flashes occur.

Joss and Li[28] argued that if the magnetic field is sufficiently strong to funnel the accretion onto the magnetic polar caps of the neutron star, then the effective accretion rate in the polar cap regions is enhanced by a factor of $\sim 10^2$ (for a fixed total accretion rate $\dot{m}$) and the instability of the nuclear burning shells should be strongly suppressed (see also Taam and Picklum[26] and Joss[27]). The value of B required to funnel the accretion is not well determined, but available estimates[33] yield $B \sim 10^{12}$ G.

Taam and Picklum[26] have studied the thermal evolution of a carbon-burning shell in an accreting neutron star but have not yet carried their computations through a complete thermonuclear flash. Their models assume low core temperatures for the neutron star ($\lesssim 10^7$ K); such cores would not be in thermal equilibrium with the surface nuclear-burning shells, but they might represent the properties of an older neutron star that has begun to accrete matter during the past $\sim 10^3$ yr.

COMPARISON WITH OBSERVATIONAL PHENOMENA

X-Ray Burst Sources

The results of the numerical calculations of helium-burning flashes (TABLE 1 and FIGURES 4–9) strongly support the conjecture that the bursts from most observed x-ray burst sources result from such flashes. In particular, the typical burst rise times, decay time scales, peak luminosities, total emitted energies, spectral properties, low-energy "tails," and recurrence intervals[1,2] are reproduced remarkably well by such flashes.

However, there are some difficulties with this model. It does not predict the observed complex burst structures and recurrence patterns, which vary from one burst source to another and often vary with time in a single source. It is quite possible that these complexities will be better understood when some of the approximations of the present model calculations are relaxed; I shall return to this point in my concluding remarks.

Another embarrassment for the helium-flash model is the ratios α of time-averaged persistent x-ray luminosity to time-averaged burst luminosity from the observed burst sources. In this model, α should just be the ratio of the gravitational energy released by accretion (more or less continuously) to the nuclear energy released in the flashes. The numerical value of α should thus be[24,25]

$$\alpha \simeq 1 \times 10^2 \left(\frac{M}{M_\odot}\right)\left(\frac{R}{10\ \text{km}}\right)^{-1}. \qquad \textbf{(5)}$$

However, some burst sources have reported[1] values of α significantly less than 1×10^2; the worst case is MXB 1659 − 29, which was observed to have an *upper limit* of $\alpha < 25$ during one of its burst-active states.[2] Moreover, in some sources the recurrence intervals have been observed to increase (decrease) when the persistent luminosity increased (decreased), which is opposite to the trend expected from this model.

If the helium-flash model is correct, the smaller values of α might simply be telling us that some nuclear-flashing neutron stars have masses moderately lower than 1 $M_\odot$ and radii moderately larger than 10 km. (However, this explanation seems to fail for MXB 1659 − 29, which appears to have the smallest effective radius among the well-observed sources.)[3,4] Alternatively, it is possible that the accretion-driven luminosity is emitted anisotropically and/or largely shifted to photon energies $\lesssim 1$ keV; Milgrom[35] has proposed a specific mechanism (an accretion disk that extends all the way to the surface of a weakly magnetized neutron star) that could produce both these effects and thereby reduce the observed values of α by up to a factor of ~ 2. It has also been suggested[25] that the observed values of α might be reduced by the storage of nuclear fuel during burst-inactive phases; this idea is now rendered less likely by the results of the evolutionary computations, which suggest that each flash is sufficiently strong to consume virtually all the available fuel.

The Rapid Burster

It is important to realize that bursts from the "rapid burster," MXB 1730 − 335, are at present unique and almost certainly cannot be the result of thermonuclear flashes. The recurrence intervals between bursts are $\sim 10^{1-2}$ s, and $\alpha \lesssim 2$ for this source.[36] Hoffman, Marshall, and Lewin[37] have described these bursts as "Type II" and the bursts from other sources as "Type I." However, Hoffman, Marshall, and Lewin also found that the rapid burster occasionally emits "special" bursts whose properties much more closely resemble the Type I bursts from other sources. Moreover, the ratio of the time-averaged luminosity in the Type II bursts from the rapid burster to that in its Type I bursts is $\sim 10^2$. Hoffman, Marshall, and Lewin made the intriguing speculation that the Type I bursts from MXB 1730 − 335 are the result of thermonuclear flashes on an accreting neutron star, while the Type II bursts are the result of an unstable accretion flow onto the same object.

X-Ray Pulsars

Let us accept, for the sake of discussion, that most observed x-ray burst sources can, in fact, be understood as helium-burning flashes on accreting neutron stars. Since x-ray pulsars are also widely believed to be accreting neutron stars, it is then puzzling, at first sight, that these objects do not also display bursting behavior. However, the strong magnetic field ($\gtrsim 10^{12}$ G; reference 33) that funnels the accretion onto the magnetic polar caps of an x-ray pulsar will also enhance the efficiency of radiative and conductive heat transport above its nuclear burning shells.[28] Moreover, the heat released by accretion and hydrogen burning will have a much greater influence upon the inner burning shells if the freshly accreted matter is confined to the polar caps, rather than spread uniformly over the neutron-star surface.[26-28] These effects should strongly tend to stabilize the nuclear burning shells of an x-ray pulsar against thermonuclear flashes[26–28] (see the section entitled Numerical Models above).

With this picture, we can also understand why the persistent x-ray flux from x-ray burst sources is unpulsed: those neutron stars whose magnetic fields are too weak to funnel the accreting matter should be precisely those that can undergo thermonuclear flashes.[26–28,33] If the magnetic field was originally as strong as in an x-ray pulsar but has since decayed, then the neutron star must be quite old (probably older than 10^7 yr; references 38 and 39). The lack of x-ray eclipses in burst sources may also reflect membership in relatively old binary systems,[40] or it may result from x-ray beaming effects that set in after the neutron-star magnetic field has decayed.[35] The concentration of x-ray burst sources in the direction of the galactic center and the identification of a few of them with globular clusters[1,23] may well be other manifestations of membership in an older galactic population than the x-ray pulsars, which are distributed through the disk of the galaxy and whose binary companion stars are usually of early spectral type.

Gamma-Ray Burst Sources

It has been suggested[12] that carbon-burning flashes on accreting neutron stars result in cosmic γ-ray bursts. This idea now seems less plausible; a more detailed look at the heat transport properties of the neutron-star surface layers demonstrates that the rising convective elements above any nuclear burning shell would cool to temperatures $\ll 10^{10}$ K before reaching the photosphere. More generally, the short ($\lesssim 10^5$ s) recurrence intervals between x-ray bursts, together with the lack of any observed recurrrence in γ-ray bursts, tends to rule against any direct relation between these two phenomena.[2]

Transient X-Ray Sources

The recent calculations by Taam and Picklum,[26] as well as earlier work,[12,25] indicate that up to $\sim 10^{26}$ g of carbon could be stored on a neutron star prior to the ignition of a carbon flash. The available nuclear energy ($\sim 10^{45}$ ergs) is then far in excess of the amount released in an x-ray burst; in fact, this energy is comparable to the amount released by the spectrally "soft" transient x-ray sources, such as

A0620–00 (reference 41). Thus carbon flashes on an accreting neutron star might be a tenable model for such sources; this suggestion is a variant on the original idea put forth by Van Horn and Hansen[22] (see also the discussion by Matilsky *et al.*[41]). Complete evolutionary computations of carbon flashes are highly desirable to further investigate this possibility.

Concluding Remarks

We have just begun to grasp all of the intricacies of nuclear processes in the surface layers of accreting neutron stars. The theoretical problems are fascinating, not only as investigations in fundamental physics, but also for their possible applications to x-ray burst sources and other observational phenomena.

The only evolutionary computations of neutron-star thermonuclear flashes that have been completed to date[27,28] relied on a number of simplifying assumptions and approximations. These included (1) the assumption of spherically symmetric accretion, (2) the assumption that the neutron-star core is in thermal equilibrium, and (3) the neglect of all nuclear-burning shells except the helium-burning shell. These approximations will have to be relaxed in future studies before this phenomenon and its observational implications can be more fully understood.

Small but significant violations of spherical symmetry might result from accretion through a relatively weak ($\lesssim 10^{11}$ G) magnetic field or the residual angular momentum of the accreting matter. If thermonuclear flashes result in x-ray bursts, such violations could be the key to the observed complexities in burst structure and recurrence patterns.[1,2] For example, a thermonuclear flash that ignites on one portion of the neutron-star surface may propagate around the star, in a pattern that varies from flash to flash and from one star to another.[27] A thorough investigation of such possibilities will eventually require two- or three-dimensional numerical computations, which will be much more difficult than the computations of spherically symmetric models that have been attempted to date.

The importance of thermal equilibrium of the neutron-star core has been only tentatively explored[27] (see Figure 4). If there is a change in the average accretion rate, the thermal inertia of the core is sufficient to require the elapse of $\sim 10^3$ yr for thermal equilibrium to be reestablished.[25] Thus departures from thermal equilibrium are entirely possible, and they could have a substantial effect upon the behavior of the nuclear burning shells.

The outer hydrogen-burning shell, while almost certainly not the seat of significant thermonuclear flashes, may nonetheless play an important role in determining the behavior of the helium-burning shell. As Taam and Picklum[26] have pointed out, the decay time scales of the beta-unstable nuclei in the CNO cycle limit the hydrogen-burning rate to such an extent that, under some circumstances, the hydrogen-burning shell should be forced inward until it overlaps the helium-burning shell. This effect could significantly complicate the relevant nuclear physics and influence the stability of the two interacting shells.[26] The possible role of a carbon-burning shell interior to the helium-burning shell also remains largely unexplored.

The unique properties of the "rapid burster" suggest the fascinating possibility that its Type I bursts result from thermonuclear flashes on an accreting neutron star, while its (rapid) Type II bursts are caused by an unstable accretion flow onto the same object[37] (see the section entitled The Rapid Burster above). If this picture is correct, two fundamental and difficult questions are raised: (1) Which one among the various accretion instabilities that have been proposed[3–11] (or yet to be proposed!) is responsible for the Type II bursts? (2) Why is this mechanism operative in the "rapid burster" but not in any other known x-ray burst source or x-ray pulsar?

Many of these issues can be attacked by additional calculations in the immedicate future. Further progress, and more surprises, are bound to be forthcoming.

References

1. Lewin, W. H. G. 1979. *In* Advances in Space Exploration, Vol. 3, X-Ray Astronomy. W. A. Baity and L. E. Peterson, Eds.: 133. Pergamon Press, Oxford, England; and earlier reviews cited therein.
2. Lewin, W. H. G. & P. C. Joss. 1977. Nature (London) **270:** 211.
3. Svestka, J. 1976. Astrophys. Space Sci. **45:** 21.
4. Henriksen, R. N. 1976. Astrophys. J. **210:** L19.
5. Liang, E. P. T. 1977. Astrophys. J. **211:** L67; Astrophys. J. **218:** 243.
6. Baan, W. 1977. Astrophys. J. **214:** 245.
7. Wheeler, J. C. 1977. Astrophys. J. **214:** 560.
8. Lamb, F. K., A. C. Fabian, J. E. Pringle & D. Q. Lamb. 1977. Astrophys. J. **217:** 197.
9. Grindlay, J. E. 1978. Astrophys. J. **221:** 234.
10. Neugebauer, M. & B. T. Tsurutani. 1978. Astrophys. J. **226:** 494.
11. Apparao, K. M. V. & S. M. Chitre. Submitted to Astrophys. Space Sci.
12. Woosley, S. E. & R. E. Taam. 1976. Nature (London) **263:** 101.
13. Maraschi, L. & A. Cavaliere. 1977. *In* Highlights of Astronomy, Vol. 4. E. A. Müller, Ed. Part I: 127. D. Reidel Publishing Co. Dordrecht, the Netherlands.
14. Lamb, D. Q. & F. K. Lamb. 1977. Ann. N.Y. Acad. Sci. **302:** 261.
15. Cowie, L. L., J. P. Ostriker & A. A. Stark. 1978. Astrophys. J. **226:** 1041.
16. Joss, P. C. 1979. Comments on Astrophys. **8:** 109.
17. Giacconi, R., H. Gursky, E. Kellogg, E. Schreier & H. Tananbaum. 1971. Astrophys. J. **167:** L67.
18. Schreier, E., R. Levinson, H. Gursky, E. Kellogg, H. Tananbaum & R. Giacconi. 1972. Astrophys. J. **172:** L79.
19. Tananbaum, H., H. Gursky, E. M. Kellogg, R. Levinson, E. Schreier & R. Giacconi. 1972. Astrophys. J. **174:** L143.
20. Rosenbluth, M. N., M. Ruderman, F. Dyson, J. N. Bahcall, J. Shaham & J. Ostriker. 1973. Astrophys. J. **184:** 907.
21. Hansen, C. J. & H. M. Van Horn. 1975. Astrophys. J. **195:** 735.
22. Van Horn, H. M. & C. J. Hansen. 1974. Astrophys. J. **191:** 479.
23. Grindlay, J., H. Gursky, H. Schnopper, D. R. Parsignault, J. Heise, A. C. Brinkman & J. Schrijver. 1976. Astrophys. J. **205:** L127.
24. Joss, P. C. 1977. Nature (London) **270:** 310.
25. Lamb, D. Q. & F. K. Lamb. 1978. Astrophys J. **220:** 291.
26. Taam, R. E. & R. E. Picklum. 1978. Astrophys. J. **224:** 210.
27. Joss, P. C. 1978. Astrophys. J. **225:** L123.
28. Joss, P. C. & F. K. Li. 1979. Astrophys. J. In press.
29. Schwarzschild, M. & R. Härm. 1965. Astrophys. J. **142:** 855.
30. Giannone, P. & A. Weigert. 1967. Z. Astrophys. **67:** 41.
31. Lodenquai, J., V. Canuto, M. Ruderman & S. Tsuruta. 1974. Astrophys. J. **190:** 141.
32. Canuto, V. 1970. Astrophys. J. **159:** 641.
33. Arons, J. & S. M. Lea. Submitted to Astrophys. J.
34. van Paradijs, J. 1978. Nature (London) **274:** 650.
35. Milgrom, M. 1978. Astron. Astrophys. **67:** L25.

36. Lewin, W. H. G., J. Doty, G. W. Clark, S. A. Rappaport, H. V. D. Bradt, R. Doxsey, D. R. Hearn, J. A. Hoffman, J. G. Jernigan, F. K. Li, W. Mayer, J. McClintock, F. Primini & J. Richardson. 1976. Astrophys. J. **207:** L95.
37. Hoffman, J. A., H. L. Marshall & W. H. G. Lewin. 1978. Nature (London) **271:** 630.
38. Ruderman, M. 1972. Ann. Rev. Astron. Astrophys. **10:** 427; and references therein.
39. Flowers, E. & M. Ruderman. 1977. Astrophys. J. **215:** 302.
40. Joss, P. C. & S. A. Rappaport. 1979. Astron. Astrophys. **71:** 217.
41. Matilsky, T., H. Bradt, J. Buff, G. W. Clark, G. Jernigan, P. C. Joss, B. Laufer, J. McClintock & D. Zubrod. 1976. Astrophys. J. **210:** L127; and references therein.
42. Lewin, W. H. G. & G. W. Clark. Ann. N.Y. Acad. Sci. This volume.

EXTRAGALACTIC X-RAY SOURCES

K. A. Pounds

X-Ray Astronomy Group
University of Leicester
Leicester LE1 7RH, England

A. C. Fabian

Institute of Astronomy
University of Cambridge
Cambridge CB3 0HA, England

INTRODUCTION

The first complete surveys of the extragalactic x-ray sky ($|b| > 10°$) are now available from the Uhuru and Ariel V satellites, in the form of the 4U[1] and 2A[2] catalogues. Both surveys are based on data obtained from scanning, fan-beam proportional counter systems, sensitive over the energy band 2–20 keV, and both reached a limiting source detection of ~1.5 Uhuru ct s^{-1} (~0.5 Ariel counts s^{-1}, ~2.5 $\times 10^{-11}$ erg cm^{-2} s^{-1} or 2μJy at ~10^{18} Hz). Comparison of 2A and 4U source positions with optical and radio catalogues has led to the establishment of two major classes of extragalactic x-ray emitter, namely, galaxy clusters and active galaxies. Some fifty examples of each class are now known and additional observations, from both satellite and rocket experiments, have contributed to a growing wealth of information on the spectra, spatial distribution, and time variability of the x-ray emission.

Extension of the Ariel V survey (in the forthcoming 3A catalogue) and the new observations from the larger, scanning experiments on the HEAO-1 spacecraft, will further increase the numbers of known extragalactic sources, but a *substantially* deeper all sky survey, demanding high angular resolution as well as detection sensitivity, to avoid source confusion, appears unlikely before the mid-1980s. The present symposium seems, then, an opportune moment for an overview of the extragalactic x-ray sky, particularly on the threshold of the HEAO-2 (Einstein Observatory) programme, which will provide much more detailed observations of the classes of sources discussed herein as well as others, yet unknown.

X-RAY EMISSION FROM CLUSTERS OF GALAXIES

Identification of X-Ray Clusters

TABLE 1 lists 41 rich clusters of galaxies that are reliably identified with 2A or 4U sources. They are all of Abell distance class ≤4, or the equivalent for the southern clusters, with the exception of A1146 (D=5, R=4, BM I, 2A error box 0.37 deg^2), A2204 (D=5, R=3, BM II, revised error box 0.6 deg^2), and SC 0316−44 (D=5, R=1, BM I, 2A error box 0.2 deg^2). The richness and centrally condensed nature of these three exceptions provides strong support for their inclusion in the light of the subsequent discussion of the properties of x-ray clusters. The random occurrence of a

0077–8923/80/0336–0496 $01.75/1 © 1980, NYAS

cluster of $D \leq 4$ in *any* x-ray error box in the 2A catalogue (total area for 107 sources $\sim$15 deg^2) is only 0.3, indicating few, if any, chance coincidences. A second requirement for a source to be listed in TABLE 1 is that the 2A x-ray flux should be compatible with a nonvarying source. This criterion has ruled out two rich southern clusters, CA 0340−538 (D=3, R=2), and Sersic 40/6 (D=3, R=3), both of which appear good on positional grounds alone. Possibly a field source or an active galaxy in each of these clusters is contributing a substantial fraction of the recorded x-ray flux. The distribution of clusters in TABLE 1 shows 16 clusters now identified with 3U sources, including several confirmed by later, improved positions, and 17 x-ray clusters first reported in the 2A catalogue. More recent additions to the list have come from the 4U

TABLE 1

X-RAY SOURCES IDENTIFIED WITH CLUSTERS OF GALAXIES

Catalogue	Number	Sources
3U catalogue identifications	11	Virgo, Coma, Perseus, Centaurus, Cyg A, 3C 129, A 262, 401, 1367, 2256, 2319
3 U identifications confirmed by 2A position	5	A478, 754, 1060,* 2147, 2199
Kellogg list confirmed by 2A position	1	A 496
New identifications in 2A catalogue†	17	A 85, 119, 133, 644, 1146, 1795, 1904, 2029, 2142, 2312, 2440, 2589, SC 0316−44, 0626−54, 1344−32, 2009−57, K1 44
Post 2A catalogue identifications		
(i) SSI observations‡	3	A 999, 1644, 2204
(ii) 4U catalogue§	4	A 539, 2065, 2657, 2666
Total =	41	X-ray clusters identified with SSI and Uhuru sources

*First confirmed by Ariel V CPC expt. (Ives and Sanford)[84]

†Exclude SC 1251−29 and SC 1326−31, now re-identified with EX Hya (Watson *et al*)[9] and MCG-6-30-15[85]

‡McHardy[7] and Ricketts[4]

§Forman *et al*[1]

catalogue[3] and from a search of a wide list of clusters[4] using the whole Ariel V data base and a point-summation technique (PST), in which some additional sensitivity is achieved by summing all data regardless of scan position angle about a chosen point in the sky. In total 41 x-ray clusters are listed, 38 of which are in common with the list (also based on 2A and 4U) recently published by Jones and Forman.[3]

Cluster Morphology

Several authors have discussed the correlation of detected x-ray emission with the morphology of a cluster of galaxies. The most recent and careful analyses have been by Bahcall[5,6] and McHardy[7] using 2A catalogue sources, and by Jones and Forman[3] using

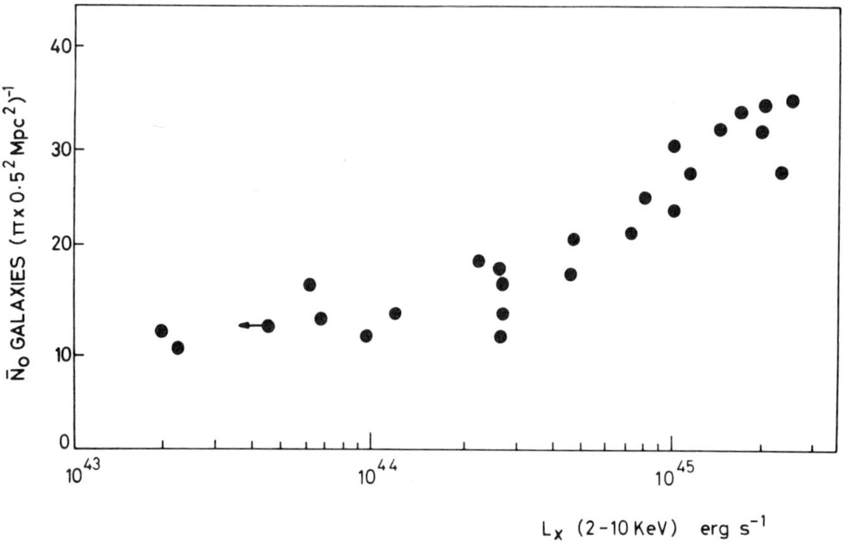

FIGURE 1. Central galaxy density $\overline{N}_0$ within 0.5 Mpc radius of the cluster center plotted against L_x for all the x-ray clusters of TABLE 1 with measured red shifts (after Bahcall[5]).

2A and 4U identifications. Bahcall and McHardy both find a correlation of x-ray luminosity with cluster richness, enhanced if a dominant central galaxy is present, and with a scarcity of spirals in the cluster. McHardy finds a positive correlation of x-ray emission with richness *and* with Bautz–Morgan type, while Jones and Forman believe cluster richness is the dominating factor. Bahcall argues persuasively that the "central galaxy density" is a better measure than cluster richness, being more closely related to the core region of a cluster, where most of the x-radiation arises.

In FIGURE 1 (after Bahcall[5]) we show the x-ray luminosities L_x of the x-ray clusters listed in TABLE 1 for which the red shifts are known, plotted against the central galaxy density $\overline{N}_0$ within 0.5 Mpc of the cluster center. For greater consistency the x-ray luminosities are all derived from 2A catalogue fluxes. Flux values in the 4U catalogue are averages of positive detections only and thus overstate fluxes, particularly of the fainter sources. The correlation of L_x with $\overline{N}_0$ is good over the range of two orders in luminosity and the scatter is small, particularly since all Bautz–Morgan (and Rood–Sastry) types of clusters have been plotted together. Conversely the small scatter demonstrates the *major* dependence of L_x is on $\overline{N}_0$. This is, of course, entirely consistent with the now favored model for cluster x-ray emission as thermal bremsstrahlung from a hot intracluster gas, the central density of which is proportional to the central density of galaxies.

A second impressive correlation is that of cluster x-ray luminosity with the fraction of galaxies F(Sp) which are spirals (FIGURE 2). Again the values of L_x are from 2A fluxes, while measurements of F(Sp) are taken from Bahcall,[6] McHardy[7] and—for several rich southern clusters—from Tytler and Vidal.[8] The latter authors pointed out the exceptionally high F(Sp) of SC 1252−28 for its reported x-ray intensity, and it is interesting to note that the corresponding 2A source has now been reidentified[9] with the dwarf nova EX Hydrae. The clear relation between spiral fraction and x-ray luminosity is interesting in that it is perhaps the clearest evidence yet that the evolution of galaxies is affected by their membership of a rich cluster.

The X-Ray Luminosity Function of Clusters

Using a statistically complete sample of x-ray clusters (Abell + Virgo), McHardy[7] has calculated the x-ray luminosity function of clusters, by the standard maximum volume method and the known 2A catalogue sky coverage correction. The resulting luminosity function is shown in FIGURE 3. The upper limit at the high-luminosity end of the function corresponds to the fact that no sources of power greater than $10^{45.75}$ erg s^{-1} are seen with a flux greater than 1 Ariel counts s^{-1}; the survey is complete to this limit over 90% of the sky. (It is not clear that the luminosity function actually truncates here, since at their implied densities, any cluster is likely to be at a significant red shift, with the result that cosmological effects will have reduced their x-ray fluxes below the Ariel V detection limit[10]). It can be seen that the total density of x-ray clusters approaches that of all Abell clusters, ρ_A, at a luminosity of $\sim 10^{43}$ erg s^{-1}. The slight excess above ρ_A is caused by the fact that the maximum accessible volume of space in which sources of low luminosity can be seen is small compared with the

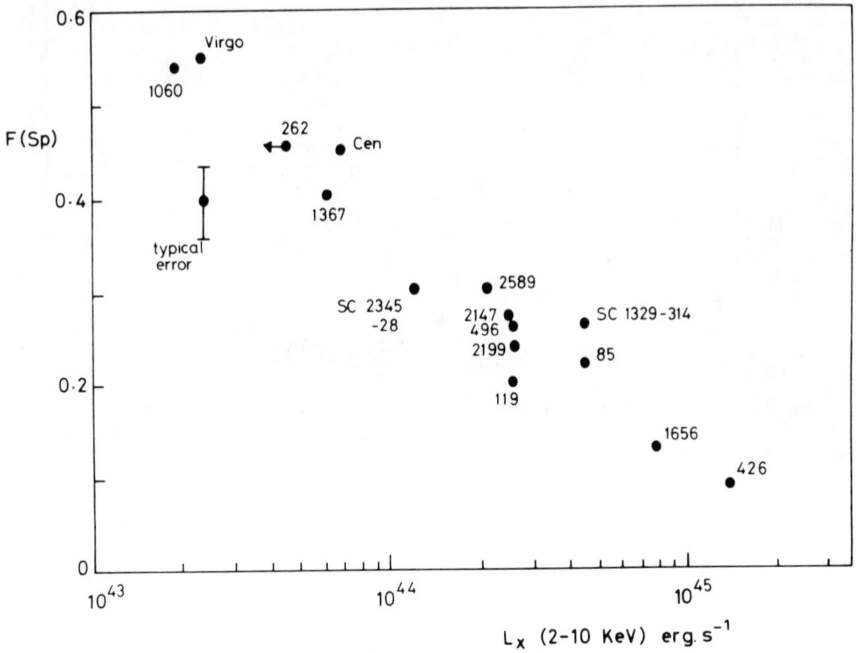

FIGURE 2. Fraction of spiral galaxies F(Sp) plotted against L_x for the clusters listed in TABLE 1. $H_0 = 50$ km s^{-1} Mpc^{-1} as throughout paper.

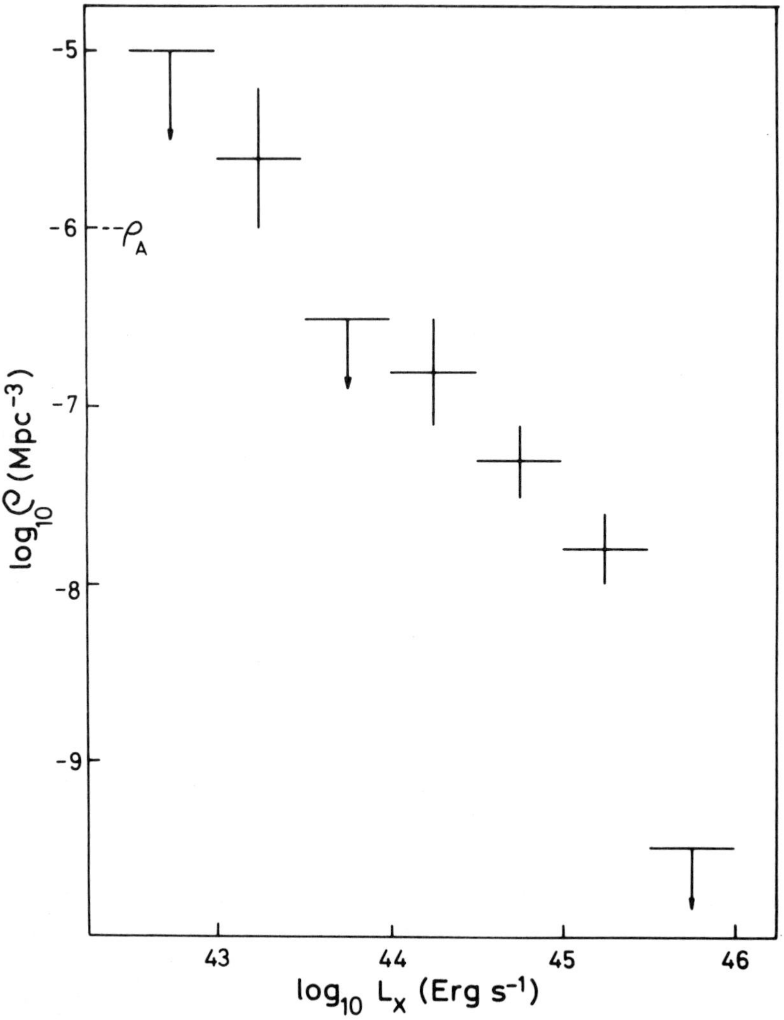

FIGURE 3. X-ray luminosity function for clusters of galaxies (after McHardy[7]). Only Abell clusters have been included for completeness and the dashed line indicates their nominal space density of 10^{-6} Mpc^{-3}.

inverse of the space density of such sources. It is ,therefore, predicted that all clusters will emit x-rays at a luminosity greater than 10^{43} erg s^{-1} in the 2–10-keV range, so that no distinction should be made between x-ray and non–x-ray clusters in the future. The only allowable distinction is between powerful and weak x-ray sources.

The contribution of clusters to the x-ray diffuse background, calculated from the luminosity function of FIGURE 3 (i.e., neglecting evolutionary effects) is ~7% at 2–10keV.[83] However, it has also been noted[10] that rich clusters of galaxies are not numerous enough to produce *all* of the x-ray background.

X-Ray Spectra of Clusters

Remarkably, considering the low spectral resolving power of current proportional counter detectors, the improved statistical precision provided by nonscanning collimated detectors on Ariel V and OSO-8 has yielded x-ray spectra of considerable significance in the study of x-ray clusters. The breakthrough came with the MSSL Ariel V measurement of the Perseus cluster in 1975, showing clear evidence for a monoenergetic feature near 7 keV.[11] The interpretation of this feature (FIGURE 4) as

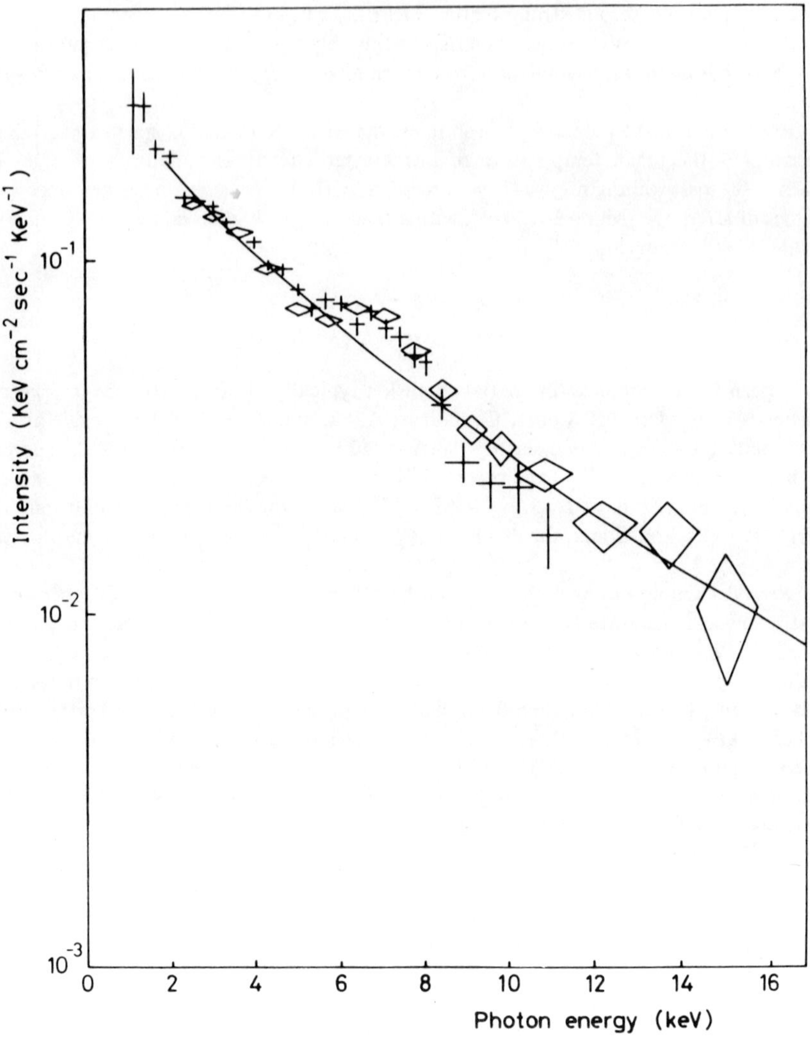

FIGURE 4. Ariel V Collimated Proportional Counter (CPC) spectrum of the Perseus cluster from 2–15 keV. The solid line is the best fit continuum spectrum ($kT \sim 6$ keV), above which the feature attributed to iron line emission near 7 keV may be clearly seen.

thermal line emission from highly stripped iron nuclei (Fe XXIV–VI) provided, at once, the strongest evidence for hot plasma ($T \sim 10^7$–10^8 K) in the cluster, together with a unique determination of the abundance of iron in the intracluster medium. The MSSL and GSFC groups have since found evidence for similar iron line emission from several other clusters of galaxies (Coma, Centaurus, Virgo, and SC 0626−54 are the best established) while upper limits from the spectra of fainter x-ray clusters are all generally consistent with a similar iron abundance, of about 0.5 that of the cosmic values.[12]

Recent soft x-ray spectra from HEAO–1[13] and rocket data from the CFA group[14] have both shown a second strong feature, near 1.1 keV, in the x-ray spectrum of M87 which dominates the Virgo cluster emission at this energy. The initial interpretation of this feature is again as line emission from thermalized iron, this time from electrons in the *L* shell.

In addition to the detection of iron lines, the MSSL and GSFC spectra provide a measure of the mean temperature of intracluster gas. In the majority of cases a thermal bremsstrahlung fit gives best agreement with the measured 2–20-keV spectra. Values of kT range from ~2–12 keV, with a tendency for higher values for clusters of greater x-ray luminosity.[12,15]

Structure of Nearby Rich Clusters

Uhuru found evidence for spatial extension typically of 30–50 arc min in 6 x-ray clusters (Virgo, Perseus, Coma, Centaurus, A262, and A2256).[16] This result needs confirmation for the last two clusters, since A262 is very faint (the Ariel V sky survey 3σ upper limit is only 0.4 of the reported 3U intensity) while the linear size of x-ray emission from A2256, at $D=3$, is ~4 M pc. However, for the nearby rich clusters in Virgo, Perseus, and Coma, the extension is in no doubt and recently crude maps have been obtained.

Perseus. Observations with the small field, nonimaging telescope on Copernicus first resolved a separate component of x-radiation from the massive central galaxy NGC 1275,[17] together with the core of the cluster emission. The Berkeley and Columbia groups made several scans across Perseus with narrow fan-beam detectors and confirmed discrete emission from NGC 1275 (25 ± 10% of the total Perseus flux at 0.5–3 keV) together with an extended source lying along the well-known line of galaxies, from NGC 1275 to IC 310.[18] The most recent and complete map to date is reproduced in FIGURE 5. This is from a 120-s exposure with a Kirkpatrick–Baez imaging telescope, flown in December 1975 on an Aerobee 350 rocket by the CFA group.[19] The energy range is 0.5–1.5 keV and the pixels of 1.4 × 1.4 arc min (although the telescope resolution was ~4 arc min). A bright, unresolved source accounting for 25% of the total flux is found to be coincident with NGC 1275 (within the rocket aspect uncertainty of 3 arc min), superimposed on a diffuse emission extending over the whole field (1° in diameter). No discrete emission from NGC 1265 is seen and no evidence is found for the enhanced emission along the line of galaxies. The most straightforward interpretation of FIGURE 5 is that the Perseus cluster emission is symmetric about the cluster center with a measured "core radius" of the hot gas ~17 arc min. From observations with the SAS-3 instruments, Ulmer and Jernigan[20] find a similar value at 1.5–5.5 keV, increasing to 40^{+6}_{-16} arc min at 5.5–15.5 keV.

Virgo. It was never clear from the Uhuru data how much, if any, of the extended x-ray emission was associated with the cluster rather than with the giant elliptical galaxy M87. This question has recently been clarified by Lawrence[21] using Ariel V SSI data. He finds evidence for two components at 2–20 keV, one being centered on M87 and the other ~0.5° to the NW, close to the cluster center as defined by the distribution of elliptical galaxies.[22] FIGURE 6 shows the resulting SSI map of the Virgo

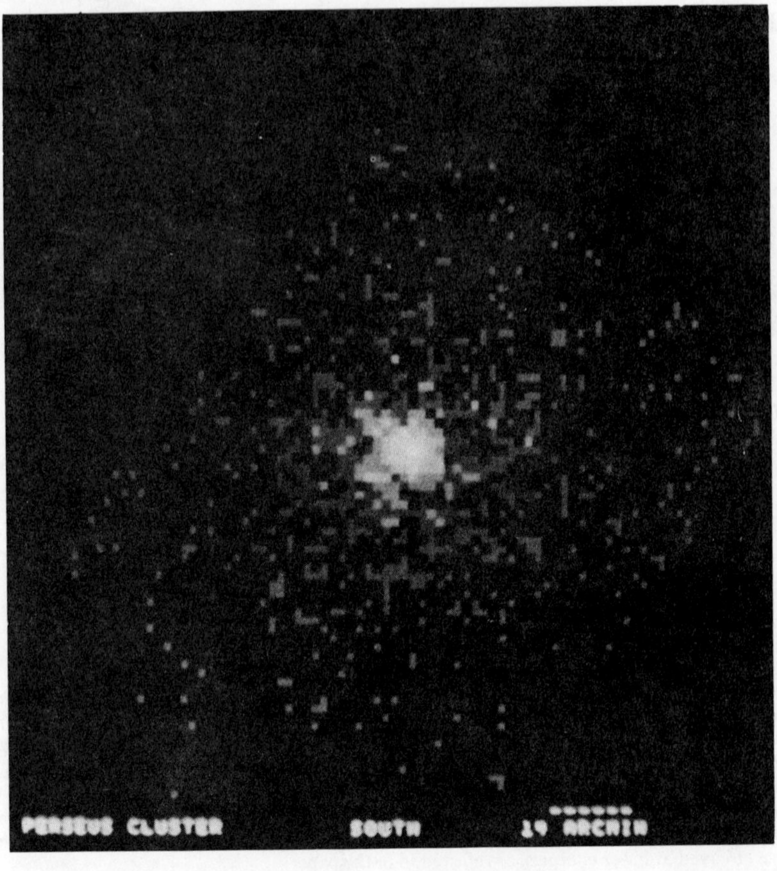

FIGURE 5. X-ray image of the Perseus cluster (0.5–1.5 keV) in bins of 1.4 × 1.4 arc min. The bright spot coincides with NGC 1275. Data points are quantized in 5 grey levels of 1–2, 3–6, 7–14, 15–30, and 42 counts, the total image containing 1200 counts in all, only ~50 of which are background events.

region, with a source (≤ 1° in diameter) centered on M87 and an extended source associated with the cluster as a whole. Independent support for this conclusion is provided by spectral studies of Virgo with the Ariel V CPC. Davison[23] finds a best fit spectrum with two temperature components of 1 and 7 keV, with the harder emission arising from a region to the NW of M87.

An x-ray image of Virgo, taken with the CFA Kirkpatrick–Baez telescope, shows[24] an extended, diffuse emission at 0.15–1.5 keV, centered on M87 with a "core radius" of ~15 arc min. Thus it seems that the Virgo region contains a hard, extended source associated with the cluster as a whole, together with a softer source, also extended, centered on M87.

Coma. This cluster, the archetypal evolved cluster, differs from Perseus and Virgo in having no active central galaxy. Copernicus observations[25] indicated a 0.5–1.5 keV "core radius" of 40 arc min, slightly larger than the Uhuru 2–6-keV value. A recent map obtained with the CFA rocket-borne telescope[26] shows a featureless emission,

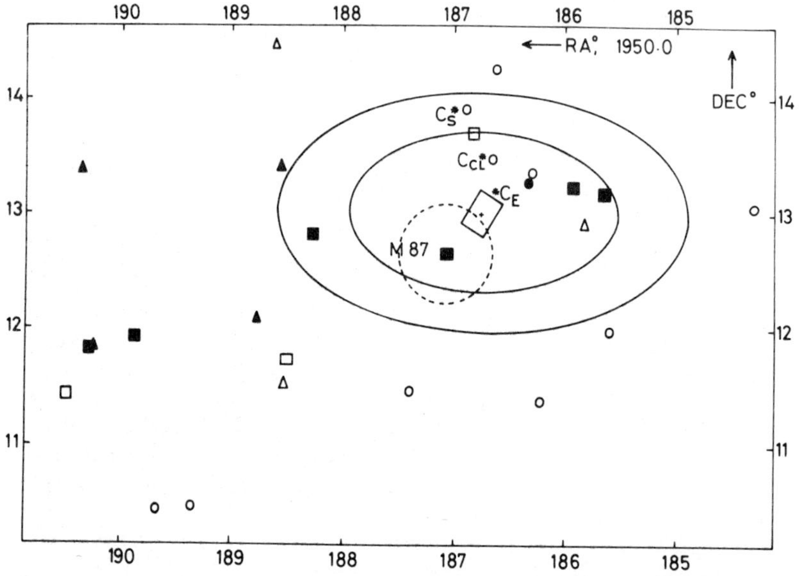

FIGURE 6. Ariel V map of the Virgo cluster at 2–10 keV. The rectangle is the 90% confidence error box for the centroid of the extended (cluster) source, the three stars marking the center of the cluster of determined from C_E (elliptical galaxies), C_s (spiral galaxies), and C_{cl} (all galaxies in the cluster). The main galaxies in the central cluster region are also shown; ellipticals as squares, spirals as triangles, and lenticulars as circles. The inner and outer ellipses represent the 50% and 20% peak power contours of the extended cluster x-ray source. The dashed circle indicates the extent of the optical halo of M87 which is also approximately the size of the x-ray source observed at lower energies, as referred to in the paper.

well-fitted by an isothermal gas sphere model with no evident emission from the prominent galaxies NGC 4874 and NGC 4869.

General Discussion of Cluster X-Radiation

The current observational status of cluster x-ray sources allows several facts to be inferred. First, the bulk of the x-rays below ~10 keV result from thermal bremsstrahlung, predominantly from the cores of the clusters. The electron temperature is close to the virial temperature, so that the speed of sound in the intracluster gas is similar to the velocity dispersion of the galaxies. Second, the mass of gas in the core is only a few percent of the total core binding mass. Third, the x-ray data are consistent with an iron abundance in the hot gas that is about half the "cosmic" (or solar) value. The cooling time in the observed gas is close to or greater than, the age of the universe, so that to a first approximation the present state of the intracluster gas can be fitted by a quasistatic model, evolving slowly due to infall, cooling, conduction, mass loss, and energy input from galaxies. Our ignorance of the role of these last processes is usually summarized by employing a polytropic equation of state for the cluster gas (pressure $\propto$ density$^{\gamma}$). Many models based on this assumption, together with that of hydrostatic equilibrium, have been constructed in the past.[27-29] The attraction of these models is that the x-ray data (both spectral and spatial) from the scanning satellites merits fitting by only a few parameters. Caution should, however, be employed in extrapolating these models beyond the observed gas in the cluster cores and in making inferences about the outer regions of clusters (total mass of gas, etc.). Careful studies of the cluster x-ray emission away from the core, and indirect information from the Sunyaev–Zel'dovich microwave dip[30] in the vicinity of clusters[31-33] will be invaluable here.

Interpretation of the several observed correlations between x-ray and other properties of rich clusters is complicated by a lack of understanding of the mass and energy balance in the intracluster gas. Ram-pressure stripping[34] may well be responsible for the lack of spirals in the cores of rich clusters, but the cluster richness may of itself inhibit their formation.[35] Also, the importance of relativistic particles on the energy balance, and thus of the relevance of radio emission is unclear. A considerable extrapolation of the inferred electron spectra is required in order to heat the intracluster gas.[36] It seems more likely that electrons will make a contribution to the observed emission above 10 keV, by Compton scattering the microwave background.

The HEAO-2 satellite should provide far more detailed images of the cores of rich clusters than have been glimpsed from earlier data. For example, the inflow of gas due to cooling may be important in the cores of the more luminous clusters,[37-39] and such thermally unstable flows could dominate the appearance of the central regions of clusters such as Perseus.

As yet the data do not allow firm statements on the source of the gas or of the mechanism of heating, although the presence of large amounts of iron argues that a substantial fraction of the emitting gas is not primordial. It will be important to discover whether the iron line emission is widely dispersed through the cluster, is concentrated in higher density regions in the cluster core, or lies nearby individual galaxies.[40]

The evolution of clusters of galaxies is potentially within the grasp of x-ray astronomers, given the deep x-ray survey capabilities of the imaging x-ray telescope on HEAO-2 and later satellites. It seems unlikely that strong cooling ever occurred at the present x-ray temperatures ($T \gtrsim 3 \times 10^7$ K) in clusters of galaxies without contradicting observations of present day clusters, or of the x-ray background.[41] This means that an origin of clusters via the cooling and fragmentation of protoclusters[42] must have occurred at lower temperatures and with a high efficiency in gas turning to "dark matter." Such clusters would be expected to appear more luminous in the past, i.e., to show positive evolution. However, an origin for clusters of galaxies via the aggregation of already condensed subunits suggests that clusters could have been less x-ray luminous in the past, i.e., showing negative (or weakly positive) evolution.[43,44] The sense of the evolution in this case depends upon the balance between the gas supply (from primordial gas and/or mass loss from galaxies) and the deepening of the cluster potential.

SEYFERT GALAXIES

X-Ray Seyferts as a Class

Prior to the appearance of the 2A catalogue, data on the x-ray emission from active galaxies was scanty. This is now no longer so, with the number of identified sources of this type comparable to that of clusters. The most notable increase in a particular astronomical class is for type-1 Seyfert galaxies. TABLE 2 summarizes the present position, with 29 Seyferts now reliably identified with 2A, Uhuru, SAS-3, or HEAO-1 x-ray sources. As seen from TABLE 2, the majority of the Seyfert identifications are based on 2A catalogue sources and the establishment of this second main class of extragalactic x-ray emitter is one of the most important achievements of the Ariel V survey. Now, as with clusters, there are sufficient identified Seyferts to establish their x-ray luminosity function and study the relation of x-ray emission with other Seyfert properties. Spectral data on the x-ray Seyferts is still patchy, but NGC

TABLE 2

SUMMARY OF X-RAY SEYFERTS

Catalogue	Number	Sources
3U catalogue identifications	3	NGC 4151, 1275,* 3C 390.3†
2A catalogue identifications	10	NGC 3783, 5548, 6814, 7469 IC 4329A Mk 79, 279, 376, 509 MCG 8-11-11‡
2A source identifications (post catalogue)	4	MCG-2-58-22‡ ESO 113-IG 45‡ ESO 141-G55‡ MCG-6-30-15¶
4U sources	5	Mk 335, 506, 541, 1040‡ NGC 4593
Ariel V SSI PST search	1	NGC 3227§
SAS-3 sources	2	III Zw 2,** 3C 120††
HEAO-1 sources	4	3C 382, NGC 7213, 3C 111 ESO 108-G35‡
Total =	29	X-ray Seyferts

*Resolved from Perseus cluster by Copernicus.[17]
†Confirmed by Copernicus position.[55]
‡Seyfert discovered after x-ray detection.[51–53, 85, 86]
¶Identified from improved SAS-3 position.[85]
§Elvis *et al.*[50]
**Schnopper *et al.*[87]
††Schnopper *et al.*[88]

4151 is notable in having the hardest spectrum, now detected beyond 1 MeV,[45] of any cosmic x-ray source. This bright x-ray Seyfert also shows a substantial low-energy cutoff[46] and evidence of time variability on a time scale of several days,[46,47] with one report of a flare lasting only 730 s.[48]

The x-ray luminosities of the Seyferts in TABLE 2 extends from $10^{42.1}$ erg s^{-1} (NGC 3227) to $10^{45.2}$ erg s^{-1} (III Zw 2), a range closely comparable with that of clusters. FIGURE 7 shows the distribution of L_x for detected Seyferts against red shift z, and it is interesting to note the range of luminosities covered by the Seyferts extends more than four-fifths of the way from NGC 3227 to the quasar 3C 273. A similar demonstration of Hβ and $L\alpha$ intensities has been given by Weedman,[49] who argues that this supports both the contention that QSO's are (simply) distant Seyferts, at $z > 0.1$, *and* a cosmological interpretation of the red shifts. Optically the first distinguishing property of a Seyfert galaxy is its extremely bright nucleus, followed by the breadth of its emission lines. It now seems clear that a third feature of Seyfert galaxies is their powerful x-ray emission and it would be surprising if it were found that this was not associated with the Seyfert nucleus. The observed variability for several Seyferts supports this view.

Elvis *et al.*[50] have made a detailed comparison of 15 x-ray Seyferts with their optical, infrared and radio emission. They find the x-ray luminosity to be correlated with the infrared power (at 3.5μm and 10μm), the optical continuum power, the FWZI of the broad emission lines, and with the luminosity in the Hα emission line, but not significantly with the radio power or the luminosity in the forbidden optical emission lines. Elvis *et al.* suggest that these correlations strongly support the view that the x-ray emission originates in the dense inner nuclear region, of extent <0.1 pc, which is peculiar to type-1 Seyferts. In passing, it seems clear that the high success rate of identifying x-ray Seyferts in the 2A catalogue is due, in part, to the x-ray to optical continuum correlation, in that the detected x-ray Seyferts are also among the brightest known.

Of the 29 x-ray Seyferts listed in TABLE 2, it is interesting that six have been found to be type-1 Seyfert galaxies as a result of an optical search of the source error boxes. The first new Seyfert to be discovered in this way, MCG 8-11-11, previously listed as a barred spiral galaxy, was found by Ward *et al.*[51] to be by far the brightest extragalactic object in the 0.093-deg^2 error box of 2A 0551+446, and a type-1 Seyfert. A continuing search of the error boxes of unidentified 2A sources by the same group, at the Anglo–Australian Observatory, the Royal Greenwich Observatory, and Leicester, quickly led to the discovery[52] of three further type-1 Seyferts, viz., ESO 113-IG45, MCG–2-58-22, and ESO 141-G55. Notably, ESO 113-IG45 is optically the most luminous (and the brightest) in the southern hemisphere, comparable with a low-luminosity QSO, while MCG–2-58-22 has Balmer lines broader than any other known Seyfert, except 3C 382. The identification of ESO 113-IG45 with 2A 0120−591 was also proposed, independently, by Ricker[53] on the basis of positional coincidence and the Seyfertlike properties of this galaxy reported by Fairall.[54]

Over the past year, several additional x-ray Seyferts have been identified, five with refined source positions in the 4U catalogue, four from HEAO-1 observations, and one further 2A source identification (2A 1326−310 ≡ MCG–6-30-15) on the basis of a substantially improved error box obtained by SAS-3. An improved SAS-3 x-ray position has also confirmed the 2A identification of IC 4329 A, distinguishing this Seyfert from the nearby cluster of galaxies. A rapidly increasing number of precise (~

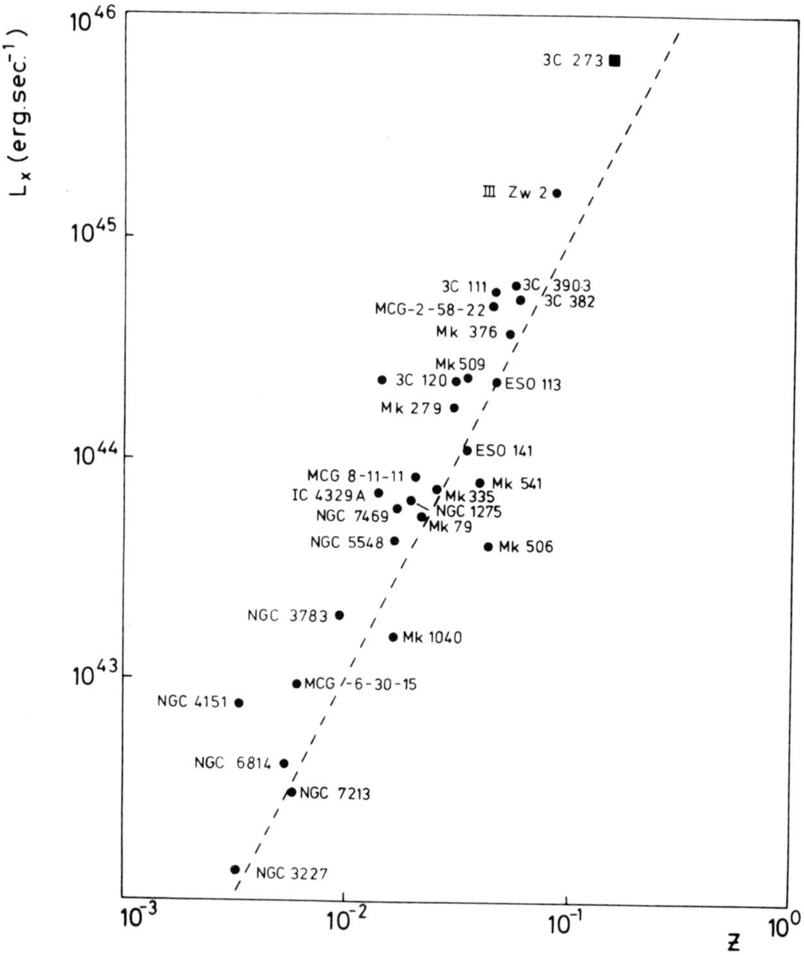

FIGURE 7. Luminosity/red-shift diagram for Seyfert type-1 x-ray sources and for the QSO 3C 273. The dashed line shows, approximately, the sensitivity limit of the 2A Sky Survey at 0.5 Ariel count s^{-1}.

arc min) positions, obtained with the SAS-3 or HEAO-1 modulation collimator experiments, has confirmed the great majority of these x-ray Seyferts; indeed it is remarkable that none of the 15 original 2A x-ray Seyferts have been disallowed by the more precise positions.

Variability in Seyfert X-Radiation

Establishing variability in the intensity of a faint x-ray source is difficult, particularly when such evidence is based on a comparison of data from different instruments. Nevertheless, there is now no doubt that some x-ray Seyferts do vary on quite short time scales and it can be anticipated that over the next few years observations of greater precision will show this to be a feature common to the class. Charles *et al.*[55] first showed an apparent factor of four variation for 3C 390.3 between Uhuru (1971) and Copernicus (1973) measurements. In the same way Ives *et al.*[46] found evidence for a factor 2.5 increase in the 2–6-keV emission of NGC 4151 from 1971 to 1974. More recently Elvis[47] reported in Ariel V SSI observations of NGC 4151 a flare with a factor 1.7 ± 0.2 increase in less than 3 days during August 1975. Also from the SSI data, Ward *et al.*[51] have found MCG 8-11-11 to vary by at least a factor 3 on a time scale of 30 days or less. A comparison of Uhuru and Ariel V SSI data on Mk 279[48] further indicates variablility on a time scale < 3 years.

The best studied x-ray Seyfert remains NGC 4151, the first to be discovered, and also the brightest. Tananbaum *et al.*,[48] from Uhuru, have reported flares in the x-radiation from NGC 4151 at an implied rate of 1 or 2 per day, with one event showing an order-of-magnitude change in ~730 s. Extended observations with the Ariel V SSI have not confirmed such large-scale variability, but do show flares on a 1–2-day time scale to be a feature of the NGC 4151 emission. FIGURE 8 shows a sample of SSI observations. Lawrence[56] has recently discussed these and similar data in terms of a shot noise emission model for NGC 4151.

OTHER ACTIVE GALAXIES

Before 1975 very few active galaxies of any type had been identified as an x-ray source. The 3U catalogue[57] lists a handful of other active galaxies, including the nearby radio galaxy Cen A, the bright quasar 3C 273, and (albeit in a very large error

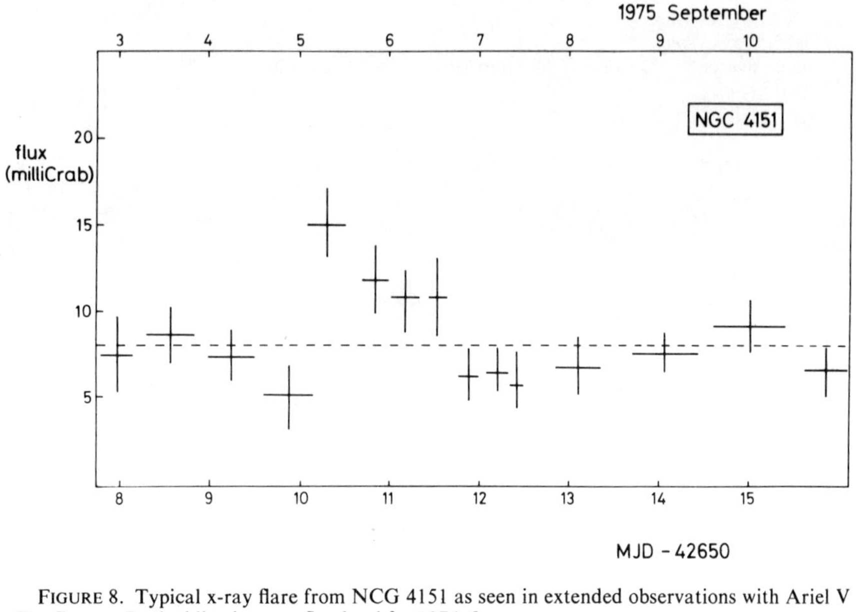

FIGURE 8. Typical x-ray flare from NCG 4151 as seen in extended observations with Ariel V Sky Survey. Dashed line is mean flux level for 1974–8.

TABLE 3
SUMMARY OF OTHER X-RAY/ACTIVE GALAXY IDENTIFICATIONS

Type	Number	Sources
Radio galaxies (?)	4	Cen A (NGC 5128) M 87* Pic A 3C 445
QSO's	3	3C 273 MR 2251 − 178† QSO 0241 + 622‡
BL Lacertids	7	Mk 421, 501 3C 371¶ I Zw 186§ 2A 1219 + 305** PKS 0548 − 322†† PKS 2155 − 304‡‡
High-excitation emission line galaxies	10	NGC 526A,¶¶ 1365,§§ 1685, 2110, 2992, 5506, 7582 M 82 MCG-5-3-16 Mk 334
Total =	24	

*Separation of M 87 from Virgo cluster by Ariel V SSI.[21]
†Identification of 2A source based on SAS-3 position.[63, 89]
‡Identification of 4U source from improved SAS-3 position.[90]
¶HEAO-1 A2 source.[91]
§HEAO-1 A1 source.[66]
**Optical search of 2A error box.[92]
††HEAO-1 A2 source.[93]
‡‡HEAO-1 A3 source.[94]
¶¶Identification of 2A source from improved HEAO-1 A3 position.[95]
§§Ward *et al.*[52]

box) the irregular galaxy M82. Again, subsequent observations have considerably extended this list. TABLE 3 summarizes the present nonSeyfert identifications in this active galaxy category.

Radio Galaxies

Although several active galaxies combine powerful x-ray and radio emission (e.g., NGC 1275, 3C 390.3, 3C 273) the majority of extragalactic x-ray sources are, at most, weak radio emitters. In fact, only for one object, Cen A, is a physical relation between x-radiation and radio emission established to date. Although Cen A is tentatively listed in TABLE 3 under the subheading "radio galaxies," the major x-ray emission arises from the nucleus of the central, giant elliptical galaxy NGC 5128.[58] X-ray observations spanning almost a decade—a more complete "light curve" than at any other wavelength—have shown the ~2–10-keV emission to vary by an order of magnitude, with approximately factor-of-two changes within a few days. A compilation of data up to 1977 is given by Lawrence *et al.*,[59] and additional OSO-8 observations are reported by Mushotsky *et al.*[60] The latter authors also report radio flare changes in phase with the x-ray variability and argue for a dominant x-ray production by Compton scattering of the radio photons by a single relativistic electron population. X-radiation has recently been detected also from the large radio lobes of Cen A.[61] Again, the emission is likely to be by the so-called inverse Compton process, though in these extended regions the scattered photons are more likely to be of the microwave background. The derived magnetic field of 0.7 μG is close to the equipartition value and indicates Cen A is merely the first crude example of an x-ray map of extended radio sources that will be obtained over the next few years.

X-Radiation from QSO's

Until recently 3C 273 was distinguished as the nearest and brightest QSO and also the only certain QSO identified as an x-ray source. A PST search of 65 radio-bright QSO's using the Ariel V sky survey data produced only one marginal additional detection.[62] Upper limits to the x-ray luminosity ranged from ~0.15 to ~50 times that of 3C 273.

Now, a refined SAS-3 position of the source 2A 2251−179 has led to its unambiguous identification with a bright (B ~ 16) quasi-stellar object.[63] Given the preliminary designation MR 2251−178, this object—at a red shift of 0.068—has an x-ray luminosity varying (at 2–20 keV) from 5×10^{44} to 1.6×10^{45} erg s^{-1}, which exceeds the optical luminosity by a factor of ~10. Most recently, a third x-ray QSO has been reported, again from the optical search of a small SAS-3 RMC error box. As yet unnamed, the quasi-stellar object identified with the x-ray source 2S 0241+622 (=4U 0241+61) is, at a red shift of 0.0438, the nearest known QSO and represents a further example of the increasing number of active galaxies whose outstanding properties have been recognized only after the discovery of an adjacent x-ray source.

Although the number of identified x-ray QSO's is still insufficient to allow a useful study of the relation of the x-ray emission to other properties—or to provide a useful limit to the likely evolutionary effects—the increased sensitivity of the HEAO-1 sky survey may do so, and certainly HEAO-2 will.

BL Lacertids

The first positive detection of x-radiation from an object of this class came with the identification of the bright BL Lac-type object, MK 421, with the x-ray source 2A 1101+38. The identification was first proposed by Ricketts *et al.*[64] on the basis of positional coincidence, albeit with a x-ray error box 0.15 deg^2, and highly variable x-ray emission. Subsequent Ariel V observations have shown the x-ray source to persist at a generally faint level (~0.8 Ariel counts s^{-1}) with a further flaring in late December 1977. No simultaneous optical observations have yet been obtained, but the rapidly improving location has subsequently strengthened the identification with MK 421. The revised Uhuru (4U) catalogue contains a source, 4U 1651+39, identified with a second BL Lac-type object, MK 501. It is interesting to note that MK 421 and 501 are in the brightest in 3 or 4 (in V) of the 32 BL Lacertae objects listed by Stein *et al.*[65] Early analysis of HEAO-1 data has produced a third candidate, I Zw 186.[66] The ill-understood nature of BL Lac-type objects, with their compact appearance, featureless blue continuum, and high optical variability, possibly lying somewhere between type-1 Seyfert galaxies and QSO's, makes the establishment of power x-ray emission especially interesting.

High-Excitation Narrow Emission Line Galaxies

Also of interest is the rapidly increasing number of x-ray identifications with active galaxies having high-excitation forbidden line spectra, but without the broad permitted lines characteristic of type-1 Seyferts. This number, now 10 if M 82 is included, suggests strongly that powerful x-ray emission is also a feature of these objects. Perhaps the prototype of this kind of galaxy is NGC 5506.[67] This galaxy has long been a candidate identification for the x-ray source 3U 1410−03 (=2A 1410−029),[68] and this has now been confirmed by SAS-3 RMC observations[69] which produced a 1.1 arc min radius error circle (90% confidence) centered only ~12 arc sec north of the galaxy nucleus (FIGURE 9).

The optical search of 2A error boxes by Ward *et al.*[52] produced two other emission line galaxies, NGC 7582 and NGC 2992, and gave the first indication of a new class of x-ray emitter being established. A resulting PST search of the Ariel V sky survey data in the vicinity of 25 other bright emission line galaxies revealed a further positive detection, NGC 1365. The nearby galaxy M82 is morphologically very similar to

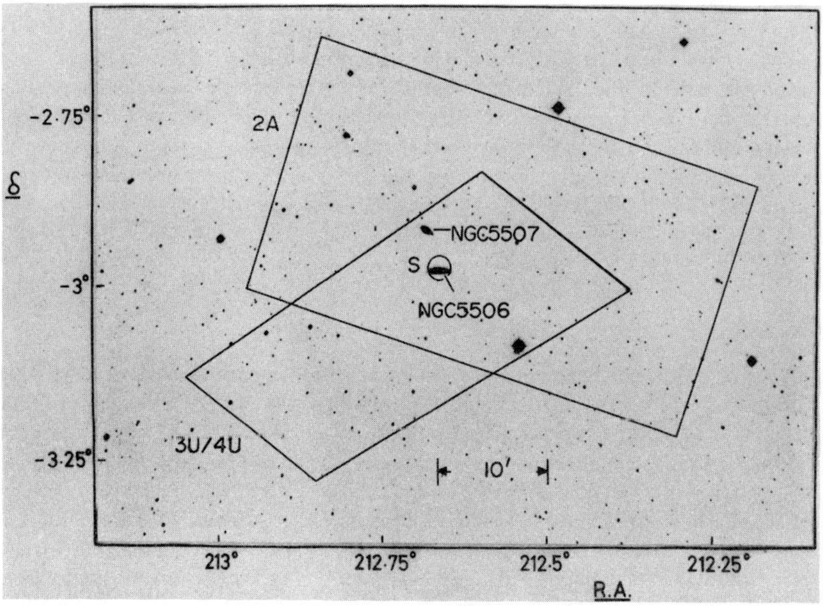

FIGURE 9. Various error boxes for the source 2A 1410−029 are shown superimposed on the blue Palomar Sky Survey photograph. The active galaxy NGC 5506, provisionally identified in the 3U catalogue, and supported by 2A, is confirmed by the much improved SAS-3 position.

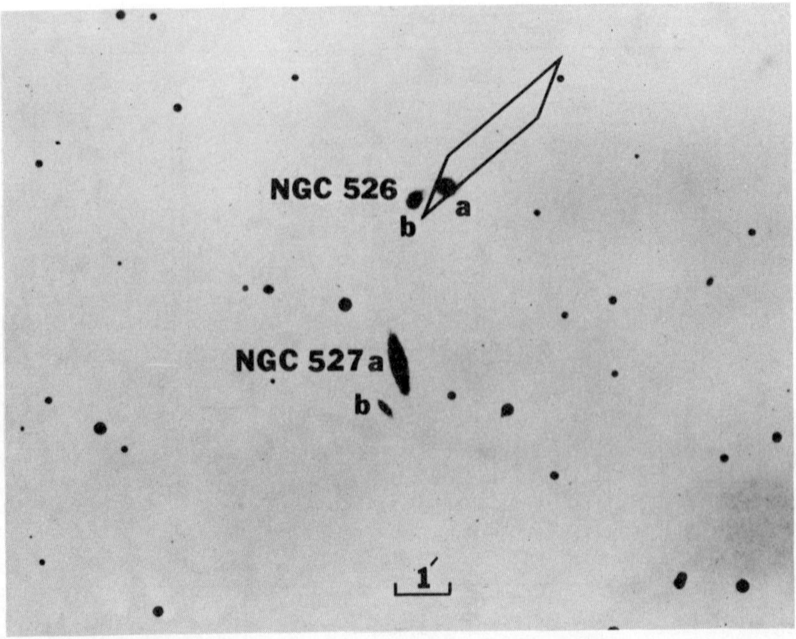

FIGURE 10. HEAO-1 scanning modulation collimator error box pinpoints the emission line galaxy NGC 526A as the counterpart of the x-ray source 2A 0120−353.

NGC 5506[67] and shows emission lines in places, although its nucleus is obscured by dust. The SSI error box of 2A 0904+700 includes M82 and makes it a reasonable additional candidate in this class.

Four most recent additions have resulted from SAS-3 or HEAO-1 observations. MCG 5-23-16 and NGC 2110 were found by optical spectroscopy of bright galaxies in the arc min SAS-3 error boxes of 2A 0946−310 and S 0549−746,[70,71] respectively, while a further precise location, with the HEAO-1 modulation collimator experiment, allowed the emission line galaxy NGC 526A to be identified with the Ariel source 2A 0120-353 (FIGURE 10).

In summary, it is now clear that strong x-ray emission is a characteristic of high-excitation narrow emission line galaxies, with luminosities from $\sim 10^{40.3}$ (M82) to $\sim 10^{43.7}$ (NGC 526A), lying to the lower end of the Seyfert type-1 range. Further, the fact that several are variable x-ray emitters on a time scale of days (FIGURE 11 shows the best example to date, from Ariel V SSI observations) provides the first clear evidence that these galaxies, too, possess a compact "quasarlike" region deep in their nucleus.

DISCUSSION

The 2–10-keV x-ray luminosities of galactic nuclei range from $L_x \sim 10^{36}$ erg s^{-1} for our own galactic nucleus to $L_x \sim 10^{46}$ erg s^{-1} for quasars. The x-ray spectra often

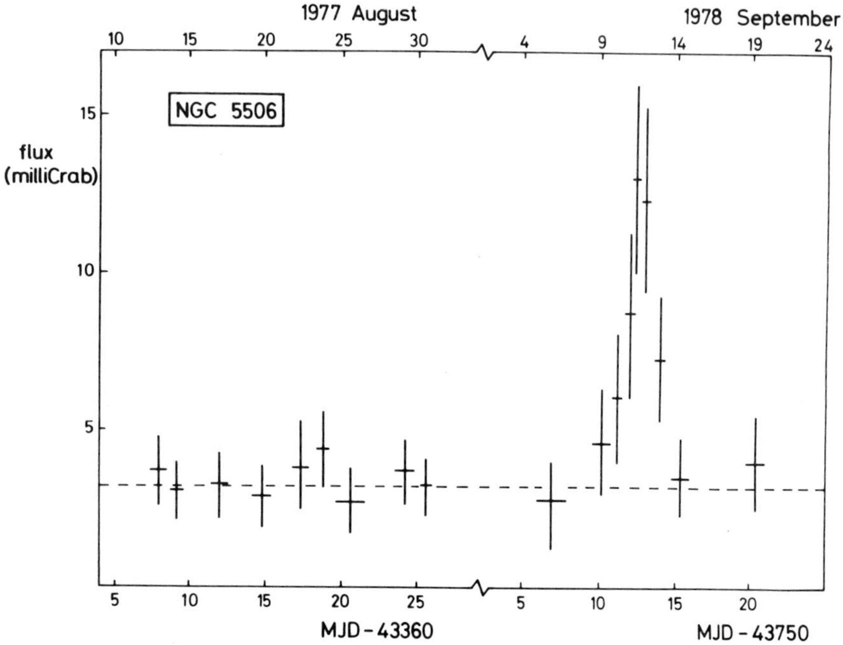

FIGURE 11. X-ray flare from NGC 5506 detected during extended Ariel V SSI observations. Dashed line is mean flux level for 1974–8.

appear hard and may be considered as quasi-power-law continua, similar to the continuum spectra observed in the radio and optical wavebands. The hard x-ray and γ-ray luminosities may exceed the total emission from the nucleus at other wavebands, as is the case with the nearby galaxy Centaurus A (NGC 5128) in which the spectrum has been confirmed above 1 MeV[72] and also NGC 4151.[73,74,45] These high luminosities, coupled with rapid variations, strongly suggest that the x-rays originate close to the power source in these galactic nuclei, as noted in the previous sections.

Luminous outbursts of radiation at any wavelength, of L erg s^{-1}, cannot occur on time scales Δt much shorter than the light crossing time of the emitting region. Moreover, if the electron scattering optical depth τ_{es} through this region exceeds unity, the photons diffuse out on a timescale τ_{es} times larger than the light crossing time. Considering a spherical region of radius R, containing particle density n, we see that

$$\tau_{es} \simeq n\sigma_T R \quad (1)$$

and

$$\Delta t \geqslant (\tau_{es} + 1)R/c. \quad (2)$$

Cavallo and Rees[75] have discussed a limitation on n due to the efficiency η with which matter is converted to energy ($\eta < 1$), i.e.,

$$L\Delta t \simeq \eta Mc^2, \quad (3)$$

where M is the mass directly involved in the energy production, or

$$n \geqslant \frac{3L\Delta t}{4\pi R^3 \eta m_p c^2} \tag{4}$$

(the inequality resulting when more matter resides in the source than is actually involved in the outburst). Δt is a minimum when the source has a radius such that $\tau_{es} = 1$. Thus

$$L \leqslant 2 \times 10^{41}(\eta/0.1)\Delta t \quad \text{erg s}^{-1}. \tag{5}$$

Storing the energy in, for example, magnetic fields and then releasing it does not overcome this limit unless the processes take place in a vacuum. Matter is needed to generate and confine the magnetic fields and this itself will present some opacity. Limit (5), displayed in FIGURE 12, is such that direct variability on time scales much

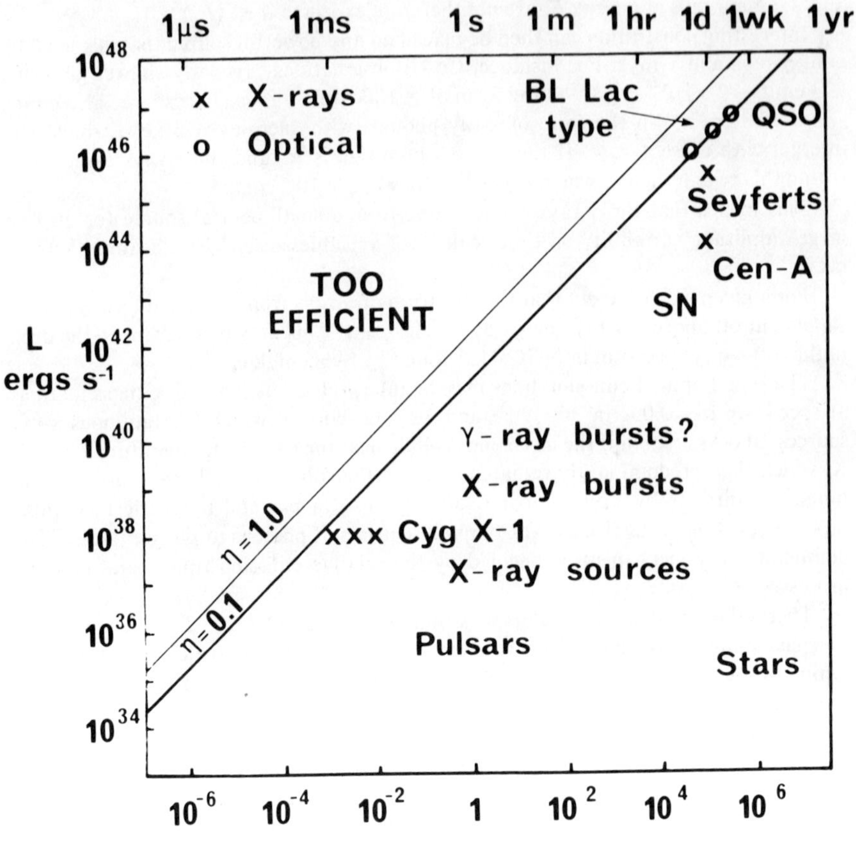

FIGURE 12. Constraints on luminosity variations on a time scale Δt seconds. See text for definition of η and general discussion.

less than a day ought not to be observed in the most luminous quasars, unless relativistic outflow occurs. (Note that the $L^{-2.5}$ shape of the x-ray Seyfert luminosity function implies that this limit ought to be observed most readily in the most luminous sources.)

A second limit applies to the hard x-ray and γ radiation due to photon–photon collisions.[76,77] Photons above a threshold energy E_s may collide and produce electron–positron pairs with harder photons of energy E_γ provided that

$$\left(\frac{E_s}{m_e c^2}\right)\left(\frac{E_\gamma}{m_e c^2}\right) \gtrsim 2. \tag{6}$$

The optical depth to photon–photon collisions,

$$\tau_{\gamma\gamma} \simeq n_x \sigma_T R, \tag{7}$$

where n_x is the number density of photons within a bandwidth $\sim 2\ E_s$ centered around 2 E_s.[77] Gamma rays of energy E_γ are not therefore expected if $\tau_{\gamma\gamma}(E_\gamma) \gtrsim 1$.

Interesting constraints can then be placed on any powerful source that is known to emit γ-rays and x-rays. For instance, Cos-B observations[78] recently showed that 3C 273 emits $\sim 2 \times 10^{46}$ erg s^{-1} in the form of $\gtrsim$ 100-MeV γ rays. This source also emits $\sim 5 \times 10^{45}$ erg s^{-1} in the form of x-ray photons with energies $\geqslant$ 5 keV, which can interact with these γ-rays. If the γ-ray source size is R_γ and the x-rays came from within this region then τ_γ would exceed 1 unless $R_\gamma \geqslant 10^{18}$ cm.

This means that the γ rays cannot come from a small central source (or display large-amplitude variability on time scales < 1 yr) unless relativistic beaming effects occur.

For a given r_γ, we expect that the spectrum of γ-rays from a compact x-ray source should cut off above a γ-ray energy at which $\tau_\gamma \simeq 1$. This may be relevant to the drop in the soft γ-ray spectrum in NGC 4151 found by Schoenfelder *et al.*[45]

The broad optical emission lines may be interpreted[79] as indicating mass motions at speeds up to $\sim 0.03c$ in 3C 390.3 and 3C 382, both of which are luminous x-ray sources. Shocks created if the gas clouds collide heat the gas to a temperature $T_s \gtrsim 10^9$ K, which then predominantly radiates x-rays, either directly by thermal bremsstrahlung, or indirectly by Comptonizing soft photons. Shocks and magnetic fields may produce relativistic electrons that Compton scatter soft photons to x-ray energies. The dominant x-ray mechanism is decided by the relative emission time scales of these processes.

Thermal bremsstrahlung occurs on a time scale of $t_{br} \simeq 2 \times 10^{11} T^{1/2}/n$ s where T is the gas temperature and n the particle density. Short-time-scale variations in a luminous source imply high particle densities and consequently high optical depths to electron scattering, $\tau_{es} \simeq 30 L_{42} t_3^{-1}$. Lightman, Giacconi, and Tananbaum[80] have pointed out that Comptonization of the bremsstrahlung photons will then occur. The observed spectra are, however, quasi-power-law spectra, not Wien spectra, consistent with complete Comptonization. This may be due to ranges of temperatures, and perhaps, more important, to Comptonization of soft photons incident in the outer regions of hot clouds. These would be reflected out at higher energies. The electrons cool on a time scale, $t_c \simeq 2 \times 10^7/\epsilon_{rad}$ s, where ϵ_{rad} is the energy density in soft photons.

Relativistic (inverse) Compton scattering cools electrons with Lorentz factor γ on a time scale of $t_{ic} \simeq 2 \times 10^7/\gamma\epsilon_{rad}$ s. Multiple scattering, in the situation where $\epsilon_x \gg \epsilon_{rad}$[81] can lead to much of the energy in the source emerging as γ rays. [This requires a higher η in Equation 3 if L (x-ray) is used in place of L (total)]. The synchro-self-Compton process, in which electrons Compton scatter the synchrotron photons they produce on an ambient magnetic field, has attractions when observationally comparing the radio and x-ray fluxes from the same galactic nuclei. The reason that rough agreement may usually be found for such a hypothesis is mostly due to the enormous sensitivity of the x-ray flux to the radio parameters (source size, self-absorption turnover, and radio flux raised to powers exceeding five). Any reasonable uncertainty in those parameters then allows an x-ray flux of the observed magnitude (or one several orders of magnitude brighter or fainter).

The actual powerhouse responsible for the activity in these galactic nuclei is not yet defined by the observations. (A review of likely candidates had been given by Rees.[82] It seems clear, however, that much of the energy is radiated as x- and γ-rays and it is a reasonable hope that some features (temporal or spectral) will emerge from x-ray or γ-ray observations to give an insight into the mechanisms involved.

Finally, a brief discussion of the implication of the extragalactic x-ray observations reported herein on the nature of the x-ray background. Warwick has shown[83] how inclusion of the emission line galaxies and QSO's, respectively, at the low- and high-luminosity ends, extends the x-ray luminosity function of type-1 Seyferts to give a total "active galaxy" luminosity function, covering the range $L_x \sim 10^{42}$–10^{46} erg s^{-1}. Nevertheless, he finds the contribution from all known active galaxies to the (2–10 keV) diffuse background to still not exceed 10–15%. Thus, the sum of both major groups of extragalactic sources, clusters and active galaxies, account for <30% of the background.

An upturn in the low-luminosity end of the active galaxy luminosity function (e.g., many more weaker emission line galaxies) could only contribute a substantial fraction of the x-ray background if a similarly large fraction of *all* galaxies ($M_{pg} \leqslant -19$) are active in x rays ($\geqslant 10^2$ times more luminous than the galaxies in the local group). A modest cosmological evolution of x-ray active galaxies could, however, easily explain the remainder of the x-ray background after accounting for unevolved sources. Thus, there remains no strong case for a substantial diffuse component of the (2–10-keV) x-ray background. The recently launched HEAO-2 could well decide.

References

1. FORMAN, W. *et al.* 1978. Astrophys. J. Suppl. **38:** 357.
2. COOKE, B. A. *et al.* 1978. Mon. Not. R. Astron. Soc. **182:** 455.
3. JONES, C. & W. FORMAN. 1978. Astrophys. J. **224:** 1.
4. RICKETTS, M. J. 1978. Mon. Not. R. Astron. Soc. **183:** 51P.
5. BAHCALL, N. A. 1977. Astrophys. J. **217:** L77.
6. BAHCALL, N. A. 1977. Astrophys. J. **218:** L93.
7. MCHARDY, I. 1978. Mon. Not. R. Astron. Soc. **184:** 783.
8. TYTLER, D. & N. V. VIDAL. 1978. Mon. Not. R. Astron. Soc. **182:** 33P.
9. WATSON, M. G. *et al.* 1978. Mon. Not. R. Astron. Soc. **184:** 79P.
10. FABIAN, A. C. & M. J. REES. 1978. Mon. Not. R. Astron. Soc. **185:** 109.
11. MITCHELL, R. J. *et al.* 1976. Mon. Not. R. Astron. Soc. **176:** 29P.
12. MUSHOTZKY, R. F. *et al.* 1978. Astrophys. J. **225:** 21.

13. LEA, S. M. *et al.* Proc. IAU/COSPAR Symposium, Innsbruck, June 1978.
14. GORENSTEIN, P. *et al.* Proc. IAU/COSPAR Symposium, Innsbruck, June 1978.
15. MITCHELL, R. J. *et al.* 1977. Mon. Not. R. Astron. Soc. **181:** 25P.
16. KELLOGG, E. & S. Murray. 1974. Astrophys. J. **193:** 2.
17. FABIAN A. C. *et al.* 1974. Astrophys. J. **189:** L59.
18. CASH, W. *et al.* 1976. Astrophys. J. **209:** L111.
19. GORENSTEIN, P. *et al.* 1978. Astrophys. J **224:** 718.
20. ULMER, M. P. & J. G. JERNIGAN. 1978. Astrophys. J. **222:** L85.
21. LAWRENCE, A. 1978. Mon. Not. R. Astron. Soc. **185:** 423.
22. DE VAUCOULEURS, G. 1961. Astrophys. J. Suppl. **6:** 213.
23. DAVISON, P. 1978. Mon. Not. R. Astron. Soc. **183:** 39P.
24. GORENSTEIN, P. *et al.* 1977. Astrophys. J. **216:** L95.
25. GRIFFITHS, R. E. *et al.* 1974. Nature (London) **250:** 471.
26. GORENSTEIN, P. Proc. IAU/COSPAR Symposium, Innsbruck, June 1978.
27. LEA, S. M. 1976. Astrophys. J. **203,** 569.
28. GULL, S. F. & K. J. E. NORTHOVER. 1975. Mon. Not. R. Astron. Soc. **173:** 585.
29. CAVALIERE, A. & R. FUSCO-FEMIANO. 1976. Astron. & Astrophys. **49:** 137.
30. SUNYAEV, R. A. & Y. B. ZEL'DOVICH. 1972. Comments Astrophys. Space Phys. **4:** 173.
31. GULL, S. F. & K. J. E. NORTHOVER. 1976. Nature (London) **263:** 572.
32. LAKE, G. & R. P. PARTRIDGE. 1977. Nature (London) **270:** 502.
33. BIRKINSHAW, M. *et al.* 1978. Nature (London) **275:** 40.
34. GUNN, J. E. & J. R. GOTT. 1972. Astrophys. J. **176:** 1.
35. GOTT, J. R. & T. X. THUAN. 1976. Astrophys. J. **204:** 649.
36. LEA, S. M. & G. D. HOLMAN. 1978. Astrophys. J. **222:** 29.
37. COWIE, L. L. & J. BINNEY. 1977. Astrophys. J. **215:** 723.
38. FABIAN, A. C. & P. E. J. NULSEN. 1977. Mon. Not. R. Astron. Soc. **180:** 479.
39. MATHEWS, W. G. & J. N. BREGMAN. 1978. Astrophys. J. **224:** 308.
40. FABIAN, A. C. & J. E. PRINGLE. 1977. Mon. Not. R. Astron. Soc. **181:** 5P.
41. FABIAN, A. C. & P. E. J. NULSEN. 1978. Mon. Not. R. Astron. Soc. **186:** 783.
42. DOROSHKEVITCH, A. C. *et al.* (1974). *In* Confrontation of Cosmological Theories and Observation, Longair, Ed: 213. D. Reidel Publishing Co. Dordrecht, The Netherlands.
43. COWIE, L. L. & S. C. PERRENOD. 1978. Astrophys. J. **219:** 354.
44. PERRENOD, S. C. 1978. Astrophys. J. **226:** 566.
45. SCHÖNFELDER, V. 1978. Nature (London) **274:** 344.
46. IVES, J. C. *et al.* 1976. Astrophys. J. **207:** L159.
47. ELVIS, M. S. 1976. Mon. Not. R. Astron. Soc. **177:** 7P.
48. TANANBAUM H. *et al.* 1978. Astrophys. J. **223:** 74.
49. WEEDMAN, D. W. 1976. Astrophys. J. **208:** 30.
50. ELVIS, M. S. *et al.* 1978. Mon. Not. R. Astron. Soc. **183:** 129.
51. WARD, M. J. *et al.* 1977. Astron. Astrophys. **56:** 219.
52. WARD, M. J. *et al.* 1978. Astrophys. J. **223:** 788.
53. RICKER, G. R. 1978. Nature (London) **271:** 334.
54. FAIRALL, A. P. 1977. Mon. Not. R. Astron. Soc. **180:** 391.
55. CHARLES, P. A. *et al.* 1975. Mon. Not. R. Astron. Soc. **170:** 17P.
56. LAWRENCE, A. *et al.* 1979. Mon. Not. R. Astron. Soc. To be published.
57. GIACCONI R. *et al.* 1974. Astrophys. J. Suppl. **27:** 37.
58. DELVAILLE, J. P. *et al.* 1978. Astrophys. J. **219:** L81.
59. LAWRENCE, A. *et al.* 1977. Mon. Not. R. Astron. Soc. **181:** 43P.
60. MUSHOTZKY, R. F. *et al.* 1978. Astrophys. J. **220:** 790.
61. COOKE, B. A. *et al.* 1978. Mon. Not. R. Astron. Soc. **182:** 661.
62. WHITE, G. J. & M. J. RICKETTS 1977. Mon. Not. R. Astron. Soc. **181:** 435.
63. RICKER, G. R. *et al.* 1978. Nature (London) **271:** 35.
64. RICKETTS, M. J. *et al.* 1976. Nature (London) **259:** 546.
65. STEIN, W. A. *et al.* 1976. Ann. Rev. Astron. Astrophys. **14:** 173.
66. SHULMAN, S. Proc. IAU/COSPAR Symposium, Innsbruck, June 1978.
67. WILSON, A. S. *et al.* 1976. Mon. Not. R. Astron. Soc. **177:** 673.
68. BAHCALL, J. N. *et al.* 1975. Astrophys. J. **199:** L9.

69. Ricker, G. R. *et al.* 1979. Astrophys. J. To be published.
70. Schnopper, H. W. *et al.* 1978. Nature (London) **275:** 719.
71. Bradt, H. V. *et al.* 1978. Astrophys. J. **226:** L111.
72. Hall, R. D. *et al.* Astrophys. J. **210:** 631.
73. Di Cocco, G. *et al.* 1978. Nature (London) **270:** 319.
74. Auriemma, G. *et al.* 1978. Astrophys. J. **221:** L7.
75. Cavallo, G. & M. J. Rees 1978. Mon. Not. R. Astron. Soc. **183:** 359.
76. Jelley, J. 1966. Nature (London) **211:** 472.
77. Herterich, K. 1974. Nature (London) **250:** 311.
78. Swanenburg, B. N. *et al.* 1978. Nature (London) **275:** 298.
79. Ostriker, J. P. *et al.* 1977. Astrophys. J. **208:** L61.
80. Lightman, A. P. *et al.* 1978. Astrophys. J. **224:** 375.
81. Rees, M. J. 1967. Mon. Not. R. Astron. Soc. **137:** 444.
82. Rees, M. J. 1977. Ann. N. Y. Acad. Sci. **302:** 613.
83. Warwick, R. S. 1979. Proc. R. Soc. London. **366:** 391.
84. Ives, J. C. & P. W. Sanford 1976. Mon. Not. R. Astron. Soc. **176:** 13P.
85. Schnopper, H. W. *et al.* 1978. I.A.U. Circ. No. 3202.
86. Ward, M. J. *et al.* 1978. Astron. Astrophys. **70:** L79.
87. Schnopper, H. W. *et al.* 1978. Astrophys. J. **222:** L91.
88. Schnopper, H. W. *et al.* 1977. Astrophys. J. **215:** L7.
89. Canizares, C. R. *et al.* 1978. Astrophys. J. **226:** L1.
90. Apparao, K. M. V. *et al.* 1978. 1978. Nature (London) **273:** 450.
91. Marshall, F. E. *et al.* 1978. Nature (London) **275:** 624.
92. Wilson, A. S. *et al.* 1979. Mon. Not. R. Astron. Soc. **187:** 109.
93. Reigler, G. R.. *et al.* 1978. I.A.U. Circ. No. 3261.
94. Griffiths, R. E. *et al.;* R. A. Wade *et al.* 1978. I.A.U. Circ. No. 3279.
95. Griffiths, R. E. *et al.* 1979. Astrophys. J. **230:** L21.

X-RAY SPECTRA OF ACCRETING DEGENERATE STARS*

N. D. Kylafis, D. Q. Lamb†‡, A. R. Masters, and G. J. Weast†

Department of Physics
University of Illinois at Urbana-Champaign
Urbana, Illinois 61801

INTRODUCTION

The nature of the vast majority of compact galactic x-ray sources is still unknown. Only in the case of the ~10% from which x-ray pulsations have been detected can the nature of the source be deduced thus far with any degree of confidence. The rotation periods and spin-up behavior found from timing observations imply that these sources are accreting magnetic neutron stars.[40,60] The x-ray and UV spectrum is a second source of clues about the nature of the underlying star, as well as about the extreme conditions of temperature and magnetic field strength that may exist in the emission region. While the x-ray spectra of many sources have been extensively observed (see, e.g., references 1, 81, 45, 32), the lack of systematic theoretical studies of x-ray emission from degenerate stars has so far prevented the exploitation of these observations.

We report here exciting new results that are changing the situation. Early theoretical studies focused on x-ray emission from neutron stars. Here we primarily discuss recent work on x-ray emission from degenerate dwarfs. These studies hold out the promise that the nature of x-ray sources and their physical properties (such as mass, magnetic field strength, intrinsic luminosity, and distance) may be determined from their x-ray spectra alone. An ability to gain such information from x-ray observations is important because most x-ray sources are optically faint, and are therefore difficult and time consuming to study optically.

NONMAGNETIC DEGENERATE DWARFS

Studies of x-ray emission from accreting nonmagnetic degenerate dwarfs include those by Hoshi,[30] Aizu,[2] Hayakawa,[22] Hayakawa and Hoshi,[23] Fabian, Pringle and Rees,[19] and Katz.[33] Recently Kylafis and Lamb[36] (see also Kylafis[35]) have carried out a comprehensive study that spans the entire range of accretion rates and stellar masses, and that includes the important, but previously unexplored, regime at moderate and high accretion rates. Their detailed numerical calculations, which will be the primary focus of our attention here, show that the characteristics of x radiation from nonmagnetic degenerate dwarfs are qualitatively different than heretofore stated. In fact, a number of x-ray sources, including both Sco X-1 and Cyg X-2, have

*This research was supported in part by NSF Grant PHY78-044404 and NASA contract NAS5-11450.

†Present address: Department of Physics, Massachusetts Institute of Technology, Cambridge, Massachusetts 02139.

‡John Simon Guggenheim Memorial Fellow.

0077–8923/80/0336–0520 $01.75/1 © 1980, NYAS

spectra, and show correlations between spectral temperature and intensity, remarkably like that of degenerate dwarfs with hard x-ray luminosities $L_h \sim 0.5–5 \times 10^{36}$ erg s^{-1}.

Assumptions

For simplicity and brevity, the calculations we will discuss assume: (1) steady, spherically symmetric accretion, (2) no magnetic field, (3) complete ionization of the accreting matter, and (4) no nuclear burning. However, we make the following remarks.

If spherically symmetric accretion occurs, the results we discuss are valid even if a strong magnetic field is present as long as bremsstrahlung cooling dominates cyclotron cooling. Bremsstrahlung cooling will dominate if the stellar magnetic field $B \lesssim 6 \times 10^6 (L/10^{36})^{2/5}$ G where L is the total accretion luminosity.[48,38] Even if accretion occurs via a disk, cyclotron cooling may not be important yet the magnetic field may still be strong enough to channel the flow of matter, so that accretion occurs approximately radially over a fraction of the stellar surface.[19] In these cases the results may still be used, together with knowledge of the overall geometry, to determine the character of the emitted radiation. If accretion occurs via a disk and the magnetic field is not strong enough to channel the flow of accreting matter, only soft, or no, x-rays are expected.[19,56]

Additional calculations,[35] which include the ionization structure around the x-ray source and the effects of absorption,[22,19] show that the accreting matter is strongly photoionized and that, except for low mass ($M \lesssim 0.5\ M_\odot$) stars or high accretion rates, the assumption of complete ionization is an excellent first approximation. Even in the latter cases, absorption generally occurs far away from the x-ray source at distances ($\sim 10^{11}$cm) comparable to or greater than the separation of low mass binaries. Absorption therefore may not occur if the density of gas surrounding the binary system is low.

Nuclear burning of accreting matter does not occur in the hot x-ray emission region, but may occur deeper in the envelope of the star. Over extended periods of time, there may be either little or no nuclear burning, presumably ending eventually with a violent nuclear outburst (cf. reference 64) or, possibly, steady nuclear burning at the accretion rate.[52,62,63] If nuclear burning occurs, the resulting intense soft x-ray blackbody radiation significantly increases Compton cooling in the emission region.[33] The effects are more extreme than, but analogous to, those encountered in the high mass ($M \gtrsim 1\ M_\odot$) stars described in the present paper.

Parameter Regimes

With the above assumptions, the following picture of x-ray emission by nonmagnetic degenerate dwarfs emerges. As accreting matter flows toward the star, a strong standoff shock forms far enough above the star for the hot, post-shock matter to cool and come to rest at the stellar surface.[30,2] The standoff distance $d \equiv r_s - R \simeq 1/4\ v_{ff}(r_s)\ t_{cool}(r_s)$, where r_s is the shock radius, R is the stellar radius, v_{ff} is the free fall velocity, and t_{cool} is the time scale for cooling, due usually to bremsstrahlung. Roughly half of the bremsstrahlung photons are emitted outward and contribute toward the observed x-ray flux; the other half are emitted inward and intercept the stellar surface, where they are reflected or absorbed. The blackbody flux resulting from the energy lost by the absorbed and reflected photons appears as UV and soft x radiation. If the mass of the degenerate dwarf is high ($M \gtrsim 1\ M_\odot$), this blackbody flux from the stellar photosphere plays a significant role in cooling matter in the hot, post-shock emission region.[33]

X and UV radiation from nonmagnetic degenerate dwarfs is thus a function of both stellar mass M and mass accretion rate $\dot{M}$, and the parameter regimes encountered are conveniently displayed on an $(M, \dot{M})$ plane as shown in FIGURE 1. A natural unit for the accretion rate is the Eddington rate $\dot{M}_E = 4\pi cR/\kappa_{es}$ at which gravitational and radiation forces balance, assuming Thomson scattering. The corresponding unit of luminosity is $L_E = (GM/R)\dot{M}_E$, so that $L/L_E = \dot{M}/\dot{M}_E$. At low accretion rates, bremsstrahlung and Compton cooling are inefficient in removing the gravitational potential energy released by the infall of accreting matter and the shock stands at a large distance above the stellar surface ($d \gg R$). Inward of the shock, the inflowing matter forms a hot, settling atmosphere in which most of the energy is liberated near the stellar surface.[19,33] As the mass accretion rate is increased, the shock moves in until eventually $d \ll R$. The transition between the two regimes occurs when the free-fall time scale $t_{ff}(R) \simeq t_{cool}(R)$ where t_{cool} is the smaller of t_{br} and t_C, the bremsstrahlung and Compton cooling time scales. As the accretion rate is increased still further, the electron scattering optical depth τ_{es} from the emission region to infinity reaches unity. At higher accretion rates, the spectrum is severely degraded by Compton scattering. Finally, if the accretion rate is sufficiently high, a bremsstrahlung fit* to the observed spectrum gives a temperature $T_{obs} \simeq 2\, m_e c^2/\tau_{es}{}^2$ (reference 31) that is less than 2 keV, and the star ceases to be a hard x-ray source. The solid curves labeled $\tau_{es} = 1$ and $d = R$ divide accretion onto nonmagnetic degenerate dwarfs into three parameter regimes, labeled I, II, and III.[33]

We note that much previous work has focused on regime II, utilizing the implied condition $d \ll R$ to simplify the calculations[30,2] (see also references 19 and 33). However, detailed calculations show that the line $d = 0.1\ R$ lies close to or, for $M \gtrsim 0.8\ M_\odot$, above the line $\tau_{es} = 1$ (see FIGURE 1). Thus for all practical purposes regime II does not exist.

Spectrum

The x-ray and UV spectrum produced by accretion onto nonmagnetic degenerate dwarfs generally has three components: (1) a hard x-ray bremsstrahlung component produced by the hot, post-shock emission region, (2) a soft x-ray blackbody component produced by bremsstrahlung photons that are absorbed by the stellar surface and reemitted, and (3) secondary radiation produced by Compton heating of infalling matter above the shock.

These three components are clearly visible in FIGURE 2, which shows the x-ray and UV spectrum produced by accretion onto a 1.0 $M_\odot$ star at a rate 0.15 $\dot{M}_E$, corresponding to $\tau_{es} = 14$. This spectrum is typical of that found in regime I. At such accretion

*Since the temperatures we encounter are often high, we will always quote temperatures obtained using a single-temperature thin-bremsstrahlung spectrum *with* frequency-dependent Gaunt factor.

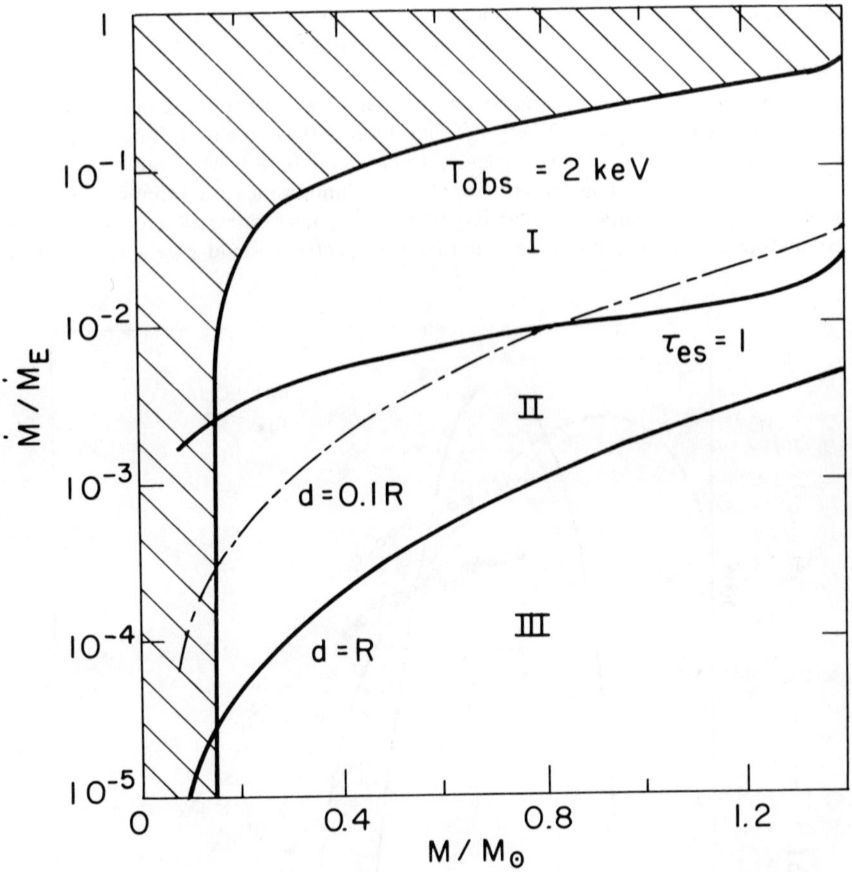

FIGURE 1. Parameter regimes in the $(M, \dot{M})$ plane. Below the solid curve labeled $d = R$, cooling of the hot, post-shock matter as it reaches the stellar surface requires a shock standoff distance $d > R$; above the curve, $d < R$. Along the dot-dashed curve, $d = 0.1\ R$ as determined from detailed calculations. Above the solid curve labeled $\tau_{es} = 1$ the electron scattering optical depth from the stellar surface to infinity exceeds unity. Above the solid curve labeled $T_{obs} = 2$ keV, a bremsstrahlung fit to the observed x-ray spectrum gives a temperature less than 2 keV and the star ceases to be a hard x-ray source (shaded region). The solid curves labeled $\tau_{es} = 1$ and $d = R$ divide accretion onto magnetic degenerate dwarfs into three regimes, labeled I, II, and III. (From Kylafis and Lamb.[36] By permission of *The Astrophysical Journal*.)

rates, the electron scattering optical depth τ_{es} from the emission region to infinity is large. The bremsstrahlung component is then degraded by Compton scattering, and the blackbody component contains an additional contribution from bremsstrahlung photons that are initially emitted outward, but that scatter, intercept the stellar surface, and are absorbed there. The secondary radiation, which is a consequence of such Compton scattering, is important only if degradation of the bremsstrahlung spectrum is substantial. From the spectrum illustrated in FIGURE 2, we draw three important conclusions: (1) degenerate dwarfs can produce larger luminosities than earlier estimates indicated, (2) even high-mass stars can produce low-temperature ($\lesssim$10 keV) x-ray spectra, and (3), for moderate and high accretion rates and high ($M \gtrsim 1.2\ M_\odot$) stellar masses, the spectra exhibit a high-energy tail.

The degraded bremsstrahlung spectrum observed at infinity has a hard (>2 keV) x-ray luminosity $L_h = 1.3 \times 10^{36}$ erg s^{-1}. This luminosity is close to the maximum hard x-ray luminosity $L_h^{max} = 4 \times 10^{36}$ erg s^{-1}, which occurs for a 1.3 $M_\odot$ star, that we find for spherically accreting degenerate dwarfs. Such luminosities are 4 times larger, and occur for accretion rates 10 times larger, than previous estimates suggested.[15,19] A bremsstrahlung fit to the degraded hard x-ray spectrum would give a temperature

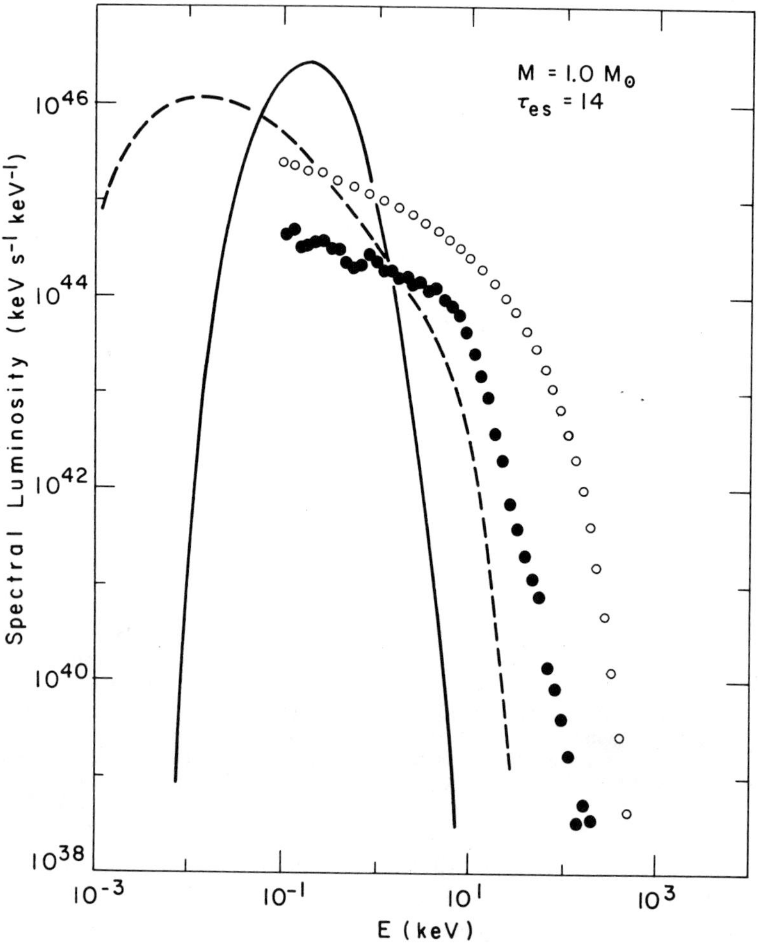

FIGURE 2. X-ray and UV spectrum produced by accretion onto a 1.0 $M_\odot$ degenerate dwarf at a rate $\dot{M} = 0.15\ \dot{M}_E$, corresponding to $\tau_{es} = 14$. Key: ○○○ spectrum produced by the emission region, ●●● degraded spectrum observed at infinity, — blackbody spectrum emitted from the stellar surface, --- secondary radiation produced by Compton heating. (From Kylafis and Lamb.[36] By permission of *The Astrophysical Journal.*)

$T_{obs} = 7$ keV, far smaller than the shock temperature $T_s = 48$ keV.† *Note, however, that* T_{obs} *does not correspond to any physical temperature in the source.* Above the knee at $\simeq 10$ keV, the hard x-ray spectrum is a power law of index $\alpha = -3.6$, with a slight high-energy tail. If the stellar mass is increased, holding the accretion rate constant, the high-energy tail becomes more and more pronounced. It is a result of the Klein–Nishina cross section, which decreases at energies $\sim m_e c^2$ and therefore leaves unscattered high-energy photons that would otherwise be degraded. We note that heretofore only neutron stars[84,3,61] were thought able to produce high-energy tails (see, e.g., reference 33). Below the knee, secondary radiation fills in the spectrum, producing a shoulder between the blackbody and hard x-ray components.

The blackbody component, which is mildly Comptonized as it passes through the hot emission region, has a temperature $T_{bb} = 0.046$ keV and a luminosity $L_{bb} = 2.0 \times 10^{37}$ erg s^{-1}. The total soft x-ray luminosity L_s, which we define by $L_s \equiv L - L_h$, is equal to 2.1×10^{37} erg s^{-1}, an order of magnitude larger than the hard x-ray luminosity. In contrast, an intense soft x-ray component is not expected for nonmagnetic neutron stars since the blackbody limit imposes a Rayleigh–Jeans falloff of the spectral flux below ~ 1 keV, even if accretion occurs over the entire surface of the neutron star (cf. reference 3). Detection of such a component would therefore imply that the observed source is a degenerate dwarf. Both the blackbody component and the secondary radiation have Rayleigh–Jeans spectra that are cut off at low energies, due to free–free absorption in the cooler infalling matter above the shock. The visual characteristics of such sources are thus likely dominated by emission from a disk, or from the heated face of the companion star (cf. references 49 and 50).

Correlation Between Spectral Temperature and Luminosity

A fourth conclusion is that at moderate and high accretion rates the spectral temperature varies dramatically and the star exhibits a pronounced correlation between x-ray spectral temperature and luminosity. FIGURE 3 shows the variation in the x-ray spectrum of a 1.0 $M_\odot$ star produced by changes in the accretion rate. The six spectra shown span the entire range of accretion rates from regime III through I. At low accretion rates, $\tau_{es} < 1$ and the observed hard x-ray spectrum is essentially that produced in the emission region. However, as the accretion rate is increased, τ_{es} exceeds unity and Compton scattering begins to degrade the spectrum.[31] As the accretion rate is increased further, degradation of the spectrum becomes ever more severe. Finally, due to the combined effects of degradation and weakening of the shock by radiation pressure, the bremsstrahlung component disappears altogether. The star then ceases to be a hard (i.e., $T_{obs} > 2$ keV) x-ray source.

FIGURE 4 summarizes the behavior of L, L_h, L_s, and T_{obs}. At low accretion rates, roughly half the accretion luminosity L appears as hard x-rays (L_h) and half appears as soft x-rays (L_s). As the accretion rate is increased, τ_{es} exceeds unity and the blackbody component contains an increasing contribution from bremsstrahlung

†Even the initial bremsstrahlung spectrum produced by the emission region is substantially softer than qualitative arguments would suggest. The shock temperature is 48 keV, yet a bremsstrahlung fit to the initial spectrum gives $\simeq 35$ keV. This is due to the fact, pointed out by Katz,[33] that the bulk of the bremsstrahlung emission comes from denser, and therefore cooler, layers of the emission region near the stellar surface.

photons that are initially emitted outward, but that scatter, intercept the stellar surface, and are absorbed there. As a result, L_s rises rapidly. L_h rises more slowly, reaches a maximum of 1.9×10^{36} erg s^{-1} when $\tau_{es} = 7$, and then falls sharply. In contrast, the observed temperature T_{obs} is high at low accretion rates but decreases as the accretion rate is increased, due to Compton degradation.

The resulting correlation between T_{obs} and L_h is shown in FIGURE 5. Note that the accretion rate increases as one moves from upper left to lower right along the curves.

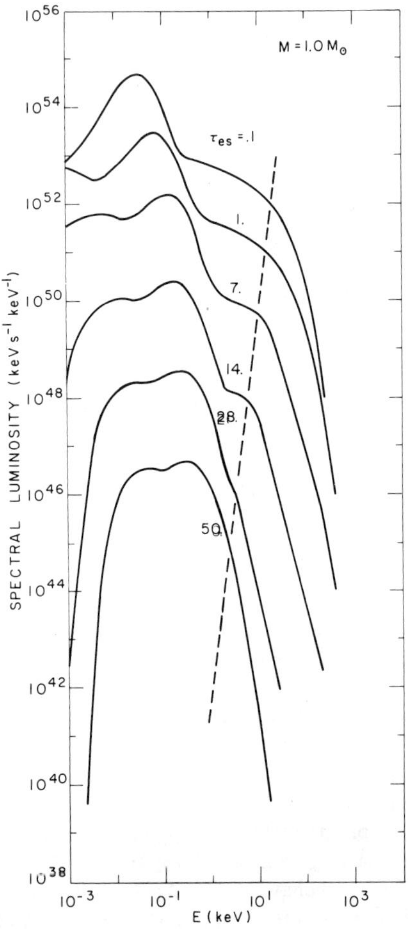

FIGURE 3. Variation in the x-ray and UV spectra of a 1.0 $M_\odot$ degenerate dwarf produced by changes in the accretion rate and, therefore, τ_{es}. The six spectra shown correspond to accretion rates $\dot{M} = 1.1 \times 10^{-3}$, 1.1×10^{-2}, 7.8×10^{-2}, 0.15, 0.27, and 0.43 $\dot{M}_E$. The units on the vertical axis correspond to the 0.43 $\dot{M}_E$ ($\tau_{es} = 50$) curve. The other spectra have been multiplied by successively larger powers of 100 for clarity of presentation; thus, for example, the $\tau_{es} = 0.1$ spectrum has a true spectral luminosity $\simeq 8 \times 10^{42}$ keV s^{-1} keV^{-1} at 1 eV. The dashed line indicates the changing position of the hard x-ray cutoff produced by Compton scattering. (From Kylafis and Lamb.[36] By permission of *The Astrophysical Journal*.)

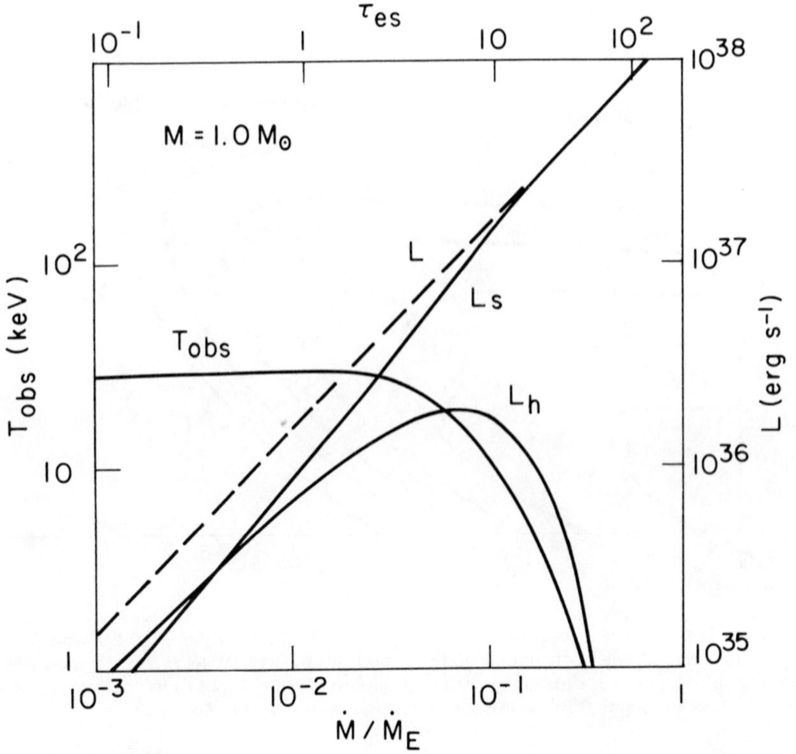

FIGURE 4. Behavior of L, L_h, L_s, and T_{obs} as a function of mass accretion rate for a 1.0 $M_\odot$ degenerate dwarf. The luminosity units are given along the right axis and the temperature units along the left. (From Kylafis and Lamb.[36] By permission of *The Astrophysical Journal.*)

For sources found in the lower right of FIGURE 5, *an increase in T_{obs} and L_h therefore corresponds to a decrease in the accretion rate;* T_{obs} and L_h increase since the smaller accretion rate lessens Compton degradation of the hard x-ray spectrum. A correlation like that in FIGURE 5 is not expected for nonmagnetic neutron stars (see below) and may be an important signature of degenerate dwarf x-ray sources.

Comparison with Nonmagnetic Neutron Stars

Studies of x-ray emission from radially accreting nonmagnetic neutron stars‡ include those by Zel'dovich and Shakura,[84] Alme and Wilson,[3] and Shapiro and Salpeter.[61] If no collisionless shock forms, the accreting matter behaves as a beam of charged particles decelerated by Coulomb collisions in the dense surface layers of the neutron star. As shown in FIGURE 6, the resulting spectrum is essentially a Compton-

‡The interesting, but difficult, case of disk accretion onto a nonmagnetic neutron star has so far received little attention. If the disk is optically thick, we expect the x-ray spectrum to resemble that of a blackbody having a temperature ≃ few keV.

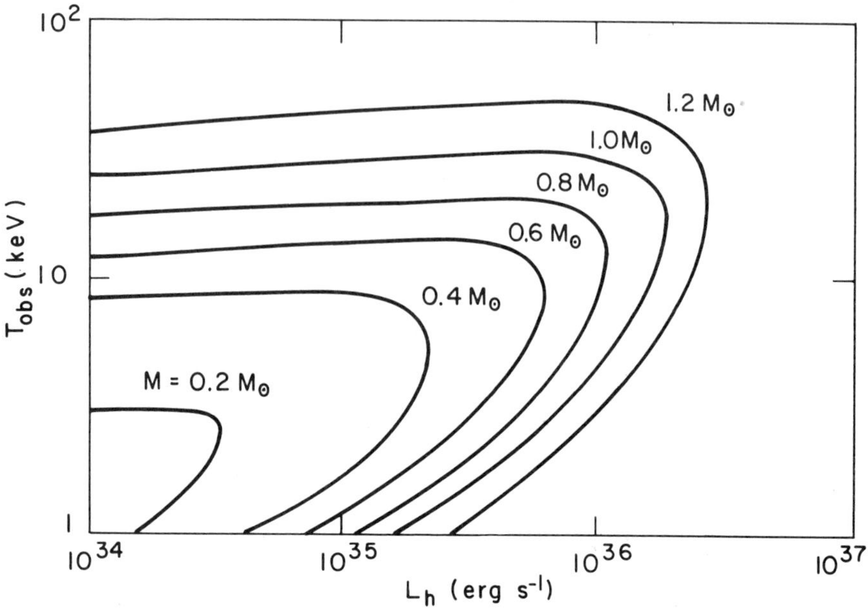

FIGURE 5. Correlation between T_{obs} and L_h for $M = 0.2$, 0.4, 0.6, 0.8, 1.0, and 1.2 $M_\odot$ degenerate dwarfs. The contribution of the blackbody component to L_h, which increases L_h somewhat along the lower-right-hand portion of the curve for $M = 1.2\ M_\odot$, has not been included. (From Kylafis and Lamb.[36] By permission of *The Astrophysical Journal.*)

ized blackbody spectrum that may have a high-energy tail if the mass of the star is low.[84,3] If a collisionless shock does form,§ the hot, post-shock emission region is cooled primarily by inverse Compton scattering of blackbody photons emitted from the stellar surface. As shown in FIGURE 7, the resulting spectrum is again essentially a Comptonized blackbody spectrum having, possibly, a high-energy tail.[61]

Because of the neutron star's deep potential well, and great efficiency in converting the infall of matter into x radiation, $\tau_{es} < 1$ until $L \simeq L_E$. Thus, Compton scattering is unimportant until $L \simeq L_E$ and, in contrast to degenerate dwarfs, the spectral temperature is expected to be a weak function of luminosity, varying roughly as $L_h^{1/4}$ at moderate and even high luminosities. When $L \simeq 0.9\ L_E$, radiation pressure allows the density of accreting matter to increase, and τ_{es} exceeds unity. Because v/c of the infalling matter is near unity, Compton scattering causes the observed spectral temperature T_{obs} to *increase*[82] and produces a power law tail[79] (see also reference 51). As L approaches L_E even more closely, absorption causes the effective radius of the stellar surface to increase and the spectral temperature to decrease. Eventually the effective radius becomes so large that T_{obs} falls below 2 keV, and the star ceases to be a hard x-ray source. From this discussion we expect the correlation between T_{obs} and L_h

§A collisionless shock is not expected unless a magnetic field is present. But the required magnetic field is sufficiently weak that it need not channel the flow of accreting matter nor affect the cooling rate in the emission region.

for nonmagnetic neutron stars to be quite different from that shown in FIGURE 5 for degenerate dwarfs.

These results and the detailed calculations of x-ray emission from degenerate dwarfs described above provide criteria for distinguishing between degenerate dwarf and neutron star x-ray sources. For example: (1) Spherically accreting degenerate dwarfs have a maximum hard x-ray luminosity $L_h^{max} = 4 \times 10^{36}$ erg s^{-1}. Therefore the brightest galactic x-ray sources cannot be degenerate dwarfs (see also reference 19). (2) The correlation between spectral temperature T_{obs} and hard x-ray luminosity L_h is expected to be quite different for degenerate dwarfs and neutron stars. Therefore, the pronounced correlation shown in FIGURE 5 may be an important signature of

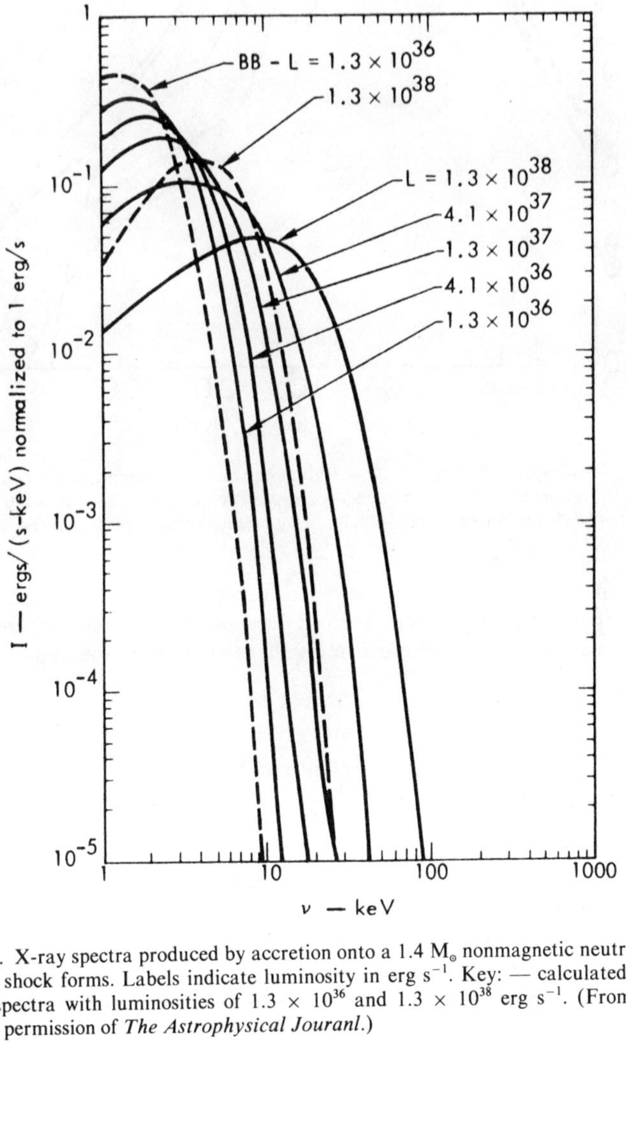

FIGURE 6. X-ray spectra produced by accretion onto a 1.4 $M_\odot$ nonmagnetic neutron star if no collisionless shock forms. Labels indicate luminosity in erg s^{-1}. Key: — calculated spectra; --- blackbody spectra with luminosities of 1.3×10^{36} and 1.3×10^{38} erg s^{-1}. (From Alme and Wilson.[3] By permission of *The Astrophysical Jouranl.*)

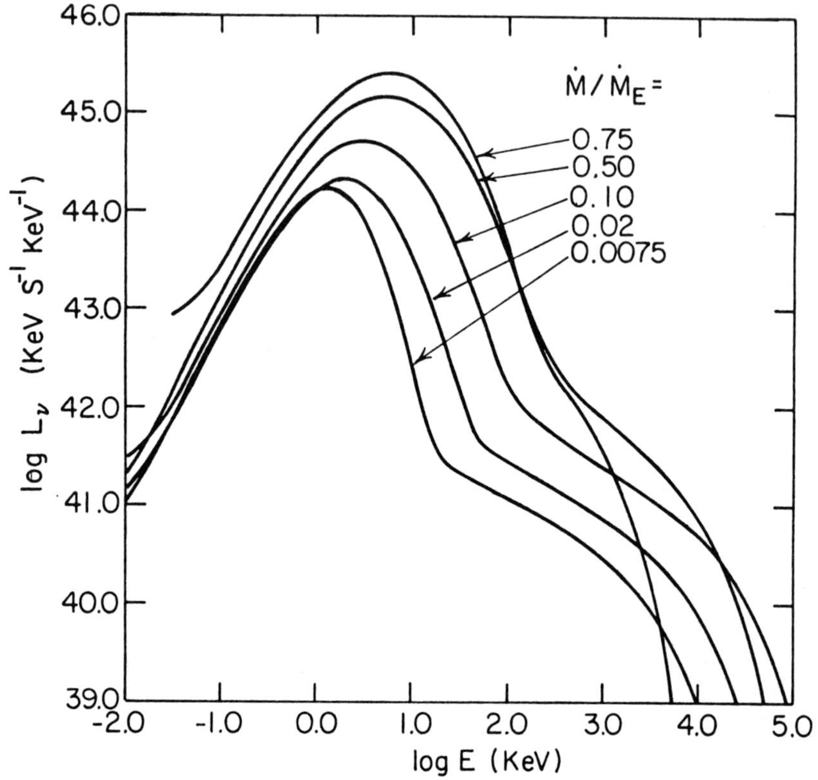

FIGURE 7. X-ray spectra producted by accretion onto a 1.1 $M_\odot$ nonmagnetic neutron star if a collisionless shock forms. The spectra shown correspond to accretion rates $\dot{M}$ = 0.075, 0.02, 0.1, 0.5, and 0.75 $\dot{M}_E$. (From Shapiro and Salpeter.[61] By permission of *The Astrophysical Journal.*)

degenerate dwarf x-ray sources. (3) Degenerate dwarf spectra all have an intense UV or soft x-ray component that is not expected for nonmagnetic neutron stars. Detection of such a component would therefore imply that the observed source is a degenerate dwarf.

Implications for Observed X-Ray Sources

Recent observations of Cyg X-2 by the Copernicus satellite[8] show variations in count rate accompanied by spectral changes, as shown in FIGURE 8, that are in very good agreement with the curves in FIGURE 5. This agreement suggests that Cyg X-2 is an accreting degenerate dwarf with a mass of ~0.4 $M_\odot$ at a distance ~300 pc.[9]

The nature of the so-called "galactic bulge" or Sco X-1-like x-ray sources is unknown. As a class, these sources cannot be accreting nonmagnetic degenerate dwarfs for the following reason. If they lie symmetrically about the galactic center, as their distribution in galactic coordinates suggests, many of them have x-ray luminosities near L_E. The calculations discussed above show that such x-ray luminosities cannot be attained by nonmagnetic degenerate dwarfs. If we place these sources nearer to us, so that their luminosities are not so large, their space density is then so great that the resulting x-ray emission exceeds the x-ray background observed, for example, in the direction of the galactic bulge.

Although the Sco X-1-like sources, as a class, cannot be accreting nonmagnetic degenerate dwarfs, some individual members of the class can be. FIGURE 9 shows that sources such as 3U1617 − 15 (Sco X-1), 3U0614 + 09, 3U1728 − 16 (GX9 + 9), and 3U1758 − 25 (GX5 − 1) exhibit a pronounced correlation between T_{obs} and L_h[45] (see also reference 53) and may be degenerate dwarf x-ray sources.[9] Verification that these x-ray sources are degenerate dwarfs would clearly have an enormous impact on galactic x-ray astronomy. And if the sources behave as calculations predict, comparison of theory and observation will allow the masses, intrinsic luminosities, and distances of these x-ray sources to be determined from their x-ray spectra alone.

MAGNETIC DEGENERATE DWARFS

Fabian, Pringle, and Rees[19] and Masters, Fabian, Pringle, and Rees[48] have discussed the qualitative features of accretion onto magnetic degenerate dwarfs. More recently, Lamb and Masters[38,39] (see also Masters[47]) carried out detailed numerical calculations of high-harmonic cyclotron emission from a hot plasma, from which they

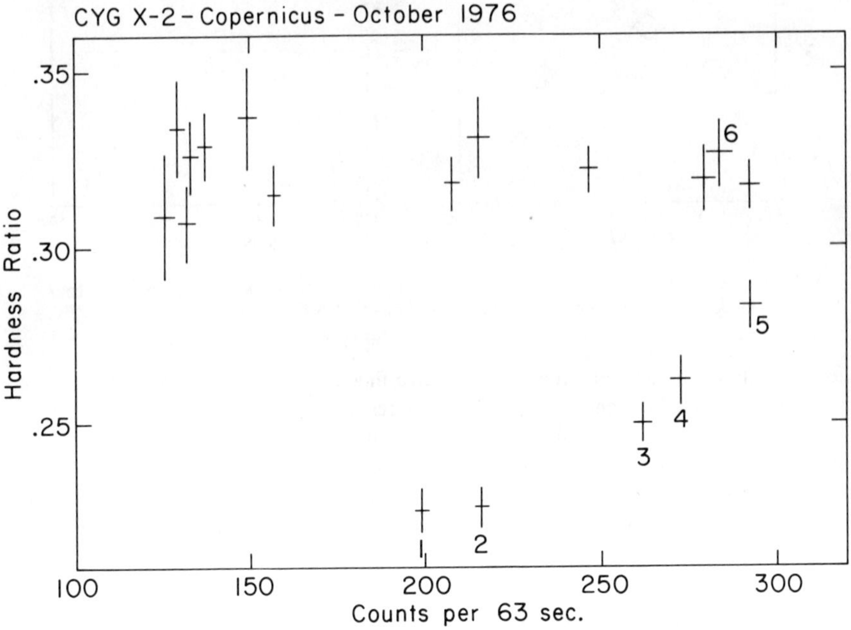

FIGURE 8. Hardness ratio versus the number of counts per 63 seconds observed by the Copernicus satellite 20–21 October 1976 for Cyg X-2. Each point represents the summation of ~2^h of data, and the first six points are numbered chronologically. (From Branduardi.[8])

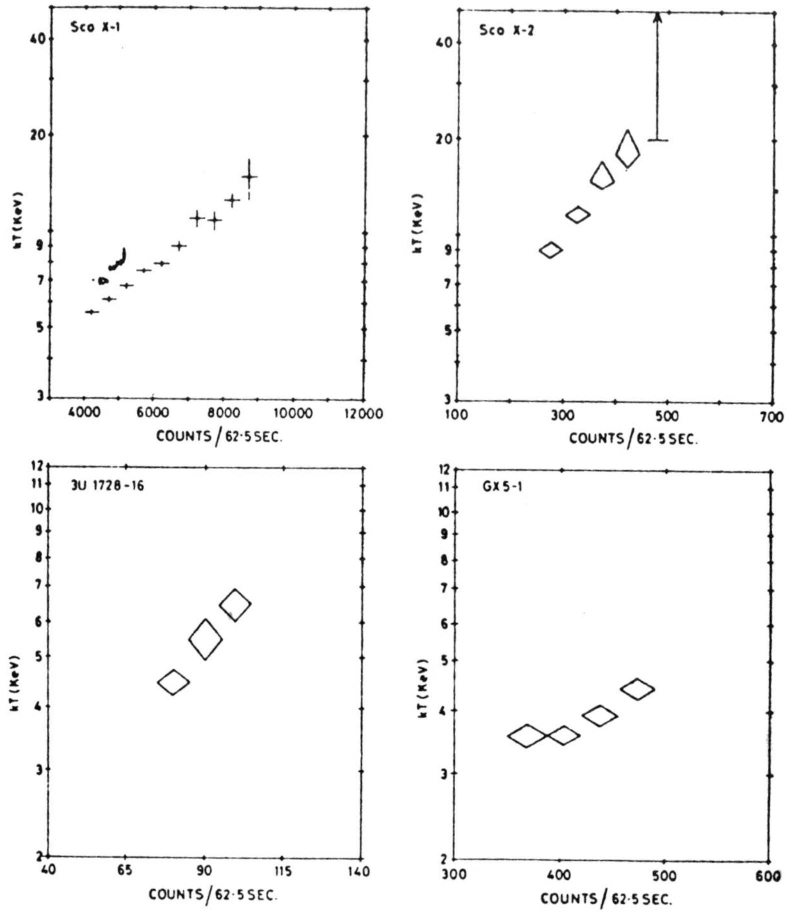

FIGURE 9. Spectral temperature versus the number of counts per 63 seconds for four Sco X-1-like sources. (From Mason *et al.*[45] By permission of *M.N.R.A.S.*)

have developed a self-consistent, quantitative model of x-ray and UV emission from accreting magnetic degenerate dwarfs. Their results, which will be the primary focus of our attention here, show that the soft x-ray and hard x-ray emission observed from AM Her is characteristic of these objects. Their results further reveal that the qualitative features of the spectrum are sufficient to determine with relative confidence the conditions in the x-ray emission region, including the value of the magnetic field strength.

High-Harmonic Cyclotron Emission from a Hot Plasma

The physical conditions of temperature ($\simeq 10$ keV), density ($\gtrsim 10^{-10}$ g cm^{-3}), and magnetic field strength ($\gtrsim 10^{6}$ G) relevant to accreting magnetic degenerate dwarfs have been encountered earlier in plasma fusion devices using magnetic confinement.

In such devices, the cyclotron spectrum is moderately self-absorbed at the fundamental and the emission is dominated by the first few harmonics[75,76,16,17,29] (see reference 7 for a lucid review of this work). One might therefore hope to be guided by such studies in attacking the problem of x-ray emission from accreting magnetic degenerate dwarfs. However, the emission regions of such stars are much larger than fusion devices (~10^5 m versus ~1 m). The self-absorption is therefore much greater, and optical depths at the fundamental may easily reach 10^9![48]

The actual emitted cyclotron spectrum is then typically like that indicated by the

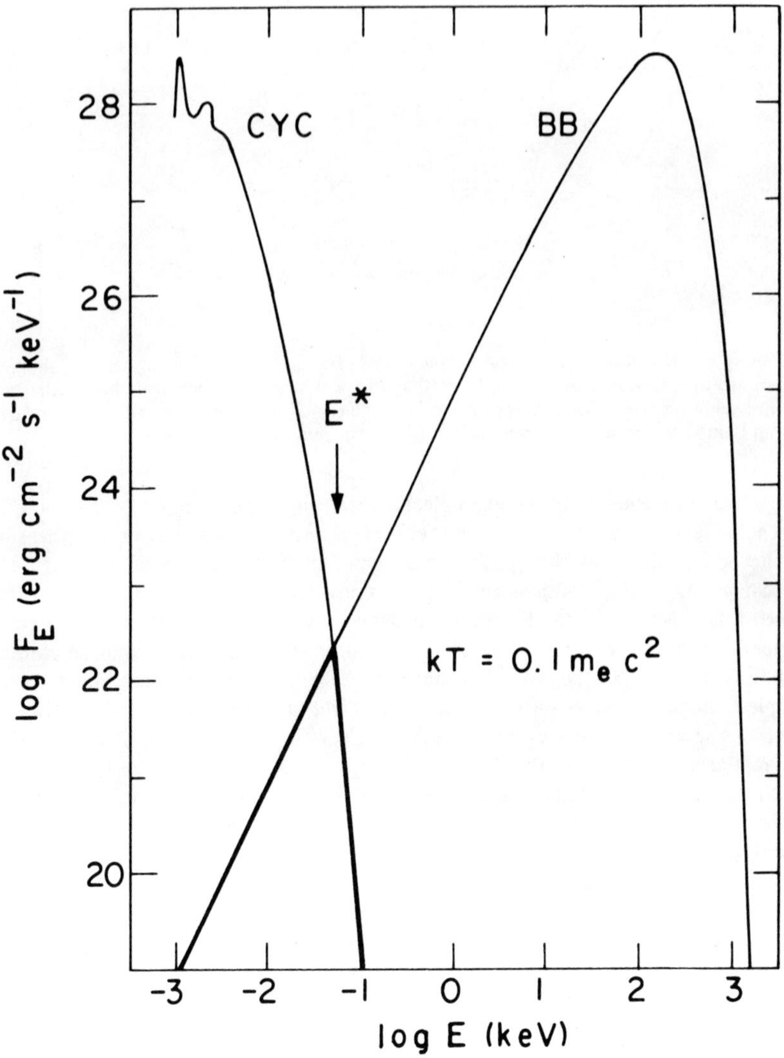

FIGURE 10. Optically thin cyclotron (left) and blackbody (right) fluxes for $kT = 0.1\ m_ec^2$, $B = 10^8$ G, $\Lambda = 10^9$. The heavy line is the cyclotron flux actually emitted.

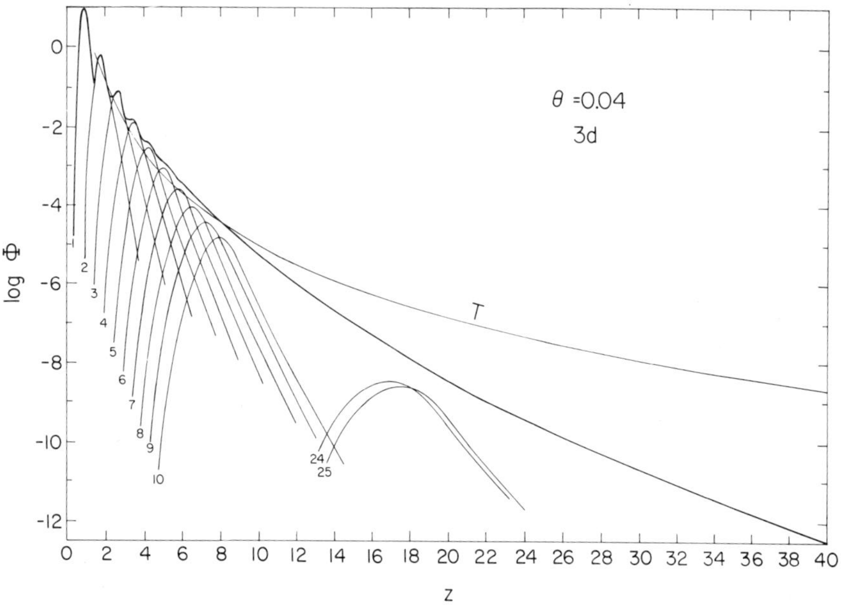

FIGURE 11. Dimensionless spectral flux $\Phi = 0.04$. The heavy curve is the total flux; the light curves are the individual harmonics. Note that the peak of the nth harmonic lies further and further below nz as n becomes large. The curve labeled "T" is the fit obtained by Trubnikov.[77] (From Lamb and Masters.[38] By permission of *The Astrophysical Journal.*)

heavy line in FIGURE 10. In this example, the electron temperature $kT_e = 0.1\ m_ec^2$, the magnetic field strength $B = 10^8$ G, and the size parameter Λ, which is a rough measure of the optical depth at the cyclotron fundamental[75,29] is 10^9. Below $E^* = 50$ eV, corresponding to the $\simeq$50th harmonic, the source is heavily self-absorbed and the emitted flux is limited to the Rayleigh–Jeans tail of a blackbody spectrum with temperature $kT_e = 0.1\ m_ec^2$. Above E* the source is optically thin, and the emitted flux falls rapidly. This behavior introduces a tremendous complication into even the simplest model of cyclotron emission from accreting magnetic degenerate dwarfs. Neither the optically thin cyclotron spectrum nor the blackbody spectrum gives an adequate representation of the spectrum actually emitted, and hence of the cyclotron cooling rate. Instead, accurate knowledge of the cyclotron emissivity at high harmonics is first required. Then the value of E^*, and hence the cooling rate, must be found taking into account frequency-dependent radiative transfer within the emission region.

As a first step in developing a self-consistent model of accreting magnetic degenerate dwarfs, Lamb and Masters[38,39] (see also Masters[47]) have carried out accurate calculations of the cyclotron emissivity at high harmonics. Examples of their results are shown in FIGURES 11 and 12. The dimensionless flux $\Phi\,(z, \theta)$ shown in these figures is the cyclotron emissivity normalized to the Rayleigh–Jeans spectral flux in a single polarization. It is a function only of $\theta = kT_e/m_ec^2$ and $z = \omega/\omega_c$, where ω_c is the cyclotron frequency. With this normalization, E^* for a particular emission region having a size parameter Λ is just given by that z for which $\Lambda\omega = 1$. All the dependence on the density and geometry of the emission region is contained in the size parameter Λ, while the magnetic field enters only in Λ and as a scale factor for the dimensionless frequency z. FIGURE 11 shows the cyclotron spectral flux from an optically thin plasma with electron temperature $T_e = 20$ keV. The heavy curve shows the total spectral flux, while the light curves indicate the contributions of individual harmonics. The radiation at high harmonics is due primarily to electrons in the high-velocity tail of the Maxwellian distribution. At the mildly relativistic temperatures of interest here, γ is appreciably greater than one for electrons in this tail. Therefore the peak of the

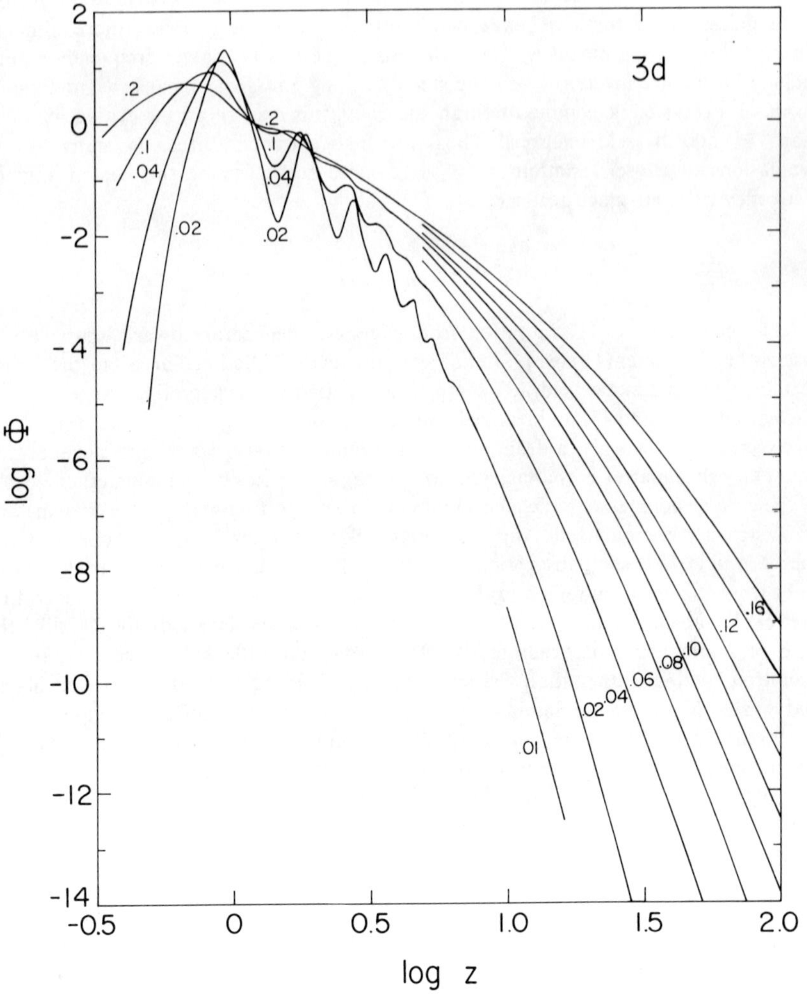

FIGURE 12. Dimensionless spectral flux Φ from a three-dimensional distribution of electrons for the indicated values of the electron temperature θ.

nth harmonic lies further and further below nz as n becomes large. The curve labeled "T" shows the fit obtained by Trubnikov[77] to calculations by Drummond and Rosenbluth[17] for the range $5 < z < 10$. The agreement is encouraging there, but the necessity of new calculations is evident from the fact that this fit already differs from the actual Φ by $\sim 10^4$ when $z \simeq 40$. FIGURE 12 summarizes the results for three-dimensional electron distributions with $\theta = 0.01$–0.02. All results are shown for $z > 5$, along with selected results for $z < 5$.

FIGURES 11 and 12 illustrate three important features of the cyclotron emissivity: (1) most of the emission occurs above the fundamental if the temperature is moderate or high, (2) the cyclotron emissivity falls steeply with frequency, and (3) thermal broadening of the individual harmonics produces a smooth continuum at high frequencies. These features make computation of the spectral flux more difficult. Thermal broadening means that calculations of the flux at a given frequency z must include harmonics up to $n \sim 2z$. The steeply falling emissivity means that individual harmonics must be determined to high accuracy; this may entail the evaluation of as many as 500 Bessel functions. Thus, for higher temperatures, as many as 10^5 evaluations of Bessel functions may be required to determine the flux at a given frequency z to four-place accuracy.

Assumptions

The calculations of x radiation from magnetic degenerate dwarfs that we will discuss here assume: (1) steady, radial accretion over a fractional area f of the stellar surface, (2) a magnetic field, (3) complete ionization of the accreting matter, (4) no nuclear burning, and (5) an idealized emission region.

Accretion over only a fraction of the stellar surface takes into account the possibility that matter is channeled onto the magnetic poles by the magnetic field. As in the case of nonmagnetic degenerate dwarfs, the accreting matter is often expected to be strongly photoionized. Any absorption will then occur far away from the x-ray source, and if $f \ll 1$ such absorption will have little importance except when the line of sight from the emission region to the observer lies directly in the column of accreting matter. If nuclear burning occurs, it will produce a large blackbody flux. Unlike the case of nonmagnetic degenerate dwarfs, it may do little else since, in general, cyclotron cooling, rather than Compton cooling resulting from the enhanced blackbody flux, will remain the dominant cooling mechanism in the emission region.

Rather than attempting to solve the full coupled hydrodynamic and frequency-dependent radiative transfer problem, which would determine completely the structure of the hot, post-shock emission region and the shape and angular dependence of the emitted flux, the emission region has been idealized as follows. The emission region is assumed to be a geometrically thin slab¶ having uniform temperature, density, and magnetic field strength. These assumptions will be roughly correct as long as $d \ll (2f)^{1/2}R$, so that the area of the slab's face is larger than that of its edges, and $d < R$. Variations and anisotropies in the emission and absorption of cyclotron

¶The calculations can be extended in a straightforward fashion to the opposite geometry of a tall, thin cylinder if the temperature, density, and magnetic field strength are assumed to be uniform in the emission region.

radiation are neglected, and a total cyclotron cooling rate is used instead. This rate corresponds to emission equal to the appropriate Rayleigh–Jeans spectrum below E^* and to an optically thin cyclotron spectrum above E^*. These assumptions allow the important parameter regimes to be delineated and approximate spectra to be calculated.

Parameter Regimes

With the above assumptions, the following picture of x-ray emission by magnetic degenerate dwarfs emerges. It is analogous to that for nonmagnetic degenerate dwarfs. As accreting matter flows toward the star, a strong standoff shock forms far enough above the star for the hot, post-shock matter to cool and come to rest at the stellar surface.[19] The standoff distance $d \equiv r_s - R \simeq v_f(r_s)t_{cool}(r_s)$ where v_f is the flow velocity of the matter, t_{cool} is the time scale for cooling, due now to both bremsstrahlung and cyclotron emission, and the other quantities are defined as before. Roughly half of the cyclotron flux is emitted outward and forms a blackbody-limited component in the UV. Roughly half of the bremsstrahlung flux is emitted outward and forms a hard x-ray component. The other halves of the cyclotron and bremsstrahlung fluxes are emitted inward and are reflected or absorbed by the stellar surface. The resulting blackbody flux forms a UV or soft x-ray component with $L_{bb} \simeq L_{cyc} + L_{br}$, where L_{bb}, L_{cyc}, and L_{br} are the luminosities in the blackbody, cyclotron, and bremsstrahlung components.

The effective accretion rate of the accreting sector of a magnetic star is $\dot{M}/f$, and the corresponding luminosity is L/f. X-ray and UV radiation from magnetic degenerate dwarfs is thus a function of stellar mass M, magnetic field strength B, and effective luminosity L/f. The dependence on stellar mass is significant but is less than that of the other two variables. If we specify the mass of the star, the parameter regimes encountered are conveniently displayed on a $(B, L/f)$ plane as shown for a 1 $M_\odot$ star in FIGURE 13. The upper left of the plane corresponds to low magnetic field strengths and high effective luminosities (and thus high densities in the emission region). In this portion of the plane, bremsstrahlung cooling dominates cyclotron cooling in the hot, post-shock emission region, and the results discussed earlier for nonmagnetic degenerate dwarfs apply. As one increases B or lowers L/f, moving toward the lower right in FIGURE 13, cyclotron cooling becomes more important until eventually it dominates.[48] The dashed line shows the location at which this occurs, as determined from detailed numerical calculations equating t_{cyc} and t_{br}, the cyclotron and bremsstrahlung cooling time scales.

As long as bremsstrahlung emission dominates, electrons and ions generally exchange energy faster than the rate at which the electrons lose energy. The electrons and ions in the hot, post-shock emission region are then able to achieve a single temperature and may be treated as a single fluid. This situation, which is the usual one, is illustrated in the left portion of FIGURE 14. The curves labeled T_e and T_i illustrate schematically the temperature profiles of the electrons and the ions in the emission region. On the other hand, if cyclotron emission dominates and one goes to sufficiently large B or low L/f, the rate at which the electrons are cooled can be rapid enough to prevent their ever achieving a common temperature with the ions.[19] The electrons and ions must then be treated as two separate fluids. This situation is

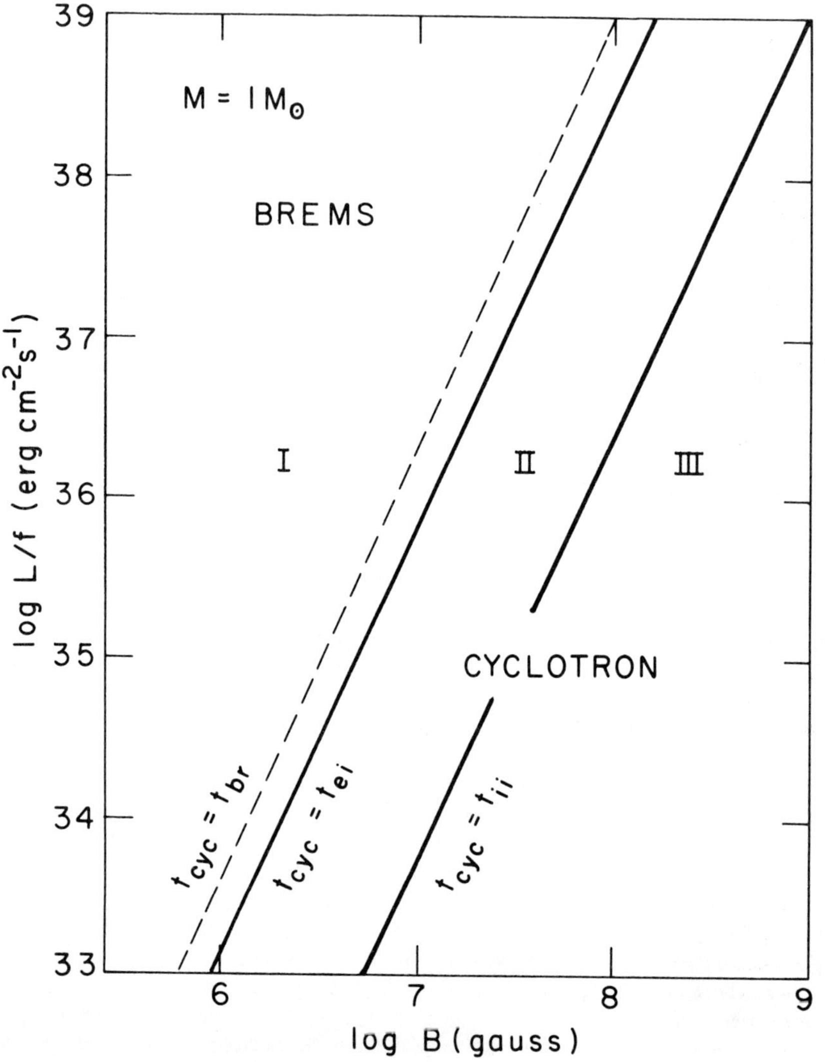

FIGURE 13. Parameter regimes in the $(B, L/f)$ plane for a $M = 1\ M_\odot$ degenerate dwarf. The one-fluid regime lies above and to the left of the solid line labeled $t_{cyc} = t_{ei}$, the two-fluid regime below and to the right. The nonhydrodynamic regime lies below and to the right of the solid line labeled $t_{cyc} = t_{ii}$. Cyclotron emission dominates below and to the right of the dashed lines, bremsstrahlung above and to the left. The solid curves labeled $t_{cyc} = t_{ei}$ and $t_{cyc} = t_{ii}$ divide accretion onto magnetic degenerate dwarfs into three parameter regimes, labeled I, II, and III. (From Lamb and Masters.[38] By permission of *The Astrophysical Journal*.)

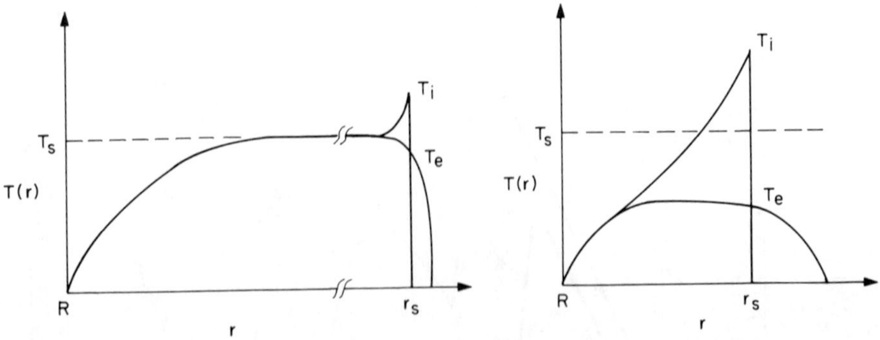

FIGURE 14. (Left) Schematic picture of electron and ion temperatures as a function of radius for the one-fluid regime (regime I). (Right) Schematic picture of electron and ion temperatures as a function of radius for the two-fluid regime (regime II).

illustrated schematically in the right portion of FIGURE 14. The transition between the one- and two-fluid regimes occurs when the cyclotron cooling time scale t_{cyc} equals the electron–ion energy exchange time scale t_{ei}, and is shown as a solid line in FIGURE 13.

As one moves further toward the lower right in FIGURE 13, the cyclotron cooling rate becomes so great that the ions lose their energy to the electrons before the ions have sufficient time to form a Maxwellian velocity distribution via Coulomb collisions among themselves. The behavior of the ions is then no longer hydrodynamic and new methods must be introduced to treat the problem. Nevertheless, we expect the rate of energy loss by the ions to remain nearly the same as that in the two-fluid regime since the rate of energy loss will still be determined primarily by the velocity of the electrons. The onset of the nonhydrodynamic regime occurs when the cyclotron cooling time scale t_{cyc} equals the ion–ion energy exchange time scale t_{ii}, and is also shown by a solid line in FIGURE 13.

As one moves still further toward the lower right in FIGURE 13, two other effects become important. First, as we have seen, the high-energy tail of the electron velocity distribution is responsible for the bulk of cyclotron emission at high harmonics. Cyclotron emission may depopulate this tail so rapidly that Coulomb collisions are unable to maintain it. The rate of cyclotron emission is then altered. Second, cyclotron emission reduces the component of electron velocity perpendicular to the magnetic field. This reduction may become so rapid that Coulomb collisions are unable to maintain an isotropic electron velocity distribution ($t_{cyc} \ll t_{ee}$). However, both of these effects become important only after the onset of nonhydrodynamic behavior by the ions.

The solid curves labeled $t_{cyc} = t_{ei}$ and $t_{cyc} = t_{ii}$ divide accretion onto magnetic degenerate dwarfs into three parameter regimes, labeled I, II, and III.

Spectrum

The x-ray and UV spectrum produced by accretion onto magnetic degenerate dwarfs generally has four components: (1) a blackbody-limited UV cyclotron component

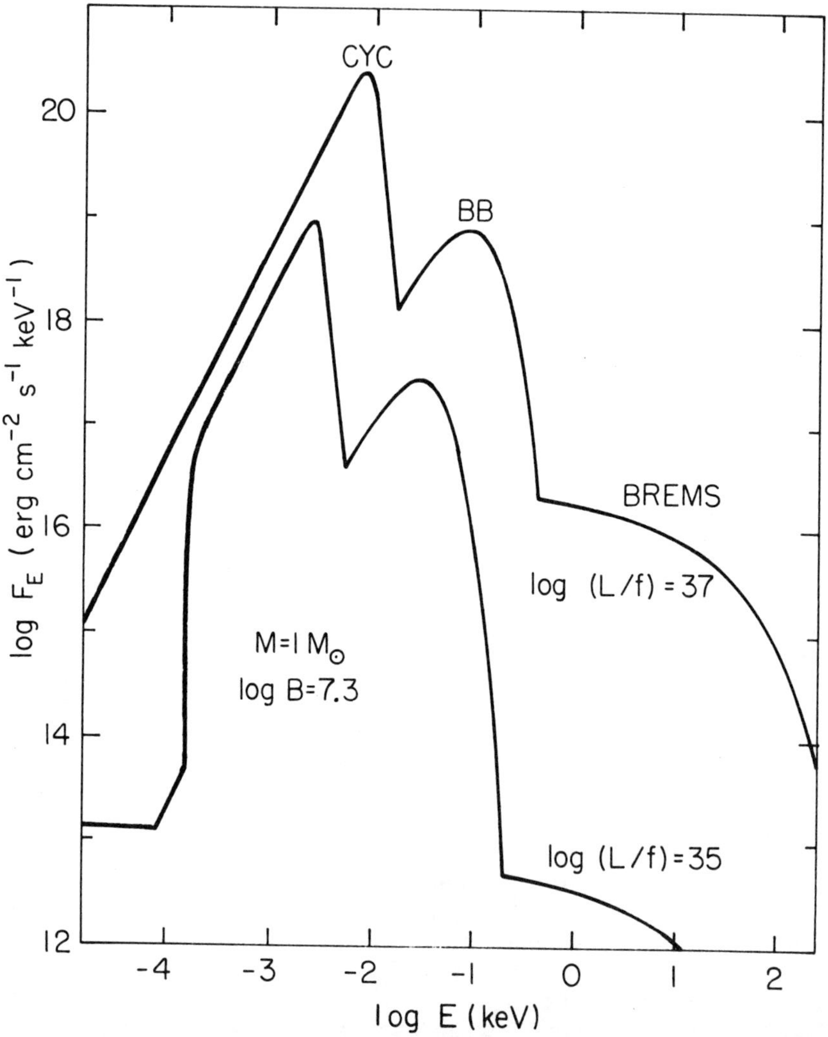

FIGURE 15. X-ray and UV spectra produced by two different accretion rates onto a 1 $M_\odot$ degenerate dwarf having a magnetic field of $B = 2 \times 10^7$ G. The portion of each spectrum below ~5 eV is less certain due to possible secondary radiation. (From Lamb and Masters.[38] By permission of *The Astrophysical Journal*.)

produced by the hot emission region, (2) a hard x-ray bremsstrahlung component also produced by the hot emission region, (3) a hard UV or soft x-ray blackbody component produced by cyclotron and bremsstrahlung photons that are absorbed by the stellar surface and reemitted, and (4) secondary radiation from infalling matter above the shock that is heated by Compton scattering or cyclotron absorption and, possibly, from the stellar surface around the emission region that is heated by radiative transfer. The first three components are clearly visible in FIGURES 15 and 16, which show spectra produced by the hot, post-shock emission region alone. Since the secondary radiation is not included, the spectra represent with relatively more uncertainty the actual observed spectrum below ~5 eV. FIGURE 15 shows the x-ray and UV spectra produced by accretion at two different rates, corresponding to $L/f = 10^{35}$ and 10^{37} erg s^{-1}, onto a 1.0 $M_\odot$ star having a magnetic field of 2×10^7 G. FIGURE 16 shows the x-ray and UV spectrum produced by accretion at a rate corresponding to

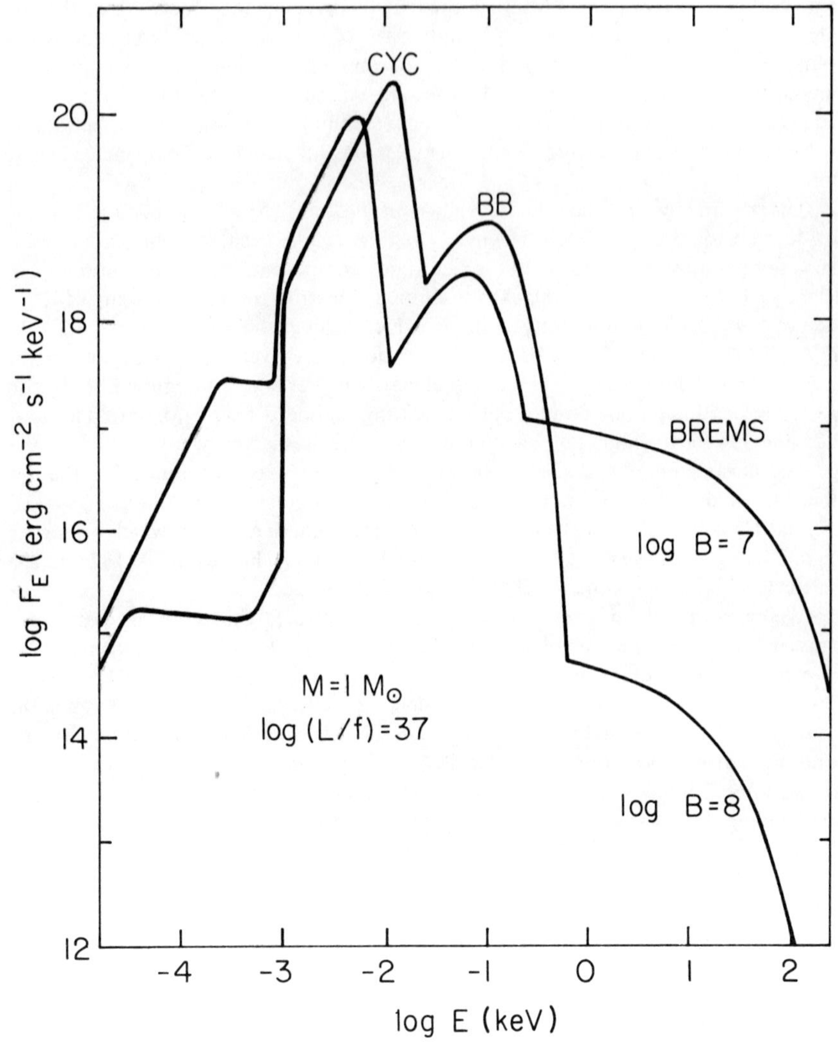

FIGURE 26. X-ray and UV spectra produced by accretion onto two different 1 $M_\odot$ degenerate dwarfs having magnetic fields of 10^7 and 10^8 G. The portion of each spectrum below ≃eV is less certain due to possible secondary radiation.

$L/f = 10^{37}$ erg s^{-1} onto two different 1.0 $M_\odot$ stars having magnetic field strengths of 10^7 and 10^8 G.

From the spectra illustrated in FIGURES 15 and 16, we draw two important conclusions. First, magnetic degenerate dwarfs are predicted to be strong UV sources with only a few percent of the total accretion luminosity ordinarily appearing as soft and hard x-rays, and therefore accessible to existing detectors. Second, the position and relative strength of the spectral components change with variations in the accretion rate, and differ for different magnetic fields. For example, with the change in accretion rate shown in FIGURE 15, the blackbody component moves from the UV into the soft x-ray region, and the luminosity of the bremsstrahlung hard x-ray component increases by nearly 4 orders of magnitude while the total accretion luminosity is increasing by only 2. As a second example, with the different magnetic field strengths shown in FIGURE 16, the bremsstrahlung hard x-ray component differs by nearly 3 orders of magnitude while the cyclotron and blackbody components hardly differ.

The parameters of the spectra shown in FIGURE 15 are as follows. For the low-luminosity example, which belongs to regime II, the bremsstrahlung component has a temperature $T_e = 21.0$ keV, the blackbody component has a temperature $T_{bb} = 10.6$ eV, and the peak of the blackbody-limited cyclotron component occurs at $E^* = 2.85$ eV. For the high luminosity example, which belongs to regime I, $T_e = 56.2$ keV, $T_{bb} = 32.2$ eV, and $E^* = 9.05$ eV. The ratio of the cyclotron luminosity to the bremsstrahlung luminosity, $q = L_{cyc}/L_{br}$, signals whether one is in regime I or II, with $q \ll 10$ indicating regime I and $q \gg 10$ indicating regime II for a 1 $M_\odot$ star. Thus $q = 2.75$ and 334 for the high- and low-luminosity spectra respectively.

The parameters of the spectra shown in FIGURE 16 are as follows. For the low magnetic field strength example, which belongs to regime I, $T_e = 56.1$ keV, $T_{bb} = 22.9$ eV, and $E^* = 5.77$ eV. For the high magnetic field strength example, which belongs to regime II, $T_e = 22.5$ keV, $T_{bb} = 33.5$ eV, and $E^* = 13.0$ eV. Finally, $q = 0.142$ and 233 for the low and high magnetic field strength spectra respectively.

We expect $f \ll 1$ for most magnetic degenerate dwarfs. Since the hot, post-shock emission region is then small and both cyclotron and blackbody components are blackbody-limited at low energies, the visual characteristics of most such stars are likely *not* dominated by flux from the emission region. Instead, the flux observed in the visual is likely dominated by secondary radiation; that is, either by emission from infalling matter, above the shock, that is heated by Compton scattering or cyclotron absorption (cf. references 11 and 67) or, possibly, by emission from the stellar surface, surrounding the emission region, that is heated by radiative transfer.

Let us note that if the geometry of the hot, post-shock emission region is such that most of the flux escapes through the face rather than through the edges of the emission region [i.e., $d \ll (2f)^{1/2}R$], then Compton degradation of the bremsstrahlung hard x-ray component will occur if L/f exceeds $\sim 10^{37}$ erg s^{-1} (see also reference 59). Such degradation is identical to that encountered in nonmagnetic degenerate dwarfs, and the shape of the resulting hard x-ray spectrum will be similar to the spectra shown in FIGURE 3. Under these circumstances, the following behavior may occur. As the accretion rate increases, the luminosities of the blackbody and bremsstrahlung components increase, much as shown in FIGURE 15. However, as the accretion rate increases still further, Compton degradation becomes severe and the luminosity of the bremsstrahlung component falls rapidly while that of the blackbody component continues to rise. At the same time, the observed spectral temperature T_{obs} of the bremsstrahlung component will decrease dramatically, while that of the blackbody component continues to increase. Secondary radiation produced by Compton heating of infalling matter above the shock may contribute to the UV and soft x-ray flux.

Correlation Between Spectral Temperature, Luminosity, and Magnetic Field Strength

Unlike the case of nonmagnetic degenerate dwarfs, where variations in the shapes and strengths of the spectral components are a function only of mass accretion rate, here both mass accretion rate and magnetic field strength are important. It is therefore necessary to display such variations by plotting the contours of quantities of interest on a $(B,L/f)$ plane. A set of such contours is shown in FIGURE 17 for a 1.0 $M_{\odot}$ star. Bremsstrahlung and cyclotron emission dominates in the same regions as in FIGURE 13. Contours of constant E^* are shown as thick solid lines and those of constant T_{bb} as dashed lines. The thin solid lines in the bremsstrahlung-dominated region show contours of constant q, while those in the cyclotron-dominated region show contours of constant T_e. To the upper right of the curve labeled "soft excess," the blackbody luminosity in soft x-rays exceeds the bremsstrahlung luminosity in hard x-rays.

Near and above $L/f = L_E = 1.4 \times 10^{38}$ erg s^{-1}, radiation pressure can be important and modify the results, but because photons can easily scatter out of the accretion column if $f \ll 1$, the Eddington luminosity does not represent the stringent upper limit to the luminosity that it does in the case of nonmagnetic degenerate dwarfs. Below and to the left of the curve $E^* = 2$ eV, the assumption $d < R$ breaks down. In both of these regions, the values shown are expected to remain qualitatively correct.

From FIGURE 17 we draw a third important conclusion, namely, that observation of the qualitative features of the spectrum is generally sufficient to determine with relative confidence the physical conditions in the x-ray emission region, including the value of the magnetic field. As we have seen, the cyclotron component is ordinarily, and the blackbody component may frequently be, inaccessible to existing detectors. Nevertheless, optical, soft x-ray, and hard x-ray observations will often be sufficient to determine the mass and magnetic field of the star, and its accretion rate. For example, observation of T_e and T_{bb} fixes a point in the $(B,L/f)$ plane of FIGURE 17, and also constrains the star to a limited range of masses. Measurement of E^*, q, or a constraint from optical observations further narrows the range of allowed physical parameters and tests the consistency, and hence validity, of the basic model. Furthermore, if an independent estimate of f exists, the intrinsic luminosity L is then known, as is the distance of the source.

Implications for Observed Sources

It will be important to carry out detailed calculations of the spectra to be expected from accreting magnetic degenerate dwarfs in which that emission region is not

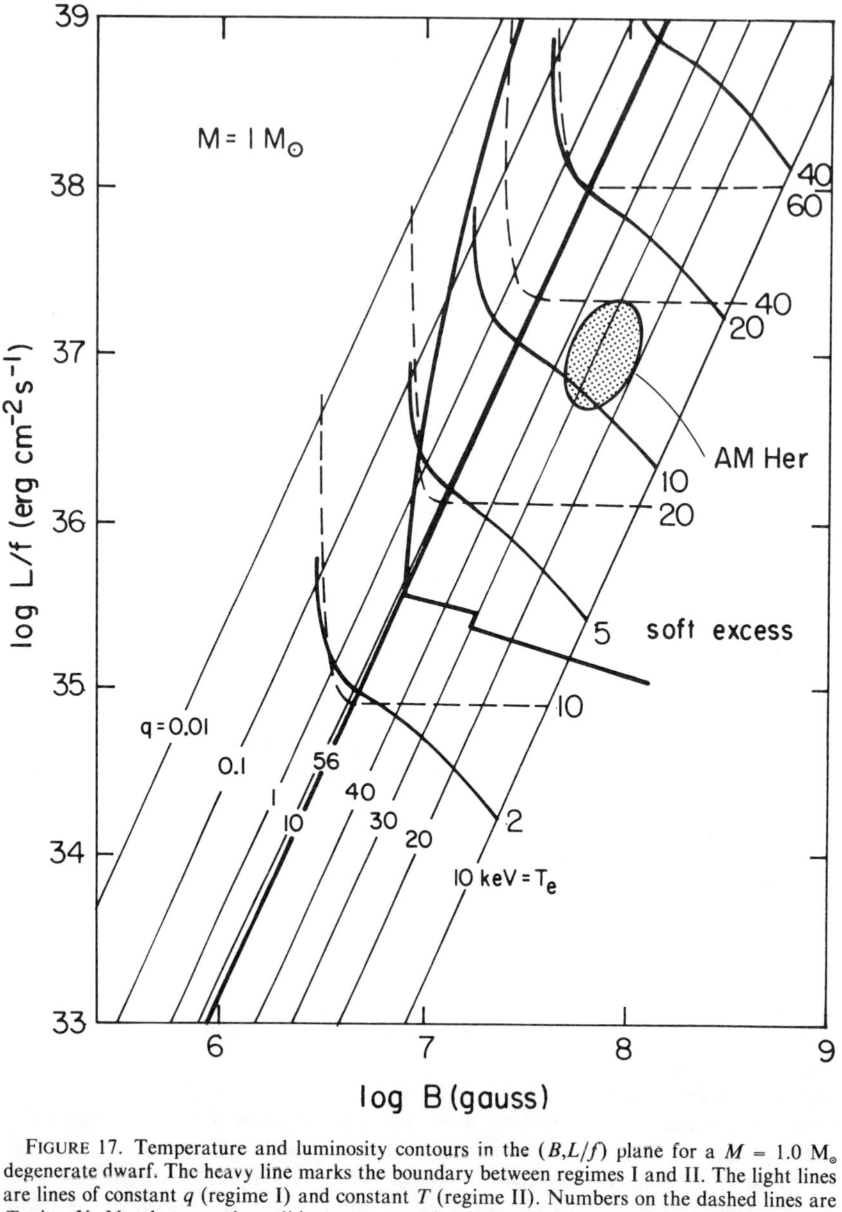

FIGURE 17. Temperature and luminosity contours in the $(B, L/f)$ plane for a $M = 1.0\ M_\odot$ degenerate dwarf. The heavy line marks the boundary between regimes I and II. The light lines are lines of constant q (regime I) and constant T (regime II). Numbers on the dashed lines are T_{bb} in eV. Numbers on the solid curves are E^* in eV. The spectral flux of the blackbody component exceeds that of the bremsstrahlung component at 0.25 keV to the right and above the heavy line labeled "soft excess." Near and above $L/f = L_E/f = 1.4 \times 10^{38}$ erg s^{-1}, radiation pressure can be important and modify the results, but because photons can easily scatter out of the accretion column if $f \ll 1$, the Eddington luminosity does not represent the stringent upper limit to the luminosity that it does in the case of nonmagnetic degenerate dwarfs. Below and to the left of the curve $E^* = 2$ eV, the assumption $d < R$ breaks down. In both of these regions, the values shown are expected to remain qualitatively correct. The shaded region labeled "AM Her" donated values of L/f and B consistent with the optical, soft x-ray, and hard x-ray data for this source. (From Lamb and Masters.[38] By permission of *The Astrophysical Journal*.)

idealized. However, it is useful and interesting to compare the present calculations with observations.

AM Her shows strong circular polarization ($\gtrsim 10\%$) at visual wavelengths[71,66] and is believed to be an accreting magnetic degenerate dwarf.[11,12,67,55] Soft x-ray observations[27,26,10,80] and hard x-ray observations[69,65] show two distinct components with temperatures $\gtrsim 30$ keV‖ and $\lesssim 40$ eV. Tuohy *et al.*[80] have shown that, if these two components are interpreted as the bremsstrahlung and blackbody components predicted in the above model of Lamb and Masters,[38] then optical observations constrain the temperature of the blackbody component to be $T_{bb} \simeq 40$ eV. This result requires AM Her to lie in the shaded ellipse in FIGURE 17 centered on $B = 10^8$ G and $L/f = 10^{37}$ erg s^{-1}. However, this interpretation should be regarded with caution since recent UV observations[74,58] find a UV flux an order of magnitude less than predicted by the above parameters.

AN UMa and VV Puppis also show strong circular polarization at visual wavelengths[72,34,43,42] and recently AN UMa had been detected in soft x-rays.[25] Furthermore, the weak hard x-ray source 2A0311 − 227 has recently been found to be a soft x-ray source[24,13] and to exhibit strong circular polarization at visual wavelengths.[73] All of these objects can be interpreted in terms of the above model.

DQ Her is also believed to be an accreting magnetic dwarf.[37,6] This system underwent a nova outburst in 1934 and shows coherent small amplitude optical pulsations at 71 s. V533 Her, which underwent a nova outburst in 1963, shows coherent small amplitude optical pulsations at 63 s[54] and is almost certainly a similar system. Unlike the AM Her systems, however, neither DQ Her nor V533 Her have been detected in soft or hard x-rays and DQ Her only shows a small amount of circular polarization at optical wavelengths.[70] Furthermore, both systems exhibit clear optical evidence of an accretion disk and neither have rotation periods synchronized with the binary orbital period. Interpreted in terms of the above model, these differences from the AM Her systems suggest that DQ Her and V533 Her have somewhat weaker magnetic fields and, concomitantly, lower values of L/f. The weaker magnetic field would be less able to enforce synchronous rotation of the degenerate dwarf and to prevent formation of a disk. Lower values of B and L/f would make the copious production of soft x-rays more difficult.

Soft and hard x-rays have been detected from several cataclysmic variables, including SS Cyg (cf. references 57,28,44) U Gem,[46,68] and EX Hya.[83,14] These sources, unlike the AM Her objects, again show strong optical evidence of a disk. Emission from the boundary layer of an optically thick accretion disk extending down to the stellar surface can, at outburst, produce soft x-rays[56] but the production of hard x-rays, as detected in several of these sources, is difficult to understand. Emission of hard x-rays may thus imply the presence of a magnetic field, which may be weak but which disrupts the disk and leads to radial inflow near the star. The above model would then apply, with the spectral shape depending on whether bremsstrahlung emission or cyclotron emission dominates. Clearly, it is important to attempt to verify this possibility, perhaps by optical polarimetry.

‖Staubert *et al.*[65] fit an 18-keV exponential spectrum to their hard x-ray data and that of Swank *et al.*[68] Note, however, that a thin bremsstrahlung (exponential) spectrum *with Gaunt factor* fits the combined data very poorly.

Magnetic Neutron Stars

Calculations of the properties of x radiation from magnetic neutron stars have proved to be extremely difficult. Aspects of the problem have been treated by Gnedin and Sunyaev,[20,21] Lamb, Pethick, and Pines,[41] Basko and Sunyaev,[4,5] and Elsner and Lamb,[18] among others. Yet no calculation to date has combined all the important

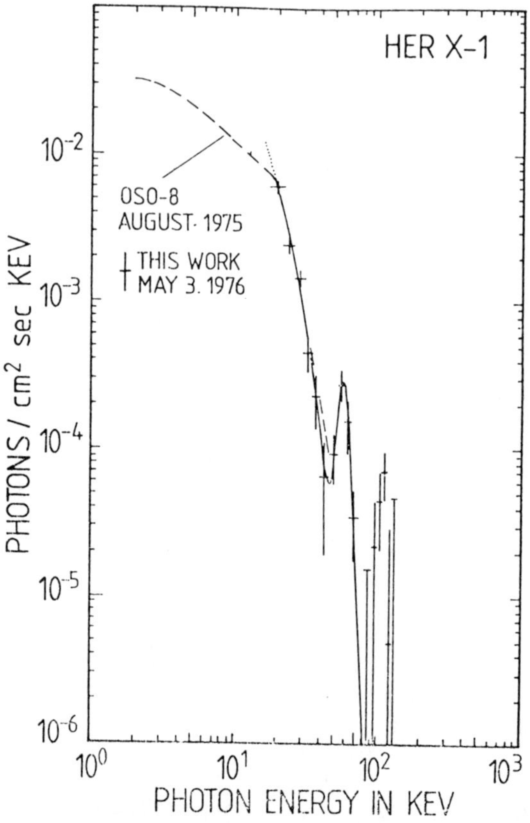

FIGURE 18. Deconvoluted x-ray spectra of the Her X-1 pulses. Key: — best-fitting exponential spectrum with a Gaussian line to the data points; (the error bars are $\pm 1\ \sigma$; the upper limits are $2\ \sigma$), ---- total x-ray spectrum of Her X-1 observed by OSO-8 during the 1975 August "on" state. (From Trumper *et al.*[78] By permission of *The Astrophysical Journal.*)

pieces of physics necessary to obtain a self-consistent solution. The lack of reliable, quantitative results constitutes a last, but large, gap in the theoretical framework necessary to carry out a systematic study of x-ray source spectra. The importance of a concerted attack on this problem has been emphasized by the exciting recent observation of a cyclotron feature near ~50 keV in the spectrum of Her X-1.[78] FIGURE

18 shows the spectrum observed. An understanding of such features should allow the direct measurement of neutron star magnetic fields.

Concluding Remarks

Comprehensive, detailed numerical studies of x-ray emission from accreting degenerate stars can provide a powerful framework for interpreting the extensive observations of x-ray spectra that now exist. Several large gaps in this theoretical framework remain; most notably, reliable, quantitative results for x-ray emission from magnetic neutron stars. Nevertheless, the new results and progress reported here hold out the promise that the nature and the physical properties (such as magnetic field strength, accretion rate, luminosity, and distance) of many of the galactic x-ray sources may be determined from observations of their x-ray spectra alone.

Acknowledgments

We are grateful to Fred Lamb for many lengthy and valuable discussions of the physics of x-ray emission from degenerate stars. We also are grateful to Graziella Branduardi, Dave Hearn, Keith Mason, Larry Petro, John Raymond, and Aldo Treves for extensive discussions of the observations. We thank Susan Black and Trisha Welch for their assistance in preparing this manuscript.

References

1. Agrawal, P. C., S. Biswas, G. S. Gokhale, V. S. Iyengar, P. K. Kunte, R. K. Manchanda & B. V. Sreekantan. 1971. Astrophys. Space Sci. **10:** 500.
2. Aizu, K. 1973. Prog. Theor. Phys. **49:** 1184.
3. Alme, M. L. & J. R. Wilson. 1973. Astrophys. J. **186:** 1015.
4. Basko, M. M. & R. A. Sunyaev. 1975. Astron. Astrophys. **42:** 311.
5. Basko, M. M. & R. A. Sunyaev. 1976. Mon. Not. R. Astron. Soc. **175:** 395.
6. Bath, G. T., W. D. Evans & J. E. Pringle. 1974. Mon. Not. R. Astron. Soc. **166:** 113.
7. Bekefi, G. 1966. Radiation Processes in Plasmas. Wiley. New York.
8. Branduardi, G. 1977. Ph.D. Thesis. University of London. Unpublished.
9. Branduardi, G., N. D. Kylafis, D. Q. Lamb & K. O. Mason. 1979. Astrophys. J. Lett. **235.** In press.
10. Bunner, A. N. 1978. Astrophys. J. **220:** 261.
11. Chanmugam, G. & R. L. Wagner. 1977. Astrophys. J. Lett. **213:** L13.
12. Chanmugam, G. & R. L. Wagner. 1978. Astrophys. J. **222:** 641.
13. Charles, P. A., K. O. Mason & S. Bowyer. 1979. IAU Circular 3329.
14. Cordova, F. A. & G. R. Riegler. 1979. Mon. Not. R. Astron. Soc. **188:** 103.
15. DeGregoria, A. J. 1974. Astrophys. J. **189:** 555.
16. Drummond. W. E. & M. N. Rosenbluth. 1961. Phys. Fluids **4:** 277.
17. Drummond, W. E. & M. N. Rosenbluth. 1963. Phys. Fluids **6:** 276.
18. Elsner, R. F. & F. K. Lamb. 1976. Nature (London) **262:** 356.
19. Fabian, A. C., J. E. Pringle & M. J. Rees. 1976. Mon. Not. R. Astron. Soc. **173:** 43.
20. Gnedin, Yu. N. & R. A. Sunyaev. 1973a. Astron. Astrophys. **25:** 333.
21. Gnedin, Yu. N. & R. A. Sunyaev. 1973b Mon. Not. R. Astron. Soc. **162:** 53.
22. Hayakawa, S. 1973. Prog. Theor. Phys. **50:** 459.
23. Hayakawa, S. & R. Hoshi. 1976. Prog. Theor. Phys. **55:** 1320.
24. Hearn, D. R. 1979. IAU Circulars 3326 & 3327.

25. HEARN, D. R. & F. J. MARSHALL. 1979. Astrophys. J. Lett. **232:** L21.
26. HEARN, D. R. & J. A. RICHARDSON. 1977. Astrophys. J. Lett. **213:** L115.
27. HEARN, D. R., J. A. RICHARDSON & G. W. CLARK. 1976. Astrophys. J. Lett. **210:** L23.
28. HEISE, J., R. MEW, A. C. BRINKMAN, E. H. B. M. GROENSCHILD, A. J. F. DEN BOGGENDE, J. SCHRIJVER & J. E. GRINDLAY. 1978. Astron. Astrophys. **63:** L1.
29. HIRSCHFIELD, J. L., D. E. BALDWIN & S. C. BROWN. 1961. Phys. Fluids. **4:** 198.
30. HOSHI, R. 1973. Prog. Theor. Phys. **49:** 776.
31. ILLARIONOV, A. F. & R. A. SUNYAEV. 1972. Astr. Zh. **49:** 58 (Eng. trans: Soviet Astron. AJ **16:** 45, 1972).
32. JONES, C. 1977. Astrophys. J. **214:** 856.
33. KATZ, J. I. 1977. Astrophys. J. **215:** 265.
34. KRZEMINSKI, W. & I. SERKOWSKI. 1977. Astrophys. J. Lett. **216:** L45.
35. KYLAFIS, N. D. 1978. Ph.D. Thesis. University of Illinois. Unpublished.
36. KYLAFIS, N. D. & D. Q. LAMB. 1979. Astrophys. J. Lett. **228:** L105.
37. LAMB, D. Q. 1974. Astrophys. J. Lett. **192:** L129.
38. LAMB, D. Q. & A. R. MASTERS. 1979. Astrophys. J. Lett. **234.** In press.
39. LAMB, D. Q. & A. R. MASTERS. 1979. In preparation.
40. LAMB, F. K. 1977. Proceedings Eighth Texas Symposium on Relativistic Astrophysics. Ann. N.Y. Acad. Sci. **302:** 482.
41. LAMB, F. K., C. J. PETHICK & D. PINES. 1973. Astrophys. J. **184:** 271.
42. LIEBERT, J. & H. S. STOCKMAN. 1979. Astrophys. J. **229:** 652.
43. LIEBERT, J. & H. S. STOCKMAN. J. R. P. ANGEL, N. J. WOOLF, K. HEGE & B. MARGON. 1979. Astrophys. J. **225:** 201.
44. MASON, K. O., F. CORDOVA & J. SWANK. 1979. IAU/COSPAR Symposium on X-ray Astronomy. Advances in Space Exploration. Vol. 3. In press.
45. MASON, K. O., P. A. CHARLES, N. E. WHITE, F. L. CULHANE, P. W. SANFORD & K. T. STRONG. 1976. Mon. Not. R. Astron. Soc. **177:** 513.
46. MASON, K. O., M. LAMPTON, P. CHARLES & S. BOWYER. 1978. Astrophys. J. Lett. **226:** L129.
47. MASTERS, A. R. 1978. Ph.D. Thesis, University of Illinois. Unpublished.
48. MASTERS, A. R., A. C. FABIAN, J. E. PRINGLE & M. J. REES. 1977. Mon. Not. R. Astron. Soc. **178:** 501.
49. MILGROM, M. 1976. Astron. Astrophys. **50:** 273.
50. MILGRAM, M. & J. I. KATZ. 1976. Astrophys. J. **205:** 545.
51. OKUDA, T. & S. SAKASHITA. 1977. Astrophys. Space Sci. **52:** 35.
52. PACZYNSKI, B. & A. ZYTKOW. 1978. Astrophys. J. **222:** 604.
53. PARSIGNAULT, D. R. & J. E. GRINDLAY. 1978. Astrophys. J. **225:** 970.
54. PATTERSON, J. 1978. IAU Circular 3283.
55. PRIEDHORSKY, W. C. & W. KRZEMINSKI. 1978. Astrophys. J. **219:** 597.
56. PRINGLE, J. 1977. Mon. Not. R. Aston. Soc. **178:** 195.
57. RAPPAPORT, S., W. CASH, R. DOXSEY, J. MCCLINTOCK & G. MOORE. 1974. Astrophys. J. Lett. **187:** L5.
58. RAYMOND, J. C., J. H. BLACK, R. J. DAVIS, A. K. DUPREE, H. GURSKY, L. HARTMANN & T. A. MATILSKY. 1979. Astrophys. J. Lett. **230:** L95.
59. RICKETTS, M. J., A. R. KING & D. J. RAINE. 1979. Mon. Not. R. Astron. Soc. **186:** 233.
60. SCHREIER, E. 1977. Proceedings Eighth Texas Symposium on Relativistic Astrophysics. Ann. N.Y. Acad. Sci. **302:** 445.
61. SHAPIRO, S. L. & E. E. SALPETER. 1975. Astrophys. J. **198:** 671.
62. SION, E. M., M. J. ACIERNO & D. A. TURNSHEK. 1978. Astrophys. J. **220:** 636.
63. SION, E. M., M. J. ACIERNO & S. TOMCZYK. 1979. Astrophys. J. **230:** 832.
64. STARRFIELD, S., W. M. SPARKS & J. W. TRURAN. 1974. Astrophys. J. Suppl. **28:** 247.
65. STAUBERT, R., E. KENDZIORRA, W. PIETSCH, C. REPPIN, J. TRUMPER & W. VOGES. 1978. Astrophys. J. Lett. **225:** L113.
66. STOCKMAN, H. S. & T. A. SARGENT. 1979. Astrophys. J. **229:** 652.
67. STOCKMAN, H. S., G. D. SCHMIDT, J. R. P. ANGEL, J. LEIBERT, S. TAPIA & E. A. BEAVER. 1977. Astrophys. J. **217:** 815.
68. SWANK, J. H., E. A. BOLDT, S. S. HOLT, R. E. ROTHSCHILD & P. J. SERLEMITSOS. 1978. Astrophys. J. Lett. **226:** L133.

69. SWANK, J., M. LAMPTON, E. M. BOLDT, S. HOLT & P. SERLEMITSOS. 1977. Astrophys. J. Lett. **216:** L71.
70. SWEDLUND, J. B., J. C. KEMP & R. D. WOLSTENCROFT. 1974. Astrophys. J. Lett. **193:** L11.
71. TAPIA, S. 1977. Astrophys. J. Lett. **212:** L125.
72. TAPIA, S. 1977. IAU Circular No. 3054.
73. TAPIA. S. 1979. IAU Circular No. 3327.
74. TREVES, A. 1978. Private communication.
75. TRUBNIKOV, B. A. 1958. Ph.D. Thesis. AEC-tr-4073 (Washington: Office of Technical Services).
76. TRUBNIKOV, B. A. 1961. Phys. Fluids. **4:** 195.
77. TRUBNIKOV, B. A. 1973. JETP Lett. **16:** 25.
78. TRUMPER, J., W. PIETSCH, C. REPPIN, W. VOGES, R. STAUBERT & E. KENDZIORRA. 1978. Astrophys. J. Lett. **219:** L105.
79. TUCHMAN, Y. & R. Z. YAHEL. 1977. Astrophys. Space Sci. **50:** 473.
80. TUOHY, I. R., F. K. LAMB, G. P. GARMIRE & K. O. MASON. 1978. Astrophys. J. Lett. **226:** L17.
81. ULMER, M. P. 1975. Astrophys. J. **196:** 827.
82. VITELLO, P. 1978. Astrophys. J. **225:** 694.
83. WATSON, M. G., M. R. SHERRINGTON & R. F. JAMESON. 1978. Mon. Not. R. Astron. Soc. **184:** 79P.
84. ZEL'DOVICH, YA. B. & N. I. SHAKURA. 1969. Astron. Zh. **46:** 225 (Eng. trans: Soviet Astron. AJ **13:** 175).

OPTICAL COUNTERPARTS OF COMPACT GALACTIC X-RAY SOURCES*

Bruce Margon†

Department of Astronomy
University of California
Los Angeles, California 90024

INTRODUCTION

Ten years ago there was one known optical counterpart of a compact galactic x-ray source: a peculiar 13th magnitude blue flickering star[1] associated with Scorpius X-1. As recently as five to six years ago, there were perhaps only a half-dozen such counterparts known, and it was straightforward for review papers on the topic[2] to be reasonably comprehensive. Today, there are of order 50 known stellar optical counterparts to galactic x-ray sources,[3] and the task of reviewing them individually has become hopeless. The reason for this recent explosion in the number of identified sources is rather easy to identify; it is due to the splendid success of the satellite-borne modulation collimator experiments aboard Ariel V, HEAO-1, and in particular, SAS-3. These instruments have permitted the derivation of positional error boxes for even fairly weak sources (10^{-4} the intensity of Sco X-1) to accuracies of order 30 arcsec, compared to positions spanning a substantial fraction of one degree previously available from mechanically-collimated detectors such as those on Uhuru, OSO-7, and Copernicus. This new series of accurate positions makes possible for the first time a "brute force" search technique. Every star in the error box can be examined photometrically or spectroscopically, in search of some telltale (even unspecified!) peculiarity, as even at low galactic latitude such positions in favorable cases may contain only a dozen candidate stars visible on deep plates. At high latitude such a box often contains only one or two stars to a limit of 20th magnitude. A perusal of the available data indicates that this method has recently had a success rate of over 50%, i.e., at least half of the sources with positions of this accuracy are ultimately optically identified. In passing, it is interesting to note that this short-lived and highly productive phase of correlative ground-based observations may rapidly be coming to an end. Positions from the HEAO-2 High Resolution Imager (HRI) have an accuracy of order 1–2 arcsec, thus eliminating the need for *any* search technique!

The successes of the past few years have lengthened the list of optical counterparts to such an extent that it seems prudent to begin by excluding certain classes of objects from this review. As sensitivities improve and the energy bandpasses of the observations continue to move toward longer x-ray wavelengths, increasing numbers of low-luminosity, soft x-ray sources are being identified.[4] The nearest of the RS CVn stars,[5] normal stellar chromospheres,[6] very hot white dwarfs,[7,8] flare stars,[9,10] and dwarf novae[11–14] are appearing in profusion, and the identifications in these categories will undoubtedly continue to increase until the source horizon imposed by interstellar

*This work was supported by National Science Foundation grant No. AST 77-27745.
†Alfred P. Sloan Foundation Research Fellow.

0077–8923/80/0336–0550 $01.75/1 © 1980, NYAS

photoelectric opacity is reached. We will exclude these soft, low-luminosity sources from further discussion here for two reasons. First, in many cases the stellar sources are not compact, relativistic stars, and thus beyond the purview of this conference. Secondly, although interesting individually, the x-ray luminositites of these sources are many orders of magnitude below those of the well-known hard x-ray binaries.[15] In fact, as shown in TABLE 1, where estimates of parameters of these classes appear, in many cases the total luminosity of the entire Milky Way galaxy due to this class of star does not equal that of one Cen X-3-type source. Thus we will initially restrict our attention to what might be called the "classic" group of sources, with x-ray luminosities in the 10^{36}–10^{38} erg/s range within the bandpass 1–10 keV.

COUNTERPARTS OF SCO-LIKE SOURCES

One could have predicted ten years ago, when only Sco X-1 was optically identified, that many or most optical counterparts of the luminous x-ray sources would be very faint. The spatial distribution of the bright sources about the galactic center[16]

TABLE 1

CHARACTERISTICS OF LOW LUMINOSITY X-RAY SOURCES

Class	X-Ray Luminosity (erg/sec)	Total Number in Galaxy	Total Galaxian Luminosity (erg/s)
RS CVn	10^{31}	10^{6}	10^{37}
α Cen	3×10^{27}	10^{9}	3×10^{36}
HZ 43/Feige 24	3×10^{34}	10^{4}:	3×10^{38}
Flare stars	5×10^{29}	10^{9}	5×10^{38}
Dwarf novae	10^{32}	10^{5}:	10^{37}

shows them to be many kiloparsecs distant, say, distance moduli of 13–14 mag. The absolute magnitude of the Sco X-1 counterpart is about $M_V = 4$. For sources at low galactic latitude, which is typical, we must also add 1–2 mag of interstellar extinction. Thus we would predict that the typical counterpart of this class will be uncomfortably faint, about 18–19th mag, based only on the assumption that Sco X-1 is typical. Such objects are now indeed showing up, so quickly that most previous attempts to tabulate them[17,18] are all obsolete. The recent MIT compendium[3] is reasonably up-to-date as of the date of this conference, and an upcoming review[19] should be complete as of mid-1979. All of these objects are rather similar (indeed, they are observationally selected to be that way!) with emission lines superposed on a usually featureless blue/ultraviolet continuum. In the best cases, such as the spectrum of AM Her shown in FIGURE 1, the emission is so unusual and enormously strong relative to the continuum that there can be no doubt concerning the validity of the identification. All too often, however, the faintness of the candidates makes them challenging spectroscopic objects even for the telescopes in the 3–4-m class, and if the emission is less than

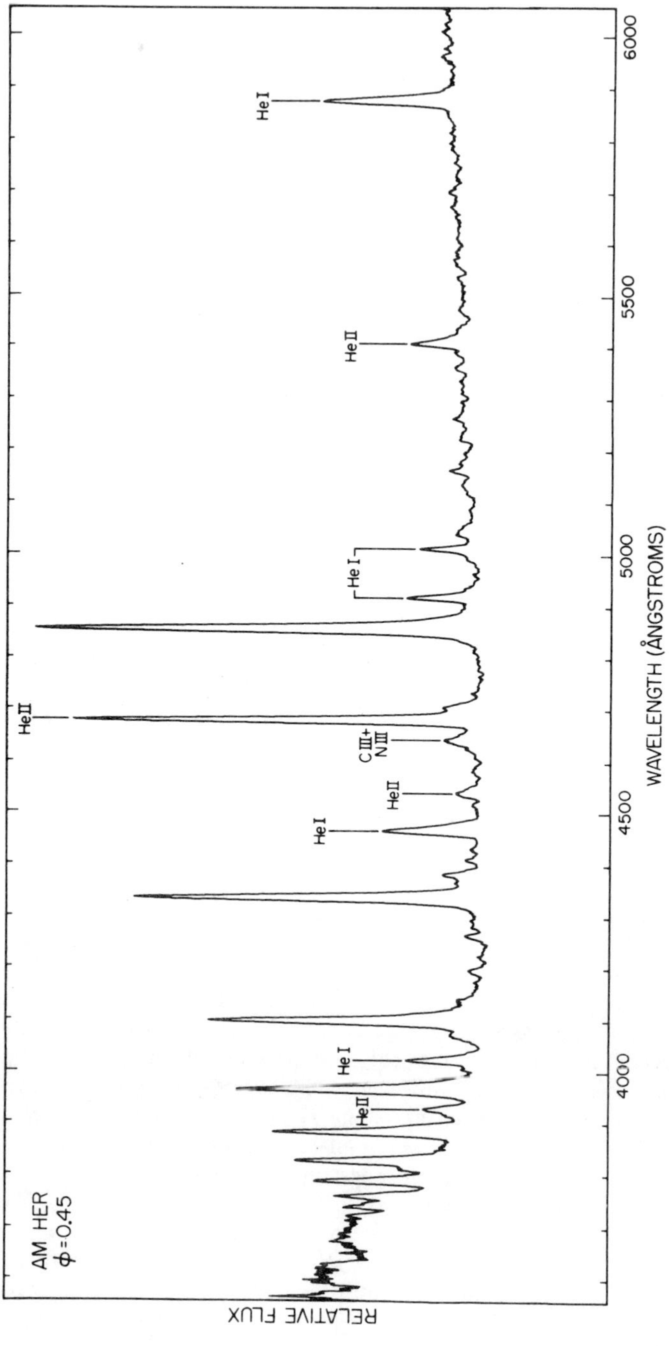

FIGURE 1. Spectrophotometry of AM Herculis, obtained with the Lick Observatory 3-m Shane reflector, with 1-nm resolution, on 1978 June 11, at photometric phase 0.45 (near minimum light). The stronger emission lines have been labeled, except that, for clarity, the numerous Balmer emissions have not been identified.

several times the continuum strength, the spectra can be ambiguous. Typical examples here are 4U0614+09 (reference 20) and 2A 0042+323 (reference 21), both objects claimed as identifications on the grounds of very weak emission lines. In the former case, the identification has been vindicated by successively smaller x-ray error boxes and improved spectroscopy.[22] If the early history of radio astronomy is an appropriate guide, we might expect that one or two of the currently-suggested counterparts will turn out to be misidentifications, but that the great majority will prove correct. Incidentally, the spectroscopic verifications may in many cases come from ultraviolet satellite observations, where many of the strong resonance transitions are accessible. A comparison of the faint optical emission lines in the spectrum of HZ Her with the strong UV emission[23] is an impressive example. Many of the x-ray star counterparts are too faint for observation with the International Ultraviolet Explorer (IUE), but will be straightforward for the Space Telescope Faint Object Spectrograph.

In addition to Balmer and He I emission, the lines most often seen in the optical counterparts are He II 4686 and the C III/N III multiplet at 4640/4650. The high excitation of these features leaves little doubt that they are somehow connected with the presence of x-rays in the system. Sadly, the observations are generally of insufficient signal-to-noise and velocity resolution to specify unambiguously in a given case whether the lines arise in hot gas immediately surrounding a compact x-ray star, or instead are excited on the sub-x-ray source point in the atmosphere of a normal stellar companion, due to the intense incident ionizing flux. It seems likely at this time that both cases are applicable in different systems. In at least one system, Her X-1/HZ Her, the exact physical process giving rise to the 4640 emission is understood (even if the location of the emitting region is not) to be the Bowen fluorescence mechanism,[24,25] as the near-ultraviolet spectrum shows other Bowen-pumped lines.[26]

To summarize, then, there are well over a dozen of these Sco-like sources now identified, yielding some good news and some bad news. The good news is that the very considerable observational and theoretical work that has gone into trying to understand Sco X-1 is almost certainly applicable to other objects as well. This list will probably continue to expand, since there are as yet unidentified sources which have x-ray properties similar to Sco X-1 (reference 27). The bad news is twofold. First, these are very optically faint objects, and telescopes in the 2.5-m class and above will be required to study them spectroscopically. Second, to be blunt—what is Sco X-1? After many years of frustrating study, we are still really not nearly as sure of the appropriate model as we are for some of the systems with more luminous counterparts, such as Cen X-3 and Vel X-1. It seems reasonably conservative to state that all the evidence available is consistent with this class of x-ray sources (with low-luminosity primary stars) all being binaries, although this is certainly not universally accepted.[28] In some of the systems, synchronous spectroscopic and photometric variations are seen, and the periods range from about 0.1 through 10 days. Most theories invoke Roche-lobe overflow to power these systems (as the normal star in the system cannot have much of a stellar wind), although in the longer period systems, such as that proposed for Cyg X-2 (reference 29), this assumption drives other system parameters to perhaps unreasonable extremes. The chief problem is that for most of the currently identified objects in this class, there is as yet simply no observational evidence one way or another bearing on the binary nature.

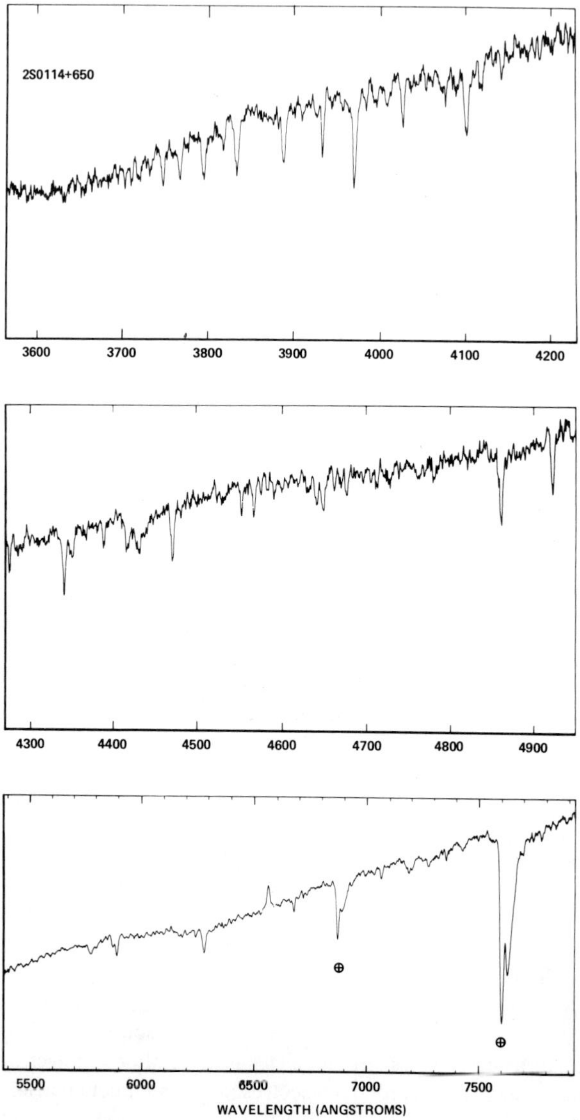

FIGURE 2. The spectrum of LS I + 65° 010, the optical counterpart of the weak x-ray source 2S 0114+650 (reference 30 and 33). The upper and middle panels were obtained with the Intensified Image Dissector Scanner at the 2.1-m reflector of the Kitt Peak National Observatory; the lower panel data are from the Robinson–Wampler Image Tube Scanner at the Lick Observatory 3-m Shane reflector. With the exception of the H-alpha emission, the spectrum is that of a normal early-type star; we classify it as B0.5III. The telluric absorption bands are also indicated.

Luminous Counterparts

It would be remiss not to at least briefly mention the recent progress on optical identifications and studies of galactic x-ray sources with luminous ($M_V < 0$) visible counterparts. These sources are obviously much easier to locate optically, so it is little wonder that identifications of objects like Cyg X-1, Vela X-1, and 4U0900−40 all came rather early. Most of the currently known examples of these luminous counterparts are within a few kiloparsecs distance, so in this case we can be relatively certain that identifications of more distant examples of this class will require only more accurate positions. Except in cases of very severe interstellar extinction, these objects should be easily visible throughout the galaxy. Examples of the fainter but intrinsically luminous counterparts are just beginning to appear, and the HEAO-2 data will doubtless extend this particular group especially quickly. Figure 2 is an example of one such system, the x-ray source 2S 0114+650 (reference 30). The identification with the 11th magnitude reddened early-type star LS I + 65° 010 (references 31 and 32) has been made by myself and co-workers[33] based on the prominent H-alpha emission and similarity to the other members of the class (see Figures 2 and 3). This spectrum is typical of the group, in that the large ratio of optical to x-ray luminosities in these sources tends to make the visible spectra unspectacular. On the other hand, this increase in optical luminosity means that these inevitably brighter systems are far better studied than the Sco-like group, and so we are observationally on much firmer ground in understanding the nature of these stars.

Many of the systems with luminous primaries are known to be single-lined spectroscopic binaries, with periods of order a few days; many are also synchronous photometric variables, where the variation is due to distortion of the star by the x-ray companion. In roughly a dozen cases, the associated x-ray sources are also x-ray pulsars, with periods in the range 1–10^3 s. In the most favorable systems, where all of these effects are seen simultaneously, the velocity of the x-ray component is available from the measured orbital Doppler shift in the pulse, and the system can be solved as a classical double-lined spectroscopic binary, yielding masses for both components.[34] Further, longer term observations of changes in the pulse period due to transfer of angular momentum between the two stars yields data that accurately estimate the moment of inertia of the compact component, and quite unambiguously show these objects to be neutron stars.[35,36]

Stubborn Identifications

It would be projecting an overly optimistic picture not to admit that there are some "stubborn" cases where no identifications of optical counterparts are forthcoming. For some sources, such as 1758−205 and 1822−000, excellent positions are available, but the fields prove to be completely blank to the limit of the *Palomar Observatory Sky Survey* plates.[37,38] For at least a half-dozen other sources, only a few candidate stars are visible in the x-ray error boxes, and spectroscopic and photometric examination of *all* of them yields no obvious candidates. This problem is particularly endemic among the bright "bulge" sources, those located within 30 degrees longitude of the galactic center. For example, GX340+0, GX349+2, GX354+0, GX3+1, and GX13+1 are fields that have been quite exhaustively studied by myself and colleagues at other

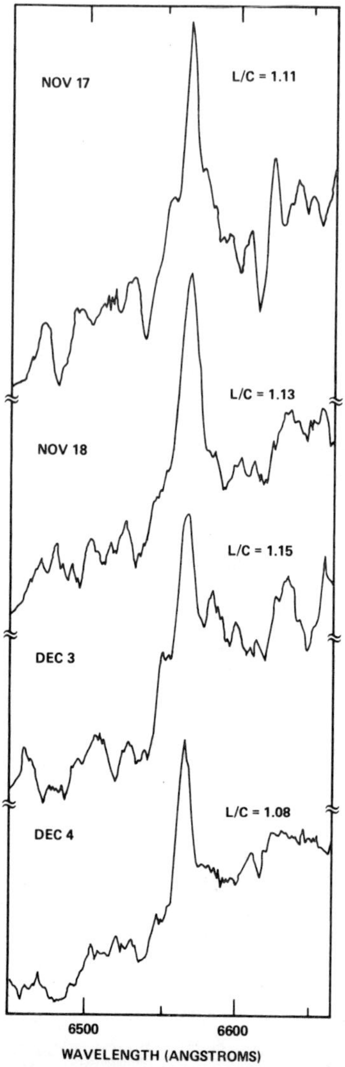

FIGURE 3. The H-alpha emission profiles in LS I + 65° 010 on four different nights in 1977. The line-to-continuum ratio of each emission feature is also indicated, showing possible evidence of variability.

institutions, and still await identification. Finally, the fascinating sources located at or very near the cores of globular clusters[39] still have no specific optical counterparts within the clusters.

The potential thus exists that these unidentified sources may actually belong to a class or classes completely separate from the identified group. There is some sketchy x-ray evidence that this may be so. For example, none of the globular cluster or bright bulge sources show x-ray eclipses, as is common in the "field" binary sources. Also, the incidence of x-ray sources, i.e., number of sources per solar mass of material, is very much larger in globular clusters than the field.[40] On the other hand, equally preliminary evidence may be used to argue that the bulge and globular cluster sources are intimately related to the field binaries: their x-ray luminosities are very similar, for example. Furthermore, at least three of the bulge sources, GX9+9, GX17+2, and GX1+4, are optically identified (Davidsen *et al.*[41] and references therein). The first two of these have stellar counterparts that are not extraordinary. The third, GX1+4, is an interesting case in that it is an x-ray pulsar, and the optical spectrum is definitely composite,[42] yielding proof that at least this source is a binary. On the other hand, the x-ray spectrum of GX1+4 is rather atypical for the bulge sources, and so this may not be the optimal example.

The nature of the globular cluster sources is a particularly interesting issue, as massive black holes have been suggested as possible explanations of these sources.[43,44] Although stellar candidates have already been suggested in several clusters, e.g., NGC 1851 (references 45 and 46), NGC 6712 (reference 47), and NGC 7078 (reference 48), no compelling evidence exists for any of the associations as yet. The HEAO-2 Einstein Observatory positions for these sources, which should have accuracies of order 2 arcsec, will yield an invaluable contribution to this problem, and these data should be available very shortly. Although a derived position very near the center of any one cluster will only imply that the source is the most massive star in the cluster, strong statistical arguments will be possible[39] if this is true for all of the cluster sources as a group, and the requirements on the minimum source masses may be substantial.

Optical Counterparts of X-Ray Bursters

It is a curious accident of history that the first x-ray bursts were detected from globular cluster sources.[49] There are now four or five bursting sources identified with stellar counterparts, objects that are almost surely not in globular clusters. The best studied of these optical counterparts is 4U 1735−444. The remarkable spectrum of this object obtained by McClintock *et al.*[50] is reproduced in Figure 4, together with a spectrum of Sco X-1 for comparison. The detailed resemblance of both the continuum slope (and thus broadband colors) as well as emission lines is very marked. The optical spectrum[71] of the x-ray burster Ser X-1, although poorly observed, is also similar to Sco X-1. Still another source with a Sco-like optical spectrum, 4U 0614+09 (reference 20) is also probably a burster.[51] There are two interpretations of these coincidences, unfortunately in conflict with each other. The first inference might be that these optical spectra are so remarkably similar because the sources are quite similar, and bursters are simply a modification upon the normal "field" binary source behavior. Alternatively one could argue that the distinctive colors and emission spectrum are due simply to the irradiation of a gas by an intense x-ray ionizing flux, and these properties say nothing about the nature of the object underlying the irradiated layer. Certainly the bursters are at least somewhat different from the Sco-like sources, if for no other reason than Sco X-1 is known not to burst! There are also more detailed differences; for example, the optical counterpart of 4U 1735−444 does not show the distinctive photometric flickering well known in Sco X-1.[17,52] At least partial resolution of this issue will undoubtedly come soon. Not only do we now have

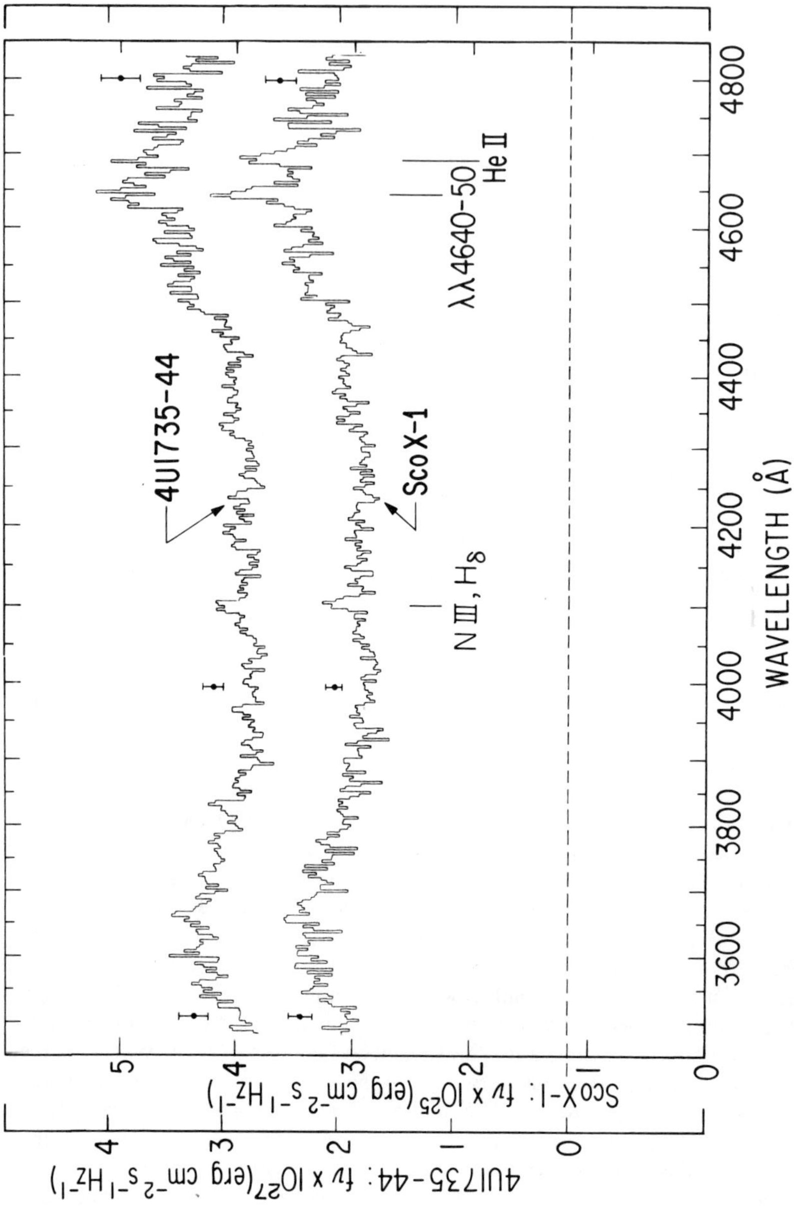

FIGURE 4. Spectrophotometry of the x-ray burster 4U 1735–444, together with a spectrum of Sco X-1 as a comparison (from reference 50). Note the strong resemblance of both the continuum and emission spectrum of the two stars. (Copyright by the American Astronomical Society; reproduced with permission.)

the optical counterparts of several stellar-like bursters available for intensive study, but in at least one of these objects, again 4U 1735−444, optical bursts simultaneous with the x-ray bursts have been reported.[52] The detailed time-history and spectrum of the optical bursts should yield very important new insights into the nature of these objects.

Photometry of the Optical Counterparts of X-Ray Pulsars

There has been important recent progress, after a brief period of stagnation, in the area of fast photometry of the optical counterparts of x-ray pulsars. The motivation for this work, at least as I view it, is as follows. One of the most important observations that the x-ray astronomer can contribute to the theoretical physicist is rather directly measured data on the equation of state of matter at ultrahigh densities, via inferences on the mass of neutron stars. To obtain these masses, as mentioned previously, velocity curves are needed for both components of the x-ray binary. For the x-ray pulsars, the Doppler shift in the period does yield these data to a very high level of accuracy (far better than available from visible spectroscopy). However, despite the above-mentioned successes in a few systems, for many of the x-ray pulsars it is difficult or impossible to accurately derive the velocity curve of the visible component. The reasons for these difficulties are several: the systems are often faint, making them spectroscopically very awkward, especially at high dispersion. In addition, the absorption line profiles are often broad and difficult to measure, and their velocities may be distorted by emission. An alternative method of obtaining the necessary data is to attempt to detect the "reflected" optical pulsations from the normal companion of the x-ray pulsar. The periodic beam of ionizing flux must cause a burst of optical recombination radiation on the surface of the cooler star, if the recombination time scales are short compared to the pulse period, which is thought to be the case. Thus detection of these optical pulses and an orbital Doppler shift therein could yield a velocity curve of very high accuracy. An elegant analysis of this type was performed for the HZ Her/Her X-1 system several years ago.[53] There exist a dozen more such optically identified x-ray pulsars where a similar analysis is in principle possible, but the results of searches for the optical analogs of the x-ray pulses have been uniformly discouraging.[54–59] Recently, however, there has been a positive detection of 7.7-s optical pulsations in 1627–673 (reference 60), which has been confirmed by an independent group[61] and thus seems observationally secure. Although the Sco X-1-like counterpart of this system[62] is quite faint (V = 18.5), the amplitude of the optical pulsations is rather large (about 4%) and so an extensive study of the phenomenon should be possible.

Future Prospects: The Exotic Optical Counterpart of A1909+04

I will conclude by briefly mentioning some very recent data of my own, in collaboration with my colleagues at the University of California,[63] on a newly identified optical counterpart that perhaps may presage some of the future surprises in this field. The optical counterpart of A1909+04, a weak and rather undistinguished x-ray source,[64] has recently been identified[65] with a peculiar emission line star that is

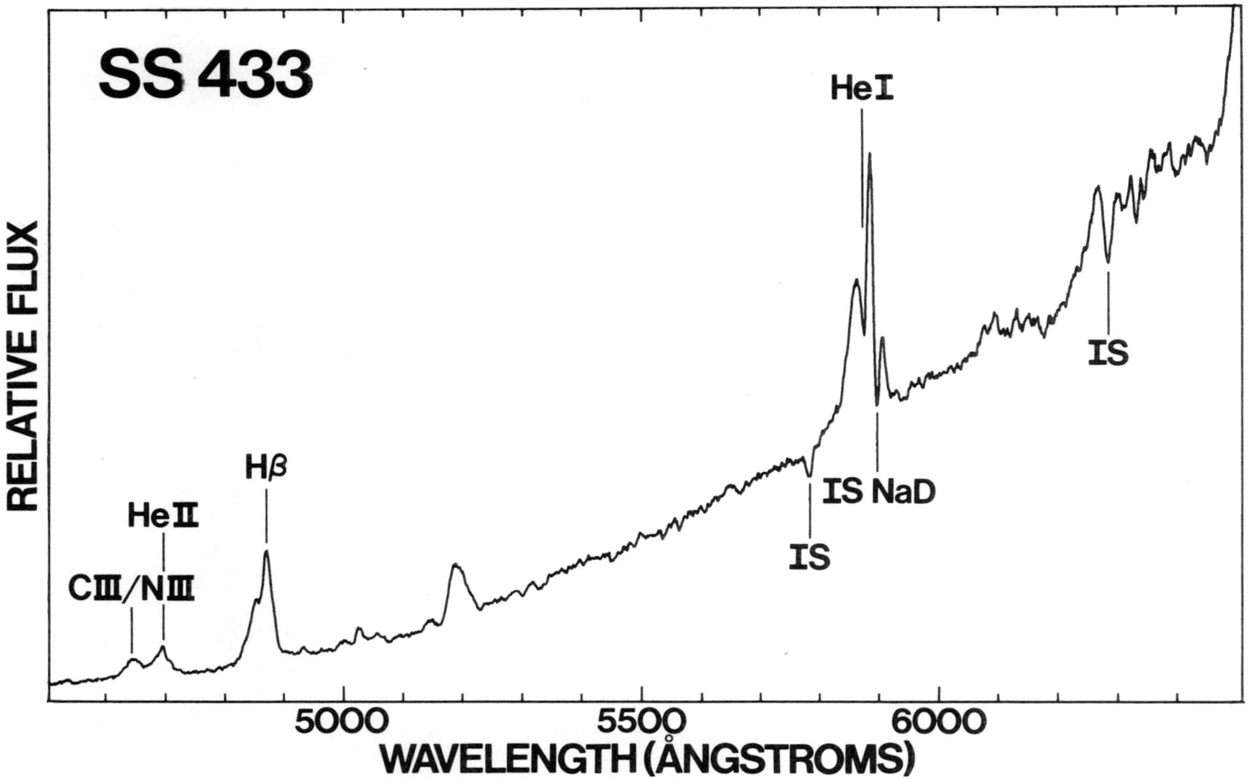

FIGURE 5. The blue/green spectrum of SS 433, the bizarre optical counterpart of the x-ray source A1909+04, obtained with 0.8-mm resolution of the Lick Observatory 3-m Shane reflector on 1978 September 30 (reference 63). The enormous H-alpha emission line has been truncated for convenience in scaling, and is responsible for the precipitous rise in flux near 645 nm. Several interstellar absorption lines and diffuse bands are marked "IS." The unidentified intense emission line at 520 nm is a typical example of the variable green feature, and He I 587.6 nm is most probably contaminated by the variable red feature, also discussed in the text.

also a variable nonthermal radio source.[66–68] The object has been previously cataloged as SS 433. We have obtained extensive spectrophotometric observations of this rather bright, red star (V = 14, B – V = 2) during 1978 September through December, using the Hale 5-m, Kitt Peak 4-m, Lick 3- and 0.6-m, and Mt. Wilson 2.5-m reflectors. In agreement with the discovery observations,[65] we find the spectrum to be dominated by intense Balmer and He I emission, variable on a time scale of days. The familiar He II 4686 and C III/N III 4640/50 emission is also present and variable. A typical spectrum of the object in the blue/green region is shown in FIGURE 5. The Balmer line profiles, when observed at Coude resolution, are complex, composite, and show night-to-night structural changes (FIGURE 6). The most bizarre of all spectral features, however, are readily apparent in FIGURE 7. There are very strong emission lines at unfamiliar wavelengths in the red (600 nm) and near-infrared (740 nm). Inspection of FIGURE 6 also shows a similarly strange line in the green, near 519 nm. The most perplexing property of these emission lines is seen in FIGURE 7, where spectra obtained on three of four consecutive nights are displayed sequentially. The features change their wavelengths daily by very substantial amounts! In the four nights spanned by the figure, the infrared line has moved to longer wavelengths by 15 nm, while the red line has moved in the opposite sense, to shorter wavelengths by 7 nm. If these motions are interpreted as due to the Doppler effect, they correspond to velocity changes of +6000 and –3500 km/s, respectively. We have observed these anomalous features to traverse an even larger range of wavelengths. Our spectra on 25 separate nights yield strong evidence that the infrared feature has moved from about 704 nm to 762 nm in roughly a month, corresponding to velocity changes of 25 000 km/s! Our data on the red and green features are less complete, but indicate similarly large motions. There is also evidence for minor but correlated sign reversals in the motions of the features. We have found no obvious periodicities in the system, but at least the red feature may have made a roundtrip traverse between 585 and 612 nm in about 60 days.

It is, to phrase it mildly, very difficult to interpret these drastic spectral changes in the context of any previously known stellar spectrum. Two models have already been advanced that attribute our observations to Doppler shifts of Balmer emission lines, one where the central object is a miniature radio galaxy spraying out two collimated plasma jets,[69] and another where the emission originates from a disk surrounding a massive black hole.[70] At the moment we favor perhaps less speculative explanations such as Zeeman splitting to explain the features, achieving the line motions by postulating that the emission region moves slightly with respect to the compact magnetic star, thus changing the magnetic field strength in the region. Obviously, only extensive further observations will decide amongst the models.

Aside from the intrinsic interest in this unusual object for its own sake, it is also hopefully indicative of what some future x-ray observations may yield. From measurements of the interstellar absorption lines in the spectrum, we know that the object is at least 3.5 kpc distant, and is thus optically very luminous. The x-ray emission is in fact only a minor perturbation in the system, but is the vital attribute that called attention to the otherwise relatively anonymous SS 433. There previously seemed to exist at least a small hazard that as x-ray observations became more and more sensitive (especially with the overwhelming advance of HEAO-2), we might see simply the same type of object (e.g., Sco X-1) at greater distances. We can see now with the

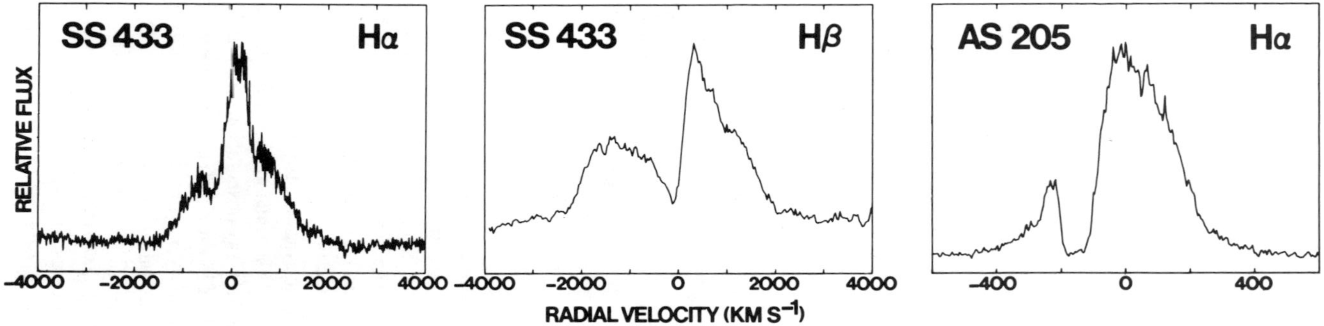

FIGURE 6. High resolution Balmer line profiles of SS 433 and a comparison T Tauri star, AS 205. All wavelengths have been converted for convenience to the local standard of rest velocity. (Left) The H-alpha line is SS 433, obtained at the Coude focus of the Hale 5-m reflector with 0.06-nm resolution on 1978 October 19. Note the superposition of sharp upon broad emission, indicating the probable presence of a disk. (Center) The SS 433 H-beta profile, obtained with 0.4-nm resolution at the Lick 3-m Shane reflector on 1978 October 1. Compare this complex structure with the same line in FIGURE 5, observed just one day earlier. (Right) A similar profile, seen at H-alpha in the T Tauri star AS 205. Such profiles have been interpreted as due to accretion of infalling matter.[72] These data were obtained with 0.08-nm resolution at the Mt. Wilson 2.5-m Hooker reflector (reference 63).

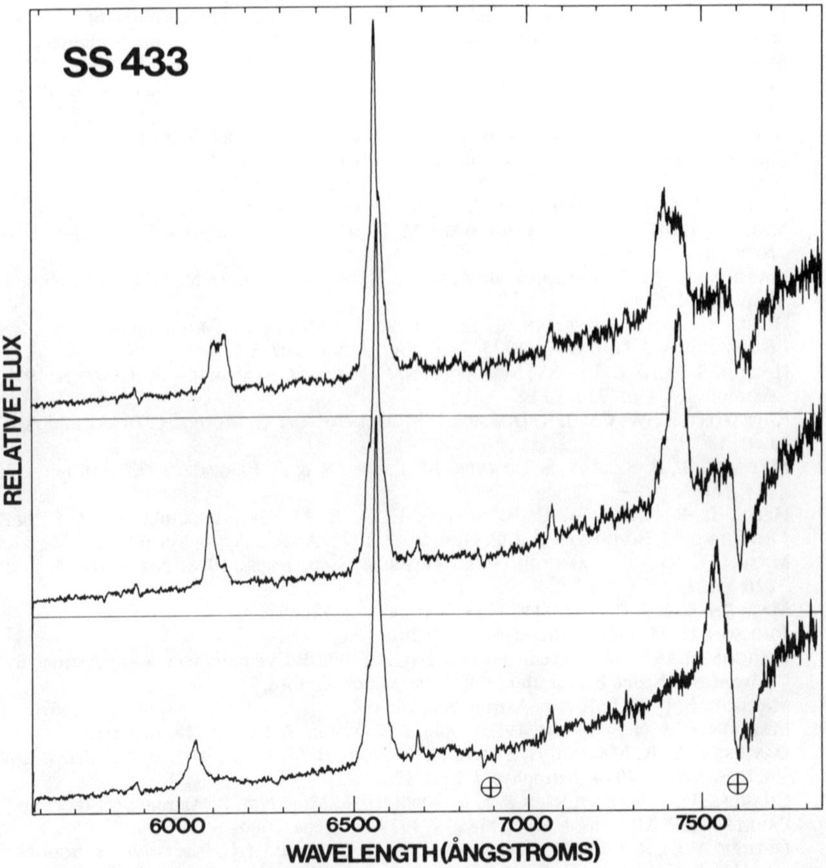

FIGURE 7. The red/infrared spectrum of SS 433 on three of four consecutive nights, obtained with 1-nm resolution on the Lick 0.6-m reflector.[63] The dramatic changes in both wavelength and profile of the two unidentified emission features flanking H-alpha are well illustrated in this example. He I 587.6-, 667.8-, and 706.8-nm emission is visible, as are the unresolved interstellar NaD lines and 628.4-nm band. The telluric *A* and *B* bands have been indicated. The upper, center, and lower panels were observed on 1978 October 23, 24, and 26, respectively.

detection of SS 433, something entirely different, that this is certainly not the case. Thus x-ray astronomy will undoubtedly continue to draw our attention to some of the most peculiar and exotic stars in the galaxy.

REFERENCES

1. SANDAGE, A., P. OSMER, R. GIACCONI, P. GORENSTEIN, H. GURSKY, J. WATERS, H. BRADT, G. GARMIRE, B. V. SREEKANTAN, M. ODA, K. OSAWA & J. JUGAKU. 1966. Astrophys. J. **146:** 316.

2. Tananbaum, H. 1973. *In* X- and Gamma-Ray Astronomy, Proceedings of the IAU Symposium No. 55. H. Bradt & Giacconi, Eds.: 9. Reidel. Dordrecht, the Netherlands.
3. Bradt, H. V., R. E. Doxsey & J. G. Jernigan. 1979. *In* Proceedings of the IAU/COSPAR Symposium on X-ray Astronomy, Advances in Space Exploration. Vol. 3. Pergamon. Oxford.
4. Garmire, G. 1979. *In* Proceedings of the IAU/COSPAR Symposium on X-ray Astronomy, Advances in Space Exploration. Vol. 3. Pergamon. Oxford. In press.
5. Walter, F., P. Charles & S. Bowyer. 1978. Astron. J. **83:** 1539.
6. Nugent, J. & G. Garmire. 1978. Astrophys. J. Lett. **226:** L83.
7. Margon, B., J. Liebert, G. Gatewood, M. Lampton, H. Spinrad & S. Bowyer. 1976. Astrophys. J. **209:** 525.
8. Margon, B., M. Lampton, S. Bowyer, R. Stern & F. Paresce. 1976. Astrophys. J. Lett. **210:** L79.
9. Heise, J., A. C. Brinkman, J. Schrijver, R. Mewe, E. Gronenschild, A. den Boggende & J. Grindlay. 1975. Astrophys. J. Lett. **202:** L73.
10. Haisch, B. M., J. L. Linsky, M. Lampton, F. Paresce, B. Margon & R. Stern. 1977. Astrophys. J. Lett. **213:** L119.
11. Rappaport, S., W. Cash, R. Doxsey, J. McClintock & G. Moore. 1974. Astrophys. J. Lett. **187:** L5.
12. Margon, B., P. Szkody, S. Bowyer, M. Lampton & F. Paresce. 1978. Astrophys. J. **224:** 167.
13. Heise, J., R. Mewe, A. C. Brinkman, E. H. B. M. Gronenschild, A. J. F. den Boggende, J. Schrijver & J. E. Grindlay. 1978. Astron. Astrophys. **63:** L1.
14. Mason, K. O., M. Lampton, P. Charles & S. Bowyer. 1978. Astrophys. J. Lett. **226:** L129.
15. Margon, B. & J. P. Ostriker. 1973. Astrophys. J. **186:** 91.
16. Johnson, H. M. 1966. Astrophys. J. **143:** 261.
17. Margon, B. 1979. *In* Proceedings of the IAU/COSPAR Symposium on X-ray Astronomy, Advances in Space Exploration. Vol. 3. Pergamon. Oxford.
18. Margon, B. 1978. Bull. Am. Astron. Soc. **10:** 499.
19. Margon, B. & H. V. Bradt. 1980. Annu. Rev. Astron. Astrophys. **18.** In press.
20. Davidsen, A., R. Malina, H. Smith, H. Spinrad, B. Margon, K. Mason, F. Hawkins & P. Sanford. 1974. Astrophys. J. Lett. **193:** L25.
21. Charles, P., J. Thorstensen & S. Bowyer. 1978. Mon. Not. R. Astron. Soc. **183:** 29p.
22. Parkes, G., P. Murdin & K. O. Mason. 1979. In preparation.
23. Dupree, A. K., R. J. Davis, H. Gursky, L. W. Hartmann, J. C. Raymond, A. Boggess, A. Holm, Y. Kondo, C. C. Wu, F. Macchetto, M. C. W. Sanford, A. J. Willis, R. Wilson, F. Ciatti, J. B. Hutchings, H. M. Johnson, J. Jugaku, D. C. Morton, A. Treves & E. P. J. van den Heuvel. 1978. Nature (London) **275:** 30.
24. McCintock, J. E., C. R. Canizares & C. B. Tarter. 1975. Astrophys. J. **198:** 641.
25. Hatchett, S., J. Buff & R. McCray. 1976. Astrophys. J. **206:** 847.
26. Margon, B. & J. G. Cohen. 1978. Astrophys. J. Lett. **222:** L33.
27. Mason, K. O., P. A. Charles, N. E. White, J. L. Culhane, P. W. Sanford & K. T. Strong. 1976. Mon. Not. R. Astron. Soc. **277:** 513.
28. Paczynski, B. 1979. *In* Proceedings of the IAU/COSPAR Symposium on X-ray Astronomy, Advances in Space Exploration. Vol. 3. Pergamon. Oxford.
29. Cowley, A. P., D. Crampton & J. B. Hutchings. 1979. Astrophys. J. **231:** 539.
30. Dower, R. G., K. M. V. Apparao, H. V. Bradt, R. E. Doxsey, J. G. Jernigan & J. Kulik. 1978. Nature (London) **273:** 364.
31. Hardorp, J., K. Rohlfs, A. Slettebak & J. Stock. 1959. Luminous Stars in the Northern Milky Way. Vol. 1. Warner and Swasey Observatory. Hamburg.
32. Gonzalez, G. & G. Gonzalez. 1954. Bull. Obs. Ton. Tac. **9:** 21.
33. Margon, B., K. B. Kwitter, H. V. Bradt & R. Dower. 1979. In preparation.
34. Bahcall, J. N. 1978. Annu. Rev. Astron. Astrophys. **16:** 241.
35. Mason, K. O. 1977. Mon. Not. R. Astron. Soc. **178:** 81p.
36. Rappaport, S. & P. C. Joss. 1977. Nature (London) **266:** 683.
37. Doxsey, R. E., K. M. V. Apparao, H. V. Bradt, R. G. Dower & J. G. Jernigan. 1977. Nature (London) **269:** 112.

38. Doxsey, R. E., K. M. V. Apparao, H. V. Bradt, R. G. Dower & J. G. Jernigan. 1977. Nature (London) **270:** 586.
39. Jernigan, J. G. & G. Clark. 1979. Astrophys. J. Lett. **231:** L125.
40. Katz, J. I. 1975. Nature **253:** 698.
41. Davidsen, A., R. Malina & S. Bowyer. 1976. Astrophys. J. **203:** 448.
42. Davidsen, A., R. Malina & S. Bowyer. 1977. Astrophys. J. **211:** 866.
43. Silk, J. & J. Arons. 1975. Astrophys. J. Lett. **200:** L131.
44. Bahcall, J. N. & J. P. Ostriker. 1975. Nature (London) **256:** 23.
45. Vidal, N. V. & K. C. Freeman. 1975. Astrophys. J. Lett. **200:** L9.
46. Bolton, A. J. & E. A. Malia. 1977. Astron. Astrophys. **59:** L23.
47. Liller, W. 1978. Bull. Am. Astron. Soc. **10:** 641.
48. Ford, H., B. Margon, D. Jenner & A. Klemola. 1978. IAU Circular 3306.
49. Grindlay, J., H. Gursky, H. Schnopper, D. R. Parsignault, J. Heise, A. C. Brinkman & J. Schrijver. 1976. Astrophys. J. Lett. **205:** L127.
50. McClintock, J. E., C. R. Canizares & D. E. Backman. 1978. Astrophys. J. Lett. **223:** L75.
51. Swank, J. H., R. H. Becker, E. A. Boldt, S. S. Holt & P. J. Serlemitsos. 1978. Mon. Not. R. Astron. Soc. **182:** 349.
52. Grindlay, J. E., J. E. McClintock, C. R. Canizares, J. van Paradijs, L. Cominsky, F. K. Li & W. H. G. Lewin. 1978. Nature (London) **274:** 567.
53. Middleditch, J. & J. Nelson. 1976. Astrophys. J. **208:** 567.
54. Margon, B., J. Nelson, G. Chanan, J. R. Thorstensen & S. Bowyer. 1977. Astrophys. J. **216:** 811.
55. Margon, B., J. R. Thorstensen, S. Bowyer, K. O. Mason, N. E. White, P. W. Sanford, G. Parkes, R. P. S. Stone & J. Bailey. 1977. Astrophys. J. **218:** 504.
56. Margon, B. 1978. Astrophys. J. **219:** 613.
57. Krzeminski, S. & W. C. Priedhorsky. 1978. Publ. Astron. Soc. Pac. **90:** 434.
58. Nelson, J., F. Cordova & J. Middleditch. 1979. Astrophys. J. **229:** 294.
59. Peterson, B. A., J. Middleditch & J. Nelson. 1975. Astrophys. J. Lett. **195:** L131.
60. Ilovaisky, S., C. Motch & C. Chevalier. 1978. Astron. Astrophys. **70:** L19.
61. Canizares, C. R., J. E. McClintock, F. K. Li & J. E. Grindlay. 1978. Bull. Am. Astron. Soc. **10:** 509.
62. McClintock, J. E., C. R. Canizares, H. V. Bradt, R. E. Doxsey, J. G. Jernigan & W. A. Hiltner. 1977. Nature (London) **270:** 320.
63. Margon, B., H. C. Ford, J. I. Katz, K. B. Kwitter, R. K. Ulrich, R. P. S. Stone & A. Klemola. 1979. Astrophys. J. Lett. **230:** L41.
64. Seward, F. D., C. G. Page, M. J. L. Turner & K. A. Pounds. 1976. Mon. Not. R. Astron. Soc. **175:** 39p.
65. Clark, D. H. & P. Murdin. 1978. Nature (London) **276:** 44.
66. Ryle, M., J. L. Caswell, G. Hine & J. Shakeshaft. 1978. Nature (London) **276:** 571.
67. Seaquist, E. R., P. C. Gregory & P. C. Crane. 1978. IAU Circular 3256.
68. Feldman, P. A., C. R. Purton, T. Stiff & S. Kwok. 1978. IAU Circular 3258.
69. Fabian, A. C. & M. Rees. 1979. Mon. Not. R. Astron. Soc. **187:** 13p.
70. Terlevich, R. J. & J. E. Pringle. 1979. Nature (London) **278:** 719.
71. Thorstensen, J., P. Charles & S. Bowyer. 1978. IAU Circular 3253.
72. Ulrich, R. K. 1976. Astrophys. J. **210:** 377.

IRON LINES IN THE SPECTRA OF X-RAY BINARIES

M. M. Basko

Space Research Institute
Academy of Sciences of the USSR
Moscow 117810 USSR

INTRODUCTION

The discovery of an iron line emission at $\epsilon = 6$–7 keV is the first prominent achievement of cosmic x-ray spectroscopy. Now this line is found in the spectra of almost all types of cosmic x-ray sources, including x-ray binaries,[1–7] supernova remnants,[8–11] single galaxies,[12] and clusters.[13–15] For many of them a theoretical interpretation of the iron line emission presents no difficulty. In the case of supernova remnants and clusters of galaxies the x-ray flux both in lines and continua comes as a thermal emission of hot ($T > 10^7$ K) optically thin gas clouds—a model thoroughly explored by theoreticians. The line is excited by electron collisions with Fe^{+25} and Fe^{+25} ions. A different mechanism operates in the active galaxy Cen A, where the iron line emission is due to K_α fluorescence of a cold ($T < 10^7$ K) gas surrounding the central x-ray source with a continuous spectrum. Here the excitation mechanism is the K-shell photoionization of low ions of iron Fe I–XVII. Such a model is characterized by a specific relation between the K_α-line ($\epsilon = 6.40$ keV) equivalent width and the magnitude of the K jump at $\epsilon = 7.11$ keV : the number of line photons emitted is three times less (the fluorescence yield $\omega_K = 0.34$) than the number of photons absorbed in the continuum above the K edge.

In the case of x-ray binaries, however, the attempts to explain the observed properties of iron line emission meet with serious difficulties. I shall try to illustrate this with an example of the best studied x-ray binary Her X-1. We know that the x-ray spectrum of this source is formed in hot regions at the magnetic poles, where the accreting material collides with the surface of the neutron star. If one assumes that the hot spots occupy $\sim 1/100$ of the total surface area of the neutron star (a typical value in the theory of x-ray pulsars), then the photoionization rate for Fe^{+25} in the atmosphere is $\gamma^{+25} \approx 7.3 \times 10^{12} L_{37}$ s^{-1}, where L_{37} is the x-ray luminosity in units of 10^{37} erg s^{-1}. From the shape of the x-ray spectrum of Her X-1 one can say that the electron temperature in the atmosphere of hot spots $T \gtrsim 10^8$ K, and the scale height is $H \approx 10^2$ cm. The recombination coefficient at such a temperature is $\alpha_{tot}{}^{+25} \approx 1.3 \times 10^{-12}$ cm^3 s^{-1}, and for the electron density $n_e = (\delta_T H)^{-1} \approx 1.5 \times 10^{22}$cm^{-3} at the level with Thomson optical depth $\tau_T - 1$ we arrive at the value of Fe^{+25} concentration $\chi^{+25} \approx 10^{-3}$. Under these conditions the principal mechanism of populating the $2p$ level of Fe^{+25} is recombination, and according to Equations 9 and 10a (in Equation 9 one must set $\Omega_s = 4\pi/100$, $r_A = R \approx 10^6$ cm, $\tau_T = 1$) the equivalent width of the iron line $W \approx 1$ eV, if one assumes normal chemical composition. The above value of W is at least two orders of magnitude less than the observed one, $W \approx 300$ eV.[3] Mészáros[16] obtained similar values, $W < 10$ eV, in the case of disk accretion onto black holes.

Thus, a natural suggestion comes to mind that in x-ray binaries the iron line is formed away from the x-ray star, and the excitation is due to fluorescence. A

0077–8923/80/0336–0566 $01.75/1 © 1980, NYAS

fluorescent K_α line of iron, originating in the atmosphere of the normal component of the binary, had been predicted[17] more than four years ago. The detailed calculations[18,19] have shown, however, that this line is not what is observed because it should (i) have significantly smaller values of the equivalent width ($W \approx 15$ eV for Her X-1); (ii) show strong correlation with the orbital phase, which contradicts the observations of Her X-1[3] and Cyg X-3[6]; (iii) be centered at $\epsilon = 6.40$ keV and have a small width $\Delta\epsilon \sim 30$ eV, which is in a disagreement with reference 7.

And even when one adds the fluorescence of the accretion disk and that of the stellar wind, the theoretical values of W for the HZ Her /Her X-1 system[19] are still 3–5 times less than the observed ones.

But though K_α fluorescence of weakly ionized iron fails to account for the observed emission feature, I should like to emphasize that the narrow K_α line *has to be found* in the spectra of x-ray binaries at the level predicted by the above mentioned calculations[18,19] (if our understanding of these objects is any good). The discovery and systematic observations of this line will render us much new information about that type of cosmic x-ray source. Possibly, it is this line that has been detected[20] in the "off" state of Her X-1.

Below I shall consider a model that is an attempt to explain the observed powerful iron line emission as a fluorescence of the gaseous shell at the Alfvén surface. It has been realized recently, when trying to interpret the observed pulse profile and soft x-ray emission of Her X-1,[21-23] that the Alfvén shell may be an important constituent of the models of binary x-ray pulsars. Calculations of the iron line emission presented below give some more support to this idea. In my opinion, the model under discussion is the only one for the time being that can claim to be able to account for the observed properties of the iron line in x-ray binaries, but it has its difficulties too.

Parameters of the Gas in the Alfvén Shell

The location of the Alfvén shell r_A to a great extent depends on the accretion geometry in the region $r > r_A$. In a spherically symmetric case[24-26]

$$r_A \gtrsim r_A{}^0 = 2^{-3/7}\mu^{4/7}(LR)^{-2/7}(GM)^{1/7}$$
$$= 2.89 \times 10^8\, \mu_{30}{}^{4/7}(L_{37}R_6)^{-2/7}(M/M_\odot)^{1/7}\ \text{cm}, \qquad (1)$$

where μ_{30} is the dipole magnetic moment of the neutron star in units of 10^{30} G cm^3, M is its mass, R_6 is its radius in units of 10^6 cm, and L_{37} is the x-ray luminosity in units of 10^{37} erg s^{-1}. In the case of disk accretion the radius of the Alfvén shell is much more uncertain, although the theoretical estimates[27,28] show it to be of the same order of magnitude as $r_A{}^0$. Keeping in mind this uncertainty and the fact that the HZ Her/Her X-1 system disk accretion takes place, I shall treat r_A as *a free parameter* that is to be determined from comparison of the observed value of the iron line equivalent width with the calculated one.

It will be only natural to assume that the Alfvén shell subtends a substantial solid angle $\Omega_s \gtrsim 1$ as seen from the neutron star, since otherwise its contribution to the iron line emission will not be dominating. But in order for the accreting material to come out of the disk plane and spread along the magnetic field lines over a substantial solid

TABLE 1

THE VALUES OF PHYSICAL PARAMETERS IN THE ALFVÉN SHELL AND THE EQUIVALENT WIDTH OF THE IRON LINE FOR $\mu = 10^{30}$ G cm^3

R_A (cm)	10^8	7×10^7	5×10^7	4×10^7	3×10^7	2.5×10^7	2.34×10^7	2×10^7
ξ	3.53×10^4	1.21×10^4	4.41×10^3	2.26×10^3	953	552	458	282
n_e(cm^{-3})	8.50×10^{16}	5.06×10^{17}	2.72×10^{18}	8.30×10^{18}	3.50×10^{19}	8.70×10^{19}	1.19×10^{20}	2.66×10^{20}
T(K)	4.01×10^7	2.83×10^7	1.33×10^7	6.79×10^6	2.82×10^6	1.60×10^6	1.34×10^6	1.52×10^6
x^{+25}	0.0118	0.04426	0.1835	0.3817	0.4023	0.2402	0.1908	0.1387
x^{+24}	4.57×10^{-5}	6.75×10^{-4}	0.01432	0.09868	0.4729	0.7305	0.7920	0.8529
x^{+7}	3.44×10^{-5}	1.33×10^{-4}	7.29×10^{-4}	2.71×10^{-3}	1.43×10^{-2}	3.83×10^{-2}	5.22×10^{-2}	7.46×10^{-2}
W^{+25}(eV)	0.49	2.3	12.6	32	43	29	24	19
W^{+24}(eV)	5.2×10^{-3}	0.095	2.5	21	125	217	245	285
$\tau_{ff}\|_{\epsilon=kT}$	4.9×10^{-9}	1.2×10^{-7}	1.0×10^{-5}	3.8×10^{-4}	0.040	0.79	2.1	3.2
L_{ff}/L	7.0×10^{-4}	2.0×10^{-3}	4.6×10^{-3}	7.1×10^{-3}	1.3×10^{-2}	1.8×10^{-2}	2.0×10^{-2}	2.4×10^{-2}

angle, the condition

$$\rho v^2 \leq H^2/8\pi \tag{2}$$

must be fulfilled in the Alfvén shell. And since from the very meaning of the Alfvén surface the strong inequality cannot obtain in (**2**), we can use Equation 2 to estimate the electron density in the shell. The calculations of the gas flow in the Alfvén shell[26] show that over the major part of the shell one can, with a sufficient degree of accuracy, adopt $V \approx 0.3 V_{ff} = 0.3\ (2GM/r_A)^{1/2}$. Assuming $N_{He} = 0.1\ N_H$, from (**2**) we get

$$N_e = 8.5 \times 10^{16}\ \mu_{30}{}^2\ r_8{}^{-5}\ (M/M_\odot)^{-1}\ \mathrm{cm}^{-3}. \tag{3}$$

Here r_8 is the Alfvén radius r_A in units of 10^8cm.

The Thomson optical depth $\tau_T = \sigma_T N_e \delta$ of the shell can be estimated from the condition that all material falling into the neutron star, $\dot{M} = LR/GM$, passes through the cross section $S \approx \Omega_s r_A \delta$ of the shell,

$$\tau_T = 0.52 L_{37} r_8{}^{-1/2} \Omega_s{}^{-1} R_6 (M/M_\odot)^{-3/2}. \tag{4}$$

The formula (**4**) is valid for $\delta \ll r_A$, which is equivalent to the condition $r_A < (\Omega_s/3.8)^{2/7} r_A{}^0$ that always holds in cases of interest here.

All estimates below are performed for Her X-1, for which within a factor $\lesssim 2$, $L_{37} = 3$, $R_6 = M/M_\odot = 1$, and $\Omega_s = 3$. The value of the dipole magnetic moment is much more uncertain, and the calculations were done for $\mu_{30} = 1$ (TABLE 1) and $\mu_{30} = 10$ (TABLE 2).

THERMAL BALANCE AND IONIZATION EQUILIBRIUM

Since according to Equation 4 the Thomson optical depth of the shell $\tau_T \lesssim 1$, the temperature and the ionization balance can be evaluated in the optically thin approximation. In this approximation they depend (for a fixed x-ray spectrum) on one parameter only,

$$\zeta = L/n_e r_A{}^2 = 1.18 \times 10^4\ L_{37} \mu_{30}{}^{-2} r_8{}^3 (M/M_\odot). \tag{5}$$

In the range of interest here, $\xi > 10^2$, it is sufficient to include in the equation of thermal balance the following processes:[29] Compton heating, bremsstrahlung cooling, heating and cooling by photoionization, and recombination of oxygen ($O^{+8} \rightleftarrows O^{+7}$) and iron ($Fe^{+26} \rightleftarrows Fe^{+25} \leftrightarrows Fe^{+24}$) ions. The cooling in the resonance line of He^+ can be neglected here because the values of $\xi \lesssim 300$, at which it might become important, are attained at such a high density in the shell that the electron deexcitation of the $2p$ level of He^+ completely dominates its radiative decay, while the optical depth in the line $\tau \gtrsim 10^3$. Under such conditions only a small fraction of L_α photons of He^+ created in the shell escapes.

If such a high density is adopted in the shell that it becomes opaque for the photons $\epsilon \sim kT$, the free–free cooling occurs through the surface emission rather than through the emission out of the whole volume. For more details on thermal balance see reference 30. The plot of the temperature $T\ (\xi)$ evaluated as described above is shown in FIGURE 1.

TABLE 2

THE VALUES OF PHYSICAL PARAMETERS IN THE ALFVÉN SHELL AND THE EQUIVALENT WIDTH OF THE IRON LINE FOR $\gamma = 10^{31}$ G cm^3

R_A(cm)	4×10^8	3×10^8	2.5×10^8	2×10^8	1.5×10^8	1.0×10^8	8.24×10^7	7.5×10^7
ξ	2.26×10^4	9.53×10^3	5.52×10^3	2.82×10^3	1.19×10^3	353	197	149
n_e(cm^{-3})	8.30×10^{15}	3.50×10^{16}	8.70×10^{16}	2.66×10^{17}	1.12×10^{18}	8.50×10^{18}	2.24×10^{19}	3.58×10^{19}
T(K)	3.61×10^7	2.47×10^7	1.64×10^7	8.52×10^6	3.56×10^6	1.06×10^6	6.98×10^5	7.57×10^5
x^{+25}	0.0200	0.0615	0.134	0.134	0.445	0.136	0.0624	0.0504
x^{+24}	1.32×10^{-4}	1.34×10^{-3}	7.09×10^{-3}	0.0546	0.353	0.856	0.936	0.949
x^{+7}	5.82×10^{-5}	1.90×10^{-4}	4.80×10^{-4}	1.75×10^{-3}	9.35×10^{-3}	0.0783	0.171	0.205
W^{+25}(eV)	0.42	1.58	4.0	11.5	20.6	8.5	4.44	3.74
W^{+24}(eV)	7.6×10^{-3}	0.092	0.56	5.1	41	130	159	168
$\tau_{ff}\|_{\epsilon=kT}$	3.46×10^{-10}	6.31×10^{-9}	7.27×10^{-8}	2.44×10^{-6}	2.54×10^{-4}	0.163	2.04	2.58
L_{ff}/L	5.19×10^{-4}	1.18×10^{-3}	1.81×10^{-3}	2.85×10^{-3}	5.04×10^{-3}	1.14×10^{-2}	1.82×10^{-2}	2.09×10^{-2}

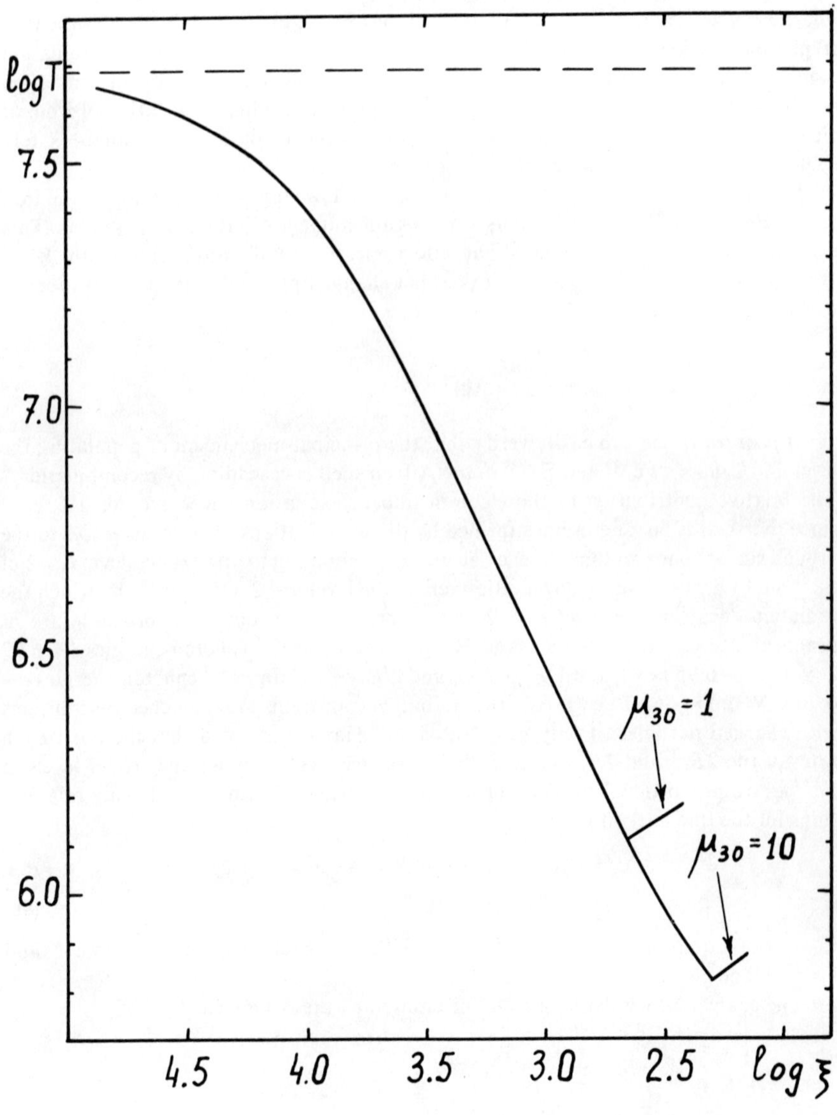

FIGURE 1. The electron temperature T in the Alfvén shell as a function of parameter $\xi = L/n_e r_A^2$.

The ionization equilibrium was calculated according to the following equations:

$$\gamma^{+25} x^{+25} = \alpha_{\text{tot}}{}^{+25} n_e (1 - x^{+25} - x^{+24}), \qquad \textbf{(6a)}$$

$$\gamma^{+24} x^{+24} = \alpha_{\text{tot}}{}^{+24} n_e x^{+25}, \qquad \textbf{(6b)}$$

$$(\gamma^{+7} + C^{+7} n_e) x^{+7} = \alpha_{\text{tot}}{}^{+7} n_e (1 - x^{+7}), \qquad (7)$$

where x^{+24}, x^{+25}, and x^{+7} are the fractions of ions Fe^{+24}, Fe^{+25}, and O^{+7}; γ^{+n} is the rate of photoionization by x-rays of the corresponding ion; $\alpha_{tot}{}^{+n}$ is the recombination coefficient summed over all excited levels; C^{+7} is the electron impact ionization coefficient for the ion O^{+7}. The contributions of electron impact ionization of iron, of dielectronic recombination onto Fe^{+25}, and of triple recombination can be safely neglected under the conditions of interest here.

I do not include in the present calculations the low stages of ionization of the iron atoms Fe^{+23}, Fe^{+22}, . . . , which make a significant contribution at $\xi \lesssim 500$. This neglect, however, has only a slight effect on the total equivalent width $W = W^{+25} + W^{+24}$ of the line, because it does not change appreciably the total number of K-shell ionizations.

Equivalent Width of the Iron Line

A posteriori one can easily verify that the principal mechanism of populating the level $n = 2$ in ions Fe^{+25} and Fe^{+24} in the Alfvén shell is cascading by recombination.* The relative contribution of the electron impact excitation never exceeds 1%. And since the shell is opaque in lines formed by direct transitions from levels $n \geq 3$ to the ground state,[30] one can think of each electron recombining to any excited level $n > 2$ of Fe^{+25} and Fe^{+24} as coming down ultimately to the level $n = 2$. Of all ways by which the radiative decay of the level $n = 2$ can occur, only the four most probable are of practical interest: $2P_{3/2} \rightarrow 1S_{1/2}$ and $2P_{1/2} \rightarrow 1S_{1/2}$ in Fe^{+25} (photon energies $\epsilon = 6.97$ keV and $\epsilon = 6.95$ keV), and $2^1P_1 \rightarrow 1^1S_0$ and $2^3P_1 \rightarrow 1^1S_0$ in Fe^{+24} (photon energies $\epsilon = 6.70$ keV and $\epsilon = 6.67$ keV). All other radiative transitions are quenched by collisions with charged particles, if only $n_e \gtrsim 10^{18}\mathrm{cm}^{-3}$.[30] Having assumed that the population rates of the $2P_{1/2}$ and $2P_{3/2}$ levels in Fe^{+25}—as well as of singlet and triplet levels in Fe^{+24}—are proportional to their statistical weights, we obtain the following relationships for the line intensities:

$$I(1S_{1/2} - 2P_{1/2}) : I(1S_{1/2} - 2P_{3/2}) = 1{:}2, \qquad \textbf{(8a)}$$

$$I(1^1S_0 - 2^1P_1) : I(1^1S_0 - 2^3P_1) = 1{:}3. \qquad \textbf{(8b)}$$

In view of Equations 8, the total line intensities for each of the two ions Fe^{+24} and Fe^{+24} will be given below.

The equivalent width of the iron line emitted by the Alfvén shell is

$$W = \beta n_{Fe} \frac{\tau_T}{\sigma_T} \frac{\epsilon_0 \epsilon_{max}}{L} \Omega_s r_A{}^2 =$$

$$= 6.85 \times 10^{11}\, \beta \tau_1 L_{37}{}^{-1} \mu_{30}{}^2 r_8{}^{-3} \Omega_s (M/M_\odot)^{-1} Y_{Fe}\ \mathrm{eV}, \qquad \textbf{(9)}$$

where the excitation coefficient $\beta(\mathrm{cm}^3\,\mathrm{s}^{-1})$ is given by

$$\beta^{+25} = (\alpha_{tot}{}^{+25} - \alpha_1{}^{+25})(1 - x^{+25} - x^{+24}), \qquad \textbf{(10a)}$$

$$\beta^{+24} = (\alpha_{tot}{}^{+24} - \alpha_1{}^{+24}) x^{+25}. \qquad \textbf{(10b)}$$

*When combined with the fact that iron is ionized exclusively by the x-ray continuum from the central source, this justifies the use of the term "fluorescence" for the line emission mechanism under discussion.

In deriving Equation 9, it is taken into account that all the line photons $\epsilon_0 = 6.7$ keV escape from the shell,[30] and the spectrum of the central x-ray source is assumed to be flat, $dL/d\epsilon$ = constant, up to the photon energy $\epsilon_{max} = 20$ keV. The iron abundance is written in the form

$$n_{Fe} = 3 \times 10^{-5} n_H Y_{Fe}. \tag{11}$$

Inserting τ_T from Equation 4 into Equation 9, we get

$$W = 3.56 \times 10^{11} \beta \mu_{30}{}^2 r_8{}^{-7/2} R_6 (M/M_\odot)^{-5/2} Y_{Fe} \text{ eV}. \tag{12}$$

It is difficult to judge the dependence of W on the Alfvén radius r_A and other parameters of the neutron star by appearance of Equation 12 since a complicated function β enters there. This dependence can be easily examined, however, in two limiting cases: (i) the iron is almost fully ionized, $x^{+24} \ll x^{+25} \ll 1$, which occurs at radii $r_A \gtrsim 0.2\, r_A{}^0$ where $T \gtrsim 10^7$ K; (ii) the iron is predominantly in the form of ions with the filled K shell, $x^{+25} \ll 1$, $1 - x^{+24} \ll 1$, which occurs at radii $r_A \lesssim 0.1\, r_A{}^0$ where $T \sim 10^{6K}$. When we note that $W^{+25} \gg W^{+24}$ in case (i), and $W^{+24} \gg W^{+25}$ in case (ii), and make necessary simplifications in Equations 12 and 10 with the aid of Equation 6b, we get

$$W \approx \begin{cases} 3.56 \times 10^{11} (\alpha_{tot}{}^{+25} - \alpha_1{}^{+25}) \mu_{30}{}^2 r_8{}^{-7/2} R_6 (M/M_\odot)^{-5/2} Y_{Fe} \text{ eV}, & \textbf{(13a)} \\ \qquad r_A \gtrsim 0.2\, r_A{}^0, & \\ 62.8\,(1 - \alpha_1{}^{+24}/\alpha_{tot}{}^{+24}) L_{37} z_8{}^{-1/2} R_6 (M/M_\odot)^{-3/2} Y_{Fe} \text{ eV}, & \textbf{(13b)} \\ \qquad r_A \lesssim 0.1\, r_A{}^0. & \end{cases}$$

From Equation 13a one sees that while $r_A \gtrsim 0.2 r_A{}^0$, the equivalent width of the iron line rapidly increases with the decreasing r_A—more rapidly than $r_A{}^{-7/2}$ because the decrease of r_A is accompanied with the increase in recombination rate $(\alpha_{tot}{}^{+25} - \alpha_1{}^{+25})$ due to the temperature drop. Nevertheless, the typical values of $W \lesssim 10$ eV in this region. At still lower values of $r_A \lesssim 0.1\, R_A{}^0$ the dependence $W(r_A)$ flattens and becomes so weak, $W \alpha\, r_A{}^{-1/2}$, that one can consider the value $W_{max} = W(0.1 r_A{}^0)$ given by Equation 13b as *the maximum possible value of the iron line equivalent width that can be expected from the Alfvén shell* (a weak dependence of the factor $(1 - \alpha_1{}^{+24}/\alpha_{tot}{}^{+24})$ on the temperature T is unimportant for $T \lesssim 3 \times 10^6$ K). From Equation 13b it follows that W_{max} is directly proportional to the luminosity L of the x-ray source, and increases with the increasing hardness of its spectrum, namely

$$W_{max} \propto \int_{\epsilon_{th}}^{\infty} \frac{dL}{d\epsilon} \left(\frac{\epsilon_{th}}{\epsilon}\right)^{2.84} \frac{d\epsilon}{\epsilon}, \tag{14}$$

where $\epsilon_{th} = 8.83$ keV is the ionization threshold of Fe^{+24}. The typical values of W_{max} amount to a few hundreds of electron volts—just what is observed. It is interesting to note that W_{max} does not depend on the magnetic moment μ (as long as $r_A > R$). It should be mentioned also that W is always proportional to the iron abundance Y_{Fe}, and does not depend on the value of the solid angle Ω_s until the optically thin approximation breaks down.

The values of equivalent widths W^{+25} and W^{+24}, calculated separately for ions Fe^{+25} and Fe^{+24}, are listed in TABLES 1 ($\mu_{30} = 1$) and 2 ($\mu_{30} = 10$). These tables contain also such other parameters of the shell as plasma temperature and density, fractions of iron and oxygen ions, free–free optical depth at the frequency $\epsilon = kT$, ratio of soft to hard x-ray luminosities L_{ff}/L and parameter ξ.

The Profile of the Iron Line

Locally, at each point of the Alfvén shell every component of the iron line originating from a particular atomic transition is rather narrow. The Doppler broadening (which exceeds the natural width) is small enough,

$$\Delta\epsilon_D = \epsilon(2kT/Am_pC^2)^{1/2} = 0.4\,T_6^{1/2}\ \text{eV}. \tag{15}$$

In principle, one could expect much more significant broadening due to the macroscopic motions. But if plasma in the shell is threaded by magnetic field lines (which I consider to be the most likely possibility), then the maximum gradient of the macroscopic velocity V_m that we can adopt is of the order of the gravitational potential gradient. Taking into account that in all cases of practical interest $\delta \ll r_A$, we can get

$$\Delta\epsilon_m \approx \epsilon\frac{\Delta V_m}{c} \approx \epsilon\frac{\delta}{2r_A}\frac{0.3\,v_{ff}}{c} = 4.3\times10^{-4}\xi R_6\Omega_s^{-1}\left(\frac{M_\odot}{M}\right)\text{eV}. \tag{16}$$

Due to the small profile width (**15**) and (**16**), the Alfvén shell at each point is optically thick for the line photons. The detailed analysis[30] shows, however, that the processes of "true absorption" are not effective enough to prevent their escape from the shell.

To a distant observer the iron line will have a much broader profile due to the Compton scattering and superposition of macroscopic velocities from different parts of the Alfvén shell. Since the Thomson optical thickness of the shell $\tau_T \approx 0.5$–1 (see Equation 4), while the optical depth with respect to the resonance scattering greatly exceeds 1, every line photon undergoes at least one scattering by an electron. After the very first such scattering a photon with the energy $\epsilon = 6.7$ keV spreads due to the Doppler effect over the energy interval[31]

$$\Delta\epsilon = \epsilon(4kT/m_eC^2)^{1/2} \approx 180\,T_6^{1/2}\ \text{eV} \tag{17}$$

and comes out of resonance. The Compton recoil has the same order of magnitude,

$$\Delta\epsilon \approx -\epsilon^2/m_eC^2 \approx -90\ \text{eV}. \tag{18}$$

The flow of gas along the Alfvén surface broadens the line profile by

$$\Delta\epsilon \sim \epsilon\frac{0.3\,V_{ff}}{C} \approx 110\,r_8^{-1/2}(M/M_\odot)^{1/2}\ \text{eV}. \tag{19}$$

Thus, *the observed line profile should have FWHM ~ 200–400 eV*. At such width different transitions in one and the same ion blend, and one might hope to resolve the observed iron line in no more than two components corresponding to ions Fe^{+25} and Fe^{+24}.

A salient feature of the line emission mechanism under discussion is that the line photons are created at the expense of photons from the ionizing continuum. To the corresponding suppression of continuum one must add the emission resulting from direct recombinations to the ground state. Figure 2 illustrates the shape of the x-ray spectrum in the 6–10-keV region that one could expect from Her X-1 for $r_8 = 0.2$ and $\mu_{30} = 1$. The spectrum is averaged over all sight directions. In evaluating the line profile, only the Doppler broadening by scattering on thermal electrons was accounted for. Note that the Fe^{+25} emission comprising some 6% of the total equivalent width is absolutely undistinguishable against the Fe^{+25} line.

In reality, the recombination line of Fe^{+24} should be much less conspicuous than in FIGURE 2 because of (i) the reduction of Fe^{+25} abundance after Fe^{+23} and other lower ions of iron are included into the ionization equilibrium, and (ii) smearing effect of Compton scattering and macroscopic motions not accounted for in FIGURE 2. A small (~10%) decrease of the continuum level at $\epsilon \gtrsim 9$ keV predicted in this model is unlikely to be detected with proportional counters, but must be discovered of course

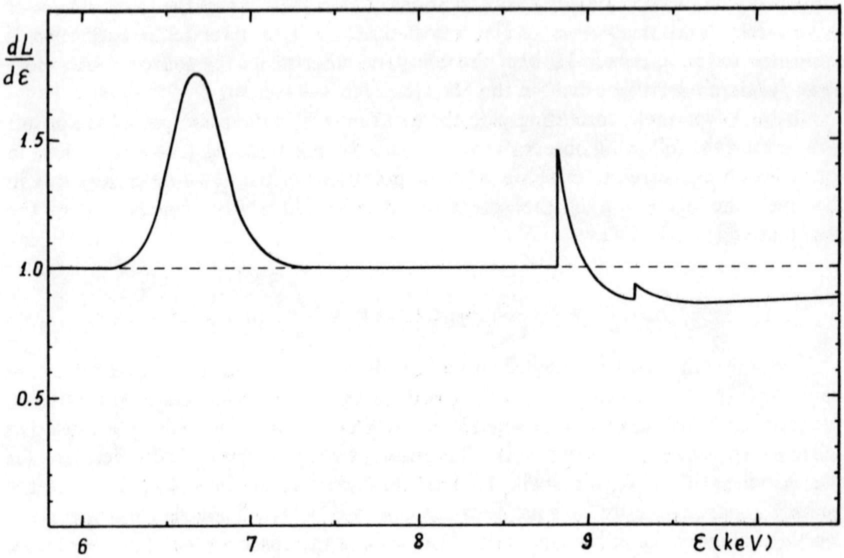

FIGURE 2. The approximate shape of the x-ray spectrum in the vicinity of the iron line that one would expect from Her X-1 for $r_A \approx 2 \times 10^7$ cm and $\mu = 10^{30}$ G cm^3. The spectrum is averaged over a period of 1.24-s pulsations.

with more sensitive spectral instruments. If shielding by the Alfvén shell is the principal cause of pulsing behavior of Her X-1[21,22] then the K-edge absorption at $\epsilon \geq 8.83$ keV should be most pronounced between the pulses and may amount to ~30–40% there.

SOFT X-RAY EMISSION

The plasma in the Alfvén shell has been already proposed[21-23] as the most likely source of the strong emission in a soft x-ray band $\epsilon = 0.1$–0.5 keV discovered in Her

X-1.[32] The results of the line emission calculations have some important implications for such an interpretation.

If one demands that, in accordance with observations, the equivalent width of the iron line emitted by the Alfvén shell in Her X-1 be $W \approx$ 200–300 eV, then, as calculated above, the plasma temperature in the shell $T \approx 10^6$ K, while its optical thickness for free–free absorption at $\epsilon \lesssim kT$ is greater than 1. This implies that the spectrum of the bremsstrahlung from the shell must be close to that of a blackbody with temperature $T \approx 10^6$ K, which is in excellent agreement with the observations.[32] But the radius of the Alfvén shell r_A in this case turns out to be significantly (by an order of magnitude) less than the conventional value $r_A \approx r_A^0$ (see Equation 1), which results in a rather low value of the total soft x-ray luminosity L_{ff}; see TABLES 1 and 2. The ratios $L_{ff}/L \lesssim 0.02$–0.03, predicted by the present model, are at least an order of magnitude too small to match the observations of Her X-1.[32] From the point of view of this model, the contradiction can be resolved either if an interstellar extinction is overestimated in reference 32, or if there is some other soft x-ray source, much more powerful than the Alfvén shell, in the HZ Her/Her X-1 system.

If the Alfvén shell, corotating with the neutron star, is the principal source of soft x-rays, then the following observational test can be proposed: the 1.24-s pulsations in soft x-rays must correlate in phase with the pulsations of the *absolute* energy flux in the line—the maximum of the soft x-ray flux should always correspond to the maximum flux in the line.

IRON LINE EMISSION IN OTHER X-RAY BINARIES

If one assumes that the model under consideration is general enough and can be applied to the whole class of binary x-ray sources, then immediately a conclusion suggests itself that strong iron line emission is to be found in the spectra of pulsating sources only, since it is x-ray pulsars that possess strong magnetic fields necessary for the existence of the Alfvén shell. To this one must add an empirical fact that the pulsating sources usually have harder x-ray continua, which is not the least important for high values of the equivalent width. And though the observational data, published up to now, are rather scarce, the above conclusion seems to agree with them. Among the best studied x-ray binaries a strong iron line has been detected in the spectra of Her X-1,[3] Cen X-3,[33] and 4U090040040,[5] well known x-ray pulsars, while all the efforts to find this line in the spectra of the much brighter sources, Sco X-1 and Cyg X-1,[34] which do not pulsate, have been unsuccessful.

Special attention should be paid to the x-ray source Cyg X-3, which produces the most powerful iron line emission of all x-ray binaries, $W \approx$ 1–2 keV.[2,6] This source has not yet proven to be an x-ray pulsar of the same type as Her X-1 and Cen X-3. Only one period, $p = 4.8^h$, has been firmly established in its flux variations, which may conceivably represent the orbital motion as well as the spinning of the neutron star.[35] Below I show that the above model can be applied to this x-ray source too, without leading to any obvious contradiction with the observations.

The principle mechanism of line excitation in the above model is fluorescence, which, due to a high density of the ionizing x-ray flux, occurs at the frequency ϵ = 6.7–6.9 keV, as contrasted to the fluorescence of cold iron ϵ = 6.40 keV.[19] There are two observational evidences that favor such an interpretation: (i) the photon energy in the line as measured with the crystal spectrometer[7] is $\epsilon = 6.84 \pm 0.17$ keV, and (ii) the absolute energy flux in the line seems to remain constant[36] when Cyg X-3 undergoes transitions from a low to the high state, and the x-ray flux at $\epsilon \sim 4$ keV increases by an order of magnitude.[2] The latter property becomes a natural consequence of the fluorescence mechanism, if one notes that the x-ray flux at $\epsilon > 8.8$ keV, responsible for the line excitation, does not show any noticeable change in such transitions.[2] Relatively large values of the equivalent width, $W \gtrsim$ keV, can be explained in terms of large x-ray luminosity (cf. Equation 14)—even in the low state $L > 4 \times 10^{37}$ $(d/10$ kpc$)^2$ erg s^{-1}.[2] As for the shift of the line center from 6.7 to 6.5 keV with the binary phase,[6] one apparently has to construct a more detailed model of Cyg X-3 to account for this effect.

Main Conclusions

The above calculations convince us that in principle the powerful iron line emission, discovered in the spectra of several x-ray binaries, can be interpreted as the reradiation of the x-ray continuum, emerging from the surface of the neutron star, by the plasma shell on the Alfvén surface. If one requires $W \approx 300$ eV in the case of Her X-1, then for $n_{\mathrm{Fe}} = 3 \times 10^{-5} n_H$, $\mu = 10^{30}$ G cm^3, and $L = 3 \times 10^{37}$ erg s^{-1} one arrives at the following values of the shell parameters: $r_{\mathrm{A}} \approx 2 \times 10^7$ cm $\approx 0.1\, r_{\mathrm{A}}{}^0$, $n_e \approx 3 \times 10^{20}$ cm^{-3}, and $T \approx 1.5 \times 10^6$ K. The most striking is an unusually low value of the Alfvén radius r_{A}, which is in conflict with theoretical estimates[28] and too small to account for the observed soft x-ray flux.[32] In view of this, any observational test that could support or refute the above model becomes of special interest.

The model presented here predicts an appreciable spectral width of the observed iron line, $\epsilon/\Delta\epsilon \lesssim 15$–$30$, centered at the energy $\epsilon = 6.70$ keV ($2^3P_1 \rightarrow 1^1S_0$ transition in heliumlike iron), which makes it clearly distinct from the K_α emission of low ionized iron with $\epsilon = 6.40$ keV and $\epsilon/\Delta\epsilon \sim 200$–$300$. The latter result implies that, in order to obtain the shape of the iron line profile in x-ray binaries, one does not necessarily have to build detectors with a very high spectral resolution $\epsilon/\Delta\epsilon \sim 10^3$–$10^4$.

References

1. Sanford, P. W., K. O. Mason & J. Ives. 1975. Mon. Not. R. Astron. Soc. **173:** 9P.
2. Serlemitsos, P. J., E. A. Boldt, S. S. Holt, R. E. Rothschild & J. L. R. Saba. 1975. Astrophys. J. Lett. **201:** L9.
3. Pravdo, S. H., R. H. Becker, E. A. Boldt, S. S. Holt, P. J. Serlemitsos & J. H. Swank. 1977. Astrophys. J. Lett. **215:** L61.
4. Pravdo, S. H., E. A. Boldt, S. S. Holt & P. J. Serlemitsos. 1977. Astrophys. J. Lett. **216:** L23.
5. Becker, R. H., R. E. Rothschild, E. A. Boldt, S. S. Holt, S. H. Pravdo, P. J. Serlemitsos & J. H. Swank. 1978. Astrophys. J. **221:** 912.
6. Becker, R. H., J. K. Robinson-Saba, E. A. Boldt, S. S. Holt, S. H. Pravdo, P. J. Serlemitsos & J. H. Swank. 1978. Astrophys. J. Lett. **224:** L113.
7. Kestenbaum, H. L., K. S. Long, R. Novick, M. C. Weisskopf & R. S. Wolff. 1977. Astrophys. J. Lett. **216:** L19.

8. Serlemitsos, P. J., E. A. Boldt, S. S. Holt, R. Ramaty & A. F. Brisken. 1973. Astrophys. J. Lett. **184:** L1.
9. Davison, P. J. N., J. L. Culhane & R. J. Mitchell. 1976. Astrophys. J. Lett. **206:** L37.
10. Pravdo, S. H., R. H. Becker, E. A. Boldt, S. S. Holt, R. E. Rothschild, P. J. Serlemitsos & J. H. Swank, 1976. Astrophys. J. Lett. **206:** L41.
11. Becker, R. H., E. A. Boldt, S. S. Holt, S. H. Pravdo, R. E. Rothschild, P. J. Serlemitsos & J. H. Swank. 1976. Astrophys. J. Lett. **209:** L65.
12. Mushotzky, R. F., P. J. Serlemitsos, R. H. Becker, E. A. Boldt & S. S. Holt. 1978. Astrophys. J. **220:** 790.
13. Mitchell, R. J., J. L. Culhane, P. J. N. Davison & J. C. Ives. 1976. Mon. Not. R. Astron. Soc. **175:** 29.
14. Serlemitsos, P. J., B. W. Smith, E. A. Boldt. S. S. Holt & J. H. Swank, 1977. Astrophys. J. Lett. **211:** L63.
15. Mushotzky, R. F., P. J. Serlemitsos, B. W. Smith, E. A. Boldt & S. S. Holt. 1978. Astrophys. J. **225:** 21.
16. Meszaros, P. 1974. Astron. Astrophys. **35:** 171.
17. Basko, M. M., R. A. Sunyaev & L. G. Titarchuk. 1974. Astron. Astrophys. **31:** 249.
18. Hatchett, S. & R. Weaver. 1977. Astrophys. J. **215:** 285.
19. Basko, M. M. 1978. Astrophys. J. **223:** 268.
20. Pravdo, S. H., E. A. Boldt, S. S. Holt, R. E. Rothschild & P. J. Serlemitsos. 1978. Astrophys. J. **225:** L53.
21. McCray, R. & F. K. Lamb. 1976. Astrophys. J. Lett. **204:** L115.
22. Basko, M. M. & R. A. Sunyaev. 1976. Astron. Zh. **53:** 950.
23. Sunyaev, R. A., 1976. Pisma Astron. Zh. **2:** 287.
24. Arons, J. & S. M. Lea. 1976. Astrophys. J. **207:** 914.
25. Elsner, R. F. & F. K. Lamb. 1977. Astrophys. J. **215**: 897.
26. Basko, M. M. 1977. Astron. Zh. **54:** 1051.
27. Scharlemann, E. 1978. Astrophys. J. **219:** 617.
28. Ghosh, P. & F. K. Lamb. 1978. Astrophys. J. Lett. **223:** L83.
29. Buff, J. & R. McCray. 1974. Astrophys. J. **189:** 147.
30. Basko, M. M. 1978. Preprint of Space Research Institute No. 450.
31. Chandrasekhar, S. 1960. Radiative Transfer. Dover, New York.
32. Shulman, S., H. Friedman, G. Fritz, R. C. Henry & D. J. Yentis. 1975. Astrophys. J. Lett. **199:** L101.
33. Swank, J. H. 1977. Private communication.
34. Boldt, E. A. 1977. Ann. N.Y. Acad. Sci. **302:** 329.
35. Sunyaev, R. A. 1976. Pisma Astron. Zh. **2:** 334.
36. Pravdo, S. H. 1978. Preprint.

THEORETICAL ASPECTS OF X-RAY BINARIES

R. Kippenhahn and H.-C. Thomas

Max-Planck-Institut für Physik und Astrophysik
8000 Munich 40
Federal Republic of Germany

X-ray binaries, cataclysmic variables, and other systems of which at least one component is highly evolved, such as the binary pulsar, are the final product of a rather complicated evolution of two stars. This evolution probably involves mass exchange within the system and mass loss from the system. The three primary processes of loss and exchange are stellar winds, Roche lobe overflow, and explosive ejection of material. Although it is very difficult, there have been bold attempts to sketch the evolutionary history leading from two innocent main sequence stars to the exotic final products mentioned above. Van den Heuvel,[1] for instance, has shown how a detached initial system consisting of two stars of 20 and 8 solar masses, having a period of 4.7 days, ends up as an x-ray binary consisting of a 2 solar mass neutron star and a 22.6 solar mass star, after a rather complicated metamorphosis. The x-ray production is nursed along by a stellar wind from the more massive of the two. Such a "scenario" of the birth of an x-ray binary is based on many assumptions, for instance, that there be no mass loss from the system during the Roche overflow.

Recently, Drs. Meyer-Hofmeister and Ritter and the authors became interested in the technique of constructing such "scenarios." In van den Heuvel's scenario as in others one combines our understanding of the evolution of single stars of constant mass with our knowledge of the different possibilities for mass exchange. In addition, the conditions under which supernova explosion occurs have to be specified.

We therefore tried to write a program which we have called a *scenario machine*. It uses the known facts about stellar evolution and well-known rules of thumb for the outcome of certain events. Starting with the masses of two main sequence stars and an orbital period together with the fractional mass and angular momentum losses as initial parameters, the machine computes the fate of the system. This fate is governed by the normal single-star evolution of the two components, which is, however, interrupted by what we call *events* such as mass exchange or a supernova explosion of one component. More specifically, one can think of two simple types of events that are illustrated in FIGURE 1. The first event, mass exchange, comprises the three well-known cases *A*, *B*, and *C*. The second event is a supernova explosion. In all three of the cases *A*, *B*, and *C* the initially more massive star becomes the less massive component. The differences among the three cases are defined by the stage of evolution at which Roche lobe overflow begins. In case *A*, the primary becomes a helium-enriched object that still burns hydrogen in its central region.[2] In case *B*, the primary becomes practically a helium star with central helium burning, unless its mass is so small that it becomes a white dwarf. In case *C*, the highly evolved carbon-oxygen core with a thin helium envelope is all that is left from the originally more massive component.[3]

The second of the basic events is a supernova explosion. In our scenario machine we have assumed that each star whose mass is above the Chandrasekhar limit, and

0077-8923/80/0336-0579 $01.75/1 © 1980, NYAS

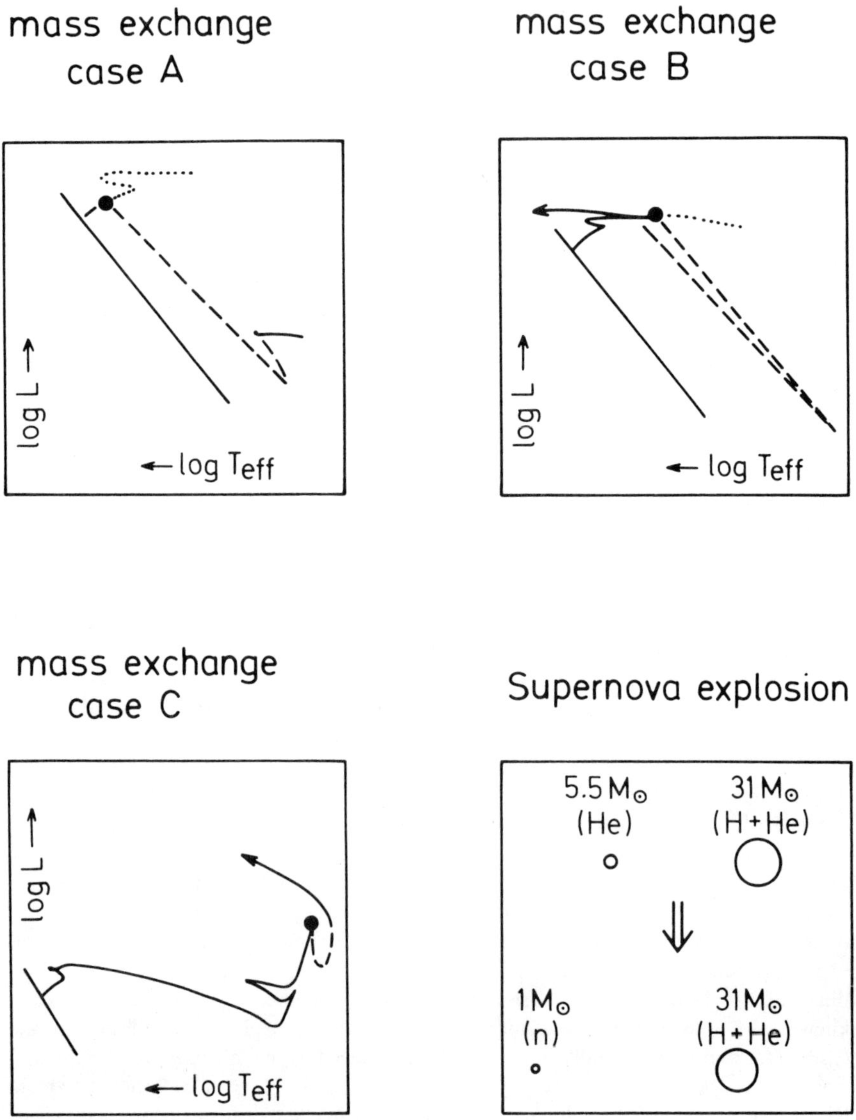

FIGURE 1. Basic events to be included in the scenario machine. The typical evolution of the mass-losing component in the Hertzsprung–Russell diagram is plotted for the three cases of mass exchange. The supernova explosion is assumed to leave the companion star unharmed and to produce a remnant of 1 $M_{\odot}$.

which does not undergo mass exchange or drastic mass loss, will end up in due course as a supernova. We assume further that a neutron star of one solar mass is left.

The ultimate fate of the system as a whole is not uniquely defined by cases *A, B,* and *C,* but depends upon the amount of mass and the amount of angular momentum that are lost from the system. The minimum set of free parameters is the following:

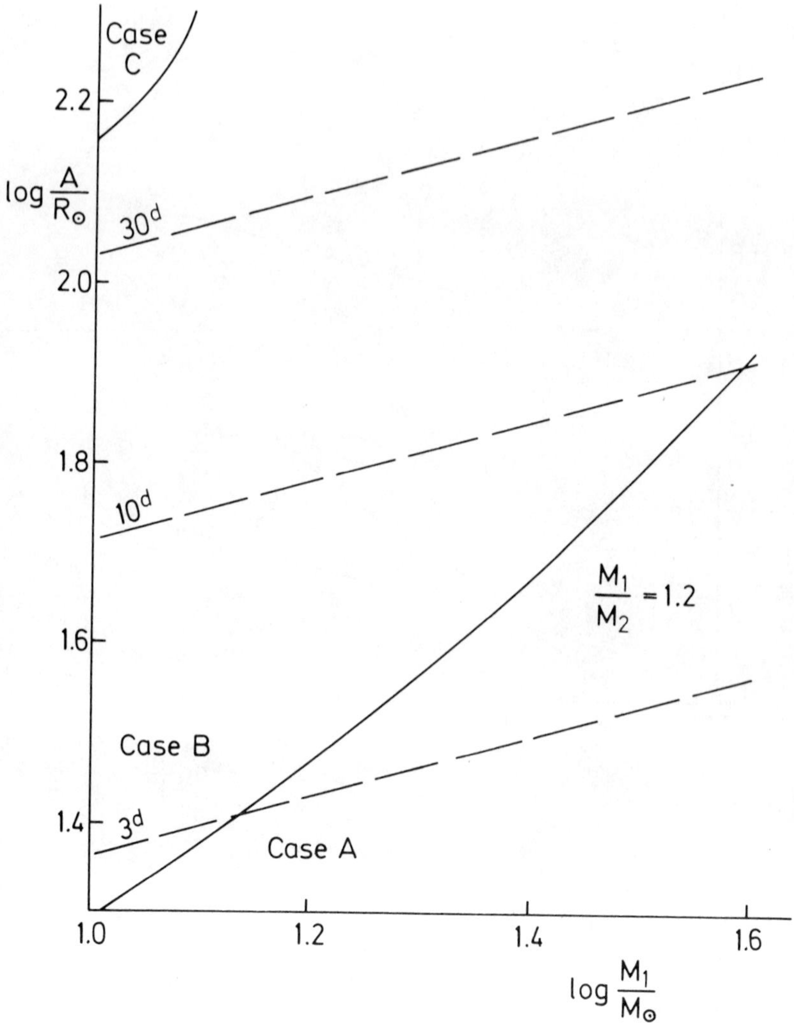

FIGURE 2. Boundaries between the three different kinds of mass exchange are drawn for primary masses varying from 10 to 40 $M_\odot$, for distances of the two components between 30 and 300 $R_\odot$, and a mass ratio of 1.2:1. Dashed lines give the period in days of the binary system before mass exchange takes place.

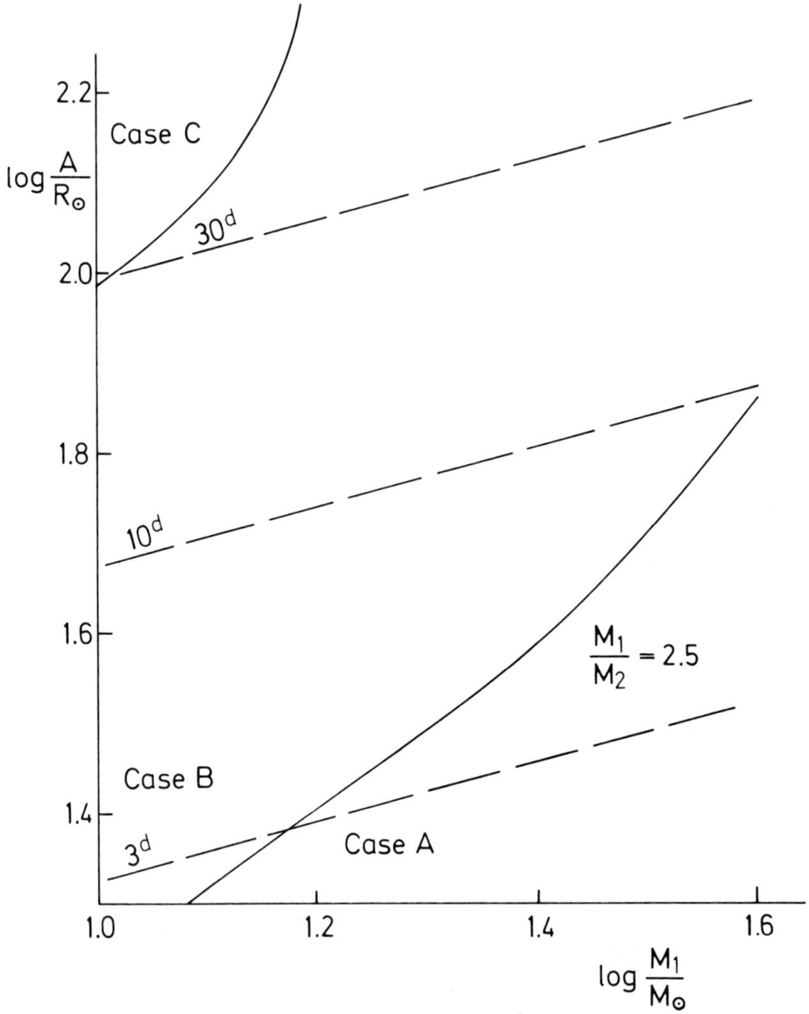

FIGURE 3. Same as FIGURE 2 but with a mass ratio of 2.5:1.

M_1/M_2, M_1, P, f, g. The first three, mass ratio, mass of the primary component, and period of the system, are obvious. The fourth parameter takes into account mass loss from the system. If a mass ΔM leaves the primary, then the fraction $f\Delta M$ falls onto the secondary, while the fraction $(1-f)\ \Delta M$ leaves the system. $f = 1$ therefore gives the conservative case in which no mass leaves the system during the mass exchange. The parameter g defines the fraction of angular momentum that, in the nonconservative case, is carried by the mass leaving the system. If one gram of material leaves the system, then the angular momentum carried by this mass is g times the average angular momentum per gram in the system. Results of evolutionary computations for

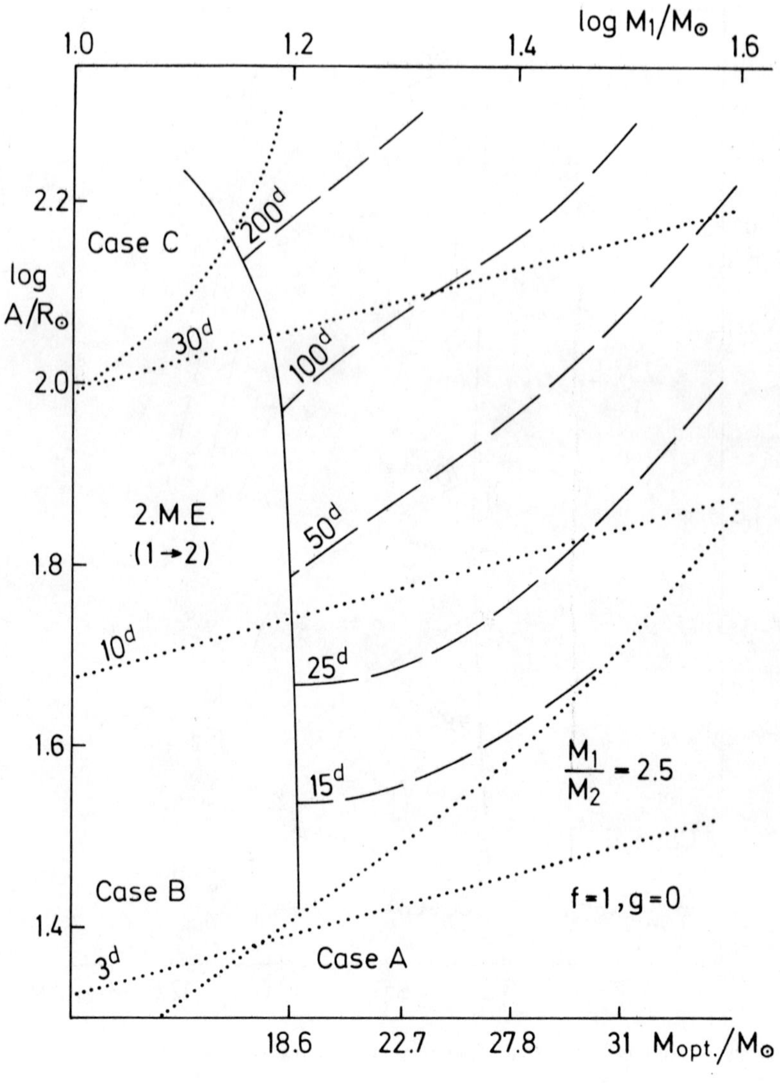

FIGURE 4. Results of the evolution including the events built into the scenario machine for the parameter field of FIGURE 3. Mass exchange (case *B* only) is assumed to be conservative ($f=1$, $g=0$). For initial values to the right of the solid line, bound systems of a neutron star circling a massive companion are obtained after a supernova explosion. The dotted lines here correspond to the lines of FIGURE 2. The final periods indicated by dashed lines and the mass of the companion are given in the lower horizontal scale.

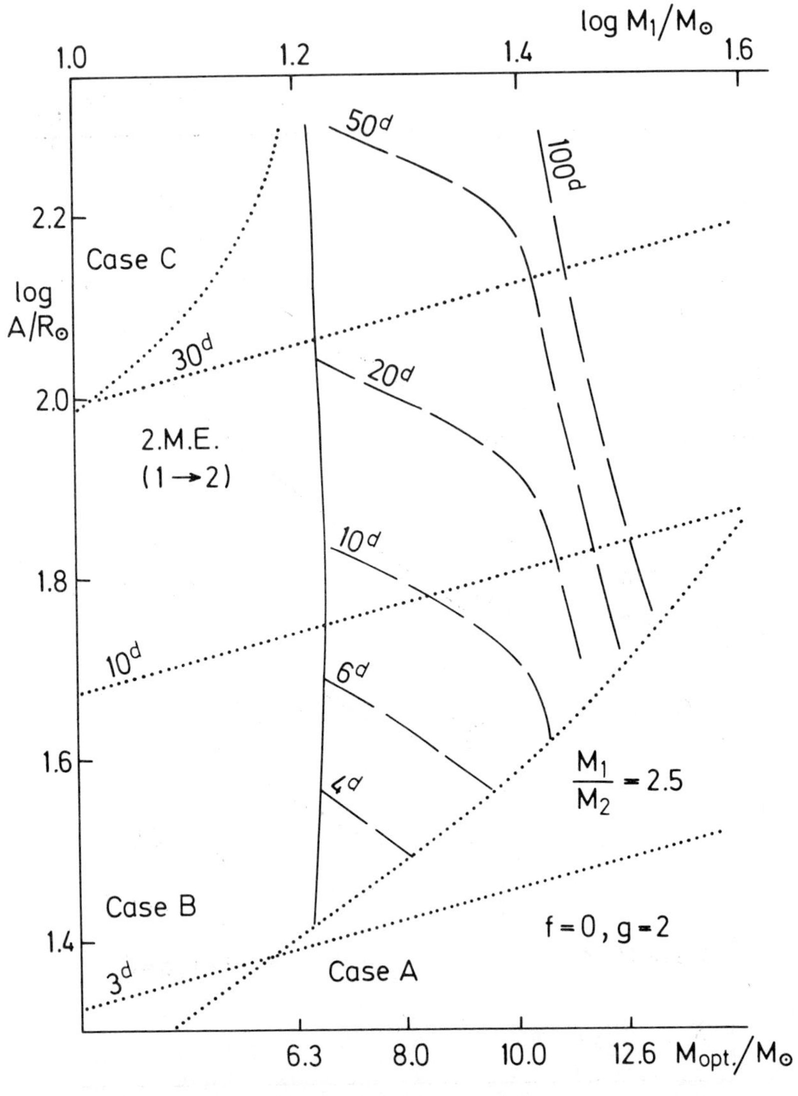

FIGURE 5. Same as FIGURE 4 but for extreme nonconservative mass exchange (f=0, g=2).

a sequence of discrete mass values are stored in the scenario machine. An interpolation program makes it possible to deal with masses in between. Given an initial parameter set, the machine computes the momentary Roche and star volumes. For each subsequent time step these volumes are compared in order to find out whether or not mass exchange starts. In addition it checks whether or not the age of the system is consistent with the assumption that neither of the two components has undergone a supernova explosion. The first three elements of our parameter set define the character of the initial mass exchange. This is shown in FIGURES 2 and 3 for two different values for the initial mass ratio. In the following we mainly deal with case *B*. Case *A* is rather complicated and has not yet been incorporated into the machine. The two figures show which initial conditions give mass exchange according to case *B*. In particular, we can look at van den Heuvel's initial system. It is a case that undergoes mass exchange according to case *B*, but as one can see, a slight change in the initial parameters would give a mass exchange according to case *A*.

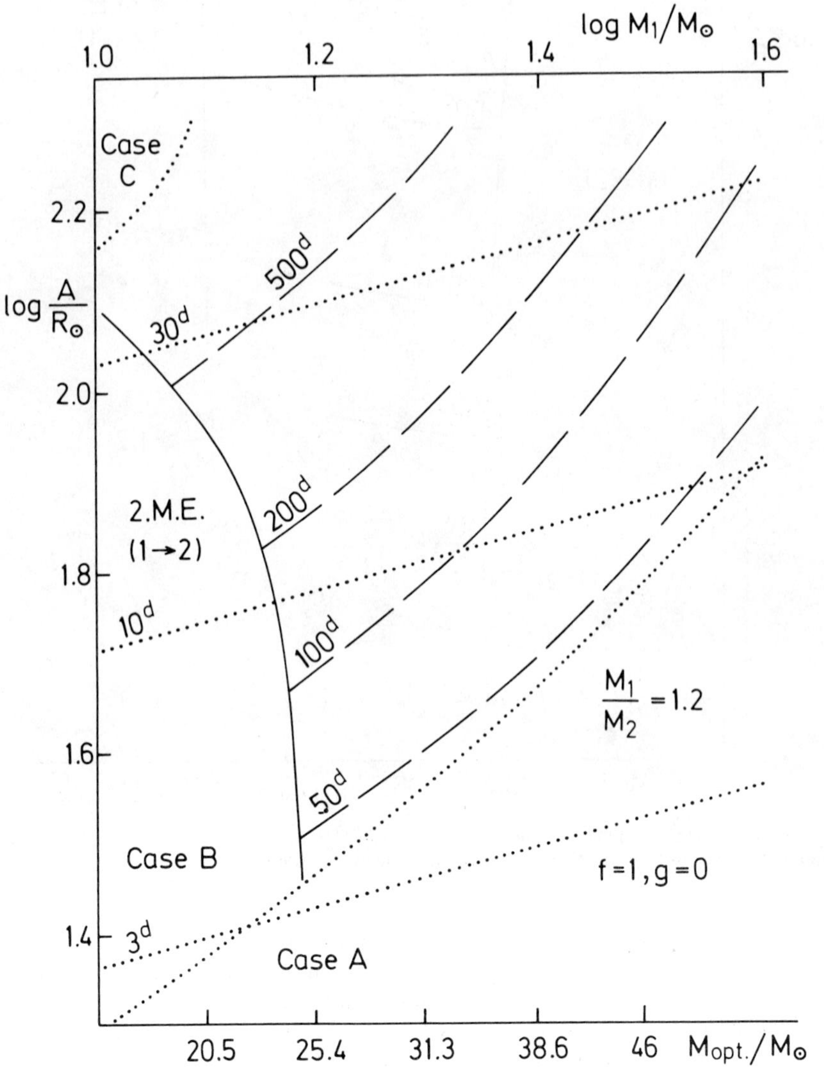

FIGURE 6. Same as FIGURE 4 but for a mass ratio of 1.2:1.

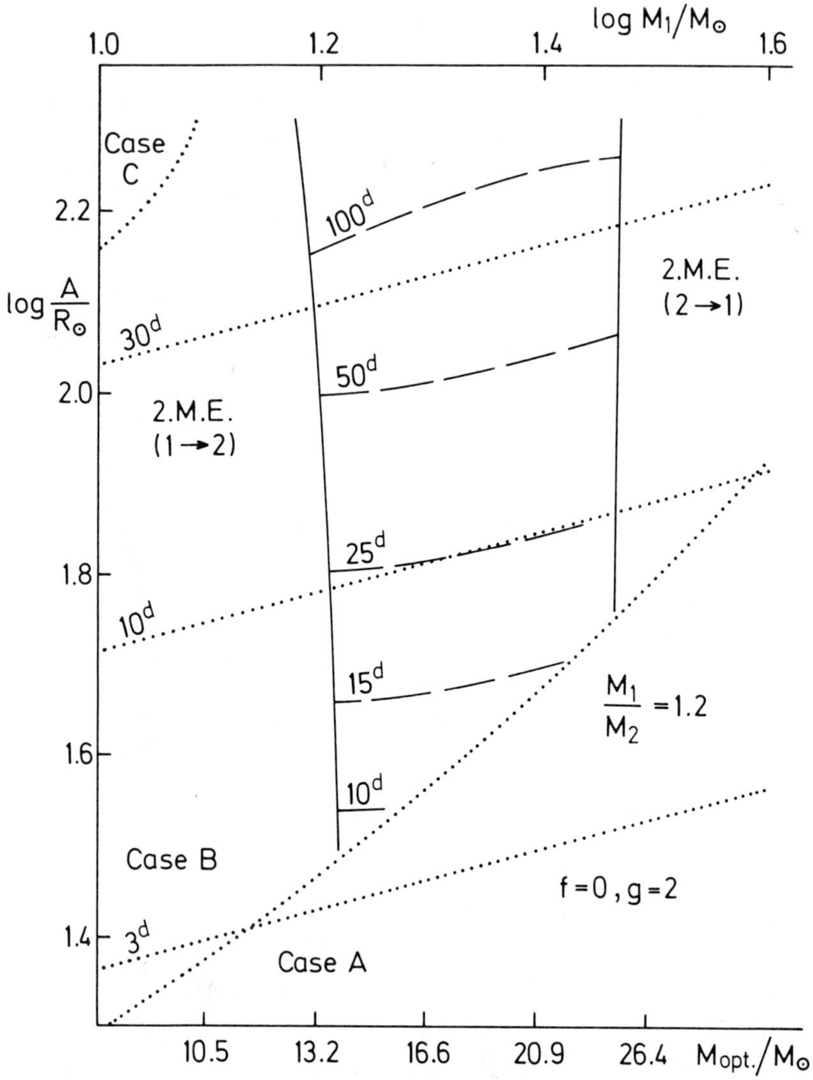

FIGURE 7. Same as FIGURE 5 but for a mass ratio of 1.2:1. Here the systems following the standard scenarios for x-ray binaries are confined by two solid lines.

Four permutations of two different values of M_1/M_2 and of f, g, the final period, final mass of the optical component, and the type of initial mass exchange as functions of initial mass and initial separation are shown in FIGURE 4–7. For each initial data set only the results for case B mass exchange in which a 1 $M_\odot$ neutron star remains after a supernova explosion are given.

Since the scenario described by van den Heuvel corresponds to $M_1/M_2 = 2.5, f = 1$, and $g = 0$, we can compare it with the results of our computations given in FIGURE 4.

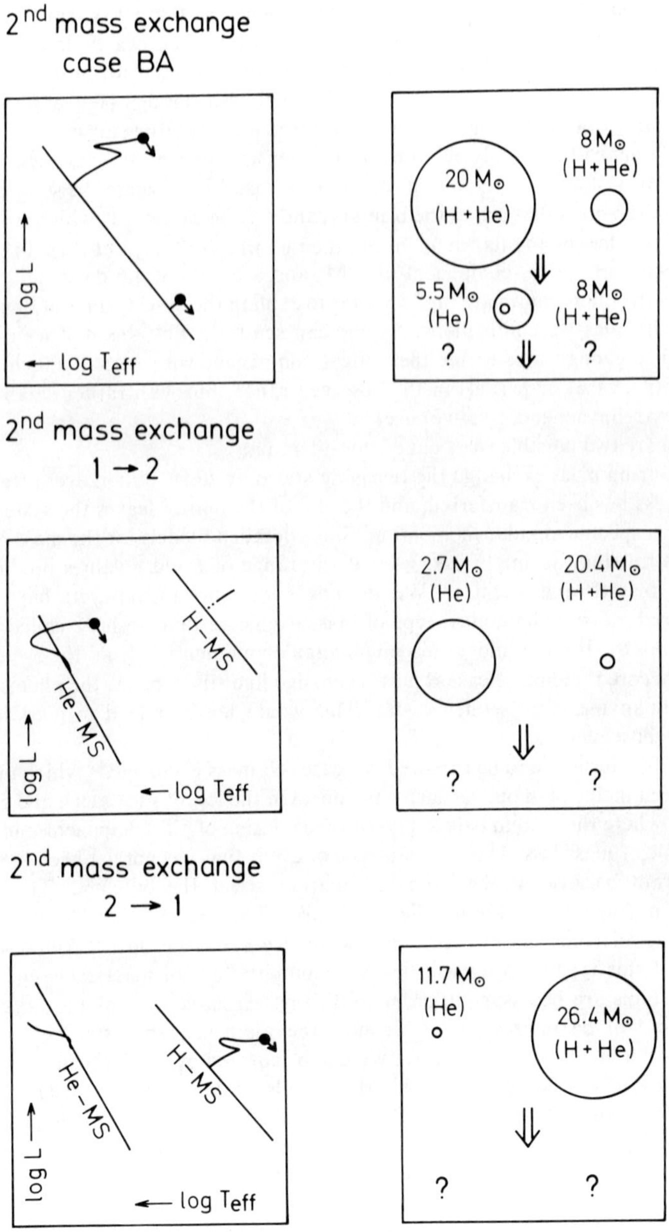

FIGURE 8. Three examples for as yet unexplored types of events encountered in our calculations. (Top) Large losses of angular momentum in a (extreme nonconservative) mass exchange bring the system into contact even if the secondary component is not receiving any matter. (Middle) The He star produced in case *B* mass exchange fills its Roche lobe a second time. (Bottom) The original secondary star fills its Roche lobe before the He star has had enough time to undergo a supernova explosion.

In this conservative case the final system resembles the x-ray emitting system obtained by van den Heuvel after mass exchange and supernova explosion. The final system consists an optical component of 22.5 $M_\odot$ and a 1 $M_\odot$ neutron star. The period we predict is 14.6 days. Our result is not identical with van den Heuvel's because his mass exchange starts at a slightly different time and his neutron star has 2 $M_\odot$. But the initial system will certainly not undergo a conservative mass exchange (van den Heuvel[1]). If we follow up its fate in the extreme case $f = 0$, where the secondary does not receive the mass that leaves the primary, and for the case $g = 2$, which corresponds to a moderate loss of angular momentum, then we find from FIGURE 5 that the system ends up with an optical component of 8 $M_\odot$ and a period of 4.6 days. This example illustrates the dilemma we are in if we want to explain the observations of binary x-ray sources with massive companions. As one can see from FIGURES 4–7 we can either obtain large enough masses for the optical component with periods much too high (conservative case) or periods in the observed range, but with rather small primary masses (extreme nonconservative case).

There are two possible ways out of this dilemma.

(1) During mass exchange the receiving star only starts to boil over after a large part of mass has been transferred, and the rest of the matter leaves the system with a rather high specific angular momentum. Since detailed models for the mass exchange phase are missing, we might ask: what is the range of f and g values producing the expected masses and periods? We do not know the answer yet, but we have encountered a new hypothetical type of mass exchange that we have called case *BA* (see FIGURE 8). If one removes too much angular momentum from the system, then the helium core remnant of case *B* mass exchange finds itself below the photosphere of the accompanying main sequence star. This would however lead to mass loss with unknown consequences.

(2) The alternative to be explored is a case *AB* mass exchange,[4,5] which up to now has not been included in our scenario machine. On the left of FIGURES 4 and 5 we find two areas where the system before supernova explosion of one component undergoes a second type of mass loss. This is a new type of event that has not yet been included in our program because we have no information about the outcome. The event is described in FIGURE 8 in the middle two figures. The star that has lost its hydrogen envelope again feeds mass to the main sequence star. We do not know what the outcome of this type of event will be. New computations of mass exchange in close binary systems are necessary in order to fill the left-hand sides of FIGURES 4–7. In addition it will be interesting to see how the main sequence star behaves if its hydrogen-rich material on the surface were to be covered by a helium envelope.

In FIGURE 7 on the right-hand side there is also unknown territory. Here the star that has received mass during the phase of first mass exchange is now giving mass back to the helium star. As a result, the helium star will try to become a red giant and fill its Roche lobe again. This situation is sketched in the two drawings at the bottom of FIGURE 8. This event also is not yet included in our program.

That the events not included in our scenario machine are important in nature can be seen from Cen X-3. It is not possible to find values f and g such that a system with $P = 2^d$ and $M_1 = 25\ M_\odot$ can be obtained by case *B* mass exchange.

In planning the computer program we encountered a whole set of problems that have to be solved in order to make the scenario machine more powerful. Some of them can even be solved by well-known stellar evolution codes, but there are others for which the underlying physics must first be studied. Some of the situations that are not understood are:

1. mass exchange in a system in which the component losing mass is a helium star;
2. evolution of stars that, during their main sequence stage, have received mass and now have a chemical structure different from that of a normally evolved star;
3. mass exchange in binary systems near mass ratio unity;
4. evolution of helium stars that have gained a hydrogen envelope;
5. final outcome case *A* mass exchange;
6. mass exchange in systems in which semiconvection is important;
7. mixing in a hydrogen star that receives helium at its surface;
8. mass loss in a system in which the receiving star has no time to adjust thermally;
9. mass loss from helium stars (can helium stars that have too much mass to become white dwarfs lose enough mass to ultimately settle down as white dwarfs?)

References

1. VAN DEN HEUVEL, E. P. J. 1977. Ann. N. Y. Acad. Sci. **302:** 44.
2. KIPPENHAHN, R. & A. WEIGERT. 1967. Z. Astrophys. **65:** 251.
3. LAUTERBORN, D. 1970. Astron. Astrophys. **7:** 150.
4. ZIÓLKOWSKI, J. 1969. Astrophys. Space Sci. **3:** 14.
5. HORN, J. 1971. Bull. Astron. Inst. Czech. **22:** 37.

SUMMARY OF WORKSHOP ON ACCRETION IN ASTROPHYSICS

P. Mészáros

Max-Planck-Institut für Physik und Astrophysik
D-8000 Munich 40
Federal Republic of Germany

In the workshop on accretion in astrophysics most of the discussions centered on the theory of accreting neutron stars and black holes, as well as on some of the relevant x-ray and optical observations.

There has been lately considerable interest in the problem of the accreting magnetized neutron stars, in particular those that are observed to be x-ray pulsars, and several papers at the workshop were devoted to the question of the formation of the x-ray spectrum, which exhibits power-law breaks and, in some objects, features identified as cyclotron lines. The plasma instabilities in the accretion column and at the Alfvén surface, the proper radiative opacities to be used, and other relevant physical processes involved were hotly discussed subjects. Another interesting topic was the problem of spherical accretion onto black holes. Some of the questions dealt with here concerned the inclusion of dissipative mechanisms and the treatment of the radiation-coupled hydrodynamics, and the main thrust at this workshop was in the discussion of possible upper limits on the luminosity and accretion rates in this type of flow. Disk flows were also discussed, in particular the pure hydrodynamic instabilities that might arise in them, as well as those involving the presence of a dipole magnetic field from the compact object, and the predicted observational properties of disks in novae. At a more fundamental level, there were several presentations on the near-vacuum neighborhood of black holes, dealing with the electrodynamics, particle orbits, and N-particle simulations around rotating and nonrotating black holes. The experimental papers presented dealt with new results on the hard x-ray spectrum of the black hole candidate Cygnus X-1, and with recent HEAO-A2 observations of spectra of x-ray burst sources. A more detailed discussion of the individual papers presented is given below.

The interaction between the material falling in towards the magnetic pole of a neutron star and its stationary atmosphere was considered by Kirk,[1] allowing for collective effects in the strong field. The major such effect turned out to be the excitation of ion-sound turbulence. This excitation occurs as a result of a relative drift motion of electrons and ions in the atmosphere, which is set up as a result of the differing stopping lengths of infalling electrons and infalling ions. The interaction between an infalling ion and the turbulent atmosphere was found to be sufficiently strong to reduce considerably the stopping length below previous values in the literature. This result was then used to set upper limits on the density of an atmosphere through which material can accrete without producing a nonradiative shock.

The effect of the peculiar opacities in the strong magnetic fields of x-ray pulsars was discussed by Ventura.[2] The free–free and Thomson opacities exhibit a strong dependence on the photon frequency, on the direction of propagation, and on the

0077–8923/80/0336–0590 $01.75/1 © 1980, NYAS

polarization characteristics of the plasma normal modes. Each of the two polarization modes may thus give access to widely different layers of the emission region in the accretion tunnel of x-ray pulsars, which should strongly influence the spectrum, polarization, and beaming of the emergent radiation and therefore a careful analysis of the structure of the normal modes is required. In work done in collaboration with Mészáros, it was pointed out that the birefringent property of the magnetic vacuum is a decisive factor in determining the ellipticity of the modes, and leads to opacities that diverge widely from the usually used cold-plasma opacities. It was also indicated that, when the Thomson opacity exceeds the free–free, the two modes are effectively coupled through Thomson scattering, and this may lead to strong spectral and beaming features even in the absence of strong temperature gradients.

The continuum spectrum and cyclotron line formation via resonant Compton–cyclotron scattering in Hercules X-1 was discussed by Heyvaerts,[3] in collaboration with Bonazzola and Puget. A simplified model of the accretion column was adopted, involving a plane parallel geometry, and the transfer at the line frequency was considered to be of a resonant scattering character. It was argued that both modes must be resonant, and an angle- and polarization-averaged cross section was taken. For simplicity, velocity gradients were neglected, and it was assumed that in thermal balance, black body photons from the surface are Comptonized by cooler columnar gas. A good fit to the data (see FIGURE 1) was obtained by solving the diffusion equation in the space and frequency variables, for a Thomson optical depth of a few units.

A different investigation of the spectrum of Hercules X-1, based on a Monte Carlo calculation, was presented by Yahel.[4] The temperature of the plasma column was taken to be higher than the blackbody value for the photons emitted from the base, and infall effects were neglected. The full angular, polarization, and energy dependence of the cold-plasma scattering cross sections, as well as of the collisional processes, was included, assuming for simplicity a thermalized particle distribution. The radiative transport was solved by simulating the random walk of the photons until escape. The emergent spectral intensity was found to be a strong function of the angle between the observer and the magnetic field. For angles close to 90°, a definite feature resembling the cyclotron line appears, whose structure depends on the magnetic field strength and on the thermal conditions in the plasma.

Spherical accretion on black holes was considered by Maraschi,[5] reporting on work done with Perola and Treves, as a model for hard x-ray emitting galactic nuclei. The inclusion of dissipation effects, with equipartition between gravitational, turbulent, and magnetic energies, leads to thermal electrons of temperature 10^9–10^{10} °K inside radii of $10^{17}\ M_8$ cm. The synchrotron radiation is Comptonized then into the MeV region, yielding a power law. Depending on the hardness of the core spectrum, Compton heating of the infalling material may cause an inversion of the infall at intermediate radii of $\sim 10^{18}$ cm. Therefore, either the whole accretion rate has to be supplied by mass sources inside this radius, or else hard spectra cannot be a persistent feature of active galactic nuclei, the activity being sporadic. The heating may also be important in determining the mechanism of formation of line-emitting clumps.

The question of how near to the Eddington luminosity may a spherically accreting black hole emit was discussed by Begelman.[6] The inclusion of dissipative mechanisms leads to an emissivity $L(r) < GM\dot{M}/r$, and if $\dot{M}$ exceeds a critical value $\dot{M}_E$, the total

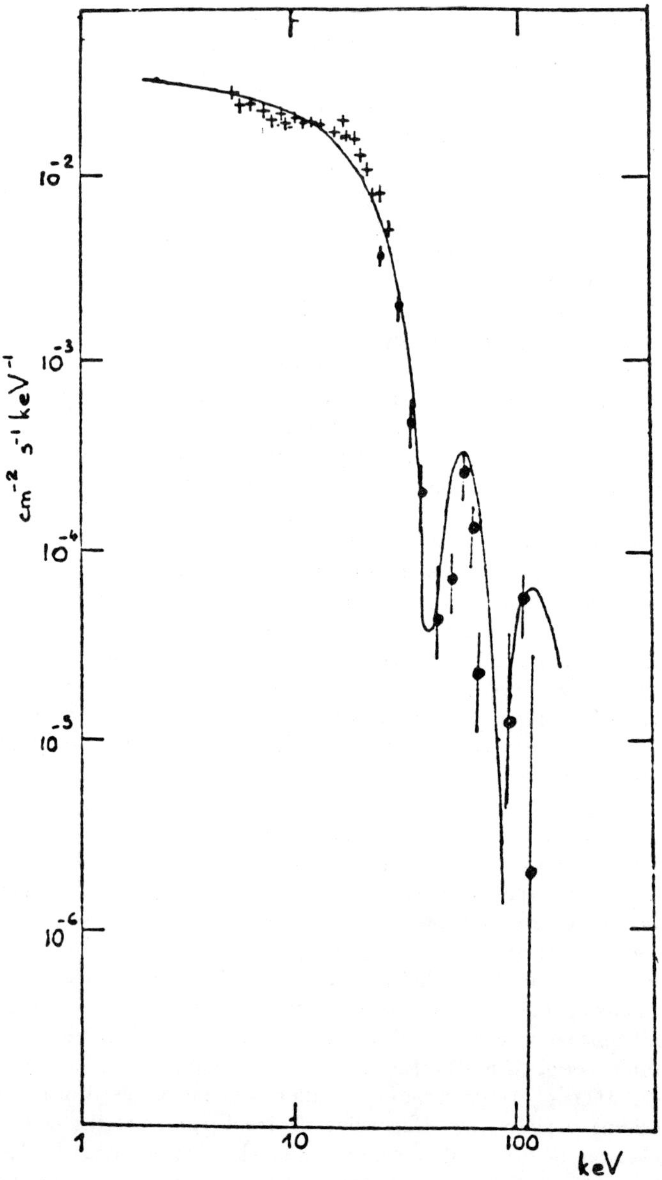

FIGURE 1. Hercules X-data from Pravdo *et al.* (Astrophys. J. **216:** L23, 1977) (crosses, $E <$ 20 keV) and from Trümper *et al.* (Ap. J. **219:** L105, 1978, and references therein) (filled circles, $E <$ 20 keV) are shown and compared to the theoretical spectrum (solid line) for $\omega_H/m_ec^2 = 0.065$, $kT/m_ec^2 = 0.06$ and $\tau = 1.82$. The absolute value of the flux has been adjusted to give the best fit. As a distance of 6 kpc, these parameters give the right flux for a column diameter of 3×10^3 cm.

flow luminosity may try to exceed the Eddington luminosity L_E. For these conditions however photon trapping sets in, so that L is always $< L_E$. Treating the trapping surface as a sink of radiation leads to $L < 0.65\ L_E$, and somewhat more detailed calculations, taking into account various qualitative features of the flow such as breaks in the density run, shocks, etc., leads to $L < 0.6\ L_E$, in confirmation of the previous simpler estimate. The conclusion drawn is that radiation pressure cannot prevent the spherically symmetric accretion rate $\dot{M}$ on a black hole from attaining arbitrarily large values.

Also addressed to the question of what the maximum spherical accretion rate of a black hole is was an interesting contribution by Wheeler,[7] in collaboration with Gilden. They used a fully relativistic hydrodynamic code, which incorporated diffuse radiation transport and also nonadiabatic heating due to magnetic field line reconnection, to study time-dependent, optically thick flows. The internal energy generated destabilizes the material in a purely relativistic way, ensuring collapse of the inflowing matter. In all cases, accretion proceeded at the free-fall rate, regardless of whether the diffusion time scale was shorter or longer than the dynamical time scale. Any envelope within the accretion radius of the black hole will thus be swallowed on a dynamical time scale with relatively small net release of energy. The inclusion of angular momentum would not necessarily affect this conclusion.

The mathematical problem of the spherical accretion flow onto a black hole was the topic of a contribution by Allan.[8] By means of analytical techniques, considering polytropic flows of variable indices to allow for the presence of other components besides thermal pressure, e.g., radiation, it was shown that in a relativistic analysis—unlike in the classical case—there is more than one critical point, and it was concluded that the accretion rate can be uniquely determined.

The time-dependent spherically symmetric nonmagnetic supercritical accretion onto a neutron star was discussed by Newman,[9] in collaboration with Winkler. An implicit technique with variable Eddington factor was used, the opacities being due to Thomson and free–free processes. For a solar-mass star with initial surface temperature of 10^6 K, the imposition of an accretion rate of 5×10^{19} g s^{-1} and the assumption that all the kinetic energy of the material is converted into radiation at the surface leads to an increase of the surface temperature up to about 10^8 K, while the total luminosity emitted approaches the Eddington value. The inner velocity is greatly reduced as this limiting luminosity is approached. The efficiency of conversion of gravitational energy into radiation was about 0.5% in this case.

The interaction of an accretion disk with the magnetosphere of a neutron star was the topic of a paper by Anzer,[10] working in collaboration with Börner. The neutron star's axis of rotation was taken perpendicular, and the magnetic axis was taken parallel to the plane of the disk, so that the disk was able to penetrate diamagnetically into the atmosphere. The relative velocity difference leads to a Kelvin–Helmholtz instability of the interface over a narrow region around the corotation radius. Turbulent diffusion then causes the matter to enter the magnetosphere, being eventually channeled towards the polar caps. There is an approximate balance near the corotation radius between the amount of matter being flung out and that being accreted, which is reflected upon the corresponding amounts of braking and accelerating torques acting on the star. Application of the model to Hercules X-1 leads to substantial agreement between the predicted and observed temporal variation of the pulse period.

The hydrodynamic stability of axisymmetric, inviscid flows was investigated by Drury.[11] Steady differential rotation was assumed, and after linearizing and Fourier decomposing in time and the nonradial coordinates, a system of two coupled linear ordinary differential equations is obtained. For incompressible disturbances, with uniform densities and power law rotation curves ($\Omega = r^{-\lambda}$, $0<\lambda<2$), one can prove the stability, while for oscillatory disturbances there is a generic mechanism that can cause instability. There is a regular singularity at the point where the phase angular velocity of the perturbation equals that of the basic flow. A wave disturbance incident on this point gives rise to transmitted and reflected waves, the amplitude of which latter may exceed that of the incident. Reflection back and forth of this amplified disturbance leads to unstable behavior.

The theoretical optical continua and line spectra of accretion disks for novalike objects were the subject of Mayo's[12] paper, in work done with Wickramasinghe and Whelan. UBV colors and hydrogen Balmer absorption profiles were computed for steady state disks, as a function of the mass flow rate, inclination angle, and central mass. In the UBV plane, the disks lie between the blackbody line and the main sequence. A comparison was made with observations of dwarf novae and novalike variables, and it is found that in the outburst state, dwarf novae are in the disk region, but not during quiescence, and the same appears true for the novalike variables (UX Uma's), suggesting that these latter are also dwarf novae stuck in outburst. The broad absorption lines of the many dwarf novae and UX Uma's can be explained as arising from the optically thick part of the disk.

Measurements of the hard x-ray spectrum of the black hole candidate Cygnus X-1 were reported by Trümper,[13] in work done with Steinle, Voges, Reppin, Kendziorra, and Staubert. They cover the range 15–150 keV, and pin down the spectrum in the "low" state to an unprecedented accuracy, based on three flights which agree with each other within 30%, and with previous rocket data within 40%, both in slope and absolute intensity. A steepening of the power law below 100 keV is evident, and the whole normal spectrum can be taken to be better defined (to $\lesssim$ 40%) than the previously thought factor of 2. A good fit has been achieved by Sunyaev and Trümper with a model involving Comptonization of cooler disk photons by a gas of $kT_c = 27$ keV and a Thomson optical depth $\tau_0 = 5$ (see FIGURE 2).

The spectra of x-ray burst sources were discussed by Swank,[14] based on four Type I and about two hundred rapid bursts observed by the HEAO-A2 experiment, for which 12 energy bands are available between 2 and 60 keV every 1.28 s. Spectral fits to a blackbody model are reasonably acceptable, while bremsstrahlung fits can be ruled out. The quality of the blackbody fit is somewhat worse but still acceptable for the rapid bursts. The peak luminosities are not less than, and probably exceed, the Eddington luminosity. Knowledge of a relationship between peak luminosity and the Eddington limit would yield the relation $M(R)$ of the bursting source, and it would appear from the data for four sources that they have the same mass and radius. The effective radius for the largest rapid bursts is twice that of the special bursts, and it is suggested that steady emission from the rapid burster may originate in the infall from this radius to the surface.

A novel way of discussing the general nonequilibrium states of black holes, or in general of any "horizon," was described by Damour,[15] in collaboration with Carter. Defining new concepts of velocity, surface density of momentum, surface pressure, etc., the Einstein equations have the form of a Navier–Stokes equation, and one can

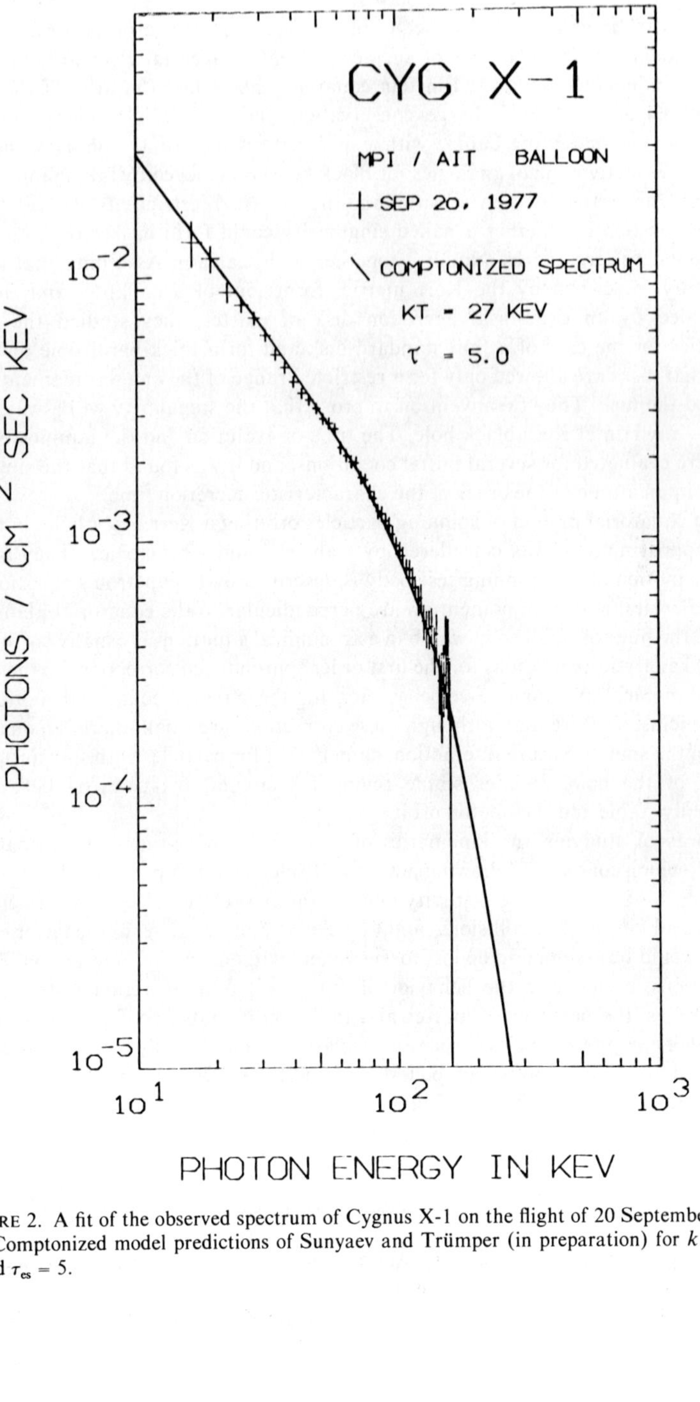

FIGURE 2. A fit of the observed spectrum of Cygnus X-1 on the flight of 20 September 1977, to the Comptonized model predictions of Sunyaev and Trümper (in preparation) for $kT_c = 27$ keV and $\tau_{es} = 5$.

say that the mechanical behavior of the surface of a black hole is analogous to that of a viscous fluid bubble endowed with both shear and bulk surface viscosities. The same approach can be used for the nonequilibrium electromagnetic properties of a black hole. By defining new concepts of surface current, tangential electric field, normal magnetic inductance, etc., the Einstein equations then adopt the form of the laws of Ohm, of Faraday, and of charge conservation, the surface of a black hole being analogous to a conducting bubble with a surface resistivity of 377 ohms. Consistency with the thermodynamical analogies for black holes is achieved by taking into account the "heat" dissipated by the viscosities and the electrical restrictivity.

The question of whether a naked singularity could form in nature from regular initial data was treated by Nobili,[16] together with Calvani. Assuming that a naked singularity, described by the Kerr metric, forms out of a collapse, and that it is surrounded by an equatorial accretion disk of matter, they studied the general properties for the case of a thin standard disk and for a thick barotropic disk. They found that disks are allowed only for a restricted range of the angular momenta of the disk and the hole. They then went on to prove that the singularity will always evolve towards a maximal Kerr black hole. The time of evolution and the luminosity of the disk were evaluated for several initial conditions, and it was found that the singularity dresses up in a time of the order of the characteristic accretion time.

The equatorial motion of spinning particles orbiting a Kerr black hole, in the first order approximation, was considered by Calvani[17] and Abramovicz. They assumed that the motion of the spinning test body is described by Papapetrou's equations, and that the orbital and spin momentum are perpendicular to the equatorial plane of the hole. In the limit of small spin, which in astronomical situations is usually correct, they obtained analytic expressions for the first order spin-induced corrections to the energy, angular momentum, angular velocity, etc., for the circular equatorial motion. The main conclusions were that, although these corrections are small, there is a qualitative effect of the spin-curvature interaction, namely that for particles with spin antiparallel to that of the hole, another stable region for circular orbits exists between the marginally stable and the photon orbit.

A way of studying the kinematics of black hole space-times was discussed by Sharp,[18] which consists of following numerically clouds of test particles, either massive or massless, of negligible self-gravity and interacting solely by collisions, in a certain background metric. The collisions might be elastic or dissipative, in the latter case the energy would be assumed to be lost to the system without further interaction. Possible applications might be to the behavior of dust and gas in accretion systems, stellar distributions, the behavior of matter and radiation in stable configurations, and the removal of energy via Penrose processes. Isolated or continuously fed systems could be mimicked through appropriate boundary conditions. One might also study some hydrodynamic processes by adapting to this situation the particle in a cell techniques.

List of Contributors to the Workshop on Accretion in Astrophysics

1. Kirk, J., Max-Planck-Institut für Physik und Astrophysik, 8000 München 40, Föhringer Ring 6, West Germany.
2. Ventura, J., Max-Planck-Institut für extraterrestrische Physik, 8046 Garching

bei München, West Germany, and P. MÉSZÁROS Max-Planck-Institut für Physik und Astrophysik, 8000 München 40, Föhringer Ring 6, West Germany.

3. BONAZZOLA, S. & J. HEYVAERTS, Observatoire de Meudon, 92190 Meudon, France, and J. G. PUGET, Institut d'Astrophysique, 98 bis Bld. Arago, 75014 Paris, France.
4. YAHEL, R. Z., Max-Planck-Institut für extraterrestrische Physik, 8046 Garching, West Germany, and University of Jerusalem, Israel.
5. MARASCHI, L., G. C. PEROLA & A. TREVES, Instituto di Fisica dell'Università Milano, Via Celoria 16, 20133 Milano, Italy.
6. BEGELMAN, M. C., Institute of Astronomy, Madingley Road, Cambridge, CB3 OHA, U.K.
7. WHEELER, J. C. & D. L. GILDEN, J.I.L.A., University of Colorado, Boulder, CO 80309, USA.
8. ALLAN, P., Department of Astronomy, University of Manchester, Manchester M13 9PL, U.K.
9. NEWMAN, M. J., Los Alamos Scientific Laboratory, T-DOT 228, P.O. Box 1663, Los Alamos, NM 87545, USA, and K.-H. WINKLER, Max-Planck-Institut für Physik und Astrophysik, Föringer Ring 6, 8000 München 40, West Germany.
10. ANZER, U. & G. BÖRNER, Max-Planck-Institut für Physik und Astrophysik, 8000 Müchen 40, Föhringer Ring 6, West Germany.
11. DRURY, L., Institute of Astronomy, Madingley Road, Cambridge CB3 OHA, U.K.
12. MAYO, S. K. & J. A. J., WHELAN, Institute of Astronomy, Madingley Road, Cambridge CB3 OHA, U.K. and D. T. WICKRAMASINGHE, Australian National University, Canberra, Australia.
13. STEINLE, R., W. VOGES, W. PIETSCH, C. REPPIN & J. TRÜMPER, Max-Planck-Institut für extraterrestrische Physik, 8046 Garching, West Germany, and E. KENDZIORRA, & R. STAUBERT, Astronomisches Institut der Universität Tübingen, Tübingen, West Germany.
14. SWANK, J. H., Code 661, Laboratory for High Energy Astrophysics, NASA-GSFC, Greenbelt, MD 20771, USA.
15. DAMOUR, T. & B. CARTER, Groupe d'Astrophysique Relativiste, Observatoire de Meudon, 92190 Meudon, France.
16. CALVANI, M. Institute of Astronomy, and L. NOBILI, Institute of Physics, University of Padova, Vicolo dell'osservatorio 5, Padova, Italy.
17. ABRAMOWICZ, M. A., Copernicus Astronomical Center, Warsaw, Poland, and M. CALVANI, Institute of Astronomy, University of Padova, Vicolo dell'osservatorio 5, Padova, Italy.
18. SHARP, N. A., Institute of Astronomy, Madingley Road, Cambridge CB3 OHA, U.K.

SUMMARY OF WORKSHOP ON HUBBLE DIAGRAM AND EXPANSION OF THE UNIVERSE*

G. A. Tammann

European Southern Observatory
1211 Geneva 23, Switzerland

I.E. Segal (M.I.T.) reported on "The Redshift-Magnitude Relation" and on seemingly strong deviations from a linear relation between the distances and recession velocities of galaxies.[1] Somewhat milder deviations from a linear relation in the local expansion field were suggested in a paper "Extragalactic Distance Scale and Hubble Constant" distributed by G. de Vaucouleurs (University of Texas at Austin.)[2]

R.M. Soneira (Institute of Advanced Study, Princeton) had a preprint distributed "A Cosmological Redshift–Distance Square–Law?" in which he challenged the results of Segal and showed that after proper statistical treatment the data in the *Second Reference Catalogue of Bright Galaxies* favor a linear Hubble expansion law.[3] Yahil (State University of New York at Stony Brook) reported on "The Linearity of the Local Expansion Field": the data in the *Revised Shapley-Ames Catalogue,* which is the largest available, magnitude-limited sample of galaxies with known redshifts, occupy in the plane of the magnitude–log *cz* diagram a region which is exactly predicted by the luminosity function of galaxies (taken from large clusters), the cut-off function of the *Catalogue* (determined from Zwicky's fainter *Catalogue of Galaxies*) and an *assumed* linear expansion law. Any claims for strong deviations from a linear expansion field are therefore due to a Malmquist-like selection bias, and the actual deviations from linearity must be small. The only detectable systematic peculiar motion is a local infall velocity of ~ 150 km s^{-1} toward the Virgo cluster center. This velocity allows one to determine the excess mean mass density within a sphere around the Virgo cluster with our Galaxy on its periphery. If the clustering of galaxies is representative for the clumpiness of the total mass, then galaxy counts indicate that the large-scale mean mass density is four times smaller than within the test volume. The resulting mass density is strongly deficient to close the universe, i.e., $q_0 \approx 0.02$.[4]

Philippe Crane (ESO, Geneva) tested "The Hubble Diagram for First-Ranked Cluster Galaxies" if there could exist any possible exception to the Hubble law which relates redshifts linearly to distance. The result indicates conclusively that redshift is directly proportional to distance.[5]

Fang Li-zhi (University of Science and Technology, China) analyzed "The Redshift-Magnitude Relation for Quasars with Resolved Radio Components" on the assumption of zero-diameter evolution.[6]

A. Cavaliere (Unità di Ricerca di Asiago-Padova) proposed in a talk "Cosmic Distances from X-ray and Microwave Observations of Clusters of Galaxies" to use the Zeldovich-Sunyaev effect (i.e., the depression of the Rayleigh-Jeans part of the cosmic background radiation by Thomas scattering in x-ray clusters) to determine distances.[7] The same effect was invoked by F. Melchiorri (Istituto di Fisica Superiore,

*Supported by the Swiss National Science Foundation.

Florence) for "A Measurement of the Cosmological Deceleration Parameter through the Sunyaev-Zeldovich Effect."[8]

D.A. Schwartz (Cambridge, Mass.) discussed the "Prospects of the Measurement of q_0 by the Apparent Volume Density of X-Ray Objects". J.K. Katgert-Merkelijn (Sterrewacht Leiden) in collaboration with L. Padrielli and C. Lari showed that present data on the apparent diameter-redshift relation of radio galaxies are compatible with $q_0 = 0$ and no size evolution.

J.R. Gott (Princeton Univ.) addressed the question "The 3K Microwave Background: Is it Hawking Radiation Associated with the Big Bang Singularity?",[9] and R.D. Sherman (Jet Propulsion Laboratory, Pasadena) presented "Models of a Hot Intercluster Medium."[10] J. Chela-Flores (University Simon Bolivar, Caracas) discussed the "Pairing in the Cosmic Neutrino Background."[11]

References

1. Segal, I. E. 1976. Mathematical Cosmology and Extragalactic Astronomy. Academic Press, New York.
2. De Vaucouleurs, G. 1979. Astrophys. J. **233:** 433. References therein.
3. Soneira, R. M. 1979. Astrophys. J. Lett. **230:** L63.
4. Sandage, A., G. A. Tammann & A. Yahil. 1979. Astrophys. J. **232:** 352; G. A. Tammann, A. Yahil & A. Sandage. 1979. ESO Sci. Preprint No. 67; A. Yahil, A. Sandage & G. A. Tammann. 1980. Astrophys. J. In press; A. Yahil, A. Sandage & G. A. Tammann. 1980. Physica Scripta. In press; G. A. Tammann, A. Sandage & A. Yahil. 1980. Physica Scripta. In press; A. Yahil, A. Sandage & G. A. Tammann. 1980. Lecture Notes of the Les Houches Summer Course 1979. In press.
5. Crane, P. 1979. Astrophys. J. Lett. In press.
6. Fang Li-zhi, Zhou You-yuan, Cheng Fu-zhen & Chu Yao-quan. 1978. Preprint.
7. Cavaliere, A., L. Danese & G. De Zotti. 1979. Astron. Astrophys. **75:** 322.
8. Fabbri, R., F. Melchiorri, F. Mencaraglia & V. Natale. 1979. Astron. Astrophys. **74:** L20.
9. Gott, J. R. III. 1978. Preprint.
10. Sherman R. D. & J. Silk. 1979. Astrophys. J. Lett. **231:** L61.
11. Alonso V. & J. Chela-Flores. 1978. Preprint.